Elements of General
and
Biological Chemistry

Eighth Edition

John Wiley and Sons, Inc.

New York   Chichester   Brisbane   Toronto   Singapore

Production supervised by Lucille Buonocore
Illustrations by John Balbalis with the assistance of
the Wiley Illustrations Department
Photo research by Hilary Newman
Manuscript edited by Jeannette Stiefel under the
supervision of Deborah Herbert
Cover Illustration by Roy Wiemann
Design by Kevin Murphy

Library of Congress Cataloging in Publication Date:

Holum, John R.
    Elements of general and biological chemistry / John R. Holum. —
  8th ed.
      p. cm.
    Includes bibliographical references and index.
    ISBN 0-471-51757-7
    1. Biochemistry.   I. Title.
  QP514.2.H64   1991
    574.19′2—dc20

                  90-38202
                  CIP

Printed in the United States of America

10 9 8 7 6 5 4 3 2 1

About the Author

JOHN HOLUM   is on the faculty of Augsburg College, Minneapolis, MN. He did his undergraduate work at St. Olaf College and earned his Ph.D. (organic chemistry) at the University of Minnesota. Additional studies were taken as sabbatical leaves at California Institute of Technology and Harvard University. In 1974 he was given the Distinguished Teaching Award of the Minnesota Section of the American Chemical Society. He is a member of Phi Beta Kappa, Phi Lambda Upsilon, Sigma Xi, and Sigma Pi Sigma. The National Science Foundation has awarded him several research grants and a Science Faculty Fellowship. He is the author or coauthor of several texts in chemistry, all published by John Wiley & Sons, and a reference work, *Topics and Terms in Environmental Problems* (Wiley – Interscience). He has also authored papers for the *Journal of the American Chemical Society,* the *Journal of Organic Chemistry,* and the *Journal of Chemical Education.* He has been active on the Examinations Committee and the Committee on Chemistry for Professional Health Care Students of the Division of Chemical Education of the ACS, and has spoken often at Divisional and Regional meetings, as well as at conferences of the Two-Year College Chemistry Association. His textbooks in chemistry for professional health care students have been widely used in America and abroad for over 25 years.

TABLE OF ATOMIC WEIGHTS AND NUMBERS

Based on the 1987 Report of the Commission on Atomic Weights and Isotopic Abundances of the International Union of Pure and Applied Chemistry and for the elements as they exist naturally on earth. Scaled to the relative atomic mass of carbon-12. The uncertainties in values, between ± 1 and ± 9 units in the last digit of an atomic weight, are in parentheses after the atomic weight. (From *Journal of Physics and Chemistry Reference Data*, Vol. 17(1988), pp. 1791.)

Element	Symbol	Atomic Number	Atomic Weight		Element	Symbol	Number	Atomic Weight	
Actinium	Ac	89	227.0278	(L)	Neon	Ne	10	20.1797(6)	(g, m)
Aluminum	Al	13	26.981539(5)		Neptunium	Np	93	237.0482	(L)
Americium	Am	95	243.0614	(L)	Nickel	Ni	28	58.69(1)	
Antimony	Sb	51	121.75(3)		Niobium	Nb	41	92.90638(2)	
Argon	Ar	18	39.948(1)	(g, r)	Nitrogen	N	7	14.00674(7)	(g, r)
Arsenic	As	33	74.92159(2)		Nobelium	No	102	259.1009	(L)
Astatine	At	85	209.9871	(L)	Osmium	Os	76	190.2(1)	(g)
Barium	Ba	56	137.327(7)		Oxygen	O	8	15.9994(3)	(g, r)
Berkelium	Bk	97	247.0703	(L)	Palladium	Pd	46	106.42(1)	(g)
Beryllium	Be	4	9.012182(3)		Phosphorus	P	15	30.973762(4)	
Bismuth	Bi	83	208.98037(3)		Platinum	Pt	78	195.08(3)	
Boron	B	5	10.811(5)	(g, m, r)	Plutonium	Pu	94	244.0642	(L)
Bromine	Br	35	79.904(1)		Polonium	Po	84	208.9824	(L)
Cadmium	Cd	48	112.411(8)	(g)	Potassium	K	19	39.0983(1)	
Calcium	Ca	20	40.078(4)	(g)	Praseodymium	Pr	59	140.90765(3)	
Californium	Cf	98	251.0796	(L)	Promethium	Pm	61	144.9127	(L)
Carbon	C	6	12.011(1)	(r)	Protactinium	Pa	91	231.0359(2)	(Z)
Cerium	Ce	58	140.115(4)	(g)	Radium	Ra	88	226.0254	(L)
Cesium	Cs	55	132.90543(5)		Radon	Rn	86	222.0176	(L)
Chlorine	Cl	17	35.4527(9)		Rhenium	Re	75	186.207(1)	
Chromium	Cr	24	51.9961(6)		Rhodium	Rh	45	102.90550(3)	
Cobalt	Co	27	58.93320(1)		Rubidium	Rb	37	85.4678(3)	(g)
Copper	Cu	29	63.546(3)	(r)	Ruthenium	Ru	44	101.07(2)	(g)
Curium	Cm	96	247.07003	(L)	Samarium	Sm	62	150.36(3)	(g)
Dysprosium	Dy	66	162.50(3)	(g)	Scandium	Sc	21	44.955910(9)	
Einsteinium	Es	99	252.083	(L)	Selenium	Se	34	78.96(3)	
Erbium	Er	68	167.26(3)	(g)	Silicon	Si	14	28.0855(3)	
Europium	Eu	63	151.965(9)	(g)	Silver	Ag	47	107.8682(2)	(g)
Fermium	Fm	100	257.0951	(L)	Sodium	Na	11	22.989768(6)	
Fluorine	F	9	18.9984032(9)		Strontium	Sr	38	87.62(1)	(g, r)
Francium	Fr	87	223.0197	(L)	Sulfur	S	16	32.066(6)	(r)
Gadolinium	Gd	64	157.25(3)	(g)	Tantalum	Ta	73	180.9479(1)	
Gallium	Ga	31	69.723(1)		Technetium	Tc	43	98.9072	(L)
Germanium	Ge	32	72.61(2)		Tellurium	Te	52	127.60(3)	(g)
Gold	Au	79	196.96654(3)		Terbium	Tb	65	158.92534(3)	
Hafnium	Hf	72	178.49(2)		Thallium	Tl	81	204.3833(2)	
Helium	He	2	4.002602(2)	(g, r)	Thorium	Th	90	232.0381(1)	(g, r, Z)
Holmium	Ho	67	164.93032(3)		Thulium	Tm	69	168.93421(3)	
Hydrogen	H	1	1.00794(7)	(g, m, r)	Tin	Sn	50	118.710(7)	(g)
Indium	In	49	114.82(1)		Titanium	Ti	22	47.88(3)	
Iodine	I	53	126.90447(3)		Tungsten	W	74	183.85(3)	
Iridium	Ir	77	192.22(3)		(Unnilennium)	(Une)	109	273	(n, Q)
Iron	Fe	26	55.847(3)		(Unnilhexium)	(Unh)	106	263.118	(L, n)
Krypton	Kr	36	83.80(1)	(g, m)	(Unniloctium)	(Uno)	108	265	(n, Q)
Lanthanum	La	57	138.9055(2)	(g)	(Unnilpentium)	(Unp)	105	262.114	(L, n)
Lawrencium	Lr	103	260.150	(L)	(Unnilquadium)	(Unq)	104	266.11	(L, n)
Lead	Pb	82	207.2(1)	(g, r)	(Unnilseptium)	(Uns)	107	262.12	(L, n)
Lithium	Li	3	6.941(2)	(g, m, r)	Uranium	U	92	238.0289(1)	(g, m, Z)
Lutetium	Lu	71	174.967(1)	(g)	Vanadium	V	23	50.9415(1)	
Magnesium	Mg	12	24.3050(6)		Xenon	Xe	54	131.29(2)	(g, m)
Manganese	Mn	25	54.93805(1)		Ytterbium	Yb	70	173.04(3)	(g)
Mendelevium	Md	101	258.10	(L)	Yttrium	Y	39	88.90585(2)	
Mercury	Hg	80	200.59(3)		Zinc	Zn	30	65.39(2)	
Molybdenum	Mo	42	95.94(1)		Zirconium	Zr	40	91.224(2)	(g)
Neodymium	Nd	60	144.24(3)	(g)					

(g) Geologically exceptional specimens of this element are known that have different isotopic compositions. For such samples, the atomic weight given here may not apply as precisely as indicated.

(L) This atomic weight is for the relative mass of the isotope of longest half-life. The element has no stable isotopes.

(m) Modified isotopic compositions can occur in commercially available materials that have been processed in undisclosed ways, and the atomic weight given here might be quite different for such samples.

(n) Name and symbol are assigned according to rules developed by the IUPAC.

(Q) Mass number of first isotope observed.

(r) Ranges in isotopic compositions of normal samples obtained on earth do not permit a more precise atomic weight for this element, but the tabulated value should apply to any normal sample of the element.

(Z) Despite having no stable isotopes, the terrestrial compositions of samples of the long-lived isotopes allow a meaningful atomic weight.

Elements of General
and
Biological Chemistry

Preface

This book is a shorter version of the fourth edition of *Fundamentals of General, Organic, and Biological Chemistry* (1990), also published by John Wiley & Sons. Its content incorporates the recommendations of the Task Force on Chemical Education for Health Professions (M. Treblow, Chairman), sponsored by the Division of Chemical Education of the American Chemical Society and described in the *Journal of Chemical Education,* July 1984, page 620 ("A Syllabus for a One-Semester Chemistry Course for Health Professions"). Hence, this text meets the needs for a basic text in a one-term course in chemistry for students aiming for careers in professional health care fields.

There is more in this book than can reasonably be taught in just one term. But although general agreement exists about topics, there is not a similar agreement about emphasis. Moreover, the incoming preparations of students in this kind of course varies widely, ranging from classes in which all studied chemistry in high school to classes resulting from open admissions policies. This explains why many schools have used earlier editions for two-term courses.

The theme of the previous editions, the molecular basis of life, continues. Topics in general and organic chemistry are therefore included only if they are either essential background for the study of biochemistry or are directly related to the chemistry of living processes. Thus the first eight chapters, largely general chemistry, conclude with an emphasis on acids, bases, and buffers. These involve concepts directly related to perhaps the most important single topic for future nurses (if, indeed, just one can be named), the chemistry of the acid–base balance of blood.

One change in these chapters is to discontinue the use of the terms "Arrhenius acid" or "Arrhenius base." In aqueous media it all comes down to proton transfers, and the hydronium ion is only one proton donor.

In keeping with society's interest and concern about the environment, we now have a special topic on acid rain.

Another change is to discontinue the teaching of the concept of acid or base normalities.

Acid–base titration problems are solved simply as ordinary problems in stoichiometry. The concept of an *equivalent* is retained only in connection with equivalents of ions as they relate to ionic charges. Aqueous solutions of clinical interest still use the units of equivalent/liter or meq/mL, so this must be taught. It is used in a special topic on the anion gap and its significance for diagnosis. The study of how to do the calculations for making dilute solutions from concentrated solutions is now in a special topic.

A third change in the first section of the book is the reorganization of the chapter on solutions and colloids to place all aspects of colloids after the study of solutions.

The next five chapters, organic chemistry, include only those functional groups whose nature and chemistry are essential to the biochemistry that follows. These are the groups in carbohydrates, lipids, proteins, nucleic acids, and many important drugs.

Only the barest minimum organic chemistry is included, because the available time has to be very carefully allotted. Thus, alkyl halides are barely mentioned because this system occurs nowhere among the biochemicals to be studied later. Very little is done with aromatic chemistry, because the details of aromatic electrophilic substitution reactions will not be exploited later. (What it means to be *aromatic,* and some of the characteristic reactions of the benzene ring, are studied, because this ring does occur among some amino acids and proteins.)

When the study of organic chemistry starts, frequent mention is made of the kinds of biological chemicals that happen to have the particular functional group currently being studied, because students appreciate the reminders that the theme of the course continues throughout all of the study of organic chemistry. We believe, however, that the soundest pedagogical approach is to introduce these groups, one after the other, as they occur among the *simplest,* monofunctional compounds. Large, complex structures such as glucose or hemoglobin or DNA can be sources of terror rather than wonder when they are introduced too early.

The next eight chapters take up the major kinds of compounds found in the human body, how they are fitted for their uses in metabolism, and how they react in cells. We continue the segregation of catabolism and anabolism into separate chapters. Thus Chapter 19 ("Molecular Basis of Energy for Life") begins with a broad overview of biochemical energetics (which might serve in some courses as the only coverage of this topic), and then it takes up the ways in which carbohydrates, lipids, and amino acids can be broken down for energy. This chapter closes with the problems of acidosis that can arise when the body cannot bring a balanced use of various substances to bear on its needs for chemical energy.

Chapter 20 ("Metabolism and Molecule Building") goes into the absorption, distribution, and biosynthesis of carbohydrates, lipids, and amino acids. One of the special topics in this chapter concerns recent advances in our understanding of the relationships of lipoprotein complexes, cholesterol, liver receptors, and heart disease. Earlier, in Chapter 15 ("Lipids"), a new special topic on the omega-3 fatty acids and heart disease has been added.

Chapter 21 ("Nucleic Acids") continues in its former location because teachers of this course seem increasingly to omit the topic entirely. Those who do this argue that students have encountered several of its major points in earlier courses, including high school biology. Nonetheless, this chapter has been updated, and it now includes a special topic on genetic fingerprinting.

The book then closes with a chapter on radioactivity with an emphasis on health-related aspects. There is little reason why this chapter could not be studied much earlier, but its present location does take advantage of the prior study of nucleic acids, genes, and genetic damage that radiations can cause. A new special topic on radon-222 as part of background radiation is now provided.

The design of this edition is mostly like that of the previous edition. There are **margin comments.** Some are reminders. Some restate a point. Some are small, illustrative tables to which the neighboring paragraph refers. Some are structures that need not be memorized.

There are **Special Topics** on matters of current interest, and a list is provided following the Table of Contents. All have been updated as needed, and several new special topics, mentioned above, appear. For the first time, sets of Review Exercises are now offered for the Special Topics.

A comprehensive package of instructional materials, described in detail on page xiii, is available to help the students.

A Teachers' Manual includes the answers to all of the Practice Exercises and Review Exercises.

Other Design Features That Aid Students

Chemistry is one of the disciplines in which important scientific terms can be sharply defined. We have tried to do so at the first occasion of using the term or as soon thereafter as possible, at or near the place where the **key term** is highlighted by a boldface color treatment. Then our aim has been to use these terms as carefully and consistently as possible. At the end of the book, there is a **glossary** where each of the key terms is defined. (The *Study Guide* has a Glossary for each chapter.)

Each chapter has a **Summary** that uses the key terms in a narrative survey. The main section of each chapter also begins with a **summary statement** that announces what is coming and that serves during test review periods to highlight the major topics.

Special labels identify sets of **Review Exercises** that are about a common topic. Within most chapters are several **Practice Exercises,** and most of these immediately follow a **worked example** that provides a step-by-step description of how to solve a certain kind of problem. The **factor-label** method is used for nearly all computations. The **answers** to all Practice Exercises are found at the back of the book together with the answers to selected Review Exercises.

The **Appendix on Mathematical Concepts** provides a review of what exponentials are and how to manipulate them. This Appendix also discusses how to use pocket calculators to handle exponentials and to carry out chain operations.

Continuing a long tradition, we have tried to make the **Index** the most thorough, most cross-referenced index in any text of this type.

JOHN R. HOLUM
Augsburg College

Acknowledgments

Over the many years of writing instructional materials, my family—Mary, my wife, and our daughters, Liz, Ann, and Kathryn—have been Gibralters of support. They, rather than my teaching or writing, are my career and so such teaching and writing are seen by us as one of the ways by which our family has tried to be of help to others. I am pleased to say "thank you" to them for being such nice people.

Here, at Augsburg College, I have enjoyed years and years of support from Dr. Earl Alton, Chemistry Department Chair, Dr. Ryan LaHurd, Dean, and Dr. Charles S. Anderson, President. My freedom to write stems in no small measure from the freedom that these caring people have accorded me.

Helpful people abound at John Wiley & Sons, too. They do good work. I think of the special support of my Chemistry Editor, Dennis Sawicki and his assistant, Sandra Harding. Chief Illustrator, John Balbalis, and the Wiley Illustration Department have been skillful, artistic, and faithful in handling artwork for many years. Photo Research Manager, Stella Kupferberg solves problems and makes this facet of production worry-free. The Designer, Kevin Murphy, stands in the long Wiley tradition of artistry and imagination. Editing Manager, Deborah Herbert and Copy Editor, Jeannette Stiefel, have seen to the smoothing out of stylistic and grammatical problems. Finally, the Supervisor of Production, Lucille Buonocore, has taken impressive pains to ensure that the innumerable details of production have been all handled well. It's an impressive team, and I count myself to be fortunate indeed for having become associated with John Wiley & Sons in the first place.

Special Acknowledgments to Professional Critiques. Thanks in part to the suggestions and critiques of teachers and scientists this book or its longer version, *Fundamentals of General, Organic, and Biological Chemistry* (Fourth edition, 1990), have been the first books of their kind to bring to the attention of students at this level the major developments concerning the molecular basis of life. I am particularly pleased to thank Sandra Olmsted,

University of Minnesota, for her unusual care in checking both the scientific and the typographical accuracies of this book. Others whom I wish to thank are:

Earl R. Alton, Augsburg College

Arlin Gyberg, Augsburg College

Alan Smith, University of Southern Maine, and his students in CHY 101 and 103

Gary Hemphill, Clinical Laboratory, Metropolitan Medical Center (Minneapolis)

Floyd L. James, Miami University, Oxford, Ohio

Merle K. Loken, University of Minnesota Medical School

Robert G. Martinek, Illinois Department of Health (Chicago)

Erwin Mickelberg, Augsburg College

Paul Mueller, The Johns Hopkins University Medical School

Neal Thorpe, Augsburg College

Michael Uricheck, Western Connecticut State University

John Davidson, Eastern Kentucky University

Henry Pigott, Victoria College

Martin Levine, Borough of Manhattan Community College

Dennis Sardella, Boston College

Robert Nelson, Georgia Southern University

J.R.H.

Supplementary Materials for Students and Teachers

The complete package of supplements that are available to help students to study and teachers to plan the course and operate the associated laboratory work includes the following:

Laboratory Manual for Elements of General and Biological Chemistry, eighth edition.   An instructor's manual for these experiments is a section in the general Teachers' Manual described below.

Study Guide for Elements of General and Biological Chemistry, eighth edition.   This softcover book contains chapter objectives, chapter glossaries, additional worked examples and exercises, sample examinations for each chapter, and the answers to all Exercises.

Teachers' Manual for Elements of General and Biological Chemistry, eighth edition.   This softcover supplement is available to teachers, and it contains all of the usual services for *both the text and the laboratory manual*.

Test Questions for Elements of General and Biological Chemistry, eighth edition.   Multiple choice test questions in camera-ready form. Available without charge to instructors only who adopt this book. Write to Chemistry Editor, John Wiley & Sons, Inc., 605 Third Ave., New York, NY 10158

Transparency Acetates.   Instructors who adopt this book can receive from John Wiley & Sons, without charge, a set of transparencies of 100 figures and tables in this book. Write to Chemistry Editor, John Wiley & Sons, Inc., 605 Third Ave., New York, NY 10158.

J.R.H.

Contents

Special Topics

Chapter 1
Goals, Methods, and Measurements

What this family of mute swans knows by instinct, we know by intellect: life is better if we work with nature, not against her. Knowing how nature works at the molecular level of life, the subject of this book, helps professional health care workers apply nature's gifts to the healing arts.

1.1 CHEMISTRY AND THE MOLECULAR BASIS OF LIFE

The theme of this book is the molecular basis of life.

Centuries ago, people surely noticed that many *different* animals drank at the same water holes, breathed the same air, ate the same kinds of food, and enjoyed the same salt licks. Ancient farmers knew that the droppings of animals nourished plants, and that animals prospered by eating plants. Some animals could eat weaker animals and grow.

Evidently, at some deep level of existence, living things can exchange parts. These parts are not organs and tissues but much smaller things, extremely tiny particles called molecules made of even smaller particles called atoms. All of life, whether plant or animal, has a *molecular* basis, and chemistry has been the route to its discovery. **Chemistry** is the study of that part of nature that deals with substances, their compositions and structures, and their abilities to be changed into other substances. There are so many different substances that we have to have a plan of study.

Our Strategy. Life at the molecular level involves molecules and chemical reactions that are often complicated. The symbols we use for them, however, are actually less complex than many symbol systems you have already mastered, like those used to draw maps. You learned how to read and understand dozens of maps by mastering just a few map symbols. Our symbols for molecules are like maps because the same pieces of molecules, like molecular "map signs," occur over and over again. It will be a good idea, therefore, before we study some of the most complicated molecules in nature (Chapters 14–22), to learn these "signs" among simpler substances. Our chapters on organic compounds (Chapters 9–13) do this.

As we said earlier, molecules are made of atoms. It really isn't possible to understand molecules without first learning about atoms and how their (even tinier) parts get reorganized into molecules. This study occurs mainly in the first eight chapters, together with the essential background about a variety of substances such as acids, bases, salts, and solutions. All these studies rest on experimental evidence that involved taking measurements of physical quantities. In this chapter we will learn about some of the measurements that have been useful.

1.2 PROPERTIES AND PHYSICAL QUANTITIES

A physical property differs from a chemical property by being observable without changing a substance into a different substance.

A **property** is any characteristic of something that we can use to identify and recognize it when we see it again. The observations of some properties, however, change an object or a sample of a substance into something else. For example, we can measure how much gasoline it takes to drive a car 100 miles, but this measurement uses up the gasoline. As it burns it changes into water and carbon dioxide (the fizz in soda pop). A property that, when observed, causes a substance to change into new substances is called a **chemical property,** and what is being observed is called a **chemical reaction.** A chemical property of iron, for example, is that it rusts in moist air; it changes slowly into a reddish, powdery substance, iron oxide, quite unlike metallic iron. **Chemistry** is the study of these kinds of changes in substances, how they occur, and how atoms become reorganized as they happen.

Properties such as color, height, or weight that can be observed without changing the object into something different are called **physical properties.** We usually rely on such properties to recognize and name things. For example, some physical properties of liquid water are that it is colorless and odorless; that it dissolves sugar and table salt but not butter; that it makes a thermometer read 100 °C (212 °F) when it boils (at sea level); and that if it is mixed with gasoline it will sink, not float. If you were handed a glass containing a liquid having these properties, your initial hypothesis undoubtedly would be that it is water. Think of how often each day you recognize things (and people) by simply observing physical properties.

Well over 6 million chemical substances are known.

The atoms of all of the kinds of matter are made of varying combinations of just three extremely tiny particles: electrons, protons, and neutrons.

Notice how much a description of water's properties depends on human senses, our abilities to see, taste, feel, and to sense hotness or coldness. Our senses, however, are limited, so inventors have developed instruments that extend the senses and make possible finer and sharper observations. These devices are equipped with scales or readout panels, and the data we obtain by using them are called *physical quantities*.

A **physical quantity** is a property to which we can assign both a numerical value *and a unit*. Your own height is a simple example. If it is, say, 5.5 feet, its numerical value, 5.5, and its unit, feet, together tell us at a glance how much greater your height is than an agreed-upon reference of height, the foot.

The unit in a physical quantity is just as important as the number. If you said that your height is "two," people would ask, "Two what?" If you said "two yards," they would know what you meant (provided they knew what a yard is). (But they might ask, "*Exactly* two yards?") This example shows that we cannot describe a physical property by a physical quantity without giving both a number and a unit.

$$\text{Physical quantity} = \text{number} \times \text{unit}$$

Physical Quantities Are Obtained by Measurements. A **measurement,** is an operation by which we compare an unknown physical quantity with one that is known. Maybe, as you were growing up, someone measured your height by comparing it with how many sticks, perhaps one-foot rulers, it took to equal your height. Usually the number of sticks did not match your height exactly, so fractions of sticks called inches (each with their own fractions) were also used. It's probably quite obvious to you by now that somebody has decided what an inch, a foot, or a yard is and that the rest of us have agreed to these definitions. And that's just what they are, definitions. We'll learn those that are the most useful in chemistry in the next section.

1.3 UNITS AND STANDARDS OF MEASUREMENT

The most fundamental quantities of measurement are called base quantities and each has an official standard of reference for one unit.

Mass, Length, Time, and Temperature Degree Are Base Quantities. The most fundamental measurements in chemistry are those of mass, volume, temperature, time, and quantity of chemical substance.

Mass is the measure of the inertia of an object. Anything said to have a lot of inertia such as a train engine, a massive boulder, or an ocean liner is very hard to get into motion, or if it is in motion, it is difficult to slow it down or make it change course. It is this inherent resistance to any kind of change in motion that we call **inertia,** and *mass* is our way of describing inertia quantitatively. A large inertia means a large *mass*.

A large mass doesn't always mean a large *weight*. Your mass does not depend on where you are in the universe, but your weight does. Your weight is a measure of the gravitational force of attraction that the earth makes on your body. This gravitational force is less on the moon, which is a smaller object than the earth — about one-sixth of that on earth. But the mass of an astronaut, the fundamental resistance to any change in motion, is the same on the moon as on the earth. When we use a laboratory balance to *weigh* something, we are actually measuring mass because we are comparing two weights *at the same place on the earth* and, therefore, under the same gravitational influence. One weight is the quantity being measured, and the other is a "weight" (or set of weights) built into the weighing balance. Although we commonly call the result of the measurement a "weight," we'd more properly call it the *mass* of the object or the sample. (We'll generally speak of masses, not weights, in this book.)

2 = a number
2 yards = a physical quantity

Large inertia goes with large mass.

Quantitative describes something expressible by a number and a unit.

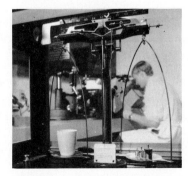

A traditional two-pan balance showing a small container on the left pan and some weights on the right pan.

Volume of a cube = (*l*)³

The **volume** of an object is the space it occupies, and space is described by means of a more basic physical quantity, length. The volume of a cube, for example, is the product of (length) $\times$ (length) $\times$ (length), or (length)³. **Length** is a physical quantity that describes how far an object extends in some direction, or it is the distance between two points.

A fundamental quantity such as mass or length is called a **base quantity** and any other quantity such as volume that can be described in terms of a base quantity is called a **derived quantity.**

Another base quantity in science is **time** — our measure of how long events last. We need this quantity to describe how rapidly the heart beats, for example, or how fast some chemical reaction occurs.

Still another important physical quantity is **temperature degree,** which we use to describe the hotness or coldness of an object.

All of these base quantities are necessary to all sciences, but chemistry has a special base quantity called the *mole* that describes a certain amount of a chemical substance. It consists of a particular (and very large) *number* of tiny particles without regard to their masses or volumes. We will not study this base quantity further until we know more about these particles.

Every Base Unit Has a Reference Standard of Measurement.   To measure and report the mass of an object, its temperature, or any of its other base or derived physical quantities, we obviously need some units and some references. By international treaties among the countries of the world, the reference units and standards are decided by a diplomatic organization called the General Conference of Weights and Measures, headquartered in Sèvres, a suburb of Paris, France. The General Conference has defined a unit called a **base unit** for each of seven base quantities, but we need units only for the five that we have already mentioned: mass, length, time, temperature, and mole. We also need some units for the derived quantities, for example, for volume, density, pressure, and heat. The standards and definitions of base and derived quantities and units together make up what is now known as the **International System of Units** or the **SI** (after the French name, *Système International-ale d'Unités*).

The other two SI base quantities are electric current and luminous intensity. Their base units are called the ampere and the candela, respectively.

Each base unit is defined in terms of a **reference standard,** a physical description or embodiment of the base unit. Long ago, the reference (such as it was) for the *inch* was ''three barleycorns, round and dry, laid end to end.'' Obviously, which three barleycorns were picked had a bearing on values of length under this ''system.'' And if the barleycorns got wet, they sprouted. You can see that a reference standard ought to have certain properties if it is to serve the needs of all countries. It should be entirely free of such risks as corrosion, fire, war, theft, or plain skulduggery, and it should be accessible at any time to scientists in any country. The improvement that the SI represents over its predecessor, the *metric system,* is not so much in base units as in their reference standards.

An alloy is a mixture of two or more metals made by stirring them together in their molten states.

The SI base unit of length is called the **meter,** abbreviated **m,** and its reference standard is called the *standard meter.* Until 1960, the standard meter was the distance separating two thin scratches on a bar of platinum–iridium alloy stored in an underground vault in Sèvres (Figure 1.1). This bar, of course, could have been lost or stolen, so the new reference for the meter is based on a property of light, something available everywhere, in all countries, and that obviously cannot be lost or damaged. This change in reference didn't change the actual length of the meter, it only changed its official reference.[1]

In the United States, older units are now legally defined in terms of the meter. For example the yard (yd), roughly nine-tenths of a meter, is defined as 0.9144 m (exactly). The foot (ft), roughly three-tenths of a meter is defined as 0.3048 m (exactly).

---

[1] The SI now defines the standard meter as how far light will travel in 1/299,792,458 of a second. It is thus based on the speed of light as measured by an ''atomic clock.''

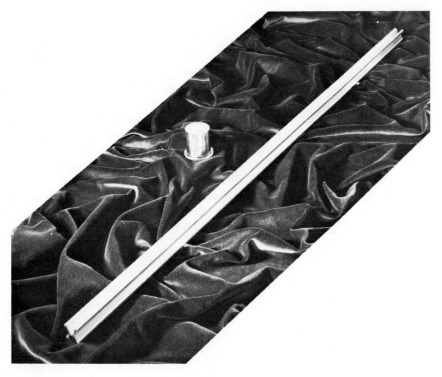

FIGURE 1.1
The SI standard mass and the former standard meter. Shown here are the U.S. copies kept at the National Bureau of Standards in Washington, D.C.

In chemistry, the meter is usually too long for convenience, and submultiples are often used, particularly the **centimeter,** or **cm,** and the **millimeter,** or **mm.** Expressed mathematically, these are defined as follows:

$$1 \text{ m} = 100 \text{ cm} \quad \text{or} \quad 1 \text{ cm} = 0.01 \text{ m}$$
$$1 \text{ m} = 1000 \text{ mm} \quad \text{or} \quad 1 \text{ mm} = 0.001 \text{ m}$$
$$1 \text{ cm} = 10 \text{ mm} \quad \text{or} \quad 1 \text{ mm} = 0.1 \text{ cm}$$

Notice that the subunits are in fractions based on 10. The millimeter, for example, is one-tenth of a centimeter. As we'll often see, this makes many calculations much easier than they were under older systems (where, e.g., the inch was one-twelfth of a foot, and the foot was one-third of a yard).

The inch (in.) is about two and a half centimeters; more exactly,

$$1 \text{ in.} = 2.54 \text{ cm} \quad \text{(exactly)}$$

Table 1.1 gives several relationships between various units of length.

1 cm   2 cm

TABLE 1.1
Some Common Measures of Length[a]

SI	U.S. Customary
1 kilometer (km) = **1000** meters (m)	1 mile (mi) = **5280** feet (ft)
1 meter = **100** centimeters (cm)	= **1760** yards (yd)
1 centimeter = **10** millimeters (mm)	1 yard = **3** feet (ft)
	1 foot = **12** inches (in.)
1 meter = 39.37 inches	1 foot = **30.48** centimeters
= 3.280 feet	= **0.3048** meter
= 1.093 yard	1 inch = **2.54** centimeters

[a] Numbers in boldface are exact.

TABLE 1.2
Some Common Measures of Mass[a]

SI

1 kilogram (kg) = **1000** grams (g)
1 gram = **1000** milligrams (mg)
1 milligram = **1000** micrograms (μg, γ, or mcg)[b]

U.S. Customary (Avoirdupois)[c]

1 short ton = **2000** pounds (lb avdp)
1 pound = **16** ounces (oz avdp)
1 ounce = **16** drams (dr avdp)
1 dram = 437.5 grains (grain)[d]

Apothecaries'

1 pound (lb ap) = **12** ounces (oz ap, or ℥) = **5760** grains
1 ounce = **8** drams (dr ap, or ℈) = **480** grains
1 dram = **60** grains

Other Relationships

1 kilogram = 2.205 lb avdp = 2.679 lb ap = 15,432 grains
= 35.27 oz avdp = 32.15 oz ap

1 lb avdp = 453.6 grams   1 lb ap = 373.2 grams
1 oz avdp = 28.35 grams   1 oz ap = 31.10 grams
1 grain = 0.0648 gram   = 64.8 milligrams
1 gram = 15.43 grains   1 dr ap = 3.887 grams

[a] Numbers in boldface are exact.

[b] The microgram is sometimes called a *gamma* in medicine and biology.

[c] These are the common units in the United States, not the apothecaries' units.

[d] The National Bureau of Standards has adopted no symbol for *grain*. Pharmacists usually symbolize it by *gr*. There is another apothecaries' weight called the *scruple* (20 grains), but it is no longer listed in the *U.S. Pharmacopoeia.*

One drop of water is about 60 mg.

1 kg of butter.

One cubic meter holds a little over 250 gallons.

The SI base unit of mass is named the **kilogram,** abbreviated **kg,** and its reference is named the *standard kilogram mass.* This is a cylindrical block of platinum–iridium alloy housed at Sèvres under the most noncorrosive conditions possible (Figure 1.1). This is the only SI reference that could still be lost or stolen, but no alternative has yet been devised. Duplicates made as much like the original as possible are stored in other countries. One kilogram has a mass roughly equal to 2.2 pounds in the U.S. customary system (the avoirdupois system). Table 1.2 gives a number of useful relationships among various units of mass, including some old apothecaries' units. Unless we state otherwise, however, we will use only the SI and the U.S. customary (avoirdupois) system in this book.

The most often used units of mass in chemistry are the SI kilogram, the **gram (g),** the **milligram (mg),** and the **microgram (μg).** These are defined as follows:

$$1 \text{ kg} = 1000 \text{ g} \quad \text{or} \quad 1 \text{ g} = 0.001 \text{ kg}$$
$$1 \text{ g} = 1000 \text{ mg} \quad \text{or} \quad 1 \text{ mg} = 0.001 \text{ g}$$
$$1 \text{ mg} = 1000 \text{ } \mu\text{g} \quad \text{or} \quad 1 \text{ } \mu\text{g} = 0.001 \text{ mg}$$

Lab experiments in chemistry usually involve grams or milligrams of a substance.

The SI unit of volume, one of the important derived units, is the cubic meter (m³) but this is much too large for convenient use in chemistry. An older unit, the **liter,** abbreviated **L,** is accepted as a *unit of convenience.* The liter occupies a volume of 0.001 m³ (exactly), and one liter is almost the same as one liquid quart; 1 quart (qt) = 0.946 L. Even the liter is often too

TABLE 1.3
Some Common Measures of Liquid Volume[a]

SI
1 cubic meter (m^3) = **1000** liters (L)
1 liter = **1000** milliliters (mL)
1 milliliter = **1000** microliters (μL, λ, or lambda)

U.S. Customary and Apothecaries'
1 gallon (gal) = **4** liquid quarts (liq qt)
1 liquid quart = **2** liquid pints (liq pt)
1 liquid pint = **16** liquid ounces (liq oz) (fluidounce, f$\mathfrak{z}$, fl oz, in the apothecaries' system)
1 liquid ounce = **8** fluidrams (f$\mathfrak{z}$) = **480** minims (m)
1 fluidram = **60** minims (m)

Other Relationships
1 cubic meter = 264.2 gallons
1 liter = 1.057 liquid quarts = 2.113 liquid pints
=33.81 liquid ounces
1 milliliter = 16.23 minims
1 liquid ounce = 29.57 milliliters
1 liquid quart = 946.4 milliliters
1 fluidram = 3.696 milliliters

Miscellaneous Approximate Equivalents (Unofficial)
1 liquid pint = 2 cups = 4 gills
1 cup = 16 tablespoonfuls = 250 mL
1 tablespoon = 3 teaspoonfuls = 15 mL
1 teaspoonful = 5 mL

[a] The apothecaries's system uses the U.S. Customary system for liquid measures. Numbers in boldface are exact.

large for convenience in chemistry, and two submultiples are used, the **milliliter (mL)** and the **microliter (μL).** These are related as follows:

$$1 \text{ L} = 1000 \text{ mL} \quad \text{or} \quad 1 \text{ mL} = 0.001 \text{ L}$$
$$1 \text{ mL} = 1000 \text{ } \mu\text{L} \quad \text{or} \quad 1 \text{ } \mu\text{L} = 0.001 \text{ mL}$$

In routine chemistry work, the milliliter is by far the most common unit you will encounter. Table 1.3 gives several other relationships among units of volume. Figure 1.2 shows apparatus used to measure volumes in the lab.

The milliliter (mL) is identical to one cubic centimeter (1 cm^3 or 1 cc) of volume, so you will sometimes see the symbols cm^3 or cc used for mL.

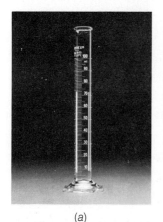

(a)

(b)

(c)

FIGURE 1.2
Apparatus for measuring liquid volumes. *(a)* Graduated cylinder—a "graduate". *(b)* Graduated pipet. *(c)* Volumetric flask. A line is etched part way up the narrow neck, and when the liquid level is at this level, its volume has the value also etched on the flask.

The SI unit of time is called the **second,** abbreviated **s.** The SI *definition,* however, involves complexities of atomic physics that are entirely beyond our needs. Fortunately, the *duration* of the SI second is the same as before, for essentially all purposes. The second is 1/86,400 of a mean solar day. Decimal-based multiples and submultiples of the second are used in science, but so are such deeply entrenched old units as minute, hour, day, week, month, and year.

The SI unit for degree of temperature is called the **kelvin, K.** (Be sure to notice that the abbreviation is K, not °K.) The kelvin is the same size as the **degree Celsius** (°C), which was once called the degree centigrade (also °C). Then it was defined as $\frac{1}{100}$ the interval between the freezing point of water (named 0 °C) and the boiling point of water (named 100 °C). The most extreme coldness possible is −273.15 °C, and this is named 0 K on the Kelvin scale. The kelvin is the name of the degree on this scale, and it is identical with the Celsius degree. Only the *numbers* assigned to points on the scale differ. See Figure 1.3, where the scales are compared.

The *kelvin* is named after William Thomson, Baron Kelvin of Largs (1842–1907), a British scientist.

Because 0 K corresponds to −273.15 °C, we have the following simple relationships between kelvins and degrees Celsius (where we follow common practice of rounding 273.15 to 273).

$$°C = K - 273$$

or

$$K = °C + 273$$

PRACTICE EXERCISE 1   Normal body temperature is 37 °C. What is this in kelvins?

The Kelvin scale is used in chemistry mostly to describe temperatures of gases. The Celsius scale is more popular for most other uses. (It is even rapidly supplanting the old, familiar Fahrenheit scale in medicine.) The **degree Fahrenheit (°F)** is five-ninths the size of the degree Celsius. To convert a Celsius temperature, t_C, to a Fahrenheit temperature, t_F, we can use either of the following equations. (Practice exercises that use these equations occur at the end of Section 1.6.)

$$t_C = \frac{5 \ °C}{9 \ °F} (t_F - 32 \ °F)$$

$$t_F = \frac{9 \ °F}{5 \ °C} \times t_C + 32 \ °F$$

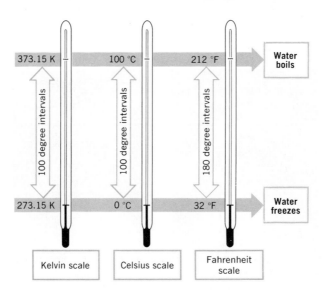

FIGURE 1.3
Relationships between the Kelvin, Celsius, and Fahrenheit scales of temperature.

TABLE 1.4
Some Common Temperature Readings in °C and °F

	°F	°C
Room temperature	68	20
Very cold day	−20	−29
Very hot day	100	38
Normal body temperature	98.6	37
Hottest temperature the hands can stand	120	49

Table 1.4 gives some common temperatures in both degrees Celsius (°C) and degrees Fahrenheit (°F).

1.4 SCIENTIFIC NOTATION

Scientific notation expresses very large or very small numbers in exponential forms to make comparisons and calculations easier.

The typical human red blood cell has a diameter of 0.000008 m. Whether we want to write it, say it, or remember it, 0.000008 m is an awkward number, and to make life easier scientists have developed a method called **scientific notation** for recording very small or very large numbers. In scientific notation (sometimes called exponential notation), a number is written as the product of two numbers. The first is a decimal number with a value usually between 1 and 10, although sometimes a wider range is used. Following this number is a times (×) sign and then the number 10 with an exponent or power. For example, we can write the number 4000 as follows:

$$4000 = 4 \times 1000 = 4 \times 10 \times 10 \times 10$$
$$= 4 \times 10^3$$

Appendix I has a review of exponential numbers.

When the decimal point is omitted, we assume that it is after the last digit in the number.

Notice that the exponent 3 is the number of places to the left that we have to move the decimal point in 4000 to get to 4, which is a number in the desirable range.

$$4\ \underset{3}{0}\ \underset{2}{0}\ \underset{1}{0}.$$

If our large number is 42,195, the number of meters in a marathon distance, we can rewrite it as follows after figuring out that we have to move the decimal point four places to the left to get a decimal number between 1 and 10.

$$42,195\ \text{m} = 4.2195 \times 10^4\ \text{m}$$

$$4\ \underset{4}{2}\ \underset{3}{1}\ \underset{2}{9}\ \underset{1}{5}.$$

In rewriting numbers smaller than 1 in scientific notation, we have to move the decimal point to the *right* to get a number in the acceptable range of 1 to 10. This number of moves is the value of the *negative* exponent of 10. For example, we can rewrite the number 0.000008 as

$$0.000008 = 8 \times 10^{-6}$$

$$0.\underset{1}{0}\ \underset{2}{0}\ \underset{3}{0}\ \underset{4}{0}\ \underset{5}{0}\ \underset{6}{8}$$

You should not continue until you are satisfied that you can change large or small numbers into scientific notation. For practice, do the following exercises.

ble.

TABLE 1.5
SI Prefixes for Multiples and Submultiples of Base Units[a]

	Prefix	Symbol
$1\,000\,000\,000\,000\,000\,000 = 10^{18}$	exa	E
$1\,000\,000\,000\,000\,000 = 10^{15}$	peta	P
$1\,000\,000\,000\,000 = 10^{12}$	tera	T
$1\,000\,000\,000 = 10^{9}$	giga	G
$1\,000\,000 = 10^{6}$	**mega**	**M**
$1\,000 = 10^{3}$	**kilo**	**k**
$100 = 10^{2}$	hecto	h
$10 = 10^{1}$	deka	da
$0.1 = 10^{-1}$	**deci**	**d**
$0.01 = 10^{-2}$	**centi**	**c**
$0.001 = 10^{-3}$	**milli**	**m**
$0.000\,001 = 10^{-6}$	**micro**	**μ**
$0.000\,000\,001 = 10^{-9}$	nano	n
$0.000\,000\,000\,001 = 10^{-12}$	pico	p
$0.000\,000\,000\,000\,001 = 10^{-15}$	femto	f
$0.000\,000\,000\,000\,000\,001 = 10^{-18}$	atto	a

[a] The most commonly used prefixes and their symbols are in boldface. Thin spaces instead of commas are used to separate groups of three zeros to illustrate the format being urged by the SI (but not yet widely adopted in the United States).

Express each number in scientific notation. Let the decimal part be a number between 1 and 10.

(a) 545,000,000   (b) 5,670,000,000,000   (c) 6454
(d) 25   (e) 0.0000398   (f) 0.00426
(g) 0.168   (h) 0.00000000000987 (see footnote 2)

Prefixes to the Names of SI Base Units Are Used to Specify Fractions or Multiples of These Units. If we rewrite 3000 m as 3×10^3 m and try to pronounce the result, we have to say "three times ten to the third meters." There's nothing wrong with this, but it's clumsy. This is why the SI has names for several exponential expressions — not independent names but prefixes that can be attached to the name of any unit. For example, 10^3 has been assigned a prefix name of *kilo-*, abbreviated *k-*. Thus 1000 or 10^3 meters can be called 1 kilometer. Abbreviated, this becomes 10^3 m = 1 km.

With just a few exceptions, the prefixes set by the SI go with exponentials in which the power is 3, 6, 9, 12, 15, and 18 or with $-3, -6, -9, -12, -15,$ and -18. These are all divisible by 3. Table 1.5 has a list of the SI prefixes and their symbols. Those given in boldface are so often encountered in chemistry that they should be learned now.

Notice that there are four prefixes that do not go with powers divisible by 3. The SI hopes their usage will gradually fade away, but this hasn't happened yet. The two in boldface have to be learned. However, *centi* is used almost entirely in just one physical quantity, the centime-

---

[2] Some of the numbers in this exercise illustrate a small problem that the SI is trying to get all scientists to handle in a uniform way. In part (h), for example, you might have gotten a bit dizzy trying to count closely spaced zeros. The SI recommends — and most European scientists have accepted the suggestion — that in numbers having four or more digits the digits be grouped in threes separated by thin spaces. For large numbers, just omit the commas. Thus 545,000,000 would be written as 545 000 000. The number 0.00000000000987 becomes 0.000 000 000 009 87. It will be a while before you see this usage very often in the United States, but when you do you'll now know what it means. Incidentally, European scientists use a comma instead of a period to locate the decimal point. You might see this yourself soon when you first weigh anything in the lab. If the weighing balance was made in Europe, a reading such as 1,045 g means 1.045 g.

ter. *Deci* is limited almost completely to another physical quantity, the deciliter (100 mL or $\frac{1}{10}$ L), and you won't see it often in strictly chemical situations. (Clinical chemists often use it because it saves space on clinical report sheets to abbreviate 100 mL to just 1 dL.)

To take advantage of the SI prefixes, we sometimes have to modify a rule used in converting a large or small number into scientific notation. The goal in this conversion will now be to get the exponential part of the number to match one with an SI prefix even if the decimal part of the number isn't between 1 and 10. For example, we know that the number 545,000 can be rewritten as 5.45×10^5, but 5 isn't divisible by 3, and there isn't an SI prefix to go with 10^5. If we counted 6 spaces to the left, however, we could use 10^6 as the exponential part.

$$5\,4\,5\,0\,0\,0. = 0.545 \times 10^6$$
$$\quad 6\ 5\ 4\ 3\ 2\ 1$$

Now we could rewrite 545,000 m as 0.545×10^6 m or 0.545 Mm (megameter), because the prefix *mega*, abbreviated M, goes with 10^6. We also could have rewritten 545,000 as 545×10^3, and then 545,000 m could have been written as 545 km (kilometers) because *kilo* goes with 10^3.

1 dL $= \frac{1}{10}$ liter

But $\frac{1}{10}$ liter $=$ 100 mL

Therefore,

1 dL = 100 mL

We usually put a zero in front of a decimal point in numbers that are less than 1, such as in 0.545. This zero just helps us remember the decimal point, and it doesn't count as a significant figure.

EXAMPLE 1.1   REWRITING PHYSICAL QUANTITIES USING SI PREFIXES

Problem: Bacteria that cause pneumonia have diameters roughly equal to 0.0000009 m. Rewrite this using the SI prefix that goes with 10^{-6}.

Solution: In straight exponential notation, 0.0000009 m is 9×10^{-7} m, but -7 is not divisible by 3 and no SI prefix goes with 10^{-7}. If we move the decimal six places instead of seven to the right, however, we get 0.9×10^{-6} m. The prefix for 10^{-6} is *micro* with the symbol μ, so

$$0.0000009\ m = 0.9 \times 10^{-6}\ m = 0.9\ \mu m$$

The diameter of one of these bacteria is 0.9 micrometers (0.9 μm).

PRACTICE EXERCISE 3   Complete the following conversions to exponential notation by supplying the exponential parts of the numbers.

(a) $0.0000398 = 39.8 \times 10^5$ (b) $0.000000798 = 798 \times 10^8$
(c) $0.000000798 = 0.798 \times$ ____   (d) $16500 = 16.5 \times$ ____

PRACTICE EXERCISE 4   Write the abbreviation of each of the following:

(a) milliliter   (b) microliter   (c) deciliter
(d) millimeter   (e) centimeter   (f) kilogram
(g) microgram   (h) milligram

PRACTICE EXERCISE 5   Write the full name that goes with each of the following abbreviations.

(a) kg   (b) cm   (c) dL   (d) μg
(e) mL   (f) mg   (g) mm   (h) μL

PRACTICE EXERCISE 6   Rewrite the following physical quantities using the standard SI abbreviated forms to incorporate the exponential parts of the numbers.

(a) 1.5×10^6 g   (b) 3.45×10^{-6} L   (c) 3.6×10^{-3} g
(d) 6.2×10^{-3} L   (e) 1.68×10^3 g   (f) 5.4×10^{-1} m

PRACTICE EXERCISE 7   Express each of the following physical quantities in a way that uses an SI prefix.
(a) 275,000 g   (b) 0.0000625 L   (c) 0.000000082 m

1.5 ACCURACY AND PRECISION

The way in which the number part of a physical quantity is expressed says something about the precision of the measurement but nothing about its accuracy.

Most people use the terms *accuracy* and *precision* as if they meant the same thing, but they don't. **Accuracy** refers to the closeness of a measurement (or the average of several measurements) to the true value. In an accurate measurement, the instrument is faithful and its user knows how to use it. **Precision** means the degree to which successive measurements agree with each other. It also means the fineness of the measurement when only one is made.

Figure 1.4 illustrates the difference between accuracy and precision in the measurement of someone's height. Each dot represents one measurement. In the first set of results, the dots are tightly clustered close to or exactly at the true value, and obviously a skilled person was at work with a carefully manufactured meter stick. This set illustrates both high precision and great accuracy. In the second set, a skilled person, without realizing it, evidently used a faulty meter stick, one that was mislabeled by a few centimeters. The precision is as great as that shown by the first set, because the successive measurements agree well with each other. But they're all untrue, so the accuracy is poor. In the third set of measurements, someone with a good meter stick did careless work. Only by accident do the values average to the true value,

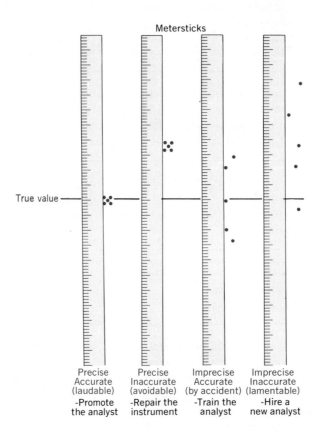

FIGURE 1.4
Accuracy and precision.

so the accuracy, in terms of the average, turned out to be high, but the precision is terrible and no one would really trust the average. The last set displays no accuracy and no precision.

No matter what physical quantities we use, we want to be able to judge how accurately and precisely they were measured. Concerning accuracy, we have a problem, because when we read the value of some physical quantity in a report or a table, we have no way of telling *from it alone* if it is the result of an accurate measurement. Someone might write, for example, "4.5678 mg of antibiotic," but in spite of all its digits we cannot tell from this report alone if the balance was working or if the person using it knew how to handle it and read it correctly. A skilled and careful experimenter frequently checks the instruments against references of known accuracy. Thus the question of *accuracy* is a human problem. We learn to trust the *accuracy* of data by employing trained people, giving them good instruments, requiring that they prove they are doing consistently accurate work, and rewarding consistently good results.

The Digits Known To Be Correct in a Quantity Are All But One of Its Significant Figures. We can indicate something about the precision of a measurement by the way we write its value. We do this by the way we round off the value to leave it with a certain number of *significant figures*. The number of **significant figures** in a physical quantity is the number of digits known with complete certainty to be correct plus one more. For example, the quantity "4.56 mg of antibiotic" has three significant figures. The first two, the 4 and the 5, are known to be correct, but the analyst is acknowledging a small uncertainty in the last digit. Unless otherwise stated, the uncertainty is assumed to be *one* unit of the last digit, so this report means that the mass of the sample is closer to 4.56 mg than to 4.55 mg or 4.57 mg. If the mass had been reported as 4.560 mg, then it has four significant figures. The 4, 5, and 6 are certainly true, but there is some uncertainty in the last digit, 0. The true mass is closer to 4.560 mg than to 4.559 mg or 4.561 mg. Thus "4.560 mg" discloses a greater precision or fineness or measurement than 4.56 mg. Sometimes you will see a report with a value such as 4.560 ± 0.001 mg or 4.560 ± 0.005. The symbol ± stands for "plus or minus," and what follows it, like 0.001 or 0.005, indicates how much uncertainty is carried in the last digit.

The ± sign in 4.560 ± 0.001 mg means that the physical quantity is in the range of 4.559–4.561 mg.

Figuring out how many significant figures there are in a number is easy, provided we have an agreement on how to treat zeros. Are all the zeros counted as *significant* in such quantities as 4,500,000 people, or 0.0004500 L, or 400,005 m? We will use the following rules to decide.

1. **Zeros sandwiched between nonzero digits are always counted as significant.** Thus both 400,005 and 400.005 have six significant figures.

   The quantities 4056 g and 4506 g both have four significant figures.

2. **Zeros that do no more than set off the decimal point on their *left* are never counted as significant figures.** Although such zeros are necessary to convey the general *size* of a quantity, they do not say anything about the *precision* of the measurement. Thus such quantities as 0.045 mL, 0.0045 mL, and 0.00045 mL all have only two significant figures.

   The other zeros are needed to locate the decimal points in these numbers, and they definitely are important in this sense. They just have nothing to do with precision.

3. **Trailing zeros to the *right* of the decimal point are always significant.** Trailing zeros are any that come to the right of a decimal point at the very end of the number, as in 4.56000. This number has three trailing zeros, and because they are to the right of the decimal point, all are significant. The number 4.5600 has six significant figures and represents considerable precision or fineness of measurement.

4. **Trailing zeros that are to the *left* of the decimal point are counted as significant only if the author of the book or article has somewhere said so.** The zeros in 4,500,000 are trailing zeros, but are they significant? Suppose this number stands for the population of a city. A city's population changes constantly as people are born and die, and as they move in and out. No one could claim to know a population is *exactly* 4,500,000 — not 4,499,999 and not 4,500,001, but 4,500,000. Most scientists handle this problem by restating the number in scientific notation so that any desired trailing zeros can be placed *after* the decimal point. By doing this, as many or as few of

such zeros can be given to convey the proper degree of precision. If the census bureau feels that the population is known to be closer to 4,500,000 than to 4,400,000 or to 4,600,000 people, and that no better precision than this is possible, then it should show only two significant figures in the result. It should report the population as 4.5×10^6 people. Giving the population as 4.50×10^6 people indicates greater precision — to three significant figures. There are four significant figures in 4.500×10^6.

Not everyone agrees with this way of handling trailing zeros that stand to the *left* of the decimal point, so you have to be careful. Some say that they aren't significant unless the decimal point is actually given, as in 45,000. L. When the decimal point is the last item in a number, however, it is easily forgotten at the time of making the record. This problem is avoided by switching to scientific notation so that all trailing zeros come after the decimal point. This is the practice we will usually follow in this book, unless noted otherwise, or unless the context makes the intent very clear.

A Few Rules Govern the Rounding Off of Calculated Physical Quantities. When we mathematically combine the values of two or more measurements, we usually have to round the result so that it has no more significant figures than allowed by the original data. Normally, such rounding is done at the *end* of a calculation (unless specified otherwise) to minimize the errors introduced by rounding. There are four simple rules for rounding.

1. When we multiply or divide quantities, the result can have no more significant figures than carried by the least precise quantity (the one with the fewest significant figures).

2. When we add or subtract numbers, the result can have no more decimal places than are in the number having the fewest decimal places.

3. When the first of the digits to be removed by rounding is 5 or higher, round the digit to its left *upward* by one unit. Otherwise, drop it and all others after it.

4. Treat exact numbers as having an infinite number of significant figures.

An *exact number* is any that we define to be so, and we usually encounter exact numbers in statements relating units. For example, all of the numbers in the following expressions are exact and, for purposes of rounding calculated results, have an infinite number of significant figures.

We use the period in the abbreviation of inch (in.) to avoid any confusion with the preposition *in,* which has the same spelling.

$$1 \text{ in.} = 2.54 \text{ cm (exactly, as defined by law)}$$
$$1 \text{ L} = 1000 \text{ mL (exactly, by the definition of mL)}$$

The significance of having an infinite number of significant figures lies in our not letting such numbers affect how we round results. It would be silly to say that the "1" in "1 L" has just one significant figure when we intend, by definition, that it be an exact number.

EXAMPLE 1.2   ROUNDING THE RESULT OF A MULTIPLICATION OR A DIVISION

Problem: A floor is measured as 11.75 m long and 9.25 m wide. What is its area, correctly rounded?

Solution:

$$\text{Area} = \text{length} \times \text{width}$$
$$= 11.75 \text{ m} \times 9.25 \text{ m}$$
$$= 108.6875 \text{ m}^2 \quad \text{(not rounded)}$$

Resist the impulse that some owners of new calculators have of keeping all the digits they paid for.

But the measured width, 9.25 m, has only three significant figures, whereas the length, 11.75 m, has four. We have to round the calculated area to three significant figures.

$$\text{Area} = 109 \text{ m}^2 \quad \text{(correctly rounded)}$$

EXAMPLE 1.3 ROUNDING THE RESULT OF AN ADDITION (OR A SUBTRACTION)

Problem: Samples of a medication having masses of 1.12 g, 5.1 g, and 0.1657 g are mixed. How should the total mass of the resulting sample be reported?

Solution: The sum of the three values, obtained with a calculator, is 6.3857 g, which shows four places following the decimal point. However, one mass is precise only to the first decimal place, so we have to round to this place. The final mass should be reported as 6.4 g. Notice that the value of the second sample mixed, 5.1 g, says nothing about the third or fourth decimal places. We don't know whether the mass is 5.06 or 5.14 g or any value in between. We only know that the mass if between 5.05 and 5.15 g. The sample just wasn't measured precisely. This is why we can't know anything beyond the first decimal place in the sum.

PRACTICE EXERCISE 8   The following numbers are the numerical parts of physical quantities. After the indicated mathematical operations are carried out, how must the results be expressed?

(a) 16.4×5.8

(b) $5.346 \div 6.01$

(c) 0.00467×5.6324

(d) $2.3000 - 1.00003$

(e) $16.1 + 0.004$

(f) $(1.2 \times 10^2) \times 3.14$

(g) $9.31 - 0.00009$

(h) $\dfrac{1.0010}{0.0011}$

1.6 THE FACTOR-LABEL METHOD IN CALCULATIONS

In calculations involving physical quantities, the units are multiplied or canceled as if they were numbers.

Many people have developed a mental block about any subject that requires the use of mathematics. They know perfectly well how to multiply, divide, add, and subtract, but the problem is in knowing *when,* and no pocket calculator tells this. We said earlier that the inch is defined by the relationship, 1 in. = 2.54 cm. This fact has to be used when a problem asks for the number of centimeters in some given number of inches, but for some people the problem arises in knowing whether to divide or multiply.

Science teachers have worked out a method called the *factor-label* method for correctly setting up such a calculation and *knowing* that it is correct. The **factor-label method** takes a relationship between units stated as an equation (such as 1 in. = 2.54 cm), expresses the relationship in the form of a fraction, called a **conversion factor,** and then multiplies some given quantity by this conversion factor. In this multiplication, identical units (the "labels") are multiplied or canceled as if they were numbers. If the remaining units for the answer are correct, then the calculation was correctly set up. We can learn how this works by doing an example, but first let's see how to construct conversion factors.

The relationship, 1 in. = 2.54 cm, can be restated in either of the following two ways and both are examples of conversion factors:

$$\frac{2.54 \text{ cm}}{1 \text{ in.}} \quad \text{or} \quad \frac{1 \text{ in.}}{2.54 \text{ cm}}$$

If we read the divisor line as "per," then the first conversion factor says "2.54 cm per 1 in." and the second says "1 in. per 2.54 cm." These are merely alternative ways of saying that "1 in. equals 2.54 cm." Any relationship between two units can be restated as two conversion factors. For example,

Some call the factor-label method the cancel-unit or the factor-unit method.

When we divide both sides of the equation 2.54 cm = 1 in. by 2.54 cm, we get:

$$\frac{2.54 \text{ cm}}{2.54 \text{ cm}} = \frac{1 \text{ in.}}{2.54 \text{ cm}}$$

This only restates the relationship of the centimeter and the inch, it doesn't change it. The use of a conversion factor just changes units, not actual quantities.

$$1 \text{ L} = 1000 \text{ mL} \quad \frac{1000 \text{ mL}}{1 \text{ L}} \quad \text{or} \quad \frac{1 \text{ L}}{1000 \text{ mL}}$$

$$1 \text{ lb} = 453.6 \text{ g} \quad \frac{453.6 \text{ g}}{1 \text{ lb}} \quad \text{or} \quad \frac{1 \text{ lb}}{453.6 \text{ g}}$$

PRACTICE EXERCISE 9   Restate each of the following relationships in the forms of their two possible conversion factors.

(a) 1 g = 1000 mg   (b) 1 kg = 2.205 lb

Suppose we want to convert 5.65 in. into centimeters. The first step is to write down what has been given, 5.65 in. Then we multiply this by the one conversion factor relating inches to centimeters that lets us cancel the unit no longer wanted and leaves the unit we want. We therefore choose the converson factor with "in." in the denominator because we must cancel "in." in the numerator.

$$5.65 \cancel{\text{ in.}} \times \frac{2.54 \text{ cm}}{1 \cancel{\text{ in.}}} = 14.4 \text{ cm} \quad \text{(rounded correctly from 14.351 cm)}$$

Notice how the units of "in." cancel. Only "cm" remains, and it is on top in the numerator where it has to be. Suppose we had used the wrong conversion factor.

The arithmetic is correct, but the result is still all wrong.

$$5.65 \text{ in.} \times \frac{1 \text{ in.}}{2.54 \text{ cm}} = 2.22 \frac{(\text{in.})^2}{\text{cm}} \quad \text{(correctly rounded)}$$

That's right. We *must* do to the units exactly what the times sign and the divisor line tell us, and (in.) times (in.) equals (in.)^2 just as $2 \times 2 = 2^2$. Of course, the units in the answer, (in.)^2/cm, make no sense, so we know with certainty that we cannot set up the solution this way. The reliability of the factor-label method lies in this use of the units (the "labels") as a guide to setting up the solution. Now let's work an example.

EXAMPLE 1.4   USING THE FACTOR-LABEL METHOD

Problem: How many grams are in 0.230 lb?

Solution: From Table 1.2, we find that 1 lb = 453.6 g, so we have our pick of the following conversion factors.

$$\frac{453.6 \text{ g}}{1 \text{ lb}} \quad \text{or} \quad \frac{1 \text{ lb}}{453.6 \text{ g}}$$

To change 0.230 lb into grams, we want "lb" to cancel and we want "g" in its place in the numerator. Therefore, we pick the first conversion factor; it's the only one that can give this result.

$$0.230 \cancel{\text{ lb}} \times \frac{453.6 \text{ g}}{1 \cancel{\text{ lb}}} = 104 \text{ g} \quad \text{(correctly rounded)}$$

There are 104 g in 0.230 lb. (We rounded from 104.328 g to 104 g because the given value, 0.230 lb, has only three significant figures. Remember that the "1" in "1 lb" has to be treated as an exact number because it's in a definition.)

PRACTICE EXERCISE 10   The *grain* is an old unit of mass still used by some pharmacists and physicians, and 1 grain = 0.0648 g. How many grams of aspirin are in an aspirin tablet containing 5.00 grain of aspirin?

On page 8, equations were given relating degrees Celsius and degrees Fahrenheit. The use of these equations illustrates further how units no longer wanted cancel each other. This can be demonstrated by using these equations to work the following practice exercises.

PRACTICE EXERCISE 11   A child has a temperature of 104 °F. What is this in degrees Celsius?

PRACTICE EXERCISE 12   If the water at a beach is reported at 15 °C, what is this in degrees Fahrenheit? (Would you care to swim in it?)

1.7 HEAT ENERGY

Heat is a factor in most chemical changes, and one common unit of heat is the calorie.

If you place two objects with different temperatures together, their temperatures eventually equalize. The colder object becomes warmer and the warmer becomes colder. Something flows between the two to make their temperatures equal, and what transfers is not matter but one of several forms of energy called **heat.**

Energy of Motion Is Called Kinetic Energy.   We define **energy** as a capacity for causing change. For example, a falling rock has energy because it can make other objects in its way move aside or break apart. The energy associated with motion is called **kinetic energy,** and its amount depends on both the mass and the velocity of what is in motion. The more massive is the rock or the more rapidly it is moving, the more it is able to cause changes, and its kinetic energy is related to its mass (m) and its velocity (v) by the following equation:

Kinetic is from the Greek *kinetikos,* meaning ''of motion.''

$$\text{kinetic energy} = \tfrac{1}{2}mv^2$$

If the rock is a piece of hot, newly solidified volcanic lava, it has the capacity to cause any cooler object nearby to become warmer. It has a temperature-changing capacity, and we say that it has thermal energy or heat.

If the rock is so hot that it glows, it has an illumination-changing capacity. It emits light, still another form of energy.

During its fall, the rock might also make a sound, especially if it cracks apart as it cools. Now the rock has a capacity to change the noice level, and it emits sound energy.

The rock might land on a narrow ledge part way down the face of a cliff. Although it is motionless, soundless, and soon cool, no one would linger long beneath the ledge because the rock still has energy. It still can cause change, although this capacity is inactive; it's only a *potential.* Energy in storage as in the rock on a ledge is called **potential energy,** an inactive form of energy related to location or, as we'll see next, to chemical makeup.

Chemical Energy Is One Kind of Potential Energy.   The rock that we have endowed with so many kinds of energy has still another. If the rock were an iron meteorite, and it fell into a vat of concentrated acid instead of onto the ledge, we would see all sorts of interesting changes. The acid would become hotter, a gas would bubble away from the iron's surface, and the iron would dissolve. If the rock, however, were a gold nugget, none of these changes would happen. Thus, the iron meteorite has a potential for causing change simply because it is iron. Potential energy that exists because of composition is called **chemical energy.**

The Calorie Is a Common Unit of Heat Energy.   Our foods have chemical energy, and our bodies can convert them into other forms such as electrical energy (when the nervous system works), kinetic energy (when muscles flex), sound energy (when we speak), and into the chemical energy of other substances, which the body makes from foods.

One way to measure the chemical energy in a food is to burn a weighed sample and see by how much the evolving heat changes the temperature of a weighed sample of water. This works because heat is the one form of energy into which all other forms can be entirely converted. This is why scientists developed a general unit of energy related specifically to heat, the **calorie.** One calorie of energy is the energy that can change the temperature of one gram of water by one degree Celsius (specifically, the one degree between 14.5 and 15.5 °C). This isn't much energy, so a multiple, the kilocalorie (kcal), is often used.

$$1000 \text{ cal} = 1 \text{ kcal}$$

Each Substance Has a Thermal Property Called Its Specific Heat. You probably know that our bodies have to control their temperatures carefully, yet they must function both in very hot climates, when heat flows into the body, and very cold conditions, when the body loses heat. Substances differ widely in their abilities to absorb (or lose) heat when the temperature changes, and to compare these abilities we use the concept of specific heat. The **specific heat** of a substance is the amount of heat that changes the temperature of one gram of the material by one degree Celsius. Since, by definition, one calorie changes the temperature of one gram of water by one degree Celsius, the specific heat of water is one calorie per gram per degree.

$$\text{specific heat of water} = \frac{1 \text{ calorie}}{\text{gram } °C}$$

The specific heat of water is higher than for almost any other known material. For example, the heat that changes the temperature of 1 g of water by 1 °C changes that of 1 g of iron by 10 °C. The specific heat of iron is $0.10 \dfrac{\text{cal}}{\text{g } °C}$. Thus, on an equal mass basis, iron has one-tenth the ability of water to absorb heat for the same change in temperature.

Water's High Specific Heat Is Important to Us. The adult body is about 60% water, and because of water's high specific heat, we can absorb heat or lose it with little life-threatening changes in temperature. Water's high specific heat also explains why cold water is such a good coolant, which we know by experience when we reach for a cooling drink or plunge into a pool on a hot day. The management of the body's heat budget is a vital topic at the molecular level of life, and we will return to it in a later chapter.

Specific Heats of Common Substances

Substance	Specific Heat (cal/g °C)
Alcohol	0.58
Gold	0.03
Granite	0.19
Iron	0.10
Olive oil	0.47
Water	1.0

1.8 DENSITY

One of the important physical properties of a liquid is its density, its amount of mass per unit volume.

An Object's Density Is the Ratio of Its Mass to Its Volume. One useful intensive property of a substance, particularly if it is a fluid, is its density. **Density** is the mass per unit volume of a substance.

$$\text{Density} = \frac{\text{mass}}{\text{volume}}$$

The density of mercury, the silvery liquid used in most thermometers, is 13.60 g/mL, which makes mercury one of the most dense substances known. The density of liquid water is 1.0 g/mL.

Don't make the mistake of confusing *heaviness* with *denseness*. A pound of mercury is just as heavy as a pound of water or a pound of feathers, because a pound is a pound. But a pound of mercury occupies only $\frac{1}{13.6}$ the volume of a pound of water.

The density of a substance varies with temperature, because for samples of most substances the volume but not the mass changes with temperature. Most substances expand in volume when warmed and contract when cooled. The effect isn't great if the substance is a liquid or a solid. The density of mercury, for example, changes only from 13.60 to 13.35 g/mL when its temperature changes from 0 to 100 °C, a density change of only about 2%. Notice from the data in the margin how the density of water, to two significant figures, is 1.0 g/mL in the (liquid) range of 0 to 30 °C (32–86 °F).

Water is unusual in that its density *decreases* when it is cooled in the range between 3.98 and 0 °C, where it freezes. This decrease occurs because of a small increase in volume, the term in the denominator of the equation defining density. Then when liquid water freezes, it expands another 10% in volume. Ice, therefore, has a density less than liquid water, so ice floats on water. If this did not happen, the entire ecology of northern lakes and the Artic ocean would be different. If their waters froze and the ice sank instead of floated, the ice would likely never melt.

One of the uses of density is in calculating what volume of a liquid to take when the problem or experiment specifies a certain mass. Often it is easier (and sometimes safer) to measure a volume than a mass, as we will note in the next example.

Density of Water at Various Temperatures

Temperature (°C)	Density (g/mL)
0	0.99987
3.98	1.00000
10	0.99973
20	0.99823
30	0.99567
40	0.99224
60	0.98324
80	0.97183
100	0.95838

EXAMPLE 1.5   USING DENSITY TO CALCULATE VOLUME FROM MASS

Problem: Concentrated sulfuric acid is a thick, oily, and very corrosive liquid that no one would want to spill on the pan of an expensive balance to say nothing of the skin. It is an example of a liquid that is usually measured by volume instead of by mass, but suppose an experiment called for 25.0 g of sulfuric acid. What volume (in milliliters) should be taken to obtain this mass? The density of sulfuric acid is 1.84 g/mL.

Solution: The given value of density means that 1.84 g acid = 1.00 mL acid. This gives two possible conversion factors:

$$\frac{1.84 \text{ g acid}}{1 \text{ mL acid}} \quad \text{or} \quad \frac{1 \text{ mL acid}}{1.84 \text{ g acid}}$$

The "given" in our problem, 2.50 g of acid, should be multiplied by the second of these conversion factors to get the unit we want, mL.

$$25.0 \text{ g acid} \times \frac{1 \text{ mL acid}}{1.84 \text{ g acid}} = 13.6 \text{ mL acid}$$

Thus if we measure 13.6 mL of acid, we will obtain 25.0 g of acid. (The pocket calculator result is 13.58695652, but we have to round to three significant figures.)

PRACTICE EXERCISE 13   An experiment calls for 16.8 g of methyl alcohol, the fuel for fondue burners, but it is easier to measure this by volume than by mass. The density of methyl alcohol is 0.810 g/mL, so how many milliliters have to be taken to obtain 16.8 g of methyl alcohol?

PRACTICE EXERCISE 14   After pouring out 35.0 mL of corn oil for an experiment, a student realized that the mass of the sample also had to be recorded. The density of the corn oil is 0.918 g/mL. How may grams are in the 35.0 mL?

The **specific gravity** of a liquid is the ratio of the mass contained in a given volume to the mass of the identical volume of water at the same temperature. If we arbitrarily say that the "given volume" is 1.0 mL, then the water sample has a mass of 1.0 g (or extremely close to this over a wide temperature range). This means that dividing the mass of some liquid that occupies 1.0 mL by the mass of an equal volume of water is like dividing by 1, but all the units cancel. Specific gravity has no units, and a value of specific gravity is numerically so close to its density that we usually say they are numerically the same. This fact has resulted in a rather limited use of the concept of specific gravity, but one use occurs in medicine.

In clinical work, the idea of a specific gravity surfaces most commonly in connection with urine specimens. Normal urine has a specific gravity in the range of 1.010–1.030. It's slightly higher than water because the addition of wastes to water usually increases its mass more rapidly than its volume. Thus the more wastes in 1 mL of urine the higher is its specific gravity.

Figure 1.5 shows the traditional method to measure the specific gravity of a urine specimen, by using a urinometer. Its use, however, has largely been supplanted by a method, the use of a refractometer, that needs only one or two drops of urine for the measurement. The refractometer is an instrument that measures the ratio of the speed of light through air to its speed through the sample being tested. This ratio can be correlated with the concentration of dissolved substances in the urine. (*How* the refractometer does this is beyond the scope of our study.)

One of the important functions of the kidneys is to remove chemical wastes from the blood stream and put them into the urine being made. The kidneys' mechanism for doing this does not remove those substances from the blood that ought to remain in the blood. The clinical significance, therefore, of a change in the concentration of substances dissolved in the urine is that it indicates a change in the activity of the kidneys. This might be the result of kidney disease so that substances that should stay in the blood leak into the urine being made. Or it might indicate that, somewhere else in the body, wastes are being generated more rapidly than the kidneys can remove them.

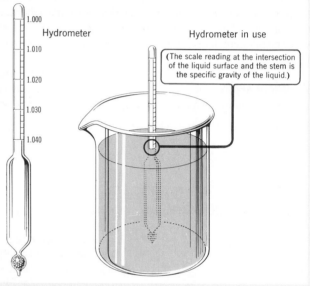

FIGURE 1.5
A hydrometer designed to serve as a urinometer.

Specific gravity is a property of a fluid that is very similar to density. It is not used very often in chemistry, but in clinical work the specific gravity of liquid specimens (such as urine) helps to reveal the nature of an illness, as described in Special Topic 1.1.

SUMMARY

Chemistry and the molecular basis of life   Down at the level of nature's tiniest particles, we find the "parts"—molecules—that nature shuffles from organism to organism in the living world. One of the many ways of looking at life is to examine its molecular basis, the way in which well-being depends on chemicals and their properties.

Physical quantities   Physical properties are those that can be studied without changing the substance into something else. They include mass, volume, time, temperature, color, and density. For our purposes, the important base quantities are (with the names of the SI base units given in parentheses) mass (kilogram), length (meter), time (second), temperature degree (kelvin), and quantity of chemical substance (mole).

Special prefixes can be attached to the names of the base units to express multiples or submultiples of these units. To select a prefix we have to be able to put very large or very small numbers into scientific notation.

Precision and accuracy   Whether we obtain data from direct measurements or by calculations, we have to be careful not to imply too much precision by using the incorrect number of significant figures. When we add or subtract numbers, the decimal places in the result can be no more than the least number of decimal places among the original numbers. When we multiply or divide, we have to round the result to show the same number of significant figures as are in the least precise original number.

Factor-label method The units of the physical quantities involved in a calculation are multiplied or canceled as if they were numbers. To convert a physical quantity into its equivalent in other units, we multiply the quantity by a conversion factor that permits the final units to be correct. The conversion factor is obtained from a defined relationship between the units.

Heat energy Heat is the energy that transfers from one object to another when there is a difference in temperature between them. When the heat raises the temperature of one gram of water one degree Celsius, one calorie of energy has moved. Each substance has the thermal property of specific heat, the number of calories per gram of the substance that transfer when the temperature changes by one degree Celsius. The high specific heat of water helps living systems maintain steady temperatures.

Density Density is the ratio of mass to volume. Since volumes of substances change with temperature, density is temperature dependent. The change, however, is usually quite small.

REVIEW EXERCISES

The answers to review exercises that require a calculation and whose numbers are marked with an asterisk are given in Appendix V. The answers to all other review exercises are given in the *Study Guide to Elements of General and Biological Chemistry,* 8th edition.

Molecular Basis of Life

1.1 Give some common experiences that illustrate that at some deep level of existence we can use the same "parts" as, say, a kitten.

1.2 Life, besides having a molecular basis, can be studied in terms of other bases. What are some? (One, for example, is the psychological basis.)

Physical Quantities, Properties, and Measurements

1.3 What is meant by the word *property?*

1.4 How does a *physical property* differ from a *chemical property?*

1.5 How does a physical quantity differ from a number?

1.6 What is meant by the *inertia* of some object, and how is its inertia related to its mass?

1.7 Why is *volume* considered to be a less basic quantity than *length?*

1.8 What general name do we give to any fundamental quantity in terms of which less fundamental quantities are defined?

1.9 Name the five base quantities to be used in this book.

1.10 What is the name of the base unit for each of the following quantities?
   (a) time   (b) mass
   (c) temperature degree   (d) length
   (e) quantity of chemical substance

1.11 What relationship does a *reference standard* have to a *base quantity?*

1.12 What are some of the properties that a reference standard should have, ideally?

1.13 Which reference standard in the SI is least ideal in terms of its being secure from any kind of physical or chemical loss?

1.14 What are the names and symbols for the common submultiples that we will use in this book for each of the following base units?
   (a) meter   (b) liter   (c) kilogram

1.15 How many centimeters make 1 m?

1.16 How many millimeters make 1 cm?

1.17 How many grams are in 1 kg?

1.18 How many milligrams are in 1 g?

1.19 Is the yard slightly shorter or slightly longer than the meter?

1.20 Is the liquid quart slightly smaller or slightly larger than the liter?

Temperature Degrees and Temperature Scales

1.21 The value $-273.15\ °C$ is a peculiar number to pick to be equivalent to 0 K. Why was it selected?

1.22 Suppose you took an unmarked, unetched thermometer and immersed it in a slush of ice and water until the mercury level stopped changing and then you marked the mercury level with a wax crayon. What is the name of this line on the Celsius scale? On the Fahrenheit scale? On the Kelvin scale?

1.23 Suppose that you repeated the experiment of Exercise 1.22 only this time immersed the unmarked thermometer in boiling water (at sea level), and you marked the mercury level after it stopped changing. What is the name given to this mark in each of the three temperature scales?

1.24 How many scale divisions are there between the two marks created by the experiments described in Exercises 1.22 and 1.23 when the distances are subdivided into (a) Fahrenheit degrees, (b) Celsius degrees, and (c) kelvins?

1.25 How do the kelvin and the degree Celsius compare in size?

1.26 How do the degree Celsius and the degree Fahrenheit compare in size?

*1.27 Water freezes at 0 °C and it boils at 100 °C. What are these values in kelvins?

1.28 Many people would find a temperature of 294 K to be comfortable. What is this in degrees Celsius?

*1.29 A German recipe calls for baking batter at 210 °C. What Fahrenheit setting should you use?

1.30 A weather report from a station in Alaska gave the outside temperature as -40 °F. What is this in degrees Celsius?

*1.31 The pool temperature at an Austrian health spa is 31 °C. What is this in degrees Fahrenheit?

1.32 An infant's temperature is 38.5 °C. Is this normal? (Do a calculation. Normal temperature is considered to be 98.6 °F.)

*1.33 Wishing to make an outside temperature of 109 °F seem less hot to a friend visiting from Europe, you tell him what it is in °C. What number do you report? (Will this gesture work?)

1.34 Anesthetic ether is a vapor above 94.3 °F. What is this in degrees Celsius?

SI Prefixes and Scientific Notation

1.35 Rewrite the following physical quantities with their units abbreviated.
  (a) 26 micrograms of vitamin E
  (b) 28 millimeters wide
  (c) 5.0 deciliters of solution
  (d) 55 kilometers in distance
  (e) 46 microliters of solution
  (f) 64 centimeters long

1.36 Rewrite the following physical quantities with their units written out in full.
  (a) 125 mg of water   (b) 25.5 mL of coffee
  (c) 15 kg of salt   (d) 12 dL of iced tea
  (e) 2.5 μg of ozone   (f) 16 μL of fluid

1.37 Rewrite the following data in scientific notation in which the decimal part of the number is between 1 and 10.
  (a) 0.013 L   (b) 0.000006 g
  (c) 0.0045 m   (d) 1455 s

1.38 Express each of these quantities in scientific notation and limit the decimal part of the number to a number between 1 and 10.
  (a) 24,605 m   (b) 654,115 g
  (c) 0.0000000095 L   (d) 0.00000568 s

1.39 Use a suitable SI prefix to express each of the quantities in Exercise 1.37.

1.40 Reformulate the numbers in the quantities of Exercise 1.38 so that they can be expressed in units that employ suitable SI prefixes.

Accuracy, Precision, and Significant Figures

1.41 According to one almanac, the population of the United States in 1790 was 3,939,214. Rewrite this figure in scientific notation retaining only three significant figures.

1.42 When a meter stick was used to take five successive measurements of a person's height, the following data were recorded: 172.7 cm, 172.9 cm, 172.6 cm, 172.6 cm, 172.8 cm. The meter stick had earlier been checked against an official reference standard and found to be good. The true value of the height was verified as 172.7 cm.
  (a) Can the measurements be described as *accurate?* Why?
  (b) Can they be described as *precise?* Why?

1.43 Examine the following numbers.
  (A) 3.7200×10^3 (B) 3720   (C) 3.720
  (D) 0.03720   (E) 37,200   (F) 0.00372
  (G) 3.720×10^3 (H) 0.0372   (I) 3.72×10^9

  (a) Which of these numbers has three significant figures? Identify them by their letters.
  (b) Which has four significant figures?
  (c) Which of them has five significant figures?

1.44 If we multiply $3.4462 \times 55.1 \times 10^8$, how many significant figures can we permit the answer to have?

1.45 If we add 0.00014 to 1.36, how many places after the decimal point can we permit to stand in the answer?

1.46 Rewrite the following number according to the number of significant figures specified by each part.

$$144,549.09$$

  (a) seven   (b) five   (c) four
  (b) three   (e) one   (f) two

1.47 The relationship between the gram and the microgram is given by

$$1 \text{ g} = 1,000,000 \ \mu\text{g}$$

How many significant figures are considered to be in each number?

Converting between Units

(*Note:* In all problems unless noted otherwise, *lb* or *pound* refer to the avoirdupois system, *lb avdp.*)

1.48 Write each of the following relationships between units for physical quantities in the forms of two conversion factors.
  (a) 1 mile = 5280 ft
  (b) 1 dram = 60 grains
  (c) 1 lb = 453.6 g
  (d) 1 ounce = 480 grains
  (e) 1 m = 39.37 in.
  (f) 1 liquid ounce = 480 minims

1.49 Given the relationships of Exercise 1.48, which of the two following calculated answers is more likely to be correct in each part. You should be able to make this kind of judgment without actually doing a calculation.
  (a) 250 minims = 0.520 liquid ounce or 1.20×10^4 liquid ounce
  (b) 3.50 ounce = 0.00729 grain or 1.68×10^3 grain
  (c) 0.350 lb = 159 g or 0.000772 g

1.50 What are the units in the result of the following calculation?

$$1.0 \text{ kg} \times 1 \frac{\text{m}}{\text{s}} \times 1 \frac{\text{m}}{\text{s}}$$

(The resulting units are the SI units for energy, a derived quantity.)

*1.51 Convert each of the following physical quantities into the units specified. Use tables in this chapter to find relationships between units.
  (a) 75.5 in. into centimeters (the height of an adult male)
  (b) 50.5 kg into pounds (the mass of an adult female)

1.52 Using relationships between units found in this chapter, convert each of the following quantities into the new units specified.
(a) 70.0 kg into pounds (the mass of an adult male)
(b) 64.0 in. into centimeters (the height of an adult female)

*1.53 A 500-mL bottle of soda contains how many liquid ounces (to three significant figures)?

1.54 If a gas tank holds 16.0 U.S. gallons, how many liters does it hold?

*1.55 Driving a car with a mass of 4.6×10^3 lb, you come to a bridge with a sign that reads, "Closed to all vehicles weighing more than 1.5×10^3 kg." Should you cross? (Do the calculation.)

1.56 A foreign car has a mass of 915 kg. What is this in pounds?

*1.57 If you can get 32 equal-sized butter pats from a one-quarter-pound (0.25 lb) stick of butter, what is the mass of each pat in grams?

1.58 The *carat* is a measurement jewelers use to describe the mass of a precious stone. 1 carat = 200 mg. What is the mass in milligrams of a diamond rated as 0.750 carat?

*1.59 Mount Everest in Nepal is the highest mountain in the world — 29,028 ft. What is this in meters? In kilometers?

1.60 The highest mountain in the United States is Alaska's Mount McKinley at 6194 m. How high is this in feet? In miles?

Heat Energy

1.61 What condition is necessary in order for us to say that heat flows from some object into another?

1.62 What is the significance of the unit of *heat* energy when it comes to the measurement of other forms of energy?

1.63 How many kilocalories will raise the temperature of 1.0 kg of water from 14.5 to 15.5 °C?

1.64 What is the specific heat of water, and what is the significance of this value being unusually high when compared with other substances?

Density

1.65 If dissolving something in water increased the mass of the system by 0.5 g for each 0.5 mL increase in its volume, how would the density be affected?

1.66 Within limits, dissolving a solid in water increases the masses of the system more rapidly than the volume. What does this do to the density of the system, cause it to increase, decrease, or remain the same?

*1.67 Taking the density of lead to be 11.35 g/cm³, how many pounds of lead fill a milk container with a volume of 1.00 qt?

1.68 Aluminum has a density of 2.70 g/cm³. A milk carton with a volume of 1.00 qt could hold how many grams of aluminum? How many kilograms? How many pounds?

*1.69 To avoid spilling any liquid onto an expensive balance, a student obtained 30.0 g of acetic acid by measuring a corresponding volume. The density of acetic acid is 1.06 g/mL. What volume in milliliters was taken?

1.70 In an experiment to test how well methyl alcohol works as an antifreeze, a technician took 275 mL of this liquid. Its density is 0.810 g/mL. How many grams of methyl alcohol were taken?

Specific Gravity (Special Topic 1.1)

1.71 What is the difference between density and specific gravity?

1.72 What fact about water makes the density of some object and its specific gravity the same (or very nearly so, depending on the number of significant figures used)?

1.73 In which fluid does a urinometer float sink farther, one of low density or one of high density?

1.74 It is possible to dissolve 2.0 g of salt in 100 mL of water with hardly any change in the total volume. Calculate the specific gravity of the resulting salt solution if no change in volume occurs.

1.75 An abnormally high specific gravity of a urine specimen tells us what about the specimen?

Chapter 2
The Nature of Matter: The Atomic Theory

In the popular imagination, the atom is seen as a miniature solar system with electrons whizzing around an atomic nucleus. A better model is studied in this chapter.

2.1 MATTER, ITS KINDS AND STATES

Elements and compounds have definite compositions; mixtures do not.

Matter is anything that occupies space and has mass, and there are three kinds of matter called elements, compounds, and mixtures.

Elements Are the Building Blocks of Compounds. **Elements** are substances that cannot be broken down into anything simpler, yet stable. Familiar examples are aluminum, iron, copper, silver, and gold. **Compounds** are invariably made from two or more elements, and they obey the **law of definite proportions,** an important law of chemical combination.

Stable here means that it can be stored at room temperature.

> Law of Definite Proportions In a given chemical compound, the elements are invariably combined in the same proportion by mass.

Water is a typical compound. Regardless of its source, when water is broken down into its elements, hydrogen and oxygen, they are always obtained in a mass ratio of 1.01 g of hydrogen to 8.00 g of oxygen.

Substance	Freezing point
Water	0°C
Oxygen	−218 °C
Hydrogen	−259 °C

Mixtures Do Not Obey the Law of Definite Proportions. Using great care to exclude sparks and ultraviolet (UV) light (the kind that gives a sunburn), hydrogen and oxygen can be blended into a mixture without producing water. Like the blending of perfume with air, **mixtures** are invariably made by the blending of two or more compounds or elements in any relative amounts. This blending is just a **physical change,** which means any change unaccompanied by a chemical reaction. The separation of a mixture into its original parts is likewise done just by physical changes.

If we made a mixture of hydrogen and oxygen in a mass ratio of 1.01 g of hydrogen to 8.00 g of oxygen and then exposed it to a spark, the mixture would detonate. When everything cooled down, only droplets of water would be present. No oxygen or hydrogen would remain. (Had we used different mass ratios, we would have still obtained water, but either some hydrogen or some oxygen would have been left over.) The spark set off the change of the two elements into a compound, and this change is an example of a *chemical reaction.* As we learned in Chapter 1, this is an event in which substances change into other substances. The substances that change, like hydrogen and oxygen, are called the **reactants** and the substances that form are called the **products.**

A chemical reaction is the *only* way that a compound can be made from its elements, and it's the only way that a compound can be broken down into its elements. There are many kinds of chemical reactions each with hundreds of examples, and we'll begin our study of a selected few in the next chapter.

The reaction of hydrogen with oxygen that gives water also generates heat, so it's described as an **exothermic reaction.** Many reactions do not happen unless the reactants are continuously heated. In other words, they absorb heat, and are called **endothermic reactions.** Some of the chemical reactions that convert cake batter into a cake are endothermic.

One of the major features of all chemical reactions is the law of conservation of mass.

> Law of Conservation of Mass Mass is neither gained nor lost in a chemical reaction; mass is conserved.

When 1.01 g of hydrogen combines, for example, with 8.00 g of oxygen, exactly 9.01 g of water forms. There is no loss in mass as the reactants change into the product.

Kinds and *states* are different. The kinds of matter are elements, compounds, and mixtures. The states are solid, liquid, and gas.

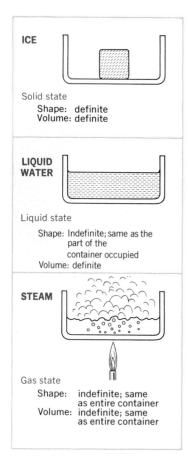

FIGURE 2.1
The three states of matter, as illustrated by water.

The Three Kinds of Matter Can Occur in Any One of Three Physical States. Everyone is familiar with ice, liquid water, and steam (Figure 2.1), and these illustrate the three possible physical **states of matter.** Matter in the **gaseous state** depends entirely on its container for both the shape and the volume of the sample. In the **liquid state,** matter takes up whatever shape its container has, but the sample has a definite volume that does not depend on the container. In the **solid state,** matter has both a definite shape and a definite volume regardless of what holds it.

The changes that convert water into steam or ice are physical, not chemical changes, because they don't make water into a compound that isn't water.

We've introduced a number of fundamental terms in this section, which we will use throughout our study. They apply to substances in bulk, to samples that we can handle and see. We will now shift levels, and study atoms, the smallest particles of matter. Remember that we're interested in the molecular basis of life. Molecules are made of atoms, and to understand molecules, we need to know atoms.

2.2   ATOMIC THEORY

Atoms of the same element have the same average mass; those of different elements have different average masses.

Centuries ago some Greek philosophers wondered whether matter, such as a piece of gold, could be broken into an infinite number of pieces. Or is there a limit? Is there some small piece that, if broken, would not give simply smaller pieces *of the same substance?* They decided that there is a limit, that gold, for example, consists of tiny invisible pieces that cannot be cut further, even in the imagination, and still give *gold* pieces. The Greek word for "not cuttable" is *atomos,* and from it we get the modern word, *atom.*

The Laws of Chemical Combination Support Our Belief in Atoms.   The Greek atom was only an idea. Scientific evidence came centuries later, emerging slowly as the laws of definite proportion and conservation of mass became established. John Dalton (1766–1844), an English chemist, reasoned that these laws virtually compel us to believe in the existence of the tiny unbreakable particles that the Greeks called atoms. He said that each kind of atom must have an unchangeable mass, and that none of this mass is lost as atoms form compounds. Thus if compounds are made from a definite proportion *by atoms,* they cannot help but have a *definite* proportion *by mass.* Here is the full list of Dalton's postulates. We'll use them as a starting point for our discussion of atoms and then see what modifications were required by later discoveries.

John Dalton

Dalton's Atomic Theory

1. Matter consists of definite particles called atoms.
2. Atoms are indestructible. They can become relocated in chemical reactions, but not broken up.
3. All atoms of a particular element are identical in mass.
4. Atoms of different elements are different in mass.
5. By joining together in different ratios as whole particles, atoms form different compounds.

Keep in mind that what we are working toward here is the structure of the atom, because we can better understand the chemical properties of both elements and compounds with this knowledge.

Atoms Are Made of Subatomic Particles.   Contrary to Dalton, atoms are not indestructible. To break an atom apart, however, takes the enormous energies available only in atom-smashing machines and not in chemical reactions. When atoms are broken apart, the debris includes *particles found in the atoms of all elements.* This remarkable fact makes the study of atomic structure vastly simpler.

Electrons, Protons, and Neutrons are Subatomic Particles.   Three of the **subatomic particles** produced by atom-smashing experiments are especially important to the chemistry of the elements: the **electron,** the **proton,** and the **neutron.** All three have mass. Two, the electron and the proton, are also electrically charged. This means that they can exert pushes and pulls, forces of repulsion and attraction, on each other without physically touching. The charge on the proton has the same *intensity* as that of the electron but is opposite in *character.*

Objects whose electrical charges are opposite in character experience a mutual attraction for each other. It's a law of nature: *unlike charges attract.* An electron and a proton thus attract each other. But another law of nature is that *like charges repel.* A proton therefore repels another proton, and an electron repels another electron. We'll use these two laws about charged particles often as we explain how atoms can join to make compounds.

The *amount* (intensity) of charge on a proton or an electron is defined as one unit, and so to symbolize their equal but opposite electrical charges we use the symbols $1+$ and $1-$. The proton has a charge of $1+$ and the electron's charge is $1-$.

Table 2.1 gives the masses of the three subatomic particles. Their values in grams, have small and awkward numbers, so there's an advantage to defining a new unit of mass, the **atomic mass unit** or **amu.**

$$1 \text{ amu} = 1.6605665 \times 10^{-24} \text{ g}$$

In amu, the masses of the proton and the neutron, when rounded to two significant figures, are both 1.0 amu. For example, for the proton, we have

$$1.67262305 \times 10^{-24} \frac{\text{g}}{\text{proton}} \times \frac{1 \text{ amu}}{1.6605665 \times 10^{-24} \text{ g}} = 1.0072605 \text{ amu/proton}$$

This rounds to 1.0 amu for the mass of one proton, a much easier number to use. The mass of the electron, a much lighter particle, is 1/1836 that of the proton. The practical result of the much smaller mass of an electron is that we can ignore the contributions that electrons make to the masses of whole atoms.

An Atom's Protons and Neutrons Make Up Its Atomic Nucleus.   Early in this century, British scientists led by Lord Rutherford (1871–1937) discovered that atoms are mostly empty space, that essentially all of the atom's mass is in an extremely dense central core, a particle that they named the **nucleus.** It has all of the atom's heavy, subatomic particles, its protons and neutrons. Because the atom's protons are in its nucleus, the nucleus also has all of the atom's positive charge. Since the charge on one proton is $1+$, the total nuclear charge equals the number of protons.

Charge on an atomic nucleus = number of protons

Physicists are presently studying the structure of the proton.

Like charges repel. The comb and the hair have like charges.

Even the 108 electrons in the largest known atom contribute only 0.02% to the atom's mass.

The 1908 Nobel prize in chemistry went to Ernest Rutherford.

TABLE 2.1
Properties of Three Subatomic Particles

Name	Mass		Electric Charge	Common Symbols
	In grams	In amu		
Electron	$9.1093897 \times 10^{-28}$ g	0.0005485712 amu	$1-$	e^-
Proton	$1.67262305 \times 10^{-24}$ g	1.00727605 amu	$1+$	p^+ or p
Neutron	1.674954×10^{-24} g	1.008665 amu	0	n

Each Element Has a Unique Atomic Number. We can now identify something special about a given element. All atoms of the same element have identical nuclear charges, meaning identical numbers of protons. Atoms of different elements have different nuclear charges, or different numbers of protons. Thus each element owns a unique number, called its **atomic number,** the number of protons in one of its atoms.

You are not to expect to memorize atomic numbers or mass numbers for the various elements.

> Atomic number of element = positive charge on its atomic nuclei
> = number of protons per atom

The Number of An Atom's Electrons Also Equals the Atomic Number. All atoms are electrically neutral. This means that the positive charge on the nucleus is exactly balanced by the negative charge contributed by the atom's electrons. The number of protons, therefore, must equal the number of electrons so that each charge of $1+$ is neutralized in an electrical sense by each charge of $1-$. The atomic number, therefore, also tells us how many electrons an atom has.

> Atomic number of an element = number of protons
> = number of electrons

Each Kind of Atom Has a Unique Mass. The sum of an atom's neutrons and protons is its **mass number.**

> Mass number = protons + neutrons

Since each neutron and proton has a mass of 1.0 amu (rounded), the mass number is the same as the mass of the atom in amu. (Only when the highest precision requiring several significant figures is needed would we have to modify this statement.)

To summarize what we have learned about atoms thus far we can say that atoms are tiny neutral particles that consist of nuclei and electrons, and their nuclei consist of protons and neutrons. Each element has its own atomic number, which equals its protons. Each kind of atom has its own mass number, which equals its protons plus neutrons.

PRACTICE EXERCISE 1   What are the mass numbers of atoms that have the following nuclear compositions?

(a) 7 protons and 8 neutrons   (b) 12 protons and 12 neutrons
(c) 11 protons and 13 neutrons

PRACTICE EXERCISE 2   How many neutrons are in each of these atoms?

(a) Atomic number 4, mass number 9   (b) Atomic number 17, mass number 35
(c) Atomic number 17, mass number 37

2.3 ELECTRON CONFIGURATIONS OF ATOMS

Each element has a unique electron configuration, and this determines its chemical properties.

The space taken by an atomic nucleus is a small fraction of the space occupied by the entire atom. If an atomic nucleus were a tennis ball, the outer edge of the atom would be at least 30 football fields away. Existing in this space are the electrons.

Electrons Are Confined to Particular Energy Levels.   The electrons do not have unlimited freedom. They normally do not escape, for example, because they are attracted toward the oppositely charged nucleus. But they don't fall into the nucleus either, because they are in rapid motion of a type that tends to hurl them outward. They are also largely confined to various spaces; they are much more likely to be in certain regions of space near the nucleus than others.

Niels Bohr (1885 – 1962), a Danish scientist, was the first to suggest this, and he called the regions the *allowed energy states* or the **energy levels** of an atom. He also postulated that as long as electrons remain in the allowed levels *of lowest energy,* the atom neither emits nor absorbs energy. The atom enjoys its greatest stability as long as its electrons stay in their lowest energy states. When they don't, we can have chemical reactions, which is why we're studying where an atom's electrons are.

To help the public visualize this picture of the atom, Bohr likened the energy levels to the orbits that planets follow around the sun. His picture of the atom was therefore dubbed the *solar system model* (Figure 2.2). This view has long since been discarded, but Bohr's postulates remain valid: An atom has only certain allowed energy states for electrons, and an atom is stable as long as the electrons stay in their lowest states.

The energy level nearest the nucleus is called the *first level,* or level 1. It corresponds to the lowest energy that an electron in an atom can have. An electron in level 1 is held more firmly and is more stable than an electron anywhere else. Additional levels are numbered in order as they are found farther from the nucleus: level 2, level 3, and so forth.

Each numbered level has its own capacity for electrons. Level 1 can hold up to 2 electrons; level 2 can have 8; and level 3 can have 18. Table 2.2 summarizes these limits.

Electrons Can Change Energy Levels When Atoms Absorb or Emit Energy.   If light of the right energy strikes an atom, the atom absorbs it. The mechanism for accepting this energy is that an electron in a lower level jumps to a higher level. The difference between the two levels in energy terms exactly matches the energy absorbed from the light. The atom is now an *excited atom,* and it is not in its most stable state. In time, the promoted electron falls back to the lower level, and the atom then emits energy.

Another way to make excited atoms is by heat. If you heat an iron rod hot enough, it glows. The heat promotes electrons of the iron atoms in huge numbers to higher energy states. As the electrons drop back down, as the excited atoms "relax," they emit light. By studying the kinds of light emitted and their relative intensities, scientists learned about energy levels.

Niels Bohr (1885 – 1962) won the 1922 Nobel prize in physics.

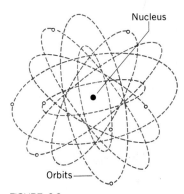

FIGURE 2.2
This solar system model of the atom, after Niels Bohr, is commonly used in popular print to depict an atom.

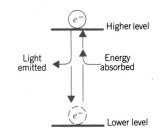

Electrons shift to higher levels when the correct amount of energy is absorbed, and they emit this energy when they drop back again.

TABLE 2.2
The Principal Energy Levels

Principal level number	1	2	3	4	5	6	7
Maximum number of electrons observed in nature[a]	2	8	18	32	32	18	8

[a] In theory, levels 5, 6, and 7 could accommodate 50, 72, and 98 electrons, respectively. Levels 1, 2, 3, and 4 are filled in theory as well as among certain elements by the numbers of electrons given.

Electrons do not move in simple orbits, like planets about the sun. Their motions are too complicated even to know with exactness, so we give attention to the regions where they most probably reside instead, regardless of how they move. These individual regions are called **atomic orbitals.**

The principal energy levels that are discussed in Section 2.3 have one or more sublevels, and each sublevel is made up of one or more atomic orbitals. The first energy level has only one sublevel. (It is its own sublevel, in other words.) And this sublevel has just one orbital, called the 1s orbital. The 1 refers to main level 1 and the s denotes the shape of the orbital. It looks like a sphere with the atomic nucleus in the center, as seen in the accompanying figure.

The second energy level has two sublevels, designated 2s and 2p. The 2s sublevel, like the 1s, has just one orbital, so we can call it the 2s orbital. It looks like the 1s

orbital from the outside—spherical. the 2p sublevel has three orbitals, each with a figure 8 cross-sectional shape, seen in the accompanying drawing. The axes of these three are mutually perpendicular, like the x, y, and z axes, so the 2p orbitals are individually named the $2p_x$, $2p_y$, and the $2p_z$ orbitals. These differ only in orientation, not in energy. An electron can be in any of these three orbitals and have the same energy.

Energy level 3 has three sublevels and three kinds of orbitals, one of the s type, three of the p, and five of still another type, the d orbitals (whose shapes we'll not discuss because they would have no bearing on the properties of the first 20 elements. There is a fourth kind of orbital, the f orbital, at level 4 and higher, which we'll also leave to more comprehensive books.)

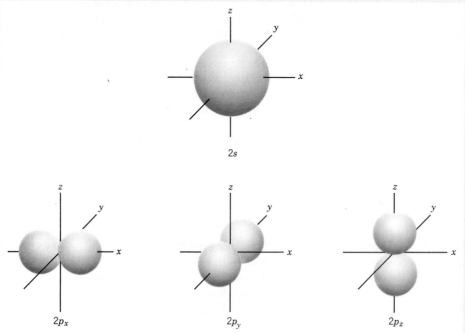

Two kinds of orbitals, the s and the p. All main energy levels have one s orbital. Levels 2 and higher all have three p orbitals.

A Few Rules Give the Pattern of the Electron Configurations of Elements 1 through 20. The distribution of electrons among the available levels is called the **electron configuration** of the atom. We'll study those of just the 20 simplest kinds of atoms. These include nearly all of the elements found in living system, so we do not have to say too much about atoms of the higher atomic numbers.

Given the atomic number of any of the 20 simplest atoms, we can figure out the electron configuration by using the following rules:

1. Use the atomic number as the number of electrons to arrange.

2. Place electrons one by one into the energy levels, filling the lowest level first and then moving out. Let a maximum of 2 electrons go into level 1 and a maximum of 8 go into level 2.

SPECIAL TOPIC *Continued*

Each atomic orbital can hold up to two electrons, no more. When two are present, they must spin in opposite directions. The spin of an electron is like the spin of the earth about its axis. When two electrons spin in opposite directions they behave like two magnets oriented to attract each other. This attraction helps to compensate for the electrical repulsion that exists between two electrons.

The accompanying table gives the electron configurations, orbital by orbital, of the first 20 elements. Each electron is represented by an arrow, and oppositely pointing arrows denote oppositely spinning electrons. Notice at atomic numbers 6 and 7 or 14 and 15 that two electrons do not go into the same orbital if another *of identical energy* is available.

Electron configurations by orbitals are written as follows, using nitrogen (atomic number 7) as an example:

$$1s^2 2s^2 2p_x{}^1 2p_y{}^1 2p_z{}^1$$

The superscripts give the number of electrons in each orbital. You can see that level 2 of the nitrogen atom has a total of 5 electrons, the number predicted by the use of the rules given in Section 2.2. Level 2 of the neon atom (atomic number 10) has 8 electrons and all orbitals of levels 1 and 2 are full.

Electron Configurations by Orbitals of Elements 1–20

Atomic Number	Element	1s	2s	$2p_x$	$2p_y$	$2p_z$	3s	$3p_x$	$3p_y$	$3p_z$	4s
1	H	↑									
2	He	↑↓									
3	Li	↑↓	↑								
4	Be	↑↓	↑↓								
5	B	↑↓	↑↓	↑							
6	C	↑↓	↑↓	↑	↑						
7	N	↑↓	↑↓	↑	↑	↑					
8	O	↑↓	↑↓	↑↓	↑	↑					
9	F	↑↓	↑↓	↑↓	↑↓	↑					
10	Ne	↑↓	↑↓	↑↓	↑↓	↑↓					
11	Na	↑↓	↑↓	↑↓	↑↓	↑↓	↑				
12	Mg	↑↓	↑↓	↑↓	↑↓	↑↓	↑↓				
13	Al	↑↓	↑↓	↑↓	↑↓	↑↓	↑↓	↑			
14	Si	↑↓	↑↓	↑↓	↑↓	↑↓	↑↓	↑	↑		
15	P	↑↓	↑↓	↑↓	↑↓	↑↓	↑↓	↑	↑	↑	
16	S	↑↓	↑↓	↑↓	↑↓	↑↓	↑↓	↑↓	↑	↑	
17	Cl	↑↓	↑↓	↑↓	↑↓	↑↓	↑↓	↑↓	↑↓	↑	
18	Ar	↑↓	↑↓	↑↓	↑↓	↑↓	↑↓	↑↓	↑↓	↑↓	
19	K	↑↓	↑↓	↑↓	↑↓	↑↓	↑↓	↑↓	↑↓	↑↓	↑
20	Ca	↑↓	↑↓	↑↓	↑↓	↑↓	↑↓	↑↓	↑↓	↑↓	↑↓

3. At level 3, stop at 8 electrons. It can hold more, but more do not enter it until we reach atomic numbers above 20, which are beyond our present interest. (We leave the explanation to more comprehensive books.)

Level 3 will resume filling at element 21, and it becomes filled at element 29 (copper).

4. Use level 4 for the 19th and 20th electrons.

Figure 2.3 shows two ways we could represent an electron configuration. But the dots of part (*a*) are tedious to count, so we'll use the type given in part (*b*). What we're really after is the situation in the highest *occupied* level, the **outside level,** or outer level. When we know this, we can make some sense of the chemistry of the corresponding element. Table 2.3 gives the electron configurations for the first 20 elements. For additional details about these configurations, see Special Topic 2.1.

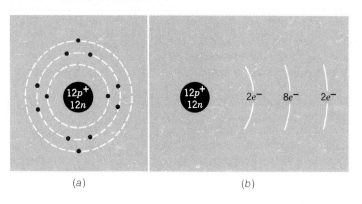

FIGURE 2.3
Electron configurations — two ways to display them. (*a*) The electron-dot symbol of an atom of atomic number 12 and mass number 24. (*b*) A concise display of the same electron configuration.

(*a*)   (*b*)

TABLE 2.3
The Electron Configurations of Elements 1–20, Main Energy Levels

Element	Symbol	Atomic Number	Level Number			
			1	2	3	4
Hydrogen	H	1	1			
Helium	He	2	2			
Lithium	Li	3	2	1		
Beryllium	Be	4	2	2		
Boron	B	5	2	3		
Carbon	C	6	2	4		
Nitrogen	N	7	2	5		
Oxygen	O	8	2	6		
Fluorine	F	9	2	7		
Neon	Ne	10	2	8		
Sodium	Na	11	2	8	1	
Magnesium	Mg	12	2	8	2	
Aluminum	Al	13	2	8	3	
Silicon	Si	14	2	8	4	
Phosphorus	P	15	2	8	5	
Sulfur	S	16	2	8	6	
Chlorine	Cl	17	2	8	7	
Argon	Ar	18	2	8	8	
Potassium	K	19	2	8	8	1
Calcium	Ca	20	2	8	8	2

EXAMPLE 2.1   WRITING AN ELECTRON CONFIGURATION

Problem: An atom has an atomic number of 19 and a mass number of 39. What is the composition of its nucleus and its electron configuration?

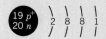

Solution: The nucleus has 19 protons (the atomic number) and 20 neutrons (mass number — atomic number). There are 19 electrons. Two electrons go into level 1 and 8 go into level 2. This leaves 9 electrons. Only eight of these go into level 3 (rules 3 and 4). The last electrons is in level 4. Using the kind of display of Figure 2.3*b*, we write the electron configuration as shown in the margin.

PRACTICE EXERCISE 3   What is the composition of the nucleus and the electron configuration of each of the following atoms? (Use this exercise to learn the rules; do not use the table.)

(a) Atomic number 7, mass number 14   (b) Atomic number 13, mass number 27
(c) Atomic number 20, mass number 40

2.4 ELEMENTS

Atoms of the same element have identical atomic numbers and electron configurations.

As we move more deeply into the nature of atoms and elements, we can make our definitions sharper. We said earlier, for example, that an element cannot be broken into anything that is simpler, yet stable. We now add that an **element** is a substance whose atoms all have identical nuclear charges and electron configurations. An **atom,** we can now say, is the smallest representative sample of an element; it has only one nucleus; and it is always an electrically neutral particle.

The atomic numbers go as high as 108, and this many different elements are known. Of these 90 occur naturally, and the rest have been prepared using high-energy devices.

Most of the Elements Are Metals. **Metals** are substances that conduct electricity and that can be polished, drawn into wires, and hammered into sheets. About 20 elements are **nonmetals.** They are poor conductors of electricity, and the solid nonmetals shatter when struck. All of the gaseous elements and one liquid element are nonmetals. A few elements are difficult to classify. Carbon, for example, is brittle, but in the form of graphite it conducts electricity.

Each Element Has an Atomic Symbol. Each element has a shorthand symbol consisting of either one or two letters. The first is always capitalized and the second letter, if any, is always lower case. Table 2.4 gives several, and these should be learned now.

Virtually Every Element Consists of a Mixture of a Small Number of Its Isotopes.
Although all of the atoms of an element have identical atomic numbers and electron configurations, nearly all elements consist of a mixture of atoms that differ slightly from each other in mass number. Such atoms, identical in atomic number and electron configuration, but different in mass number are called **isotopes.**

Be sure to distinguish between substances and particles. Elements, compounds, and mixtures are substances. Atoms, protons, neutrons, and electrons are particles of which substances are made.

The gaseous elements are

Hydrogen	Helium
Nitrogen	Neon
Oxygen	Argon
Fluorine	Krypton
Chlorine	Xenon
Radon	

TABLE 2.4
Symbols for Common Elements[a]

C	Carbon	Al	Aluminum	Cl	Chlorine	Ag	Silver (*argentum*)
F	Fluorine	Ba	Barium	Mg	Magnesium	Cu	Copper (*cuprum*)
H	Hydrogen	Br	Bromine	Mn	Manganese	Fe	Iron (*ferrum*)
I	Iodine	Ca	Calcium	Pt	Platinum	Pd	Lead (*plumbum*)
N	Nitrogen	Li	Lithium	Zn	Zinc	Hg	Mercury (*hydrargyrum*)
O	Oxygen	Ra	Radium			K	Potassium (*kalium*)
P	Phosphorus					Na	Sodium (*natrium*)
S	Sulfur						

[a] Each atomic symbol in the first column is the capitalized first letter of the name of the element. Each symbol in the second group consists of the first two letters of the name of the element. Each symbol in the third group has the same first letter as the name of the element, but the second letter in the symbol isn't the second letter of the name. The symbols in the fourth group are from the Latin names of the elements.

Isotopes are often symbolized as follows, where the superscripts are mass numbers and the subscripts are atomic numbers:

$$^{35}_{17}Cl \qquad ^{37}_{17}Cl$$

The difference between the isotopes of an element is in their numbers of neutrons. For example, chlorine (atomic number 17) consists chiefly of two isotopes called chlorine-35 and chlorine-37, where the numbers are mass numbers. An atom of chlorine-35 has 18 neutrons $(35 - 17)$ and one of chlorine-37 has 20 neutrons $(37 - 17)$.

It is very important to realize that the atoms of both of the chlorine isotopes have the same atomic number, 17. They have, therefore, identical numbers of protons and electrons, 17, and they have identical electron configurations. With the same electron configuration, the two chlorine isotopes have identical *chemical* properties. The only difference between an element's isotopes is in mass numbers, and we need to know about isotopes to understand the nature of atomic weights.

The Atomic Weight of an Element Reflects the Relative Abundances of Its Isotopes.
The two chlorine isotopes occur *everywhere,* all over the earth, in the same ratio, roughly $3:1$. For every four chlorine atoms in nature, three are chlorine-35 and one is chlorine-37. To find the average mass of these four atoms, we'll assume that their ratio is exactly $3:1$.

Mass of 3 atoms of chlorine-35	3×35 amu $= 105$ amu
Mass of 1 atom of chlorine-37	1×37 amu $= \underline{\ 37 \text{ amu}}$
	Mass of these four atoms $= 142$ amu

Dividing 142 amu by 4 gives us 35.5 amu as the average mass of the chlorine atoms as this element occurs naturally, anywhere on the earth. (The actual average is 35.457 amu; the ratio is not exactly $3:1$.)

The average mass of the atoms of the various isotopes of any given element, as they occur in their natural proportions, is called the **atomic weight** of the element. A Table of Atomic Weights and Numbers may be found inside the front cover of this book. The existence of isotopes explains why atomic weights, unlike mass numbers, are not simple, whole numbers. Most of the elements are like chlorine; the proportions of their isotopes are the same regardless of their origins, geologically. (Footnotes to this table describe the exceptions. They will not concern us.)

It is important to realize that isotopes are *substances,* not particles. It isn't easy, but the chlorine isotopes can be separated from each other and placed into different containers. The chemical properties of the two isotopes, as we said, are the same. Only the masses of their atoms differ. Thus an element, which is one of the three kinds of *matter,* is usually made of two or a few more *substances* called its *isotopes* (The element xenon, for example, has nine stable isotopes, the most of any element.) About 250 isotopes occur in nature. More than a thousand have been made by nuclear transformations that we will study in Chapter 22. Many are useful in medicine, as we'll also see.

PRACTICE EXERCISE 4   Suppose that element number 25 consists of two isotopes in a $50:50$ ratio. One has 25 neutrons and the other has 30. What is the atomic weight of this element?

2.5   THE PERIODIC LAW AND THE PERIODIC TABLE

In a given family of representative elements, the atoms of all members have the same outside-level electron configurations.

If each of the 108 elements were completely unlike the others, the study of chemistry would be extremely difficult. Fortunately, the elements can be sorted into a small number of families whose members have much in common. Dimitri Mendeleev, a Russian scientist, was the first to notice this as he wrote a chemistry textbook for his students, which was published in 1869.

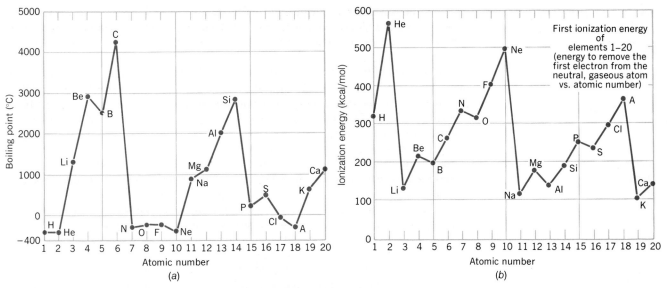

FIGURE 2.4
Periodicity and two properties of elements 1–20. (*a*) Boiling points versus atomic numbers. (*b*) Ionization energies versus atomic numbers.

The Properties of the Elements Reoccur Periodically.   When Mendeleev organized some of the physical and chemical properties of the known elements, he observed that the properties seemed to go through cycles from the elements of lowest atomic weight to the highest. The cycles weren't geometrically perfect, but there were definite rises and falls. One improvement, which came with the discovery of atomic numbers after Mendeleev's time, was to use atomic numbers instead of atomic weights to rank the elements. As seen in Figure 2.4*a*, the temperatures at which the first 20 elements boil do not rise ever higher as the atomic numbers increase. Instead, they rise and fall. They vary periodically, not linearly with increasing atomic number.

The ionization energies of the elements display a similar rising and falling, as seen in Figure 2.4*b*. This energy is the energy needed to make one electron leave each atom in a sample whose mass (in grams) equals the element's atomic weight.

Enough other properties of elements also vary periodically with atomic number to establish a general law of nature, the **periodic law.**

The mass of an element in grams equal to its atomic weight is called one mole of the element, and the mole (abbreviated *mol*) is an SI base unit.

> Periodic Law   The properties of the elements are a periodic function of their atomic numbers.

This natural law makes possible a useful way to organize and study the elements, as we'll see next.

The Periodic Table Organizes the Elements to Show Off Their Periodic Properties.
Notice that in the plot of boiling points versus atomic numbers, helium, neon, and argon are at the bottoms of cycles, and they are at the tops in Figure 2.4*b*. They seem to form a set of elements that behave alike. Let us now start to line up the elements horizontally by atomic number, but start new rows so as to make a set of similar elements fall into the same *vertical column*. When we do this, the rest of the columns *automatically* are made up of other sets of similar elements. The resulting display of the elements is called the **periodic table,** shown inside the front cover of this book.

If one element in a group or chemical family forms a compound with, say, chlorine, then all the others in the family do, too, and all the compounds have the same atom ratios or formulas.

Each horizontal row in the periodic table is called a **period** and each vertical column is called a **group** or a **chemical family.** What makes this so helpful is that a study of chemistry can focus first on just a few families rather than dozens of individual elements.

Some Families of Elements Are Representative Elements.   The periods in the periodic table are not all of the same length, and several are broken. This is necessary if the highest priority is to have the vertical columns contain elements with similar properties, particularly *chemical* properties. Thus period 1, with only hydrogen and helium, is very short, and like the next two periods is separated into two parts.

The groups have both numbers and letters. One set makes up an A-series: IA, IIA, IIIA, and so forth up to VIIA. This set plus Group 0 are called the **representative elements.** The groups in the B-series, clustered near the middle of the periodic table, are called the **transition elements.**

The transition elements represent successive fillings of *inner* energy levels while their outside levels hold one or two electrons.

The two rows of elements placed outside the table are the **inner transition elements.** (The table would not fit well on the page if the inner transition elements were not handled this way.) Elements 58 through 71 constitute the *lanthanide* series, named after element 57, which just precedes this series. The series of elements 90 to 103 is the *actinide series.*[1]

The lanthanides are also called the *rare earth* elements, and some are essential in electronic devices.

Several of the groups of representative elements have family names. Those in Group IA, for example, are the **alkali metals,** because they all react with water to give an alkaline or caustic (skin-burning) solution. The elements in Group IIA are the **alkaline earth metals,** because their compounds are commonly found in "earthy" substances, like limestone.

The elements in Group VIIA are called the **halogens** after a Greek word signifying their salt-forming ability. For example, chlorine in Group VIIA is present in table salt (in a chemically combined form with sodium).

The elements in Group 0, all gases, were discovered after Mendeleev's work. Except for a few compounds that xenon and krypton form with fluorine and oxygen, the Group 0 elements react with nothing. Hence, they are called the **noble gases** (*noble* signifying limited activity).

Other groups of representative elements are named after their first members, for example, the **carbon family** (Group IVA), the **nitrogen family** (Group VA), and the **oxygen family** (Group VIA).

PRACTICE EXERCISE 5   Referring to the periodic table, pick out the symbols of the elements as specified.

(a) A member of the carbon family: Sr, Sn, Sm, S
(b) A member of the halogen family: C, Ca, Cl, Co
(c) A member of the alkali metals: Rn, Ra, Ru, Rb
(d) A member of the alkaline earth metals: Mg, Mn, Mo, Md
(e) A member of the noble gas family: Ac, Al, Am, Ar

The periodic table was established before electron configurations were known.

Electron Configurations Are Behind the Arrangement of the Periodic Table.   The numbers of electrons housed in the principal energy levels of several representative elements are given in Table 2.5. When we look at the outside levels, family by family, we see two very striking facts. First, the atoms of each family all have identical numbers of electrons in their outside levels and, second, the numbers in the outside levels differ from family to family.

All atoms of Group IA metals have one electron in the outside level. In Group IIA, all atoms have two outside level electrons. In fact, *among the representative elements, the group number (I, II, and so forth, through VII) is equal to the number of electrons in the outside level.*

All atoms of the noble gases (except those of helium) have eight outer level electrons, an *outer octet.* Helium atoms have just two outer level electrons, but this level (number 1) cannot

[1] There is an effort being made by some chemists to do away with the A and B organization. Until this movement succeeds in changing the standard references and the standard chemistry examinations in the United States, we feel bound to stay with the A and B system.

TABLE 2.5
Electron Configurations Among Four Families of Representative Elements

Family	Element	Atomic Number	Principal Level Number						
			1	2	3	4	5	6	7
Group IA	Lithium	3	2	1					
The alkali metals	Sodium	11	2	8	1				
	Potassium	19	2	8	8	1			
	Rubidium	37	2	8	18	8	1		
	Cesium	55	2	8	18	18	8	1	
	Francium	87	2	8	18	32	18	8	1
Group IIA	Beryllium	4	2	2					
The alkaline earth metals	Magnesium	12	2	8	2				
	Calcium	20	2	8	8	2			
	Strontium	38	2	8	18	8	2		
	Barium	56	2	8	18	18	8	2	
	Radium	88	2	8	18	32	18	8	2
Group VIIA	Fluorine	9	2	7					
The halogens	Chlorine	17	2	8	7				
	Bromine	35	2	8	18	7			
	Iodine	53	2	8	18	18	7		
	Astatine	85	2	8	18	32	18	7	
Group 0	Helium	2	2						
The noble gases	Neon	10	2	8					
	Argon	18	2	8	8				
	Krypton	36	2	8	18	8			
	Xenon	54	2	8	18	18	8		
	Radon	86	2	8	18	32	18	8	

hold more than two electrons anyway. The noble gas elements, as we said, are the least chemically reactive of all elements, so we might suspect that having eight electrons in an outer electron level somehow relates to chemical stability. As we'll see in the next chapter, the outer octet will be one of the most significant features of atomic structure that we will study.

The electron configurations of the series B elements, such as the various transition elements, have their own fairly regular patterns, but our study will not require knowledge of them.

EXAMPLE 2.2 FINDING INFORMATION IN THE PERIODIC TABLE

Problem: How many electrons are in the outside level of an atom of iodine?

Solution: Until you become more familiar with the locations of certain elements in the periodic table, you'll have to use the Table of Atomic Weights and Numbers (inside the front cover of this book) to find the atomic number of a given element. Doing this, we find that the atomic number of iodine is 53. Now we use the periodic table and find that iodine is in Group VIIA. Being one of the A-type elements, we know that iodine is a *representative* element, which means that its group number is the same as the number of outside level electrons—7.

PRACTICE EXERCISE 6   How many electrons are in the outside level of an atom of each of the following elements?
(a) Potassium   (b) Oxygen   (c) Phosphorus   (d) Chlorine

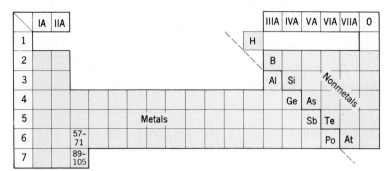

FIGURE 2.5
Locations of metals, nonmetals, and metalloids in the periodic table.

Metals and Nonmetals Are Segregated in the Periodic Table.   Another use of the periodic table is to tell which elements are metals and which are not. As seen in Figure 2.5, the nonmetals are all in the upper-right-hand corner. All of the noble gases (Group 0) and the halogens (Group VIIA) are nonmetals. The few elements along the borderline between metals and nonmetals are sometimes called **metalloids,** and they have properties that are partly metallic and partly nonmetallic.

If you compare the locations of the elements in Figure 2.5 with the electron configurations given in Table 2.5, you will see that all of the nonmetals, except hydrogen and helium, have four-to-eight outer electrons. Except for hydrogen and helium, all atoms with one, two, or three outer electrons belong to metals. Only among some elements with high atomic numbers do we find metals with more than three outside-level electrons. Tin and lead of Group IVA are common examples.

SUMMARY

Matter   Matter, anything with mass that occupies space, can exist in three physical states: solid, liquid, and gas. Broadly, the three kinds of matter are elements, compounds, and mixtures. Elements and compounds obey the law of definite proportions. Mixtures do not. A chemical reaction, an event in which substances change into different substances, occurs without any loss of mass.

Dalton's atomic theory   The law of definite proportions and the law of conservation of mass in chemical reactions led John Dalton to the theory that all matter consists of discrete, noncuttable particles called atoms. When atoms of different elements combine to form compounds, they combine as *whole* atoms; they do not break apart.

Atomic structure   Atoms are tiny electrically neutral particles that are the smallest representatives of an element that can display the element's chemical properties. Each atom has one nucleus, a dense inner core, surrounded by enough electrons to balance the positive nuclear charge. All of the atom's protons and neutrons are in the nucleus, and each of these subatomic particles has a mass of 1.0 amu. The mass of the electron is only 1/1836 the mass of a proton. The proton has a charge of $1+$, the electron's charge is $1-$, and the neutron is electrically neutral.

Atomic numbers and atomic weights   The atomic number of an element is the number of protons in each atom. It also equals the number of electrons in one atom. Nearly all elements occur as a mixture of a small number of isotopes. The isotopes of an element share the same atomic number, but their atoms have small differences in their numbers of neutrons. To fully characterize an isotope, we have to specify both the atomic number and the mass number, the number of protons plus neutrons. The atomic weight of an element is numerically equal to the average mass (in amu) of the atoms of all its isotopes as they occur naturally together.

Electron configurations   To write an electron configuration of an atom, we place its electrons one by one into the available energy levels. We start with the level of lowest energy, level 1, which can hold up to 2 electrons. Level 2 can hold 8 electrons, and level 3 can handle 18 (but is limited to 8 until atomic number 21 is reached). As long as electrons remain in their levels, an atom neither absorbs nor radiates energy.

Periodic properties   Because many properties of the elements are periodic functions of atomic numbers, the elements fall naturally into groups or families that we can organize into vertical columns in a periodic table. Atoms of the representative elements that belong to the same family have the same number of outside-level electrons. This number equals the A-group number itself. (Helium does not fit this generalization.)

The horizontal rows of the periodic table are called periods. The nonmetallic elements are in the upper-right-hand corner of the periodic table. The metals, the great majority of the elements, make up the rest of the table. At the border between metals and nonmetals occur the elements — the metalloids — that have both metallic and nonmetallic properties.

REVIEW EXERCISES

The answers to the review exercises that require a calculation and whose numbers are marked by an asterisk are given in Appendix V. The answers to all other review exercises appear in the *Study Guide* that accompanies this book.

Matter

2.1 If we think about a warm room, bathed in sunlight, we might imagine that energy itself occupies space. Why don't we call it matter?

2.2 What are the names of the three states of matter?

2.3 In terms of shapes and volumes, how do solids differ from gases?

2.4 What is the essential difference between a chemical reaction and a physical change?

2.5 Which of the following events are physical changes?
(a) The splintering of rock into gravel.
(b) The change in the color of leaves during the fall season.
(c) The decay of plant remains on a forest floor.
(d) The eruption of an oil well just "brought in" by drillers.
(e) The seeming disappearance of sugar as it is stirred into coffee or tea.
(f) The swing of a compass needle near another magnet.
(g) The formation of a raindrop in a cloud.

2.6 What is the general name given to the event by which elements are changed into compounds?

2.7 Hydrogen reacts with oxygen to give water and heat evolves.
(a) What are the *reactant(s)* and the *product(s)*?
(b) Is the reaction endothermic or exothermic?

2.8 What important facts about compounds distinguish them from either elements or mixtures?

2.9 What important mass relationship between reactants and products in a chemical reaction is always observed?

Atomic Theory

2.10 What are the names, electrical charges, and masses of the three subatomic particles? (Use the amu as the unit of mass, and omit the mass of the lightest of the three.)

2.11 Name two subatomic particles that attract each other.

2.12 What subatomic particles have forces of repulsion between them?

2.13 We will learn in the next chapter that it is possible for particles of an atomic size to have unequal numbers of electrons and protons. If particle M has 13 protons and 10 electrons, what is the electrical charge on M?

2.14 It costs energy to make an electron leave an atom, but which of the following changes would cost the *least* energy, and why?

(a) Removal of an electron from a particle having 12 protons and 12 electrons.
(b) Removal of an electron from a particle having 12 protons and 11 electrons.

2.15 Rutherford found that essentially all of the mass of an atom is in its nucleus. The following calculations will give you an idea of what this means.
(a) Calculate the density in g/cm^3 of the nucleus of a hydrogen atom from the following data. Assume that the nucleus is a perfect sphere so that you can calculate its volume by the standard equation, volume of sphere $= \frac{4}{3}\pi r^3$, where $r =$ radius and $\pi = 3.14$. The radius of the hydrogen nucleus is 5×10^{-14} cm and the mass of the nucleus is 1.67×10^{-24} g.
(b) Convert the answer to part (a) into units of metric tons per cubic centimeter. 1 metric ton = 1000 kg.

2.16 Particle M has 13 protons and 10 electrons, and particle Y has 8 protons and 10 electrons. Do they attract each other, repel each other, or leave each other alone? Explain.

2.17 The atomic mass unit is roughly on the order of what size, 10^{-230} g, 10^{-23} g, 10^{-3} g, or 10^{23} g?

2.18 To five significant figures, calculate the mass in grams of a sample of 6.0220×10^{23} hydrogen atoms, each one having just one proton. Ignore the mass of the electron. How does the result compare with the atomic weight of hydrogen? (To two significant figures, are the results the same or different?)

Electron Configurations

2.19 Briefly describe the picture of atomic structure that emerged from Niels Bohr's theory.

2.20 What features of Bohr's theory are still valid?

2.21 What is the maximum number of electrons that can be in each?
(a) Principal level 2   (b) Principal level 3

2.22 Write the atomic number, the mass number, and the electron configuration of an atom with six protons and seven neutrons.

2.23 What are the atomic number, mass number, and electron configuration of an atom with 18 protons and 21 neutrons?

2.24 Write the electron configuration of silicon (atomic number 14).

2.25 The iron atom has 2 electrons in the first level, 8 in the second, 14 in the third, and 2 in the fourth. Using only this information, and general facts about electron configurations, answer the following questions.
(a) What is the atomic number of iron?
(b) Is principal level 3 completely filled?

2.26 Write the electron configurations of the elements that have the following atomic numbers, using only these numbers.
(a) 11   (b) 8   (c) 17   (d) 20

2.27 Write the electron configurations of the elements that have the following atomic numbers.
(a) 12   (b) 19   (c) 9   (d) 3

Elements and Their Isotopes

2.28 Roughly, how many elements are known: 80, 100, 200, or 2000?

2.29 Roughly, how many elements are not solids under ordinary room conditions, but are either liquids or gases: 2, 13, 44, 75, 180, or 1885?

2.30 Indium is a substance that can be given a shiny finish, and it conducts electricity well. Is indium more likely to be a metal or a nonmetal?

2.31 Phosphorous is a dark reddish, powdery substance that doesn't conduct electricity. Is it more likely to be a metal or a nonmetal?

2.32 Elements and isotopes are *substances,* not tiny particles — although they consist of such particles. What is the distinction between the terms *element* and *isotope of an element?*

2.33 Which of the following are pairs of isotopes? (Use the hypothetical symbols for your answer.)
M has 12 protons and 13 neutrons
Q has 13 protons and 13 neutrons
X has 12 protons and 12 neutrons
Z has 13 protons and 12 neutrons

2.34 The members of which of the following pairs of elements would be expected to have identical chemical properties and why?
Pair 1. *A* (10 protons and 9 neutrons)
         B (10 protons and 10 neutrons)
Pair 2. *C* (7 protons and 6 neutrons)
         D (6 protons and 6 neutrons)

2.35 Carbon, atomic number 6, has three isotopes that are found in nature. The most abundant (98.89%) has a mass number of 12. The isotope with a mass number of 13 makes up 1.11% of naturally occurring carbon. The third isotope, carbon-14 — obviously present in the merest trace, because 98.89% + 1.11% = 100.00% — makes possible the dating of ancient artifacts.
(a) What is the same about these isotopes?
(b) In what specific feature of atomic structure do they differ?

2.36 Later, when we want to describe some *chemical* reaction of sulfur, which consists principally of two isotopes (mass numbers 32 and 34), we will use the symbol S. Why don't we have to specify which isotope?

Atomic Symbols

2.37 The symbol BN stands for a compound (boron nitride), not an element. How can we tell this *from the symbol itself?*

2.38 What are the symbols of the following elements?
(a) Iodine   (b) Lithium   (c) Zinc
(d) Lead   (e) Nitrogen (f) Barium

2.39 Write the symbols for the following elements.
(a) Carbon   (b) Chlorine   (c) Copper
(d) Calcium   (e) Fluorine   (f) Iron

2.40 What are the symbols of the following elements?
(a) Hydrogen   (b) Aluminum   (c) Manganese
(d) Mangesium   (e) Mercury   (f) Sodium

2.41 Write the symbols for the following elements.
(a) Oxygen   (b) Bromine   (c) Potassium
(d) Silver   (e) Phosphorus (f) Platinum

2.42 What are the names of the elements represented by the following symbols?
(a) P   (b) Pt   (c) Pb   (d) K
(e) Ca   (f) C   (g) Hg   (h) H
(i) Br   (j) Ba   (k) F   (l) Fe

2.43 Give the name of each element represented by the following symbols.
(a) S   (b) Na   (c) N   (d) Zn
(e) I   (f) Cu   (g) O   (h) Li
(i) Mn   (j) Mg   (k) Ag   (l) Cl

Atomic Weights

2.44 Mass numbers are always whole numbers. Atomic weights almost never are. Explain.

2.45 Iodine (atomic number 53), which is needed in the diet to have a healthy thyroid gland, occurs in nature as only one isotope, iodine-127. Using just this information, what is the atomic weight of iodine (to three significant figures)?

2.46 Bromine, in the same chemical family as iodine, occurs in nature as a nearly 1 : 1 mixture of two isotopes, bromine-79 and bromine-81. Using just this information, which is the atomic weight of bromine (to two significant figures)?

2.47 An element consists of two isotopes, *X* and *Y*, in a ratio of 4 : 1. Atoms of *X* have 30 protons and 34 neutrons. Those of *Y* have 30 protons and 38 neutrons. What is the atomic weight of this element (to three significant figures)?

2.48 What is the atomic weight of an element that consists of three isotopes in a ratio of 1 : 1 : 1 if the atomic number is 30, and one isotope has 30 neutrons, the second has 36, and the third has 38?

2.49 To two significant figures, the atomic weight of carbon is 12 and the atomic weight of magnesium is 24.
(a) How many times heavier are magnesium atoms than carbon atoms?
(b) If we counted out in separate piles 10^{23} atoms of carbon and the identical number of magnesium atoms, which pile of atoms would have the larger mass, and by what factor?
(c) If we weighed out 2.0 g of carbon and we wanted a sample of magnesium that had the identical number of atoms, how many grams of magnesium should we weigh out?

*2.50 How many times heavier are carbon atoms than hydrogen atoms (on average)? Calculate the answer to two significant figures. Use data from the Table of Atomic Weights and Numbers inside the front cover.

*2.51 Carbon and oxygen atoms can chemically combine to form particles consisting of these atoms in a ratio of $1:1$. (The substance that forms is the poisonous gas, carbon monoxide.)
  (a) How much heavier are oxygen atoms than carbon atoms (to three significant figures)?
  (b) If all of the atoms in a sample of 12.0 g of carbon are to combine entirely with oxygen atoms, how many grams of oxygen are needed?

The Periodic Table

2.52 What general fact about the chemical elements makes possible the stacking of these elements in the kind of array seen in the periodic table?

2.53 What do the terms *group* and *period* refer to in the periodic table?

2.54 Without consulting the periodic table, is an element in Group VA a representative or a transition element?

2.55 Without consulting the periodic table, is an element in Period 2 a representative or a transition element?

2.56 Give the group number and the chemical family name of the set of elements to which each of the following belongs.
  (a) Sodium (b) Bromine (c) Sulfur (d) Calcium

2.57 For each of the following elements give the group number and the name of family to which it belongs.
  (a) Iodine (b) Phosphorus
  (c) Magnesium (d) Lithium

2.58 The electron configuration of a representative element is

$$2 \quad 8 \quad 7$$

Answer the following questions without looking at a periodic table.
  (a) How may electrons are in level 2, and is it filled?
  (b) Is the element more likely to be a metal or a nonmetal? Why?
  (c) What is the group number and the family name associated with this element?

2.59 The electron configuration of a representative element is

$$2 \quad 8 \quad 18 \quad 18 \quad 8 \quad 2$$

Answer the following questions without consulting the periodic table.

(a) Is this element more likely a metal or a nonmetal? How can you tell?
(b) What is the group number of this element?

2.60 The accompanying diagram shows a section of the periodic table, except that hypothetical atomic symbols are used. The numbers are atomic numbers but not all of the numbers are given symbols. Without referring to an actual periodic table, answer the following questions.

7	8	9	10
15	16	17	18
W	X	Y	Z
33	34	35	36

(a) What elements, if any, are shown as belonging to the same family as Y? Write the number(s).
(b) What elements, if any, are in the same period as W? Write the number(s).
(c) Above the first horizontal row, write the group numbers for the elements of this row.
(d) If the outside level of an atom of number 33 has five electrons, how many electrons are in the outside level of an atom of number 15? Of number 16? Of number 36?
(e) If number 10 is one of the noble gases, which is also a noble gas, 9, 11, or 36?
(f) What is the atomic number of the least metallike element in the same family as element W?
(g) If number 18 is a nonmetal, and number 33 is a metalloid, are the elements represented here metals or nonmetals?
(h) If the outer level of element W has five electrons, how many electrons are in the outer levels of the elements whose atomic numbers are 33, 34, and 35?

Atomic Orbitals (Special Topic 2.1)

2.61 Principal energy level 2 has how many sublevels? How many orbitals are in each sublevel, and how are they named?

2.62 If an orbital is to hold two electrons, what must be true about them?

2.63 The three 2-p orbitals are alike in what ways? How are they different?

2.64 Write the electron configurations by atomic orbitals for carbon (atomic number 6) and sodium (atomic number 11).

Chapter 3
The Nature of Matter: Compounds and Bonds

Ionic Compounds

Names and Formulas of Ionic Compounds

The Octet Rule

Molecular Compounds

Polar Molecules

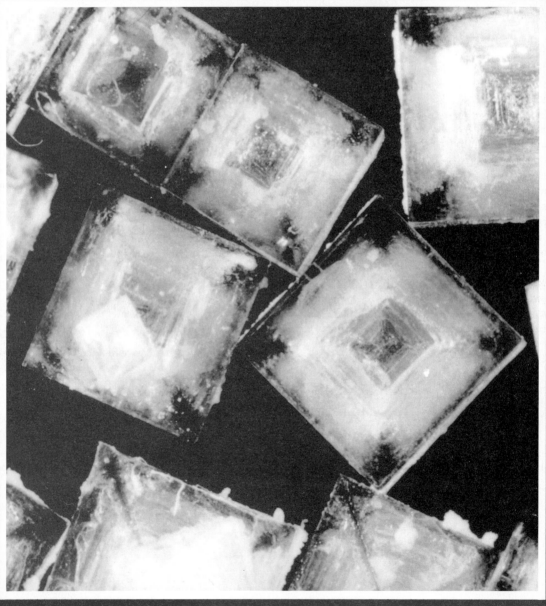

When carefully grown, crystals of table salt, sodium chloride, are perfect cubes. What makes this happen will be learned in this chapter.

3.1 IONIC COMPOUNDS

Oppositely charged particles called ions, made from metal and nonmetal atoms, attract each other in ionic compounds.

Whatever holds us together — skin and bones or cell membranes and blood vessels — must handle stresses and strains, both physical and chemical. We now ask: What must be true at the atomic level to explain how some kinds of matter stick together in bulk, and other kinds do not?

The law that unlike charges attract is our first clue. Electrical forces of attraction are at the heart of what holds matter together. But atoms are electrically *neutral,* so how can they attract each other? They can't, not *as atoms.* Many combinations of atoms, however, can reorganize their electron configurations to form new particles able to attract each other electrically. Strong forces of attraction, called **chemical bonds,** can exist between such reorganized atoms. Our broad goal in this chapter is to learn the nature of these bonds.

Two Broad Families of Substances Are Molecular and Ionic Compounds. The electrons and nuclei of atoms can become reorganized in two common ways. One kind, particularly common among combinations of nonmetal atoms, gives a new kind of small particle called a molecule. A **molecule** is an electrically neutral particle made up of two or more atomic nuclei surrounded by enough electrons to make the tiny package neutral. Compounds consisting of molecules are called **molecular compounds,** which we will study in Section 3.4. Sugar, vitamin C, cholesterol, and aspirin are examples, and all are made out of atoms of carbon, hydrogen, and oxygen, nonmetal elements. Water, made of hydrogen and oxygen, is also a molecular compound.

Molecule is from a Greek term meaning *little mass.*

The other way to reorganize the electrons and nuclei of atoms into compounds occurs when metals react with nonmetals. Tiny particles called **ions** that have opposite electrical charges form because their neutral atoms have gained or lost electrons. Since unlike charges attract, oppositely charged ions attract each other, very strongly in fact. Compounds made of oppositely charged ions are called **ionic compounds** which we will study next.

Ion is from the Greek *ienai,* meaning "to go." In solution ions can move under the influence of an electric current, but molecules can't.

Ions Form When Electrons Transfer between Atoms. Sodium chloride (table salt) is a typical ionic compound. Its parent elements, sodium and chlorine, cannot be stored together because they react violently. The following changes in electron configurations happen among billions of billions of sodium and chlorine atoms.

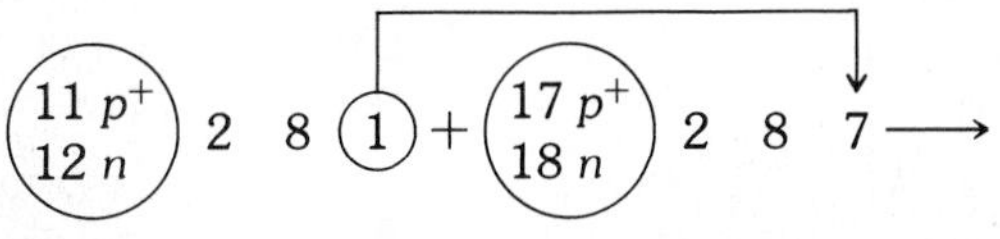

One sodium atom — Na   One chlorine atom — Cl

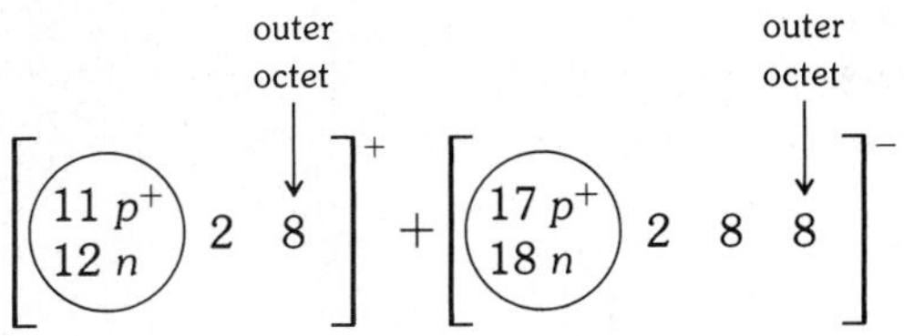

One sodium ion — Na^+ One chloride ion — Cl^-

For purposes of illustration, we have picked the sodium-23 and chlorine-35 isotopes. Sodium is a soft, silvery metal that has to be stored out of contact with air or water. Chlorine is a greenish-yellow poison gas.

The new particle with the sodium nucleus is no longer an atom, because it no longer is neutral. It has a charge of $1+$. (The $11+$ of its 11 nuclear protons is balanced by only $10-$ from its 10 remaining electrons.) Similarly, the new particle with a chlorine nucleus isn't an atom either. It carries a charge of $1-$. (It has 17 protons for $17+$ but 18 electrons for $18-$, so the net charge is $1-$.) Charged particles of atomic size are called **ions.**

Electron-Transfer Reactions Are Examples of Redox Reactions.   Because electron transfers are such important chemical events, they have a special vocabulary. When an atom loses an electron, we call the change an **oxidation.** If an atom gains an electron, we call the event a **reduction.** An electron lost from one atom cannot simply disappear. It has to have a place to go, an electron-acceptor atom. Electrons always transfer from something to something. Thus, it takes two substances for an electron-transfer reaction.

Reactions that involve electron transfers are **redox reactions,** a contraction of reduction – oxidation reaction. In every such reaction, one substance is the **oxidizing agent,** because it causes the oxidation of the other substance by accepting electrons from it. Chlorine is the oxidizing agent in the reaction of sodium and chlorine. By accepting electrons from sodium atoms, it oxidizes them to sodium ions, Na^+.

Another species in a redox reaction is a **reducing agent,** because it causes the reduction of something by giving it electrons. Sodium is the reducing agent in our reaction. It reduces chlorine by giving electrons to chlorine atoms.

In the reaction of sodium with chlorine we can assign these new terms as follows:

Oxidizing agent   Chlorine   Substance oxidized   Sodium

Reducing agent   Sodium   Substance reduced   Chlorine

The oxidizing agent always is itself reduced; the reducing agent is always itself oxidized.

At the molecular level of life, the entire chemical process of using oxygen depends on a series of redox reactions called the *respiratory chain.*

Chlorine bleaches such as Clorox work by oxidizing dyes to uncolored products.

TABLE 3.1
Some Important Ions[a]

Group	Element	Symbol for the Neutral Atom	Symbol for Its Common Ion	Name of Ion
IA	Lithium	Li	Li^+	Lithium ion
	Sodium	Na	Na^+	Sodium ion
	Potassium	K	K^+	Potassium ion
IIA	Magnesium	Mg	Mg^{2+}	Magnesium ion
	Calcium	Ca	Ca^{2+}	Calcium ion
	Barium	Ba	Ba^{2+}	Barium ion
IIIA	Aluminum	Al	Al^{3+}	Aluminum ion
VIA	Oxygen	O	O^{2-}	Oxide ion
	Sulfur	S	S^{2-}	Sulfide ion
VIIA	Fluorine	F	F^-	Fluoride ion
	Chlorine	Cl	Cl^-	Chloride ion
	Bromine	Br	Br^-	Bromide ion
	Iodine	I	I^-	Iodide ion
Transition elements	Silver	Ag	Ag^+	Silver ion
	Zinc	Zn	Zn^{2+}	Zinc ion
	Copper	Cu	Cu^+	Copper(I) ion (cuprous ion)[b]
			Cu^{2+}	Copper(II) ion (cupric ion)
	Iron	Fe	Fe^{2+}	Iron(II) ion (ferrous ion)
			Fe^{3+}	Iron(III) ion (ferric ion)

[a] Other common ions are listed in Table 3.4.

[b] The names in parentheses are older names, but still often used.

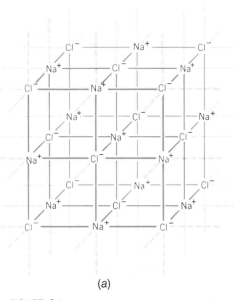

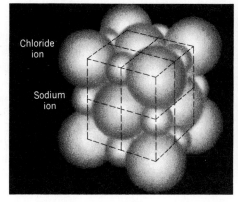

(a)   (b)   (c)

FIGURE 3.1
The structure of a sodium chloride crystal. *(a)* This schematic shows the alternating pattern of Na^+ and Cl^- ions. *(b)* Sodium ions have much smaller diameters than chloride ions. As seen here the sodium ions are surrounded by chloride ions as nearest neighbors, and the like-charged ions are just a little farther apart. *(c)* The regular shape of a sodium chloride crystal comes from the regular arrangement of its ions.

The Symbols and Names of Monatomic Ions Are Like the Names of Their Elements.
When an ion has just one nucleus, it's called a *monatomic ion,* and it's named after its parent element, either the same name or something close to it. All ions derived from metals have the same name as the element (plus the word *ion*). Monatomic ions from nonmetals have names that end in *-ide,* as in *chloride ion,* whose parent element is chlorine. The names, symbols, and electrical charges of several common ions are in Table 3.1, and they must be learned now.

When a physician or nurse refers to the *sodium level* of the blood, the sodium *ion* level is always meant.

The Force of Attraction between Oppositely Charged Ions Is Called an Ionic Bond.
The reaction between sodium and chlorine is typical of those between any element in Group IA and any in Group VIIA. Electron transfers occur from the metal to the nonmetal, and ions form. Of course atoms are so small that any visible sample, even the tiniest speck, has billions of billions of them. When actual samples are mixed, therefore, a storm of electron transfers occurs, and countless numbers of oppositely charged ions form.

Even a speck-sized sample of sodium has at least 10^{19} atoms.

Because like charges repel, the new sodium ions repel each other. The chloride ions repel each other also. But because unlike charges attract, sodium ions and chloride ions attract each other. Out of all these attractions and repulsions, the storm subsides to produce firm, hard, and regularly shaped crystals of sodium chloride. Spontaneously, the unlike-charged ions, Na^+ and Cl^-, nestle together as closest neighbors and like-charged ions stay just a little farther apart, as seen in Figure 3.1. The Na^+ ions have Cl^- ions as closest neighbors, and Cl^- ions have Na^+ ions as nearest neighbors. As close neighbors, they cannot help but attract each other.

The forces of repulsions within the crystal of sodium chloride, Na^+ from Na^+ and Cl^- from Cl^-, are still there, of course. But because they act over slightly longer distances, they cannot overcome the attractions of Na^+ to Cl^- ions. The net force of attraction between oppositely charged ions is called the **ionic bond.**

Crystals of Ionic Compounds Have Regular Shapes. As you can see in Figure 3.1, the ions end up evenly spaced in the crystal. This is why the crystal has a regular shape, as seen in the photo of Figure 3.1c. Such crystals can be shattered, of course. A hammer blow makes one layer of ions shift so that like-charged ions momentarily become closest neighbors. Now, at least along this layer, the net force is one of repulsion, not attraction, and the crystal splits.

Many familiar substances besides sodium chloride are ionic. Sodium bicarbonate in baking soda, barium sulfate in "barium X-ray cocktails," sodium hydroxide in lye and drain cleaners, and calcium sulfate in plaster of paris are other examples of ionic compounds.

Ionic Compounds Are Electrolytes. People in professional health care fields speak of the "electrolyte balance" of this or that body fluid, where **electrolyte** refers to any substance that can furnish ions in a solution in water. Every fluid in every living thing contains dissolved ions. Even among the large molecules in living organisms, molecules of proteins, for example, there generally are several electrically charged sites. Thus, at the molecular level of life, electrolytes and ions are everywhere. As we will see later, the term *electrolyte* comes from the fact that solutions of electrolytes can conduct electricity.

> Clinical chemists can measure the concentrations of several ions in the blood, and changes in concentrations can signify particular health problems.

3.2 NAMES AND FORMULAS OF IONIC COMPOUNDS

The ions in an ionic compound must assemble in whatever ratio lets the compound be electrically neutral.

Shorthand symbols for compounds are called **chemical formulas,** and they are made from the symbols of their elements to show the ratios, *by atoms,* present in the compound.

Because compounds are electrically neutral, the formula of an ionic compound must have its ions in a ratio that balances plus and minus charges. The formula of sodium chloride, for example, must show its Na^+ and Cl^- ions in a 1 : 1 ratio to ensure neutrality, because then each $1+$ charge is balanced by a $1-$ charge.

The formula of an ionic compound is made up of the symbols of its ions, but the electrical charges are omitted. They are "understood." The formula of sodium chloride is thus written as NaCl, not as Na^+Cl^-. By convention, the positive ion is placed first. The formula Na_2Cl_2 also has a 1 : 1 ratio of ions, but by convention, the ratio is usually expressed in the smallest whole numbers.

The formula of a compound between calcium ions, Ca^{2+}, and oxide ions, O^{2-}, must also have them in a ratio of 1 : 1, because only this ratio balances the $2+$ of each calcium ion with the $2-$ of each oxide ion. The formula is CaO.

> The *net* charge on any compound is zero.

> A formula that uses the *smallest* whole numbers to give the ratio of atoms is called an **empirical formula.**

Subscripts Are Used in Formulas When Atom Ratios Are Not 1 : 1. A compound made of calcium ions and chloride ions must have *two* Cl^- ions to one Ca^{2+} ion, because two minus charges are needed for *every* particle with a $2+$ charge. To show this ratio, the formula includes numbers called **subscripts.** These follow and are placed one-half of a line below their associated atomic symbols. The subscript 1 is never used because just writing a symbol means that you are taking at least one of it. Remembering that the positive ion goes first, the formula of the compound of the calcium ion and the chloride ion, therefore, is $CaCl_2$ (and not Cl_2Ca nor $Ca^{2+}Cl_2^-$ nor as Ca_2Cl_4).

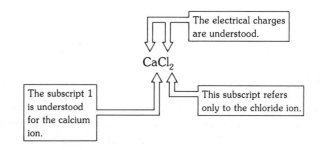

Ionic Compounds Are Named after Their Ions. To name an ionic compound, we assemble the names of the ions in the order in which they appear in the formula (except the word *ion* is left out). For example,

NaCl   Sodium chloride (sodium *ion* plus chloride *ion*)

CaO   Calcium oxide (calcium *ion* plus oxide *ion*)

$CaCl_2$ Calcium chloride (calcium *ion* plus two chloride *ions*)

Four metal ions at the bottom of Table 3.1 are given two names, for example, copper(I) ion or cuprous ion. The first name is the modern name, but the second is still widely used. Both have to be learned. The roman numeral in "copper(I)" or "iron(III)" stands for the positive charge on the ion, Cu^+ or Fe^{3+}. [Notice there is no space between "copper" and "(I)" in "copper(I)."] In the common names of the ions of copper and iron, the *-ous* endings goes with the ion of lower charge and the *-ic* ending goes with the ion of higher charge.

There are two kinds of operations involving names and formulas that you must learn: How to write formulas from names and names from formulas. We will limit the next examples and exercises to those that involve the ions in Table 3.1. (Appendix II has further information.)

EXAMPLE 3.1   WRITING FORMULAS FROM NAMES OF IONIC COMPOUNDS

Problem: Write the formula of aluminum oxide.

Solution: We first use the name to identify the ions, and then we recall their electrical charges, Al^{3+} and O^{2-}. We know that 3+ isn't canceled by 2−, so we can't simply write AlO. But the lowest common multiple of 2 and 3 is 6 ($2 \times 3 = 6$). So to balance charges we pick the smallest number of aluminum ions that give a total charge of 6+ and the smallest number of oxide ions that give a total charge of 6−. We need 2 Al^{3+}, because [$2 \times (3+) = 6+$], and 3 O^{2-}, because [$3 \times (2-) = 6-$]. The ratio, therefore, must be 2 Al^{3+} to 3 O^{2-}. Writing Al first, with a subscript of 2, and O next, with a subscript of 3, the formula is

Notice that the charges on the ions are omitted in the formula.

$$Al_2O_3 \quad \text{(the answer)}$$

The strategy in Example 3.1 to find and use the lowest common multiple of the ionic charges always works. Try Practice Exercise 1 to develop experience.

PRACTICE EXERCISE 1   Write the formulas of the following compounds.

(a) Silver bromide (a light-sensitive chemical used in photographic film).

(b) Sodium oxide (a very caustic substance that changes to lye in water).

(c) Ferric oxide (the chief component in iron rust).

(d) Copper(II) chloride (an ingredient in some laundry-marking inks).

EXAMPLE 3.2   WRITING NAMES FROM FORMULAS OF IONIC COMPOUNDS

Problem: Write the name of $FeCl_3$, using both the modern and the older forms.

Solution: The symbol Fe stands for one of the two ions of iron, but which one, Fe^{2+} or Fe^{3+}? This situation makes us use our knowledge of the charge on one ion, Cl^-, to figure out the charge on the other. The Cl_3 part of $FeCl_3$ tells us there must be a total negative charge of 3−, because $3 \times (1-) = 3-$. The Fe part of $FeCl_3$ has to neutralize this, so we must be dealing with the Fe^{3+} ion, named either the iron(III) ion or the ferric ion. Thus the modern name of $FeCl_3$ is iron(III) chloride, and the older name is ferric chloride.

Notice that there is no space between "iron" and "(III)" in iron(III).

PRACTICE EXERCISE 2   Write the names of each of the following compounds. When a name can be given in either a modern form or an older form, write both.

(a) CuS   (b) NaF   (c) FeI_2 (d) $ZnBr_2$ (e) Cu_2O

The technique used in Example 3.2 to find the electrical charge of an unfamiliar ion from the known charge of another in an ionic compound greatly reduces the number of ions and their charges that have to be memorized. Practice Exercise 3 provides further examples. It involves unfamiliar ions not in Table 3.1.

PRACTICE EXERCISE 3   What are the charges on the *metal* ions in each of these substances?

(a) Cr_2O_3 (a green pigment is stained glass).
(b) HgS ("Chinese red," a bright, scarlet-red pigment).
(c) $CoCl_2$ (an ingredient in invisible ink).

A Small Particle Made from the Atoms Shown in a Formula Is Called a Formula Unit.   It is not possible to have a sample of NaCl that we could manipulate in the lab that consists of just one sodium ion and one chloride ion. Yet, it is still useful to have a name for such a particle. We call it a *formula unit,* and we define this term so that it can apply to any kind of substance, an element, an ionic compound, or a molecular compound. A **formula unit** is that very small particle made of the atoms specified by the formula regardless of how the atoms are held together. One formula unit of NaCl thus consists of one sodium ion and one chloride ion. One formula unit of sodium metal, Na, consists of one atom of sodium. One formula unit of water, H_2O, consists of one molecule of water.

3.3   THE OCTET RULE

Atoms and ions whose outside energy levels hold eight electrons are substantially more stable than those that do not.

Only krypton and xenon of the noble gases form a few compounds, and these are limited to compounds with fluorine and oxygen.

The Outer Octet Is a Condition of Unusual Stability.   The Group 0 elements have almost no reactions. Evidently there is something special and stable about the two kinds of electron configurations that occur among them. One is the **outer octet,** eight electrons in the outside level. The other is a *filled* level 1 when it is the outside level (as in helium). Thus, *noble gas configurations are conditions of unusual chemical stability.* This fact will help us understand how elements in other families form the ions they do, and why all ions of the same family have the same charge.

A sodium atom, in Group IA, has one electron in its outside level, level 3, and sodium is a reactive element. After a reaction, however, in which it loses its level 3 electron to form the Na^+ ion, level 2 with its eight electrons becomes the new outside level. Thus Na^+, the sodium *ion,* has an outer octet and the noble gas configuration of neon. Whatever makes an outer octet give stability to the neon *atom* evidently works in the sodium *ion,* because this ion is also chemically very stable. (It is stable provided something of opposite charge, like a chloride ion, is near it. We can never speak of a *charged* particle as particularly stable unless it's in a charge-balanced environment.)

A chlorine atom, in Group VIIA, has seven electrons in its highest occupied energy level, number 3. When it accepts one electron to become Cl^-, this outer level acquires an octet, and the chloride *ion* has the same configuration as an atom of a noble gas, argon. Like argon, the chloride ion is unusually stable (provided that something of opposite charge is nearby).

When we put samples of sodium and chlorine together, the ions that are able to form by electron transfers are only those with noble gas configurations. The spontaneous reaction of sodium with chlorine thus leads to greater stability for the atoms of both. They change from particles that do not have noble gas configurations to those that do. The atoms sacrifice neutrality for stability. It doesn't matter that the new particles are charged, because they are *oppositely* charged. Once formed, the ions attract each other as ionic bonds develop.

We haven't explained *why* a noble gas configuration is stable. We only point out that for some reason it is. (No one knows why.) The pattern we noted for sodium and chlorine is so consistently observed among the reactions of the representative metals and nonmetals that it's virtually a law of nature. It's called the **octet rule.**

> Octet Rule The atoms of the reactive representative elements tend to undergo those chemical reactions that most directly give them electron configurations of the nearest noble gas.

Group Number	Oxidation Number
IA	1+
IIA	2+
IIIA	3+
VIA	2−
VIIA	1−
0	None

Table 3.2 shows how the ions of the representative elements have noble gas configurations. The charges of these ions are called their **oxidation numbers,** and the octet rule lets us predict the oxidation numbers of the most likely ions of the representative elements. The next three examples illustrate this.

TABLE 3.2
Electron Configurations of Ions and Comparable Noble Gases

Group	Common Element	Atomic Number	Atoms Electron Configurations (main levels only) 1	2	3	4	5	6	Ion	Ions Electron Configurations (main levels only) 1	2	3	4	5	6	Nearest Noble Gas
IA Alkali metals	Li	3	2	1					Li^+	2						Helium
	Na	11	2	8	1				Na^+	2	8					Neon
	K	19	2	8	8	1			K^+	2	8	8				Argon
IIA Alkaline earth metals	Mg	12	2	8	2				Mg^{2+}	2	8					Neon
	Ca	20	2	8	8	2			Ca^{2+}	2	8	8				Argon
	Ba	56	2	8	18	18	8	2	Ba^{2+}	2	8	18	18	8		Xenon
VIA Oxygen family	O	8	2	6					O^{2-}	2	8					Neon
	S	16	2	8	6				S^{2-}	2	8	8				Argon
VIIA Halogens	F	9	2	7					F^-	2	8					Neon
	Cl	17	2	8	7				Cl^-	2	8	8				Argon
	Br	35	2	8	18	7			Br^-	2	8	18	8			Krypton
	I	53	2	8	18	18	7		I^-	2	8	18	18	8		Xenon
0 Noble gases	He	2	2						The noble gases do not form stable ions							
	Ne	10	2	8												
	Ar	18	2	8	8											
	Kr	36	2	8	18	8										
	Xe	54	2	8	18	18	8									

EXAMPLE 3.3   USING THE OCTET RULE

Problem: When nutritionists speak of the calcium requirement of the body, they always mean the calcium *ion* requirement. Calcium has atomic number 20. What charge does the calcium ion have, and what is the symbol of this ion?

Solution: There are two methods to solve this kind of problem, and you should learn both. The first is to work with the electron configuration of the atom, and the second is to use the periodic table.

Using the rules learned in Chapter 2, we write the electron configuration of element 20, calcium, as follows:

$$2 \quad 8 \quad 8 \quad 2$$
$$\text{level} \quad 1 \quad 2 \quad 3 \quad 4$$

Only *very* rarely are there situations in which atoms gain or lose more than three electrons.

This gives us what we need to know; the calcium atom has two electrons in the outside level and eight in the next inner level. Only by losing *both* of the outer electrons can this atom get a new outside level with an octet. Losing one electron won't do. Neither will losing three or more. The only stable ion that calcium can form, therefore, is one with the following electron configuration.

$$2 \quad 8 \quad 8$$
$$\text{level} \quad 1 \quad 2 \quad 3$$

By losing two electrons the net charge on the particle becomes 2+, so the symbol for the calcium ion is Ca^{2+}.

The second way to arrive at this answer is to find calcium in the periodic table; it's in Group IIA. Because all of the Group IIA elements have two outside-level electrons and all can acquire configurations of the nearest noble gases by losing these, the Group IIA ions all bear charges of 2+. Therefore the Group IIA ions are Be^{2+}, Mg^{2+}, Ca^{2+}, Sr^{2+}, Ba^{2+}, and Rn^{2+}.

EXAMPLE 3.4   USING THE OCTET RULE

Problem: Oxygen can exist as the oxide ion in such substances as calcium oxide, an ingredient in cement. What is the symbol for the oxide ion, including its electrical charge?

Solution: As in Example 3.3, we will solve this in two ways. First, the electron configuration of an oxygen atom (atomic number 8) is

$$2 \quad 6$$
$$\text{level} \quad 1 \quad 2$$

This shows us what we need to know; the outside level of the oxygen atom has six electrons, just two short of an octet and a neon configuration. We might have also noted that the oxygen atom has six too many electrons to have the helium configuration. However, gaining two electrons to become like neon is much easier than losing six, so oxygen achieves a noble gas configuration most directly by accepting two electrons from some metal atom donor. The configuration of a stable ion of oxygen, therefore, is

$$2 \quad 8$$
$$\text{level} \quad 1 \quad 2$$

The two extra electrons make the ionic charge 2−, so the symbol for the oxide ion is O^{2-}.

Using the periodic table, we note that oxygen is in Group VIA, so this tells us that its outside level has six electrons. It must pick up two electrons—not just one and not more than two—to have a noble gas configuration. These two extra electrons give the particle a charge of 2−, so we can write O^{2-} directly.

The next worked example introduces an important rule about the elements in Groups IVA and VA. *They generally do not form ions.*

EXAMPLE 3.5   USING THE OCTET RULE

Problem: Hardly any element is involved in more compounds than carbon (Group IVA). Roughly 6 million are known. Can carbon atoms change to ions? If so, what is the symbol of the ion?

Solution: We will use the shorter method of solving this. Because carbon is in Group IVA, its atoms have four outside-level electrons. To achieve a noble gas configuration, a carbon atom either must lose these four (and become helium like) or gain four (and become neon like). In one or two rare situations, carbon does the latter—becomes the C^{4-} ion—but this is so rare that we ignore it. Almost equally rare is the formation of ions of Group VA elements. Their atoms would have to acquire three electrons or lose five to have outer octers. Therefore, we can state a rule: **Any *nonmetal* atoms in either Groups IVA and VA do not form ions.** (Exceptions can be made as needed.)

PRACTICE EXERCISE 4   Write the electron configuration of an atom of each of the following elements, and from this deduce the charge on the corresponding ion. If the atom isn't expected to have a corresponding ion, state so. The numbers in parentheses are atomic numbers. Do not use the periodic table for this practice exercise.

(a) Potassium (19)   (b) Sulfur (16)   (c) Silicon (14)

PRACTICE EXERCISE 5   Write the electron configurations of the *ions* of the elements in Practice Exercise 4 that can form ions.

PRACTICE EXERCISE 6   Relying on their locations in the periodic table, write the symbols of the ions of each of the following elements. Always remember that no symbol of an ion is complete without its electrical charge. (The numbers in parentheses are atomic numbers.)

(a) Cesium (55)   (b) Fluorine (9)
(c) Phosphorus (15)   (d) Strontium (38)

3.4   MOLECULAR COMPOUNDS

In molecules, pairs of atoms that could not become ions by electron-transfer share pairs of electrons in covalent bonds.

The nonmetals of Groups IVA and VA rarely become ions. Yet one of them, carbon, probably occurs in more different compounds that any other element. Evidently nature forms other kinds of bonds than the ionic bond. They occur often among compounds made entirely of nonmetals, and such compounds consist of molecules, not ions. A **molecule,** as we said, is a small, electrically neutral particle made of at least two nuclei and enough electrons to make the system neutral. A compound that consists of molecules is called a **molecular compound.** Molecules are the *formula units* of molecular compounds. Some elements are also molecular, as we will see next.

The three fundamental *chemical* particles are the atom, the ion, and the molecule.

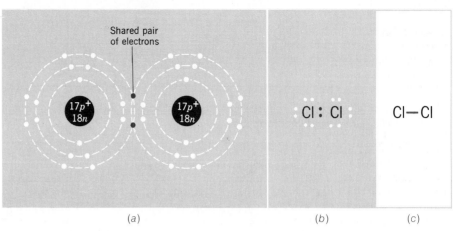

FIGURE 3.2
The covalent bond in the Cl—Cl molecule. *(a)* An electron-dot representation that shows which pair of electrons is shared. *(b)* A condensed electron-dot symbol where only outer-level electrons are shown. *(c)* The bond-line structure, the structural formula, of a chlorine molecule.

Co- from cooperative; *-valent* from the Latin *valere,* to be strong, signifying strong binding.

Electron Density becomes Concentrated between Two Nuclei in the Covalent Bond.
Unlike the noble gas elements (Group 0), the halogens (Group VIIA) consist of diatomic molecules with the formulas F_2, Cl_2, Br_2, and I_2. The other diatomic elements are oxygen, O_2, hydrogen, H_2, and nitrogen, N_2. The diatomic particles of these elements cannot be explained in terms of ions, because we would need *positively* charged as well as negatively charged ions. No nonmetal atom forms a positively charged ion. If a fluorine atom (Group VIIA), for example, with seven outer electron became F^+, it would have six outside-level electrons, not an outer octet. Fluorine, therefore, cannot form F^+ that could then attract F^- and so form an ionic F_2 formula unit. Yet, we know that to fashion any kind of chemical bond we must have an attraction between opposite charges. Thus if we can't use F^+ and F^- to make F_2, there must be another way to get an attraction between these atoms.

The source of such an attraction was first proposed by an American chemist, G. N. Lewis (1875–1946). He suggested that a pair of outer-level electrons, one from each atom, can be shared between two nuclei. Being *shared* means that *the pair spends most of its time between the nuclei.* The pair thus gives to this region a relatively high density of negative charge, a high **electron density.** The nuclei are naturally attracted to this region, because unlike charges attract. This electrical force of attraction created by sharing a pair of electrons is called the **covalent bond.** As each gives up an electron for the shared pair, the two initially separated atoms lose their individual identities, their nuclei stay permanently near each other, and a molecule results.

Figure 3.2 shows three ways to depict the shared pair of electrons in the chlorine molecule. Part *(a)* is an electron-dot representation; part *(b)* strips away all inner-level electrons (leaving their presence understood), and part *(c)* shows the shared pair represented by a line. The line is a particularly useful symbol for a covalent bond because it so easly lets us depict molecules with many bonds.

The Octet Rule Applies to Covalent Bonds.
If we count both electrons of a shared pair for each atom, the octet rule can be used for molecules. We can see this in the chlorine molecule, for example,

(The circles enclose octets when the
shared pair is counted for each atom)

The hydrogen molecule, Figure 3.3, illustrates how the shared pair gives each hydrogen atom the helium configuration. For a discussion in greater depth of the covalent bond, based on atomic orbitals that merge to form molecular orbitals, see Special Topic 3.1.

SPECIAL TOPIC 3.1   MOLECULAR ORBITALS

In Special Topic 2.1 we learned about atomic orbitals, the homes of individual electrons in atoms. Here we'll learn about the homes of the shared electron pairs of covalent bonds in molecules, molecular orbitals.

Suppose we have two atoms approaching each other on a collision course. Each atom has an atomic orbital with just one electron in it. In this circumstance as the two orbitals hit, they partly merge to form a new kind of orbital, a **molecular orbital,** which encloses the nuclei of both atoms. The two electrons, one from each individual atom's merging orbital, now move within this space. They experience an attraction toward two nuclei now. This attraction toward two nuclei is what we mean by the sharing of a pair of electrons between the covalently bonded atoms.

The shared electrons are naturally between the two nuclei most of the time, so this is where the electron density

is the greatest. This electron density is what attracts and holds the nuclei and creates the bond.

The figure on the left, below, illustrates how the molecular orbital of a hydrogen molecule forms. Two $1s$ atomic orbitals, each with an electron, partially merge to give a space within which the two hydrogen nuclei now are found. This partial merging is sometimes called the *overlapping* of atomic orbitals.

The figure on the right shows how two p orbitals can overlap to create the molecular orbital for the covalent bond in the fluorine molecule.

In other situations (not illustrated), an s orbital can overlap with one lobe of a p orbital.

In double bonds, two pairs of atomic orbitals overlap. And in triple bonds, three pairs of orbitals overlap.

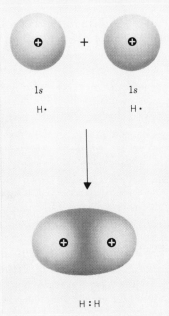

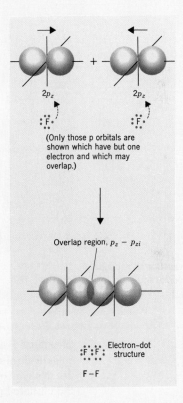

Above: The molecular orbital in the hydrogen molecule is created by the partial overlapping of the $1s$ orbitals of two hydrogen atoms. The shared pair of electrons spends most of the time between the two nuclei, so these nuclei are attracted toward each other.

Right: When p orbitals of separated fluorine atoms partially overlap, end to end, a molecular orbital forms in which a shared pair of electrons can reside and create a covalent bond between the fluorine atoms.

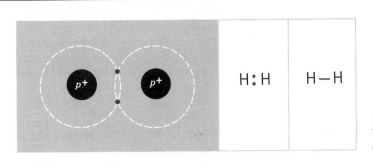

FIGURE 3.3
The covalent bond in H—H represented three ways.

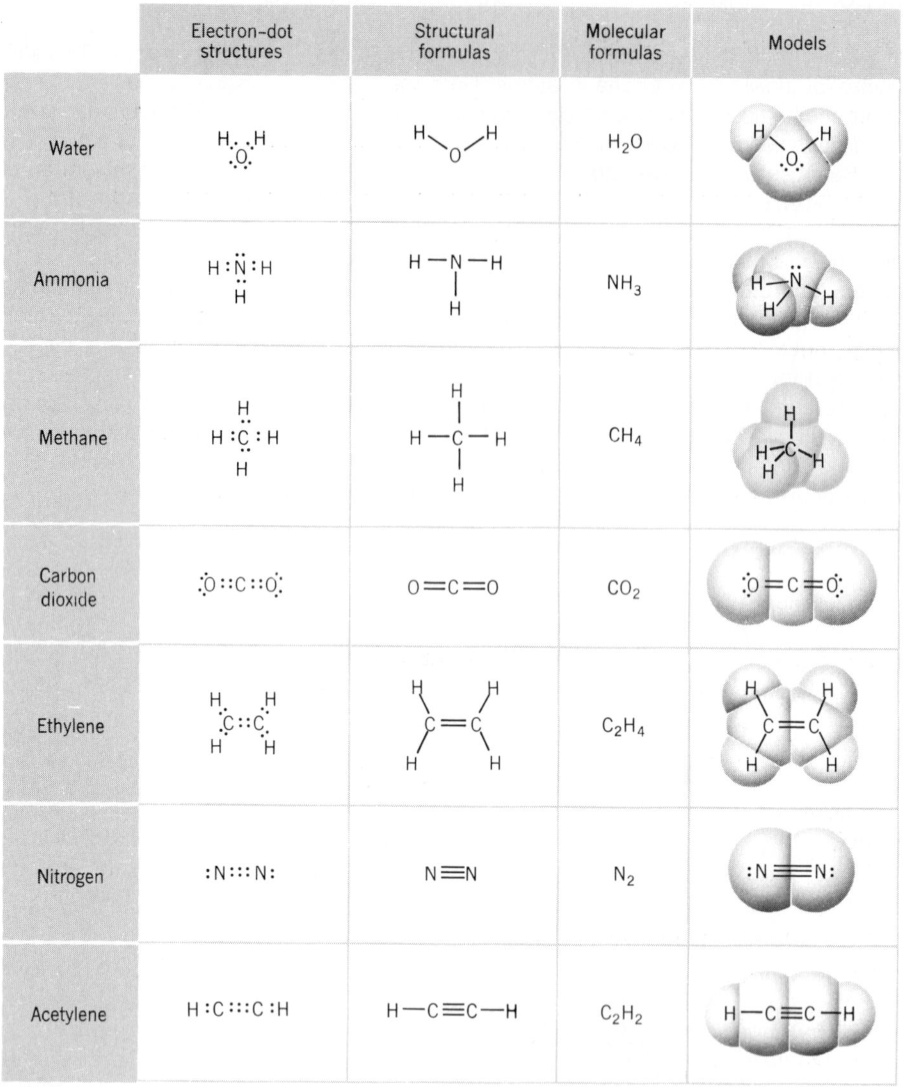

	Electron–dot structures	Structural formulas	Molecular formulas	Models
Water	H:O:H (with O below)	H—O—H	H_2O	
Ammonia	H:N:H, H	H—N—H, H	NH_3	
Methane	H:C:H, H top/bottom	H—C—H with H top/bottom	CH_4	
Carbon dioxide	:O::C::O:	O=C=O	CO_2	
Ethylene	H C::C H (with H's)	H₂C=CH₂ structure	C_2H_4	
Nitrogen	:N:::N:	N≡N	N_2	
Acetylene	H:C:::C:H	H—C≡C—H	C_2H_2	

FIGURE 3.4
Some common molecular compounds in which single, double, and triple bonds occur.

Structural Formulas Show the Patterns of Connections in Molecules. Because molecules made of several atoms would require us to write a large number of dots, chemists seldom use this symbolism. Instead, each shared pair of electrons is represented by a straight line and the unshared pairs (those not between two atoms) are normally omitted. The symbol of a molecule in which lines between atomic symbols represent covalent bonds is called a **structural formula** or, simply a **structure.** A structure depicts an atom-to-atom sequence in a molecule. Figure 3.4 shows several.

Every structural formula has a **molecular formula,** which gives only the composition of one molecule with no information about internal sequences. When a molecule has three or more atoms, however, more than one structure can be written, so a structural formula gives more information than a molecular formula. The molecular formula of water, for example, is H_2O, and with no additional instructions we might imagine that the structure is either H—H—O or H—O—H. The correct structure is H—O—H and, as we'll learn soon, the octet rule helps us reject H—H—O.

Two or Three Pairs of Electrons Can be Shared. When one pair of electrons is shared, the covalent bond is a **single bond.** But many molecules involve two or three shared pairs. When

TABLE 3.3
Covalences of Some Common Elements

Element	Atomic Number	Electron Configuration (by main levels)					Covalence[a]
		1	2	3	4	5	
Carbon	6	2	4				4
Nitrogen	7	2	5				3
Oxygen	8	2	6				2
Sulfur	16	2	8	6			2
Fluorine	9	2	7				1
Chlorine	17	2	8	7			1
Bromine	35	2	8	18	7		1
Iodine	53	2	8	18	18	7	1
Hydrogen	1	1					1

[a] Only the most common covalences are given. Some of these elements have multiple covalences, which we will study later.

two pairs of electrons are shared, we have a **double bond,** and with three shared pairs, we have a **triple bond.** As seen in Figure 3.4, carbon dioxide has two double bonds, ethylene has one double bond, and both nitrogen and acetylene have one triple bond. In all cases, as the electron-dot structures show, the octet rule is obeyed.

Double bonds between carbon atoms occur often among the compounds at the molecular level of life.

The Number of Covalent Bonds That an Atom Can Have in a Molecule Is Called Its Covalence.
As Figure 3.4 shows, atoms of nonmetals differ in their electron-sharing abilities. Oxygen can form two bonds; nitrogen can form three; and carbon, four. Halogen atoms and hydrogen atoms can each form one covalent bond. The covalent combining ability of an element, called its **covalence,** equals the number of electrons its atoms must obtain by sharing to have a noble gas configuration. Thus a hydrogen atom has to get a share of one more electron to achieve a helium configuration, so its covalence is 1. Several common covalences are given in Table 3.3. These must be learned.

The Group VIIA elements (the halogens) all have seven outer-level electrons and all need a share in one more electron to have outer octets. Therefore, their covalences are 1 also. Notice, now, that the covalence equals [8 − (group number)], or with the halogens, 1 = 8 − 7.

The Group VIA elements (oxygen family) have six outer-level electrons; they need at least a share in two more for octets; and so they have covalences of 2. And this covalence, 2, equals (8 − 6), where 6 is the group number.

Group VA elements (nitrogen family) have five outer-level electrons and so they need a share in three more electrons (8 − 5) for octets. These elements, therefore, have covalences of 3, as seen in the ammonia molecule, Figure 3.4.

The Group IVA elements (carbon family) have four outer-level electrons. Their need is for four more electrons (by sharing) to achieve octets, so their covalences are 4, which is 8 − 4. Thus the methane molecule has four bonds from carbon, all single. Carbon also has four bonds (two double bonds) in carbon dioxide (Figure 3.4).

Think of the covalence as the number of lines (bonds) that must radiate from the atom in a structure. Thus a nitrogen atom, with a covalence of 3, must have three lines. They can be three single bonds, a single plus a double bond, or one triple bond, as long as there are three. Examine Figure 3.4 for additional examples. We can now see why we cannot accept H—H—O as the structure for water, H_2O. One of the H atoms has two lines (bonds) from it (one too many), and the O atom has only one line (one bond too few). But the structure H—O—H satisfies the covalences.

PRACTICE EXERCISE 7   Convert the following molecular formulas into structures using the covalences of Table 3.3. (*Hint:* Draw in the bonds on the atom of highest covalence first and use those of covalence 1 last.)

(a) CCl_4 (b) H_2S (c) NCl_3 (d) $CHBr_3$

The Atoms in Polyatomic Ions Are Held Together by Covalent Bonds.   We didn't mention earlier that several ions, the **polyatomic ions,** consist of many atoms because we needed to know about the covalent bond to explain them. These ions are clusters of atomic nuclei and electrons held together by covalent bonds, and all atoms have octets. But in a polyatomic ion, the total plus charge on all nuclei is not perfectly balanced by the total number of electrons. There are either one, two, or three too many electrons, or (rarely) one or two too few. A simple example is the hydroxide ion (OH^-), whose structure is given by **1.**

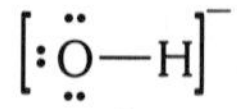

$$\left[:\overset{\cdot\cdot}{\underset{\cdot\cdot}{O}}-H\right]^-$$
1

In **1,** oxygen has an octet, and hydrogen has the helium configuration, so the conditions of stability are met. The OH^- ion, however, has 10 electrons (2 in oxygen's inner level plus 2 shared electrons plus the 6 shown as dots); and to balance the $10-$ charge, OH^- has only 9 protons (8 in the oxygen nucleus and 1 in the hydrogen nucleus). The net charge is, therefore, $1-$.

Other polyatomic ions can be analyzed this way, and important examples are shown in Table 3.4. Their names, formulas, and charges must now be learned. Many are involved as electrolytes in body fluids.

TABLE 3.4
Some Important Polyatomic Ions

Name	Formula
Ammonium ion	NH_4^+
Hydronium ion[a]	H_3O^+
Hydroxide ion	OH^-
Acetate ion	$C_2H_3O_2^-$
Carbonate ion	CO_3^{2-}
Bicarbonate ion[b]	HCO_3^-
Sulfate ion	SO_4^{2-}
Hydrogen sulfate ion[c]	HSO_4^-
Phosphate ion	PO_4^{3-}
Monohydrogen phosphate ion	HPO_4^{2-}
Dihydrogen phosphate ion	$H_2PO_4^-$
Nitrate ion	NO_3^-
Nitrite ion	NO_2^-
Hydrogen sulfite ion[d]	HSO_3^-
Sulfite ion	SO_3^{2-}
Cyanide ion	CN^-
Permanganate ion	MnO_4^-
Chromate ion	CrO_4^{2-}
Dichromate ion	$Cr_2O_7^{2-}$

[a] This ion is known only in a water solution.

[b] Formal name: hydrogen carbonate ion.

[c] Common name: bisulfate ion.

[d] Common name: bisulfite ion.

Parentheses Sometimes Enclose Polyatomic Ions in Formulas. In ammonium sulfide the ratio of ions is two NH_4^+ ions to one S^{2-} ion. To show this ratio without getting subscripts confused, we place parentheses as follows.

$$(NH_4)_2S$$

Ammonium sulfide

The 4 goes with H to show a ratio of 4 H to 1 N *within* the ammonium ion, but the 2 is outside the parenthesis to show that it goes with the entire ammonium ion. Thus in $(NH_4)_2S$ there are two NH_4^+ ions giving a total of 2 N and 8 H.

PRACTICE EXERCISE 8   Using Tables 3.1 and 3.4, write formulas of the following. (Eventually, you must be able to do this without referring to the tables.)

(a) Sodium nitrate   (b) Potassium hydroxide   (c) Calcium hydroxide
(d) Magnesium carbonate   (e) Sodium sulfate   (f) Ammonium phosphate

---

PRACTICE EXERCISE 9   Write the names of the following compounds.

(a) Li_2CO_3 (b) $NaHCO_3$ (c) $KMnO_4$ (d) NaH_2PO_4 (e) $(NH_4)_2HPO_4$

3.5 POLAR MOLECULES

Even electrically neutral molecules can attract each other if they are polar.

If molecules are neutral, how can they stick together? Sugar molecules, for example, stack together naturally to make beautiful crystals that are not easy to melt. What holds sugar molecules together in such crystals? We will introduce the answer here, an answer deeply relevant to such matters as the strengths of muscle fibers and nylon threads, the resilience of skin, and the molecular basis of the action of hormones.

Shared Pairs between Unlike Atoms Are Usually Not Equally Shared. The space between two nuclei occupied by shared electrons is a region with a high density of negative charge. This place is sometimes said to have an **electron cloud,** because we can imagine the electrons swarming like a cloud of bugs. Clouds are mobile, however, and some clouds are thick while others are thin. Let us now see what can happen to electron clouds when atoms with *different* nuclear charges are held by a covalent bond.

The electron cloud of an electron pair shared between nuclei of different charge usually becomes stabilized nearer one end of the bond than the other. When the two atoms are of the same period in the periodic table, the nucleus with the *larger* positive charge pulls the electron cloud somewhat away from the other nucleus. This thins the electron cloud at one end of the bond and thickens it at the other end.

The nucleus of lower charge has the thinner cloud, so its positive charge is not entirely canceled where this nucleus is located in the molecule. A fraction of the positive charge of this atomic nucleus is still able to exert whatever influence any charge can exert. Such a fractional charge is called a *partial charge,* and we use the Greek lowercase letter delta, δ, to stand for *partial.* Thus a partial positive charge has the symbol $\delta+$.

The hydrogen fluoride molecule, for example, has a $\delta+$ charge at its hydrogen end. The hydrogen nucleus has only a charge of $1+$, but the fluorine nucleus has a charge of $9+$, nine times as great. Moreover, because fluorine's other electrons are not concentrated between the two nuclei in H—F, they cannot entirely shield the influence of fluorine's nuclear charge from the shared pair. This is why the electron cloud of the shared pair in H—F is pulled toward the fluorine end. This is the reason that the hydrogen end does not retain enough electron density

The two atoms being compared must be of the same period so that they have the same number of *inner level* electrons shielding the nuclei from those in the outer level.

The octet theory does not explain the angularity of the water molecule.

H—O—H

H   H
 \\ /
  O

Linear model   Angular model
(incorrect)   (correct)

The easiest way to understand the correct geometry is in terms of a simple theory with a very long name—the **valence-shell electron-pair repulsion theory,** or **VSEPR** for short. *Valence shell* refers to the outer-level electrons of an atom. *Electron pair* refers to the strong tendency for electrons in valence shells in molecules to occur in pairs, each pair making up one electron cloud. *Repulsion* describes the effect of each cloud on other clouds.

VSEPR theory says that the directions taken by bonds at a central atom (such as the O atom in H_2O, the N atom in NH_3, and the C atom in CH_4) are determined by the need for valence-shell electron-pair clouds to avoid each other as much as possible. Whether the electron-pair clouds are occupied by shared pairs or unshared pairs, they try to keep apart.

A simple analogy involving balloons will help. If you have four balloons of the same shape and tie them together at a common point, the balloons spontaneously take up positions shown in the accompanying photograph. Now imagine that lines—they're called *axes*—project from the center to the farthest points of the balloons' surfaces. These axes intersect at angles of 109.5°, and their ends are at the corners of a regular tetrahedron, shown in the drawing. Now imagine that each balloon represents an electron-pair cloud, and the common point is the nucleus of some central atom. The clouds, like the balloons, try to stay out of each other's way. In other words, the natural orientation for the axes of four electron-pair clouds is tetrahedral. They make angles of 109.5°.

When four balloons are tied together, their axes don't lie in a plane but point, instead, to the corners of a regular tetrahedron (below).

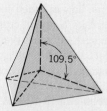

109.5°

A regular tetrahedron—a four-sided figure bounded by identical, equilateral triangles. Any two lines from corners to the center make an angle of 109.5°.

to neutralize fully the positive charge of the hydrogen nucleus where it is. This is also the reason that the hydrogen end has a $\delta+$ charge. Now let's look at the other end.

When the F end of the H—F molecule pulls some electron density toward itself, it gathers more than enough to neutralize the positive charge located there on the fluorine nucleus. The F end of the molecule, therefore, has a partial negative charge, $\delta-$.

We can't say precisely what the sizes of these fractions are, but because the molecule as a whole is electrically neutral, the algebraic sum of the $\delta+$ and the $\delta-$ must be zero.

Remember, molecules taken as a whole are neutral.

Polar Bonds Have Opposite Partial Charges at Either End. When a covalent bond has a $\delta+$ at one end and a $\delta-$ at the other, it is called a **polar bond.** We can symbolize the electrical polarity of the bond in hydrogen fluoride in either of two ways, as seen in structures **2** and **3.**

$$\overset{\delta+\ \ \ \delta-}{H—F} \qquad \overset{\longmapsto}{H—F}$$

2   **3**

When all four electron-pair clouds are involved in molecular orbitals — covalent bonds — to the central atom, as in methane, CH_4, the bond angles are exactly tetrahedral, 109.5°, as the accompanying drawing shows.

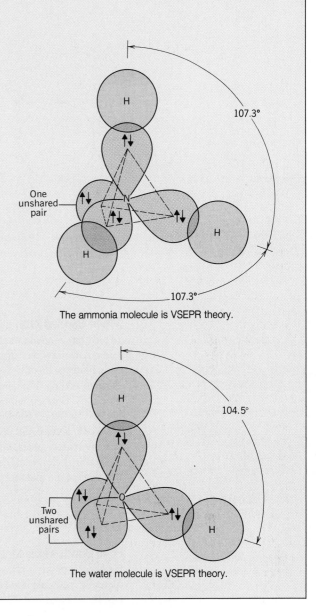

The ammonia molecule is VSEPR theory.

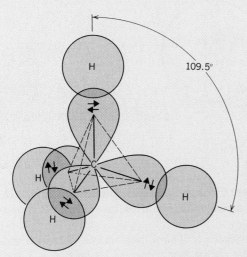

The methane molecule is VSEPR theory.

When one pair is unshared and three are in covalent bonds, as in ammonia, NH_3, the cloud of the unshared pair, not being involved with another atomic nucleus, slightly pushes the other clouds away and squeezes the bond angles from 109.5° to 107.3°, shown on the top right.

When two pairs are unshared and two are in bonds, as in H_2O, there is a slightly greater squeezing effect on the pairs in the bonds. Thus the bond angle in water is forced from 109.5° to 104.5°, shown on the right.

The water molecule is VSEPR theory.

In **3**, the arrow points toward the end of the bond that is richer in electron density. At the other end, there is a hint of the positive character by the merger of the arrow with a plus sign. Because there are two partial, opposite charges in H—F, this molecule is sometimes said to have an **electrical dipole**.

A magnet is a good analogy to an electrically polar, diatomic molecule, like H—F. A magnet has a magnetic dipole, two poles labeled north and south. Perhaps you have played with toy magnets and know that two magnets can stick to each other *if they are lined up properly*. In fact, if you have a great many magnets, and line them up correctly, you can make all of them cling together. You just have to make sure that poles of opposite kind are nearest neighbors and that poles that are alike are as far apart as possible.

Molecules that are electrically polar can stick to each other just like this. Given the freedom to move, they will line up automatically the way magnets can (see Figure 3.5). This is how neutral molecules can stick to each other. How tightly they stick depends on the sizes of the partial charges and the shapes of the molecules. Shapes might interfere with molecules getting close enough for the forces of attraction between opposite partial charges to work. Special Topic 3.2 describes how several simple molecules maintain unique shapes.

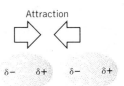

Attraction

Two polar molecules

Two polar molecules can attract each other.

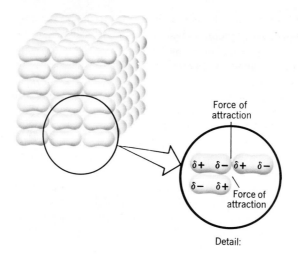

FIGURE 3.5
Polar molecules attract each other in
crystal of a molecular compound.

Detail:

When Bonded Atoms Have Different Electronegativities, the Bond Is Polar.   The ability
of an atom of an element to draw electron density toward itself in a covalent bond is called the
electronegativity of the element.

Fluorine has the highest electronegativity of all of the elements. This is the result of the
fluorine atom's high nuclear positive charge being shielded by only level 1 and level 2
electrons. Oxygen, which stands just to the left of fluorine in the periodic table, has the next
highest electronegativity. Its atoms have one less positive charge than fluorine atoms, but still
have only level 1 and level 2 electrons. The element with the third highest electronegativity,
you might now guess, lies just to the left of oxygen. It is nitrogen with atoms that have one less
charge on their nuclei than oxygen atoms. We won't need to know the relative electronegativities of all of the elements, but among those of particular importance in our study, the order is,

This order of electronegativities
should be learned.

$$O > N > C > H$$

The difference between C and H is very small.

Polar Bonds Make Molecules Polar If the Bond Polarities Do Not Cancel.   It's easy to tell
if a *bond* is polar; it always is if the atoms at each end have different electronegativities. For a
diatomic molecule like H—F, when the bond is polar so is the molecule. Of course diatomics
like H—H and F—F, which involve identical atoms, cannot be polar.

PRACTICE EXERCISE 10   Examine the bonds in each structure. Write $\delta+$ and $\delta-$ signs at the correct ends of each bond.

       H
       |
(a) H—N—H   (b) H—F   (c) H—O—O—H   (hydrogen peroxide)

Whether larger molecules are polar in an overall sense depends not just on the presence
of polar bonds but also on the geometry of the molecule. The polarities of individual bonds can
cancel each other if the molecular geometry is right. Consider, for example, the carbon
dioxide molecule, **4**.

O=C=O   H
              \\
                O—H

4   **5**

Because oxygen is more electronegative than carbon, each carbon–oxygen double bond in carbon dioxide must be polar. But these two dipoles point in exactly opposite directions, so in an overall sense they cancel each other. This leaves the molecule as a whole nonpolar. The water molecule, **5**, on the other hand, is angular. Its two individual O—H bond polarities cannot cancel, and the water molecule is polar.

Another useful way to think about the polarity of a molecule uses the idea of a *center of density of charge,* something like a "balance point" for electrical charge. The symmetry of the carbon dioxide molecule, **4**, tells us that all the positive charges on the three nuclei balance around the center of the carbon nucleus. Similarly, all of the negative charges contributed by all the electrons must also balance *around the identical point.* Because these two centers are in the same place, the molecule is nonpolar.

In a **polar molecule,** the centers of density of positive and negative charge are not at the same place. We do not have to be able to pinpoint these centers exactly to know if they are in the same place or not. Wherever they are in the water molecule, **5**, we know that because of the angularity of the molecule they can't be at the same point. Hence, we know that this molecule must be polar.

A center of electrical charge is like a center of gravity, a single point from which all charge (or mass, in the case of gravity) can be thought of as acting.

SUMMARY

Ionic bonds and ionic compounds A reaction between a metal and a nonmetal usually occurs by a transfer of an electron from the metal atom to the nonmetal atom. The metal atom changes into a positively charged ion and the nonmetal atom becomes a negatively charged ion. The oppositely charged ions aggregate in whatever whole-number ratio ensures that the product is electrically neutral. The electrical force of attraction between the ions is called the ionic bond, and compounds made of ions are called ionic compounds. They are also electrolytes.

Octet rule Among the representative elements, a useful guide to predict how many electrons transfer is that the resulting ions must have electron configurations of the nearest noble gases. Metal atoms lose one, two, or three outer-level electrons to achieve this, and nonmetal atoms gain enough new outer-level electrons to acquire such configurations. However, the nonmetal elements of Groups IVA and VA seldom become ions.

Formulas The formulas of ionic compounds begin with the symbol of the metal ion (without the sign). Subscripts are used to give the ratios of the ions. When two or more polyatomic ions are in the formula, parentheses must enclose their symbols. The names of ionic compounds are based on the names of the ions (except that the word *ion* is omitted). The name of the positively charged ion precedes that of the other ion.

Redox reactions Electron transfers are called redox reactions. An atom that loses electrons is oxidized, and one that gains electrons is reduced. Anything that causes an oxidation is an oxidizing agent; a reducing agent is anything that causes a reduction. Metals tend to be reducing agents and nonmetals are oxidizing agents. Simple, monatomic ions have oxidation numbers that equal their electrical charges. Elements have oxidation numbers of zero.

Molecular compounds Atoms of nonmetals can form molecules by sharing pairs of outer-level electrons. Each shared pair constitutes one covalent bond, and each pair is counted as the joint property of both atoms when octets are checked. The shared pair creates a region of relatively high electron density between the two atoms toward which the nuclei are electrically attracted, and this attraction is called a covalent bond.

Covalence numbers The minimum number of covalent bonds a nonmetal atom can have equals the number of electrons it must get by sharing to achieve an octet. Sometimes two or three pairs of electrons are shared, which means that double or triple bonds occur. In polyatomic ions, the atoms are joined by covalent bonds, but the overall numbers of electrons and protons do not balance.

Polar molecules If individual bond polarities, caused by electronegativity differences between the joined atoms, do not cancel, the otherwise neutral molecule is polar and it can stick to adjacent molecules much as magnets can stick together.

REVIEW EXERCISES

The *Study Guide* that accompanies this book contains answers to these review exercises.

Ions and Ionic Compounds

3.1 The term *chemical bond* is another name for what kind of force?

3.2 An atom and an ion are both small particles. What is the fundamental difference between the two?

3.3 What subatomic particle in an atom must increase or decrease in number in order for the atom to become an ion?

3.4 Lithium (Group IA) and fluorine (Group VIIA) can react to form ionic bonds. Use electron configurations (including the composition of the nuclei) as we did in Section 3.1 for the reaction of sodium and chlorine to show how a lithium atom and a fluorine atom can change to ions.

3.5   Show how magnesium atoms and fluorine atoms can cooperate to form ions that will aggregate in the correct ratio. (Follow the directions given in Review Exercise 3.4.)

3.6   Write diagrams to show how oxygen atoms and sodium atoms can cooperate to form ions that will aggregate in the correct ratio. (Follow the directions given in Review Exercise 3.4.)

3.7   If M is the symbol of some representative metal, and the symbol of its ion is M^{3+}, in what group in the periodic table is M?

3.8   An atom of the representative nonmetal X can accept two electrons and become an ion. In what group in the periodic table is X?

3.9   An element is in Group IIA. What charges can its ions have?

3.10 Two kinds of electron configurations are exceptionally stable, chemically. Describe them in your own words.

3.11 Using only what can be deduced from electron configurations built from the atomic numbers, write electron configurations of each of the following *ions*. The atomic numbers are given in parentheses.
(a) Calcium ion (20)   (b) Aluminum ion (13)

3.12 In certain compounds, the hydrogen atom can exist as a negatively charged ion called the *hydride ion*. What is likely to be the electron configuration of this ion? Write a symbol for the ion.

3.13 Some substances are described as *electrolytes*. What kinds of compounds are most likely to be electrolytes?

3.14 It is common in studies of heart conditions to hear scientists speak of the *sodium level* of the blood. *Level* refers to the concentration, the ratio of substance to volume. To what specifically does *sodium* refer?

Oxidation and Reduction

3.15 If the number of electrons of a particle increases, has it been oxidized or reduced?

3.16 In a reaction in which an oxidation happens, what other event must also happen?

3.17 What is the oxidation number of the sodium ion?

3.18 What is the oxidation number of the Pb(IV) ion?

3.19 What are the oxidation numbers of the metals in each of the following compounds? Write also the chemical symbol of the metal *ion*.
(a) BiF_3 (b) CdI_2 (c) Cr_2S_3
(d) Gd_2O_3 (e) $SnCl_2$ (f) TiF_3

3.20 If the oxidation number of oxygen is $2-$, what are the oxidation numbers of the other elements in the following compounds?
(a) Mn_2O_7 (b) TiO_4 (c) W_2O_5 (d) Rb_2O

Names and Formulas Involving Monatomic Ions

3.21 If Sr is strontium, what is the name of Sr^{2+}?

3.22 If Te is tellurium (group VIA), what is the name of Te^{2-}?

3.23 Give the correct symbols, including the charges, for the following ions.
(a) Potassium ion   (b) Aluminum ion
(c) Iodide ion   (d) Copper(I) ion
(e) Barium ion   (f) Sulfide ion
(g) Sodium ion   (h) Ferric ion
(i) Oxide ion   (j) Lithium ion
(k) Silver ion   (l) Magnesium ion
(m) Cupric ion   (n) Bromide ion
(o) Calcium ion   (p) Iron(II) ion
(q) Fluoride ion   (r) Chloride ion
(s) Zinc ion   (t) Barium ion

3.24 Give the names of the following ions. When an ion has more than one name, one older and the other a modern name, write both.
(a) Na^+ (b) Fe^{3+} (c) Li^+ (d) O^{2-}
(e) S^{2-} (f) Ba^{2+} (g) Cu^+ (h) I^-
(i) Al^{3+} (j) K^+ (k) Zn^{2+} (l) F^-
(m) Fe^{2+} (n) Cl^- (o) Ca^{2+} (p) Cu^{2+}
(q) Br^- (r) Mg^{2+} (s) Ag^+

3.25 If the older name of the Hg^{2+} ion is *mercuric ion,* what is the most likely name of Hg^+? (Actually, this ion exists doubled up as Hg_2^{2+}, but the charge is still $1+$ per Hg.)

3.26 An older name for the Sn^{4+} ion is *stannic ion.* Which is the more likely formula for the *stannous ion,* Sn^{2+} or Sn^{5+}?

3.27 The ion Pb^{2+} was once called the *plumbous ion.* What is its modern name?

3.28 The ion Au^{3+} has the older name of *auric ion.* What is its modern name?

3.29 The formula of calcium fluoride is CaF_2. What law of nature is violated, if any, by writing the formula as F_2Ca?

3.30 Write the formula of each of the following compounds.
(a) Lithium chloride   (b) Barium oxide
(c) Aluminum sulfide   (d) Sodium bromide
(e) Cupric oxide   (f) Ferric chloride

3.31 What are the formulas of the following compounds?
(a) Cuprous sulfide   (b) Potassium fluoride
(c) Sodium sulfide   (d) Calcium iodide
(e) Magnesium chloride   (f) Ferrous bromide

3.32 Write the names of the following compounds. Wherever two names are possible, a modern name and an older name, write both.
(a) $FeBr_3$ (b) $MgCl_2$ (c) NaF
(d) ZnO   (e) $CuBr_2$ (f) Li_2O

3.33 What are the names of the following compounds? (If both a modern and an older name are possible, give both names.)
(a) KI   (b) CaS   (c) $BaCl_2$
(d) Al_2O_3 (e) $FeCl_2$ (f) AgI

Octet Rule

3.34 The chemical properties of which family of elements is most closely associated with the octet rule? Explain.

3.35 State the octet rule.

3.36 Do the atoms of the elements in Groups VIA and VIIA tend to gain or lose electrons to form ions? Explain.

Molecules and Molecular Compounds

3.37 In general terms, what is a *molecule,* and in what way or ways is a molecule different from an atom? From an ion?

3.38 Draw figures to illustrate a brief discussion of how two fluorine atoms develop a covalent bond in F_2.

3.39 Discuss how the hydrogen atom and a chlorine atom interact to make the H—Cl molecule. Draw figures to illustrate your discussion.

3.40 Hydrazine, N_2H_4, has been used as a rocket fuel. Which of the following structures is correct for hydrazine, A, B, or C?

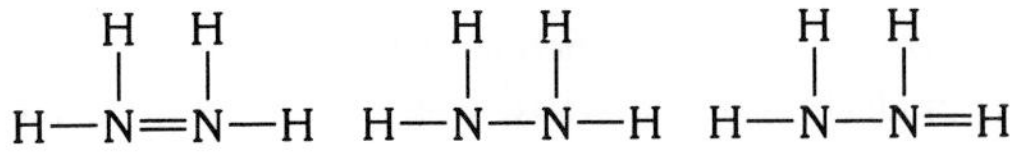

3.41 The raw material for polypropylene, widely used in making indoor — outdoor carpeting, is propylene, C_3H_6. Which of the following structures for propylene is correct, A or B?

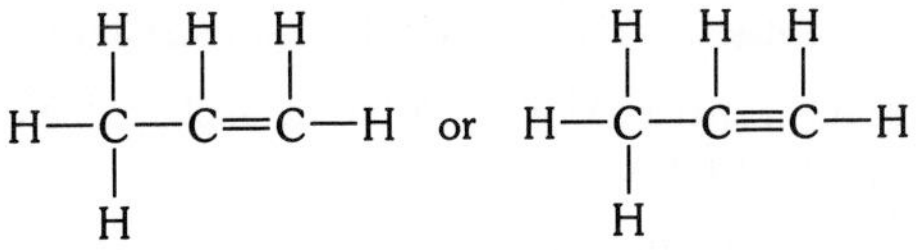

3.42 Without referring to the periodic table or any other table, write the electron configuration of an atom of atomic number 14, and predict its covalence.

3.43 A molecular compound between germanium and hydrogen has the following structure.

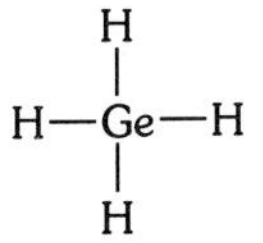

(a) What is the covalence of Ge in this compound?
(b) Germanium is one of the representative elements. Without referring to the periodic table, using only the answer to part (a), to what group in the periodic table does germanium belong?

3.44 The molecular formula of phosphine is PH_3. Write a structural formula for this compound.

3.45 The molecular formula of carbon disulfide is CS_2. Write a structural formula that is consistent with the covalences of carbon and sulfur.

3.46 Antimony, atomic number 51, is a representative element. It forms a compound with hydrogen called stibine. Using the location of antimony in the periodic table as a clue, write the molecular and structural formulas of stibine.

3.47 One of the components of a certain brand of chlorine bleach has the molecular formula HClO, hypochlorous acid. Write a structural formula of this compound that is consistent with the covalences of H, Cl, and O, which were given in this chapter.

Polyatomic Ions and Formulas Involving Them

3.48 What are the names of the following ions?
(a) HCO_3^- (b) SO_4^{2-} (c) NO_3^-
(d) OH^- (e) NH_4^+ (f) CN^-
(g) HPO_4^{2-} (h) CrO_4^{2-} (i) CO_3^{2-}
(j) MnO_4^- (k) HSO_4^- (l) HSO_3^-
(m) NO_2^- (n) PO_4^{3-} (o) $Cr_2O_7^{2-}$
(p) $H_2PO_4^-$ (q) H_3O^+ (r) $C_2H_3O_2^-$

3.49 Write the formulas of the following ions.
(a) Carbonate ion (b) Nitrate ion
(c) Hydroxide ion (d) Ammonium ion
(e) Phosphate ion (f) Cyanide ion
(g) Hydronium ion (h) Monohydrogen phosphate ion
(i) Nitrite ion (j) Bicarbonate ion
(k) Bisulfate ion (l) Dihydrogen phosphate ion
(m) Dichromate ion (n) Bisulfite ion
(o) Chromate ion (p) Sulfate ion
(q) Sulfite ion (r) Acetate ion

3.50 Write the formulas of the following compounds.
(a) Ammonium phosphate
(b) Potassium monohydrogen phosphate
(c) Magnesium sulfate
(d) Calcium carbonate
(e) Lithium bicarbonate
(f) Potassium dichromate
(g) Ammonium bromide
(h) Iron(III) nitrate

3.51 What are the formulas of the following compounds?
(a) Sodium dihydrogen phosphate
(b) Copper(II) carbonate
(c) Silver nitrate
(d) Zinc bicarbonate
(e) Potassium bisulfate
(f) Ammonium chromate
(g) Calcium acetate
(h) Ferric sulfate

3.52 Write the names of the following compounds.
(a) $NaNO_3$ (b) $CaSO_4$
(c) KOH (d) Li_2CO_3
(e) NH_4CN (f) Na_3PO_4
(g) $KMnO_4$ (h) $Mg(H_2PO_4)_2$

3.53 What are the names of the following compounds?
(a) K_2HPO_4 (b) $NaHCO_3$
(c) NH_4NO_2 (d) $ZnCrO_4$
(e) $LiHSO_4$ (f) $Ca(C_2H_3O_2)_2$
(g) $K_2Cr_2O_7$ (h) $NaHSO_4$

3.54 What is the total number of atoms of all kinds in one formula unit of each of the following substances?

(a) $(NH_4)_2SO_4$ (b) $Al_2(HPO_4)_3$ (c) $Ca(HCO_3)_2$

3.55 One formula unit of each of the following substances has how many atoms of all kinds?

(a) $(NH_4)_3PO_4$ (b) $Fe_2(SO_4)_3$ (c) $Ca(C_2H_3O_2)_2$

3.56 Iron(III) glycerophosphate, $Fe_2[C_3H_5(OH)_2PO_4]_3$, has sometimes been used to treat iron deficiency anemia. How many atoms of all kinds are present in one formula unit of this substance?

Polar Molecules

3.57 What has happened to the electron cloud involving a shared pair of electrons between two atoms X and Y if Y is more electronegative than X?

3.58 Referring to the preceding question, if X and Y are in the same period in the periodic table, which most likely has the higher atomic number? Explain.

3.59 Chlorine, atomic number 17, is not quite as electronegative as oxygen, atomic number 8, yet chlorine has a much larger positive charge on its nucleus. What feature about a chlorine atom explains why its nucleus is not as effective in making chlorine more electronegative than oxygen?

3.60 Suppose that the difference in electronegativity between X and Y in compound XY is so great that the electron cloud is pulled toward one end effectively to an extent of 100%. Can we call the bond between X and Y a covalent bond anymore? If not, then what is a better name for the bond?

3.61 Suppose that X and Y can form a diatomic molecule, $X—Y$, and that Y is less electronegative than X.

(a) Is the molecule $X—Y$ polar?

(b) If so, where are the $\delta+$ and the $\delta-$ partial charges located?

3.62 Study the following molecules. For those that are polar, write in $\delta+$ and $\delta-$ where they are properly located.

(a) $H—H$ (b) $H—F$ (c) $F—F$
(d) $Cl—Cl$ (e) $N\equiv N$ (f) $H—I$

3.63 Suppose that X and Y are elements in the same group of nonmetals in the periodic table, but X stands above Y. Which of the two has the higher electronegativity? Explain.

3.64 Suppose that M and Z are in the same period in the periodic table, but M precedes Z. Which of these elements has the higher electronegativity? Explain.

Molecular Orbitals (Special Topic 3.1)

3.65 What event brings a molecular orbital into existence? Give an example.

3.66 In molecular orbital theory, what does *sharing* of a bonding pair of electrons mean? How does this create the covalent *bond*?

3.67 What atomic orbitals overlap to make the molecular orbital for the bond in the following molecules?

(a) $H—H$ (b) $F—F$ (c) $H—F$

VSEPR Theory (Special Topic 3.2)

3.68 What do the letters VSEPR stand for, and what feature of molecular structure is VSEPR theory meant to explain?

3.69 Briefly explain how VSEPR theory explains the bond angle in CH_4 (methane).

3.70 The water molecule does not have a linear geometry. Why?

Chapter 4
Chemical Reactions: Equations and Mass Relationships

Chemical Equations
Avogadro's Number
Formula Weights and Molecular Weights

The Mole
Reactions in Solution
Molar Concentration

Measurements of mass and volume let us relate experimental quantities of reactants to numbers of molecules. How this works is studied in this chapter.

4.1 CHEMICAL EQUATIONS

The coefficients of a balanced equation give the proportions of the formula units in the reaction.

The technical term for the study of weight relationships in chemistry is *stoichiometry* after the Greek *stoicheion* (element) and *metron* (measure).

There is neither basic understanding nor practical application of any chemical event in nature apart from a fundamental grasp of weight relationships in chemical reactions. These are at the heart of almost anything done with substances in medicine, nutrition and dietetics, and respiratory care, for example, whether the work is with medications, diets and recipes, or respiratory therapy.

Our own study of weight relationships began in Chapter 2 when we learned that atoms combine only in whole-number ratios to form compounds. The formula units of compounds behave in the same way; they react only in whole-number units. The equations that symbolize reactions give the ratios of these units, so we have to learn to write chemical equations before we can study weight relationships.

All Reactant Atoms Are Somewhere among the Products in a Balanced Chemical Equation. A **chemical equation** is a condensed description of a reaction that uses chemical formulas of the reactants and the products. The formulas of the reactants, separated by plus signs, are on one side of an arrow and the formulas of the products, also separated by plus signs, are on the arrowhead side of the arrow. For example, the formation of iron(II) sulfide from iron and sulfur is described by the following equation.

$$Fe \quad + \quad S \quad \longrightarrow \quad FeS$$

Translation:

Iron   reacts with   sulfur
in a ratio of one atom of
iron to one atom of sulfur . . . to give . . . iron(II) sulfide

A more complicated example is the equation for the formation of aluminum sulfide, Al_2S_3, from aluminum and sulfur. The reactants here are not in a simple 1 : 1 ratio, so numbers called **coefficients** have to be used. These are placed in front of the appropriate formulas to show the proportions of formula units that are involved.

$$2Al \quad + \quad 3S \quad \longrightarrow \quad Al_2S_3$$

Translation:

Aluminum   reacts with   sulfur
in a ratio of 2 aluminum atoms
to 3 sulfur atoms   . . . to give . . . 1 formula unit of aluminum sulfide

Coefficients are multipliers for the associated formulas. The symbol $2C_3H_8$, for example, occurs in the equation for the combustion (burning) of propane to water and carbon dioxide.

$$2C_3H_8 + 10O_2 \longrightarrow 8H_2O + 6CO_2$$
Propane

In $2C_3H_8$, the numbers 3 and 8 are subscripts and the 2 is a coefficient, a multiplier for the whole formula. In $2C_3H_8$, therefore, there are 6C and 16H atoms. As with subscripts, whenever a coefficient is 1, the 1 isn't written; it is understood.

We are never allowed to change subscripts, once we have the right formulas, just to get an equation to balance. If we change H_2O to H_2O_2, for example, we change the formula for water to the formula for hydrogen peroxide, an entirely different substance. To write a balanced equation we must use correct formulas and then adjust only the *coefficients*, never the subscripts.

EXAMPLE 4.1   BALANCING A CHEMICAL EQUATION

Remember, the formula unit of chlorine is a molecule, Cl_2, not an atom.

Problem: Sodium, Na, reacts with chlorine, Cl_2, to give sodium chloride, NaCl. Write the balanced equation for this reaction.

Solution: The first step is to set down the correct formulas in the format of an equation. (Never worry about the coefficients until the correct formulas are down, then never change the formulas.)

$$Na + Cl_2 \rightarrow NaCl \quad \text{(unbalanced)}$$

$NaCl_2$ doesn't even exist.

So far, we see two chlorine atoms on the left (in Cl_2) but only 1 on the right. We can't fix this by writing $NaCl_2$, because this isn't the correct formula for sodium chloride. The only way we are allowed to get two Cl atoms on the right is to put a coefficient of 2 in front of NaCl.

$$Na + Cl_2 \rightarrow 2NaCl \quad \text{(unbalanced)}$$

Of course, writing 2 in front of NaCl makes it a multiplier for both Na and Cl, so now we have two Na atoms on the right and just one on the left. To fix this, we write a 2 before the Na on the left:

$$2Na + Cl_2 \rightarrow 2NaCl \quad \text{(balanced)}$$

Notice particularly how we used a subscript, the 2 in Cl_2, to suggest a coefficient for another formula on the other side of the arrow. This is standard strategy in balancing equations.

EXAMPLE 4.2   BALANCING A CHEMICAL EQUATION

Problem: Iron, Fe, can be made to react with oxygen, O_2, to form an oxide with the formula Fe_2O_3. Write the balanced equation for this reaction.

Solution: We first write down the correct formulas in the format of an equation.

Oxygen, remember, exists as molecules, O_2, not as atoms.

$$Fe + O_2 \longrightarrow Fe_2O_3 \quad \text{(unbalanced)}$$

Next we exploit subscripts to suggest coefficients. Oxygen has a subscript of 2 in O_2 and a subscript of 3 in Fe_2O_3. To get a balance, we use the 3 as a coefficient for O_2, so that in $3O_2$ we will have 6O atoms on the left side of the equation. Then we use the 2 (suggested by O_2) as a coefficient for Fe_2O_3 so that in $2Fe_2O_3$ we will get 6O atoms on the right.

$$Fe + 3O_2 \rightarrow 2Fe_2O_3 \quad \text{(unbalanced)}$$

Now there are six O atoms on the left (in $3O_2$) and six on the right (in $2Fe_2O_3$). Of course, the coefficient of 2 in the formula on the right also means that there are 4Fe atoms on the right. To fix this, we simply use a coefficient of 4 on the left, for Fe.

$$4Fe + 3O_2 \longrightarrow 2Fe_2O_3 \quad \text{(balanced)}$$

PRACTICE EXERCISE 1   In the presence of an electrical discharge, oxygen, O_2, can be changed into ozone, O_3. Write the balanced equation for this reaction.

PRACTICE EXERCISE 2   Aluminum, Al, reacts with oxygen to give aluminum oxide, Al_2O_3. Write the balanced equation for this change.

Sometimes, the adjustments to coefficients produce balanced equations whose coefficients can be divided by a common divisor. For example, we might have obtained the following equation for the reaction of sodium with chlorine in Example 4.1.

$$4Na + 2Cl_2 \longrightarrow 4NaCl$$

This is a balanced equation and all of its formulas are correct, so there is nothing fundamentally wrong with it. However, chemists generally (but not always) write balanced equations that use the set of smallest whole numbers possible for the coefficients. We will follow this rule unless a special situation warrants something else.

When the formulas in an equation include polyatomic ions, and when it is obvious that they do not themselves change, then treat them as whole units in balancing equations.

EXAMPLE 4.3   BALANCING EQUATIONS INVOLVING POLYATOMIC IONS

Problem: When water solutions of ammonium sulfate, $(NH_4)_2SO_4$, and lead nitrate, $Pb(NO_3)_2$, are mixed, a white solid separates that has the formula $PbSO_4$ (lead sulfate). Ammonium nitrate, NH_4NO_3, is the other product, but it remains dissolved. Represent this reaction by a balanced equation.

Solution: As usual, we start by simply writing the correct formulas in the format of an equation.

$$(NH_4)_2SO_4(aq) + Pb(NO_3)_2(aq) \longrightarrow PbSO_4(s) + NH_4NO_3(aq)$$

The symbol (*aq*) stands for aqueous solution, meaning a solution in which water is the solvent. The symbol (*s*) means a solid, one that forms directly and is not in solution.

Because polyatomic ions are involved, we next examine the formulas to see if any of them change or if they all appear to react as whole units; and we can see that they do react as whole units. The subscript of 2 in $(NH_4)_2SO_4$ suggests that we use 2 as the coefficient in the formula on the right where (NH_4) occurs.

$$(NH_4)_2SO_4(aq) + Pb(NO_3)_2(aq) \longrightarrow PbSO_4(s) + 2NH_4NO_3(aq)$$

This automatically brought into balance the units of (NO_3) on each side of the arrow. The equation is now balanced.

PRACTICE EXERCISE 3   Balance each of the following equations.

(a) $Ca + O_2 \rightarrow CaO$
(b) $KOH + H_2SO_4 \rightarrow H_2O + K_2SO_4$
(c) $Cu(NO_3)_2 + Na_2S \rightarrow CuS + NaNO_3$
(d) $AgNO_3 + CaCl_2 \rightarrow AgCl + Ca(NO_3)_2$
(e) $Al + H_2SO_4 \rightarrow Al_2(SO_4)_3 + H_2$
(f) $CH_4 + O_2 \rightarrow H_2O + CO_2$

4.2 AVOGADRO'S NUMBER

Avogadro's number, 6.02×10^{23}, is the number of formula units in a sample of a pure substance with a mass numerically equal to the formula weight in grams.

The coefficients of a balanced equation tell us the proportions of the formula units involved. Thus the equation for the reaction of sodium with chlorine,

$$2Na + Cl_2 \longrightarrow 2NaCl$$

can be interpreted as follows:

$$2 \text{ atoms of } Na + 1 \text{ molecule of } Cl_2 \longrightarrow 2 \text{ formula units of } NaCl$$

We cannot, however, go into the lab and carry out a reaction on such a small scale. We have to use much larger numbers of particles to obtain samples that we can manipulate. We could pick almost any number we pleased, provided it was large enough to give us workable samples of formula units. The question is "What number would be the most useful as a standard number of formula units?"

The number to which chemists have long agreed is called *Avogadro's number,* after Amadeo Avogadro, an Italian chemist. Its value (to three significant figures) is 6.02×10^{23}.

$$\textbf{Avogadro's number} = 6.02 \times 10^{23}$$

Avogadro's number is a pure number with a special name, like the name *dozen* for 12. In fact, Avogadro's number is sometimes called the chemist's "dozen." Just as *dozen* can be used to signify 12 of anything, so *Avogadro's number* can be used to signify 6.02×10^{23} of anything, electrons, protons, atoms, virus particles, anything.

The usefulness of a number this complicated lies in its relationship to atomic weights. We can show this by doing a calculation using sodium's atomic weight, 22.99. This means that one atom of sodium has a mass of 22.99 amu. We learned earlier that 1 amu = 1.6606×10^{-24} g, so one atom of sodium has a mass in grams given by:

$$\frac{22.99 \text{ amu}}{1 \text{ atom Na}} \times \frac{1.6606 \times 10^{-24} \text{ g}}{1 \text{ amu}} = \frac{3.818 \times 10^{-23} \text{ g Na}}{1 \text{ atom Na}}$$

If one atom of sodium has a mass of 3.818×10^{-23} g, what is the mass of Avogadro's number of sodium atoms? We find this by the following:

$$\frac{3.818 \times 10^{-23} \text{ g Na}}{1 \text{ atom Na}} \times 6.02 \times 10^{23} \text{ atoms Na} = 22.98 \text{ g Na}$$

Thus Avogadro's number of sodium atoms weigh 22.98 g, which (ignoring the difference caused by rounding) is numerically equal to the atomic weight of sodium, a number that naturally goes with this element. The logic of using a number as complicated as Avogadro's number is that it delivers a mass of atoms numerically equal to something familiar about the element, its atomic weight in grams. The same kinds of calculations could be applied to the other elements.

EXAMPLE 4.4   UNDERSTANDING AVOGADRO'S NUMBER

Problem: How many carbon atoms are in 6.00 g of carbon? The atomic weight of carbon is 12.0.

Solution: We solve this by working with the basic meaning of Avogadro's number when it is applied to the element carbon. It's the number of carbon atoms in the grams of carbon numerically equal to the formula weight of carbon, 12.0. Thus 12.0 g of carbon = 6.02×10^{23} atoms of carbon. This gives us two conversion factors:

$$\frac{6.02 \times 10^{23} \text{ atoms C}}{12.0 \text{ g C}} \quad \text{or} \quad \frac{12.0 \text{ g C}}{6.02 \times 10^{23} \text{ atoms C}}$$

If we multiply what is given, 6.00 g of C, by the first conversion factor, the units "g C" will cancel, and our answer will be in atoms of carbon.

$$6.00 \text{ g C} \times \frac{6.02 \times 10^{23} \text{ atoms C}}{12.0 \text{ g C}} = 3.01 \times 10^{23} \text{ atoms C}$$

Thus 6.00 g of carbon contains 3.01×10^{23} atoms of carbon.

PRACTICE EXERCISE 4   How many atoms of gold are in 1.00 oz of gold? Note, 1.00 oz = 28.4 g, and the atomic weight of gold is 197.0.

4.3   FORMULA WEIGHTS AND MOLECULAR WEIGHTS

The sum of the atomic weights of all of the atoms in a chemical formula is the formula weight of the substance.

Because atoms lose no mass when they form compounds, we can extend the idea of atomic weight to any formula unit and calculate a formula weight for any substance. The **formula weight** of a compound is simply the sum of the atomic weights of all of the atoms in one formula unit. The formula weight of NaCl, for example, is calculated as follows, where we round atomic weights to the first decimal place *before* we use them in a calculation, and we follow a common practice that lets the unit of *amu* be left understood.

For the rest of this book, our policy is to round values of atomic weights to their first decimal point before starting any calculations.

$$\begin{array}{ll} \text{1 atom of Na in NaCl gives} & 23.0 \\ \text{1 atom of Cl in NaCl gives} & \underline{35.5} \\ \text{Total} & 58.5 \end{array}$$

The formula weight of NaCl is 58.5. Thus one formula unit of NaCl has a mass of 58.5 amu. Or we can say that Avogadro's number of these formula units has a mass of 58.5 g. Either meaning can be taken, depending on the situation.

The idea of a formula weight is general; it applies to anything with a definite formula, including elements as well as compounds. The formula of sodium, for example, is Na, so we can just as well say that 23.0 is its formula weight as to call this its atomic weight. The element chlorine occurs as a diatomic molecule, Cl_2, so its formula weight is twice its atomic weight or $2 \times 35.5 = 71.0$.

You should learn, before we continue, that a synonym for formula weight, namely, **molecular weight,** is used by many scientists. However, *formula weight* is a more general term, and it's the one that we will use.

EXAMPLE 4.5   CALCULATING A FORMULA WEIGHT

Problem: Some baking powders contain ammonium carbonate, $(NH_4)_2CO_3$. Calculate its formula weight.

Solution: First, look up and write down the atomic weights of all of the elements present, rounding each to the first decimal place.

N, 14.0   H, 1.0   C, 12.0   O, 16.0

Notice that in each formula unit of $(NH_4)_2CO_3$, N occurs 2 times, H occurs 8 times, C occurs 1 time, and O occurs 3 times. Therefore,

$$2N + 8H + 1C + 3O = (NH_4)_2CO_3$$
$$2 \times 14.0 + 8 \times 1.0 + 1 \times 12.0 + 3 \times 16.0 = 96.0 \quad \text{(correctly rounded)}$$

The formula weight of ammonium carbonate is 96.0. This means that one formula unit has a mass of 96.0 amu, and it means that Avogadro's number of these units has a total mass of 96.0 g.

PRACTICE EXERCISE 5   Calculate the formula weights of the following compounds.

(a) $C_9H_8O_4$ (aspirin)   (b) $Mg(OH)_2$ (milk of magnesia)
(c) $Fe_4[Fe(CN)_6]_3$ (ferric ferrocyanide or Prussian blue, an ink pigment)

4.4 THE MOLE

The *mole*, the name of the SI base unit for *amount of chemical substance*, is the formula weight in grams of a substance.

When we weigh out the formula weight in grams of any chemical, element, or compound, we obtain one **mole** (abbreviated **mol**) of it. This is the SI base unit for *quantity of substance.* Because each substance has its own formula weight, *the actual mass that equals one mole varies from substance to substance.* What does not vary is the number of formula units; regardless of the chemical, if you have 1 mol of it, you have Avogadro's number of its formula units. Thus 1 mol of Na has a mass of 23.0 g and it consists of 6.02×10^{23} Na atoms; 1 mol of Cl_2 has a mass of 71.0 g, but it also consists of 6.02×10^{23} Cl_2 molecules, and 1 mol of NaCl has a mass of 58.5 g, but it likewise consists of 6.02×10^{23} formula units of NaCl.

Think of the *mole* as the lab-sized unit of a chemical, a quantity that can be manipulated experimentally. Like any physical quantity, it can be taken in fractions or multiples. For example, the formula weight of H_2O is 18.0, so 1 mol of H_2O weighs 18.0 g. If we wished, we could weigh out a smaller sample, say 1.80 g, and then we would have 0.100 mol of water, because 1.80 is $\frac{1}{10}$ of 18.0. Or we could take 36.0 g of H_2O, and then have 2.00 mol, because 36.0 is 2 times 18.0.

Mol stands for both the plural and the singular.

What is *constant* about one mol of any substance is not the mass but the number of formula units.

An Equation's Coefficients Tell Us the Mole Proportions.   The mole concept lets us think about the coefficients in an equation in two ways at the same time. In the following equation for the reaction of Fe with S, each formula has a coefficient of 1. We can now interpret these coefficients in any of the ways given beneath the formulas in the equation:

Fe	+ S	⟶ FeS
1 atom of Fe	+ 1 atom of S	⟶ 1 formula unit of FeS
1-dozen atoms of Fe	+ 1-doz atoms of S	⟶ 1-doz formula units FeS
6.02×10^{23} atoms of Fe	+ 6.02×10^{23} atoms of S	⟶ 6.02×10^{23} formula units of FeS
1 mol of Fe	+ 1 mol of S	⟶ 1 mol of FeS

Notice that the *proportions* of the formula units all remain the same. All that changes is the *scale* of the reaction, the actual numbers of formula units used. The bottom line is what is now most important to our future work; *the coefficients in a balanced equation give us the proportions of substances in moles.* We will use this interpretation almost exclusively from here on. Thus to use once more an equation that we have employed before:

$$2Na + Cl_2 \longrightarrow 2NaCl$$

We can now interpret this to mean that for every 2 *mol* of Na that reacts, 1 mol of Cl_2 also reacts and 2 mol of NaCl forms.

There are three kinds of calculations involving moles that have to be learned. One is to use an equation's coefficients to find how many moles of one substance must be involved if a certain number of moles of another are used. We do this by using the equation's coefficients to set up conversion factors, as we will see in the next example.

EXAMPLE 4.6   USING THE MOLE CONCEPT

Problem: How many moles of oxygen are needed to combine with 0.500 mol of hydrogen in the reaction that produces water by the following equation?

$$2H_2 + O_2 \longrightarrow 2H_2O$$

Solution: The coefficients tell us that 2 mol of H_2 combine with 1 mol of O_2. This relation lets us select between the following conversion factors, which pertain just to this particular reaction.

$$\frac{2 \text{ mol } H_2}{1 \text{ mol } O_2} \quad \text{or} \quad \frac{1 \text{ mol } O_2}{2 \text{ mol } H_2}$$

For this reaction 2 mol of H_2 is chemically equivalent to 1 mol of O_2. We have to choose one of these ratios to multiply by the given quantity, 0.500 mol of H_2, to find out how much O_2 is needed. The correct ratio is the second one.

$$0.500 \text{ mol } H_2 \times \frac{1 \text{ mol } O_2}{2 \text{ mol } H_2} = 0.250 \text{ mol } O_2$$

In other words, 0.500 mol of H_2 requires 0.250 mol of O_2 for this reaction.

PRACTICE EXERCISE 6   How many moles of H_2O are made from the 0.250 mol of O_2 in Example 4.6?

PRACTICE EXERCISE 7   Nitrogen and oxygen combine at high temperature in an automobile engine to produce nitrogen monoxide, NO, an air pollutant. The equation is $N_2 + O_2 \rightarrow 2NO$. To make 8.40 mol of NO, how many moles of N_2 are needed? How many moles of O_2 are also needed?

PRACTICE EXERCISE 8   Ammonia, an important nitrogen fertilizer, is made by the following reaction: $3H_2 + N_2 \longrightarrow 2NH_3$. In order to make 300 mol of NH_3, how many moles of H_2 and how many moles of N_2 are needed?

The Molar Mass of a Substance Is the Number of Grams per Mole.   The next kind of calculation we have to learn is to convert moles to grams. This is necessary because laboratory balances do not read in moles. (If they did, we'd have to have a separate balance, marked for moles, for each and every possible formula weight!) Instead, the balances read in grams, so after we calculate how many moles to use, we have to translate the moles into grams before we can weigh out the substance.

In this calculation we use a formula weight in still another way. We write the units of grams per mole (g/mol) after a formula weight, and when we do we have the **molar mass** of the substance, the number of grams per mole. Thus the molar mass of sodium, which has an atomic weight of 23.0, is 23.0 g Na/mol Na. We can rewrite this fact as an equation:

$$1 \text{ mol Na} = 23.0 \text{ g Na}$$

We now can devise two conversion factors, as we will see in the next worked example.

EXAMPLE 4.7   CONVERTING MOLES TO GRAMS

Problem: In Example 4.6 we found that 0.250 mol of O_2 was needed. How many grams of O_2 are in 0.250 mol of O_2?

Solution: We first must calculate the formula weight of oxygen, which is two times its atomic weight (16.0) or $2 \times 16.0 = 32.0$. Therefore the molar mass of oxygen is 32.0 g O_2/mol O_2. This fact gives us the following conversion factors that relate mass of O_2 to moles of O_2.

$$\frac{32.0 \text{ g } O_2}{1 \text{ mol } O_2} \quad \text{or} \quad \frac{1 \text{ mol } O_2}{32.0 \text{ g } O_2}$$

We next multiply what was given, 0.250 mol of O_2, by whichever conversion factor lets us cancel "mol O_2" and leaves us with "g O_2," the desired final unit. This means that we use the first conversion factor.

$$0.250 \text{ mol } O_2 \times \frac{32.0 \text{ g } O_2}{1 \text{ mol } O_2} = 8.00 \text{ g } O_2$$

Thus 0.250 mol of O_2 has a mass of 8.00 g of O_2.

PRACTICE EXERCISE 9   An experiment calls for 24.0 mol of NH_3. How many grams is this?

The third kind of calculation is to convert grams to moles. The next worked example shows why this is necessary and how to do it. It's very similar to the moles-to-grams conversion. We will again use a molar mass to find the right conversion factor.

EXAMPLE 4.8   CONVERTING GRAMS TO MOLES

Problem: A student wanted to use 12.5 g of NaCl in a reaction. How many moles is this?

Solution: The relationship between grams and moles of NaCl is given by the molar mass of NaCl. We have already calculated that the formula weight of NaCl is 58.5, so its molar mass is simply 58.5 g NaCl/mol NaCl. In other words,

$$1 \text{ mol NaCl} = 58.5 \text{ g NaCl}$$

Therefore we have two possible conversion factors.

$$\frac{58.5 \text{ g NaCl}}{1 \text{ mol NaCl}} \quad \text{or} \quad \frac{1 \text{ mol NaCl}}{58.5 \text{ g NaCl}}$$

If we multiply the second factor by the given, 12.5 g NaCl, the units will cancel properly and the result will be the moles of NaCl in this sample.

$$12.5 \text{ g NaCl} \times \frac{1 \text{ mol NaCl}}{58.5 \text{ g NaCl}} = 0.214 \text{ mol NaCl} \quad \text{(rounded from 0.2136752137)}$$

Thus 12.5 g of NaCl consists of 0.214 mol of NaCl.

PRACTICE EXERCISE 10   A student was asked to prepare 6.84 g of aspirin, $C_9H_8O_4$. How many moles is this?

We can now put these kinds of calculations together in an example of a very common laboratory situation: How many grams of one substance are needed to make a given mass of another according to some equation? We'll see how in the next example.

EXAMPLE 4.9   MOLE CALCULATIONS USING BALANCED EQUATIONS

Problem: Aluminum oxide can be used as a white filler for paints. How many grams of aluminum are needed to make 24.4 g of Al_2O_3 by the following equation?

$$4Al + 3O_2 \longrightarrow 2Al_2O_3$$

Solution: It is usually a good idea at the start of a problem such as this to compute any needed formula weights and write them down for reference. When we do this, we have

$$Al, 27.0 \quad Al_2O_3, 102.0$$

Since the equation's coefficients refer to *moles,* we must first find out how many moles are in 24.4 g of Al_2O_3. We learned to do this in the previous example. The formula weight of Al_2O_3, 102.0, tells us that 1 mol Al_2O_3 = 102.0 g Al_2O_3. We use this fact to make and use the proper conversion factor.

$$24.4 \text{ g } Al_2O_3 \times \frac{1 \text{ mol } Al_2O_3}{102.0 \text{ g } Al_2O_3} = 0.239 \text{ mol } Al_2O_3$$

Thus the 24.4 g of Al_2O_3 consists of 0.239 mol of Al_2O_3.

Now we can use the coefficients of the equation to find the moles of Al needed to make 0.239 mol of Al_2O_3. The equation tells us that 4 mol of Al gives 2 mol of Al_2O_3. This fact lets us prepare the following conversion factors.

$$\frac{4 \text{ mol } Al}{2 \text{ mol } Al_2O_3} \quad or \quad \frac{2 \text{ mol } Al_2O_3}{4 \text{ mol } Al}$$

To find the moles of Al needed to make 0.239 mol of Al_2O_3, we multiply 0.239 mol of Al_2O_3 by the first factor.

$$0.239 \text{ mol } Al_2O_3 \times \frac{4 \text{ mol } Al}{2 \text{ mol } Al_2O_3} = 0.478 \text{ mol } Al$$

Thus 0.478 mol of Al will make 0.239 mol (or 24.4 g) of Al_2O_3. The problem asks for the answer in grams of Al, not moles of Al, however, so we next have to convert 0.478 mol of Al into grams of Al. We studied how to do this in Example 4.7. We use the formula weight of Al to make the correct conversion factor.

$$0.478 \text{ mol } Al \times \frac{27.0 \text{ g } Al}{1 \text{ mol } Al} = 12.9 \text{ g } Al$$

This is the answer; it takes 12.9 g of Al to prepare 24.4 g of Al_2O_3 according to the given equation.

Before working the next practice exercise, think about the overall strategy used in this example. As diagrammed in Figure 4.1, we first moved the calculations from the grams level to the moles level. We had to do this because the balanced equation's coefficients gave us the

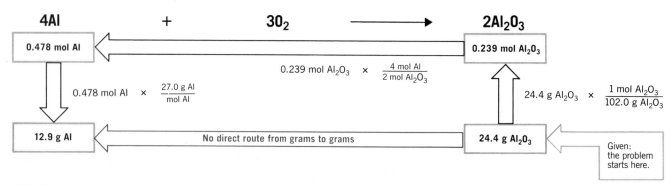

FIGURE 4.1
All calculations involving masses of reactants and products that participate in a chemical reaction must be worked out at the mole level. There is no direct route from grams of one substance to grams of another.

important relationship between Al_2O_3 and Al in moles, not grams. Then we used these coefficients to relate the moles of one substance to the moles of another. Finally, we moved back to the grams level so that we could use laboratory balances marked in grams, not moles.

PRACTICE EXERCISE 11   How many grams of oxygen are needed for the experiment described in Example 4.9? Use a diagram of the solution in the style of Figure 4.1 as you work out the answer.

PRACTICE EXERCISE 12   If 28.4 g of Cl_2 are used up in the following reaction, how many grams of Na are also used up, and how many grams of NaCl form?

$$2Na + Cl_2 \longrightarrow 2NaCl$$

4.5 REACTIONS IN SOLUTION

Virtually all of the chemical reactions studied in the lab and those that occur in living systems take place in an aqueous solution.

If the atoms, ions, or molecules of one substance are to react with those of another, they must have enough freedom to move about and find and hit each other. Such freedom exists in gases and liquids, but not in solids. Normally, to get solids to react, we put them into solution first. Then their formula units can move around. To study mass relationships when the reactants are in solution (next section), we will first learn some common terms used to describe solutions.

Solutions Consist of Solutes Dissolved in Solvents.   A **solution** is a uniform mixture of particles that are of atomic, ionic, or molecular size. A minimum of two substances are needed. One is called the solvent and the others are called the solutes. The **solvent** is the medium into which the other substances are mixed or dissolved. Normally, the solvent is a liquid. When it is water, we have an **aqueous solution.** (Unless we state otherwise, we'll always mean *aqueous* solution when we talk about solutions.)

A **solute** is anything that is dissolved by the solvent. In an aqueous solution of sugar, the solute is sugar and the solvent is water. The solute can be a gas. Club soda is a solution of carbon dioxide in water. The solute can be a liquid. Some brands of antifreeze, for example, are mostly solutions of the liquid propylene glycol in water.

Solutions Can Be Dilute, Concentrated, Unsaturated, and Saturated.   Several terms are used to describe a solution. A **dilute solution,** for example, is one in which the ratio of solute to solvent is very small, like a few crystals of sugar dissolved in a glass of water. Most of the aqueous solutions in living systems have more than two solutes, and they are dilute in each of them.

In a **concentrated solution,** the ratio of solute to solvent is large. Syrup, for example, is a concentrated solution of sugar in water.

Some solutions are **saturated solutions,** which means that it isn't possible to dissolve more of the solute at the given temperature. (The maximum ratio of solute to solvent is different at different temperatures.) If we add more solute to a saturated solution, what we add will just stay separate. If it's a solid, it will generally sink to the bottom and remain there.

An **unsaturated solution** is one in which more solute could be dissolved without having to change the temperature. The ratio of solute to solvent is thus lower in an unsaturated solution than in the corresponding saturated solution.

TABLE 4.1
Solubilities of Some Substances in Water

Solute	Solubilities (g/100 g H_2O)			
	0 °C	20 °C	50 °C	100 °C
Solids				
Sodium chloride, NaCl	35.7	36.0	37.0	39.8
Sodium hydroxide, NaOH	42	109	145	347
Barium sulfate, $BaSO_4$	0.000115	0.00024	0.00034	0.00041
Calcium hydroxide, $Ca(OH)_2$	0.185	0.165	0.128	0.077
Gases				
Oxygen, O_2	0.0069	0.0043	0.0027	0
Carbon dioxide, CO_2	0.335	0.169	0.076	0
Nitrogen, N_2	0.0029	0.0019	0.0012	0
Sulfur dioxide, SO_2	22.8	10.6	4.3	1.8 (at 90 °C)
Ammonia, NH_3	89.9	51.8	28.4	7.4 (at 96 °C)

A Solubility Is Usually Expressed as the Maximum Grams of Solute in 100 Grams of Solvent.   The amount of solute needed to give a saturated solution in a given quantity of solvent at a specific temperature is called the **solubility** of the solute. The units are usually grams of solute per 100 grams of solvent. Table 4.1 gives some examples that show how widely solubilities can vary. Notice particularly that a saturated solution can still be quite dilute. Less than a milligram of barium sulfate, for example, can dissolve in 100 g of water. Notice also that the solubilities of solids generally increase with temperature. Gases, on the other hand, become less and less soluble in water as the temperature increases.

Supersaturated Solutions Are Unstable.   Most solids, as we said, become less soluble as the temperature of a solution is reduced. When we cool saturated solutions of such solids, therefore, we can expect the now excess solute to leave the solution. This separation of a solid from a solution is called **precipitation,** and the solid is called the **precipitate.**

If we are very careful, if we use a saturated solution absolutely free of dirt, lint, or specks of solute, when we cool it we sometimes will *not* see the excess solute precipitate. This cooled

(a)   (b)   (c)   (d)

FIGURE 4.2
Supersaturation. (*a*) This solution is supersaturated. (*b*) Some seed crystals are added to it. (*c*), (*d*) Excess solute rapidly separates from the solution.

solution now has more solute than it's supposed to be able to hold. A solution whose concentration exceeds that of its saturated system is called a **supersaturated solution.** It's an unstable system.

If we now scratch the inner wall of the container with a glass rod, or if we add a "seed" crystal of the pure solute, the excess solute will usually separate immediately. This can be dramatic and pretty to see (Figure 4.2). The seed crystal provides a starting surface on which excess solute particles can form more precipitate.

4.6 MOLAR CONCENTRATION

The ratio of *moles per liter* is the most useful way to describe a solution's concentration.

We now turn to the question: How are mass relationships handled when a chemical reaction occurs in a solution? The answer requires information about the solution's concentration, which we will study next.

The Molarity of a Solution Is the Moles of Solute per Liter of Solution.   The **concentration** of a solution is the ratio of the quantity of solute to some given unit of the solution. The units in this ratio can be anything we wish, but the most useful are those of *moles of solute per liter of solution,* a ratio called the solution's **molar concentration,** or **molarity,** abbreviated M

$$M = \frac{\text{mol solute}}{\text{L solution}} = \frac{\text{mol solute}}{1000 \text{ mL solution}}$$

M = moles per liter
mol = moles

A solution, for example, might have the label "0.10 M NaCl." If so it has a concentration of 0.10 mol of NaCl per liter of solution (or per 1000 mL of solution). Notice, however, that molarity is a *ratio*. It isn't necessary to have a whole liter of solution to speak of some molar concentration. The important fact about molarity is that a given value, with its proper units, gives us two conversion factors. In our example, by using 1000 mL for 1 L, they are

$$\frac{0.10 \text{ mol NaCl}}{1000 \text{ mL NaCl solution}} \quad \text{and} \quad \frac{1000 \text{ mL NaCl solution}}{0.10 \text{ mol NaCl}}$$

To prepare a solution of known molarity, we use a special piece of laboratory glassware called a *volumetric flask*. Figure 4.3 shows how we could make one liter of a 1 M solution of

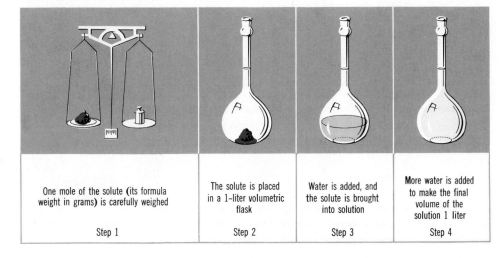

One mole of the solute (its formula weight in grams) is carefully weighed	The solute is placed in a 1-liter volumetric flask	Water is added, and the solute is brought into solution	More water is added to make the final volume of the solution 1 liter
Step 1	Step 2	Step 3	Step 4

FIGURE 4.3
The preparation of 1 L of a 1 M solution.

some solute. The grams of the solute that equals 1 mol is weighed and placed in the flask. Enough water is then added to dissolve the solute. Finally, more water is added until the level of the well-mixed solution reaches the mark that tells us the flask now holds 1 L of the solution. But, as we said, it isn't necessary to make up a whole liter of solution. So volumetric flasks of a number of capacities are available, ranging from 1 mL to 5 L and several intermediate sizes.

The concept of molarity will become clearer by studying how to do some of the calculations associated with it. In the next example we'll see what kinds of calculations have to be done in order to prepare a given volume of a solution with a specified molar concentration.

EXAMPLE 4.10   PREPARING A SOLUTION OF KNOWN MOLAR CONCENTRATION

Problem: How much sodium bicarbonate, $NaHCO_3$, is needed to prepare 500 mL of 0.125 M $NaHCO_3$? We can assume that a 500-mL volumetric flask is available.

Solution: "How much . . . ?" has to mean *grams,* because lab balances do not measure moles. We thus have to find out how many *grams* of sodium bicarbonate are in 500 mL of 0.125 M $NaHCO_3$, But before we can calculate grams we have to calculate the *moles* of $NaHCO_3$ in this much solution. Here is where the concentration gives us the needed conversion factor, because 0.125 M $NaHCO_3$, in effect, means:

$$\frac{0.125 \text{ mol } NaHCO_3}{1000 \text{ mL } NaHCO_3 \text{ solution}} \quad \text{or} \quad \frac{1000 \text{ mL } NaHCO_3 \text{ solution}}{0.125 \text{ mol } NaHCO_3}$$

If we multiply the given volume, 500 mL of $NaHCO_3$ solution, by the first conversion factor, the volume units will cancel and we will learn how many moles we need.

$$500 \text{ mL } NaHCO_3 \text{ solution} \times \frac{0.125 \text{ mol } NaHCO_3}{1000 \text{ mL } NaHCO_3 \text{ solution}} = 0.0625 \text{ mol } NaHCO_3$$

In other words, our 500 mL of 0.125 M $NaHCO_3$ solution has to contain 0.0625 mol of $NaHCO_3$.

Next, we have to convert 0.0625 mol of $NaHCO_3$ into grams of $NaHCO_3$, and for this we have the usual choice of a conversion factors given by the formula weight of $NaHCO_3$, 84.0.

$$\frac{84.0 \text{ g } NaHCO_3}{1 \text{ mol } NaHCO_3} \quad \text{or} \quad \frac{1 \text{ mol } NaHCO_3}{84.0 \text{ g } NaHCO_3}$$

If we multiply 0.0625 mol of $NaHCO_3$ by the first of these factors, then the units of mol $NaHCO_3$ will cancel and our answer will be in what we want, grams.

$$0.0625 \text{ mol } NaHCO_3 \times \frac{84.0 \text{ g } NaHCO_3}{1 \text{ mol } NaHCO_3} = 5.25 \text{ g } NaHCO_3$$

Thus to prepare 500 mL of 0.125 M $NaHCO_3$, we have to weigh out 5.25 g of $NaHCO_3$, dissolve it in some water in a 500-mL volumetric flask, and then carefully add water until its level reaches the mark, making sure that the contents become well mixed.

PRACTICE EXERCISE 13   How many grams of each solute are needed to prepare the following solutions?
(a) 250 mL of 0.100 M H_2SO_4 (b) 100 mL of 0.500 M glucose ($C_6H_{12}O_6$)

Another calculation that must sometimes be made when a reactant is available only in a solution of known molar concentration is to find the volume of the solution that holds a desired quantity of its solute. The next worked example shows how to do this.

EXAMPLE 4.11   USING SOLUTIONS OF KNOWN MOLAR CONCENTRATION

Problem: In an experiment to see if mouth bacteria can live on mannitol ($C_6H_{14}O_6$), the sweetening agent used in some sugarless gums, a student needed 0.100 mol of mannitol. It was available as a 0.750 M solution. How many milliliters of this solution must be used in order to obtain 0.100 mol of mannitol?

Solution: The conversion factors provided by the given concentration are

$$\frac{0.750 \text{ mol mannitol}}{1000 \text{ mL mannitol solution}} \quad \text{and} \quad \frac{1000 \text{ mL mannitol solution}}{0.750 \text{ mol mannitol}}$$

Therefore,

$$0.100 \text{ mol mannitol} \times \frac{1000 \text{ mL mannitol solution}}{0.750 \text{ mol mannitol}} = 133 \text{ mL mannitol solution}$$

Thus 133 mL of 0.750 M mannitol solution holds 0.100 mol of mannitol.

PRACTICE EXERCISE 14   To test sodium carbonate, Na_2CO_3, as an antacid, a scientist needed 0.125 mol of Na_2CO_3. It was available as 0.800 M Na_2CO_3. How many milliliters of this solution are needed for 0.125 mol of Na_2CO_3?

Once solutions of known molar concentration have been prepared, then the most common kind of calculation involves the weight relationships in some reaction when at least one reactant is in solution.

EXAMPLE 4.12   CALCULATIONS THAT INVOLVE MOLAR CONCENTRATIONS

Problem: Potassium hydroxide, KOH, reacts with hydrochloric acid, HCl, as follows:

$$\text{HCl}(aq) \quad + \text{KOH}(aq) \longrightarrow \text{KCl}(aq) \quad + \text{H}_2\text{O}$$

Hydrochloric   Potassium   Potassium
acid   hydroxide   chloride

How many milliliters of 0.100 M KOH are needed to react with the acid in 25.0 mL of 0.0800 M HCl?

Solution: We first have to find out how many moles of acid are present, because for each mole of acid we'll need a mole of the hydroxide (as the coefficients tell us). The molar concentration of the acid gives us these two conversion factors:

$$\frac{0.0800 \text{ mol HCl}}{1000 \text{ mL HCl solution}} \quad \text{or} \quad \frac{1000 \text{ mL HCl solution}}{0.0800 \text{ mol HCl}}$$

Therefore we multiply the given, 25.0 mL of HCl solution, by the first factor:

$$25.0 \text{ mL HCl solution} \times \frac{0.0800 \text{ mol HCl}}{1000 \text{ mL HCl solution}} = 0.00200 \text{ mol HCl}$$

The coefficients of the balanced equation tell us that 1 mol of HCl is equivalent chemically (in this equation) to 1 mol of KOH, so 0.00200 mol of HCl requires 0.00200 mol of KOH.

Finally, we have to find out what volume of the KOH solution contains 0.00200 mol of KOH. The molarity of the KOH solution gives us the option of the following conversion factors:

$$\frac{0.100 \text{ mol KOH}}{1000 \text{ mL KOH solution}} \quad \text{or} \quad \frac{1000 \text{ mL KOH solution}}{0.100 \text{ mol KOH}}$$

We can see that if we multiply 0.00200 mol of KOH by the second conversion factor, we'll have the right units:

$$0.00200 \text{ mol KOH} \times \frac{1000 \text{ mL KOH solution}}{0.100 \text{ mol KOH}} = 20.0 \text{ mL KOH solution}$$

Thus, 20.0 mL of 0.100 M KOH solution exactly neutralizes the acid in 25.0 mL of 0.0800 M HCl.

The next worked example shows how to solve a problem in which the mole-to-mole ratio of reactants in the equation is not 1 : 1.

EXAMPLE 4.13   CALCULATIONS THAT INVOLVE MOLAR CONCENTRATIONS

Problem: Sodium hydroxide, NaOH, can react with sulfuric acid by the following equation:

$$H_2SO_4(aq) + 2NaOH(aq) \longrightarrow Na_2SO_4(aq) + 2H_2O$$

| Sulfuric | Sodium | Sodium |
| acid | hydroxide | sulfate |

How many milliliters of 0.125 M NaOH provide enough NaOH to react completely with the sulfuric acid in 16.8 mL of 0.118 M H_2SO_4 by the given equation?

Solution: We start by asking how many moles of sulfuric acid are in 16.8 mL of 0.118 M H_2SO_4. Then, we will relate this amount of acid to the moles of NaOH that match it according to the coefficients. Finally, we will find out what volume of the 0.125 M NaOH solution holds this much NaOH.

First, then, the moles of H_2SO_4 that react:

The standard abbreviation for solution is soln.

$$16.8 \text{ mL } H_2SO_4 \text{ soln} \times \frac{0.118 \text{ mol } H_2SO_4}{1000 \text{ mL } H_2SO_4 \text{ soln}} = 0.00198 \text{ mol } H_2SO_4$$

Next, we find the moles of NaOH that chemically match 0.00198 mol of H_2SO_4 according to the coefficients of the equation. They tell us that 2 mol of NaOH is chemically equivalent (in this equation) to 1 mol of H_2SO_4, so we can select between the following conversion factors to match moles of NaOH to moles of H_2SO_4.

$$\frac{2 \text{ mol NaOH}}{1 \text{ mol } H_2SO_4} \quad \text{or} \quad \frac{1 \text{ mol } H_2SO_4}{2 \text{ mol NaOH}}$$

Hence, we convert 0.00198 mol H_2SO_4 to its equivalent in moles of NaOH (in this reaction) as follows:

$$0.00198 \text{ mol } H_2SO_4 \times \frac{2 \text{ mol NaOH}}{1 \text{ mol } H_2SO_4} = 0.00396 \text{ mol NaOH}$$

Finally, we find the volume of the 0.125 M NaOH solution that holds 0.00396 mol of NaOH:

$$0.00396 \text{ mol NaOH} \times \frac{1000 \text{ mL NaOH soln}}{0.125 \text{ mol NaOH}} = 31.7 \text{ mL NaOH soln}$$

Thus 31.7 mL of 0.125 M NaOH solution is needed to neutralize all of the sulfuric acid in 16.8 mL of 0.118 M H_2SO_4.

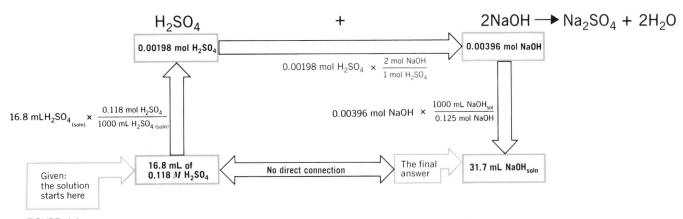

FIGURE 4.4
The calculation flow diagram for Example 4.13.

Figure 4.4 provides a pictorial summary, a calculation flow chart of the steps used to solve the problem of Example 4.13.

PRACTICE EXERCISE 15   Blood isn't supposed to be acidic, but in some medical emergencies it tends to become so. To stop and reverse this trend, the emergency care specialist might administer a dilute solution of sodium bicarbonate intravenously. Sodium bicarbonate destroys (neutralizes) acids. For example, it reacts with sulfuric acid (which is *not* present in blood) as follows:

$$2NaHCO_3(aq) + H_2SO_4(aq) \longrightarrow Na_2SO_4(aq) + 2CO_2(g) + 2H_2O$$

How many milliliters of 0.112 M H_2SO_4 will react with 21.6 mL of 0.102 M $NaHCO_3$ *according to this equation?*

SUMMARY

Equations   Chemical equations use the formulas of reactants (given before an arrow) and the products (after the arrow) to describe a chemical reaction. Coefficients in front of formulas ensure that all atoms among the reactants occur among the products, and they give the mole proportions of the substances involved. Subscripts within formulas must never be changed just to get an equation to balance.

Weight relationships   A quantity of a substance equal to its formula weight taken in grams is one mole of the substance, so to calculate a molar mass just find the formula weight and attach the units grams per mole (g/mol). One mole of any pure substance, element, or compound, consists of 6.02×10^{23} of its formula units. This number is named Avogadro's number. In working problems involving balanced equations and quantities of substances, be sure to solve them at the mole level where the coefficients can be used. Then, as needed, convert moles to grams.

Solutions   A solution is made up of a solvent and one or more solutes, and the ratio of quantity of solute to some unit quantity of solvent or of solution is called the concentration of the solution. A solution can be described as dilute or concentrated according to its ratio of solute to solvent being small or large.

A solution can also be described as unsaturated, saturated, or supersaturated according to whether it can be made to dissolve any more solute (at the same temperature). Each substance has a particular solubility in a given solvent at a specified temperature, and this is often expressed as the grams of solute that can be dissolved in 100 g of the solvent.

Molar concentration   The most useful quantitative description of the concentration of a solution is the ratio of the moles of solute per liter (or 1000 mL) of solution. This is the molar concentration or the molarity of the solution.

REVIEW EXERCISES

Answers to the review exercises whose numbers are marked by an asterisk are in Appendix V. The *Study Guide* that accompanies this book contains answers to the other review exercises. Remember to round atomic weights to their first decimal points *before* using them in any calculations.

Balanced Equations

4.1   Write in your own words what the following equation says.

$$2C + O_2 \longrightarrow 2CO \quad \text{(carbon monoxide)}$$

4.2   What does the following equation state? Write it in your own words.

$$N_2 + 3H_2 \longrightarrow 2NH_3 \quad \text{(ammonia)}$$

4.3   What would be a more acceptable way of writing the following balanced equation?

$$8H_3PO_4 + 16NaOH \longrightarrow 8Na_2HPO_4 + 16H_2O$$

4.4   Balance the following equations.
(a) $SO_2 + O_2 \rightarrow SO_3$
(b) $CaO + HNO_3 \rightarrow Ca(NO_3)_2 + H_2O$
(c) $AgNO_3 + MgCl_2 \rightarrow AgCl + Mg(NO_3)_2$
(d) $HCl + Ca(OH)_2 \rightarrow CaCl_2 + H_2O$
(e) $C_2H_6 + O_2 \rightarrow CO_2 + H_2O$

4.5   Balance each of the following equations.
(a) $NaHCO_3 + H_2SO_4 \rightarrow Na_2SO_4 + H_2O + CO_2$
(b) $Fe_2O_3 + H_2 \rightarrow Fe + H_2O$
(c) $Ca(OH)_2 + HNO_3 \rightarrow Ca(NO_3)_2 + H_2O$
(d) $NO + O_2 \rightarrow NO_2$
(e) $Al_2O_3 + H_2SO_4 \rightarrow Al_2(SO_4)_3 + H_2O$

Avogadro's Number

4.6   What is Avogadro's number?

4.7   Why did scientists select Avogadro's number and not some less complicated number to use in defining chemical unis for substances?

*4.8   How many molecules are in 6.00 g of H_2O?

4.9   A sample of aspirin ($C_9H_8O_4$) with a mass of 0.180 g (roughly the amount of aspirin in a typical 5-grain tablet) consists of how many molecules of aspirin?

*4.10   One dose of a particular medication contains 6.02×10^{20} molecules. How many grams are in this dose? The formula weight of the medication is 150.

4.11   A sample of impure water contains 3.01×10^{18} molcules of the impurity. How many milligrams of the impurity are present if its formula weight is 240?

Formula Weights

4.12   What law of nature makes it possible to compute formula weights simply by adding all of the atomic weights of the atoms given in a formula?

*4.13   Calculate the formula weight of each of the following substances.
(a) NaOH   (b) $CaCO_3$ (c) H_2SO_4
(d) Na_2CO_3 (e) $KMnO_4$ (f) Na_3BO_3

4.14   Calculate the formula weight of each of the following substances.
(a) $Mg(HCO_3)_2$ (b) Na_2HPO_4 (c) HNO_3
(d) $KC_2H_3O_2$ (e) $(NH_4)_2SO_4$ (f) $Ca_3(PO_4)_2$

Moles

4.15   How do we calculate the quantity of mass present in 1 mol of any substance?

4.16   What is the relationship between Avogadro's number and 1 mol of any substance?

4.17   What would be the practical difficulty in the lab of having balances read in moles instead of in mass units?

4.18   In converting grams to moles or moles to grams, what units are given to the formula weight of, say, H_2O, a molecular compound with a formula weight of 18.0?

4.19   The formula weight of bromine, Br_2, is 159.8. What are the two conversion factors that we can write that use this information?

*4.20   How many grams are in 2.50 mol of each of the compounds listed in Review Exercise 4.13?

4.21   How many grams are in 0.575 mol of each of the compounds given in Review Exercise 4.14?

*4.22   Calculate the number of moles that are in 75.0 g of each of the compounds listed in Review Exercise 4.13.

4.23   How many moles are in 28.6 g of each of the compounds given in Review Exercise 4.14?

Weight Relationships Involving Balanced Equations

*4.24   In the equation for the reaction of iron with oxygen,

$$4Fe + 3O_2 \longrightarrow 2Fe_2O_3$$

what two conversion factors express the *mole* relationship between each of the following pairs of substances?
(a) Fe and O_2 (b) Fe and Fe_2O_3 (c) O_2 and Fe_2O_3

4.25   Ethane, C_2H_6, burns according to the following equation.

$$2C_2H_6 + 7O_2 \longrightarrow 4CO_2 + 6H_2O$$

What two conversion factors express the *mole* relationship between each of the following pairs of compounds?
(a) C_2H_6 and O_2 (b) CO_2 and C_2H_6
(c) H_2O and O_2 (d) C_2H_6 and H_2O

*4.26   Butane, C_4H_{10}, is the fluid used in cigarette lighters. It burns according to the following equation.

$$2C_4H_{10} + 13O_2 \longrightarrow 8CO_2 + 10H_2O$$

(a) How many moles of oxygen are needed to react completely with 4.0 mol of butane?

(b) How many moles of water form from the burning of 10 mol of butane?

(c) To make 16 mol of carbon dioxide by this reaction, how many moles of oxygen are needed?

4.27 In the last step of the most commonly used method to convert iron ore into iron, the following reaction occurs.

$$Fe_2O_3 + 3CO \longrightarrow 2Fe + 3CO_2$$

(a) To make 350 mol of Fe, how many moles of CO are needed?

(b) If one batch begins with 35 mol of Fe_2O_3, how many moles of iron are made?

(c) How many moles of CO are required to react with 125 mol of Fe_2O_3?

*4.28 The Synthane process for making methane, CH_4, from coal is as follows, where we use carbon, C, to represent coal.

$$C + 2H_2 \longrightarrow CH_4$$

(a) How many moles of hydrogen are needed to combine with 37.5 mol of C?

(b) How many moles of methane can be made from 86 mol of H_2?

4.29 Gasohol contains ethyl alcohol, C_2H_6O, which burns according to the following equation.

$$C_2H_6O + 3O_2 \longrightarrow 2CO_2 + 3H_2O$$

(a) If 475 g of ethyl alcohol burn this way, how many grams of oxygen are needed?

(b) If 326 g of CO_2 form in one test involving this reaction, how many grams of ethyl alcohol burned?

(c) If 92.6 g of O_2 are consumed by this reaction, how many grams of water form?

*4.30 An industrial synthesis of chlorine is carried out by passing an electric current through a solution of NaCl in water. Other commercially valuable products are sodium hydroxide and hydrogen.

$$2NaCl + 2H_2O \xrightarrow{\text{electric current}} 2NaOH + H_2 + Cl_2$$

(a) How many grams of NaCl are needed to make 775 g of Cl_2?

(b) How many grams of NaOH are also produced?

(c) How many grams of hydrogen are made as well?

4.31 Phosphoric acid, H_3PO_4, is needed to convert phosphate rock into a fertilizer called *triple phosphate*. One way to make phosphoric acid is by the following reaction.

$$P_4O_{10} + 6H_2O \longrightarrow 4H_3PO_4$$

(a) To make 1.00×10^3 kg of phosphoric acid (one metric ton), how many kilograms of P_4O_{16} are needed?

(b) How many kilograms of water are also required?

*4.32 One method that can be used to neutralize an acid spill in the lab is to sprinkle it with powdered sodium carbonate. For example, sulfuric acid, H_2SO_4, can be neutralized by the following reaction.

$$H_2SO_4 + Na_2CO_3 \longrightarrow Na_2SO_4 + CO_2 + H_2O$$

If 45.0 g of H_2SO_4 are spilled, what is the minimum number of grams of sodium carbonate that have to be added to it to complete this reaction?

Solutions

4.33 A solution of sodium chloride at 50 °C was found to have a concentration of 36.5 g NaCl/100 g H_2O. Using information in a table in this chapter, determine whether this solution was saturated, unsaturated, or supersaturated.

4.34 A solution of barium sulfate at 20 °C has a concentration of 0.00024 g $BaSO_4$/100 g H_2O. Is it saturated, unsaturated, or supersaturated? Would it be described as dilute or concentrated?

4.35 Suppose you do not know and do not have access to a reference in which to look up the solubility of potassium chloride, KCl, in water at room temperature. Yet you need a solution that you know beyond doubt is saturated. How could such a solution be made?

4.36 Suppose you have a saturated solution of vitamin C in water at 10 °C. Without changing either the quantity of solvent or the quantity of solute, what could you do to make this solution unsaturated? What could you do to find out if this solution could be made supersaturated?

Molar Concentration

4.37 What are the differences among *molecule, mole,* and *molarity*?

4.38 What is another term for *molarity*?

4.39 When a volumetric flask is used to prepare a solution having some specified molarity, does the one who makes this solution know precisely how much *solvent* is used? Why is this information unnecessary for the uses to which the solution might be put?

*4.40 Calculate the number of grams of solute that would be needed to make each of the following solutions.
(a) 500 mL of 0.200 *M* NaCl
(b) 250 mL of 0.125 *M* $C_6H_{12}O_6$ (glucose)
(c) 100 mL of 0.100 *M* H_2SO_4
(d) 500 mL of 0.400 *M* KOH

4.41 How many grams of solute are needed to make each of the following solutions?
(a) 250 mL of 0.120 *M* Na_2CO_3
(b) 500 mL of 0.100 *M* NaOH
(c) 100 mL of 0.750 *M* $KHCO_3$
(d) 250 mL of 0.100 *M* $C_{12}H_{22}O_{11}$ (sucrose)

*4.42 How many milliliters of 0.10 *M* HCl contain 0.025 mol of HCl?

4.43 How many milliliters of 1.0 *M* H_2SO_4 would have to be taken to obtain 0.0025 mol of H_2SO_4?

*4.44 If you need 0.0010 mol of $NaHCO_3$ for an experiment, how many milliliters of 0.010 *M* $NaHCO_3$ would you have to measure out?

4.45 The stockroom has a supply of 0.10 M H_2SO_4. If you have to have 0.075 mol of H_2SO_4, how many milliliters of this stock solution do you have to take?

*4.46 There is a supply of 1.00 M NaOH in the lab. How many milliliters of this solution have to be taken in order to obtain 10.0 g of NaOH?

4.47 The stock supply of sulfuric acid is 0.50 M H_2SO_4. How many milliliters of this solution contain 5.0 g of H_2SO_4?

Weight Relationships Involving Reactions in Solution

*4.48 Barium sulfate, $BaSO_4$, is very insoluble in water, and a slurry of this compound is the "barium cocktail" given to patients prior to taking X-rays of the intestinal tract. It can be made by the following reaction.

$$Ba(NO_3)_2(aq) + Na_2SO_4(aq) \longrightarrow BaSO_4(s) + 2NaNO_3(aq)$$

In one use of this reaction to make barium sulfate—it is collected by filtering the final solution—a chemist used 250 mL of 0.100 M $Ba(NO_3)_2$. How many milliliters of 0.150 M Na_2SO_4 were needed to supply enough solute for this reaction?

4.49 How many milliliters of 0.100 M HCl are required to react completely with (and be neutralized by) 25.4 mL of 0.158 M Na_2CO_3 if they react according to the following equation?

$$2HCl(aq) + Na_2CO_3(aq) \longrightarrow 2NaCl(aq) + CO_2(g) + H_2O$$
Hydrochloric   Sodium
acid   carbonate

*4.50 Calcium hydroxide, $Ca(OH)_2$, is an ingredient in one brand of antacid tablets. It reacts with and neutralizes the acid in gastric juice, hydrochloric acid, as follows:

$$Ca(OH)_2(s) + 2HCl(aq) \longrightarrow CaCl_2(aq) + 2H_2O$$

A tablet that contains 2.00 g of $Ca(OH)_2$ can neutralize how many milliliters of 0.100 M HCl?

4.51 A nitric acid spill can be neutralized by sprinkling solid sodium carbonate on it. The reaction is

$$Na_2CO_3(s) + 2HNO_3(aq) \longrightarrow 2NaNO_3(aq) + CO_2(g) + H_2O$$

In one accident, 25.0 mL of 16.0 M HNO_3 (concentrated nitric acid, a dangerous chemical) spilled onto a stone desk top. Will 40.0 g of Na_2CO_3 be enough to neutralize this acid by the given equation? How many grams are needed?

Chapter 5
Kinetic Theory and Chemical Reactions

You have to be a believer to be a hot air balloonist — believe that nature always works in the same way. Some of the reliable properties of air important at the molecular level of life are studied in this chapter.

5.1 THE GASEOUS STATE AND PRESSURE

The earth's atmosphere exerts a pressure on the earth's surface that gives us a unit for measuring the pressure of any gas.

Because air is so vital to life, and because it is a mixture of gases, our ability to understand the molecular basis of respiration will depend on a knowledge of the gaseous state. Unlike liquids and solids, all gases obey a number of the same natural laws regardless of their chemical identities. These laws are given in terms of four physical quantities: pressure, volume, temperature, and moles. We've already learned about the last three of these, so what we have to study now is pressure. Then we'll look at the laws of gases.

Pressure, volume, temperature, and moles are called the *variables* of the gaseous state because they can be varied according to the experiment.

Pressure Is Defined as Force per Unit Area. Air is matter, so it has mass and like anything with mass is subject to the earth's gravitational attraction. This attraction makes air exert a force on each unit area of the earth, and force per unit area is called **pressure.**

A column of air 1 in.^{2} in cross section that extends from sea level to the edge of outer space exerts a force that averages 14.7 lb, so the air pressure at sea level is 14.7 lb/in.^{2}. At higher elevations less air is in the column, so the air pressure is less. On the top of Mt. Everest, for example, the earth's highest mountain, the air pressure is about one-third that at sea level.

By *outer space* we mean space beyond the earth's atmosphere where the atmosphere is so thin as to be almost nonexistent.

One way to observe air pressure is with a Torricelli barometer, a device illustrated and explained in Figure 5.1. At 0 °C and sea level, the pressure of the atmosphere supports a column of mercury in this device that averages 760 mm high, and this pressure is now defined as one **standard atmosphere,** abbreviated **atm.**

$$1 \text{ atm} = 760 \text{ mm Hg} \quad \text{(at 0 °C)}$$

Outside of health-related fields, **torr** is used as a synonym for *mm Hg*. This avoids the use of a unit of length to describe something (pressure) that isn't a length.

In scientific work and medicine, gas pressures are often expressed in a smaller unit, the **millimeter of mercury,** abbreviated **mm Hg,** defined by the following equation:

$$1 \text{ mm Hg} = \frac{1}{760} \text{ atm}$$

The air pressure in the lab might, for example, be 740 mm Hg (0.974 atm) one day and 747 mm Hg (0.983 atm) on another, depending on the weather. At the top of Mt. Everest, the pressure is close to 250 mm Hg.[1]

In a Mixture of Gases, Each Gas Exerts Its Partial Pressure Independently of the Others. Air is a mixture of gases, principally nitrogen and oxygen but with small traces of a few others. It's possible to separate 100 L of dry air into 21 L of oxygen and 79 L of nitrogen. Thus on a volume basis air is 21% O_2 and 79% N_2. *These percentages are true at all altitudes of the earth's atmosphere.* What changes with altitude is not the *volume* ratio of oxygen to nitrogen but their individual pressures called their partial pressures.

The **partial pressure** of a gas in a gas mixture such as air is the contribution the gas makes to the total pressure. It's the pressure that the gas would exert if all other gases were removed and it were all alone in the same container (at the same temperature). When dry air is at a total pressure of 760 mm Hg, the partial pressure of oxygen is 160 mm Hg and that of nitrogen is 600 mm Hg. Notice that the sum, 160 mm Hg + 600 mm Hg, is 760 mm Hg.

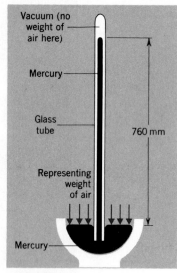

FIGURE 5.1
The Torricelli barometer (named after Evangelista Torricelli, 1608–1647, an Italian physicist). The long tube, sealed at one end, is filled with mercury and inverted into a dish of mercury. Some mercury runs out of the tube, but the remainder is held up by air pressure. No air is above the mercury inside the tube, so no air pressure acts inside to oppose the air pressure outside. The column of mercury in the tube stands 760 mm high at sea level and 0 °C.

Labels in figure: Vacuum (no weight of air here); Mercury; Glass tube; 760 mm; Representing weight of air; Mercury

[1] The SI unit of pressure is the pascal (Pa).

$$1 \text{ mm Hg} = 133.3224 \text{ Pa}$$
$$1 \text{ atm} = 101,325.024 \text{ Pa}$$

This makes the standard atmosphere about 101 kilopascals (kPa). Although most professional organizations urge health care specialists to switch from millimeters of mercury to kilopascals to report pressures of respiratory gases, it is our understanding that few have yet made the change. The mm Hg unit still appears to be the most widely used unit in medicine.

(We can also note that 160 mm Hg is 21% and 600 mm Hg is 79% of 760 mm Hg. Thus the partial pressure is directly proportional to the volume percentages, 21% O_2 and 79% N_2.)

John Dalton (of atomic theory fame) was the first to notice that partial pressures add up to the total pressure of a gas mixture. This is one of the laws of nature, now called **Dalton's Law** or the **law of partial pressures**:

Dalton's Law of Partial Pressures The total pressure exerted by a mixture of gases is the sum of the individual partial pressures.

$$P_{total} = P_a + P_b + P_c + \text{etc.}$$

In the equation for Dalton's law, P_a, P_b, and so on, refer to partial pressures of individual gases. When the gases are known, then their formulas can be substituted for the small letters. Thus P_{O_2} refers to the partial pressure of oxygen.

At High Altitudes, Oxygen-Enriched Air Is Required for Breathing. We said that when the total pressure of dry air is 760 mm Hg, the value of P_{O_2} is 160 mm Hg. At an altitude of 5.5 km (3.4 miles; 18,000 ft), the total pressure is one-half as much, 380 mm Hg, and so the partial pressure of oxygen is also one-half as much, one-half of 160 mm Hg or 80 mm Hg. This partial pressure is not high enough to force atmospheric oxygen out of the lungs and into the blood stream at a rate fast enough to support human life. Nearly all people have to use pressure tanks that deliver oxygen-enriched air to the lungs when they operate above 14,000 ft. At 18,000 ft, for example, air has to be enriched to 42% O_2 to give its P_{O_2} the same value as it has at sea level.

The Air We Exhale Includes Carbon Dioxide. The air we inhale is generally not dry air, and its water vapor makes a small contribution to the total pressure. Typical values for the partial pressures of the gases in inhaled air of about 20% relative humidity (5% water vapor at room temperature) are given in Table 5.1.

Table 5.1 also gives partial pressure data for exhaled air and for air in the lung's air sacs (alveoli). Notice the decrease in P_{O_2} between inhaled and exhaled air and the relatively large increase in P_{CO_2}. The P_{O_2} decreases, of course, because oxygen is absorbed by the blood. The

You will also see the symbol we use here for the partial pressure of oxygen, P_{O_2}, written as PO_2.

Oxygen-enriched air is air with an increased ratio of oxygen to nitrogen.

TABLE 5.1
The Composition of Air During Breathing

Gas	Partial Pressure (in mm Hg)		
	Inhaled Air	Exhaled Air	Alveolar Air[a]
Nitrogen	594.70	569	570
Oxygen	160.00	116	103
Carbon dioxide	0.30	28	40
Water vapor	5.00[b]	47	47
Totals	760.00	760	760

[a] Alveolar air is air within the alveoli, thin-walled air sacs enmeshed in beds of fine blood capillaries. Little more than bubbles of tissue, these sacs are the terminals of the successively branching tubes that make up the lungs. We have about 300 million alveoli in our lungs.

[b] A partial pressure of water vapor of 5.00 mm Hg corresponds to air with a relative humidity of about 20%, a familiar weather report term that we will not define further here.

P_{CO_2} increases because carbon dioxide, a gaseous waste product of metabolism, is removed from the blood at the lungs. The warm, moist environment in the alveoli loads the exhaled air with water vapor, as the data in Table 5.1 also show. But notice that the total pressure, regardless of the location, is still the sum of the partial pressures.

5.2   THE GAS LAWS

The pressure, volume, and temperature of 1 mol of a gas affect each other nearly identically for all gases.

The symbols are

P = pressure

V = volume

T = Kelvin temperature

n = number of moles

Gas Pressure Varies Inversely with Volume at Constant Temperature.   Robert Boyle (1627–1691), an English scientist, discovered the following relationship, now called the **pressure–volume law** or **Boyle's law.**

> **Pressure–Volume Law (Boyle's Law)** The volume of a fixed number of moles (n) of gas is inversely proportional to its pressure at a constant temperature.

Put in mathematical form, this law states that

The symbol $\propto$ stands for *is proportional to.*

$$V \propto \frac{1}{P} \quad (T \text{ and } n \text{ constant}) \tag{5.1}$$

Or after rearranging terms and introducing a constant of proportionality:

$$PV = \text{a constant}$$

This can be expressed in the following form, which is easier to use in calculations. Subscripts 1 and 2 represent initial and final states.

$$P_1 V_1 = P_2 V_2 \tag{5.2}$$

Equation 5.2 is true because each side equals the same constant.

EXAMPLE 5.1   DOING PRESSURE–VOLUME LAW CALCULATIONS

Problem: A given mass of oxygen occupies 500 mL at 760 mm Hg at 20 °C. At what pressure will it occupy 450 mL at the same temperature?

Solution: We have the following data:

$$P_1 = 760 \text{ mm Hg} \qquad P_2 = ?$$
$$V_1 = 500 \text{ mL} \qquad V_2 = 450 \text{ mL}$$

From Equation 5.2, we know that $P_1 V_1 = P_2 V_2$. Because we want to calculate P_2, we rearrange this equation to

$$P_2 = P_1 V_1 / V_2$$

Next, we insert our data into this equation.

$$P_2 = \frac{(760 \text{ mm Hg}) (500 \text{ mL})}{(450 \text{ mL})}$$
$$P_2 = 844 \text{ mm Hg} \quad \text{(correctly rounded)}$$

Thus the new pressure must be 844 mm Hg.

PRACTICE EXERCISE 1   If 660 mL of helium (He) at 20 °C is under a pressure of 745 mm Hg, what volume will the sample occupy (at the same temperature) if the pressure is changed to 375 mm Hg?

Gas Volume Varies Directly with the Kelvin Temperature at Constant Pressure.

Jacques Charles (1746–1823), a French scientist, discovered how a gas changes its volume when its temperature is raised while both the pressure and number of moles are kept constant. His discovery is now called the **temperature–volume law** or **Charles' law.**

Charles' interest stemmed from a curiosity about hot air balloons, which had just been developed.

Temperature–Volume Law (Charles' Law) The volume of a fixed mass of any gas is directly proportional to its Kelvin temperature, if the gas pressure is kept constant.

$$V \propto T \quad \text{(constant } P \text{ and mass)}$$

T must be in kelvins, K.
K = °C + 273

By rearranging terms and introducing a constant of proportionality we have:

$$\frac{V}{T} = \text{a constant}$$

A useful way to express this, where we use the subscripts 1 and 2 as we did in Equation 5.2, is

$$\frac{V_1}{T_1} = \frac{V_2}{T_2} \quad \text{(constant } P \text{ and } n) \qquad (5.3)$$

Charles' law calculations are handled much as we worked those of Boyle's law. Remember that temperatures given in degrees Celsius must be changed to kelvins.

PRACTICE EXERCISE 2   A sample of cyclopropane, an anesthetic, with a volume of 575 mL at a temperature of 30 °C, was cooled to 15 °C at the same pressure. What was the new volume?

Gas Pressure Varies Directly with Kelvin Temperature.

Joseph Gay-Lussac (1778–1850), another French scientist, discovered the way in which an increase in the temperature of a confined gas changes its pressure, a relationship now called the **pressure–temperature law** or **Gay-Lussac's law.**

Pressure–Temperature Law (Gay-Lussac's Law) The pressure of a gas is directly proportional to its Kelvin temperature if V and n are constant.

$$P \propto T \quad \text{or} \quad \frac{P}{T} = \text{a constant} \qquad (V \text{ and } n \text{ are constant})$$

This property of a gas is behind the warning not to put an aerosol can into an incinerator. The can is sealed, and it still holds residual gas. Hence, as its temperature rises so does its internal pressure. Eventually, the can explodes, which might damage the incinerator or possibly hurt bystanders.

When a chemical explosive (e.g., TNT) is detonated, a solid of small volume suddenly changes into a large-volume mixture of hot gases at high temperature and pressure.

Under Identical Pressure and Temperature, Equal Volumes of Gases Have the Same Moles.

The gas laws so far studied assume a constant size in moles of gas, n. If we want to increase the number of moles, something else has to change. Thus, if we want to keep the pressure and temperature constant as we increase the moles, we need more volume. In fact, at constant pressure and temperature, the volume of a gas is directly proportional to the number of moles:

$$V \propto n \quad \text{(at constant } P \text{ and } T)$$

This relationship becomes an equation when we insert a proportionality constant C:

$$V = Cn \quad \text{(at constant } P \text{ and } T)$$

Amadeo Avogadro discovered an important fact about this equation: C is the same for all gases. A 45-L sample of oxygen, for example, has the same number of moles as a 45-L sample of nitrogen or any other gas (under the same temperature and pressure). Thus we have another gas law, the volume–mole law or **Avogadro's law.**

> Volume–Mole Relationship (Avogadro's Law) Equal volumes of gases have equal numbers of moles when compared at the same pressure and temperature.

At 273 K and 1 Atm, One Mole of a Gas Occupies 22.4 L.   The actual volume occupied by one mole of any gas varies, of course, with pressure and temperature. To make comparisons scientists have therefore agreed to reference conditions called the **standard conditions of temperature and pressure** or **STP.** Standard pressure is 1 atm and standard temperature is 273 K (0 °C). At STP, one mole of any gas occupies 22.4 L, a value called the **standard molar volume** of a gas.

We now have a basis for learning about a much more general equation that combines the four gas laws that we have studied.

The Ratio, PV/nT Is the Same for All Gases.   As scientists worked more and more with the laws of Boyle, Charles, Gay-Lussac, and Avogadro, they eventually discovered a more general relationship. For any gas, the result of multiplying P and V and dividing by T is proportional to the moles of the gas, n:

$$\frac{PV}{T} \propto n$$

Using a proportionality constant, R, we have:

$$\frac{PV}{T} = nR \tag{5.4}$$

This equation, usually rewritten as $PV = nRT$, is the **universal gas law,** and the constant, **R**, is the **universal gas constant.** Its value (together with its units) is

For this value of R to be true, P must be in (mm Hg), V in mL, T in kelvins, and n in moles.

$$R = \frac{PV}{nT} = 6.24 \times 10^4 \, \frac{\text{mm Hg mL}}{\text{mol K}}$$

In nearly all experiments with gases, the experimenter holds the value of n constant. (No gas is added and none is permitted to escape.) Under this condition, the right side of Equation 5.4 is a constant:

Since both n and R are constants, their product has to be a constant, too.

$$\frac{PV}{T} = nR = \text{another constant} \tag{5.5}$$

Using subscripts 1 and 2 in the usual way, the form of Equation 5.5 that is particularly convenient for calculations is an equation called the **general gas law:**

$$\frac{P_1V_1}{T_1} = \frac{P_2V_2}{T_2} \qquad (5.6)$$

Both sides of Equation 5.6 equal $n \times R$, so they have to equal each other.

Equation 5.6 includes the laws of Boyle, Charles, and Gay-Lussac. Notice, for example, when both n and T are constant (Boyle's law conditions), the temperatures can be canceled in the equation, leaving Equation 5.2, $P_1V_1 = P_2V_2$, an expression of Boyle's law. (You should show how Equation 5.6 similarly incorporates the laws of Charles and Gay-Lussac.)

EXAMPLE 5.2 USING THE GENERAL GAS LAW

Problem: A sample of an anesthetic gas with a volume of 925 mL at 20.0 °C and 750-mm Hg pressure is warmed to 37.0 °C at a pressure of 745 mm Hg. What is the final volume of the gas? (Retain three significant figures in the answer.)

When temperatures are given in degrees Celsius, they must always be converted to kelvins before working any gas law problem.

Solution: It's best to collect all of the data first.

$$V_1 = 925 \text{ mL} \qquad V_2 = ?$$
$$P_1 = 750 \text{ mm Hg} \qquad P_2 = 745 \text{ mm Hg}$$
$$T_1 = 293 \text{ K} \qquad T_2 = 310 \text{ K}$$

(Don't forget that the gas laws require that all temperatures be expressed in kelvins.) Now we can use Equation 5.6.

$$\frac{(750 \text{ mm Hg})(925 \text{ mL})}{(293 \text{ K})} = \frac{(745 \text{ mm Hg})(V_2)}{(310 \text{ K})}$$

Solving for V_2, we find:

$$V_2 = \frac{(750 \text{ mm Hg})(925 \text{ mL})(310 \text{ K})}{(745 \text{ mm Hg})(293 \text{ K})}$$
$$= 985 \text{ mL}$$

PRACTICE EXERCISE 3 A fire occurred in a storage room where a steel cylinder of oxygen-enriched air was kept. The pressure of the gas in the cylinder was 300 atm at 20.0 °C. To what value does the pressure change if the temperature rises to 200 °C but the volume of the cylinder does not change?

5.3 THE KINETIC THEORY OF GASES

All of the gas laws can be explained in terms of a model of a gas described by the kinetic theory.

As the gas laws unfolded over the decades, more and more scientists asked: What must gases be like to make these laws true? What is so remarkable about the gas laws is that they hold for all gases, particularly when a gas is not close to condensing to a liquid (that is, it isn't under high pressure or isn't very cold). There are no general laws like the gas laws for liquids and solids.

An Ideal Gas Would Obey the Gas Laws Exactly. When measurements of accuracy and high precision are made, no gas obeys the gas laws *exactly* over all ranges of pressure and temperature. Many come so close, however, that it was natural for scientists to theorize about a hypothetical gas that would obey the gas laws exactly under all conditions. Such a gas is called an **ideal gas,** and scientists used the behavior of real gases to postulate the following characteristics of the ideal gas.

The noble gases behave most nearly as ideal gases. Polar gases such as water vapor or ammonia are least ideal.

Model of an Ideal Gas

1. The ideal gas consists of a large number of extremely tiny particles in a state of chaotic, utterly random motion.
2. The particles are perfectly hard, and when they collide they lose no energy because of friction.
3. The particles neither attract nor repel each other.
4. The particles move in accordance with the known laws of motion.

Because this model said that gases consist of tiny particles *in motion,* it came to be called the **kinetic theory of gases.** Let's *see* how it explains the gas laws.

Collisions of Moving Gas Molecules with the Container Create Gas Pressure.   The pressure of a gas comes from the innumerable collisions its atoms or molecules make with the walls of the container (Figure 5.2). If the walls are squeezed in to make the volume smaller, the number of collisions *per unit wall area* will increase, so the pressure must then rise. But this is just what Boyle's law says, a smaller volume must create a greater pressure (at constant T and n).

Since the atoms or molecules of a gas generally do not attract or repel each other (the third postulate of the kinetic theory), they act independently. Hence, they must make *independent* contributions to the total pressure, just as Dalton's law of partial pressure describes. (Indeed, this law suggested the third postulate.)

Heating a Gas Makes Its Molecules Move Faster.   One result of the calculations based on the model of an ideal gas is a mechanism for how a gas absorbs energy when heated. *Heat causes an increase in the average kinetic energy of the molecules of the gas.* As we've learned, the kinetic energy of a moving object is given by:

$$\text{Kinetic energy} = \tfrac{1}{2}mv^2$$

where m = mass and v = velocity. The mass of each gas molecule cannot change, so a

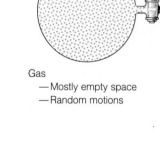

Gas
— Mostly empty space
— Random motions

FIGURE 5.2
The kinetic theory and the pressure–volume relationship (Boyle's law). The pressure of the gas is proportional to the frequency of collisions per unit area. When the gas volume is made smaller in going from (a) to (b), the frequency of the collisions per unit area of the container's walls increases. This is how the pressure increase occurs.

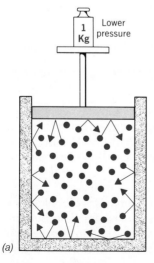
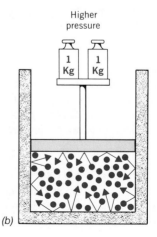

change in temperature alters the average kinetic energy only by changing the average velocity of the molecules. Thus when a gas is heated, its now more energetic molecules move with a greater average velocity. If a gas is cooled, the molecules move with a lower average velocity. At −273 C, all molecular motions in a gas stop, which is the reason for calling this temperature *absolute zero* (and making it equal to 0 K on the Kelvin scale). We will see that the relationship between temperature and average molecular velocity bears strongly on the effect of temperature on the rates of chemical reactions

With a Higher Collision Energy, Molecules of a Hotter Gas Cause a Higher Pressure. When a gas is made hotter, its molecules have a greater average energy, as we just saw. The molecules in the hotter gas, therefore, must cause an increase in the force they exert when they strike a unit area of the container's walls. If the walls cannot move, the result is an increase in gas pressure. This is behind Gay-Lussac's law: Gas pressure is proportional to gas temperature at constant volume.

With a Higher Collision Energy, Molecules of a Hotter Gas Can Move Walls. The harder hitting molecules of a hotter gas ordinarily would increase the pressure, as we just learned, but when a wall can give way, the hotter gas instead takes up more room. The gas pressure thus stays constant. At constant pressure, in other words, a gas volume must increase with temperature, as Charles' law says.

The success of the kinetic theory in explaining the gas laws was considered so outstanding that some historians have considered it one of the greatest triumphs of the human mind. It understandably encouraged scientists to see how the theory might assist our understanding of liquids and solids, too.

5.4   THE LIQUID AND SOLID STATES AND KINETIC THEORY

The molecules in liquids move about randomly, and the tendency of some to escape creates the vapor pressure of a liquid.

At the molecular level, the chief physical difference between liquids and gases is that the molecules of a liquid are nearly always in contact with neighbor molecules (Figure 5.3). Otherwise, molecules in a liquid move around randomly, like those of a gas. Gas molecules, in contrast, touch only when they collide. In fact, the reason why all gases obey the same laws is that there is so much empty space in a gas.

A Liquid's Evaporation Creates a Vapor Pressure. A fraction of the molecules at the surface of a liquid happen to be moving upward. With enough velocity they can escape into the space above. If they do not return to the liquid in equal numbers, the liquid evaporates.

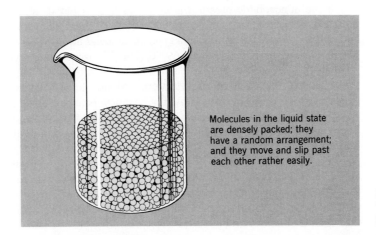

Molecules in the liquid state are densely packed; they have a random arrangement; and they move and slip past each other rather easily.

FIGURE 5.3
The liquid state as viewed by the kinetic model.

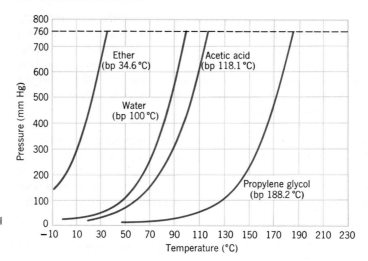

FIGURE 5.4
Vapor pressure versus temperature. (Ether was once widely used as an anesthetic. Acetic acid is the sour component in vinegar. Propylene glycol is in several brands of antifreeze mixtures.)

It's helpful to think of a liquid's vapor pressure as its *escaping tendency*.

Evaporation is the gradual change of a liquid to its gas state. The gaseous form of a liquid is sometimes called a **vapor.** A liquid that readily evaporates is called a *volatile* liquid. Ether is an example. One that does not evaporate at any noticeable rate is said to be *nonvolatile*. Salad oil is an example.

Evaporation ensures that the space above a liquid's surface contains molecules of vapor. These behave as a gas and so generate their own partial pressure. Thus the liquid's vapor contributes to the total pressure just above the liquid, in accordance with Dalton's law of partial pressure. This partial pressure is called the liquid's **vapor pressure.** It is measured by quickly introducing a liquid into a sealed space connected to a device that measures pressure. The pressure then increases to a constant new value, and the increase is the vapor pressure.

The higher the temperature of a liquid, the higher its vapor pressure, as the plots in Figure 5.4 show. At body temperature (37 °C), the vapor pressure of water is 47 mm Hg. This is the partial pressure of water vapor in both alveolar and exhaled air, as we saw in Table 5.1, so alveolar and exhaled air are saturated in water.

Vapor Pressure of Water

Temperature (°C)	Vapor Pressure (mm Hg)
20	17.5
30	31.8
37	47.1
40	55.3

In a Boiling Liquid, the Vapor Pressure Is at a Maximum.   When a liquid has been heated to a temperature high enough to make its vapor pressure equal to the atmospheric pressure, the liquid boils. Vapor pockets and bubbles now form *beneath* the liquid's surface. As they rise they create the familiar turbulence we see in a boiling liquid. The temperature at which a liquid's vapor pressure equals one atmosphere is the **normal boiling point.**

If the atmospheric pressure is lower, as it is at higher altitudes, the liquid's vapor pressure becomes equal to the outside atmospheric pressure at a lower temperature. Thus liquids have lower than normal boiling points at lower atmospheric pressures. At the top of Mount Everest, water boils at only 69 °C (156 °F) because the pressure is so low there.

In the mile-high city of Denver, Colorado, water boils at about 95 °C.

Water Has a High Heat of Vaporization.   When a liquid boils, its temperature no longer increases as it absorbs more heat. All the heat now being absorbed is used just to cause the change in state. The heat needed to convert a liquid to its vapor is called its **heat of vaporization.** Water, for example, has a heat of vaporization at 100 °C and 1 atm of 540 cal/g, one of the highest of all known values for any liquid.

Water, of course, can change from its liquid to vapor form (evaporate) at temperatures below the boiling point, and a certain heat of vaporization is needed for this at any temperature. At body temperature, for example, about 500 cal/g is needed to vaporize water. As we'll soon see, the vaporization of water from the skin and the lungs is a major mechanism for the removal of heat from the body.

SPECIAL TOPIC 5.1   HYPOTHERMIA

The body responds to a fall in its temperature by trying to increase its rate of metabolism so that more heat is generated internally. Uncontrollable shivering is the outward sign of this response, and it begins after the body temperature has decreased 2–3 °F (measured rectally). If the temperature continues to drop, the shivering will be violent for a period of time. Then loss of memory—amnesia—sets in at about 95–91 °F. The muscles become more rigid as the core temperature drops to the range of 90–86 °F. The individual must now have outside help immediately, because mental ability to take life-saving steps is now gone. The heartbeat becomes erratic, unconsciousness sets in (87–78 °F), and below 78 °F death occurs by heart failure or pulmonary edema.

Death by hypothermia, often called death by *exposure,* can happen even when the air temperature is above freezing. If you become soaked by perspiration or rain and the wind comes up, an outside temperature of 40 °F is dangerous. Those who fall overboard in cold water (32–35 °F) seldom live longer than 15–30 min.

The legendary St. Bernard dogs who brought little casks of brandy to blizzard victims in the Swiss Alps were more agents of death than life to anyone who drank the brandy. A shot of brandy in a hypothermic individual worsens the situation. Alcohol dilates (enlarges) blood capillaries. When the capillaries near the skin's surface, which are loaded with the most chilled blood in the hypothermic body, suddenly enlarge, the chilled blood moves quickly to the body's core. This rapid drop in *core* temperature is particularly life threatening.

If a victim of hypothermia is conscious and able to swallow food or drink, give warm, nonalcoholic fluids and sweet foods. As quickly as possible, get the victim dry and out of the wind. Get into a dry sleeping bag with the victim, so that your own body warmth can be used. It is a genuine medical emergency, and prompt aid is vital.

Ions and Molecules in Solids Vibrate About Fixed Points.   The ions and molecules in a solid do not move around. They are fixed relative to each other, but they still vibrate. When we heat a solid, these vibrations increase. Eventually, at a particular temperature unique for each solid, the vibrations are vigorous enough to overcome the forces of attraction that keep the crystal rigid. The particles now slip and slide, abandon old neighbors, and get new ones. Such activity occurs when a solid melts. The minimum temperature that makes this happen is called the **melting point** of the solid.

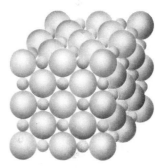

Solid (ionic)
—Densely and orderly packed
—Vibrations about fixed points

The Heat of Fusion of Water Is Unusually High.   A quantity of heat called the **heat of fusion** is required to make a solid melt. If we want to refreeze the solid, we have to take this energy out again by cooling the liquid. The heat of fusion of water is 80 cal/g, one of the highest of all known heats of fusion. This makes ice a particularly efficient coolant. To melt just one gram of ice takes 80 cal, so when a little ice is put into an icepack, it can remove as it melts a great quantity of heat from an inflammed area.

Heat Generated by Metabolism Is Removed by Evaporation, Radiation, Conduction, and Convection.   The thermal properties of water are as vital to life as any other properties because they help the body manage the heat produced by metabolism. **Metabolism** is the continuing chemical activity of all kinds that occurs in the body. Not only does it provide the energy for operating nerves, muscles, vital organs, and the synthesis of biochemicals, metabolism also tends to raise the body temperature. If this is not kept from increasing, the body experiences **hyperthermia.** In the opposite condition, **hypothermia,** the body temperature is too low. See Special Topic 5.1 for more details. Both hyperthermia and hypothermia are life-threatening.

The body loses heat by four mechanisms: the evaporation of water, and by radiation, conduction, and convection. Evaporation occurs by **perspiration** at the skin or from the inner surfaces of the lungs, the latter giving moistness to exhaled air. Perspiration can be *sensible* or *insensible;* that is, it can be obvious and noticeable in beads of sweat, or it can go unnoticed. Either form is effective, because each gram of water that evaporates and leaves the body carries with it about 500 cal (0.5 kcal) of heat, water's heat of vaporization. We ingest each day about 2.5 L of water. We lose about 1.0 L by perspiration, so this removes about 20–25% of the calories produced daily by metabolism.

Hyper, over or above
therm, heat
hypo, lower

Nearly half of the heat loss from the body occurs through the uncovered head. Outdoors in winter, when your feet are cold, put on a hat.

Radiation refers to energy like light energy. Thermal radiation (''heat rays'') is called infrared radiation because it is just a little less energetic than red light. It is invisible to the eye but special camera film can detect it. Any warm object, including the body, radiates this form of energy, and its loss tends to cool the body.

Heat loss by **conduction** occurs whenever a warmer object is in contact with one that is colder as when cold machinery, for example, is handled or when the body is in contact with cold air.

Convection is the loss by wind or draft of the warm, thin layer of air next to the skin. When this loss is prevented, the air layer is an excellent insulation. Waffle-weave underclothing is designed, for example, to trap this layer in place. Woolen fabrics also work well.

5.5   THE KINETIC THEORY AND RATES OF CHEMICAL REACTIONS

The rate of a chemical reaction increases with temperature and with reactant concentrations.

In chemical reactions, electrons and nuclei become reorganized. Old bonds change and different bonds form. The electron configurations of the reactants switch over to those of the products. If a gentle touching of the reactant particles were all it took for this to happen, no reactant would be stable in the presence of anything else. Yet, many substances are stable and can be stored in the presence of air, moisture, glass, people, and other potential reactants.

The kinetic theory helps us understand why some combinations of reactants do nothing to each other, why others can stand each other until the temperature increases too much, and why still other combinations can't be stored under any circumstances. The field of chemistry that deals with the rates of chemical reactions is called **kinetics.**

The Conversion of Kinetic to Potential Energy Makes Chemicals React.   For the particles of two reactants to change each other chemically, they must collide. Only in this nearness, however brief, can the energy of the collision make electrons of the reactant particles relocate relative to their nuclei. Generally, very light collisions will not work.

We know that the law of conservation of energy operates in nature. We now ask: What happens to their kinetic energy when colliding particles slow down or stop? Is it lost? If so, what of the law of conservation of energy?

The *total* energy—kinetic plus potential—stays constant throughout the change, but it becomes apportioned differently.

The energy that existed as *kinetic* energy is not lost; it is transformed into potential energy. In the realm of colliding molecules, the kinetic energy existing at the moment of collision usually changes stable electron–nuclei arrangements temporarily to less stable arrangements. Now chemical things can happen.

The particles, of course, might simply revert to their original configurations. This happens often, and the particles bounce away from each other without permanent change. The potential energy in the temporary and unstable arrangement at the instant of collision reconverts to kinetic energy of motion. This is how a bouncing ball can hit a sidewalk, momentarily stop, be temporarily deformed, and then bounce away, still as a ball and not something else. In like manner, many collisions of molecules lead to nothing, chemically.

In other situations, however, particles deformed by collision might undergo a reorganization of electrons and nuclei to give different chemical species. Thus the conversion of kinetic energy into potential energy during a collision can make a chemical reaction possible. The collisions that give the products are called *successful* collisions. The **rate of a reaction** is the number of successful collisions that occur each second in each unit of volume.

A Certain Minimum Collision Energy Is Needed to Make a Reaction Happen.   Almost no reactions are instantaneously rapid (explosive). Each reaction has its own minimum potential energy that must accumulate when reactant molecules collide before electrons can relocate and form new bonds. This minimum value of energy is called the reaction's **energy of activation.** Figure 5.5a shows what it means, and you can see why the energy of activation is sometimes called the ''energy hill'' or the energy barrier of a reaction.

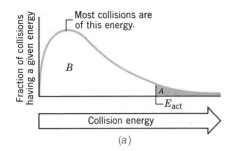
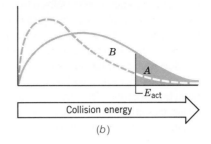

FIGURE 5.5
Energy of activation. (*a*) Only a small fraction of all of the collisions, represented by the ratio of areas, $A/(A + B)$, has enough energy for reaction. (*b*) This fraction greatly increases when the temperature of the reacting mixture is increased.

The vertical axis in Figure 5.5*a* represents changes in the *fraction* of the collisions that occur as the collision energy changes. These energies are given by the horizontal axis. They range from zero collision energy (bare taps) on the left to very large values on the right, values approaching infinity. They vary because reactant particles move over a large range of speeds, from very low values (even zero for some, for a moment) to very high speeds.

Figure 5.5*a* shows that some collisions will be such slight taps that virtually no kinetic energy changes into potential energy. However, the fraction of such collisions is essentially zero. As the value of the collision energy increases (as we move to the right in Figure 5.5*a*), the fraction of the collisions with a particular collision energy also increases until it reaches a maximum. Then, as we continue to move to the right, collisions that have increasingly higher energies become less and less likely, so the fractions with such high energies decline and the curve moves back down. Eventually, we reach a value of collision energy that provides the exact energy needed for the electron–nuclei rearrangement, the chemical reaction, to occur. This is the energy of activation, symbolized as E_{act} in the figure.

When collisions have this much energy, or any higher value, the reaction involves enough energy for it to take place. Colliding particles that do not achieve the energy of activation just bounce away, chemically unchanged.

Of course, it isn't enough to have sufficient energy. The colliding particles must hit each other just right, much as the runners in a relay race have to pass the baton correctly regardless of how fast or slowly they are moving at this critical moment.

Only a small fraction of all collisions have enough energy to be successful (result in a reaction). We can represent this fraction as $A/(A + B)$ the ratio of the shaded area, A, in Figure 5.5*a* to the total area $(A + B)$. If the energy of activation were much higher, meaning a high energy barrier, the shaded area, A, would be even smaller. Then the fraction $A/(A + B)$ would also be much smaller; the reaction would be very slow. On the other hand, if the energy of activation were very low, the fraction $A/(A + B)$ would be large, and the reaction would be rapid. In the extreme, there would be no E_{act}, A would equal $(A + B)$, the fraction would equal 1, and every well-oriented collision would be successful. In practical terms, such a reaction would be very rapid, and it would occur at the instant of mixing the reactants.

At the other extreme, E_{act} would be so high that $A/(A + B)$ would be virtually zero. Then no reaction occurs, and the "reactants" are eternally stable in each other's presence.

An Increase in the Temperature of the Reactants Makes the Fraction of Successful Collisions Higher.
It is well known that increasing the temperature of the reactants increases the rate of their reaction. As the reactant molecules acquire a higher average kinetic energy, the average collision energy also increases. See Figure 5.5*b*. The curve thus flattens out, and its maximum point shifts to the right. *The energy of activation, however, stays essentially unchanged.* Therefore, as higher temperatures shift the curve to the right, a larger fraction of the area under the curve moves to the *right* of E_{act}. The fraction of collisions given by $A/(A + B)$ thus increases, and so the reaction rate increases.

The higher the temperature, the flatter the curve becomes and the greater is the percentage of collisions that are effective.

The effect of temperature is large. As a rule of thumb, an increase of only 10 °C doubles or triples most rates.

Sometimes a patient with a high fever is partly immersed in very cold water to bring the body temperature down rapidly.

Metabolism Accelerates When Body Temperature Increases.   A large rate acceleration for a small temperature change has serious implications for health. An increase in body temperature of only 1 °F raises the rate of metabolism so much that the oxygen requirement of tissue increases by 7%. This places an extra strain on the heart, because it has to speed up the delivery of oxygen from the lungs.

All reactions in a healthy human body occur at a constant temperature of 37 °C, normal body temperature. Chemicals that do not react outside the body at this temperature, such as sugar and oxygen, do react inside. The next question is Why? We'll answer this in the next section after we have learned a little more essential background.

Each Reaction Has a Characteristic *Heat of Reaction*.   Figure 5.6 helps us make an important distinction between a reaction's energy of activation and the heat of reaction. The figure shows a kind of plot new to our study, a *progress of reaction diagram*. Its vertical axis gives *relative* values of the potential energies of the substances, either the reactants or the products, depending on where in the plot we are. The horizontal axis indicates the direction of the chemical change. The symbols of the species are included to keep the nature of the reaction in view.

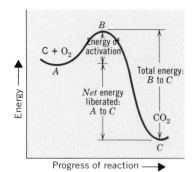

FIGURE 5.6
Progress of reaction diagram for the exothermic reaction of carbon with oxygen that produces carbon dioxide.

To follow the progress of the reaction diagrammed in Figure 5.6, begin on the left at site A with the reactants, carbon and oxygen in our example. We know that they are quite stable in each other's presence at or near room temperature. Coal (mostly carbon), after all, can be stored with no trouble in air (with its 21% O_2). As you know, to make carbon and oxygen react, we have to heat them; we ignite them, in other words. This gives their particles higher kinetic energies, and as the temperature rises, more and more collisions become closer to being successful. We are now moving in the diagram both upward toward higher potential energies of the reactants and rightward in the direction of the products. We are climbing the energy hill.

Eventually we provide the energy of activation, and we are at the top of the energy barrier at B. The reactants' electrons and nuclei can now rearrange to give product molecules, carbon dioxide. A great deal of the potential energy in the complex of electrons and nuclei at the top of the barrier now transforms into the kinetic energy of the newly forming molecules of carbon dioxide. There is quite a drop in potential energy now as the reaction progresses to products at site C. Some of this potential energy goes to repay the cost of climbing the energy hill, but there is a net excess that is liberated from the mixture. This net energy difference (A to C) between the reactants and the products is called the **heat of reaction.**

As we know, once the reaction of carbon and oxygen starts, it continues spontaneously. The reaction is exothermic, and some of the energy represented in Figure 5.6 by the distance from B to C activates still unchanged particles of the reactants. The heat of reaction in this case represents the conversion of some of the chemical energy in carbon and oxygen into the kinetic energy of the molecules of CO_2.

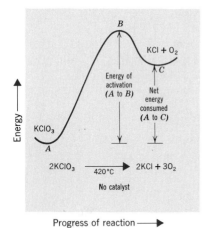

FIGURE 5.7
Progress of reaction diagram for the endothermic conversion of potassium chlorate ($KClO_3$) into potassium chloride (KCl) and oxygen (O_2).

Not all reactions liberate energy. Many chemical changes won't occur unless there is a continuous input of energy. One example is the conversion of potassium chlorate into potassium chloride and oxygen, as shown in the progress of reaction diagram of Figure 5.7. In this example, a good share of the energy of activation (A to B in this figure) is permanently retained by the product molecules as their own internal or potential energy. This net retained energy is represented in Figure 5.7 by the distance between levels A and C. The reaction is endothermic, because there is a conversion of energy supplied by the heat into the potential (chemical) energy of the products. Thus you can see that both exothermic and endothermic reactions have energies of activation, but in the exothermic reaction there is still a net release of energy, whereas in the endothermic reaction there is a net absorption of energy.

Reactions Usually Go Faster When the Reactants Are Concentrated.   Another way besides heat to increase the frequency of successful collisions (the reaction rate) is simply to increase the frequency of all collisions. Even if we do not increase their average violence, by making collisions of *all* kinds occur more often we will make successful collisions happen more frequently. We can do this by increasing the concentrations of the reactants. If the molar concentration of one reactant is doubled, the frequence of all collisions must double because there are twice as many of its particles *in the same volume*. It's like going from a stroll down a lonely country lane to an aisle of a very crowded store. An increase in the concentration of moving people increases the "excuse-me" kind of bumps and collisions.

One of the spectacular results of increasing the concentration of a reactant can be seen in the contrast between something burning in pure oxygen instead of in air. Red hot steel wool that only glows and gives off sparks when held in a bunsen burner flame bursts into flame when thrust into pure oxygen.

Someone has estimated that if the air we breathe were 30% O_2, instead of 21%, no forest fire could ever be put out, and eventually all of the world's forests would disappear. Obviously, if you are ever where oxygen-enriched air is being used, you will want all flames excluded.

People shouldn't even wear shoes with cleats or carry metal cigarette lighters (which might be accidentally dropped and made to light) in rooms where oxygen-enriched air is in use.

5.6 CATALYSTS AND REACTION RATES

Catalysts let reactions go faster by permitting lower energies of activation.

Catalysts Accelerate Reaction Rates. Now that we have the background of the previous section, we can deal with the question of why chemicals that do not react at body temperature outside the body react readily inside. One of the interesting and most important phenomena in all of nature is the acceleration of a reaction rate by a trace amount of some chemical that does not permanently change as the reaction proceeds. This phenomenon is called **catalysis,** the chemical responsible for it is called a **catalyst,** and the verb is *to catalyze.*

Enzymes Function as Catalysts. The catalysts in living systems are called **enzymes,** and a special enzyme is involved in virtually every reaction in living things. Virtually all enzymes are in a large family of biochemicals called proteins. You can easily see the catalytic effect of an enzyme by dipping a small piece of raw liver into a dilute solution of hydrogen peroxide (H_2O_2). Hydrogen peroxide decomposes as follows:

$$2H_2O_2 \longrightarrow 2H_2O + O_2$$

This is a slow reaction at room temperature or body temperature, so if you look at a sample of hydrogen peroxide, you won't notice any bubbling action. But, if you add a tiny slice of liver, an enzyme in liver called *catalase* sets off the decomposition of hydrogen peroxide and you will see a vigorous evolution of oxygen. The frothing seen when hydrogen peroxide is used to disinfect a wound is the same reaction. Hydrogen peroxide forms in certain reactions of metabolism. Since it is toxic, it has to be broken down quickly. The catalase in the liver catalyzes this reaction.

Dilute solutions of hydrogen peroxide are sold in drugstores as a bleach and disinfectant.

Catalysts Also Let Reactions Happen Under Milder Conditions. Catalase makes a reaction occur much faster at the same temperature than it does in the absence of a catalyst. A catalyst can also make a reaction take place at a much lower temperature than otherwise. A classic example is the decomposition of potassium chlorate ($KClO_3$) into potassium chloride (KCl) and oxygen that we mentioned in the previous section. Notice in the following equations how the temperature at which the reaction occurs varies with the presence of manganese dioxide (MnO_2).

Without MnO_2, the temperature has to be 420 °C
$$2KClO_3 + heat \xrightarrow[420\,°C]{} 2KCl + 3O_2$$

With MnO_2, the temperature can be only 270 °C
$$2KClO_3 + heat \xrightarrow[270\,°C]{MnO_2} 2KCl + 3O_2$$

Sometimes the special conditions for a reaction are written above or below the arrow in the equation.

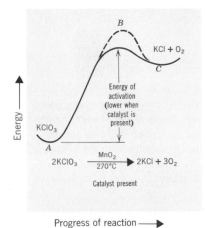

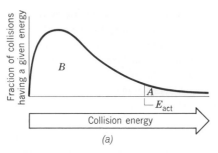

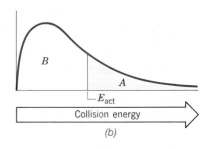

FIGURE 5.9
The effect of lowering the energy of activation on the frequency of successful collisions — the rate of the reaction. (a) When the energy of activation is high, the ratio of areas in the drawing that represents the fraction of successful collisions, $A/(A + B)$, is small and the rate of reaction is slow. (b) The ratio is much larger — the rate is much faster — when the energy of activation is less. Catalysts lower energies of activation and so increase reaction rates.

FIGURE 5.8
Progress of reaction diagram for the endothermic, catalyzed conversion of potassium chlorate into potassium chloride and oxygen. The dashed-line curve shows where the energy barrier went in the uncatalyzed reaction sketched in Figure 5.7. Notice that the net energy consumed, the heat of reaction, is identical to that of the uncatalyzed reaction, but the energy of activation is lower.

The rates of the evolution of oxygen are equal under the sets of conditions given here. Thus the rate at 420 °C in the absence of MnO_2, is observed at a substantially lower temperature in the presence of MnO_2.

With or without the catalyst, the decomposition of potassium chlorate is endothermic, as we saw in Figure 5.7. Figure 5.8 shows the progress of reaction diagram when the catalyst is present. The catalyst does not change the heat of reaction; instead it lowers the energy of activation. This is why the reaction happens faster. This is how all catalysts work. They lower the energy barrier so the fraction of all collisions with enough energy is larger than without the catalyst. See also Figure 5.9 and compare it with Figure 5.5 to see how a reduction of E_{act} increases the fraction of collisions with energies equal to or higher than E_{act}.

In summary, a catalyst either makes a reaction go faster at the same temperature or permits the reaction to go at the same rate at a lower temperature. It does this by reducing the energy of activation of a reaction but it does not affect the heat of reaction. Thus we add the effect of a *catalyst* to the effects of *temperature* and *concentration* as the important factors that govern the reaction rates.

None of the preceding discussion explained *how* a catalyst works. Each works by a different mechanism, but all we have discussed thus far is what catalysts do. When you study the subject of enzymes later, you will learn something about their mechanism of action.

SUMMARY

Gas properties   The four important variables for describing the physical properties of gases are numbers of moles (n), temperature T (in kelvins), volume (V), and pressure (P). We express pressure — force per unit area — in atmospheres (atm), or millimeters of mercury (mm Hg). Other important physical quantities in the study of gases are partial pressures, standard pressure and temperature (STP = 273 K and 1 atm), and the molar volume at STP (22.4 L).

Gas laws   All real gases obey, more or less, some important laws. Gas pressure is inversely proportional to volume (when n and T are fixed) — the pressure-volume law (Boyle's law). Gas volume is directly proportional to the Kelvin temperature (when n and P are fixed) — the temperature-volume law (Charles' law). Gas pressure is directly proportional to the Kelvin temperature (when n and V are fixed) — the pressure–temperature law (Gay-Lussac's law). Gas volume is also directly proportional to the number of moles (when T and P are fixed) — Avogadro's law. In fact, 1 mol of any gas under the same conditions of T and P has as many particles as 1 mol of any

other gas. Another gas law is that the total pressure of a mixture of gases equals the sum of their partial pressures (Dalton's law).

Boyle's, Charles', and Avogadro's laws combine in the universal gas law, $PV = nRT$, where R, the universal gas constant, holds for all gases. A variation of this, the general gas law, is the equation

$$P_1 V_1 / T_1 = P_2 V_2 / T_2.$$

Kinetic theory   An ideal gas consists of a huge number of very tiny, very hard particles in random, chaotic motion that do not attract or repel each other. The Kelvin temperature of a gas is directly proportional to the average kinetic energy of the gas particles.

Liquid state   Liquids do not follow general laws like gases, because essentially no space separates liquid particles from each other. Liquids can evaporate, and this "escaping tendency" causes vapor pressure, which varies with temperature. When its vapor pressure equals the external pressure, the liquid boils. When this external pressure is

1 atm, the boiling temperature is called the normal boiling point. A specific amount of heat, the heat of vaporization, is needed to convert a gram of liquid into its vapor state.

Solid state   The particles in a solid vibrate about fixed points, but if the solid is heated, these vibrations eventually become so violent that the particles enter the liquid state. A specific amount of heat, the heat of fusion, is needed to convert a gram of solid into its liquid state.

Managing the heat of metabolism   The body keeps from becoming hyperthermic by using radiation, conduction, convection, and perspiration to lose some of the heat generated from metabolism. The evaporation of water at the skin and via moist exhaled air uses up some of the body's heat as heat of vaporization. Both hyperthermia and hypothermia are serious conditions.

Kinetic theory and chemical reactions   Virtually all chemical reactions have an energy of activation. Reactant particles more frequently surmount this barrier and the reaction happens faster, the more concentrated they are, and the more energetically they collide. Raising the temperature increases the frequency of collisions energetic enough to be successful.

Catalysts   A catalyst (e.g., an enzyme) accelerates a reaction by lowering the energy of activation, but it does not affect the heat of reaction.

REVIEW EXERCISES

The answers to review exercises that require a calculation and whose numbers are marked with an asterisk are given in Appendix V. The answers to the other review exercises are given in the *Study Guide* that accompanies this book.

Pressure and Other Variables

5.1   What are the four principal physical quantities used to describe the physical state of any gas?

5.2   What is the difference between pressure and force?

5.3   What causes the atmosphere to exert a force on the earth?

5.4   What is one standard atmosphere of pressure?

5.5   How many mm Hg are in 1 atm?

5.6   Why doesn't all of the mercury run out of a Torricelli barometer?

*5.7   At the summit of Mt. McKinley (20,320 ft, 6194 m), the atmospheric pressure is 0.460 atm. What is this pressure in mm Hg?

5.8   The highest altitude at which a pilot or plane passenger could survive without a pressurized cabin but while breathing oxygen-enriched air is about 40,000 ft (8 miles, 13 km). The air pressure there is about 150 mm Hg. What is this in atm?

Specific Gas Laws

5.9   What fact about gases did Boyle discover?

5.10   In order for P_1V_1 to equal P_2V_2 what conditions must be true about the measurements at the times designated by the subscripts 1 and 2?

5.11   What is the law of partial pressures?

5.12   What four gases are in exhaled air?

5.13   Of the four important variables in the study of the physical properties of gases, which are assumed to be held at constant values in each of the following gas laws?
(a) Boyle's law
(b) Charles' law
(c) Law of partial pressures
(d) Avogadro's law
(e) Gay-Lussac's law

5.14   State the volume–temperature law and the physical conditions under which it holds.

5.15   What is Avogadro's law?

*5.16   A sample of oxygen with a volume of 525 mL and a pressure of 750 mm Hg has to be given a volume of 475 mL. What pressure is needed if the temperature is to be kept constant?

5.17   What is the new pressure on a sample of helium with an initial volume of 1.50 L and a pressure of 745 mm Hg if its volume becomes 2.10 L at the same temperature?

*5.18   The value of P_{N_2} at the summit of Mt. McKinley (Review Exercise 5.7) is 277 mm Hg on a day when the atmospheric pressure there is 350 mm Hg. Assuming that the air is made up only of nitrogen and oxygen, what is the partial pressure of oxygen in mm Hg up there?

5.19   At an elevation of 40,000 ft (Review Exercise 5.8), the partial pressure of nitrogen is 119 mm Hg on a day when the air pressure at this elevation is 150 mm Hg. What is the value of P_{O_2}?

*5.20   A sample of oxygen was warmed from 15 to 30 °C at constant pressure. Its initial volume was 1.75 L. What is its final volume in liters?

5.21   In order to change a 400-mL sample of nitrogen at 25 °C to a 200-mL sample with the same pressure, what must become of the temperature? (Give your answer in degrees Celsius.)

Universal Gas Law

5.22   What are the standard conditions of temperature and pressure?

5.23   What is meant by a *molar volume*?

5.24   Under what circumstances is the molar volume equal to 22.4 L?

5.25   What is the equation for the universal gas law?

5.26   To use the value of R that we calculated in this chapter, in what specific units must P, V, n, and T be?

5.27 At STP how many molecules of hydrogen are in 22.4 L?

*5.28 When an electric current is passed through water under suitable conditions, the water breaks down into hydrogen and oxygen according to the following equation:

$$2H_2O \longrightarrow 2H_2 + O_2$$

In one experiment, a dry sample of one of these gases was collected. Its volume at 748 mm Hg and 23.0 °C was 875 mL.
(a) How many moles of this gas were obtained?
(b) This sample of gas had a mass of 1.136 g. What is the formula weight of this gas? (Remember that a formula weight is numerically equal to the *ratio* of grams to moles.)
(c) Which gas was it, oxygen or hydrogen?

5.29 Consider the experiment described in Review Exercise 5.28.
(a) How may moles of hydrogen were obtained?
(b) What volume was occupied by this sample of hydrogen at 748 mm Hg and 23.0 °C?

*5.30 One source of industrial hydrogen is methane. At a high temperature, methane (CH_4) decomposes ("cracks") as follows into carbon and hydrogen.

$$CH_4 \xrightarrow[heat]{} C + 2H_2$$

(a) If 1.00 mol of CH_4 is used, how many moles of hydrogen are produced?
(b) If 100 L of methane gas, initially at 740 mm Hg and 20 °C, are used, how many liters of hydrogen will be obtained when they are measured under the same conditions of temperature and pressure?
(c) If a sample of methane with a mass of 50.0 g is cracked, what volume of hydrogen is produced when measured at 750 mm Hg and 25 °C?

5.31 Carbon dioxide can be removed from exhaled air by making the air pass through granulated sodium hydroxide. The reaction is

$$\underset{\substack{\text{Sodium} \\ \text{hydroxide}}}{NaOH} + \underset{\substack{\text{Carbon} \\ \text{dioxide}}}{CO_2} \longrightarrow \underset{\substack{\text{Sodium} \\ \text{bicarbonate}}}{NaHCO_3}$$

After this system had operated for some time, it was found that 12.4 g of NaOH had been used up.
(a) How many moles of CO_2 were responsible for this amount of change?
(b) If 11.6 g of NaOH were used up in a separate operation, how many milliliters of CO_2 gas caused this change if the gas volume were measured at 740 mm Hg and 25 °C?

Kinetic Theory of Gases

5.32 Scientists asked: What must gases be like for the gas laws to be true? What was their answer?

5.33 What is true about an ideal gas that is not strictly true about any real gas?

5.34 Dalton's law of partial pressures implies that gas molecules from different gases actually leave each other alone in the mixture, both physically and chemically (except at moments of collisions, when they push each other around). Which one of the three postulates in the model of an ideal gas is based on Dalton's law?

5.35 How does the kinetic theory of gases explain the phenomenon of gas pressure?

5.36 How does the kinetic theory of gases account for Boyle's law (in general terms)?

5.37 For 1 mol of an ideal gas, those working out the kinetic theory found that the product of pressure and volume is proportional to the average kinetic energy of the ideal gas particles.
(a) Using the universal gas law (which makes no mention of kinetic energy), to which of the four physical quantities used to describe a gas is the product of pressure and volume for 1 mol of a gas also proportional?
(b) If the product of P and V is proportional both to the average kinetic energy of the ideal gas particles and to the Kelvin temperature, what does this say about the relationship between the average kinetic energy and this temperature?

5.38 How does the kinetic theory explain (in general terms) the volume–temperature law?

5.39 The pressure–temperature law (Gay-Lussac's law) can be explained in terms of the kinetic theory in what way (in general terms)?

The Liquid State and Vapor Pressure

5.40 Why aren't there universal laws for the physical behavior of liquids (or solids) as there are for gases?

5.41 How does the kinetic theory explain
(a) How vapor pressure arises?
(b) Why vapor pressure rises with increasing liquid temperature?

Kinetic Theory and the Liquid and Solid States

5.42 The following are some common observations. Using the kinetic theory, explain how each occurs in terms of what molecules are doing.
(a) Moisture evaporates faster in a breeze than in still air.
(b) Ice melts much faster if it is crushed than if it is left in one large block.
(c) Even if hung out to dry in below-freezing weather, wet clothes will become completely dry even though they freeze first.

5.43 At room temperature, nitrogen is a gas, water is a liquid, and sodium chloride is a solid. What do these facts tell us about the relative strengths of electrical forces of attraction in these substances?

Metabolism and the Body's Heat Budget

5.44 What constitutes the body's *metabolism*, in general terms?

5.45 What name is given to the loss of body water by evaporation that does not involve the sweat glands?

5.46 Name the body's three mechanisms for losing heat that do not involve evaporation directly.

5.47 What is the difference between radiation and conduction as means for losing heat from the body?

5.48 Wearing woolen clothing minimizes heat loss from the body by what mechanism?

5.49 What is hypothermia, and why is it dangerous to life?

5.50 What is hyperthermia, and how can it be life threatening?

Kinetic Theory and Rates of Reactions

5.51 In terms of what we visualize as happening when two molecules interact to form products, how do we explain the existence of an energy barrier to the reaction—an energy of activation?

5.52 Study the accompanying progress of reaction diagram for the conversion of carbon monoxide and oxygen to carbon dioxide, and then answer the questions. The equation for the reaction is

$$2CO(g) + O_2(g) \longrightarrow 2CO_2(g)$$

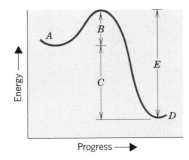

(a) What substance or substances occur at position A?
(b) What substance or substances occur at position D?
(c) Which letter labels the arrow that represents the heat of reaction?
(d) Which letter labels the arrow that stands for the energy of activation?
(e) Is the reaction exothermic or endothermic?
(f) Which letter labels the arrow that would correspond to the energy of activation if the reaction could go in reverse?

5.53 Suppose that the following hypothetical reaction occurs.

$$A + B \longrightarrow C + D$$

Suppose further that this reaction is endothermic and that the energy of activation is numerically twice as large as the heat of reaction. Draw a progress of reaction diagram for this reaction, and draw and label arrows that correspond to the energy of activation and the heat of reaction.

5.54 The reaction of X and Y to form Z is exothermic. For every mole of Z produced, 10 kcal of heat are generated. The energy of activation is 3 kcal. Sketch the energy relationships on a progress of reaction diagram.

5.55 How do we explain the rate-increasing effect of a rise in temperature?

5.56 As a rule of thumb, how much of a temperature increase doubles or triples the rates of most reactions?

5.57 Explain how a rise in body temperature can lead to a strain on the heart.

5.58 How can we increase the frequency of all collisions in a reacting mixture without raising the temperature?

5.59 When an increase in the concentration of one or more reactants causes an increase in the rate of a reaction, how do we explain this?

5.60 When an increase in the rate of a reaction has been caused by an increase in the concentration of one of the reactants, which of the following factors has been changed? (Identify them by letter.)
(a) The energy of activation.
(b) The heat of reaction.
(c) The frequency of collisions.
(d) The frequency of successful collisions.

Catalysts

5.61 In what way, if any, does a catalyst affect the following factors of a chemical reaction?
(a) The heat of reaction.
(b) The energy of activation.
(c) The frequency of collisions.
(d) The frequency of successful collisions.

5.62 What is the name for the catalysts found in living systems?

5.63 Once we have selected a particular reaction, we have to accept whatever energy of activation and heat of reaction goes with it. However, there are three things that we might try to help speed up the reaction. What are they?

Hypothermia (Special Topic 5.1)

5.64 What is hypothermia?

5.65 With a decrease of about how many degrees in the body's core temperature does hypothermia set in? What are some early signs?

5.66 Why should one not give, say, brandy to a victim of hypothermia?

Chapter 6
Water, Solutions, and Colloids

You can be sure that the people of this Algerian village who live in the shadow of water shortages never take water for granted. In this chapter we'll learn about some properties of water that are vital at the molecular level of life.

6.1 WATER

Many physical properties of water relate to its polarity and to hydrogen bonds between its molecules.

We take in more water than all other materials combined. We use it as the fluid in all cells, as a heat-exchange agent, and as the carrier in the bloodstream for distributing oxygen and all molecules from food, all hormones, minerals, and vitamins, and all disease-fighting agents.

Water is a superb solvent. It can dissolve at least trace amounts of almost anything, including rock. It is particularly good at dissolving ionic substances and the more polar molecular compounds.

In this chapter we will focus on water and some of the physical properties of aqueous solutions. To understand many aspects of life at the molecular level, we need to know why water dissolves some things well but not others, and to achieve this goal we must study the polarity of water in greater detail.

Of the 2500 mL of water we take in each day, 1200 mL is in the liquid we drink, 1000 mL is in the food we eat, and 300 mL is made by the reactions of metabolism.

Water's Boiling Point Is Unusually High. As a rule of thumb, the lower the formula weight of a liquid, the lower is its boiling point. Three simple substances with low formula weights, water, ammonia, and hydrogen fluoride, are striking exceptions, however. We can see this in Figure 6.1, where boiling points of the hydrides of the elements in Groups IVA – VIIA are plotted against their formula weights. (Hydrides are compounds of hydrogen with some other element. The hydride of oxygen, for example, is water, H_2O.)

These are *binary* hydrides because they are made of just two elements.

Look first at the plot in the figure for the Group IVA hydrides, those of carbon (CH_4), silicon (SiH_4), and germanium (GeH_4). Their boiling points nicely follow the rule of thumb we mentioned. They increase regularly with increasing formula weight. The plot just above this, however, does not follow the rule as well. Ammonia, NH_3, is badly off the straight line on which the other Group VA hydrides fall. It boils far higher than it "should." Similarly, hydrogen fluoride, HF, does not fit well with the plot of the boiling points of the Group VIIA hydrides, those of the halogens.

Water is the worst actor of all. Water "should" boil at about -100 °C, as the dashed line extension of the plot for the Group VIA hydrides indicates. But it actually boils at $+100$ °C. The high boiling point of water indicates that water molecules attract each other. Energy, therefore, is required to get them to stay apart in the vapor phase. We'll look next at the nature of the attraction of water molecules for each other.

Hydrogen Bonds Exist between Water Molecules. Because oxygen is much more electronegative than hydrogen, each of the two H—O bonds is very polar. Because the water

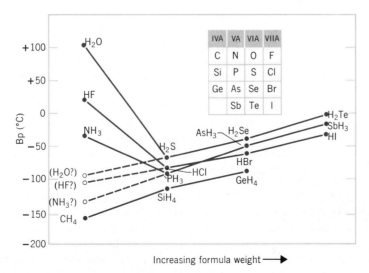

FIGURE 6.1
Boiling points versus formula weights for the binary, nonmetal hydrides of the elements in Groups IVA, VA, VIA, and VIIA.

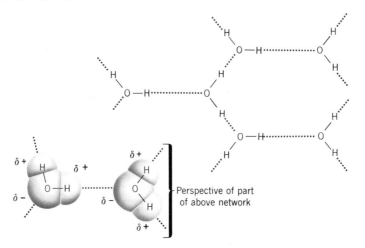

FIGURE 6.2
Hydrogen bonds in water (. . .)

The VSEPR theory (Special Topic 3.2) explains why there is an angle in the water molecule.

molecule is angular, these individual polarities do not cancel. Thus the water molecule is polar, so polar, in fact, that between water molecules there is a strong enough force of attraction to be called a bond. It's not a covalent bond or an ionic bond, so it has its own name, *hydrogen bond*. A **hydrogen bond** is the force of attraction between the $\delta+$ on H, when H is bonded to O, N, or F and the $\delta-$ on some other O, N, or F. Only when both $\delta+$ and $\delta-$ are relatively large can this special dipole–dipole attraction be great enough to justify saying that a bond, a hydrogen bond, exists. The $\delta+$ on H turns out to be large enough only when H is bonded to one of the three most electronegative atoms, F, O, or N. The high electronegativity of these, of course, ensures also that they will bear large $\delta-$ charges.

The hydrogen bond is a bridging bond between molecules, as illustrated in Figure 6.2. Like all bonds, it is a force of attraction. However, it is not nearly as strong as a covalent bond, only 5% as strong. But this is strong enough to make a difference not just in water but in such vital substances as muscle proteins, cotton fibers, and the chemicals of genes (DNA). In fact, among the biochemicals where hydrogen bonds occur, their very weakness makes them more important than any other bond, just as the weakest link in a chain is the most important.

Largely because of the hydrogen bond, water has the unusually high heats of fusion and vaporization described in Section 5.4. An extra large input of heat per gram is necessary to melt or boil water because energy is needed to overcome its innumerable hydrogen bonds.

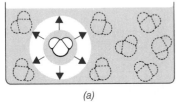

(a)

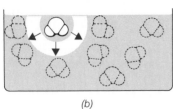

(b)

FIGURE 6.3
Surface tension. (a) In the interior of a sample of water, individual water molecules are attracted equally in all directions. (b) At the surface, nothing in the air counterbalances the downward pull that the surface molecules in water feel.

A Water Surface Acts Like a Skin Because of Hydrogen Bonding.   All liquids possess a surface tension, but that of water is unusually high. **Surface tension** is a phenomenon in which a liquid's surface acts as a thin, invisible, elastic membrane. Water's polar molecules tend to jam together where they meet the air, as shown in Figure 6.3. The polar forces that pull surface molecules inward aren't counterbalanced by forces to pull them outward. This net inward pull causes the jam-up responsible for surface tension. The same inward pull makes water form beads on a greasy or waxed surface. Wax or grease molecules, like those of air, are nonpolar and cannot attract water molecules and make the water spread out. So the water draws inward and forms a bead. Clean glass, on the other hand, has innumerable polar sites on its surface. These strongly attract water molecules, and this causes water to spread out on clean glass. When water forms beads on glass, you can be sure that the surface is unclean, that it has an invisible greasy coating.

Surface-Active Agents Reduce the Surface Tension of Water.   Like glass with innumerable polar sites, the molecules that make up the membrane of the air sacs (alveoli) in the lungs are also dotted with polar sites. But unlike glass, this membrane is flexible. The strong attraction that water molecules have for these sites should therefore be expected to pull on the whole membrane and make it cave in. Such a lung collapse, fortunately, is not normal; the membrane surface also is coated with a substance that greatly reduces water's surface tension. In some diseases, the protection is absent, and lungs do collapse.

Anything that lowers the surface tension of water is called a **surface-active agent,** or a **surfactant,** for short. All soaps and detergents are surfactants, for example. Their molecules include parts able to attract water molecules and other parts able to ''dissolve'' in oils and greases, the materials that act as glues for soil on fabrics and skin. This double action makes them excellent cleansing agents. Bile, one of the digestive juices, contains extremely powerful surfactants called bile salts. Without them, our digestive systems would be much less able to digest the fats and oils in our diets or to wash them from the particles of other kinds of food.

Bile is secreted into the upper intestinal tract from the gallbladder.

6.2 WATER AS A SOLVENT

Water dissolves best those substances whose ions or molecules can strongly attract water molecules and form solvent cages.

When crystalline table salt, NaCl, is added to water, the water molecules bombard the crystal surfaces and begin to dislodge ions (Figure 6.4). In the crystal environment, however, these

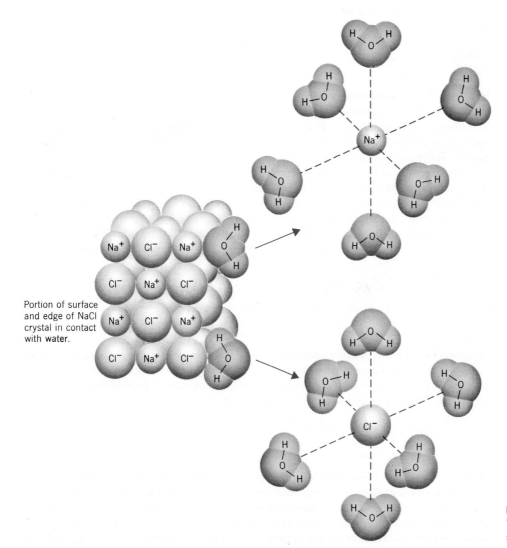

Portion of surface and edge of NaCl crystal in contact with water.

FIGURE 6.4
The hydration of ions helps ionic substances dissolve in water.

ions have oppositely charged ions as nearest neighbors. Opposite charges attract each other so strongly that Na^+ and Cl^- ions will not leave the crystal unless something else substitutes for this environment. Water molecules are polar enough to do this.

Water Molecules Can Hydrate Ions.   Because water molecules are very polar, they can surround Na^+ ions, letting the $\delta-$ sites on oxygen point toward the positively charged ion, as seen in Figure 6.4. Similarly, Cl^- ions can also attract water molecules (Figure 6.4), which become oriented so that their $\delta+$ sites point toward the negatively charged Cl^- ions.

Once in solution, Cl^- ions no longer have Na^+ ions as nearest neighbors, but they have several water molecules performing the same service. They provide each Cl^- with an environment of opposite charge. Similarly, in solution, Na^+ ions no longer have Cl^- ions as nearest neighbors, but electron-rich oxygen atoms in water molecules take their place. This phenomenon of water molecules being attracted to ions is called **hydration,** and it leads to the formation of a cage of water molecules about each ion in solution.

Hydr- is from the Greek *hydōr*, water.

Of all of the common solvents, only water has molecules both polar enough and small enough to form effective solvent cages around ions. Of course, in doing this, water molecules must give up some of their attractions for each other. Only ionic substances or compounds made of very polar molecules can break up hydrogen bonds between water molecules. Thus the formation of a solution isn't just the separation of the solute particles from each other. It is also, to some extent, the separation of solvent molecules from each other as well as the attraction between solute particles and solvent molecules.

Usually, when both ions of an ionic compound carry charges of two or three units, the compound isn't very soluble in water. The ions find more stability by remaining in the crystal than they can replace by accepting solvent cages.

Polar molecular compounds dissolve in water, too, and Figure 6.5 shows how water hydrates their molecules.

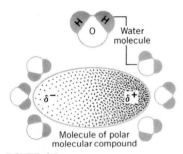

FIGURE 6.5
The hydration of a polar molecule helps polar molecular substances to dissolve in water.

Hydrates Are Compounds with Water Molecules in Their Crystals.   If we let water evaporate from an aqueous solution of any one of several substances, the crystalline residue contains intact water molecules. They are held within the crystals in *definite* proportions. Such water-containing solids are called **hydrates,** and they are true compounds because they obey the law of definite proportions.

The formulas of hydrates are written to show that intact water molecules are present. The formula of the pentahydrate of copper(II) sulfate, for example, is written as $CuSO_4 \cdot 5H_2O$, where a raised dot separates the two parts of the formula. Table 6.1 lists a number of other common hydrates.

The water present in a hydrate is called the **water of hydration**. It usually can be expelled by heat to leave a residue called the **anhydrous form** of the compound. For example,

$$CuSO_4 \cdot 5H_2O(s) \longrightarrow CuSO_4(s) \qquad + 5H_2O(g)$$

Copper(II) sulfate
pentahydrate (deep
blue crystals)

Copper(II) sulfate
(anhydrous form is
nearly white)

(as steam)

Many anhydrous forms of hydrates readily take up water and re-form the hydrates. Plaster of paris, for example, although not completely anhydrous, contains relatively less water than gypsum. When we mix plaster of paris with water, it soon sets into a hard, crystalline mass according to the following reaction.

$$(CaSO_4)_2 \cdot H_2O + 3H_2O \longrightarrow 2CaSO_4 \cdot 2H_2O$$

Plaster of
paris

Gypsum

(Notice that the first 2 in gypsum's formula is a *coefficient* for the *entire* formula.)

Some compounds in their anhydrous forms are used as drying agents or desiccants. A **desiccant** is a substance that removes moisture from air by forming a hydrate. Any substance that can do this is said to be **hygroscopic.** Anhydrous calcium chloride, $CaCl_2$, sometimes used to dehumidify damp basements, is a common desiccant. In humid air it draws

TABLE 6.1
Some Common Hydrates

Formulas	Names	Decomposition Modes and Temperatures[a]	Uses
$(CaSO_4)_2 \cdot H_2O$	Calcium sulfate hemihydrate (plaster of paris)	$-H_2O$ (163)	Casts, molds
$CaSO_4 \cdot 2H_2O$	Calcium sulfate dihydrate (gypsum)	$-2H_2O$ (163)	Casts, molds, wallboard
$CuSO_4 \cdot 5H_2O$	Copper(II) sulfate pentahydrate (blue vitriol)	$-5H_2O$ (150)	Insecticide
$MgSO_4 \cdot 7H_2O$	Magnesium sulfate heptahydrate (epsom salt)	$-6H_2O$ (150)	Cathartic in medicine
		$-7H_2O$ (200)	Used in tanning and dyeing
$Na_2B_4O_7 \cdot 10H_2O$	Sodium tetraborate decahydrate (borax)	$-8H_2O$ (60) $-10H_2O$ (320)	Laundry
$Na_2CO_3 \cdot 10H_2O$	Sodium carbonate decahydrate (washing soda)	$-H_2O$ (33.5)	Water softener
$Na_2SO_4 \cdot 10H_2O$	Sodium sulfate decahydrate (glauber's salt)	$-10H_2O$ (100)	Cathartic
$Na_2S_2O_3 \cdot 5H_2O$	Sodium thiosulfate pentahydrate (photographer's hypo)	$-5H_2O$ (100)	Photographic developing

[a] Loss of water is indicated by the minus sign before the symbol, and the loss occurs at the temperature in degrees Celsius (°C) given in parentheses.

enough water to form a liquid solution. Any substance this active as a desiccant is also said to be **deliquescent.**

6.3   DYNAMIC EQUILIBRIA IN SOLUTIONS

Equilibrium exists between the dissolved and undissolved states in a saturated solution.

A solution is saturated when it holds its maximum of solute. When excess undissolved solute is also present, the system illustrates something new to our study, a dynamic equilibrium. Dynamic equilibria occur among enough situations at the molecular level of life to make their special vocabulary one goal of this section.

Dynamic Equilibrium Is Much Activity Without Any Net Change.   Imagine that we put more solid sodium chloride into a water sample than needed to make a saturated solution. Some salt dissolves, and its ions enter the solution and move randomly about as the crystals become smaller. Some of the dissolved ions, however, find their way back. They reattach to a crystal surface and the crystal grows. The more concentrated the dissolved ions become as the salt dissolves, the more frequently will they bump into crystals and stay (reprecipitate). Eventually, the rates at which ions leave the crystals and return become equal. At this moment, in other words, the crystals are regrowing as fast as they are dissolving. This situation pertains to any saturated solution *in contact with undissolved solute* (see Figure 6.6).

In spite of considerable coming and going, the activities in our saturated sodium chloride solution oppose or cancel each other. No longer is there any net change in either the solution's concentration or in the mass of undissolved solute. Any situation with opposing activities but no net change is called a **dynamic equilibrium.** *Dynamic* signifies much activity and

At 20 °C, 36 g of NaCl in 100 g of water makes a saturated solution.

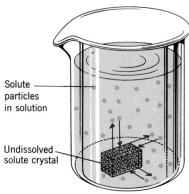

Solute particles in solution

Undissolved solute crystal

FIGURE 6.6
In a saturated solution, dynamic equilibrium exists between the ions or molecules of the solute in solution and those in the undissolved state.

Although heat often *evolves* when solids are added to *pure water,* the nearly saturated solutions require the addition of heat to make more solute dissolve.

equilibrium means no net change. In a **saturated solution** (in contact with undissolved solute) there is a dynamic equilibrium between the undissolved and the dissolved solute.

Dynamic Equilibria Have Both a Forward and a Reverse Reaction.   We represent an equilibrium by a special equation, one with half-barbed double arrows, which we can illustrate for a saturated solution.

$$\text{solute}_{\text{undissolved}} \rightleftharpoons \text{solute}_{\text{dissolved}}$$

The two oppositely pointing arrows tell us that the equation is an *equilibrium expression* or an *equilibrium equation.* They signify equilibrium between the materials on their left and those on their right. The change from left to right is called the *forward reaction,* and the opposing change is the *reverse reaction.*

Equilibria Respond to Stresses.   Once an equilibrium becomes established, no net change occurs *spontaneously.* We can still force a change, however. When we do so, we say that we *upset* the equilibrium. Anything that upsets an equilibrium is called a *stress.* Heat, for example, is a common stress. The equilibrium in a saturated solution is changed by heat. For many solid solutes, the forward reaction (dissolving) is endothermic, so heat is needed to make more solute dissolve in a saturated solution. To show this fact in the equilibrium expression, heat can be shown as if it were a reactant. For example,

$$\text{solute}_{\text{undissolved}} + \text{heat} \rightleftharpoons \text{solute}_{\text{dissolved}}$$

Assuming that undissolved solute is present, when we add heat to this saturated solution, we upset its equilibrium. The rate of the forward change (dissolving) becomes faster than the rate of the reverse change (coming out of solution). For a time, the system will not be in equilibrium, because a net change (dissolving) is happening.

Once we stop adding heat, however, and we maintain the temperature of the system at a constant but higher value, the rate at which solute returns to the crystals will catch up. The opposing rates will again be the same. Both opposing changes will be faster than they were at the lower temperature, but after both rates are the same again there is no further net change. Once again, we have dynamic equilibrium. It doesn't matter how rapidly the opposing changes occur; as long as their rates are equal, there is equilibrium.

Equilibria Shift in Whatever Direction Absorbs a Stress.   When a stress, like heat, upsets an equilibrium, we say that the equilibrium shifts. It might shift in favor of the forward reaction, as when heat applied to a saturated solution causes more solid to dissolve.

The removal of heat, cooling, can also be a stress. If we cool our saturated solution, some additional solute will join what already is undissolved. Now the rate of the reverse reaction is temporarily faster than the rate of the forward reaction. In time, the rates of the forward and reverse changes become equal. We once again have equilibrium, but now less solute is in solution and more is undissolved.

Just how a stress will shift an equilibrium can be predicted by an insight credited to Henri Louis Le Chatelier (1850–1936), a French chemist.

> Le Chatelier's Principle An equilibrium always responds to a stress by shifting in whatever direction absorbs the stress.

The stress of adding heat to our saturated solution, for example, was absorbed by the heat-consuming reaction, the forward reaction whereby more solute dissolved. The stress of removing heat or cooling the same system is absorbed by a heat-supplying reaction, the reverse change in which some solute comes out of solution.

People who work where the air pressure is high must return to normal atmospheric pressure slowly and carefully. Otherwise, they could experience the bends, or decompression sickness—severe pains in muscles and joints, fainting, and even deafness, paralysis, or death. Deep-sea divers run this risk as well as those who work in deep tunnels where air pressures are increased to help keep out water.

Under high pressure, in accordance with Henry's law, the blood dissolves more nitrogen and oxygen than at normal pressure, as the accompanying figure shows. (Because the plots are straight lines, you can tell that the solubility is *directly* proportional to increases in the partial pressures of the gases.) If blood thus enriched in nitrogen and oxygen is too quickly exposed to lower pressures, these gases suddenly come out of solution. Their microbubbles block the blood capillaries, close off the flow of blood, and lead to the symptoms we described.

If the return to normal pressure is made slowly, the gases leave the blood more slowly, and they can be re-

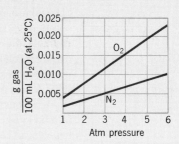

The solubilities of oxygen and nitrogen in water versus pressure.

moved as they emerge. The excess oxygen can be used by normal metabolism, and the excess nitrogen has a chance to be gathered by the lungs and removed by normal breathing. For each atmosphere of pressure above normal to which the person is exposed, about 20 min of careful decompression is usually recommended.

Whenever a system at equilibrium shifts in response to a stress, materials on one side of the double arrows increase at the expense of those on the other side. Whichever side increases is the *product* of what is now called the *favored* reaction. The forward reaction, for example, is described as favored when we heat the saturated solution, because it produces more dissolved solute.

Here's how we might use several of our new terms to summarize what we have studied about equilibria. The stress of additional heat on a saturated solution in contact with undissolved solute upsets the equilibrium and shifts it to the right, in favor of dissolved solute. Conversely, if we cool a saturated solution, the stress is now the *loss* of heat. In accordance with Le Chatelier's principle, the equilibrium shifts to the left in favor of undissolved solute.

Many chemical reactions in the body are actually shifts in chemical equilibria in response to chemical stresses.

Gas Solubility Increases with the Partial Pressure of the Gas. Gases vary widely in solubilities in water. In 100 g of water at 20 °C, for example, only 4.3 mg of oxygen dissolves but 169 mg of carbon dioxide and 10.6 g of ammonia dissolve. These figures are for a solution under a total pressure of 1 atm and in contact with only the pure gas and the water vapor that unavoidably is also present above the solution. We have to describe these conditions when we quote the solubility of a gas because the solubility of a gas is directly proportional to the partial pressure of the gas above the solution. This is the **gas solubility-pressure law,** often called **Henry's law** after its discoverer.

The increase in gas solubility with the partial pressure of the gas illustrates Le Chatelier's principle. This increase happens because pressure is a volume-reducing stress. The only way that the volume occupied by a gas above its solution can be reduced in response to an increase of pressure is for some of the gas to dissolve. This response absorbs at least some of the stress of the higher pressure.

Because the partial pressure of O_2 in air decreases with altitude, O_2 enters the blood less easily as the attitude increases.

People who work in high-pressure environments, such as deep-sea divers, run very serious risks as more air (both its nitrogen and oxygen) dissolves in their blood. Special Topic 6.1 discusses decompression sickness (the "bends"), which can disable or kill its victims.

Sometimes patients with crushing injuries or victims of carbon monoxide (CO) poisoning are placed in *hyperbaric chambers* where they receive 100% oxygen at pressures up to 2 atm (as opposed to oxygen's partial pressure of 0.2 atm in air as we ordinarily breath it). This helps their blood take up more oxygen for delivery to oxygen-starved tissue.

The more concentrated a solution is in a gas, the higher the gas's partial pressure will be above the solution at equilibrium. Sometimes this partial pressure of a gas above its solution is called the **gas tension** of the solution. In the fluid of the average human cell during rest, for example, the oxygen tension is about 30 mm Hg. In the lungs, the partial pressure of oxygen in alveolar air is slightly over 100 mm Hg. Since gases tend to diffuse from a higher to a lower pressure, oxygen for this reason alone tends to move from the lungs to the cells. (There are other forces at work, too.)

Carbon dioxide, on the other hand, has a gas tension in the fluids of cells at rest of about 50 mm Hg. But in alveolar air in the lungs its partial pressure is about 40 mm Hg. The natural tendency of carbon dioxide, therefore, is to move from cells to lungs, the direction this waste product of metabolism should move in order to be removed from the body.

Gas Solubility Decreases with Temperature. Heating an aqueous solution of a gas drives the gas out of solution. All gases dissolve exothermally, so we can represent the equilibrium as follows:

> Because hot water holds less dissolved oxygen than cold water, game fish tend to avoid it when the summer is hottest, and they move to cooler depths.

$$gas_{undissolved} \rightleftharpoons gas_{dissolved} + heat$$

When heat is added, the equilibrium changes to absorb the heat, so it must shift to the left. Gases lose essentially all of their solubility as the water temperature approaches 100 °C.

Some Gases, Like CO_2 and NH_3, Dissolve Partly by Reacting with Water. Molecules of such water-soluble gases as ammonia or carbon dioxide do not just intermingle with water molecules in solution. A significant percentage of them interacts with water in our first example of a *chemical* equilibrium. Carbon dioxide in water, for example, changes partly to carbonic acid as follows, and this reaction helps to pull CO_2 into solution:

$$CO_2(aq) + H_2O \rightleftharpoons H_2CO_3(aq)$$
$$\text{Carbonic acid}$$

This equilibrium has a vital function in the chemistry of respiration, as we will see in chapter 18.

6.4 PERCENTAGE CONCENTRATIONS

The number of grams of solute in 100 g of solution is the weight/weight percentage concentration of the solution.

Many times in the lab, test tube tests are made to detect the presence of some chemical. A solution called a **reagent** is generaly used to make the test. A few drops of the reagent might be added to the substance being tested, for example. We then look for some noticeable change that makes the test positive thus telling us that the substance we sought is present. The positive test might be a change in color or odor, the appearance of a precipitate, the generation of heat, or the evolution of a gas. You no doubt already have used reagents in the lab to make tests.

Many reagents are prepared by diluting a concentrated solution with water. A simple calculation is done first to learn how much of a concentrated solution must be taken, and Special Topic 6.2 describes this calculation.

Most tests do not require knowledge of the molar concentration of the reagent. Not that such information would hurt, but to obtain it sometimes adds to the work without returning any special gain. Nonetheless, some indication of concentration is desirable, and this is why a variety of percentage concentration expressions have been developed.

> We now use *percent* instead of *percentage* to conform to common usage.

Weight/Weight Percent Is the Kind of Percent Concentration Normally Used. The **weight/weight percent (w/w)% concentration** of a solution is the number of grams of solute in 100 g of the solution. A 10.0% (w/w) glucose solution, for example, has a concen-

Chemicals are often purchased as concentrated reagents, which then must be diluted. Sometimes the label gives directions for making these dilutions, but sometimes not. In this special topic we'll study how to do the calculations needed to prepare a dilute solution from a more concentrated solution.

One basic (and somewhat obvious) idea works here. The actual quantity of solute in the volume of concentrated solution taken is identical with the amount of solute after the dilution. We only add *solvent,* not solute, after all. If we're working with molar concentrations, the moles of solute in the desired volume of the dilute solution can be calculated from the molarity as follows, where we use the molarity (moles per liter) as a conversion factor:

$$\text{mole solute} = \text{liter}_{\text{dil soln}} \times \frac{\text{mole solute}}{\text{liter}_{\text{dil soln}}}$$

$$= \text{liter}_{\text{dil soln}} \times M_{\text{dil soln}}$$

This same number of moles of solute is to be obtained from the concentrated solution (and the purpose of this calculation is to find out this volume). The moles we want are found by a similar calculation that uses the molarity of the concentrated solution:

$$\text{mole solute} = \text{liter}_{\text{concd soln}} \times \frac{\text{mole solute}}{\text{liter}_{\text{concd soln}}}$$

$$= \text{liter}_{\text{concd soln}} \times M_{\text{concd soln}}$$

We now have two expressions for moles of solute, and they must equal each other, as we said. Therefore,

$$\text{liter}_{\text{dil soln}} \times M_{\text{dil soln}} = \text{liter}_{\text{concd soln}} \times M_{\text{concd soln}}$$

We need not use the unit of liters. We can use any volume unit we please provided it's the same unit on both sides. We normally use the milliliter unit, so our equation can be changed into the following standard working equation for doing dilution problems involving molar concentrations.

$$\boxed{\text{mL}_{\text{dil soln}} \times M_{\text{dil soln}} = \text{mL}_{\text{concd soln}} \times M_{\text{concd soln}}}$$

Work the following exercise to practice this equation.

PRACTICE EXERCISE The concentrated sulfuric acid that can be purchased from chemical supply houses is 18 M H_2SO_4. How could we use this to prepare 250 mL of 1.0 M H_2SO_4? (Answer: Dissolve 14 mL of 18 M H_2SO_4 in water and make the final volume equal to 250 mL.)

If you ever prepare a dilute solution from one that is more concentrated, especially a concentrated acid, to protect yourself *you must always wear safety glasses and always pour the concentrated solution into water as you stir the system.* Otherwise, so much heat is suddenly produced that the system will boil and spatter.

When concentrations are in percents, dilution problems use equations very similar to what we just used. For example, when the percent is a weight/weight percent, the equation is

$$g_{\text{concd soln}} \times \text{percent (w/w)}_{\text{concd soln}}$$
$$= g_{\text{dil soln}} \times \text{percent (w/w)}_{\text{dil soln}}$$

It would be easy to show by an analysis of the units that each side of this equation equals the grams of the solute.

For weight/volume percents, the equation is

$$\text{mL}_{\text{concd soln}} \times \text{percent (w/v)}_{\text{concd soln}}$$
$$= \text{mL}_{\text{dil soln}} \times \text{percent (w/v)}_{\text{dil soln}}$$

---

tration of 10.0 g of glucose in 100 g of solution. To make 100 g of this solution, you would mix 10.0 g of glucose with 90.0 g of the solvent for a total mass of 100 g.

EXAMPLE 6.1   USING WEIGHT/WEIGHT PERCENTS

Problem: How many grams of 0.900% (w/w) NaCl contain 0.250 g of NaCl?

Solution: The concentration term, 0.900% (w/w), gives us the following two conversion factors.

$$\frac{0.900 \text{ g NaCl}}{100 \text{ g NaCl soln}} \quad \text{and} \quad \frac{100 \text{ g NaCl soln}}{0.900 \text{ g NaCl}}$$

These are simply equivalent ways of understanding the concentration. Any expression of concentration in any units can be expressed as two ratios, such as we have just done. Next, we multiply the given, 0.250 g NaCl, by the second factor to make the final units g NaCl soln.

$$0.250 \text{ g NaCl} \times \frac{100 \text{ g NaCl soln}}{0.900 \text{ g NaCl}} = 27.8 \text{ g NaCl soln}$$

Thus 27.8 g of 0.900% (w/w) NaCl contains 0.250 g of NaCl. [Does the answer make sense? Be sure to ask this. If 100 g of solution holds 0.900 g of solute, we'd need *less* than 100 g of solution to hold less than 0.900 g (i.e., 0.250 g) — very roughly one-third less, as we calculated.]

EXAMPLE 6.2   PREPARING WEIGHT/WEIGHT PERCENT SOLUTIONS

Problem: A special kind of saline solution, called isotonic saline, is sometimes used in medicine. Its concentration is 0.90% NaCl (w/w). How would you prepare 750 g of such a solution?

Solution: Once again, we have to translate the concentration, 0.90% (w/w) NaCl, into conversion factors.

$$\frac{0.90 \text{ g NaCl}}{100 \text{ g NaCl soln}} \quad \text{and} \quad \frac{100 \text{ g NaCl soln}}{0.90 \text{ g NaCl}}$$

What we have to calculate is the number of grams of NaCl we must weigh out and dissolve in water to make the final mass equal to 750 g. To find this, we multiply the mass of the NaCl solution by the first conversion factor.

$$750 \text{ g NaCl soln} \times \frac{0.90 \text{ g NaCl}}{100 \text{ g NaCl soln}} = 6.8 \text{ g NaCl} \quad \text{(rounded from 6.75)}$$

Thus, if we dissolve 6.8 g of NaCl in water and add enough water to make the final mass equal to 750 g, we can prepare a label that reads 0.90% (w/w) NaCl.

PRACTICE EXERCISE 1   Sulfuric acid (H_2SO_4) can be purchased from a chemical supply house as a solution that is 96.0% (w/w) H_2SO_4. How many grams of this solution contain 9.80 g of H_2SO_4 (or 0.100 mol)?

PRACTICE EXERCISE 2   How many grams of glucose ($C_6H_{12}O_6$) and how many grams of water are needed to prepare 500 g of 0.250% (w/w) glucose?

Volume/Volume Percent Is Often Used When a Solution Is Made of Gases or Liquids.
A **volume/volume percent** or **(v/v)%,** gives us the number of volumes of one substance present in 100 volumes of the mixture. When we say, for example, that air is 21% (v/v) in O_2, we mean that there are 21 volumes of O_2 in 100 volumes of air.

A Weight/Volume "Percent" Is the Same As Grams per Deciliter.   A weight/volume percent isn't a true percent because the units do not cancel. When a concentration is given as a **weight/volume percent (w/v)%,** it means the number of grams of solute in 100 mL of the solution. Thus a 0.90% (w/v) NaCl solution has 0.90 g of NaCl in *every* 100 mL of the solution.

Sometimes the concentration units are g/dL (or g/dl). This means *grams per deciliter,* and because 1 dL = 100 mL, a concentration, for example, of 2.50 g/dL is identical to 2.50% (w/v) as we have defined it.

Weight/volume percent problems are handled through conversion factors just as we did for weight/weight percent problems.

EXAMPLE 6.3   WORKING WITH WEIGHT/VOLUME PERCENTS

Problem: How would one prepare 500 mL of a 5.00% (w/v) solution of glucose in water?

Solution: The percent gives us the following conversion factors:

$$\frac{5.00 \text{ g glucose}}{100 \text{ mL glucose soln}} \quad \text{or} \quad \frac{100 \text{ mL glucose soln}}{5.00 \text{ g glucose}}$$

We multiply the given volume by the first factor:

$$500 \text{ mL glucose soln} \times \frac{5.00 \text{ g glucose}}{100 \text{ mL glucose soln}} = 25.0 \text{ g glucose}$$

Thus to make this solution, dissolve 25.0 g of glucose in water in a volumetric flask and make the final volume equal to 500 mL.

PRACTICE EXERCISE 3   How many grams of solute are needed to prepare each of the following solutions?

(a) 100 mL of 2.00% (w/v) $KMnO_4$ (b) 25 mL of 1.0% (w/v) NaOH

Milligram Percent, Parts per Million, and Parts per Billion Are Other Concentration Units.   You will occasionally encounter some special concentration expressions that are handy when the solutions are very dilute. **Milligram percent** means the number of milligrams of solute in 100 mL (1 dL) of the solution.

For very dilute solutions, the concentration might be given in *parts per million (ppm),* which means the number of parts (in any unit) in a million parts (the same unit) of the solution. *Parts per billion (ppb)* similarly means parts per billion parts, such as grams per billion grams. This expression is used for *extremely* dilute systems.

Because of the many ways the term *percent* can be taken, there is a trend away from using it, which should be encouraged. Instead, the explicit units are given. Thus instead of referring to a concentration of, say, 10.0% (w/v) KCl, the label or the report should read 10.0 g KCl/100 mL or 10.0 g KCl/dL. When no units are given, just a percent, you have to interpret the percent as weight/weight if the solute is a solid when pure and as a volume/volume percent if the solute is a liquid when pure.

One part per billion is like 1 penny in $10 million (1 billion pennies).

6.5   COLLOIDAL DISPERSION

The sizes of the particles in a mixture, as distinct from their chemical identities, determine a number of properties.

There are three chief kinds of **homogeneous mixtures,** mixtures in which any small sample has the same composition and properties as any other sample of the same size taken anywhere else in the mixture. The solution is one kind. *Colloidal dispersions* and *suspensions* are the others, and all three types are found in the body. They differ fundamentally in the sizes of the particles involved, differences that alone cause interesting and important changes in properties. Our chief interest in this section is with colloidal dispersions, but to understand them we need to contrast them with solutions.

In Solutions, the Dispersed Particles Are Smallest.   The ions and molecules that make up solutions have formula weights of no more than a few hundred and diameters in the range of

TABLE 6.2
Solutions

Kinds	Common Examples
Gas in a liquid	Carbonated beverages (carbon dioxide in water)
Liquid in a liquid	Vinegar (acetic acid in water)
Solid in a liquid	Sugar in water
Gas in a gas	Air
Liquid in a gas[a]	
Solid in a gas[a]	
Gas in a solid	Alloy of palladium and hydrogen
Liquid in a solid	Toluene in rubber (e.g., rubber cement)
Solid in a solid	Carbon in iron (steel)

[a] True examples probably do not exist.

1 nm $= 10^{-9}$ m $= 1$ nanometer.

0.1 – 1 nm. We usually think of solutions as being liquids, but in principle the solvent can be in any state — solid, liquid, or gas – and so can the solute. Table 6.2 has a list of the several combinations that can form a solution.

Solutions are generally transparent — you can see through them — but they often are colored. Solutes do not separate from the solvent under the influence of gravity, and they can't be separated by passing them through ordinary filter paper. The blood carries many substances in solution, including the sodium ion, Na^+, and the chloride ion, Cl^-, as well as molecules of glucose, the chief sugar in blood.

In Colloidal Dispersions, the Particle Sizes Are Larger.   A **colloidal dispersion** is a homogeneous mixture in which the dispersed particles are very large clusters of ions or molecules or are macromolecules (molecules with formula weights in the thousands or hundreds of thousands). The dispersed particles have diameters in the range of 1 – 1000 nm.

Table 6.3 gives several examples of colloidal dispersions, and they include many familiar substances such as whipped cream, milk, dusty air, jellies, and pearls. The blood also carries many substances in colloidal dispersions, including proteins.

When colloidal dispersions are in a fluid state — liquid or gas — the dispersed particles, although large, are not large enough to be trapped by ordinary filter paper during filtration. They are large enough, however, to reflect and scatter light (Figure 6.7). Such light scattering by a colloidal dispersion is called the **Tyndall effect,** after British scientist John Tyndall (1820 – 1893). It is responsible for the milky, partly obscuring character of smog, or the way sunlight sometimes seems to stream through a forest canopy.

Solutions do not exhibit the Tyndall effect because the solute particles are too small.

The large, dispersed particles in a fluid colloidal dispersion eventually separate under the influence of gravity, but this can take time. Depending on the system, it can take from many hours to many decades! One factor that keeps the particles dispersed is that they are constantly buffeted by molecules of the solvent. This motion of colloidal particles is called the **Brownian movement.** Evidence for it can be seen with a good microscope. (You see the light scintillations caused as the colloidal particles move erratically about.)

Robert Brown (1773 – 1858), an English botanist, first observed this phenomenon when he saw the trembling of particles inside grains of pollen that he viewed with a microscope.

In the most stable colloidal systems, all of the particles bear like electrical charges. They repel each other, therefore, and they can't coagulate into particles so large that they must separate under the influence of gravity. Electrically charged, colloidally dispersed particles are common among proteins in living systems. Other dissolved species of opposite charge, such as small ions, make the system electrically neutral.

Some colloidal dispersions are stabilized by protective colloids. **Emulsions,** for example, are colloidal dispersions of two liquids in each other, such as oil and vinegar, which can sometimes be stabilized by protective colloids called *emulsifying agents*. Thus mayonnaise is

TABLE 6.3
Colloidal Systems

Type	Dispersed Phase[a]	Dispersing Medium[b]	Common Examples
Foam	Gas	Liquid	Suds, whipped cream
Solid foam	Gas	Solid	Pumice, marshmallow
Liquid aerosol	Liquid	Gas	Mist, fog, clouds, certain air pollutants
Emulsion	Liquid	Liquid	Cream, mayonnaise, milk
Solid emulsion	Liquid	Solid	Butter, cheese
Smoke	Solid	Gas	Dust in smog
Sol	Solid	Liquid	Starch in water, jellies,[c] paints
Solid sol	Solid	Solid	Black diamonds, pearls, opals, metal alloys

[a] The colloidal particles constitute the *dispersed phase*.

[b] The continuous matter into which the colloidal particles are scattered is called the *dispersing medium*.

[c] Sols that adopt a semisolid, semirigid form (e.g., gelatin desserts, fruit jellies) are called **gels.**

stabilized by egg yolk, whose protein molecules coat the microdroplets of the vegetable oil and prevent them from merging into drops large enough to separate.

Suspensions Are Homogeneous Only When Constantly Stirred. In **suspensions,** the dispersed or suspended particles are over 1000 nm in average diameter, and they separate under the influence of gravity. They are large enough to be trapped by filter paper. A suspension such as clay in water has to be stirred constantly to keep it from separating. A suspension, therefore, is always on the borderline between a homogeneous mixture and one that is heterogeneous (not uniform throughout). The blood, while it is moving, is a suspension, besides being a solution and a colloidal dispersion. Suspended in circulating blood are its red and white cells and its platelets.

See Table 6.4 for a summary of the chief features of solutions, colloidal dispersions, and suspensions.

FIGURE 6.7
Tyndall effect. Shown here are three beakers viewed from the top. A light source on the right directs a laser beam through all three beakers. Colloidal starch dispersions are in the outside beakers, and a salt solution is in the middle. Light is scattered by the dispersions but not by the solution.

TABLE 6.4
Characteristics of Three Homogeneous Mixtures—Solutions, Colloidal Dispersions, and Suspensions

Particle Sizes Become Larger $\longrightarrow$		
Solutions	Colloidal Dispersions	Suspensions
All particles are on the order of atoms, ions, or small molecules (0.1—1 nm)	Particles of at least one component are large clusters of atoms, ions, or small molecules, or are very large ions or molecules (1—1000 nm)	Particles of at least one component may be individually seen with a low-power microscope (over 1000 nm)
Most stable to gravity	Less stable to gravity	Unstable to gravity
Most homogeneous	Also homogeneous, but borderline	Homogeneous only if well stirred
Transparent (but often colored)	Often translucent or opaque, but may be transparent	Often opaque, but may appear translucent
No Tyndall effect	Tyndall effect	Not applicable (suspensions cannot be transparent)
No Brownian movement	Brownian movement	Particles separate unless system is stirred
Cannot be separated by filtration using filter paper	Cannot be separated by filtration	Can be separated by filtration
Homogeneous	**to**	**Heterogeneous** $\longrightarrow$

6.6   OSMOSIS AND DIALYSIS

The migration of ions and molecules through membranes is an important mechanism for getting nutrients inside cells and waste products out.

From the Greek *kolligativ*, depending on number and not on identity or nature.

Colligative properties are those properties of solutions and colloidal dispersions that depend not on the chemical identities of the solutes but on nothing more than their concentrations. Two examples of colligative properties are the depression of the freezing point and the elevation of the boiling point. Solutions have lower melting points and higher boiling points than their pure solvents. Aqueous solutions, for example, freeze not at 0 °C but at slightly lower temperatures. They also boil not at 100 °C but at slightly higher temperatures.

As we said, the amounts of the effects of colligative properties depend on concentrations, not chemical identities. Thus, if we prepare two solutions, one with 1.0 mol of NaCl in 1000 g of water and the other with 1.0 mol of KBr in 1000 g of water, each freezes at −3.4 °C (and not at 0 °C), and each boils at 101 °C (at 760 mm Hg). Both the freezing and the boiling points are the same despite the chemical difference in the solutes, because the ratios of the moles of ions to the moles of water in both solutions are identical. Otherwise, the identities of the solutes are immaterial.

The temperature at equilibrium of a mixture made from 33 g of NaCl and 100 g of ice is about −22 °C (−6 °F).

When the concentrations are very high, the effects can be quite large and important. The whole basis for the use of an antifreeze in a radiator is the large depression of the freezing point of the coolant caused by the presence of the antifreeze. A 50% (v/v) solution of antifreeze in water doesn't freeze until about −40 °F.

Colligative properties of the highest importance at the molecular level of life are osmosis and a closely related phenomenon, dialysis. As we will see, osmosis and dialysis are quite sensitive even to low concentrations of solutes.

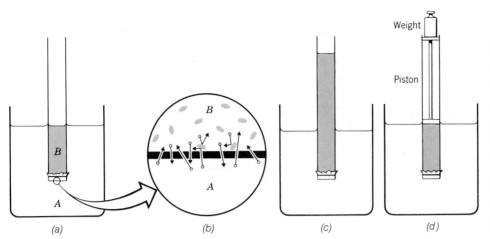

Weight

Piston

(a)   *(b)*   *(c)*   *(d)*

FIGURE 6.8
Osmosis and osmotic pressure *(a)* In the beaker, *A*, there is pure water and in the tube, *B*, there is a solution. An osmotic membrane closes the bottom of the tube. *(b)* A microscopic view at the osmotic membrane shows how solute particles interfere with the movements of water molecules from *B* to *A*, but not from *A* to *B*. *(c)* The level in *A* has fallen and that in *B* has risen because of osmosis. *(d)* To prevent osmosis, a back pressure would be needed, and the exact amount of pressure is the osmotic pressure of the solution in *B*.

Osmosis Is the Diffusion of Solvent Molecules through Membranes. Cells in the body are enclosed by cell membranes, and on both sides there are aqueous systems with substances in solution and colloidal dispersion. Materials and water have to be able to move through membranes in either direction so that nutrients can enter cells and wastes can leave.

One way that ions and molecules get through cell membranes is by **active transport.** *Active* signifies the participation of materials in the membrane itself. These accept and pass on ions and molecules by endothermic reactions. We do no more than mention this here. The other means of passing things through membranes is spontaneous dialysis, and the simplest form of dialysis is called osmosis.

Membranes of cells are **semipermeable;** they can let some but not all kinds of molecules and ions through. Cellophane, for example, is a synthetic, semipermeable membrane. In contact with an aqueous solution, cellophane lets only water molecules and other small molecules and ions through. It stops molecules of colloidal size. Evidently, cellophane has ultrafine pores just large enough for small particles but too small for colloidal particles.

Some membranes have pores so small that only water molecules can pass. Ions are hydrated, so their effective sizes are apparently too large. No other molecules can get through either. A semipermeable membrane so selective that only solvent molecules pass through is called an **osmotic membrane.**

When two solutions with different concentrations of particles are separated by an osmotic membrane, osmosis occurs. **Osmosis** is the net migration of solvent from the solution of lower concentration of solute into the solution of higher concentration. If osmosis could continue long enough, the concentrated solution would become dilute enough and the other solution (by losing solvent) would become concentrated enough to make the two concentrations equal. This way of thinking about the *direction* of the net flow of water in osmosis, incidentally, is perhaps the best way to remember how it goes. The net flow is such as to make the concentrated solution more dilute.

Figure 6.8 shows why the net osmotic flow is in one direction. It shows the special case in which pure water is on one side of the membrane. Water molecules can move in *both* directions, but the solute particles on one side interfere. Water molecules, therefore, cannot leave as frequently from this side as they can come in from the other side, where no solute particles are in the way. Thus more water molecules enter the concentrated solution than leave it, and this solution becomes increasingly diluted. Eventually, the rising column of water shown in Figure 6.8 will exert a high enough back pressure to prevent any further rise, and the osmosis will stop.

The back pressure necessary to prevent osmosis is called the **osmotic pressure** of the solution. Even dilute solutions can have high osmotic pressures. For example, a 0.100 *M* solution of sugar in water has an osmotic pressure of nearly 2000 mm Hg (2.6 atm). This solution could, in the right apparatus, support a column of water 25.3 m (83.0 ft) high.

From the Latin *permeare,* to go through.

Osmotic pressure is one factor that causes sap to rise in trees.

Separated Ions As Well As Molecules Contribute Individually to Osmotic Pressure.

The osmotic pressure of a solution has to be understood not as something that the solution is actually exerting, like some hand pushing on a surface. Instead, osmotic pressure is a potential that is directly related to the concentration of the solute particles. This potential is realized only when an osmotic membrane separates the solution from pure water.

The particles can be ions, molecules, or macromolecules. When the solute is an ionic compound, like NaCl, the particles in solution are not NaCl molecules but separated Na^+ and Cl^- ions. Two solute particles are thus in solution for each formula unit of NaCl that dissolves. The molar concentration of solute particles, therefore, is twice the molar concentration of the salt. In 0.10 M NaCl, for example, the concentration is $2 \times (0.10) = 0.20$ mol of all ions per liter. The osmotic pressure of 0.10 M NaCl is therefore twice as large as that of 0.10 M glucose, which does not break up into ions. An ionic compound like Na_2SO_4 breaks up into three ions when one formula unit dissolves, two Na^+ ions and one SO_4^{2-} ion. The concentration of particles in 0.10 M Na_2SO_4 is therefore $3 \times (0.10) = 0.30$ mol of all ions per liter.

The Osmolarity of a Solution Relates Better to Its Osmotic Pressure Than Its Molarity.

Because *molarity* does not reveal enough about a solution when we think about its osmotic pressure, scientists use the concept of **osmolarity,** abbreviated **Osm,** to express the molar concentration of all osmotically active particles in the solution. Thus a solution that has a concentration of 0.10 mol of NaCl/L has a molarity of 0.10 M but an osmolarity of 0.20 Osm. The osmolarity of 0.10 M Na_2SO_4 is 0.30 Osm.

PRACTICE EXERCISE 4   Assuming that any *ionic* solutes in this exercise break up completely into their constituent ions when they dissolve in water, what is the osmolarity of each solution?

(a) 0.010 M NH_4Cl (which ionizes as NH_4^+ and Cl^-)
(b) 0.005 M Na_2CO_3 (which ionizes as $2Na^+$ and CO_3^{2-})
(c) 0.100 M fructose (a sugar and a molecular substance)
(d) A solution that contains both fructose and NaCl with concentrations of 0.050 M fructose and 0.050 M NaCl

---

Small Solute Ions and Molecules Pass through Dialyzing Membranes.

Dialysis is like osmosis except that ordinary-sized ions and molecules can move through a dialyzing membrane. A dialyzing membrane can be thought of as having larger pores than an osmotic membrane. Cell membranes are largely dialyzing membranes but not entirely so. Cell mem-

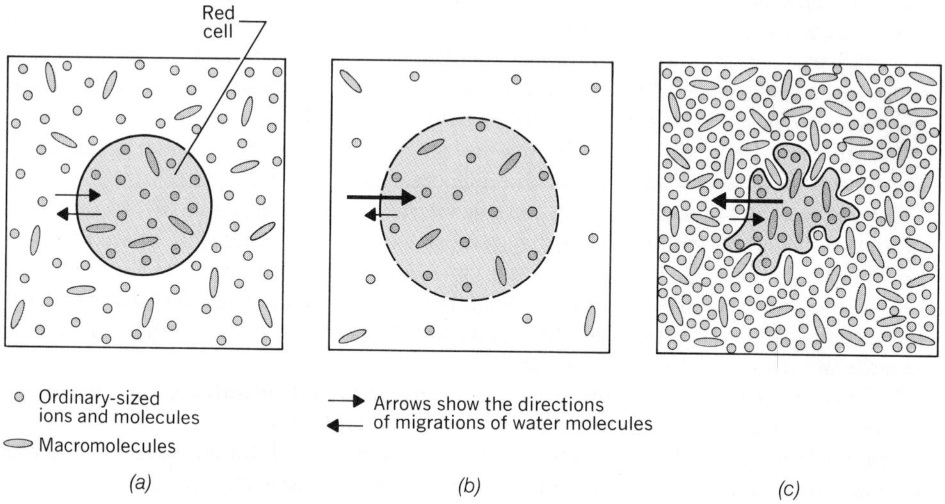

Red cell

FIGURE 6.9
Dialysis. *(a)* The red cell is in an isotonic environment. *(b)* Hemolysis is about to occur because the red cell is swollen by extra fluids brought into it from its hypotonic environment. *(c)* The red cell experiences crenation when it is in a hypertonic environment.

○ Ordinary-sized
  ions and molecules
⬭ Macromolecules

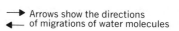
→ Arrows show the directions
← of migrations of water molecules

(a)   (b)   (c)

branes always include substances that selectively block the migrations of some small ions and molecules and allow others to pass through. Sometimes these special blocking compounds, usually proteins, are activated by hormones.

Dialysis produces a net migration of water only if the fluid on one side of the dialyzing membrane has a higher concentration in colloidal substances than the other. Colloidal-sized particles are blocked by dialyzing membranes, so they get in the way of the movements of smaller particles through the membrane. The net flow of fluid in dialysis, as in osmosis, is from the side that has the lower concentration of colloidal substances to the side with the higher concentration.

The contribution of colloidal particles to an imbalance in concentration gives the system a **colloidal osmotic pressure,** which is similar in meaning to osmotic pressure. In the next section we present some situations involving life at the molecular level where osmotic pressure relationships are very critical and depend on the colloidal osmotic pressure of blood.

Most natural membranes are dialyzing membranes. Osmotic membranes are difficult to prepare.

6.7 DIALYSIS AND THE BLOOD

When the osmotic pressure of blood varies too much, the result can be shock or harmful damage to red blood cells.

The body tries to maintain the concentrations of all of the substances in the blood within fairly narrow limits. To do this, the body has a thirst mechanism to help bring in water, and it has the machinery of diuresis and perspiration to let water leave. A number of hormones are involved in maintaining the integrity of blood, but in this section we look briefly at just two problems that occur when the osmotic pressure of the blood changes too much.

A Loss of the Blood's Macromolecules Triggers the Shock Syndrome.   When a person goes into shock, one of the many problems is that blood capillaries have become unable to prevent the loss of macromolecules from the blood. These are mostly molecules of albumin, a protein, and their loss quickly reduces the colloidal osmotic pressure of blood. With a lower colloidal osmotic pressure, water is less able to diffuse back into the blood from the surrounding spaces. Yet, water is still able to leave the blood. The result is a loss in blood volume, and this upsets the mechanisms that bring nutrients to the brain and carry wastes away. The result to the nervous system is called shock.

A *syndrome* is the whole collection of symptoms that characterize a disease.

Red Blood Cells Hemolyze in a Medium of Low Osmolarity.   In some clinical situations, body fluids need replacement or nutrients have to be given by intravenous drip. It is important that the osmolarity of the solution being added to the blood matches that of the fluid inside the red cells. Otherwise, the hemolysis or crenation of red blood cells will occur. Let's see what these mean.

Millions of red blood cells circulate in the bloodstream, and their membranes behave as dialyzing membranes. Within each red cell is an aqueous fluid containing dissolved and colloidally dispersed substances (Figure 6.9a). Although the dispersed particles are too large to dialyze, they contribute to the colloidal osmotic pressure. They help, therefore, to determine the direction in which dialysis occurs between the inside and the outside of the red cell. When red cells are placed in pure water, for example, enough fluid will migrate into the cells to make them burst open, as seen in Figure 6.9b. The rupturing of red cells is called **hemolysis,** and we say that the cells hemolyze.

When we put red blood cells in a medium with an osmolarity *higher* than their own fluid, dialysis occurs out of the cells and into the solution. Now the cells, losing fluid volume, shrivel and shrink, and this is called **crenation** (Figure 6.9c).

Two solutions of equal osmolarity are called **isotonic solutions.** If one has a lower osmotic pressure than the other, the first is said to be **hypotonic** with respect to the second. A hypotonic solution has a lower osmolarity than the one to which it is compared. Red cells hemolyze if placed in a hypotonic environment, including pure water.

SPECIAL TOPIC 6.3   HEMODIALYSIS

The kidneys cleanse the bloodstream of nitrogen waste products such as urea and other wastes. If the kidneys stop working efficiently or are removed, these wastes build up in the blood and threaten the life of the patient. The artificial kidney is one remedy for this.

The overall procedure is called *hemodialysis,* the dialysis of blood, and the figure on the lower left shows how it works. The bloodstream is diverted from the body and pumped through a long, coiled cellophane tube that serves as the dialyzing membrane. (The blood is kept from clotting by an anticlotting agent such as heparin.) A solution called the *dialysate* circulates outside of the cellophane tube. This dialysate is very carefully prepared not only to be isotonic with blood but also to have the same concentrations of all of the essential substances that should be left in solution in the blood. When these concentrations match, the rate at which such solutes migrate out of the blood equals the rate at which they return. In this way several key equilibria are maintained, and there is no net removal of essential components. The figure on the lower right shows how this works. The dialysate, however, is kept very low in the concentrations of the wastes, so the rate at which they leave the blood is greater than the rate at which they can get back in. In this manner, hemodialysis slowly removes the wastes from the blood.

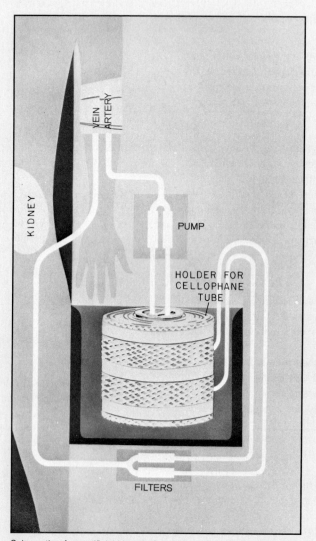

Schematic of an artificial kidney (Courtesy Artificial Organs Division, Travenol Laboratories, Inc.)

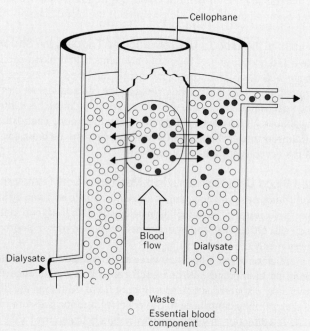

Waste molecules move out of the blood faster than they can return, but essential substances leave and return at equal rates.

If one of the two solutions has a higher osmotic pressure than the other, it is said to be **hypertonic** compared to the second. Thus 0.14 M NaCl is hypertonic with respect to 0.10 M NaCl. Red cells undergo crenation when they are in a hypertonic environment.

A 0.9% (w/w) NaCl solution, called **physiological saline solution,** is isotonic with respect to the fluid inside a red cell. Any solution that is administered into the bloodstream of a patient has to be similarly isotonic.

All of the topics we have studied in this and the preceding section are important factors in the operation of artificial kidney machines, which are discussed in Special Topic 6.3.

SUMMARY

Water   The higher electronegativity of oxygen over hydrogen and the angularity of the water molecule make water a very polar compound, so polar that hydrogen bonds exist between the molecules. The high polarity of water explains many of its unusual thermal properties, such as its high heats of fusion and vaporization, its high surface tension, and its ability to dissolve ionic and polar molecular compounds.

Hydrogen bonds   When hydrogen is covalently bonded to atoms of any of the three most electronegative elements (O, N, or F), its partial positive charge is large enough to be attracted rather strongly to the partial negative charge on an atom of O, N, or F on a nearby molecule. This force of attraction is the hydrogen bond. The most common errors are to consider this bond as the bond within a molecule of hydrogen, or to think of it as a covalent bond within some molecule. It's neither. The hydrogen bond is a force of attraction between the $\delta+$ on H in H—O, H—N, or H—F to the $\delta-$ of another O, N, or F.

Hydration   The attraction of water molecules to ions or to polar molecules leads to a loose solvent cage that shields the ions or molecules from each other. This phenomenon is called hydration. Sometimes water of hydration is present in a crystalline material in a definite proportion to the rest of the formula unit, and such a substance is a hydrate. Heat converts most hydrates to their anhydrous forms. And some anhydrous forms serve as drying agents or desiccants.

Solutions   Ions and molecules of ordinary size, if soluble in water at all, form solutions, homogeneous mixtures that neither gravity nor filtration can separate. The solubilities of most solids increase with temperature, because the process of their dissolving is usually endothermic. (More energy is needed to break up the crystal than is recovered as the solvent cages form about the ions or molecules.)

Gas solubilities   According to Henry's law, the solubility of a gas is directly proportional to its partial pressure in the space above the solution. Some gases do more than mechanically dissolve in water; part of what dissolves reacts with water to form soluble species. The availability of a gas from an aqueous solution can be expressed as the gas tension of the gas in the solution.

Percent concentration   A variety of concentration expressions have been developed to provide ways of describing a concentration without going into molar concentrations. These include weight/weight percents, volume/volume percents, and hybrid descriptions that aren't true percentages: weight/volume percent, milligram percent, parts per million, and parts per billion. When dilute solutions are made from more concentrated solutions, remember that the solute taken must all be present in the dilute solution.

Colloidal dispersions   Large clusters of ions or molecules or macromolecules do not form true solutions but colloidal dispersions. These can reflect and scatter light (Tyndall effect), experience the Brownian movement, and (in time) succumb to the force of gravity (if the medium is fluid). Protective colloids sometimes stabilize these systems. If the dispersed particles grow to an average diameter of about 1000 nm, they slip over into the category of suspended matter, and such systems must be stirred to maintain the suspension.

Osmosis and dialysis   When a semipermeable membrane separates two solutions or dispersions of unequal osmolarities, a net flow occurs in the direction that, if continued, would produce solutions of identical osmolarities. When the membrane is osmotic, only the solvent can migrate, and the phenomenon is osmosis. The back pressure needed to prevent osmosis is called the osmotic pressure, and it's directly proportional to the concentration of all particles of solute that are osmotically active — ions, molecules, and macromolecules.

When macromolecules are present, their particular contribution to the osmotic pressure is called the colloidal osmotic pressure of a solution. It is this factor that operates when the membrane is a dialyzing membrane. The permeability of blood capillaries changes temporarily when a person experiences shock, and macromolecules leave the blood. Their departure results in the loss of water, too, and the blood volume decreases.

Solutions of matched osmolarity are called isotonic. Otherwise, one is hypertonic (more concentrated) with respect to the other, and the other is hypotonic (less concentrated). Only isotonic solutions, or those that are nearly so, should be administered intravenously.

REVIEW EXERCISES

The answers to review exercises that require a calculation and whose numbers are marked with an asterisk are given in Appendix V. The answers to the remaining review exercises are given in the *Study Guide* that accompanies this book.

Hydrogen Bond

6.1   The hydrogen molecule, H—H, does not become involved in hydrogen bonding.
(a) What kind of bond occurs in a hydrogen molecule?
(b) Why can't this molecule become involved in hydrogen bonding?

6.2   The methane molecule, CH_4, does not become involved in hydrogen bonding. Why not?

6.3   Draw a structure of two water molecules. Write in $\delta+$ and $\delta-$ symbols where they belong. Then draw a correctly positioned dotted line between two molecules to symbolize a hydrogen bond.

6.4   Hydrogen bonds exist between two molecules of ammonia, NH_3. Draw the structures of two ammonia molecules. Put $\delta+$ and $\delta-$ signs where they should be located. Then draw a dotted line that correctly connects two points to represent a hydrogen bond.

6.5   The hydrogen bond between two molecules of ammonia must be much weaker than the hydrogen bond between two molecules of water.
(a) How do boiling point data suggest this?
(b) What does this suggest about the relative sizes of the $\delta+$ and the $\delta-$ sites in molecules of water and ammonia?
(c) Why are the $\delta+$ and the $\delta-$ sites different in their relative amounts of fractional electric charge when we compare molecules of ammonia and water?

6.6   If it takes roughly 100 kcal/mol to break the covalent bond between O and H in H_2O, about how many kilocalories per mole are needed to break the hydrogen bonds in a sample of liquid water?

6.7   Explain in your own words how hydrogen bonding helps us understand each of the following:
(a) The high heats of fusion and vaporization of water.
(b) The high surface tension of water.

6.8   Explain in your own words why water forms tight beads on a waxy surface but spreads out on a clean glass surface.

6.9   What does a surfactant do to water's surface tension?

6.10 What common household materials are surfactants?

6.11 What surfactant is involved in digestion? What juice supplies it? How does it aid digestion?

6.12 Bile comes from the gallbladder, and when a patient has the gallbladder removed, he or she is put on a diet that is relatively low in fats and oils. Why?

Aqueous Solutions

6.13 In a crystal of sodium chloride, the sodium ions are surrounded by oppositely charged ions (Cl^-) as nearest neighbors. What replaces this kind of electrical environment for sodium ions when sodium chloride dissolves in water?

6.14 When we say that a chloride ion in water is *hydrated,* what does this mean? (Make a drawing as part of your answer.)

6.15 Carbon tetrachloride, CCl_4, is a liquid, and its molecules are tetrahedral like those of methane, CH_4. It does not dissolve in water. Why won't water let CCl_4 molecules in?

6.16 Suppose that you do not know and do not have access to a reference in which to look up the solubility of sodium nitrate, $NaNO_3$, in water at room temperature. Yet, you need a solution that you know beyond doubt is saturated. How can you make such a saturated solution and know that it is saturated?

Hydrates

6.17 Calcium sulfate dihydrate, $CaSO_4 \cdot 2H_2O$, loses all of its water of hydration at a temperature of 163 °C. Write the equation for this reaction.

6.18 Why are hydrates classified as compounds and not as wet mixtures?

6.19 When water is added to anhydrous copper(II) sulfate, the pentahydrate of this compound forms. Write the equation.

6.20 Anhydrous sodium sulfate is hygroscopic. What does this mean? Does this property make it useful as a desiccant?

6.21 Sodium hydroxide is sold in the form of small pellets about the size and shape of split peas. It is a very deliquescent substance. What can happen if you leave the cover off of a bottle of sodium hydroxide pellets?

6.22 When 6.29 g of the hydrate of compound M was strongly heated to drive off all of the water of hydration, the residue — the anhydrous form, M — had a mass of 4.97 g. What number should x be in the formula of the hydrate, $M \cdot xH_2O$? The formula weight of M is 136.

*6.23 When all of the water of hydration was driven off of 4.25 g of a hydrate of compound Z, the residue — the anhydrous form, Z — had a mass of 2.24 g. What is the formula of the hydrate (using the symbol Z as part of it)? The formula weight of Z is 201.

Dynamic Equilibria and Le Chatelier's Principle

6.24 State Le Chatelier's principle.

6.25 Consider the following dynamic equilibrium that describes a saturated solution of sodium chloride in water.

$$NaCl(s) + heat \rightleftharpoons Na^+(aq) + Cl^-(aq)$$

(a) Write the equation for the forward reaction.
(b) Write the equation for the reverse reaction.
(c) Which reaction, forward or reverse, is endothermic? What does this mean?

(d) If a beaker containing this saturated system at 40 °C were cooled to 5 °C, which reaction, the forward or the reverse, would become more rapid than the other?

(e) After the system rested at 5 °C for a long period of time, what would become of the rates of the forward and reverse reactions? (Would they become equal or remain unequal?)

(f) Would the concentration of the NaCl solution at equilibrium at 5 °C be higher or lower than it is at 40 °C?

(g) If more solid sodium chloride were added to the solution at equilibrium at 5 °C, would the solution become more concentrated?

(h) If the solution at equilibrium at 5 °C is warmed, what specific stress is being placed on the equilibrium?

(i) The stress of part (h) causes the equilibrium to do what? (What verb is used to describe this?) This change would be in what direction?

6.26 The following expression describes the equilibrium that exists when ice and water are mixed, both at 0 °C.

$$H_2O(s) + heat \rightleftharpoons H_2O(l)$$

(a) What could be done, experimentally, to make the rate of the exothermic process greater than the rate of the endothermic change?

(b) According to Le Chatelier's principle, which change, the forward or the reverse, becomes favored by the removal of heat from this system? Does the ratio of the mass of ice to liquid water increase or decrease as a result?

(c) The addition of NaCl (which dissolves) puts Na^+ and Cl^- ions into the liquid water but not into the ice crystals. These ions slow down the rate of the reverse process but not the forward change. (The ions interfere with the return of water molecules to the crystal surface, but not with their escape from these surfaces.) In order to reestablish the equilibrium what has to be done to the rate of the forward change? Would this be done by adding heat or removing heat? In other words, would you have to raise or lower the temperature of the system to reestablish equilibrium? (Another way of asking this is "Does salty water freeze at a temperature above or below 0 °C?")

6.27 If the solubility of a compound *decreases* with increasing temperature, how would you write the equilibrium expression, as A or as B?

A   solid + heat $\rightleftharpoons$ solution

B   solid $\rightleftharpoons$ solution + heat

6.28 Ammonium chloride dissolves in water endothermically. Suppose that you have a saturated solution of this compound, that its temperature is 30 °C, and that undissolved solute is present. Write the equilibrium expression for this saturated solution, and use Le Chatelier's principle to predict what will happen if you cool the system to 20 °C.

6.29 We have to distinguish between *how fast* something dissolves in water and *how much* can dissolve to make a saturated solution. The speed with which we can dissolve a solid in water increases if we (a) crush the solid to a powder, (b) stir the mixture, or (c) heat the mixture. Use the kinetic theory as well as the concept of forward and reverse processes to explain these facts.

Gas Solubilities

6.30 What is Henry's law? How is advantage taken of it in the technology of hyperbaric chambers?

6.31 Using Le Chatelier's principle, explain why the solubility of a gas in water should increase with increasing partial pressure of the gas.

6.32 When the gas tension of CO_2 in blood is described as 30 mm Hg, what specifically does this mean?

6.33 If in one region of the body the gas tension of oxygen over blood is 80 mm Hg and in a second region it is 50 mm Hg, which region (the first or the second) has a higher concentration of oxygen in the blood itself?

6.34 Explain why carbon dioxide is more soluble in water than is oxygen.

Percent Concentrations

6.35 If a solution has a concentration of 1.2% (w/w) NaOH, what two conversion factors can we write based on this value?

6.36 A solution bears the label, 1.5% (w/v) NaCl. What two conversion factors can be written for this value?

6.37 A solution of wood alcohol in water is described as 12% (v/v). What two conversion factors are possible from this value?

6.38 How many grams of solute are needed to prepare each of the following solutions?
(a) 500 g of 0.900% (w/w) NaCl
(b) 250 g of 1.25% (w/w) $NaC_2H_3O_2$
(c) 100 g of 5.00% (w/w) NH_4Cl
(d) 500 g of 3.50% (w/w) Na_2CO_3

*6.39 Calculate the number of grams of solute needed to make each of the following solutions.
(a) 100 g of 0.500% (w/w) NaI
(b) 250 g of 0.500% (w/w) NaBr
(c) 500 g of 1.25% (w/w) $C_6H_{12}O_6$ (glucose)
(d) 750 g of 2.00% (w/w) H_2SO_4

6.40 How many grams of solute have to be weighed out to make each of the following solutions?
(a) 125 mL of 10.0% (w/v) NaCl
(b) 250 mL of 2.00% (w/v) KBr
(c) 500 mL of 1.50% (w/v) $CaCl_2$
(d) 750 mL of 0.900% (w/v) NaCl

*6.41 In order to prepare the following solutions, how many grams of solute are required?
(a) 250 mL of 5.00% (w/v) $Mg(NO_3)_2$
(b) 500 mL of 1.00% (w/v) NaBr
(c) 100 mL of 2.50% (w/v) KI
(d) 50.0 mL of 3.35% (w/v) $Ca(NO_3)_2$

6.42 How many milliliters of methyl alcohol have to be used to make 750 mL of 10.0% (v/v) aqueous methyl alcohol solution?

*6.43 A sample of 500 mL of 5.00% (v/v) aqueous ethyl alcohol contains how many milliliters of pure ethyl alcohol?

6.44 A chemical supply room has supplies of the following solutions: 5.00% (w/w) NaOH, 1.00% (w/w) Na_2CO_3, and 2.50% (w/v) glucose. If the densities of these solutions can be taken to be 1.00 g/mL, how many milliliters of the appropriate solution would you have to measure out to obtain the following quantities?
(a) 3.50 g of NaOH   (b) 0.250 g of Na_2CO_3
(c) 0.500 g of glucose   (d) 0.100 mol of NaOH
(e) 0.100 mol of glucose ($C_6H_{12}O_6$)

*6.45 The stockroom has the following solutions: 2.00% (w/w) KOH, 0.500% (w/w) HCl, and 0.900% (w/v) NaCl. Assuming that the densities of these solutions are all 1.00 g/mL, how many milliliters of the appropriate solution have to be measured out to obtain the following quantities of solutes?
(a) 0.220 g of KOH   (b) 0.150 g of HCl
(c) 0.100 mol of NaCl   (d) 0.100 mol of KOH

6.46 A student needed 50.0 mL of 10.0% (w/w) aqueous sodium acetate, $NaC_2H_3O_2$. (The density of this solution is 1.05 g/mL.) Only the trihydrate $NaC_2H_3O_2 \cdot 3H_2O$, was available, and the student knew that the water of hydration would just become part of the solvent once the solution was made. How many grams of the trihydrate would have to be weighed out to prepare the needed solution?

*6.47 How many grams of $Na_2SO_4 \cdot 10H_2O$ have to be weighed out to prepare 100 mL of 10.0% (w/w) Na_2SO_4 in water? (The density of this solution is 1.09 g/mL.)

Colloidal Dispersions

6.48 What does it mean when we describe a solution or a colloidal dispersion as *homogeneous?*

6.49 What is the basis for distinguishing among solutions, colloidal dispersions, and suspensions?

6.50 Which of the three kinds of homogeneous mixtures
(a) Can be separated into its components by filtration?
(b) Exhibits the Tyndall effect?
(c) Shows observable Brownian movements?
(d) Has the smallest particles of all kinds?
(e) Is likeliest to be the least stable at rest over time?

6.51 What kinds of particles make the most stable colloidal dispersions? Explain.

6.52 The blood is simultaneously a solution, a colloidal dispersion, and a suspension. Explain.

6.53 Why won't a solution give the Tyndall effect?

6.54 What causes the Brownian movement?

6.55 What simple test could be used to tell if a clear, colorless solution contained substances in colloidal dispersion?

6.56 What is an emulsion? Give some examples.

6.57 What is a sol? Give some examples.

6.58 What is a gel? Give an example.

Osmosis and Dialysis

6.59 If a solution that contains 1.00 mol of glucose in 1000 g of water freezes at -1.86 °C, what is the freezing point of a solution that contains 1.00 mol of glycerol in 1000 g of water? (Both are compounds that do not break up into ions when they dissolve.)

6.60 A solution that contains 1.00 mol of glucose in 1000 g of water has a normal boiling point of 100.5 °C. Another solution that contains 1.00 mol of an unknown compound in 1000 g of water has a normal boiling point of 101.0 °C. What is the likeliest explanation for the higher boiling point of the second solution?

6.61 Explain in your own words and drawings how osmosis gives a net flow of water from pure water into a solution on the other side of an osmotic membrane.

6.62 In general terms, how does an osmotic membrane differ from a dialyzing membrane?

6.63 Explain in your own words why the osmotic pressure of a solution should depend only on the concentration of its solute particles and not on their chemical properties.

6.64 Why is the osmolarity of 1 M NaCl not the same as its molarity?

6.65 Which has the higher osmolarity, 0.10 M NaCl or 0.080 M Na_2SO_4? Explain.

*6.66 Which solution has the higher osmotic pressure, 10% (w/w) NaCl or 10% (w/w) NaI? Both NaCl and NaI break up in water in the same way — two ions per formula unit.

6.67 Solution A consists of 0.5 mol of NaCl, 0.1 mol of $C_6H_{12}O_6$ (glucose, a molecular substance), and 0.05 mol of starch (a colloidal, macromolecular substance), all in 1000 g of water. Solution B is made of 0.5 mol of NaBr, 0.1 mol of $C_6H_{12}O_6$ (fructose, a molecular substance related to glucose), and 0.005 mol of starch all in 1000 g of water. Which solution, if either, has the higher osmotic pressure? Explain.

6.68 What happens to red blood cells in hemolysis?

6.69 Physiological saline solution has a concentration of 0.9% (w/w) NaCl.
(a) Is a solution that is 1.1% (w/w) NaCl described as hypertonic or hypotonic with respect to physiological saline solution?
(b) What would happen, crenation or hemolysis, if a red blood cell were placed in (1) 0.5% (w/w) NaCl? (2) In 1.5% (w/w) NaCl?

6.70 Explain how the loss of macromolecules from the blood can lead to the increased loss of water from blood and a reduction in blood volume.

Decompression Sickness (Special Topic 6.1)

6.71 The solubilities of which gases increase in blood to cause decompresson sickness? Why do they increase?

6.72 How does an increased solubility of gases in blood cause a problem when the individual returns to normal atmospheric pressure?

6.73 How does a slow decompression reduce the possibility of de-compression sickness?

6.74 What is the "rule of thumb" about the rate of decompression needed to avoid decompression sickness?

Preparing Dilute from Concentrated Solutions
(Special Topic 6.2)

6.75 Concentrated acetic acid is 17 M $HC_2H_3O_2$. How would you prepare 100 mL of 2.0 M $HC_2H_3O_2$?

*6.76 Concentrated nitric acid is 16 M HNO_3. How would you pre-pare 500 mL of 1.0 M HNO_3?

6.77 Which is the safe procedure, to add concentrated sulfuric acid to water or to add water to the concentrated acid? Why?

6.78 A student has to prepare 500 g of 1.25% (w/w) NaOH. The stock supply of NaOH is in the form of 5.00% (w/w) NaOH. How many grams of the stock solution have to be diluted to make the desired solution?

*6.79 A 10.0% (w/w) HCl solution is available from the stockroom. How many grams of this solution have to be weighed out to prepare, by dilution, 250 g of 0.500% (w/w) HCl? If the den-sity of the 10.0% solution is 1.05 g/mL, how many milliliters would provide the grams of the concentrated solution that are called for?

6.80 Concentrated hydrochloric acid is available as 11.6 M HCl. The density of this solution is 1.18 g/mL.
(a) Calculate the percent (w/w) of HCl in this solution.
(b) How many milliliters of this concentrated acid have to be taken to prepare 500 g of a solution that is 10.0% (w/w) HCl?

*6.81 Commercial nitric acid comes in a concentration of 16.0 mol/L. The density of this solution is 1.42 g/mL.
(a) Calculate the percent (w/w) of nitric acid, HNO_3, in this solution.
(b) How many milliliters of the concentrated acid have to be taken to prepare 250 g of a solution that is 10.0% (w/w) HNO_3?

Hemodialysis (Special Topic 6.3)

6.82 What does *hemodialysis* mean?

6.83 During hemodialysis, what is the *dialysate*?

6.84 With respect to the following solutes in blood, what should be the concentration of the dialysate for *effective* hemodialysis, more or less concentrated or the same concentration?
(a) Na^+ (b) Cl^- (c) urea

Chapter 7
Acids, Bases, and Salts

In their place, acids are beneficial, but that place is not in the air surrounding these Fraser firs, dead and dying partly because of acid rain. This chapter is the first of two that tells us about acids.

7.1 SOURCES OF IONS AND ELECTROLYTES

The principal ion producers in water are acids, bases, and salts.

Almost all water contains dissolved ions, whether the water is in lakes and rivers or in the fluids of living systems. Most experiments done in the lab, and nearly all clinical analyses involve ions in solution. Sometimes the tiniest imbalances in the concentrations of certain ions in cells or in the blood cause the gravest medical emergencies.

We begin here a rather extensive study of ions, a study that spreads over more than one chapter. In this chapter we will learn about the most general reactions of the principal families of ionic compounds: acids, bases, and salts.

Even the Purest Water Contains Traces of H_3O^+ and OH^- Ions. Before we study the families of compounds that supply ions in water, we have to learn about the ions that form just from water molecules. When two water molecules collide powerfully enough (Figure 7.1), a proton, H^+, transfers. Two ions form in a $1:1$ ratio, the **hydronium ion,** H_3O^+, and the **hydroxide ion,** OH^-. Any reaction in which ions form from neutral molecules is called **ionization,** and this self-ionization of water is the forward reaction in a chemical equilibrium:

$$2H_2O \rightleftharpoons H_3O^+(aq) + OH^-(aq)$$

$H\cdot$ = hydrogen atom

H^+ = hydrogen ion, a bare proton

The forward reaction is definitely not favored, which we have indicated by using arrows of unequal length. At 25 °C, the concentration of each ion is only 1.0×10^{-7} mol/L. This may seem too small to mention, but life hinges on holding the concentrations of H_3O^+ and OH^- in body fluids at about this value. In pure water their concentrations are exactly equal, but acids, bases, and some salts, change this.

We won't study the mathematical details of equilibria, so we use the device of unequal arrows to indicate which sides of equilibria are favored.

The Acidity of a Solution Depends on the Ratio of H_3O^+ to OH^-. **Acids** make the molar concentration of hydronium ions in aqueous systems higher than that of hydroxide ions. **Bases** or **alkalies** make the molar concentration of hydroxide ion greater than that of hydronium ion. In a **neutral solution,** the molar concentrations of H_3O^+ and OH^- are equal. These definitions let us define acidic, basic, and neutral solutions as follows, where we use brackets, [   ], about a formula to mean its *molar concentration*.

Svante Arrhenius (1859–1927), a Swedish chemist won the third Nobel prize in chemistry (1903) for proposing the idea of ions.

Acidic solutions	$[H_3O^+] > [OH^-]$
Neutral solutions	$[H_3O^+] = [OH^-]$
Basic solutions	$[H_3O^+] < [OH^-]$

Acidic Solutions Have Some Common Properties. When hydronium ions are in sufficient excess over hydroxide ions in water, the solutions have some common properties, all related to this excess. One is their ability to make a dye called litmus be in its red form, not its blue form. Acidic solutions turn blue litmus red, one of the oldest tests for such solutions. Typically,

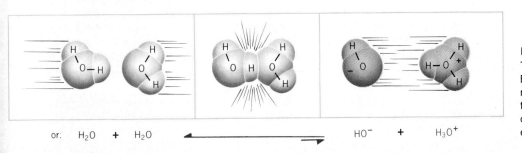

or: H_2O + H_2O $\longleftarrow\!\!\!\longrightarrow$ HO^- + H_3O^+

FIGURE 7.1
The self-ionization of water. Extremely violent collisions are necessary, and at equilibrium the fraction of all collisions occurring at any instant that are energetic enough is very small.

a solution of this dye is allowed to soak into strips of porous paper, which are then dried and sold as litmus paper. Unsurprisingly, a chemical used to indicate the presence of another is commonly called an indicator, and litmus paper is an example of an acid–base indicator.

In the lab, *never* taste a chemical unless the instructor says it's all right.

Acidic solutions also have tart or sour tastes, not that anyone should perform this test on an unknown system. (The "unknown" might be a strong poison!) But the tart tastes of citrus fruit juices are caused by citric acid. Sour milk has lactic acid. Vinegar has acetic acid. Rhubarb has oxalic acid. Sour apples have malic acid. Gastric juice contains hydrochloric acid. You can see that acids are very common in nature.

Basic Solutions Also Have Some Common Properties.   Solutions with a sufficient excess of hydroxide ion also have common properties related to this excess. They all turn red litmus blue, for example. They generally have bitter tastes and slippery or soapy textures.

The most important single reaction of acids and bases is that they neutralize each other. When mixed in the right molar proportion, acids and bases react to produce a solution that has none of the properties of either an acid or a base. The solution is a neutral solution, and the reaction is called **acid – base neutralization.**

Salts Form in Acid – Base Neutralization Reactions.   The third major family of ion-producing substances is the salts. **Salts** are ionic compounds in which the positive ion is a metal ion or any other positive ion except H^+, and the negative ion is any but OH^-. All salts are crystalline solids at room temperature because the attractions of oppositely charged ions in the crystals are very strong.

Salts are the products of acid–base neutralization reactions. Table salt or sodium chloride, for example, forms when hydrochloric acid reacts with sodium hydroxide. The base, sodium hydroxide, neutralizes the acid. (Or, with equal validity, we could say that the acid neutralizes the base.) The equation is

$$\text{HCl}(aq) \quad + \text{NaOH}(aq) \longrightarrow \text{NaCl}(aq) + \text{H}_2\text{O}$$

| Hydrochloric acid | Sodium hydroxide | Sodium chloride |

We will return to the salts at the end of this chapter after we have studied acids and bases in more detail.

Aqueous Solutions of Acids, Bases, and Salts Conduct Electricity.   The electricity in a lightning strike travels well through wet ground because soil water contains ions. Electrocardiograms can be obtained by attaching wires to the *outside* of the body only because the fluids in the skin and inside the body contain ions that can conduct very small electric currents safely.

The electricity from lightning that strikes trees often follows the root system in the ground.

Electricity in metal wires is a flow of electrons.

Unlike an electric current in a metal, a current in a solution of ions is not carried by electrons. Instead it is carried by dissolved ions, which move and transport electrons. (The word *ion* is from the Greek *ienai,* which means to move.) Let's see how ions do this.

Figure 7.2 shows a typical setup. The plates or wires that dip into the solution are called **electrodes.** The battery forces electrons to one of these electrodes, called the **cathode,** which becomes electron rich. Positive ions are attracted to the cathode, because opposite charges attract, and this is why positive ions are call **cations** (pronounced CAT-ions).

The electrons that make the cathode electron rich are pumped from the other electrode, called the **anode,** which becomes electron poor. Negative ions are attracted to the anode, and they are called **anions** (AN-ions).

When a cation arrives at the cathode, it takes an electron from its electron-rich surface. When an anion reaches the anode, it deposits an electron at the electron-poor surface. If something takes electrons from one electrode and something (else) puts them on the other, the effect is the same as if the electrons themselves were actually moving through the solution, and this is how a solution of ions carries an electric current.

Ions are chemically changed by electrolysis, and many useful substances are made this way.

The passage of an electric current through a fluid a called **electrolysis,** and substances that permit electrolysis are called **electrolytes.** For electrolysis to happen, ions must be mobile. When ions are fixed in crystals, no electrolysis occurs. Ions can be made mobile either by melting the crystals or dissolving them. Thus the term *electrolyte* refers either to a solution of ions or to the solid ionic compound. (The term does not apply to metals. Metals are called *conductors*.)

Electrolytes Can Be Strong or Weak According to Their Abilities to Supply Ions in Water.

In solutions of equal molar concentrations, electrolytes are not equally good at carrying electricity. A "good" or **strong electrolyte** is one that readily supplies ions, so it helps a strong electric current to flow. Essentially 100% of the formula units of a strong electrolyte break up or separate into ions when it dissolves. Sodium hydroxide, hydrochloric acid, and sodium chloride are all strong electrolytes as are most of the common bases, acids, and water-soluble salts.

Only a small percentage of the formula units of a **weak electrolyte** break up into ions in water even though it might be very soluble in water. Acetic acid, the acid that makes vinegar sour, is a typical weak electrolyte. Although 1 M NaCl is an excellent conductor, 1 M acetic acid is a poor conductor. Aqueous solutions of ammonia are also poor conductors, so aqueous ammonia is classified as a weak electrolyte.

Many substances are **nonelectrolytes.** They do not conduct ordinary currents of electricity (e.g., household currents) at all. Pure water is an example, and alcohol and sugar are others. As we learned, the purest of water does contain some ions, but their concentrations are too low to make it a conductor. You can see that electrolysis can be used to find out if a high percentage of the formula units of a compound break up into ions in water.

We can summarize what we have learned about strong and weak electrolytes as follows: Be sure to notice the emphasis on *percentage* ionization as the feature that dominates these definitions.

Strong electrolyte	One that is strongly ionized in water — a high percentage ionization
Weak electrolyte	One that is weakly ionized in water — a low percentage ionization
Nonelectrolyte	One that does not ionize in water — essentially zero percentage ionization

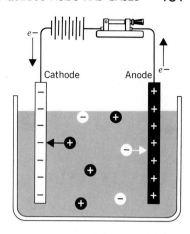

● Cation   ⊖ Anion

FIGURE 7.2
Electrolysis. Cations, positive ions, migrate to the cathode and remove electrons. Anions, negative ions, migrate to the anode and deposit electrons. The effect is a closed circuit.

7.2 THE COMMON AQUEOUS ACIDS AND BASES

Hydrochloric acid, sulfuric acid, and phosphoric acid are common acids and sodium hydroxide is a common base.

We have learned about some common properties of acidic and basic solutions. In this section we will learn about the chemicals most often used to prepare such solutions.

Hydrochloric Acid Is a Solution of Hydrogen Chloride in Water.

You can buy hydrochloric acid, HCl(aq), in hardware and plumbing supply stores (where it's likely to be called *muriatic acid*). Handle it very carefully and avoid letting either the liquid or its sharp, stinging fumes touch your skin, clothing, or eyes.

Hydrochloric acid is made by dissolving the gas hydrogen chloride, HCl(g), in water. This gas reacts promptly and essentially completely with the water to give a solution of hydronium ions and chloride ions. It is this solution to which the name *hydrochloric acid* belongs.

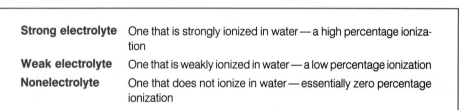

$$\text{HCl}(g) + \text{H}_2\text{O} \longrightarrow \underbrace{\text{H}_3\text{O}^+(aq) + \text{Cl}^-(aq)}$$

Hydrogen
chloride

Hydrochloric acid, HCl(aq)

The concentrated HCl(aq) of commerce is 12 M, which is 37% (w/w) in HCl, a saturated solution of HCl(g) in water. Handle it very carefully. It can cause severe chemical burns.

(g) = gas

(l) = liquid

(s) = solid

(aq) = aqueous solution

When we use the formula HCl(aq) in an equation or when we speak of hydrochloric acid, we always mean the aqueous solution of the separated ions, $H_3O^+(aq)$ and $Cl^-(aq)$.

Hydrochloric acid is called a **monoprotic acid** because the ratio of H_3O^+ ions to Cl^- ions is $1:1$. Other monoprotic acids can be made by dissolving the other hydrogen halide gases in water: hydrofluoric acid, HF(aq), hydrobromic acid, HBr(aq), and hydriodic acid, HI(aq). All except hydrofluoric acid are *strong acids*.

A **strong acid** is any that ionizes essentially 100% in solution. A **weak acid** ionizes only to a few percent or less. All strong acids are strong electrolytes. All weak acids are weak electrolytes.

> HF(aq) is highly ionized, but its ions attract each other so strongly that it behaves as a weak acid. Its relatively high concentration of fluoride ion is what makes it a very dangerous acid.

Nitric Acid Is a Strong, Monoprotic Acid.   Pure, anhydrous nitric acid, HNO_3, is known but is very rarely used. The "nitric acid" available in the laboratory is an aqueous solution in which the following chemical equilibrium exists. (We continue to use arrows of unequal lengths to indicate which side of an equilibrium is favored.) The curved arrows are meant to show how one particle hits another, forces the transfer of H^+ and makes the electron pair that held H^+ stay behind.

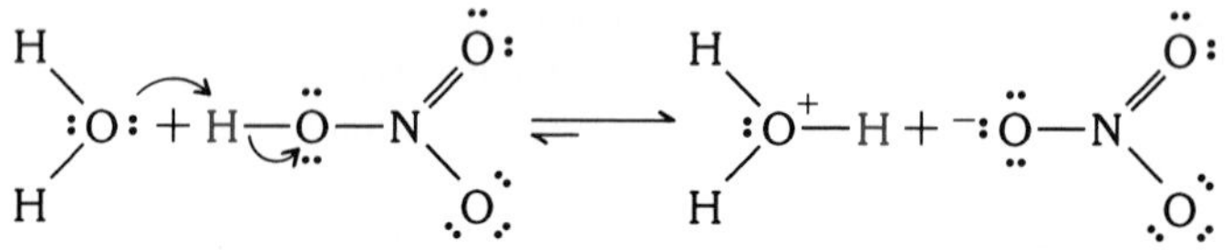

> The concentrated $HNO_3(aq)$ of commerce is 16 M, which is 71% (w/w) in HNO_3. Handle it carefully. It not only causes severe burns, it turns the skin yellow on even a brief contact.

or

$$H_2O \;+\; HNO_3 \;\rightleftharpoons\; H_3O^+ \;+\; NO_3^-$$

          Nitric   Hydronium   Nitrate
          acid   ion   ion

Acetic Acid Is a Weak Organic Acid.   Like nearly all organic acids, acetic acid is a weak acid. Although its formula, $HC_2H_3O_2$, shows four hydrogen atoms, it is a monoprotic acid. As seen in the structural formula of Figure 7.3, three H atoms are held by carbon, and only one is attached to oxygen. Only this one can transfer to a water molecule, and a powerful collision is needed (Figure 7.3). The H—O bond in acetic acid is considerably stronger than the H—Cl bond in HCl(g), so acetic acid is a weaker acid than hydrochloric acid. We can represent the equilibrium present in aqueous acetic acid as follows:

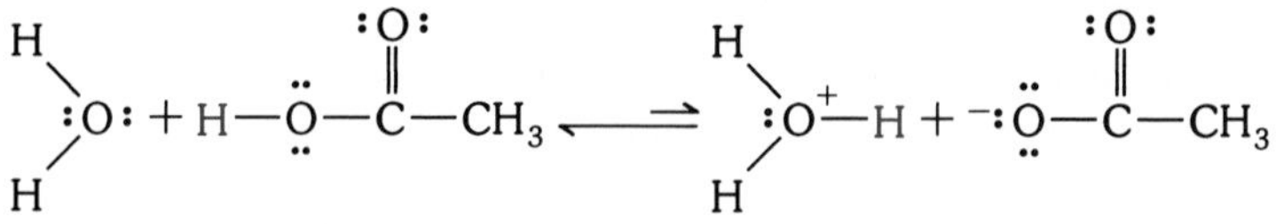

> **Organic compounds** are compounds of carbon and other nonmetals. Other compounds are **inorganic compounds.**

or,

$$H_2O \;+\; H{-}C_2H_3O_2 \;\rightleftharpoons\; H_3O^+ \;+\; C_2H_3O_2^-$$

          Acetic acid   Acetate ion

> Vinegar is about 5% (w/w) acetic acid in water.

We will continue to use $HC_2H_3O_2$ as our symbol for acetic acid in this chapter and the next. Just remember that only one hydrogen of $HC_2H_3O_2$ is active in acid–base reactions, indicated by the one H that begins the formula of the acid; acetic acid is monoprotic. (Citric acid, $H_3C_6H_5O_7$, can furnish three H^+ ions, indicated by the H_3 at the start of its formula.)

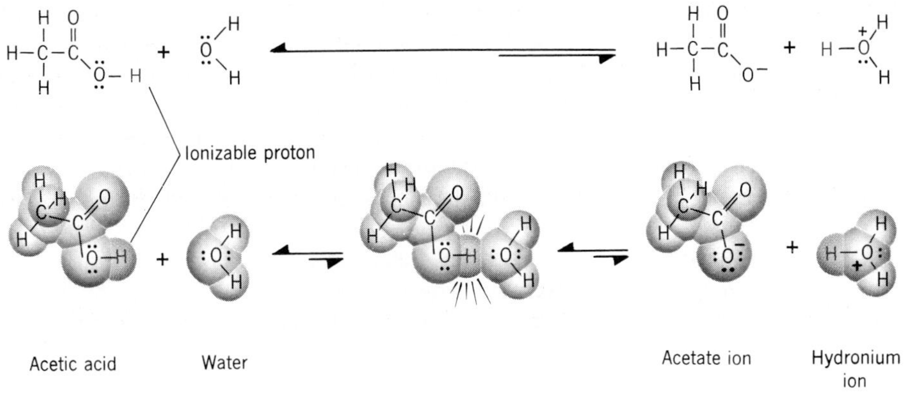

FIGURE 7.3
The ionization of acetic acid in water. Only a small fraction of the collisions are energetic enough to make a proton transfer from O in acetic acid to O in H_2O.

Sulfuric Acid Is a Major Industrial Chemical.

Sulfuric acid, $H_2SO_4(aq)$, is the only common aqueous **diprotic acid,** one that gives two hydronium ions per formula unit in solution. The ionization of the first proton is so easy that sulfuric acid is a strong acid.

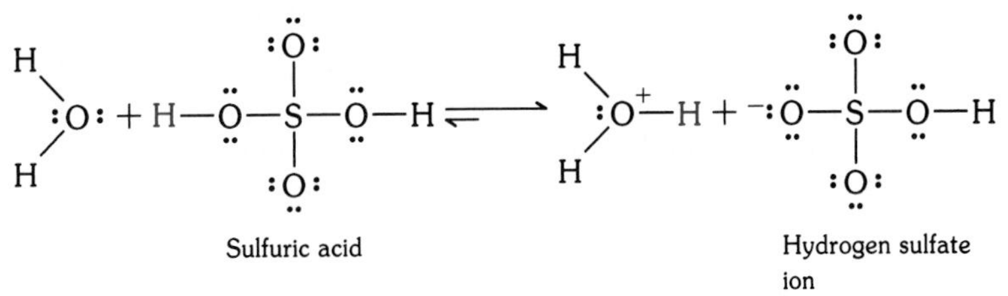

or,

$$H_2O + H_2SO_4 \rightleftharpoons H_3O^+ + HSO_4^-$$

Sulfuric acid   Hydrogen sulfate ion

The hydrogen sulfate ion is also an acid, but the transfer of H^+ away from it to a water molecule requires the movement of a positively charged particle away from one that already is oppositely charged. Although this is harder than the transfer of the first H^+ from H_2SO_4, it still happens.

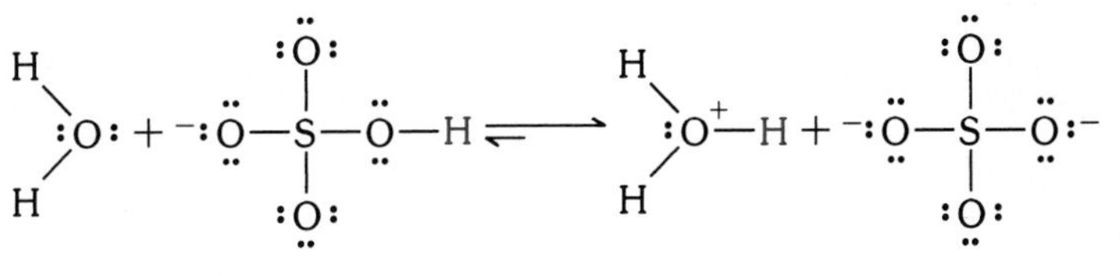

or,

$$H_2O + HSO_4^- \rightleftharpoons H_3O^+ + SO_4^{2-}$$

Hydrogen sulfate ion   Sulfate ion

Sulfuric acid forms when sulfur trioxide, SO_3, dissolves in water.

$$SO_3(g) + H_2O \longrightarrow H_2SO_4(aq)$$

Sulfur trioxide is an air pollutant that reacts with the moisture in air, so this reaction is one of many that causes the formation of acid rain (see Special Topic 8.1, page 168, on acid rain).

Sulfuric acid is sold as a 96% solution, which is nearly always diluted before being used. The 96% solution is a dense, oily liquid that must be handled with great care, because it combines with water *very* exothermically and it attacks skin and clothing voraciously. It can be safely handled, of course, and it is the "workhorse" acid for industrial uses, so much so that a country's standard of living can be judged by its annual per capita consumption of sulfuric acid.

Anions of Phosphoric Acid Occur Widely in Cell Fluids. Phosphoric acid, H_3PO_4, is the only common, inorganic, **triprotic acid,** one that can release three H^+ ions. Like the ionization of sulfuric acid, that of phosphoric acid occurs in steps, each one more difficult than the previous.

$$H_2O + H_3PO_4(aq) \rightleftharpoons H_3O^+(aq) + H_2PO_4^-(aq)$$

Phosphoric   Dihydrogen
acid   phosphate ion

$$H_2O + H_2PO_4^-(aq) \rightleftharpoons H_3O^+(aq) + HPO_4^{2-}(aq)$$

Monohydrogen
phosphate ion

$$H_2O + HPO_4^{2-}(aq) \rightleftharpoons H_3O^+(aq) + PO_4^{3-}(aq)$$

Phosphate ion

Phosphoric acid is classified as a *moderate* acid. Its percentage ionization is roughly 27% in a dilute solution, which is too low to be a strong acid but high enough to make it a good conductor of electricity in water. It is important to learn the names and formulas of the three ions available from phosphoric acid, because the phosphate ion system occurs widely in the body.

> Salts of phosphoric acid are important fertilizers. One salt, Na_3PO_4 (trisodium phosphate), is a powerful cleanser, but wear rubber gloves when using it.

TABLE 7.1
Common Acids[a]

Acid	Formula	Percentage Ionization
Strong Acids		
Hydrochloric acid	HCl	>90
Hydrobromic acid	HBr	>90
Hydriodic acid	HI	>90
Nitric acid	HNO_3	>90
Sulfuric acid[b]	H_2SO_4	>60
Moderate Acids		
Phosphoric acid	H_3PO_4	27
Sulfurous acid[c]	H_2SO_3	20
Weak Acids		
Nitrous acid[c]	HNO_2	1.5
Acetic acid	$HC_2H_3O_2$	1.3
Carbonic acid	H_2CO_3	0.2

[a] Data are for 0.1 *M* solutions of the acids in water at room temperature.

[b] In a 0.05 *M* solution. *Concentrated* sulfuric acid (99%) is particularly dangerous not only because it is a strong acid but also because it is a powerful dehydrating agent. This action generates considerable heat at the reaction site, and at higher temperatures sulfuric acid becomes even more dangerous. Moreover, concentrated sulfuric acid is a thick, viscous liquid that does not wash away from skin or fabric very quickly.

[c] Like carbonic acid, an unstable acid. A component of acid rain.

SPECIAL TOPIC 7.1   THE CARBONIC ACID SYSTEM

In the molecule of carbonic acid, H_2CO_3, the two hydrogens are held by separate oxygens. They are held less strongly, however, than the two hydrogens on oxygen in the water molecule. The reason is the extra O atom on the C=O group in the middle of the carbonic acid molecule. This oxygen atom makes the C=O group an electronegative group and therefore able to pull electron density from the oxygen atoms of the H—O groups. This weakens the H—O bonds, so the hydrogens more easily transfer to water molecules during a collision such as the following:

Carbonic acid   Bicarbonate ion

Now that there is a negative charge (on the bicarbonate ion), the second hydrogen is much harder to release, but the following collision, if it is violent enough, can bring about the transfer of a proton as indicated.

Bicarbonate ion   Carbonate ion

Carbonic acid.

Bicarbonate ion.

Carbonate ion.

Carbonic Acid Is Known Only in Solution. Carbonic acid, H_2CO_3, is a weak diprotic acid that is unusual because it is unstable. Its instability, however, is an important property when the body has to manage one of the respiratory gases, carbon dioxide, as we will see. When carbon dioxide dissolves in water, the following equilibrium forms in which roughly 1 molecule of $CO_2(aq)$ in 400 has reacted with water to form carbonic acid.

$$CO_2(aq) + H_2O \rightleftharpoons H_2CO_3(aq)$$
Carbonic acid

Then a small fraction of the carbonic acid molecules ionizes.

$$H_2CO_3(aq) + H_2O \rightleftharpoons H_3O^+ + HCO_3^-(aq)$$
Bicarbonate ion

The bicarbonate ion ionizes to a very slight extent in water.

$$HCO_3^-(aq) + H_2O \rightleftharpoons H_3O^+ + CO_3^{2-}(aq)$$
Carbonate ion

Special Topic 7.1 describes the molecular structures of carbonic acid and its anions.

Table 7.1 summarizes the common aqueous acids. You should memorize the names and formulas of all of the strong and moderate acids on this list. There are very few of them, and

Club soda and beverages such as Perrier water are little more than solutions of carbon dioxide in water. Traces of salts are also present.

once they are learned, you can be fairly certain that any unfamiliar acid you encounter will be a weak acid. It's easier to learn a few strong and moderate acids than several hundred weak acids.

The Strongest Aqueous Bases Are Hydroxides of Metals in Groups IA and IIA. The properties of bases that we mentioned in the previous section are properties of the hydroxide ion. Table 7.2 lists the most common sources of this ion.

Strong bases are those that ionize nearly 100% in water, and sodium hydroxide, NaOH, and potassium hydroxide, KOH, are the most common.

$$NaOH(s) \xrightarrow{water} Na^+(aq) + OH^-(aq)$$

$$KOH(s) \xrightarrow{water} K^+(aq) + OH^-(aq)$$

Two other strong bases are of Group IIA metals. These are magnesium hydroxide, $Mg(OH)_2$, and calcium hydroxide, $Ca(OH)_2$. They ionize essentially 100% in water, but as the data in Table 7.2 show, they are so insoluble in water that even saturated solutions provide only very dilute solutions of hydroxide ions.

$$Ca(OH)_2(s) \rightleftharpoons Ca^{2+}(aq) + 2OH^-(aq)$$

$$Mg(OH)_2(s) \rightleftharpoons Mg^{2+}(aq) + 2OH^-(aq)$$

Whereas both sodium and potassium hydroxides can be prepared in solutions concentrated enough to give severe chemical burns to the skin, calcium and magnesium hydroxide are very mild. Calcium hydroxide is a component of one commercial antacid tablet, and a slurry of magnesium hydroxide in water, called "milk of magnesia," is used as an antacid and a laxative without posing any danger of chemical burn.

The only common weak base is a solution of ammonia in water called aqueous ammonia. When ammonia dissolves in water, a small percentage of its molecules react with water.

$$NH_3(aq) + H_2O \rightleftharpoons NH_4^+(aq) + OH^-(aq)$$
Ammonia   Ammonium ion

A dilute solution (about 5%) of ammonia in water is sold as household ammonia in supermarkets. It's a good cleaning agent, but watch out for its fumes.

7.3 THE CHEMICAL PROPERTIES OF AQUEOUS ACIDS AND BASES

Because they all contribute a common ion, H_3O^+, the common aqueous acids have a number of similar chemical properties.

In this section we will study the reactions of the hydronium ion, H_3O^+. This study will be easier if we use H^+ as substitute for H_3O^+. We'll see why we can do this next.

The Terms *Hydronium Ion, Hydrogen Ion,* and *Proton* Can Be Used Interchangeably When Discussing Acids and Bases. The self-ionization of water was described as the *transfer of a proton* from one molecule of H_2O to another. The proton, of course, is a subatomic particle, but it is also a hydrogen atom that has lost its electron and so has become a *hydrogen ion,* H^+. Thus the terms *proton* and *hydrogen ion* mean the same thing.

Although protons can be *transferred* from one place to another, they have no existence as separate entities in solution. No subatomic particle has. As explained in Special Topic 7.2, H^+ is always held by a covalent bond to a water molecule or to some other species, such as a molecule of acetic acid or hydrogen chloride. In aqueous systems, a hydrogen ion or proton, however, is particularly easily available (by transfer) from a hydronium ion, as we will often see in this section. Chemists, therefore, frequently use the term "hydrogen ion" in place of "hydronium ion." We will do this, too, because it simplifies chemical equations; it lets us use H^+ to represent H_3O^+.

Household lye is sodium hydroxide. It's imperative to keep it out of reach of children.

You'll sometimes see aqueous ammonia called "ammonium hydroxide," but it's better not to use this term. The compound NH_4OH is unknown as a pure compound.

In the electron-dot symbolism of Section 3.4, the formation of the water molecule from its atoms can be expressed as follows, where each small x represents a valence-shell electron of the oxygen atom. (We'll ignore the geometrical aspects for the moment.)

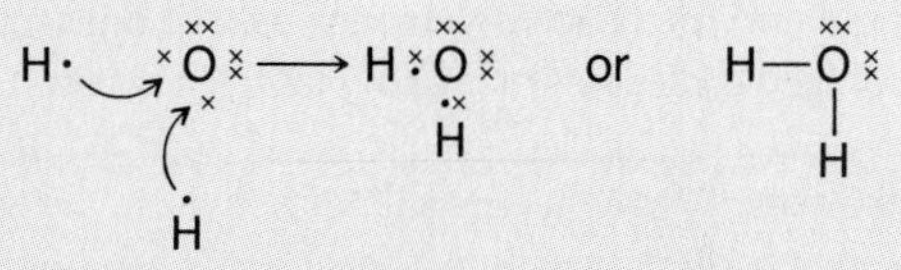

In the resulting water molecule, oxygen still has two unshared pairs of electrons. If we now imagine a bare proton, H^+, coming in to one of these pairs, we can see how a new covalent bond can form:

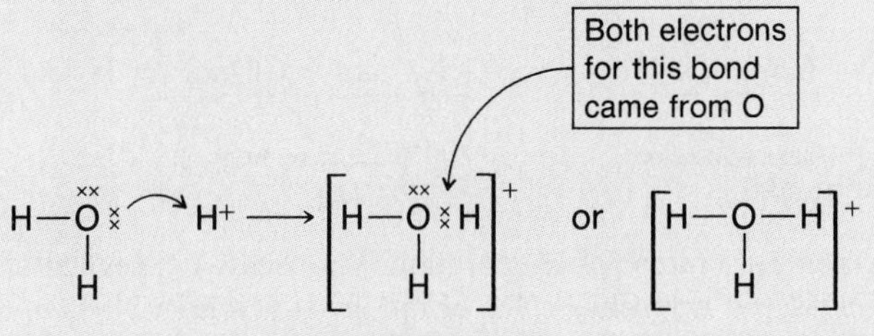

Both electrons for this bond came from O

Hydronium ion

As the accompanying figure shows, the hydronium ion assumes an approximately tetrahedral geometry in accordance with VSEPR theory (Special Topic 3.2).

When both electrons for a covalent bond are provided by one of the atoms, the bond is sometimes called a **coordinate covalent bond.** Once it forms, however, the system completely forgets the origin of the bond. It is just another covalent bond. Thus the three O—H bonds in H_3O^+ are exactly alike. Moreover, all are relatively weak. Any of the three H atoms can break away as H^+ and transfer to some proton-accepting particle with an unshared pair of electrons. Because the energy needed for this transfer is relatively small, we are justified to think of H_3O^+ as the equivalent of H^+ and to use their names and formulas interchangeably. As we will often see, this simplifies many equations.

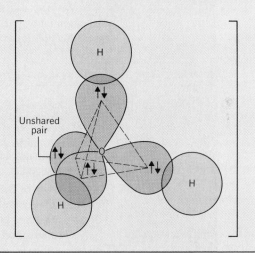

Before we study particular reactions, we will now learn how to write a special equation, the *net ionic equation,* that is helpful when we work largely with ions instead of molecules or other formula units.

TABLE 7.2
Common Bases

Base	Formula	Solubility[a]	Percentage Ionization (%)
Strong Bases			
Sodium hydroxide	NaOH	109	>90 (0.1 *M* solution)
Potassium hydroxide	KOH	112	>90 (0.1 *M* solution)
Calcium hydroxide	$Ca(OH)_2$	0.165	100 (saturated solution)
Magnesium hydroxide	$Mg(OH)_2$	0.0009	100 (saturated solution)
Weak Base			
Aqueous ammonia	NH_3	89.9	1.3 (18 °C)[b]

[a] Solubilities are in grams of solute per 1000 g of water at 20 °C except where otherwise noted.

[b] The ionization referred to here is the forward reaction of the following chemical equilibrium:

$$NH_3(aq) + H_2O \rightleftharpoons NH_4^+(aq) + OH^-(aq)$$

We say *molecular equation* even though some of the chemicals in the equation might be ionic.

A Net Ionic Equation Omits Spectator Particles. A conventional equation is called a **molecular equation;** it shows all of the substances in the full, complete chemical formulas we would need in order to plan an actual experiment. The molecular equation for the reaction of hydrochloric acid and sodium hydroxide, for example, is

$$HCl(aq) + NaOH(aq) \longrightarrow NaCl(aq) + H_2O$$

As we now know, however, $HCl(aq)$ is really $H^+(aq)$ and $Cl^-(aq)$; and $NaOH(aq)$ is actually $Na^+(aq)$ and $OH^-(aq)$. One is a *strong* acid and the other is a *strong* base, so both must be fully ionized in solution. We don't have nonionized molecules of HCl or NaOH in solution. Now let's consider the status of the products.

Nearly all salts are strong electrolytes. Virtually no salt dissolves in water without separating almost entirely in solution. One product, $NaCl(aq)$, therefore, is fully ionized, so we must show this in the ionic equation. The other product is water, a nonelectrolyte, so its molecules are not shown as separated ions in an equation.

We now use these facts about the reactants and products to expand the molecular equation into what is called the **ionic equation.** We do this by replacing the formulas of anything present as ions by the actual formulas of these ions. Thus, the ionic equation for our example is

$$\underbrace{H^+(aq) + Cl^-(aq)}_{} + \underbrace{Na^+(aq) + OH^-(aq)}_{} \longrightarrow \underbrace{Na^+(aq) + Cl^-(aq)}_{} + H_2O$$

| These came from HCl(aq) | These came from NaOH(aq) | These came from NaCl(aq) | Not ionized |

Writing an ionic equation is only a scratch-paper operation, because we next cancel all of the formulas that appear identically on opposites sides of the arrow. Neither $Na^+(aq)$ nor $Cl^-(aq)$ changes, so there is no reason to retain them; our goal is an equation that gives attention just to the actual chemical event. These ions, $Na^+(aq)$ and $Cl^-(aq)$, do serve one function; they contribute to overall electrical neutrality for their respective compounds. But in this reaction they are nothing more than *spectator particles*.

Sometimes we don't cancel, but only reduce in number. If, for example, an equation has

$$\cdots + 4H_2O \longrightarrow \cdots + 2H_2O$$

we can simplify it to

$$\cdots + 2H_2O \longrightarrow \cdots$$

We now cancel the spectator particles, whether they are ions or molecules, from the ionic equation. Finally, we rewrite what remains as the **net ionic equation,** an equation that shows only what reacts and forms:

$$H^+(aq) + OH^-(aq) \longrightarrow H_2O(aq)$$

This net ionic equation, one of the most important that we will study, describes the fundamental chemistry of the neutralization of any strong acid by any metal hydroxide in water.

Net Ionic Equations Must Be Balanced Both Electrically and Materially. We have **material balance** when the numbers of atoms of each element, regardless of how they might be chemically present, are the same on both sides of the arrow. We have **electrical balance** when the algebraic sum of the charges to the left of the arrow equals the sum of the charges to the right. No ionic equation is balanced unless both conditions exist.

EXAMPLE 7.1   WRITING A NET IONIC EQUATION

Problem: Sulfuric acid sometimes has to be neutralized by sodium hydroxide. The reaction can be carried out to produce sodium sulfate, $Na_2SO_4(aq)$, and water. Write the molecular, ionic, and net ionic equations.

Solution: We always start with a molecular equation, and to write it we put down the formulas of the reactants and products in the conventional manner, and then we balance.

The molecular equation is

$$H_2SO_4(aq) + 2NaOH(aq) \longrightarrow Na_2SO_4(aq) + 2H_2O$$

Next, we have to recall the following facts:

$H_2SO_4(aq)$ means $2H^+(aq) + SO_4^{2-}(aq)$ (This is a strong, fully ionized acid.)

$2NaOH(aq)$ means $2Na^+(aq) + 2OH^-(aq)$ (This is a strong, fully ionized metal hydroxide.)

$Na_2SO_4(aq)$ means $2Na^+(aq) + SO_4^{2-}(aq)$ (This is a salt, and (aq) tells us that it's in solution; hence, it's fully ionized.)

$2H_2O$ means $2H_2O$ (No breaking up occurs with this nonelectrolyte.)

These facts let us transform the molecular equation into the ionic equation:

$$2H^+(aq) + SO_4^{2-}(aq) + 2Na^+(aq) + 2OH^-(aq) \longrightarrow 2Na^+(aq) + SO_4^{2-}(aq) + 2H_2O$$

Formulas must be of the same physical state before they can be canceled. We could not cancel, for example, HCl(g) by HCl(aq), because their states are different.

Next we identify those particles that appear identically on opposite sides of the arrow, the spectators, and cancel them.

$$2H^+(aq) + \cancel{SO_4^{2-}(aq)} + \cancel{2Na^+(aq)} + 2OH^-(aq) \longrightarrow \cancel{2Na^+(aq)} + \cancel{SO_4^{2-}(aq)} + 2H_2O$$

This leaves us with

$$2H^+(aq) + 2OH^-(aq) \longrightarrow 2H_2O$$

We have both a material and an electrical balance, but we should note that we can divide all of the coefficients by 2 and convert them to smaller whole numbers. Thus we have reached our goal. The final net ionic equation is

$$H^+(aq) + OH^-(aq) \longrightarrow H_2O$$

In other words, the only chemical event that occurs when we mix sodium hydroxide and sulfuric acid in the ratios of the molecular equation is the reaction of $H^+(aq)$ with $OH^-(aq)$. As we said earlier, this is the neutralization of any strong acid by a metal hydroxide.

PRACTICE EXERCISE 1   Write the molecular, the ionic, and the net ionic equation for the neutralization of nitric acid by potassium hydroxide. A water-soluble salt, $KNO_3(aq)$, and water form.

Hydrogen Ions React with Bicarbonate Ions to Give CO_2 and H_2O. One of the important reactions used by the body to control the acid–base balance in the blood is the reaction of the bicarbonate ion with the hydrogen ion. All metal bicarbonates react the same way with strong, aqueous acids, and the products are carbon dioxide, water, and a salt. Sodium bicarbonate and hydrochloric acd, for example, react as follows:

$$HCl(aq) + NaHCO_3(aq) \longrightarrow CO_2(g) + H_2O + NaCl(aq)$$

Potassium bicarbonate and hydrobromic acid give a similar reaction.

$$HBr(aq) + KHCO_3(aq) \longrightarrow CO_2(g) + H_2O + KBr(aq)$$

What actually forms initially is not CO_2 and H_2O but H_2CO_3, carbonic acid. Almost all of it, however, promptly decomposes to CO_2 and H_2O.

Notice that the salt will always be a combination of the cation of the bicarbonate (Na^+ or K^+ in our examples) and the anion of the acid (Cl^- or Br^- in our examples).

EXAMPLE 7.2   WRITING EQUATIONS FOR THE REACTIONS OF BICARBONATES WITH STRONG ACIDS

Problem: What are the molecular and the net ionic equations for the reaction of potassium bicarbonate with hydriodic acid? (Assume the salt that forms is soluble in water.)

Solution: First, we figure out that the salt is KI, made of the cation from the bicarbonate and the anion from the acid. This enables us to write the molecular equation:

$$KHCO_3(aq) + HI(aq) \longrightarrow CO_2(g) + H_2O + KI(aq)$$

To prepare the ionic equation, we analyze each of the formulas in the molecular equation.

$KHCO_3(aq)$ means $K^+(aq)$ and $HCO_3^-(aq)$ (As we were told.)

$HI(aq)$ means $H^+(aq) + I^-(aq)$ (Because this is a fully ionized acid.)

$KI(aq)$ means $K^+(aq) + I^-(aq)$ (Because we treat all water-soluble salts as fully ionized.)

$CO_2(g)$ and H_2O stay the same   (Neither is ionized.)

Using these facts, we expand the molecular equation into the ionic equation.

$$[K^+(aq) + HCO_3^-(aq)] + [H^+(aq) + I^-(aq)] \longrightarrow CO_2(g) + H_2O + [K^+(aq) + I^-(aq)]$$

The $K^+(aq)$ and the $I^-(aq)$ cancel. (Draw in the cancel lines yourself.) This leaves the following net ionic equation.

$$H^+(aq) + HCO_3^-(aq) \longrightarrow CO_2(g) + H_2O$$

It is balanced both materially and electrically.

The equation produced by Example 7.2

$$H^+(aq) + HCO_3^-(aq) \longrightarrow CO_2(g) + H_2O$$

is the same net ionic equation for the reaction of all metal bicarbonates with all strong, aqueous acids.

Because this reaction destroys the hydrogen ions of the acid, it is also properly called an acid neutralization. In fact, the familiar "bicarb" used as a home remedy for acid stomach is nothing more than sodium bicarbonate. Stomach acid is roughly 0.1 M HCl, and the bicarbonate ion neutralizes this acid by the reaction we have just studied. An overdose of "bicarb" must be avoided because it can cause a medical emergency involving the respiratory gases. Another use of sodium bicarbonate is to make an isotonic solution to be given intravenously to neutralize acid in the blood. For still another use, see Special Topic 7.3.

PRACTICE EXERCISE 2   Write the molecular, ionic, and net ionic equations for the reaction of sodium bicarbonate with sulfuric acid in which sodium sulfate, $Na_2SO_4(aq)$, is one of the products.

The Hydrogen Ion Also Reacts with the Carbonate Ion to Give CO$_2$ and H$_2$O. Metal carbonates are attacked by hydrogen ions, to give carbon dioxide, water, and a salt made of the cation of the carbonate and the anion of the acid. These are the same kinds of products that bicarbonates give, but the mole relationship is different. Mole for mole, the CO_3^{2-} ion can neutralize twice as much H^+ as the HCO_3^- ion, as we'll see in the next example.

Besides aspirin, Alka-Seltzer® and similar tablets contain citric acid (a solid acid), and solid sodium bicarbonate. The response of these tablets to water illustrates the importance of water as a solvent in the reactions of ions. In the crystalline materials, ions are not free to move. As soon as these tablets hit water, however, the ions become mobile and so are free to react.

Citric acid is a triprotic acid, and we can represent it as

H_3Cit, where *Cit* stands for the citrate ion, an ion with a charge of 3−. The hydrogen ions liberated in aqueous citric acid react with the bicarbonate ions made free as sodium bicarbonate dissolves. This reaction gives the CO_2 that fizzes out of the solution.

$$H^+(aq) + HCO_3^-(aq) \longrightarrow CO_2(g) + H_2O$$

EXAMPLE 7.3   WRITING EQUATIONS FOR THE REACTIONS OF METAL CARBONATES WITH STRONG, AQUEOUS ACIDS

Problem: Sodium carbonate, Na_2CO_3, neutralizes hydrochloric acid and forms sodium chloride, carbon dioxide, and water. Write the molecular, ionic, and net ionic equations for this reaction. Assume that the reaction occurs in an aqueous solution.

Solution: We first write the formulas into a conventional, molecular equation and balance it.

$$2HCl(aq) + Na_2CO_3(aq) \longrightarrow CO_2(g) + H_2O + 2NaCl(aq)$$

Then we analyze each of the formulas in this equation to see how to use them in the ionic equation.

$2HCl(aq)$ means $2H^+(aq) + 2Cl^-(aq)$	(The acid is strong and fully ionized.)
$Na_2CO_3(aq)$ means $2Na^+(aq) + CO_3^{2-}(aq)$	(This is a salt and it is written with (aq). So it's in solution and fully ionized.)
$2NaCl(aq)$ means $2Na^+(aq) + 2Cl^-(aq)$	(This soluble salt is treated, like all salts dissolved in water, as fully ionized.)

$CO_2(g)$ and H_2O remain unchanged.

Now we can expand the molecular equation to the ionic equation.

$$[2H^+(aq) + 2Cl^-(aq)] + [2Na^+(aq) + CO_3^{2-}(aq)]$$
$$\longrightarrow CO_2(g) + H_2O + [2Na^+(aq) + 2Cl^-(aq)]$$

We can cancel the $2Na^+(aq)$ and the $2Cl^-(aq)$ from both sides of the equation, which leaves the following net ionic equation.

$$2H^+(aq) + CO_3^{2-}(aq) \longrightarrow CO_2(g) + H_2O$$

Notice in Example 7.3 that one carbonate can neutralize two hydrogen ions, twice as many as are neutralized by one bicarbonate ion. The net ionic equation that we devised in Example 7.3

$$2H^+(aq) + CO_3^{2-}(aq) \longrightarrow CO_2(g) + H_2O$$

is the same for the reactions of all of the carbonates of the Group IA metals with all strong, aqueous acids.

PRACTICE EXERCISE 3   Write the molecular, ionic, and net ionic equations for the reaction of aqueous potassium carbonate, $K_2CO_3(aq)$, with sulfuric acid to give potassium sulfate, $K_2SO_4(aq)$, a water-soluble salt, and the other usual products.

Adding a few drops of strong acid to a rock sample is a field test for carbonate rocks. A positive test is an odorless fizzing reaction.

Most carbonates are water-insoluble compounds. Calcium carbonate, $CaCO_3$, for example, is the chief substance in limestone and marble. Despite its insolubility in water, calcium carbonate reacts readily with strong, aqueous acids (with their hydrogen ions, of course).

$$CaCO_3(s) + 2HCl(aq) \longrightarrow CO_2(g) + H_2O + CaCl_2(aq)$$

When water-insoluble compounds are involved in a reaction with an aqueous solution of something, we have to write their entire formulas in net ionic equations, so the net ionic equation is

$$CaCO_3(s) + 2H^+(aq) \longrightarrow CO_2(g) + H_2O + Ca^{2+}(aq)$$

PRACTICE EXERCISE 4   Dolomite, a limestone-like rock, contains both calcium and magnesium carbonates. Magnesium carbonate is attacked by nitric acid. The salt that forms is water soluble. Write the molecular, ionic, and net ionic equations for this reaction.

The Hydrogen Ion Reacts with the Hydroxide Ion to Give Water. We already discussed the reaction of strong acids with hydroxides when we learned how to write net ionic equations. Only the Group IA hydroxides are very soluble in water, so their reactions in solution with strong, aqueous acids have net ionic equations that differ slightly from those of the water-insoluble metal hydroxides. If we let M stand for a Group IA metal, we can write the reactions of aqueous solutions of their hydroxides with a strong acid such as hydrochloric acid by the following general equation.

$$MOH(aq) + HCl(aq) \longrightarrow MCl(aq) + H_2O$$

The net ionic equation is

$$OH^-(aq) + H^+(aq) \longrightarrow H_2O$$

If we now use M for any metal in Group IIA (except beryllium), the equations are

$$M(OH)_2(s) + 2HCl(aq) \longrightarrow MCl_2(aq) + 2H_2O$$

or

$$M(OH)_2(s) + 2H^+(aq) \longrightarrow M^{2+}(aq) + 2H_2O$$

PRACTICE EXERCISE 5   When milk of magnesia is used to neutralize hydrochloric acid (stomach acid), solid magnesium hydroxide in the suspension reacts with the acid. Write the molecular and net ionic equations for this reaction.

The Hydrogen Ion Is Neutralized by Ammonia. An aqueous solution of ammonia is an excellent reagent for neutralizing acids. The unshared pair of electrons on nitrogen in ammonia can be used to make a covalent bond to H^+, furnished by an acid. When both electrons for a covalent bond are furnished by the same atom, the bond is called a **coordinate covalent bond.** The coordination of H^+ by ammonia neutralizes the acid. For example,

$$NH_3(aq) + HCl(aq) \longrightarrow NH_4Cl(aq)$$

or

$$NH_3(aq) + H^+(aq) \longrightarrow NH_4^+(aq)$$

All ammonium salts are soluble in water, so they liberate NH_4^+ ions in aqueous solutions. Many biochemicals have ammonia-like molecules that also neutralize hydrogen ions.

PRACTICE EXERCISE 6   Write the molecular and net ionic equations for the reaction of aqueous ammonia with (a) HBr(aq) and (b) $H_2SO_4(aq)$.

The Hydrogen Ion Oxidizes Many Metals.   Nearly all metals are attacked more or less readily by the hydrogen ion. Metal atoms are oxidized to metal ions, and hydrogen ions are reduced to H_2 gas. The molecular equation for the reaction will show that another product is the salt of the cation of the metal and the anion of the acid. Zinc, for example, reacts with hydrochloric acid as follows:

Remember: A loss of electrons is oxidation, and a gain of electrons is reduction.

$$Zn(s) + 2HCl(aq) \longrightarrow H_2(g) + ZnCl_2(aq)$$

The net ionic equation is

$$Zn(s) + 2H^+(aq) \longrightarrow H_2(g) + Zn^{2+}(aq)$$

Aluminum is also attacked by acids. For example, its reaction with nitric acid can be written as follows:

$$2Al(s) + 6HNO_3(aq) \longrightarrow 2Al(NO_3)_3(aq) + 3H_2(g)$$

The net ionic equation is

$$2Al(s) + 6H^+(aq) \longrightarrow 2Al^{3+}(aq) + 3H_2(g)$$

PRACTICE EXERCISE 7   Write the molecular and the net ionic equations for the reaction of magnesium with hydrochloric acid.

Metals Can Be Arranged in an Activity Series.   Metals differ greatly in their tendencies to react with aqueous hydrogen ions. The Group IA metals like sodium and potassium, for example, include the most reactive metals of all. They not only reduce protons taken from hydronium ions, they also reduce protons taken from water molecules. No acid need be present. The following reaction of sodium metal with water is extremely violent, and it should never be attempted except by an experienced chemist working with safety equipment, including a fire extinguisher. (The hydrogen gas generally ignites spontaneously as it forms.)

$$2Na(s) + 2H_2O \longrightarrow 2NaOH(aq) + H_2(g)$$

If this reaction is violent in water, it would be even more violent in aqueous acids in which the concentrations of H^+ ions are much higher.

Gold, silver, and platinum, in contrast, are stable not only toward water but also toward hydrogen ions.

With such a vast difference in the reactivities of metals toward acids, it shouldn't be surprising that the metals can be arranged in an order of reactivity. The result is called the

TABLE 7.3
The Activity Series of the Common Metals

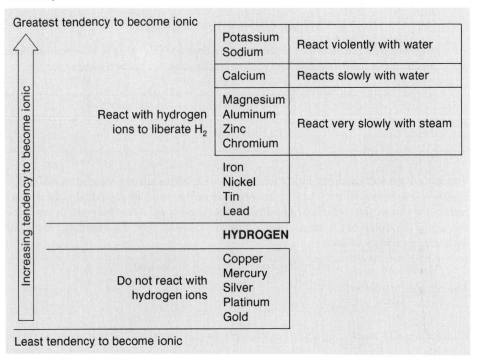

activity series of the metals, and it's given in Table 7.3. Atoms of any metal above hydrogen in the series can transfer electrons to H^+, either from H_2O or from H_3O^+, to form hydrogen gas, and the metal atoms change to metal ions. Tin and lead, however, react only very slowly. The metals below hydrogen in the activity series do not transfer electrons to H^+.

The rate of the reaction of an acid with a metal depends on the acid as well as the metal. When compared at the same molar concentrations, strong acids react far more rapidly than weak acids, as Figure 7.4 shows. These differences reflect the differences in percentage ionizations, because the actual reaction, as we have said, is with the hydrogen ion. When the concentration of hydrogen ion is high, as it can be when the acid is strong, the reaction is vigorous. In 1 M HCl, the concentration of $H^+(aq)$ is also 1 M, because for each HCl one $H^+(aq)$ is released. However, in 1 M $HC_2H_3O_2$, acetic acid (a weak acid), the actual concentration of $H^+(aq)$ is closer to 0.004 M, which is about $\frac{1}{250}$ as much. No wonder the liveliness of the reaction pictured in Figure 7.4c, the reaction of zinc with 1 M acetic acid, is much less than in Figure 7.4a, the reaction with 1 M HCl. In Figure 7.4b, the reaction is with 1 M H_3PO_4, a moderate acid, and the vigor of the reaction is somewhere in between that of the other two.

7.4   BRØNSTED ACIDS AND BASES

Acid–base reactions are proton transfers between any donors and acceptors of H^+.

In the previous section we focused on the properties of the hydronium ion as a proton donor, but in Section 7.2 we learned that H_3O^+ itself forms from still other proton donors, like HCl(g). We also saw that many chemical species can accept protons, not just OH^- but also HCO_3^- and NH_3, for example. The central feature of acid–base reactions is thus the relocation of just this tiny particle, the proton. Something donates it and something accepts it. A Danish chemist,

FIGURE 7.4
Relative hydrogen ion concentrations and the reactivity of zinc. (a) The acid is HCl(aq), a strong, fully ionized acid. (b) The acid is $H_3PO_4(aq)$, a moderate, partly ionized acid. (c) The acid is acetic acid, $HC_2H_3O_2(aq)$, a weak, poorly ionized acid. Although the molar concentrations of the acids are the same in each tube, the actual molar concentrations of their hydrogen ions are greatly different, being highest in (a), where the bubbles of hydrogen are evolving the most vigorously, next highest in (b), and lowest in (c).

Johannes Brønsted (1879–1947), was the first to take advantage of this observation to suggest the following very simple yet quite broad definitions of acids and bases.

Brønsted Acids and Bases

Acids are proton donors.
Bases are proton acceptors.

According to Brønsted, anything that donates a proton in a specific reaction is an acid, and anything that accepts one is a base. Thus, even the water molecule can be a base, because in the reaction of gaseous HCl with water, H_2O takes the proton offered by HCl(g).

$$H\text{---}Cl(g) + H_2O \rightleftharpoons Cl^-(aq) + H_3O^+$$

Brønsted   Brønsted
acid   base

The forward reaction in this equilibrium goes essentially to completion. This is why $H\text{---}Cl(g)$ is called a strong acid. Almost no proton transfer occurs as the reverse reaction. If it did, then we would have to label Cl^- as the base and H_3O^+ as the acid. The inability of Cl^- to take a proton even from H_3O^+ surely justifies our calling it a weak base, so weak in fact that we never even think of it as a base at all. It should be apparent by now that proton acceptors and donors differ widely in their abilities to give up or receive protons. It should also be evident that we need some reference of comparison before we say what is strong and what is weak.

Brønsted Acids and Bases Vary Widely in Strengths. Because water is the most common solvent used for acids and bases, it naturally provides the reference of comparison we need. In the Brønsted concept, a **strong acid** like H—Cl(g) is a strong donor of protons *to water*. A **weak acid** like acetic acid, on the other hand, is a poor proton donor *to water*. Water itself is an even weaker acid, so weak that normally we don't even think of it in such a way. Recall that only $1 \times 10^{-7}\%$ of all water molecules are at any moment self-ionized into OH^- and H_3O^+ ions.

Our reference for a strong base in water is the hydroxide ion. It binds protons very strongly (as H—OH). Thus when dissolved in water, the common strong bases, like NaOH, are called strong because they all release essentially 100% of their strongly basic hydroxide ions. The chloride ion, on the other hand, does not readily accept a proton, so it is a weak base.

Notice carefully now that the particle left behind, Cl^-, when the strong acid, HCl(g), gives up its proton is a very weak base. A strong acid thus leaves behind a weak base when the proton transfers. Notice that when a water molecule gives up a proton, the particle left behind, OH^-, is a strong base. So this weak acid, H_2O, is related to a strong base, OH^-. You can see that there are reciprocal relationships here. Let's summarize them, and then see how they can be used.

1. If an acid is strong, the particle that remains when it donates a proton is a weak base. (For example, HCl is a strong acid and Cl^- is a weak base; H_3O^+ is a strong acid and H_2O is a weak base).

2. If a base is strong, what forms when it accepts a proton will be a weak acid. (Thus OH^- is a strong base, so H_2O is a weak acid.)

3. If an acid is weak, what remains when it loses a proton will be a fairly strong base. (Thus HCO_3^- is a weak acid, and CO_3^{2-} is a strong base. But H_2O is a very weak acid, so OH^- is a very strong base.)

4. If a base is weak, what forms when it accepts a proton will be a fairly strong acid. (Thus H_2O is a weak base, and H_3O^+ is a strong Brønsted acid. The Cl^- ion is a very weak base, so H—Cl(g) is a very strong acid.)

Let us now work an example to show how these relationships plus our knowledge of which are the five strong acids (Table 7.1) enable us to figure out if a particular species is a strong or weak Brønsted acid or base.

EXAMPLE 7.4   PREDICTING IF AN ION OR MOLECULE IS A WEAK BRØNSTED ACID

Problem: Lactic acid is the acid responsible for the tart taste of sour milk. Is lactic acid a strong or a weak acid?

Solution: The list of strong acids does not include lactic acid. Hence, it is a weak acid. It's as simple as that (and we'd err very seldom).

EXAMPLE 7.5   DEDUCING IF AN ION OR MOLECULE IS A STRONG OR A WEAK BRØNSTED BASE

Problem: Is the bromide ion a strong or a weak Brønsted base?

Solution: When the question deals with a potential *base,* we find the answer in a round about fashion. We accept this because the alternative would be to memorize a rather extensive list of the stronger Brønsted bases. Here's how to do it.

Pretend that the potential base actually functions as one. Write the formula of what forms if it accepts H^+. If Br^- were a base and accepted H^+, it would have to change to HBr, which, in water, is hydrobromic acid. Now comes the crucial question. Is hydrobromic acid a strong acid? We have to know the list, and HBr is a *strong* acid. Therefore, we know it easily gives up a proton. And thus we know that what remains when the proton so readily leaves HBr, Br^-, has to be a poor proton binder. So our answer is that Br^- is a weak Brønsted base.

EXAMPLE 7.6   DEDUCING IF AN ION OR A MOLECULE IS A STRONG OR A WEAK BRØNSTED BASE

Problem: Is the phosphate ion, PO_4^{3-}, a strong or a weak Brønsted base?

Solution: Using the strategy described in Example 7.5, we pretend that this ion actually is a base, a proton acceptor. So we give it a proton, and write the result, HPO_4^{2-}. This, as a Brønsted acid, isn't on our list of strong acids, so we conclude it is a weak acid. This means that PO_4^{3-} is a good proton binder (holding the proton as HPO_4^{2-}). A good proton binder is a strong base, so our answer to the question is that PO_4^{3-} is a strong base.

PRACTICE EXERCISE 8   Classify the following particles as strong or as weak Brønsted acids.

(a) HSO_3^- (b) HCO_3^- (c) $H_2PO_4^-$

PRACTICE EXERCISE 9   Classify the following ions as strong or as weak Brønsted bases.

(a) I^- (b) NO_3^- (c) CN^- (d) NH_2^-

The relative strengths of several Brønsted acids and bases are given in Table 7.4. Above the hydronium ion, all acids are about equally strong in water. All bases above water are roughly equally weak in water. In fact, they're not really bases at all in water. And the acids listed below water are not really acids in water. They're neutral. The bases below the hydroxide ion are so strong they cannot exist in water. (They react with water, take protons, and generate hydroxide ions.)

TABLE 7.4
Relative Strengths of Some Brønsted Acids and Bases

Brønsted Acid		Brønsted Base	
Name	Formula	Name	Formula
Perchloric acid	$HClO_4$	Perchlorate ion	ClO_4^-
Hydrogen iodide	HI	Iodide ion	I^-
Hydrogen bromide	HBr	Bromide ion	Br^-
Sulfuric acid	H_2SO_4	Hydrogen sulfate ion	HSO_4^-
Hydrogen chloride	HCl	Chloride ion	Cl^-
Nitric acid	HNO_3	Nitrate ion	NO_3^-
HYDRONIUM ION	H_3O^+	WATER	H_2O
Hydrogen sulfate ion	HSO_4^-	Sulfate ion	SO_4^{2-}
Phosphoric acid	H_3PO_4	Dihydrogen phosphate ion	$H_2PO_4^-$
Acetic acid	$HC_2H_3O_2$	Acetate ion	$C_2H_3O_2^-$
Carbonic acid	H_2CO_3	Bicarbonate ion	HCO_3^-
Ammonium ion	NH_4^+	Ammonia	NH_3
Bicarbonate ion	HCO_3^-	Carbonate ion	CO_3^{2-}
WATER	H_2O	HYDROXIDE ION	OH^-
Methyl alcohol	CH_3OH	Methoxide ion	CH_3O^-
Ammonia	NH_3	Amide ion	NH_2^-
Hydrogen	H_2	Hydride ion	H^-

Increasing acid strength (upward). Increasing base strength (downward).

The Ammonium Ion Is a Brønsted Acid.   The ammonium ion occupies a special place in our study because many biochemicals such as proteins have a molecular part that is very much like this ion. Although the ammonium ion is a weak acid (Table 7.4), it still is a Brønsted acid, and it can neutralize the hydroxide ion. When we add sodium hydroxide to a solution of ammonium chloride, the following reaction occurs.

$$NaOH(aq) + NH_4Cl(aq) \longrightarrow NH_3(aq) + H_2O + NaCl(aq)$$

The net ionic equation is

$$OH^-(aq) + NH_4^+(aq) \longrightarrow NH_3(aq) + H_2O$$

This reaction neutralizes the hydroxide ion, and it leaves a solution of the weaker base, NH_3. (If the initial solution is concentrated enough, the final solution has a strong odor of ammonia.)

In some medical emergencies, when the blood has become too alkaline or too basic, an isotonic solution of ammonium chloride is administered by intravenous drip. Its ammonium ions can neutralize some of the base in the blood and bring the acid–base balance back to normal.

7.5   SALTS

A very large number of ionic reactions can be predicted from a knowledge of the solubility rules of salts.

Salts, as we said, are ionic compounds whose cations are any except H^+ and whose anions are any except OH^-. A **simple salt** is made of just *two* oppositely charged ions. Examples are $NaCl$, $MgBr_2$, and $CuSO_4$. *Mixed salts* have three or more different ions. Alum, used in water purification, is an example: $K_2SO_4 \cdot Al_2(SO_4)_3 \cdot 24H_2O$. As the formula of alum illustrates, the salt family includes hydrates. Some salts of practical value are given in Table 7.5.

TABLE 7.5
Some Salts and Their Uses

Formula and Name	Uses
$BaSO_4$ Barium sulfate	Used in the "barium cocktail" given prior to X-raying the gastrointestinal tract
$(CaSO_4)_2 \cdot H_2O$ Calcium sulfate hemihydrate (plaster of paris)	
$MgSO_4 \cdot 7H_2O$ Magnesium sulfate heptahydrate (epsom salt)	Purgative
$AgNO_3$ Silver nitrate	Antiseptic and germicide. Used in eyes of infants to prevent gonorrheal conjunctivitis. Photographic film sensitizer
$NaHCO_3$ Sodium bicarbonate (baking soda)	Baking powders. Effervescent salts. Stomach antacid. Fire extinguishers
$Na_2CO_3 \cdot 10H_2O$ Sodium carbonate decahydrate (soda ash, sal soda, washing soda)	Water softener. Soap and glass manufacture
$NaCl$ Sodium chloride	Manufacture of chlorine, sodium hydroxide. Preparation of food
$NaNO_2$ Sodium nitrite	Meat preservative

Salts Can Be Made from Acids and Hydroxides, Carbonates, and Bicarbonates. In the laboratory, salts are obtained whenever an acid is used in any of the following ways. We summarize and review these methods here and show their similarities.

$$\text{Acid} + \text{metal} \longrightarrow \text{a salt} + H_2$$
$$\text{Acid} + \text{metal hydroxide} \longrightarrow \text{a salt} + H_2O$$
$$\text{Acid} + \text{metal carbonate} \longrightarrow \text{a salt} + H_2O + CO_2$$
$$\text{Acid} + \text{metal bicarbonate} \longrightarrow \text{a salt} + H_2O + CO_2$$

If the salt is soluble in water, we have to evaporate the solution to dryness to isolate it. Sometimes, however, the salt precipitates. To predict when to expect this as well as to show still another way to make salts, we turn to the solubility rules for salts.

Solubility Rules for Salts in Water

1. All lithium, sodium, potassium, and ammonium salts are soluble regardless of the counter ion.[1]
2. All nitrates and acetates are soluble, regardless of the counter ion.
3. All chlorides, bromides, and iodides are soluble, except when the counter ion is lead, silver, or mercury(I).
4. Sulfates are soluble except when the counter ion is calcium, strontium, barium, silver, lead (Pb^{2+}), or mercury(I).
5. Salts not in the above categories are generally insoluble or, at best, only slightly soluble.

By *soluble* we mean the ability to form a solution with a concentration of at least 3–5% (w/w).

There are exceptions to these rules, but we will seldom be wrong in applying them. One of the many applications of these rules is to predict possible reactions between salts.

Salts Can Be Made from Other Salts by Double Decomposition Reactions. In addition to the reactions we have already studied to make salts, we can also make them by a "change of partners" reaction called **double decomposition.** The best way to see how this works is by a worked example.

EXAMPLE 7.7   PREDICTING DOUBLE DECOMPOSITION REACTIONS OF SALTS

Problem: What happens if we mix aqueous solutions of sodium sulfate and barium nitrate?

Solution: Because, by the solubility rules, both sodium sulfate and barium nitrate are soluble in water, their solutions contain their separated *ions.* When we pour the two solutions together, four ions experience attractions and repulsions. Hence, we have to examine each possible combination of oppositely charged ions to see which, if any, makes a water-insoluble salt. If we find one, then we can write an equation for the reaction that produces this salt. Here are the possible combinations when Ba^{2+}, NO_3^-, Na^+, and SO_4^{2-} ions intermingle in water.

$$Ba^{2+} + 2NO_3^- \xrightarrow{\ ?\ } Ba(NO_3)_2(s)$$

(This possibility is obviously out, because barium ions and nitrate ions do not precipitate together from water. "All nitrates are soluble.")

---

[1] A *counter ion* is the unnamed ion that is part of the salt. For example, in the lithium salt, LiCl, the counter ion is the chloride ion.

$$2Na^+ + SO_4{}^{2-} \xrightarrow{\ ?\ } Na_2SO_4(s)$$ (This possibility is also out. "All sodium salts are soluble.")

$$Na^+ + NO_3{}^- \xrightarrow{\ ?\ } NaNO_3(s)$$ (No. Again, "All sodium salts are soluble.")

$$Ba^{2+} + SO_4{}^{2-} \xrightarrow{\ ?\ } BaSO_4(s)$$ (Yes. Barium sulfate, $BaSO_4$, is not in any of the categories of water-soluble salts. Therefore we predict it is insoluble.)

A slurry of barium sulfate in flavored water is the barium "cocktail" a patient drinks before having an X-ray taken of the gastrointestinal tract.

Because we predicted that $BaSO_4$ can form a precipitate, we can write a molecular equation, and we'll use some connector lines to show how partners exchange—how *double* decomposition occurs.

$$Ba(NO_3)_2(aq) + Na_2SO_4(aq) \longrightarrow 2NaNO_3(aq) + BaSO_4(s)$$

The net ionic equation, however, is a better way to describe what happens.

$$Ba^{2+}(aq) + SO_4{}^{2-}(aq) \longrightarrow BaSO_4(s)$$

The sodium and nitrate ions are only spectators. To obtain the solid barium sulfate, we would filter the mixture and collect this compound on the filter. If we also wanted the sodium nitrate, we would evaporate the clear filtrate to dryness.

PRACTICE EXERCISE 10   If solutions of sodium sulfide, Na_2S, and copper(II) nitrate, $Cu(NO_3)_2$, are mixed, what, if anything, will happen chemically? Write a molecular and a net equation for any reaction.

One of the many uses of the solubility rules is to understand what it means for water to be called *hard water* and what it means to *soften* such water. These are discussed in Special Topic 7.4.

Reactions of Ions—A Summary.   Our study of the reaction of sodium sulfate and barium nitrate in Example 7.7 illustrated the power of knowing just a few facts for the sake of predicting an enormous number of others with a high probability of success. The following facts summarize those that should now be well learned.

1.   The solubility rules of the salts (because then we can assume that all of the other salts are insoluble).
2.   The five strong acids in Table 7.1 (because then we can assume that all of the other acids, including organic acids, are weak).
3.   The first two strong bases of Table 7.2 (because then we can assume that all of the rest are either weak or are too insoluble in water to matter much).

To summarize, we expect ions to react with each other if any one of the following possibilities is predicted:

1.   A gas forms that (mostly) leaves the solution—it could be hydrogen (from the action of acids on metals), or carbon dioxide (from acids reacting with carbonates or bicarbonates).
2.   An un-ionized, molecular compound forms that remains in solution— it could be water (from acid–base neutralizations), or a weak acid (by the action of H^+ on the acetate ion), or ammonia (by the reaction of OH^- on $NH_4{}^+$).
3.   A precipitate forms—some water-insoluble salt or one of the water-insoluble hydroxides.

SPECIAL TOPIC 7.4   HARD WATER

Ground water that contains magnesium, calcium, or iron ions at a high enough level to interact with ordinary soap to form scum is called **hard water.** These "hardness ions," Ca^{2+}, Mg^{2+}, Fe^{2+}, or Fe^{3+}, are either absent from **soft water** or are present in extremely low concentrations.

The anions that most frequently accompany the hardness ions are SO_4^{2-}, Cl^-, and HCO_3^-. When the bicarbonate ion is the chief anion, the hard water is **temporary hard water.** Otherwise, it is **permanent hard water.** When temporary hard water is heated near its boiling point, as in hot boilers, steam pipes, and instrument sterilizers, the bicarbonate ion breaks down to the carbonate ion.

$$2HCO_3^-(aq) \longrightarrow CO_3^{2-}(aq) + CO_2(g) + H_2O$$

The carbonate ion, unlike HCO_3^-, forms insoluble precipitates with the hardness ions. Using the calcium ion to illustrate:

$$CO_3^{2-}(aq) + Ca^{2+}(aq) \longrightarrow CaCO_3(s)$$

Such insoluble carbonate salts deposit as scaly, pipe clogging material, as the accompanying photograph illustrates.

Water Softening Is Done by Removing the Hardness Ions Hard water can be softened in various ways. Most commonly, excess soap is used. Some scum does form, but then the extra soap does the cleansing work. To avoid the scum altogether, softening agents are added before the soap is used. One common water-softening chemical is sodium carbonate decahydrate, known as washing soda. Its carbonate ions take out the hardness ions as insoluble carbonates by the kind of reaction for which we wrote the previous net ionic equation.

Another home water-softening agent is household ammonia, 5% (w/w) NH_3. We've already learned about the following equilibrium in such a solution.

$$NH_3(aq) + H_2O \rightleftharpoons NH_4^+(aq) + OH^-(aq)$$

In other words, aqueous ammonia has some OH^- ions, and the hydroxides of the hardness ions are not soluble in water. When aqueous ammonia is added to hard water,

therefore, the following reaction occurs (illustrated using the magnesium ion this time):

$$Mg^{2+}(aq) + 2OH^-(aq) \longrightarrow Mg(OH)_2(s)$$

As hydroxide ions are removed by this reaction, more are made available from the ammonia–water equilibrium. (A loss of OH^- ion from this equilibrium is a stress, and the equilibrium shifts to the right in response, as we'd predict using Le Chatelier's principle.)

Still another water-softening technique is to let the hard water trickle through zeolite, a naturally occurring porous substance rich in sodium ions. When the hard water is in contact with zeolite, sodium ions go into the water and the hardness ions leave solution and attach themselves to the zeolite. Later, the hardness ions can be flushed out by letting water that is very concentrated in sodium chloride trickle through the spent zeolite, and this restores the zeolite for reuse. Synthetic, ion-exchange materials are also used to soften water by roughly the same principle.

Perhaps the most common strategy in areas where the water is quite hard is to use synthetic detergents instead of soap. Synthetic detergents do not form scum and precipitates with the hardness ions.

Calcium carbonate deposits in a 2-in. hot water pipe after 2 years of service in northeastern New Jersey. (Courtesy of The Permutit Company, a division of Sybron Corporation.)

PRACTICE EXERCISE 11   When a solution of hydrochloric acid is mixed in the correct molar proportions with a solution of sodium acetate, $NaC_2H_3O_2$, essentially all of the hydronium ion concentration vanishes. What happens and why? Write the net ionic equation.

PRACTICE EXERCISE 12   What, if anything, happens chemically when each pair of solutions is mixed? Write net ionic equations for any reactions that occur.

(a) NaCl and $AgNO_3$ (b) $CaCO_3$ and HNO_3 (c) KBr and NaCl

TABLE 7.6
Equivalents of Ions

Ion	g/mol	g/eq
Na^+	23.0	23.0
K^+	39.1	39.1
Ca^{2+}	40.1	20.1
Mg^{2+}	24.3	12.2
Al^{3+}	27.0	9.0
Cl^-	35.5	35.5
HCO_3^-	61.0	61.0
CO_3^{2-}	60.0	30.0
SO_4^{2-}	96.1	48.1

Concentrations of Individual Ions in Solutions of Several Salts Are Often Given in Equivalents or Milliequivalents per Liter. To close this discussion of salts we will study a concentration expression often used for their individual ions. It is based not on the moles of an ion per liter but on a quantity called an *equivalent* per liter. We'll see why soon.

One **equivalent** of an ion, abbreviated **eq,** is the number of grams of the ion that corresponds to Avogadro's number, one mole, of electrical charges. For example, when the charge is unity, either $1+$ or $1-$, it takes Avogadro's number of ions to have Avogadro's number of electrical charges. Thus 1 eq for ions such as Na^+, K^+, Cl^-, or Br^- is the same as the molar mass of each ion. The molar mass of Na^+ is 23.0 g Na^+/mol, so 1 eq of $Na^+ = 23.0$ g of Na^+. This much sodium ion, 23.0 g of Na^+, contributes Avogadro's number of positive charges. Thus the equivalent weight of the sodium ion is 23.0 g Na^+/eq.

When an ion has a double charge, either $2+$ or $2-$, then the mass of one equivalent equals the molar mass divided by 2. For example, 1 mol of CO_3^{2-} ion $= 60.0$ g of CO_3^{2-}, so 1 eq of CO_3^{2-} ion $= 30.0$ g of CO_3^{2-} ion. This much carbonate ion carries Avogadro's number of negative charge. The extension of this to ions of higher charges should now be obvious. **The equivalent weight of an ion is its formula weight divided by its charge.** You can see this in Table 7.6, which gives equivalent weights for a number of ions.

The advantage of the concept of the equivalent is the simplicity of a $1:1$ ratio. Regardless of the amounts of charges on the individual ions, when cations and anions are present either in an ionic crystal or in a solution, we can always be sure that for every equivalent of positive charge there has to be one equivalent of negative charge. The condition of electrical neutrality in an ionic compound or a solution of ions is that

$$\text{eq of cations} = \text{eq of anions}$$

or,

$$\text{meq of cations} = \text{meq of anions}$$

where the **milliequivalent,** or **meq,** is related to the equivalent by the relationship, 1000 meq $= 1$ eq.

The normal ranges of values of the concentrations of several components of blood are listed on the inside back cover of this book where you will see that many are given in units of milliequivalents per liter (meq/L). There is also an application, described in Special Topic 7.5, in which the concentrations of anions in blood that are hard to determine directly can be estimated. When the level of such anions increases sufficiently, it can indicate a malfunction somewhere in the body.

SUMMARY

Ionization of water Trace concentrations of hydronium ions, H_3O^+, and hydroxide ions, OH^-, are always present in water. In neutral water, their molar concentrations are equal (and very low). In writing equations, we usually write H_3O^+ as H^+, calling the latter

The ions in blood that contribute the highest levels of concentration and charge are Na^+, Cl^-, and HCO_3^-. For example, the concentration of Na^+ normally is in the range of 135–145 meq/L; of Cl^-, 100–108 meq/L; and of HCO_3^-, 21–29 meq/L. In contrast, the levels of K^+, Ca^{2+}, and Mg^{2+} ions are on the order of only 2–5 meq/L each. The blood also carries varying concentrations of negatively charged ions of organic acids, such as the anions of acetic acid, citric acid, and many others.

The levels of the organic anions tend to rise in several metabolic disturbances such as diabetes or kidney disease, but measuring these anions is difficult. The combined levels of the organic ions, however, can be estimated by calculating a quantity known as the **anion gap,** which can be defined by the following equation:

$$\text{anion gap} = \frac{\text{meq } Na^+}{L} - \left(\frac{\text{meq } Cl^-}{L} + \frac{\text{meq } HCO_3^-}{L}\right)$$

To determine the anion gap in a patient's blood, a sample is analyzed for the concentrations in meq/L of Na^+, Cl^-, and HCO_3^-, which as we said are the most abundant ions and are also relatively easy to analyze. Then the concentration data are fed into this equation. For example, suppose that analyses found the following data: Na^+ = 137 meq/L; Cl^- = 100 meq/L; and HCO_3^- = 28 meq/L. Then the anion gap is found by

$$\text{Anion gap} = 137\,\frac{\text{meq}}{L} - \left(100\,\frac{\text{meq}}{L} + 28\,\frac{\text{meq}}{L}\right)$$
$$= 9\text{ meq/L}$$

The normal range for the anion gap is 5–14 meq/L, so in our example the anion gap of 9 meq/L falls within the normal range. This 9 meq/L is accounted for by the presence of unmeasured ions of low concentration.

If metabolic disturbances cause the levels of organic anions to rise, the body must retain cations in the blood and excrete some of the more common anions such as Cl^- and HCO_3^- to maintain the absolute requirement that the blood be electrically neutral. In other words, negative organic ions tend to expel other negative ions but retain whatever positive ions are available. This is how the anion gap widens in metabolic disturbances that generate organic anions. The anion gap routinely rises above 14 meq/L in untreated diabetes.

You can see that by using rather easily measured data on the concentrations of Na^+, Cl^-, and HCO_3^- in meq/L, and calculating the anion gap from these data, the clinical chemist can inform the health care professionals of any unusual buildups in anions which indicate possible disease. In severe exercise, the anion gap rises above normal, too, but it goes back down again in time. Thus an above normal anion gap has to be interpreted in the light of other facts.

either the hydrogen ion or the proton. In explaining these reactions, however, it's usually necessary to use the correct formula, H_3O^+.

Acidic solutions are those in which $[H_3O^+] > [OH^-]$. In basic solutions, $[OH^-] > [H_3O^+]$. Neutral solutions have these two ions in equal (and extremely low) concentrations.

Common aqueous acids and bases Acids are substances that liberate hydrogen ions in water and bases are substances that produce hydroxide ions in water. Salts are ionic compounds that involve any other ions. Strong acids or bases are those that break up into ions to the extent of a high percent, up to 90 or 100%. The five most common strong acids are hydrochloric, hydrobromic, hydriodic, sulfuric, and nitric acid. All are monoprotic except sulfuric acid, which is diprotic.

The strongest bases are the hydroxides of Groups IA and IIA metals, like NaOH, KOH, and $Ca(OH)_2$.

Electrolytes Salts in their molten states and the aqueous solutions of all soluble salts and of all strong acids and strong, soluble bases conduct electricity, and are called strong electrolytes. Solutions of weaker acids or of slightly soluble strong bases are weak electrolytes. All molecular substances, unless they change into ions by reacting with water, are nonelectrolytes. Pure water is a nonelectrolyte.

Brønsted acids and bases A Brønsted acid is any that can donate a proton, and a Brønsted base is any that can accept a proton. A strong Brønsted base is able to accept and bind a proton strongly. In water, the strongest Brønsted base is the hydroxide ion. Water is a weak Brønsted base and the chloride ion in water is an even weaker base. A strong Brønsted acid has a strong ability to give up or donate a proton. The hydronium ion is an example of a strong Brønsted acid, and gaseous hydrogen chloride is even stronger. A weak Brønsted acid has a weak or poor ability to donate a proton. It holds its proton instead. Acetic acid is a weak Brønsted acid, and the water molecule is even weaker.

Aqueous acids The strong aqueous acids react with:

Metals, to give the salt of the metal and hydrogen.

Metal hydroxides, to give a salt and water.

Metal carbonates, to give a salt, carbon dioxide, and water.

Metal bicarbonates, to give a salt, carbon dioxide, and water.

A solution of an acid is neutralized when any sufficiently strong proton-binding species is added in the correct mole proportion to make the concentration of hydrogen ion and hydroxide ion equal (and very small).

Carbonic acid and carbonates Carbonic acid, H_2CO_3, is both a weak acid and an unstable acid. When it is generated in water by the reaction of any stronger acid with a bicarbonate or a carbonate salt, virtually all of the carbonic acid decomposes to carbon dioxide and water, and most of the carbon dioxide fizzes out. The carbonate ion and the bicarbonate ion are both Brønsted bases, and the bicarbonate ion is involved in carrying waste carbon dioxide from cells, where it is made, to the lungs.

Aqueous bases   Any ion that can accept a proton from the hydronium ion in water qualifies as a base. These include the hydroxide ion, the bicarbonate ion, the carbonate ion, and ammonia plus any of the anions of weak acids. What forms when a weak acid donates a proton is a strong Brønsted base, and what forms when a strong acid gives up a proton is a weak base.

Ammonia and the ammonium ion   Ammonia is a strong base toward H_3O^+ but a weak base toward H_2O. The ammonium ion is a strong acid toward OH^- but a weak acid toward H_2O. Ammonia can neutralize strong acids and the ammonium ion can neutralize strong bases.

Salts   The chemical properties of salts in water are the properties of their individual ions. Salts can be produced by any of the reactions of strong acids that were studied (and summarized, above) as well as by double decomposition reactions. The solubility rules for salts are guides for the prediction of their reactions. If a combination of oppositely charged ions can lead to an insoluble salt, an un-ionized species that stays in solution, or a gas, then the ions react.

Equivalents of ions   An equivalent (eq) of an ion is the number of grams of the ion that carry Avogadro's number of positive or negative charges. It is calculated by dividing the molar mass of the ion by the size of the charge it carries. The concentration of an ion in a dilute solution is often given in (meq/L), where 1000 meq = 1 eq.

REVIEW EXERCISES

The answers to review exercises that require a calculation and that are marked with an asterisk are found in Appendix V. The answers to the other review exercises are found in the *Study Guide* that accompanies this book.

Atoms, Ions, and Molecules

7.1   What families of compounds are the principal sources of ions in aqueous solutions?

7.2   Review the differences between atoms and ions by answering the following questions:
  (a) Are there any atoms that have more than one nucleus? If so, give an example.
  (b) Are there any ions with more than one nucleus? If so, give an example.
  (c) Are there any ions that are electrically neutral? If so, give an example.
  (d) Are there any atoms that are electrically charged? If so, give an example.

7.3   Write the equilibrium equation for the self-ionization of water, and label the ions that are present.

Common Acids, Bases, and Salts

7.4   What features do the common aqueous acids have in common?

7.5   In the context of acid–base discussion, what are two other names that we can use for *proton*?

7.6   In terms of the molar concentrations of certain ions, what is always true in (a) an acidic solution, (b) a basic solution, and (c) a neutral solution?

7.7   Acids have a set of common reactions, and so do bases, but not salts. Explain.

7.8   Tell whether each of the following solutions is acidic, basic, or neutral.
  (a) $[H^+] = 6.2 \times 10^{-6}$ mol/L
  and $[OH^-] = 1.6 \times 10^{-9}$ mol/L
  (b) $[H^+] = 1.0 \times 10^{-7}$ mol/L
  and $[OH^-] = 1.0 \times 10^{-7}$ mol/L
  (c) $[H^+] = 1.36 \times 10^{-8}$ mol/L
  and $[OH^-] = 7.35 \times 10^{-7}$ mol/L

7.9   Salts are all crystalline solids at room temperature. Why do you suppose this is?

Electrolytes

7.10   The word *electrolyte* can be understood in two ways. What are they? Give examples.

7.11   To which electrode do cations migrate?

7.12   The anode in electrolysis has what electrical charge, positive or negative?

7.13   The electrode that is negatively charged attracts what kinds of ions, cations or anions?

7.14   Explain in your own words how the presence of cations and anions in water enables the system to conduct electricity.

7.15   When NaOH(s) is dissolved in water, the solution is an excellent conductor of electricity, but when methyl alcohol is dissolved in water, the solution won't conduct electricity at all. What does this behavior suggest about the structural natures of NaOH and methyl alcohol, whose structure is given below? (Notice that both appear to have OH groups in their formulas.)

$$
\begin{array}{c}
\text{H} \\
| \\
\text{H} - \text{C} - \text{O} - \text{H} \\
| \\
\text{H}
\end{array}
$$

Methyl alcohol

7.16   If a water-soluble compound breaks up entirely into ions as it dissolves in water, do we call it a weak or a strong electrolyte?

7.17   In the liquid state, tin(IV) chloride, $SnCl_4$, is a nonconductor. What does this suggest about the structural nature of this compound? (Is is more likely ionic or molecular?)

Aqueous Acids and Bases

7.18 What is the difference between hydrochloric acid and hydrogen chloride?

7.19 In which species is the covalent bond to hydrogen stronger, in $HCl(g)$ or in $H_3O^+(aq)$? How do we know?

7.20 We can represent an acid by the general symbol HA. Its equilibrium expression, then, is

$$HA \rightleftharpoons H^+ + A^-$$

On the basis of the way that the equilibrium arrows are drawn, is this acid weak or strong?

7.21 Is $HC_2H_3O_2$ a mono-, di-, tri, or tetraprotic acid? (What is its name?)

7.22 What are the names and the formulas of the aqueous solutions of the four hydrohalogen acids?

7.23 If we represent all diprotic acids by the symbol H_2A, write the equilibrium expressions for the two separate ionization steps.

7.24 Would the ionization of the second proton from a diprotic acid occur with greater ease or with greater difficulty than the ionization of the first proton? Explain.

7.25 Compare the structures of nitrous acid, HNO_2, and nitric acid, HNO_3.

Nitrous acid   Nitric acid

Nitrous acid is a much weaker acid than nitric acid. How does the extra oxygen in the structure of nitric acid help to explain this? (*Hint:* Study Special Topic 7.2.)

7.26 Which is the stronger acid in water, sulfurous acid or sulfuric acid?

Sulfurous acid   Sulfuric acid

Explain this along the lines of argument useful for Review Exercise 7.25.

7.27 Write the equilibrium expression for the solution of carbon dioxide in water that produces some carbonic acid.

7.28 Write the equilibrium expressions for the successive steps in the ionization of carbonic acid.

7.29 Magnesium hydroxide is practically insoluble in water, and yet it is classified as a strong base. Explain.

7.30 Ammonia is very soluble in water, and yet it is called a weak base. Explain.

7.31 What are the names and formulas of two bases that are both strong and can form relatively concentrated solutions in water?

7.32 When carbon dioxide is bubbled into pure water to form a solution, it takes only time and the help of a little warming to drive essentially all of it out of solution again. When this gas is bubbled into aqueous sodium hydroxide, however, it is completely trapped by a chemical reaction. If we assume that the reaction involves CO_2 and NaOH in a mole ratio of $1:1$, what is the molecular equation for this trapping reaction?

7.33 What is meant by *aqueous ammonia*? Why don't we call it "ammonium hydroxide"?

7.34 Write the names and the formulas of the five strong acids that we have studied.

7.35 What are the four strong bases — both the names and formulas? Which are quite soluble in water?

Net Ionic Equations

7.36 Consider the following net ionic equation.

$$2H^+(aq) + Cu(s) + NO_3^-(aq) \longrightarrow Cu^{2+}(aq) + NO_2(g) + H_2O$$

(a) Does it have material balance?
(b) Does it have electrical balance?

7.37 Complete and balance the following molecular equations, and then write the net ionic equations.
(a) $HNO_3(aq) + NaOH(aq) \rightarrow$
(b) $HCl(aq) + K_2CO_3(aq) \rightarrow$
(c) $HBr(aq) + CaCO_3(s) \rightarrow$
(d) $HNO_3(aq) + NaHCO_3(aq) \rightarrow$
(e) $HI(aq) + NH_3(aq) \rightarrow$
(f) $HNO_3(aq) + Mg(OH)_2(s) \rightarrow$
(g) $HBr(aq) + Zn(s) \rightarrow$

7.38 Complete and balance the following molecular equations, and then write the net ionic equation.
(a) $KOH(aq) + H_2SO_4(aq) \rightarrow$
(b) $Na_2CO_3(aq) + HNO_3(aq) \rightarrow$
(c) $KHCO_3(aq) + HCl(aq) \rightarrow$
(d) $MgCO_3(s) + HI(aq) \rightarrow$
(e) $NH_3(aq) + HBr(aq) \rightarrow$
(f) $Ca(OH)_2(s) + HCl(aq) \rightarrow$
(g) $Al(s) + HCl(aq) \rightarrow$

7.39 What are the net ionic equations for the following reactions of strong, aqueous acids? (Assume that all reactants and products are soluble in water.)
(a) With metal hydroxides
(b) With metal bicarbonates
(c) With metal carbonates
(d) With aqueous ammonia

7.40 Write net ionic equations for the reactions of all of the water-insoluble Group IIA carbonates, where you use $MCO_3(s)$ as their general formula, with hydrochloric acid (chosen so that all of the products are soluble in water).

7.41 If we let $M(OH)_2(s)$ represent the water-insoluble Group IIA metal hydroxides, what is the general net ionic equation for all of their reactions with nitric acid (chosen so that all of the products are soluble in water)?

7.42 If we let $M(s)$ represent either calcium or magnesium metal, what net ionic equation represents the reaction of either with hydrochloric acid?

7.43 Sodium and potassium in Group IA are higher in the activity series than calcium and magnesium in Group IIA.
(a) What does it mean to be *higher* in the activity series?
(b) If you check back to Figure 2.4b on page 35, you will see that sodium and potassium have lower ionization energies than calcium and magnesium. In what way does this fact correlate with their higher position in the activity series of the metals?

7.44 Zinc metal reacts more rapidly with which acid, 1 M nitric acid or 1 M acetic acid?

*7.45 How many moles of sodium bicarbonate can react quantitatively with 0.250 mol of HCl?

7.46 How many moles of potassium hydroxide can react quantitatively with 0.400 mol of H_2SO_4 (assuming that both H^+ in H_2SO_4 are neutralized)?

*7.47 How many grams of sodium carbonate does it take to neutralize 4.60 g of HCl?

7.48 How many grams of calcium carbonate react quantitatively with 6.88 g of HNO_3?

*7.49 How many grams of sodium bicarbonate does it take to neutralize all of the acid in 25.4 mL of 1.15 M H_2SO_4?

7.50 How many grams of potassium carbonate will neutralize all of the acid in 36.8 mL of 0.550 M HCl?

*7.51 How many milliliters of 0.246 M NaOH are needed to neutralize the acid in 32.4 mL of 0.224 M HNO_3?

7.52 How many milliliters of 0.108 M KOH are needed to neutralize the acid in 16.4 mL of 0.116 M H_2SO_4?

*7.53 For an experiment that required 12.0 L of dry CO_2 gas (as measured at 740 mm Hg and 25 °C), a student let 5.00 M HCl react with marble chips, $CaCO_3$.
(a) Write the molecular and net ionic equations for this reaction.
(b) How many grams of $CaCO_3$ and how many milliliters of the acid are needed?

7.54 How many liters of dry CO_2 gas are generated (at 750 mm Hg and 20 °C) by the reaction of $Na_2CO_3(s)$ with 250 mL of 6.00 M HCl? Write the molecular and the net ionic equations for the reaction, and calculate how many grams of Na_2CO_3 are needed.

Brønsted Acids and Bases

7.55 Acetic acid is a weak acid. (a) Write the formulas of the *two* proton-donating species in aqueous acetic acid. (Do not count H_2O.) (b) Which species has the higher molar concentration?

7.56 In aqueous ammonia, which species in equilibrium would Brønsted call the most abundant base?

7.57 Which member of each pair is the stronger Brønsted base?
(a) NH_3 or NH_2^- (b) OH^- or H_2O (c) HS^- or S^{2-}

7.58 Study each pair and decide which is the stronger Brønsted acid.
(a) $H_2PO_4^-$ or HPO_4^{2-} (b) H_2SO_3 or HSO_3^-
(c) NH_4^+ or NH_3

7.59 Which member of each pair is the stronger Brønsted acid?
(a) H_2CO_3 or HCl   (b) H_2O or OH^- (c) HSO_4^- or HSO_3^-

Salts

7.60 Write the names and formulas of three compounds that, by reacting with hydrochloric acid, give a solution of potassium chloride. Write the molecular equations for these reactions.

7.61 Write names and formulas of three compounds that will give a solution of lithium bromide when they react with hydrobromic acid. Write the molecular equations for these reactions.

7.62 Which of the following compounds are insoluble in water (as we have defined solubility)?
(a) NaOH   (b) NH_4Br (c) Hg_2Cl_2
(d) $Ca_3(PO_4)_2$.(e) KBr   (f) Li_2SO_4

7.63 Which of the following compounds are insoluble in water?
(a) $(NH_4)_2SO_4$ (b) KNO_2 (c) LiCl
(d) AgCl   (e) $Mg_3(PO_4)_2$ (f) $NaNO_3$

7.64 Identify the compounds that do not dissolve in water.
(a) NH_4NO_3 (b) $BaCO_3$ (c) $PbCl_2$
(d) K_2CO_3 (e) $LiC_2H_3O_2$ (f) Na_2SO_4

7.65 Which of the following compounds do not dissolve in water?
(a) Li_2CO_3 (b) $Na_2Cr_2O_7$ (c) NH_4I
(d) AgBr   (e) K_2CrO_4 (f) $FeCO_3$

7.66 Assume you have separate solutions of each compound in the pairs below. Predict what happens chemically when the two solutions of a pair are poured together. If no reaction occurs, state so. If there is a reaction, write its net ionic equation.
(a) LiCl and $AgNO_3$ (b) $NaNO_3$ and $CaCl_2$
(c) KOH and H_2SO_4 (d) $Pb(NO_3)_2$ and KCl
(e) NH_4Br and K_2SO_4 (f) Na_2S and $CuSO_4$
(g) K_2SO_4 and $Ba(NO_3)_2$ (h) NaOH and HI
(i) K_2S and $NiCl_2$ (j) $AgNO_3$ and NaCl
(k) $LiHCO_3$ and HBr   (l) $CaCl_2$ and KOH

7.67 If you have separate solutions of each of the compounds given below and then mix the two of each pair together, what (if anything) happens chemically? If no reaction occurs, state so, but if there is a reaction write its net ionic equation.
(a) H_2S and $CdCl_2$ (b) KOH and HBr
(c) Na_2SO_4 and $BaCl_2$ (d) $Pb(C_2H_3O_2)_2$ and Li_2SO_4
(e) $Ba(NO_3)_2$ and KCl   (f) $NaHCO_3$ and H_2SO_4
(g) Na_2S and $Ni(NO_3)_2$ (h) NaOH and HBr
(i) $Hg(NO_3)_2$ and KCl   (j) $NaHCO_3$ and HI
(k) KBr and NaCl   (l) $Pb(NO_3)_2$ and Na_2CrO_4

Equivalents and Milliequivalents of Ions

7.68 The concentration of potassium ion in blood serum is normally in the range of 0.0035 – 0.0050 mol K^+/L. Express this range in units of milliequivalents of K^+ per liter.

7.69 The concentration of calcium ion in blood serum is normally in the range of 0.0042–0.0052 eq of Ca^{2+}/L. Express this range in units of milliequivalents of Ca^{2+} per liter.

*7.70 The level of chloride ion in blood serum is normally quoted as 100–106 meq/L. How many grams and how many milligrams constitute 106 meq of Cl^-?

7.71 The sodium ion level in the blood is normally 135–145 meq/L. How many grams and how many milligrams of sodium ion constitute 135 meq of Na^+?

*7.72 The potassium ion level of blood serum normally does not exceed 0.196 g of K^+ per liter. How many milliequivalents of K^+ ion are in 0.196 g of K^+?

7.73 The magnesium ion level in plasma normally does not exceed 0.0243 g of Mg^{2+}/L. How many milliequivalents of Mg^{2+} are in 0.0243 g of Mg^{2+}?

Bonds in the Hydronium Ion (Special Topic 7.1)

7.74 One of the H—O bonds in H_3O^+ is called a *coordinate covalent bond*. What does this mean?

7.75 Once formed, are the three H—O bonds in H_3O^+ equivalent, or are they in some chemical way different?

The Carbonic Acid System (Special Topic 7.2)

7.76 Why is the C=O group an electronegative group?

7.77 Why is HCO_3^- a weaker acid than H_2CO_3?

7.78 Which is the stronger base, CO_3^{2-} or HCO_3^-? Why?

Carbonated Medications (Special Topic 7.3)

7.79 What is the name of the specific base in Alka-Seltzer® and similar preparations? What is the name of the acid?

7.80 Why doesn't the acid in these preparations cause CO_2 to form in the absence of water, since water is not an actual reactant?

7.81 Tablets from a bottle of Alka-Seltzer® that has stood open for months often have no "fizz." How do you suppose this situation arises?

Hard Water (Special Topic 7.4)

7.82 Hard water was called *hard* long before people knew about the solutes in it. Why did they call it this?

7.83 Hard water was called *temporary* hard water long before people knew the chemistry of it. Why do you suppose they called it only *temporarily* hard?

7.84 What specific cations cause the hardness of hard water? Do they cause the hardness of both permanent and temporary hard water?

7.85 What specific anion must be relatively abundant in temporary hard water? What chemical property does it have that is related to temporary hardness?

7.86 How does temporary hard water become soft simply by boiling? Give the equation(s).

7.87 How does household ammonia work as a softening agent? Give the equation(s).

7.88 What happens in zeolite as hard water trickles through it to remove the hardness ions?

The Anion Gap (Special Topic 7.5)

*7.89 The analysis of the blood from a young man recovering from polio found 137 meq of Na^+/L, 34 meq of HCO_3^-/L, and 93 meq of Cl^-/L. Calculate the anion gap. Does it suggest a serious disturbance in his metabolism?

7.90 A patient on a self-prescribed diet consisting essentially only of protein was found to have the following blood analyses after 2 weeks of the diet: Na^+, 174 meq/L; Cl^-, 135 meq/L; and HCO_3^-, 20 meq/L. Calculate the anion gap. Does it suggest a disturbance in metabolism?

Chapter 8
Acidity: Detection, Control, Measurement

The pH Concept

The Effects of Ions on pH

Buffers. Preventing Large Changes in pH

Acid–Base Titrations

It would take the flowage of Niagara Falls one hour to carry one mole of hydrogen ion past an observer if the pH of the river were 10 (which it is not). The pH concept helps us deal with ultra-low concentrations of hydrogen ions.

8.1 THE pH CONCEPT

Very low levels of H$^+$ are more easily described and compared in terms of pH values than as molar concentrations.

At the molecular level of life we must have almost perfect control over the acid–base balance of body fluids. Blood, for example, must have a molar concentration of hydrogen ion, [H$^+$], of 4.47×10^{-8} mol/L or extremely close to this. If the value of [H$^+$] goes up to 10×10^{-8} mol/L or drops to 1×10^{-8} mol/L, death is very near. Yet, we can change the acidity of pure water far more than these figures represent by adding only one drop of concentrated hydrochloric acid or one drop of concentrated sodium hydroxide to a liter.

The brackets in [H$^+$] signify that the concentration is specifically moles per liter.

In Acidosis, the Blood is Becoming More Acidic. So critical is the acid–base balance of the blood that a special vocabulary has been created to describe small shifts toward acidity or basicity. If the acidity of the blood increases, the condition is called **acidosis.** Acidosis is characteristic of untreated diabetes and emphysema, and other conditions.

If the acidity of the blood decreases, the condition is called **alkalosis.** An overdose of bicarbonate, exposure to the low partial pressure of oxygen at high altitudes, or prolonged hysteria can cause alkalosis.

In their more advanced stages, acidosis and alkalosis are medical emergencies because they interfere with respiration. **Respiration** is what brings oxygen into the body, chemically uses it, and then removes waste carbon dioxide. We need a more detailed background before we can study the molecular basis of respiration, and many of these details are studied in this chapter.

The Product of the Molar Concentrations of H$^+$ and OH$^-$ in Water Is a Constant. We now need quantitative information about the self-ionization of water. The equilibrium is

$$H_2O \rightleftharpoons H^+(aq) + OH^-(aq)$$

What is rather remarkable is that *in any dilute aqueous solution,* the result of multiplying the molar concentrations of hydrogen and hydroxide ions is a constant, *regardless of the solute.* This constant is called the **ion product constant of water,** and its symbol is K_w

K_w at Various Temperatures

$$K_w = [H^+][OH^-] \qquad (8.1)$$

Temperature (°C)	K_w
0	1.5×10^{-15}
10	3.0×10^{-15}
20	6.8×10^{-15}
30	1.5×10^{-14}
40	3.0×10^{-14}

Regardless of how we might change the value of [H$^+$] by adding either acid or base to a solution, the value of [OH$^-$] automatically and rapidly adjusts, and the *product* of the molar concentrations of these ions remains equal to K_w. The only way to change K_w is to change the temperature, as data in the margin show.

At 25 °C, both [H$^+$] and [OH$^-$] equal 1.0×10^{-7} mol/L. Therefore,

$$K_w = (1.0 \times 10^{-7})(1.0 \times 10^{-7})$$
$$= 1.0 \times 10^{-14} \text{ (at 25 °C)}$$

Units are never included in quoted values for K_w.

In most of our work, we will assume a temperature of 25 °C.

With this value of K_w, we can calculate either of the two concentrations, [H$^+$] or [OH$^-$], if we know the other.

EXAMPLE 8.1 USING THE ION PRODUCT CONSTANT OF WATER

Problem: The value of [H$^+$] of blood (when measured at 25 °C, not body temperature) is 4.5×10^{-8} mol/L. What is the value of [OH$^-$], and is the blood acidic, basic, or neutral?

A review of exponents can be found in Appendix I.

Solution: We simply use the value of $[H^+]$ in the equation for K_w.

$$K_w = 1.0 \times 10^{-14} = (4.5 \times 10^{-8}) \times [OH^-]$$
$$[OH^-] = \frac{1.0 \times 10^{-14}}{4.5 \times 10^{-8}}$$
$$= 2.2 \times 10^{-7} \text{ mol/L}$$

Because the value of $[H^+]$ is less than the value of $[OH^-]$, the blood is (very slightly) basic.

PRACTICE EXERCISE 1

For each of the following values of $[H^+]$, calculate the value of $[OH^-]$ and state whether the solution is acidic, basic, or neutral.

(a) $[H^+] = 4.0 \times 10^{-9}$ mol/L   (b) $[H^+] = 1.1 \times 10^{-7}$ mol/L   (c) $[H^+] = 9.4 \times 10^{-8}$ mol/L

The pH of an Acidic Solution Is Less Than 7.00. When we compare numbers such as we found in Example 8.1 (4.5×10^{-8} versus 2.2×10^{-7}) to see which is larger we have to look in two places for each number. We have to compare the exponents of 10 and then the numbers that appear before each 10. Yet, numbers this small (and this close together) are often encountered in dealing with the relative acidities of body fluids such as blood. To make these kinds of comparisons easier, a Danish biochemist, S. P. L. Sorenson (1868–1939), invented the concept of pH.

There are two ways to write equations to define pH, but all of our needs will be met with one, the following:

$$[H^+] = 1 \times 10^{-pH} \tag{8.2}$$

In other words, the **pH** of a solution is the negative power (the p in pH) to which the number 10 must be raised to express the molar concentration of a solution's hydrogen ions (hence, the H in pH).[1]

In pure water at 25 °C, $[H^+] = 1.0 \times 10^{-7}$ mol/L. The pH of pure water at 25 °C, therefore, is 7.00. Thus a pH of 7.00 corresponds to a neutral solution at 25 °C.[2] Because pH occurs as a *negative* exponent in Equation 8.2, a pH less than 7.00 corresponds to an acidic solution. A pH more than 7.00 corresponds to a basic solution. In pH terms, then, we have the following definitions of acidic, basic, and neutral solutions (at 25 °C).

Acidic solution   pH < 7.00
Neutral solution   pH = 7.00
Basic solution   pH > 7.00

[1] If the logs of both sides of Equation 8.2 are taken, and after a change in sign, we get the other (equivalent) way to define pH:

$$pH = -\log [H^+]$$

[2] The 7 in 7.00 comes from the *exponent* in 1.0×10^{-7}, so it actually does nothing more than set off a decimal point when we rewrite the number as 0.00000010. Hence, the 7 in the pH value of 7.00 can't be counted as a significant figure. A pH value of 7.00 therefore has just two significant figures, those that follow the decimal point, just as there are but two significant figures in the value of the molar concentration of H^+, 1.0×10^{-7} mol/L. To repeat, the number of significant figures in any value of pH is the number of figures that *follow* the decimal point.

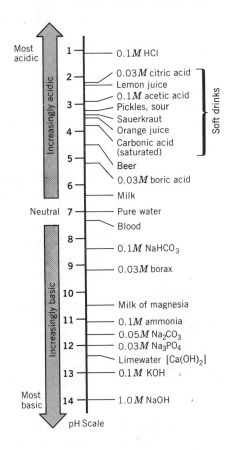

FIGURE 8.1
The pH scale and the pH values of
several common substances.

When the value of $[H^+]$ is 1 mol/L or higher, the pH concept isn't used. The exponents would no longer be negative, so there would be none of the confusion that Sorenson addressed when he invented pH. The pH values of several common substances are shown in Figure 8.1.

The definition of pH in Equation 8.2 has the number 1 before the times sign: $[H^+] = 1 \times 10^{-pH}$. What if the pH is, say, 4.56? If we insert this into Equation 8.2, this pH value means that

$$[H^+] = 1 \times 10^{-4.56} \text{ mol/L}$$

Although this is a perfectly good number, we normally reexpress exponentials so that the exponents are whole numbers. We can tell from the number as it stands that the value of $[H^+]$ is between 1.0×10^{-4} and 1.0×10^{-5} mol/L, but we can't tell exactly where between these limits without using logarithms.

Although logarithms are not hard to use, particularly when pocket calculators have both log x and 10^x functions,[3] our needs in the area of pH are not this extensive. All we need is an ability to *interpret* pH values that are not whole numbers rather than an ability to carry out logarithmic calculations. For example, a pH of 4.56 is between 4.00 and 5.00, so the value of $[H^+]$ corresponding to a pH of 4.56 has to be between 1.0×10^{-4} and 1.0×10^{-5} mol/L. Of much more importance to us is the ability to recognize that a pH of 4.56 corresponds to an acidic solution.

---

[3] For example, enter 4.56 into your calculator, change the sign by hitting the $+/-$ key (because pH is a *negative* exponent) and then hit the 10^x key. The screen should read 2.754228703 −05, meaning (after some rounding) 2.8×10^{-5} (mol/L, of course).

PRACTICE EXERCISE 2

A patient suspected of having taken an overdose of barbiturates was found to have blood with a pH of 7.10.

(a) Between what limits of values for the molar concentration of H^+ does this pH correspond to?
(b) Is the blood slightly basic, acidic, or is it neutral?
(c) If the pH of blood normally is 7.35, is the patient experiencing acidosis or alkalosis?

PRACTICE EXERCISE 3

The patient in Practice Exercise 2 had been admitted to a teaching hospital, and the attending physician asked a group of students which of two isotonic reagents, sodium bicarbonate or ammonium chloride, might be administered by intravenous drip to restore the pH of the patient's blood to a normal value of 7.35. Which solution could be used? Explain how it would work using a net ionic equation.

One drop of concentrated hydrochloric acid changes the pH of a liter of water from 7 to 4.

Small Changes in pH Correspond to Large Changes in [H⁺]. One of the very deceptive features of the pH concept is that the actual hydrogen ion concentration changes greatly — by a factor of 10 — for each change of only one unit of pH. For example, if the pH of a solution is zero (meaning that $[H^+] = 1 \times 10^0$ mol/L), only 1 L of water is needed to contain 1 mol of H^+. When the pH is 1, however, then 10 L of water (about the size of an average wastebasket) is needed to hold 1 mol of H^+. At a pH of 5, it takes a large railroad tank car full of water to include just 1 mol of H^+. If the pH of the water flowing over Niagara Falls, New York, were 10 (which, of course, it isn't), an entire 1-hour flowage would be needed for 1 mol of H^+ to pass by. And at a pH of 14, the volume that would hold 1 mol of H^+ is about one-quarter of the volume of Lake Erie, one of the Great Lakes. You can see that seemingly small changes in pH numbers signify enormous changes in real concentrations of hydrogen ions.

Lake Erie isn't alkaline. A pH of 14 is what 1 M NaOH (lye) is.

pH Is a Measure of the Acidity of a Solution, Its H_3O^+ Concentration. When the solute is a weak acid, just a small percentage of its molecules are ionized at equilibrium, so no simple correlation exists between the molar concentration of the weak acid and the pH of the solution. In 0.10 M acetic acid, for example, the pH is about 2.9, not 1.0. It would be 1.0 if acetic acid were 100% ionized, because then 0.10 mol of acetic acid would give 0.10 mol of H^+ ion in a liter of solution. And because 0.10 mol/L $= 1.0 \times 10^{-1}$ mol/L, the pH of 0.10 M acetic acid would be 1.0 *if it were a strong acid*. The pH of our acetic acid solution is thus not a measure of the molarity of the solute but of the molarity of the hydrogen ions that it liberates. When we talk about the **acidity** of a solution, we mean $[H_3O^+]$. The pH gives us a measure of the acidity of a solution, not the concentrations of all species that can neutralize OH^-.

A simple correlation does exist between solute molarity and pH in dilute solutions of strong, 100% ionized acids. In 0.010 M HCl, for example, which is the same as 1.0×10^{-2} mol HCl(aq)/L, the molar concentration of H^+ ions also has to be 1.0×10^{-2} mol/L. This is because each molecule of HCl in solution breaks up into one H^+ ion and one Cl^- ion. Because $[H^+] = 1.0 \times 10^{-2}$ mol/L, the pH of 0.010 M HCl is simply 2.0. Similarly, a solution that is 0.0001 M HNO_3 has a pH of 4.0, because 0.0001 is the same as 1×10^{-4}, and because HNO_3 is a strong acid and fully ionized. Each molecule of HNO_3 actually means one ion of H^+.

Phenolphthalein (fee-noll-THAY-lean).

Some commercial pH test papers contain several indicator dyes in the same test strip, which makes possible a whole spectrum of colors according to the pH color code on the container.

Several Dyes Indicate a Solution's pH. **Acid–base indicators** can be used to find out whether a solution is acidic, basic, or neutral. The dye phenolphthalein, for example, has a bright pink color at a pH above 10.0 and is colorless below a pH of 8.2. In the range of 8.2–10.0 phenolphthalein undergoes a gradual change from colorless to pale pink to deep pink. In Chapter 7 we introduced litmus as an indicator. It is blue above a pH of about 8.5 and red below a pH of 4.5. Each indicator has its own pH range and set of colors, and Figure 8.2 gives just a few examples.

When the solutions to be tested for pH are themselves highly colored, we cannot use indicators. Moreover, we often need more than a rough idea of pH, which is all that indicators can actually give. For such situations there are instruments called pH meters (Figure 8.3)

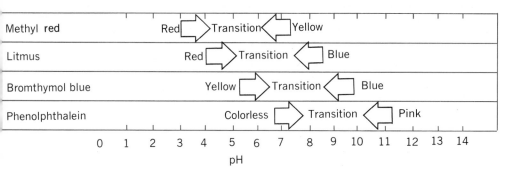

Methyl red	Red ⟹ Transition ⟸ Yellow		
Litmus	Red ⟹ Transition ⟸ Blue		
Bromthymol blue	Yellow ⟹ Transition ⟸ Blue		
Phenolphthalein	Colorless ⟹ Transition ⟸ Pink		

0 1 2 3 4 5 6 7 8 9 10 11 12 13 14
pH

FIGURE 8.2
Some common acid-base indicators.

equipped with special electrodes that can be dipped into the solution to be tested. With a good pH meter, pH values can be read to the second decimal place.

8.2 THE EFFECTS OF IONS ON pH

Aqueous solutions can be made basic or acidic by ions other than OH⁻ or H⁺.

In many laboratory situations, an aqueous solution of some salt is prepared only to have the solution turn out to be acidic or basic. If we weren't aware that this could happen, we might unknowingly prepare a salt solution that could be corrosive to metals, or could harm living things, or could make food, drink, or medications unfit. In the most general terms, salts are able to upset the 1 : 1 mole ratio of $[H^+]$ to $[OH^-]$ whenever *one* of the salt's ions, the anion or the cation, can react with water, at least to some extent. This reaction is called the **hydrolysis of ions,** and our next task is to learn a simple way to identify any ion that can hydrolyze and so affect the pH.

"Hydrolysis" is from the Greek *hydro,* water, and *lysis,* loosening or breaking—breaking or loosening by water.

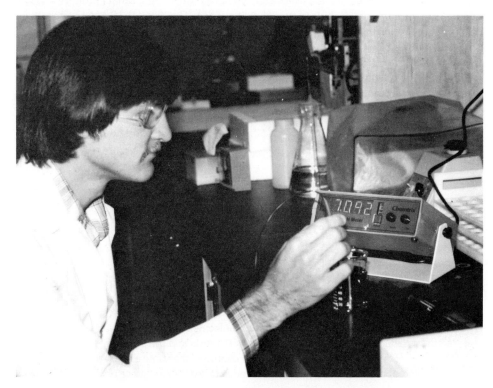

FIGURE 8.3
A pH meter.

Anions Obtained from Weak Acids Are Basic Enough to Hydrolyze. Aqueous solutions of such sodium salts as Na_3PO_4, Na_2CO_3, Na_2HPO_4, $NaHCO_3$, $NaC_2H_3O_2$, and NaH_2PO_4 test basic to litmus because their anions hydrolyze. They react with water so that an excess of OH^- over H^+ develops. The bicarbonate ion, for example, hydrolyzes to establish the following equilibrium:

$$HCO_3^-(aq) + H_2O \rightleftharpoons H_2CO_3(aq) + OH^-(aq)$$

Although the forward reaction is not strongly favored (HCO_3^- is not a strong proton acceptor, and H_2O is a very poor proton donor), the extent to which it occurs at all causes an excess of OH^- ions over H^+ ions.

Notice that all of the anions in the above list of sodium salts, PO_4^{3-}, CO_3^{2-}, HPO_4^{2-}, HCO_3^-, $C_2H_3O_2^-$, and $H_2PO_4^-$, are anions derived from *weak* acids. We learned in Chapter 7 that the anions left behind when weak acids ionize are relatively strong bases. They are strong enough bases so that all anions derived from weak acids hydrolyze and make a solution basic.

Anions such as Cl^-, Br^-, I^-, NO_3^-, and SO_4^{2-}, which are anions produced when strong acids ionize, are all extremely weak bases, too weak to react with water. Unable to hydrolyze, these anions cannot upset the $1:1$ ratio of H^+ to OH^- in water. To summarize,

---

Anions derived from all weak acids hydrolyze in water and tend to make a solution basic.

Anions of strong acids do not hydrolyze.

Most Transition Metal Cations and the Ammonium Ion Hydrolyze to Generate H_3O^+ Ions and Lower the pH of the Solution. A solution of ammonium chloride, NH_4Cl, in water turns blue litmus red, so in this solution, $[H^+] > [OH^-]$. The excess hydrogen ions come from the forward reaction of the following equilibrium involving the NH_4^+ ion.

$$NH_4^+ + H_2O \rightleftharpoons NH_3(aq) + H_3O^+(aq)$$

To be sure, the ammonium ion is a weak acid, but not so weak that it cannot produce enough hydronium ions by this hydrolysis to turn blue litmus red. The pH of $0.10\ M\ NH_4Cl$ is about 5.1, not strongly acidic, but certainly more acidic than water (roughly 100 times more acidic).

Other cations also hydrolyze. The hydrated cations of most metals, in fact, make a solution test acidic. The aluminum ion in water, for example, exists largely as the hydrate $[Al(H_2O)_6]^{3+}$. The high positive charge on the central metal ion in this hydrated ion attracts electron density from the H—O bonds of the H_2O molecules it holds. These bonds are thus weakened so $[Al(H_2O)_6]^{3+}$ can donate H^+ to water as follows:

$$[Al(H_2O)_6]^{3+}(aq) + H_2O \longrightarrow [Al(H_2O)_5(OH)]^{2+} + H_3O^+(aq)$$

By using the structure of the hydrated aluminum ion we can better see how this reaction occurs.

The arrows signify a force of attraction between $\delta-$ on O and the central, positively charged cation.

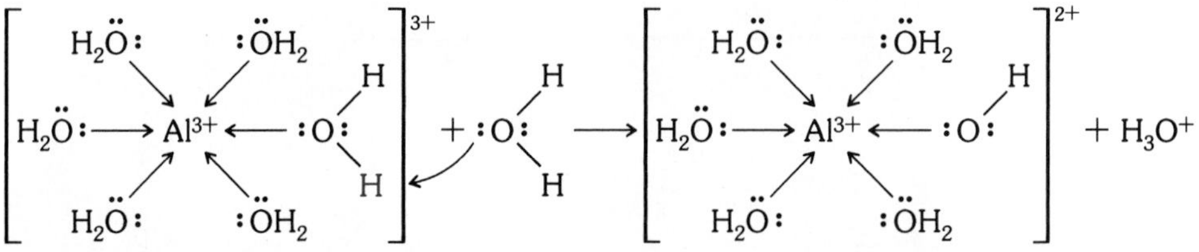

You can see how a proton transfers from a water molecule of the hydrated ion to a molecule of water in the surrounding solvent. In this way H_3O^+ forms and the solution is made acidic. A $0.1\ M$ solution of $AlCl_3$ has a pH of about 3, for example, the same as $0.1\ M$ acetic acid, but there is nothing about the formula, $AlCl_3$, to indicate that it can affect pH.

All hydrated metal ions with charges of 3+ and most with charges of 2+ on the central metal ion generate hydronium ions in water. Metal ions with such high charges are generally very small, so the *density* of their positive charge, the charge per unit volume, is high, as in the hydrated aluminum ion. A high positive charge density is able to act in a strongly electronegative way to weaken H—O bonds in the surrounding water molecules of their hydrated forms.

The only common metal ions that do *not* hydrolyze to give acidic solutions are those of Groups IA and IIA (except Be^{2+} of Group IIA), for example, Li^+, Na^+, and K^+, Mg^{2+}, Ca^{2+}, and Ba^{2+} do not hydrolyze. Evidently, except for Be^{2+}, the cations of Groups IA and IIA do not have sufficiently high positive charge densities. We can summarize what we have learned about the hydrolysis of cations as follows:

Metal ions from Groups IA or IIA (except Be^{2+}) do not hydrolyze.

Expect other metal ions as well as NH_4^+ to hydrolyze and generate H^+.

With these rules we can generally predict correctly whether a given salt will affect the pH of an aqueous solution. The exceptions would be salts where *both* cation and anion can hydrolyze, like $NH_4C_2H_3O_2$, ammonium acetate. These have to be taken on a case-by-case basis with the result hinging on the relative strengths of the cation as a proton producer and the base as a proton neutralizer. We will not work with such salts. But let's see how we can predict the hydrolysis of salts that respond to a simpler analysis.

EXAMPLE 8.2 PREDICTING HOW A SALT AFFECTS THE pH OF ITS SOLUTION

Problem: Sodium phosphate, Na_3PO_4 (trisodium phosphate), is a strong cleaning agent for walls and floors. Is its aqueous solution acidic or basic?

Solution: Na_3PO_4 involves Na^+ and PO_4^{3-}. The Na^+ ion does not hydrolyze, but PO_4^{3-} does. Its protonated form, HPO_4^{2-}, is not on our list of strong acids, so we can infer that it is a weak acid. We therefore expect PO_4^{3-} to be a relatively strong base and we expect it to hydrolyze as follows:

$$PO_4^{3-}(aq) + H_2O \rightleftharpoons HPO_4^{2-}(aq) + OH^-(aq)$$

The forward reaction in this equilibrium generates some OH^- ions, so the solution will be basic.

Notice that in Example 8.2 we did not need a table of Brønsted bases to predict that PO_4^{3-} would be basic enough to hydrolyze. We used our knowledge of just a few facts—which acids are strong acids in water and which cations do not hydrolyze—to figure out what we needed to know. We will work another example to practice using the list of strong aqueous acids to decide whether a given salt can affect the pH of its solution.

EXAMPLE 8.3 PREDICTING WHETHER A SALT AFFECTS THE ACIDITY OF ITS SOLUTION

Problem: Chromium(III) nitrate, $Cr(NO_3)_3$, is soluble in water. Does this salt make its aqueous solution acidic or basic?

Solution: This salt dissociates into $Cr^{3+}(aq)$ and three $NO_3^-(aq)$ ions. Since the nitrate ion is the anion of a strong acid, HNO_3, we know that it is unable to react with water to generate hydroxide ions. The nitrate ion does not hydrolyze. The Cr^{3+} ion, however, is not from Groups IA or IIA. Moreover, like the aluminum ion, it has a high positive charge. We therefore expect it to hydrolyze and generate some hydrogen ion, like the aluminum ion. This salt solution tests acidic.

PRACTICE EXERCISE 4   Determine without the use of tables whether each ion can hydrolyze. If so, state whether it tends to make the solution acidic or basic.

(a) CO_3^{2-} (b) S^{2-} (c) HPO_4^{2-} (d) Fe^{3+} (e) NO_2^- (f) F^-

PRACTICE EXERCISE 5   Is a solution of potassium acetate, $KC_2H_3O_2$, acidic, basic, or neutral to litmus?

PRACTICE EXERCISE 6   Is a solution of copper(II) nitrate, $Cu(NO_3)_2$, acidic, basic, or neutral?

PRACTICE EXERCISE 7   Ammonium sulfate, $(NH_4)_2SO_4$, is a nitrogen fertilizer. Could the application of an aqueous solution of this fertilizer affect the pH of the soil? If so, will it increase or decrease the pH?

8.3 BUFFERS. PREVENTING LARGE CHANGES IN pH

The pH of a solution can be held relatively constant if it contains a buffer.

There are several life-threatening situations in which the blood is forced to accept or retain either more acid or more base than normal. The pH of blood, however, cannot be allowed to vary by more than 0.2 to 0.3 pH units. Wider variations, fortunately, are prevented by the activities of buffers. Let's see what they are and how they work.

Buffers Have Components Able to Neutralize H^+ and OH^-.   Buffers are combinations of solutes that prevent large changes in pH when strong acids or bases are added to an aqueous solution. One solute of the buffer system neutralizes H^+ and another neutralizes OH^-. Fluids that contain buffers are said to be *buffered* against changes in pH.

The Phosphate Buffer Works Inside Cells.   The **phosphate buffer** consists of the pair of ions, HPO_4^{2-} and $H_2PO_4^-$, the monohydrogen and the dihydrogen phosphate ions. Of the two, $H_2PO_4^-$ has more protons, so it is the part of the buffer that neutralizes hydroxide ion. Any added OH^- shifts the following equilibrium to the right, and neutralizes the OH^-. This keeps the pH from increasing.

$$H_2PO_4^-(aq) + OH^-(aq) \rightleftharpoons HPO_4^{2-}(aq) + H_2O$$

The HPO_4^{2-} ion is the better proton acceptor of this buffer, and it neutralizes any added H^+. An increase in H^+ forces the following equilibrium to the right. This neutralizes the H^+, and the pH is kept from decreasing.

$$HPO_4^{2-}(aq) + H^+(aq) \rightleftharpoons H_2PO_4^-(aq)$$

This phosphate buffer is the principal buffer inside cells.

H_2CO_3 occurs in blood because some dissolved CO_2 reacts with water to form it.

The Carbonate Buffer Works in the Blood.   The chief buffer in blood is the **carbonate buffer**, which consists of H_2CO_3 and HCO_3^-, carbonic acid and the bicarbonate ion. Carbonic acid, by far the stronger Brønsted acid of the two, can neutralize OH^- and so prevent an increase in pH. When excess OH^- appears, the following equilibrium shifts to the right, which neutralize OH^- and prevents alkalosis.[4]

[4] It is not necessary that the CO_2 in blood be in the form of H_2CO_3 to neutralize OH^-. Here $CO_2(aq)$ reacts directly with OH^- ion as follows:

$$CO_2(aq) + OH^-(aq) \longrightarrow HCO_3^-(aq)$$

Thus the acid component of the carbonate buffer is the sum total of CO_2 and H_2CO_3. We will continue to use Equation 8.3 to represent the base-neutralizing reaction of the carbonate buffer, however, because it simplifies the discussion.

$$H_2CO_3(aq) + OH^-(aq) \rightleftharpoons HCO_3^-(aq) + H_2O \qquad (8.3)$$

The base in the carbonate buffer is HCO_3^-. When excess acid comes into the blood, it forces the following equilibrium to shift to the right, which neutralizes H^+ and so prevents acidosis.

$$HCO_3^-(aq) + H^+(aq) \rightleftharpoons H_2CO_3(aq) \qquad (8.4)$$

The Expulsion of CO₂ by Exhaling Must Also Occur to Prevent Acidosis.
When acid is neutralized by a shift of Equilibrium 8.4 to the right, the level of carbonic acid, H_2CO_3, in the blood rises. We have already learned that this is an unstable acid, and not much can exist as such in solution. Thus when the level of H_2CO_3 rises, the following equilibrium shifts to the right to produce carbon dioxide and water.

$$H_2CO_3(aq) \rightleftharpoons CO_2(aq) + H_2O \qquad (8.5)$$

We use the symbol $CO_2(aq)$ instead of $CO_2(g)$ to signify that the carbon dioxide is still in solution. When the blood moves through the capillaries of the lung's alveoli, however, where gaseous carbon dioxide can escape the body, the following change takes place, and $CO_2(g)$ now leaves. Because it leaves, we won't write the change as an equilibrium.

$$CO_2(aq) \longrightarrow CO_2(g) \qquad (8.6)$$

This important change tends to pull or shift all of the preceding equilibria of the carbonate buffer system to the right. The blood thus uses *two* mechanisms to handle excess acid. It neutralizes it by the work of the carbonate buffer, and it uses ventilation to make this neutralization permanent.

The Brain Changes the Rate of Ventilation in Response to Acidosis or Alkalosis.
Ventilation is the circulation of air into and out of the lungs. The rate of ventilation normally is controlled by a site in the brain called the respiratory center. It works first by monitoring the level of $CO_2(aq)$ in the blood. When this level rises, the brain then instructs the breathing apparatus to breathe more rapidly and deeply, a response called **hyperventilation.** This increases the rate at which CO_2 is exhaled from the lungs, and this response pulls all of the carbonate equilibria in their acid-neutralizing directions.

The successive shifts of Equilibrias 8.4 and 8.5, caused by the reaction of Equation 8.6, have a major consequence: One H^+ ion is permanently neutralized for each CO_2 molecule that leaves the body. All of these steps are summarized in Figure 8.4. Notice that the H^+ neutralized ends up in a molecule of water. The ability of this water molecule to form by

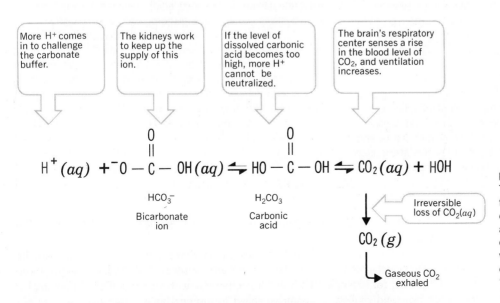

FIGURE 8.4
The irreversible neutralization of H^+ through the loss of CO_2 is one way the carbonate buffer system handles acidosis. It is the last step, the change of dissolved CO_2 into gaseous CO_2, which is exhaled, that draws all of the equilibria to the right and makes H^+ 'disappear' into H_2O.

The carbon dioxide and carbonic acid naturally present in unpolluted rain give it a pH of 5.6. Any form of precipitation with a lower pH is called **acid rain.** (Specialists in this field prefer the term *acid deposition* so that atmospheric dusts carrying acidic materials are also included.) The record low pH was 2.4, as acidic as vinegar, which occurred in rain that fell during a storm at Pitlochry, Scotland, in 1974. In the same year, rain with a pH 2.7 fell on western Norway. Over most of the eastern United States the pH of rain is less than 4.5.

The addition of acid rain to ground water, lakes, and streams lowers their pH. This change threatens aquatic plants and animals because their metabolism, like ours, is unusually sensitive to small variations in pH. As long as the pH of a lake stays between 5.5 and 6.0, fish generally are able to thrive, but when the pH drops below 4.5, they cannot live. Most of the lakes in southern Norway, a country particularly hard hit by acid rain, have no fish at all.

Not only is a low pH a direct threat to aquatic life, another threat comes from toxic metal ions leached from soil and rocks by water at a pH of 4.5 or less. The aluminum ion, for example, is often found in relatively high concentrations in fishless lakes.

Some lakes have a natural buffering capacity because they rest in basins of carbonate rock, like limestone (mostly $CaCO_3$). Carbonates neutralize hydrogen ions, as we learned in the previous chapter. For example,

$$CaCO_3(s) + 2H^+ \longrightarrow Ca^{2+}(aq) + CO_2(g) + H_2O$$

Where the underlying formations are not carbonate materials but are silicate rocks like quartzite sandstone and granite, the lakes are sensitive to acid rain. The U.S. Environmental Protection Agency classifies all of the lakes and streams in the United States as having moderate to high sensitivity, except for areas of the western Great Plains and the inland deserts of the west. Particularly sensitive are the great wilderness areas of northern Minnesota, Wisconsin, upper New York, and New England. The southern parts of Canada, which receive much of their acid rain from the United States, are also vulnerable, and the Canadian people are understandably displeased.

Oxides of Sulfur and Nitrogen Are the Chemical Agents Behind Acid Rain Acid rain is the result of the many activities of people that generate oxides of sulfur and nitrogen. We burn coal and oil to run electrical power plants and vehicles. These fuels contain traces of sulfur and sulfur compounds, and the sulfur in such fuels changes mostly to sulfur dioxide during combustion. Sulfur itself, for example, oxidizes as follows:

$$S(s) + O_2(g) \longrightarrow SO_2(g)$$

When sulfur dioxide dissolves in water it forms a hydrate, $SO_2 \cdot H_2O$, which has traditionally been written as H_2SO_3 and called sulfurous acid. Although a relatively weak acid, it is a stronger acid than carbonic acid, so when rain washes gaseous SO_2 from the atmosphere, the rainwater is acidic.

So much sulfur-bearing fuel is burned that millions of tons of SO_2 are injected into the atmosphere annually. Winds then carry it wherever they blow. The Scandinavian countries receive their acid rain largely from the industrial areas of central Europe, including England, Germany, Czechoslovakia, and Poland. The great industrial belt in the United States from Chicago to New York is a major producer of acid rain because much of the coal burned there (upwards of 60%) has relatively high levels of sulfur (2–4%). Less than 20% of western coal has this much sulfur.

The major source of the pH decreasing potential in acid rain comes from sulfur trioxide, which is made in the atmosphere from sulfur dioxide. The process is complicated, because there is no direct atmospheric reaction that occurs between SO_2 and O_2 to give SO_3. Both oxygen and the ozone in smog, however, can make SO_3 out of SO_2, particularly in the presence of sunlight and fine dust. The SO_3 reacts with water to form sulfuric acid, a strong acid, and thus is able to affect pH in smaller quantities than SO_2 and sulfurous acid. In environmental studies, the term *sulfur oxides* refers to SO_2 and SO_3 combined. There are also natural sources of these oxides, for example, volcanic eruptions. The chief source, however, is the combustion of coal and oil.

Nitrogen dioxide is another major air pollutant that dissolves in water to give acids—both nitric acid (a strong acid) and nitrous acid (a weak acid). Nitrogen dioxide forms in two steps during the high temperature combustion of fuels in the engines of vehicles and power plants. Fortunately, nitrogen and oxygen, the major gases in air, do not react with each other at ordinary temperature and atmospheric pressure. In an engine, however, the temperature becomes very high, and there are periodic moments of high pressure. Under these conditions, the following reaction is able to occur to some extent, and nitrogen monoxide (NO, "nitric oxide") forms.

$$N_2(g) + O_2(g) \longrightarrow 2NO(g)$$

Equation 8.5 and so ensure the neutralization thus depends on the loss of the CO_2 molecule from the body.

One of many lessons we can learn from these facts is that anything that interferes with the loss of CO_2 inhibits the neutralization of the H^+ ion and causes acidosis. In some people, those with emphysema, for example, the breathing apparatus works poorly. Their breathing becomes too slow and shallow, a condition called **hypoventilation.** Because they are not

Nitrogen monoxide is therefore a component of exhaust gases. When these gases diffuse and cool in the outside air, nitrogen monoxide is further oxidized to nitrogen dioxide, NO_2.

$$2NO(g) + O_2(g) \longrightarrow 2NO_2(g)$$

This is the air pollutant that largely causes the reddish-brown color of smog. Nitrogen dioxide reacts with water to give two acids.

$$2NO_2(g) + H_2O \longrightarrow HNO_3(aq) + HNO_2(aq)$$

Nitric acid   Nitrous acid

In air pollution studies, the term *nitrogen oxides* refers to NO and NO_2. (These oxides are also part of the complex atmospheric chemistry that produces the ozone, O_3, in smog, a pollutant particularly harmful to human life and to plants.)

Acid rain is not only a threat to aquatic life but also to agriculture and building materials. The accompanying photographs show the effect of air pollutants and acid rain on statuary. Limestone and marble are particularly sensitive, because they are chiefly calcium carbonate, and carbonates are dissolved by acids as we have already said. Several major cathedrals in Europe need constant repair because of the problem. Acid deposition is also corrosive to exposed metals such as railroad rails, vehicles, and machinery.

Acid Rain Is a Solvable Technical Problem; the Barriers are Economic and Therefore Political Less reliance on sulfur-containing coal and oil might be thought to be a solution to the acid deposition problem. No doubt it would, but the alternative is either drastic cutbacks in energy consumption or a major switch to nuclear power. But nuclear power presently is costlier in every way than power obtained from the burning of coal or oil and it bears its own pollution ills. Meanwhile, as coal and oil are used, the removal of most of the SO_2 from smokestack gases is possible. For example, SO_2 is absorbed by wet calcium hydroxide ("slaked lime") by the following reaction:

$$SO_2(g) + Ca(OH)_2(s) \longrightarrow CaSO_3(s) + H_2O$$

Not all SO_2 is removed, however, and given the enormous quantities of coal and oil burned worldwide, emissions of SO_2 still occur. In a technological sense, the problem is controllable. The questions remaining are mostly political, economic, and diplomatic. According to Swedish scientist Svante Oden, if SO_2 emissions are left uncontrolled, acid deposition will make the earth north of the 45th parallel a chemical desert.

A statue in 1908 (left) and the same statue in 1968 (right). Acid deposition accelerates the decay of stone, even noncarbonate stone. This statue of Baumberg sandstone at the Herten Castle in Westphalia, West Germany, lasted 206 years with little loss of features. After only 60 years of exposure to the heavily polluted air of the Rhein-Ruhr region of West Germany, virtually all features are gone.

expelling CO_2, the level of CO_2 and H_2CO_3 increases in their blood. This retention of CO_2 amounts to the retention of acid. People unable to breathe deeply enough, therefore, experience acidosis.

Any cause of shallow breathing, either because the tubes to the lungs are blocked (asthma), the lungs are fluid-filled (pneumonia), or because the brain's respiratory centers have been knocked out (by narcotics or barbiturates) means acidosis.

Too much loss of CO_2 can also be an emergency. It would mean too much loss of acid, so the problem would be alkalosis, not acidosis. Any time hyperventilation becomes uncontrolled and involuntary, as in hysterical fits or overbreathing at high altitudes, alkalosis results. Just how acidosis and alkalosis interfer with respiration are topics whose study we must postpone until we know more about the chemistry of the blood. We will return to this topic in Chapter 18.

It is not just human life that is threatened by changes in pH. All forms of life can be affected, as the discussion of acid rain in Special Topic 8.1 describes.

8.4 ACID–BASE TITRATIONS

At the end point of an acid–base titration, the number of moles of H^+ should match the number of moles of H^+ acceptor.

One of the very common kinds of chemical analysis is to determine the concentration of an acid or a base. The purpose is to find the molarity of some whole solute, like moles of acetic acid per liter of solution, not just to measure the pH of the solution.

A Solution's Capacity to Neutralize OH^- Is Its Neutralizing Capacity. As we said in Section 8.1, the pH of a solution tells us indirectly the *acidity* of a solution, what the concentration of hydronium ions is. It does not disclose the **neutralizing capacity** of the solution, its capacity to neutralize a strong base. The 1 mol of acetic acid in 1 L of 1 *M* $HC_2H_3O_2$ can neutralize 1 mol of NaOH, yet this solution has an actual quantity of H_3O^+ of only about 0.004 mol, the result of acetic acid being a weak acid.

Always remember that acids are classified as *weak* or *strong* according to their abilities to transfer a proton to one particular and very weak base, H_2O. When sodium hydroxide is added to an acid, however, we are adding a very strong base, OH^-. This base can take H^+ not only from H_3O^+ but also from $HC_2H_3O_2$. Thus the neutralizing capacity of 1 *M* acetic acid is considerably greater than its actual concentration of hydronium ion.

The Titration of an Acid with a Base Gives Data from Which Concentrations Can Be Calculated. The procedure used to measure the total acid (or base) neutralizing capacity of a solution is called **titration**. It involves comparing the volume of a solution of unknown concentration to the volume of a *standard solution* that exactly neutralizes it. A **standard solution** is simply one whose concentration is accurately known.

The apparatus for titration is shown in Figure 8.5. When a titration is used for an acid–base analysis, a carefully measured volume of the solution of unknown acidity (or basicity) is placed in a beaker or a flask. A very small amount of an acid–base indicator, like phenolphthalein, is added. Then a *standard solution* of the neutralizing reagent is added through a stopcock, portion by portion, from a special tube called a *buret,* marked in 1-mL and 0.1-mL divisions (Figure 8.5). This addition is continued until a change in color, caused by the acid–base indicator, signals that the unknown has been exactly neutralized.

End Points Ideally Occur at Equivalence Points. With a carefully selected acid–base indicator, the color change in an acid–base titration occurs when all the available hydrogen ions have reacted with all the available proton acceptors. This point in a titration is called the **equivalence point.**

A well-chosen indicator is one whose color at the equivalence point is the same as it would be in a solution made up of the *salt* that forms in the titration (and in the same concentration). When this salt has an ion that hydrolyzes, the equivalence point cannot be at pH 7.00. For example, when one mole of acetic acid has been exactly neutralized by one mole of sodium hydroxide, exactly one mole of sodium acetate has been made. Because the acetate ion hydrolyzes (but not the sodium ion), this salt gives a solution that is slightly basic to litmus,

FIGURE 8.5
The apparatus for titration. By
manipulating the stopcock, the analyst
controls the rate at which the solution
in the buret is added to the flask below.

not a solution with a pH of 7.00. So it would be poor to pick an indicator that changes color over an acidic range. (Phenolphthalein works very well in this titration.)

Whether or not the indicator has been well chosen, the analyst has little choice but to stop the titration when the indicator's color changes. This stopping point is called the **end point** of the titration. In a well-run titration, of course, the end point and the equivalence point coincide.

In Section 4.6 we studied how to do calculations involving volumes and concentrations of solutions with an emphasis on calculating volumes. Acid–base titrations, however, are usually done to determine concentrations, so we'll work through an example to see how it's done.

EXAMPLE 8.4   CALCULATING MOLARITIES FROM CONCENTRATION DATA

Problem: A student titrated 25.0 mL of sodium hydroxide solution with standard sulfuric acid. It took 13.4 mL of 0.0555 M H_2SO_4 to neutralize the sodium hydroxide in the solution. What was the molarity of the sodium hydroxide solution? The equation for the reaction is

$$2NaOH(aq) + H_2SO_4(aq) \longrightarrow Na_2SO_4(aq) + 2H_2O$$

Solution: Be sure to understand the goal first. We are to calculate the molarity of the NaOH, which means the ratio of the moles of NaOH to liters of NaOH solution. We were given (indirectly) the liters of the NaOH solution, 25.0 mL = 0.025 L. To find the moles of NaOH we need two conversion factors, one involving the molarity of the acid, and the other involving the coefficients in the equation.

The molarity of the H_2SO_4 solution, 0.0555 M, gives us the following conversion factors. We'll be using the first.

$$\frac{0.0555 \text{ mol } H_2SO_4}{1000 \text{ mL } H_2SO_4 \text{ soln}} \qquad \frac{1000 \text{ mL } H_2SO_4 \text{ soln}}{0.0555 \text{ mol } H_2SO_4}$$

The balanced equation gives us these conversion factors, and we will be using the second.

$$\frac{1 \text{ mol } H_2SO_4}{2 \text{ mol } NaOH} \qquad \frac{2 \text{ mol } NaOH}{1 \text{ mol } H_2SO_4}$$

Now let's begin with the given volume of the acid and convert it into the number of moles of NaOH it neutralized.

$$13.4 \text{ mL } H_2SO_4 \text{ soln} \times \frac{0.0555 \text{ mol } H_2SO_4}{1000 \text{ mL } H_2SO_4 \text{ soln}} \times \frac{2 \text{ mol } NaOH}{1 \text{ mol } H_2SO_4} = 0.00149 \text{ mol } NaOH$$

To convert mL acid to mol of acid

To find mols of base from mols of acid

Thus 0.00149 mol of NaOH was present in 25.0 mL or 0.025 L of NaOH solution. To find the molarity of the NaOH solution, we take the following ratio of moles to liters:

$$\frac{0.00149 \text{ mol } NaOH}{0.025 \text{ L } NaOH \text{ soln}} = 0.0596 \; M \; NaOH$$

Thus the molarity of the NaOH solution is 0.0596 M.

PRACTICE EXERCISE 8   If it takes 24.3 mL of 0.110 M HCl to neutralize 25.5 mL of freshly prepared sodium hydroxide solution, what is the molarity of the NaOH solution?

PRACTICE EXERCISE 9   If 20.0 mL of 0.125 M solution of NaOH exactly neutralized the sulfuric acid in 10.0 mL of H_2SO_4 solution, what was the molarity of the H_2SO_4 solution?

SUMMARY

pH The self-ionization of water produces an equilibrium in which the ion product constant, K_w, the product of the molar concentrations of hydrogen and hydroxide ions, is 1.0×10^{-14} (at 25 °C). If acids or bases are added, the value of K_w stays the same, but individual values of $[H^+]$ and $[OH^-]$ adjust.

A simple way to express very low values for these molar concentrations is by a value of pH, where $[H^+] = 1 \times 10^{-pH}$. When, at 25 °C, pH < 7, the solution is acidic. When pH > 7, the solution is basic. To measure the pH of a solution we use indicators, dyes whose colors change over a narrow range of pH, or we use a pH meter.

Hydrolysis of ions A dissolved salt affects the pH of the solution if one of its ions reacts more than the other with water to generate extra H^+ or OH^- ions. Cations of Groups IA and IIA metals (except Be^{2+}) do not hydrolyze. Nearly all others do. Anions derived from weak acids are relatively strong Brønsted bases and therefore hydrolyze to give basic solutions.

Buffers Solutions that contain something that can neutralize OH^- ion and something else that can neutralize H^+ ion are buffered against changes in pH when either additional base or acid is added. The phosphate buffer, present in the fluids inside cells, consists of HPO_4^{2-} (to neutralize H^+) and $H_2PO_4^-$ (to neutralize OH^-). The carbonate buffer, the chief buffer in blood, consists of HCO_3^- (to neutralize H^+) and H_2CO_3 (to neutralize OH^-). The supply of H_2CO_3 comes from the reaction of dissolved CO_2 with water.

Carbonic acid breaks down in the lungs where the CO_2 is expelled. When metabolism or some deficiency in respiration produces or retains H^+ at a rate faster than the blood buffer can neutralize it, the lungs try to remove CO_2 at a faster rate (hyperventilation). Overall, for each molecule of CO_2 exhaled, one proton is neutralized.

Acid – base titration The concentration of an acid or a base in water can be determined by titrating the unknown solution with a standard solution of what can neutralize it. The indicator is selected to have its color change occur at whatever pH the final solution would have if it were made from the salt that forms by the neutralization.

REVIEW EXERCISES

The answers to review exercises that require a calculation and that are marked with an asterisk are found in Appendix V. The answers to the other review exercises are found in the *Study Guide* that accompanies this book.

Ion Product Constant of Water

8.1   Write the equation that defines K_w.

8.2   What is the value of K_w at 25 °C?

8.3   The higher the temperature, the higher is the value of K_w. Why should there be this trend?

*8.4   At the temperature of the human body, 37 °C, the concentration of the hydrogen ion in pure water is 1.56×10^{-7} mol/L. What is the value of K_w at 37 °C?

8.5   "Heavy water" or deuterium oxide, D_2O, is used in nuclear power plants. It self-ionizes like water, and at 20 °C there is a concentration of D^+ ion of 3.0×10^{-8} mol/L. What is the value of K_w for heavy water at 20 °C?

pH

8.6   What equation defines pH in exponential terms?

8.7   The average pH of urine is about 6. Is this acidic, neutral, or basic?

8.8   The pH of gastric juice is in the range of 1.5–3.5. Is gastric juice acidic, basic, or neutral?

8.9   What is the pH of 0.01 M HCl(aq), assuming 100% ionization?

8.10 What is the pH of 0.01 M NaOH(aq), assuming 100% ionization?

8.11 Why does a pH of 7.00 correspond to a neutral solution at 25 °C?

8.12 Following surgery, a patient experienced persistent vomiting and the pH of his blood became 7.56. (Normally it is 7.35.) Has the blood become more alkaline or more acidic? Is the patient experiencing acidosis or alkalosis?

8.13 A patient brought to the emergency room following an overdose of aspirin was found to have a pH of 7.20 for the blood. (Normally the pH of blood is 7.35.) Has the blood become more acidic or more basic? Is the condition acidosis or alkalosis?

8.14 A certain brand of beer has a pH of 5.0. What is the concentration of hydrogen ion in moles per liter? Is the beer slightly acidic or basic?

8.15 The pH of a soft drink was found to be 4.5. Between what limits of molar concentration does the value of $[H^+]$ occur?

8.16 A solution of a monoprotic acid was prepared with a molar concentration of 0.10 M. Its pH was found to be 1.0. Is the acid a strong or a weak acid? Explain.

8.17 The pH of a solution of a monoprotic acid was found to be 4.56, whereas its molar concentration was 0.010 M. Is this acid a strong or a weak acid? Explain.

8.18 When a soil sample was stirred with pure water, the pH of the water changed to 7.90. Did the soil produce an acidic or a basic reaction with the water?

8.19 A patient entered the emergency room of a hospital after three weeks on a self-prescribed low-carbohydrate, high-fat diet and the regular use of the diuretic, acetazolamide. (A diuretic promotes the formation of urine and thus causes the loss of body fluid.) The pH of the patient's blood was 7.20. Was the condition acidosis or alkalosis?

Hydrolysis of Ions

8.20 Explain in your own words why a solution of sodium acetate, $NaC_2H_3O_2$, is slightly basic, not neutral.

8.21 Predict whether each of the following solutions is acidic, basic, or neutral.
(a) K_2SO_4 (b) NH_4NO_3 (c) $KHCO_3$
(d) $FeCl_3$ (e) Li_2CO_3

8.22 Predict whether each of the following solutions is acidic, neutral, or basic.
(a) KNO_3 (b) Na_2HPO_4 (c) K_3PO_4
(d) $Cr(NO_3)_3$ (e) $KC_2H_3O_2$

Buffers

8.23 When we say the blood is *buffered,* what does this mean in general terms (without specifying specific substances)?

8.24 What two chemical species make up the chief buffer in blood?

8.25 Write the equations that show how the chief buffer system in the blood works to neutralize hydroxide ion and hydrogen ion.

8.26 Explain in your own words, using equations as needed, how the loss of a molecule of CO_2 at the lungs permanently neutralizes a hydrogen ion.

8.27 In high-altitude sickness, the patient *involuntarily* overbreathes, and expels CO_2 from the body at a faster than normal rate. This results in an *increase* in the pH of the blood.
(a) Is this condition alkalosis or acidosis?
(b) Why should excessive loss of CO_2 result in an increase in the pH of the blood? (*Note:* Such a patient should be returned to lower elevations as soon as possible. It helps to rebreathe one's own air, as by breathing into a paper sack, because this helps the system retain CO_2.)

8.28 A patient with emphysema *involuntarily* hypoventilates. (In other words, the respiratory system is not working properly.) This leads to a decrease in the pH of the blood.
(a) Is this alkalosis or acidosis?
(b) Why should hypoventilation under these circumstances lead to a decrease in the pH of the blood?

8.29 What two ions constitute the phosphate buffer in cellular fluids?

8.30 Based on what should be predicted about the abilities of the anions of the phosphate buffer to hydrolyze, should one expect cellular fluids to be slightly alkaline or slightly acidic?

Acid – Base Titrations[5]

8.31 What does it mean to have a *standard* solution of, say, HCl(*aq*)?

8.32 When doing a titration, how does one know when the end point is reached?

8.33 What steps does an analyst take to ensure that the end point and the equivalence point in an acid–base titration occur together?

8.34 Give an example of a titration in which the equivalence point has a pH that is equal to 7. (Give a specific example of an acid and a base that, when titrated together, produce such a solution.)

8.35 Give a specific example of an acid and a base that, when titrated together, produce a solution with a pH greater than 7.

8.36 At the equivalence point in a titration the pH is <7. Give a specific example of an acid and a base that, when titrated together, would produce this result.

*8.37 Individual aqueous solutions were prepared that contained the following substances. Calculate the molarity of each solution.
(a) 7.292 g of HCl in 500.0 mL of solution
(b) 16.18 g of HBr in 500.0 mL of solution

8.38 What is the molarity of each of the following solutions?
(a) 25.58 g of HI in 500.0 mL of solution
(b) 9.808 g of H_2SO_4 in 500.0 mL of solution

*8.39 If 20.00 g of a monoprotic acid in 100.0 mL of solution gives a concentration of 0.5000 *M*, what is the formula weight of the acid?

8.40 A solution with a concentration of 0.2500 *M* could be made by dissolving 4.000 g of a base in 125.0 mL of solution. What is the formula weight of this base?

8.41 How many grams of each solute are needed to prepare the following solutions?
(a) 1000 mL of 0.2500 *M* HCl
(b) 750.0 mL of 0.1150 *M* HNO_3
(c) 500.0 mL of 0.01500 *M* H_2SO_4

8.42 To prepare each of the following solutions would require how many grams of the solute in each case?
(a) 100.0 mL of 0.1000 *M* HBr
(b) 250.0 mL of 0.1000 *M* H_2SO_4
(c) 500.0 mL of 1.250 *M* Na_2CO_3

*8.43 In standardizing a sodium carbonate solution, 21.45 mL of this solution was titrated to the end point with 18.78 mL of 0.1018 *M* HCl. The equation was

$$2HCl(aq) + Na_2CO_3(aq) \longrightarrow 2NaCl(aq) + CO_2(g) + H_2O$$

(a) Calculate the molarity of the sodium carbonate solution.
(b) How may grams of sodium carbonate does it contain per liter?

8.44 A freshly prepared solution of sodium hydroxide was standardized with 0.1148 *M* H_2SO_4.
(a) If 18.32 mL of the base was neutralized by 20.22 mL of the acid, what was the molarity of the base?
(b) How many grams of NaOH were in each liter of this solution?

Acid Rain (Special Topic 8.1)

8.45 What is the pH of unpolluted water in nature? Why is it less than 7?

8.46 Rain is called acid rain when its pH is less than what value?

8.47 How are some lakes buffered against decreases in pH even when they receive acid rain?

8.48 What are the names and formulas of the sulfur oxides in acid rain?

8.49 What are sources of sulfur for making the sulfur oxides in acid rain?

8.50 What is a way to remove much of the sulfur dioxide from exhaust gases at power plants? (Give an equation.)

8.51 Which sulfur oxide is the greater cause of a decrease in pH? Why?

8.52 What are the names and formulas of two nitrogen oxides in polluted air? Which causes the reddish-brown color of smog?

8.53 How do these nitrogen oxides form in the environment? (Give equations.)

8.54 Which nitrogen oxide decreases the pH of water? How does it do this? (Write the equation.)

8.55 What problem besides the decrease in pH can acid rain cause to aquatic life?

---

[5] For all of the calculations in the review exercises that follow, round atomic weights to their *second* decimal places before adding them to find formula weights.

Chapter 9
Introduction to Organic Chemistry

Organic and Inorganic Compounds   *Isomerism*
Structural Features of Organic Molecules

It's hard to believe that everything you see in this picture is made out of the electrons and nuclei of just a few elements. As our study shifts to organic chemistry in this chapter, we'll begin to see how this can be.

9.1 ORGANIC AND INORGANIC COMPOUNDS

Because most organic compounds are molecular, not ionic, they have relatively low boiling points, melting points, and solubilities in water.

Organic compounds are compounds of carbon, and there are more of these than of all the other elements combined, *except* hydrogen. The name itself, implying *organism,* arose when scientists, prior to 1828, believed that organic compounds could be made only by living organisms. They thought that a catalyst-like agency, a vital force, is essential to this synthesis, and that only living things possess it. **Inorganic compounds** were all the rest, those not dependent on a vital force.

Vita- is from a Latin root meaning ''life.''

Wöhler's Experiment Contradicted the Vital Force Theory.   In 1828, Friedrich Wöhler (1800 – 1882) succeeded in making urea, a white solid that can be isolated from urine and that everyone regarded as an organic compound. Urea was the unexpected result of his attempt to make crystalline ammonium cyanate, NH_4NCO, an inorganic compound. Wöhler prepared an aqueous solution of this salt, which consists of the ammonium ion, NH_4^+, and the cyanate ion, NCO^-, and he boiled off the water thinking, reasonably, that the residue would be the desired compound. But the product was urea. Evidently, the heat had caused the following reaction.

$$NH_4NCO \xrightarrow{\text{heat}} \begin{array}{c} H \quad O \quad H \\ | \quad\; \| \quad\; | \\ H-N-C-N-H \end{array}$$

Ammonium cyanate   Urea

Urea is the chief nitrogen waste from the body. It is also manufactured from ammonia and used as a commercial fertilizer.

There had been no living thing in the boiling solution. No vital force had been needed. The vital force theory was mortally wounded. Other syntheses of organic compounds from inorganic materials soon followed, and now the vital force theory is dead. Today, we define organic compounds without reference to organisms. They are simply the compounds of carbon other than a few types that we will mention soon. Roughly 6 million organic compounds are known, and all of them have been made, or could in principle be made, from inorganic substances.

The Differences Between Organic and Inorganic Compounds Reflect Differences in Their Chemical Bonds.   Relatively few inorganic compounds contain carbon. In those that do, mostly the carbonates, bicarbonates, and cyanides of metal ions, a carbon–carbon bond is almost never observed. The carbon atoms of organic compounds, on the other hand, are nearly always covalently bound to each other so as to form the structural ''skeletons'' of organic molecules. Atoms of several other nonmetal elements, for example, H, O, N, and S, are appended to these skeletons by covalent bonds.

Because organic compounds are made from nonmetal elements, covalent bonds are far more prevalent among them than ionic bonds, which so commonly occur among inorganic compounds, particularly the salts. Most organic compounds are thus molecular and most inorganic compounds are ionic. Organic and inorganic compounds therefore display some major differences in physical properties.

Forces of attraction between even quite polar molecules are generally much less than between oppositely charged ions. Most organic compounds, therefore, melt at temperatures well below those at which ionic compounds melt. A very large number of organic compounds, in fact, are liquids, not solids, at room temperature, but all ionic compounds are solids and melt generally well above 400 °C.

Most organic compounds also have normal boiling points below 400 °C, but most inorganic compounds have very high boiling points. The exceptions are the inorganic compounds that are not ionic, such as a large number of gases (e.g., hydrogen halides, carbon dioxide, and the oxides of sulfur) and some inorganic liquids (e.g., water).

Relatively few organic compounds dissolve in water, but many inorganic compounds are water soluble. Organic molecules are not often polar enough to be hydrated by water. However, the organic compounds that do dissolve in water are particularly important at the molecular level of life, because water is the central solvent in living things.

9.2 STRUCTURAL FEATURES OF ORGANIC MOLECULES

Organic molecules have flexible chains or rings of carbon atoms, and they nearly always have a functional group.

The uniqueness of carbon among the elements is that its atoms can bond to each other successively many times and still form equally strong bonds to atoms of other nonmetals. A typical molecule in the familiar plastic polyethylene has hundreds of carbon atoms covalently joined in succession. Each carbon binds enough hydrogen atoms to fill out its full complement of four bonds.

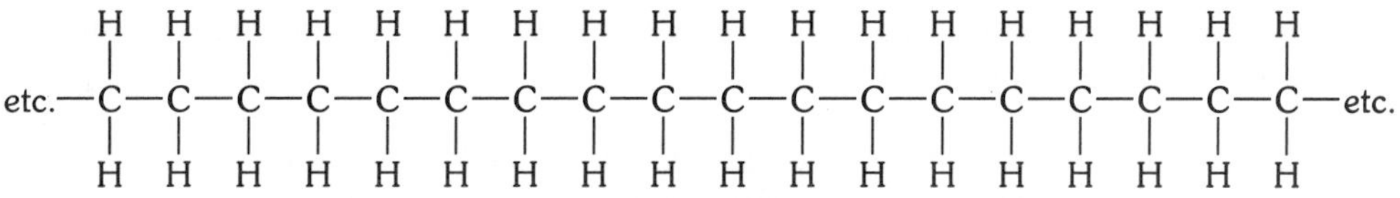

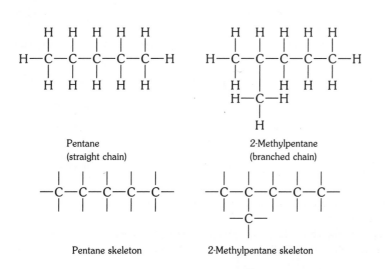

Polyethylene

Only a short segment of a typical molecule of polyethylene is shown here.

Recall that each line in a structural formula stands for one shared pair of electrons. Each carbon, therefore, has four lines (bonds), meaning four shared pairs of electrons for a total of eight outside-level electrons. In all stable, electrically neutral, organic molecules, carbon *always* has four bonds coming from it, *never* five and *never* three.

Carbon Skeletons Occur as Straight Chains, Branched Chains, and Rings. A succession of carbon atoms covalently bonded together, as in polyethylene, is called as **straight chain.** Molecules of pentane, one of many compounds in gasoline, have straight chains just five carbons long. "Straight" in this context has a very limited, technical meaning: The absence of carbon branches. "Straight chain" means that one carbon follows another with no additional carbons joined at intermediate points. All carbons, except the two at opposite ends of the chain, are joined to just two other carbons.

A molecule of 2-methylpentane illustrates a **branched chain,** which means that it has one (or more) carbon atoms joined to carbons that are *between* the ends of some parent chain.

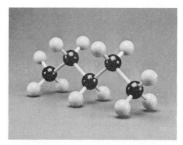

Pentane

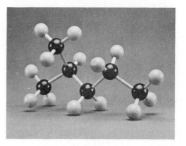

2-Methylpentane

Pentane (straight chain)

2-Methylpentane (branched chain)

Pentane skeleton

2-Methylpentane skeleton

Cyclopentane

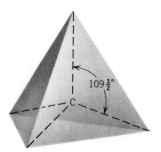

$109\frac{1}{2}°$

Tetrahedral carbon

Pentane and 2-methylpentane are sometimes called *open-chain compounds* to contrast them with compounds in which the carbon atoms form a cyclic system called a **ring**. Cyclopentane is an example of a ring compound.

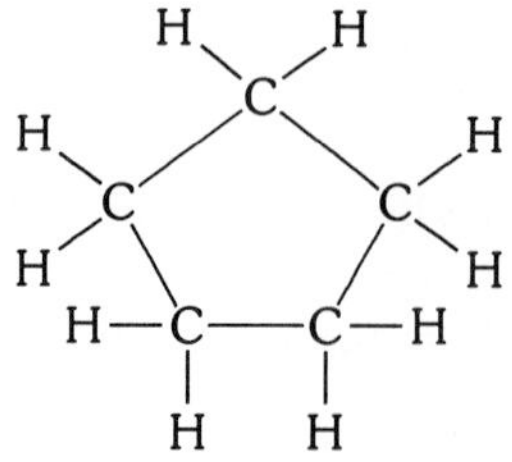

Cyclopentane

Printed Structural Formulas Usually Ignore Bond Angles. The ball-and-stick models of pentane, 2-methylpentane, and cyclopentane show the correct angle of 109.5° between any two bonds at any carbon with four single bonds (see Special Topic 3.2.). To simplify matters, the printed structures nearly always let this detail be "understood." We mentally read the correct bond angles into a printed structure.

Free Rotation Occurs about Single Bonds in Open-Chain Compounds. Another fact about molecules left to the imagination when we write their structures is their flexibility. Pieces of open-chain molecules connected by *single* bonds have a property called **free rotation.** Such pieces can rotate with respect to each other around the single bond that joins them. These rotations are generally caused by collisions between molecules, so in a sample of pentane not all of the molecules are in the fully extended form shown in the photograph. They are kinked and twisted into an almost infinite number of contorted forms, called conformations. Photographs of the models of just a few conformations of pentane are shown in Figure 9.1. Any specific property of pentane is the net effect of all of its variously twisted molecules on whatever physical agent or chemical reactant has been used to observe the property.

Molecular Structures Are Usually Displayed by Condensed Structural Formulas. So far we have shown every bond in a structural formula as a straight line, but we have also seen how it is useful to leave some molecular features to the imagination. We can also leave most of the bonds in a structural formula to the imagination. In doing this, we *must* always remember that *every carbon must have four bonds.*

With this in mind, we can group the hydrogen atoms held by a particular carbon all together to one side or the other of its symbol. Whenever a carbon holds three hydrogens, for example, we can simplify the system by writing CH_3. (We could also write H_3C, but you don't see this as often.) Just remember that these three hydrogen atoms are *individually* joined to the carbon. The simplest example of doing this occurs with the structural formula of ethane.

Ethane (C_2H_6)   H—C—C—H   becomes CH_3—CH_3

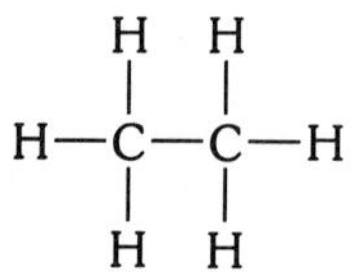

Full or expanded
structure

Condensed
structure

When a carbon holds two hydrogen atoms, we can write the system either as CH_2 or (less often seen) as H_2C.

When a carbon holds just one hydrogen, we can write it as CH or (less commonly seen) as HC.

FIGURE 9.1
Free rotation at single bonds. Photographs of models of four of the innumerable conformations of the pentane molecule are shown here followed by drawings of just their carbon skeletons. When we write the structure of pentane as $CH_3CH_2CH_2CH_2CH_3$, it stands for any of these, since some of each (and many others) exist in a sample of pentane.

When we condense a full or expanded structure following these guides, the result is called a **condensed structure,** but because this is almost always the kind used when discussing a compound, it is simply called the substance's **structure.** (We'll show how to condense ring structures to simple geometric figures in Chapter 10.)

EXAMPLE 9.1   CONDENSING A FULL STRUCTURAL FORMULA

Problem: Condense the structural formula for 2-methylpentane.

2-Methylpentane

Solution: CH₃—CH—CH₂—CH₂—CH₃
                          |
                          CH₃

PRACTICE EXERCISE 1   Condense the following expanded structural formulas.

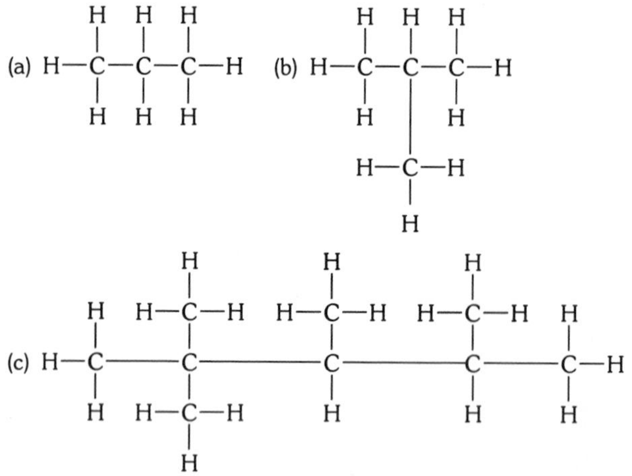

$$
\text{(a) } H-\overset{\displaystyle H}{\underset{\displaystyle H}{C}}-\overset{\displaystyle H}{\underset{\displaystyle H}{C}}-\overset{\displaystyle H}{\underset{\displaystyle H}{C}}-H
\qquad
\text{(b) } H-\overset{\displaystyle H}{\underset{\displaystyle H}{C}}-\overset{\displaystyle H}{\underset{\displaystyle \overset{|}{\underset{H}{C}}}{C}}-\overset{\displaystyle H}{\underset{\displaystyle H}{C}}-H
$$

(with H—C—H below the central carbon in (b))

(c) expanded structure with carbon chain

Because of free rotation, we have to be able to interpret zigzags. For example,

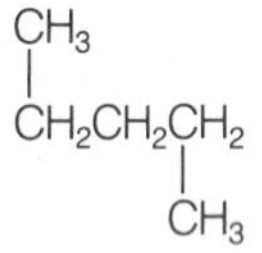

$$
\underset{\underset{CH_3}{\overset{|}{\underset{CH_2CH_2CH_2}{|}}}}{\overset{CH_3}{|}}
$$

is the same molecule as $CH_3CH_2CH_2CH_2CH_3$

There is still another useful simplification in writing condensed structures. Whenever a *single* bond appears on a *horizontal* line, we need not write a straight line to represent it; we can leave such a bond understood. We cannot do this for bonds that are not on a horizontal line. The 2-methylpentane of Example 9.1, for example, can be written as follows. Notice that the vertically oriented bond is shown by a line.

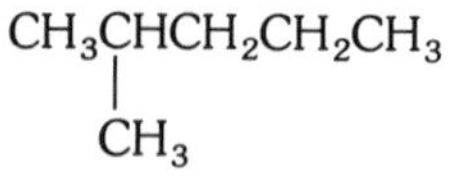

$$
\underset{\displaystyle CH_3}{CH_3\overset{|}{C}HCH_2CH_2CH_3}
\qquad \text{2-Methylpentane}
$$

PRACTICE EXERCISE 2   Rewrite the condensed structures that you drew for the answers to Practice Exercise 1 and let the appropriate carbon–carbon single bonds be left to the imagination.

PRACTICE EXERCISE 3   Just to be certain that you are comfortable with condensed structures, expand each of the following to make them full, expanded structures with no bonds left to the imagination.

(a) CH_3CH_3 (b) $CH_3\overset{\displaystyle CH_3}{\underset{\displaystyle CH_3}{\overset{|}{C}HCHCH_3}}$ (c) $CH_3CH_2\overset{\displaystyle CH_2CH_3}{\underset{\displaystyle CH_3}{\overset{|}{C}CH_2CH_2CH_3}}$

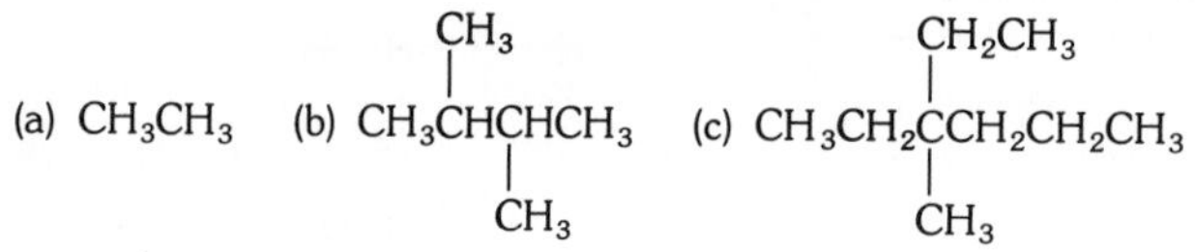

One very important skill needed to use condensed structures is the ability to recognize errors. The most common error is not having four bonds from each carbon, and showing or implying either too many or too few bonds. See if you can find the errors in the next exercise.

PRACTICE EXERCISE 4   Which of the following structures cannot represent real compounds?

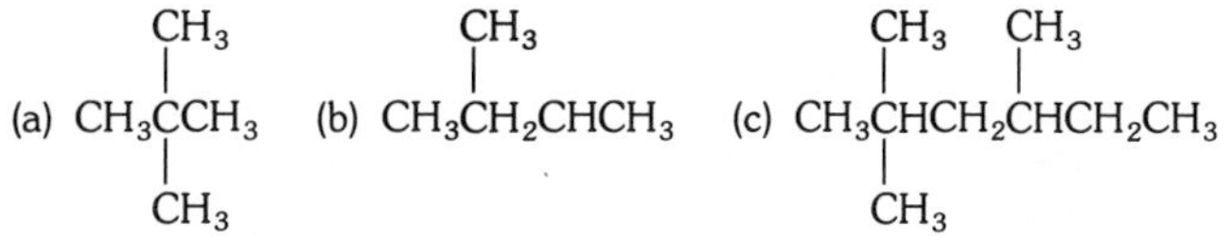

$$\text{(a)} \quad CH_3\overset{\displaystyle CH_3}{\underset{\displaystyle CH_3}{\overset{\displaystyle |}{\underset{\displaystyle |}{C}}}}CH_3 \qquad \text{(b)} \quad CH_3CH_2\overset{\displaystyle CH_3}{\overset{\displaystyle |}{C}}HCH_3 \qquad \text{(c)} \quad CH_3\overset{\displaystyle CH_3}{\overset{\displaystyle |}{C}}HCH_2\overset{\displaystyle CH_3}{\overset{\displaystyle |}{C}}HCH_2CH_3$$

Still another simplification is to group inside parentheses two or three identical groups attached to the same carbon. Although we will not do this often, here are two examples.

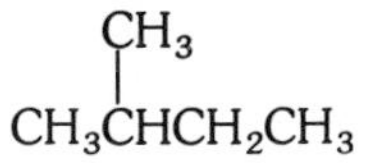

$$\overset{\displaystyle CH_3}{\overset{\displaystyle |}{C}}H_3CHCH_2CH_3 \qquad \text{can be written as} \quad (CH_3)_2CHCH_2CH_3$$

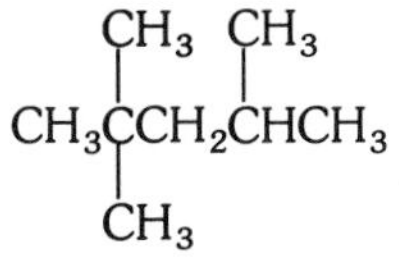

$$CH_3\overset{\displaystyle CH_3 \quad CH_3}{\underset{\displaystyle CH_3}{\overset{\displaystyle | \quad\quad |}{\underset{\displaystyle |}{C}}}}CH_2CHCH_3 \qquad \text{can be written as} \quad (CH_3)_3CCH_2CH(CH_3)_2$$

When a structure of a stable compound contains atoms other than carbon and hydrogen, the octet rule requires that *every electrically neutral oxygen or sulfur atom must have two bonds, every neutral nitrogen must have three,* and *every halogen atom must have just one.*

Another rule about condensed structures is that carbon–carbon double and triple bonds are not left to the imagination. For example,

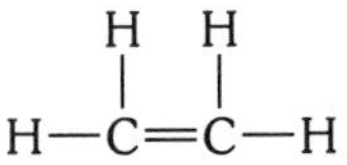

$$H\overset{\displaystyle H \quad H}{\underset{}{\overset{\displaystyle | \quad\; |}{-C=C-}}}H$$

Ethylene

O and N can have more bonds in certain *ions,* for example, H_3O^+ and NH_4^+.

condenses to $CH_2{=}CH_2$ or to $H_2C{=}CH_2$ but not to CH_2CH_2

(the raw material for making polyethylene)

Ethylene

Functional Groups Define Families of Organic Compounds. Chemicals that react with organic compounds generally attack only certain molecular parts and leave the rest alone. These parts, called **functional groups,** can be places where there are double or triple bonds or where atoms other than C or H occur. The parts of organic molecules that consist only of carbon and hydrogen and only single bonds undergo so few reactions that they are called **nonfunctional groups.**

Although over 6 million organic compounds are known, there are only a handful of functional groups, and each one serves to define a family of organic compounds. Our study of organic chemistry will be organized around just a few of these families, those outlined in Table 9.1. One important family is called the alcohol family. We have learned, for example, that ethyl alcohol is CH_3CH_2OH. Its molecules have the OH group attached to a chain of two carbons. But *chain length* is not what determines the family of a compound. It only determines the name of the specific family member, as we will see later. The chain can be any length you can think of, and the substance will be in the alcohol family provided that somewhere on the chain there is an OH group attached to a carbon from which only single bonds extend. Some simple examples of alcohols are

$$CH_3{-}OH \qquad CH_3CH_2{-}OH \qquad CH_3CH_2CH_2{-}OH \qquad CH_3\overset{\displaystyle CH_3}{\underset{\displaystyle OH}{\overset{\displaystyle |}{\underset{\displaystyle |}{C}}}}H$$

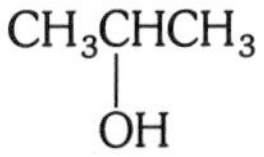

| Methyl alcohol | Ethyl alcohol | Propyl alcohol | Isopropyl alcohol |

TABLE 9.1
Some Important Families of Organic Compounds

Family	Molecular Features	Example

Hydrocarbons	Only C and H present
	Subfamilies:
	  Alkanes: only single bonds
	  Alkenes: some double bonds
	  Alkynes: some triple bonds
	
	  Aromatic: benzene ring present
		CH_3CH_3, ethane
		$CH_2=CH_2$, ethene
		$HC\equiv CH$, ethyne 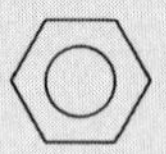
		
		⬡ , benzene

Alcohols	$-\overset{\|}{\underset{\|}{C}}-OH$ as in $R-O-H$	CH_3CH_2OH, ethyl alcohol
Ethers	$-\overset{\|}{\underset{\|}{C}}-O-\overset{\|}{\underset{\|}{C}}-$ as in $R-O-R'$	$CH_3CH_2OCH_2CH_3$, diethyl ether
Thioalcohols (mercaptans)	$-\overset{\|}{\underset{\|}{C}}-S-H$ as in $R-S-H$	CH_3SH, methyl mercaptan
Disulfides	$-\overset{\|}{\underset{\|}{C}}-S-S-\overset{\|}{\underset{\|}{C}}-$ as in $R-S-S-R'$	CH_3SSCH_3 dimethyl disulfide
Aldehydes	$-\overset{O}{\overset{\|\|}{C}}-H$ as in $R-\overset{O}{\overset{\|\|}{C}}-H$	$CH_3\overset{O}{\overset{\|\|}{C}}H$, acetaldehyde
Ketones	$-\overset{\|}{\underset{\|}{C}}-\overset{O}{\overset{\|\|}{C}}-\overset{\|}{\underset{\|}{C}}-$ as in $R-\overset{O}{\overset{\|\|}{C}}-R'$	$CH_3\overset{O}{\overset{\|\|}{C}}CH_3$, acetone
Carboxylic acids	$-\overset{O}{\overset{\|\|}{C}}-O-H$ as in $R-\overset{O}{\overset{\|\|}{C}}-O-H$	$CH_3\overset{O}{\overset{\|\|}{C}}OH$, acetic acid
Esters	$-\overset{O}{\overset{\|\|}{C}}-O-\overset{\|}{\underset{\|}{C}}-$ as in $R-\overset{O}{\overset{\|\|}{C}}-O-R'$	$CH_3\overset{O}{\overset{\|\|}{C}}OCH_2CH_3$, ethyl acetate
Phosphate esters	$-\overset{\|}{\underset{\|}{C}}-O-\underset{\underset{OH}{\|}}{\overset{\overset{O}{\|\|}}{P}}-O-H$ as in $R-O-\underset{\underset{OH}{\|}}{\overset{\overset{O}{\|\|}}{P}}-O-H$	$CH_3O\underset{\underset{OH}{\|}}{\overset{\overset{O}{\|\|}}{P}}OH$, methyl phosphate
Diphosphate esters	$R-O-\underset{\underset{OH}{\|}}{\overset{\overset{O}{\|\|}}{P}}-O-\underset{\underset{OH}{\|}}{\overset{\overset{O}{\|\|}}{P}}-O-H$	$CH_3O\underset{\underset{OH}{\|}}{\overset{\overset{O}{\|\|}}{P}}-O-\underset{\underset{OH}{\|}}{\overset{\overset{O}{\|\|}}{P}}-OH$, methyl diphosphate
Triphosphate esters	$R-O-\underset{\underset{OH}{\|}}{\overset{\overset{O}{\|\|}}{P}}-O-\underset{\underset{OH}{\|}}{\overset{\overset{O}{\|\|}}{P}}-O-\underset{\underset{OH}{\|}}{\overset{\overset{O}{\|\|}}{P}}-O-H$	$CH_3O\underset{\underset{OH}{\|}}{\overset{\overset{O}{\|\|}}{P}}-O-\underset{\underset{OH}{\|}}{\overset{\overset{O}{\|\|}}{P}}-O-\underset{\underset{OH}{\|}}{\overset{\overset{O}{\|\|}}{P}}-OH$, methyl triphosphate
Amines	$-NH_2$ as in $R-NH_2$ $-NH-$ as in $R-NH-R'$ $-\overset{\|}{N}-$ as in $R-\underset{\underset{R}{\|}}{N}-R'$	CH_3NH_2, methylamine $CH_3NHCH_2CH_3$, methylethylamine $CH_3\underset{\underset{CH_3}{\|}}{N}CH_3$, trimethylamine
Amides	$-\overset{O}{\overset{\|\|}{C}}-NH_2$ as in $R-\overset{O}{\overset{\|\|}{C}}-NH_2$	$CH_3\overset{O}{\overset{\|\|}{C}}NH_2$, acetamide

Amides can also be of the types:

$R-\overset{O}{\overset{\|\|}{C}}-NH-R'$ and $R-\overset{O}{\overset{\|\|}{C}}-N(R)_2$

Organic Families Have Family Symbols. Because all alcohols have the same functional group, they exhibit the same kinds of chemical reactions. When just one of these reactions is learned, it applies to all members of the family, literally to thousands of compounds. In fact, we will often summarize a particular reaction for an organic family by using a general family symbol. All alcohols, for example, can be symbolized by the symbol $R-OH$, where R stands for a carbon chain, any chain of whatever length. All alcohols, for instance, react with sodium metal as follows:

R is from the German word *Radikal,* which we translate here to mean *group,* as in a group of atoms.

$$2RO-H + 2Na \longrightarrow 2RONa + H_2$$

(RONa is an ionic compound, a combination of RO^- and Na^+.) If we wanted to write the specific example of this reaction that involves, say, ethyl alcohol, all we have to do is replace R by CH_3CH_2.

$$2CH_3CH_2OH + 2Na \longrightarrow 2CH_3CH_2ONa + H_2$$

Butyl alcohol, another member of the family, reacts as follows:

$$CH_3CH_2CH_2CH_2OH + 2Na \longrightarrow 2CH_3CH_2CH_2CH_2ONa + H_2$$

You can see that the reaction occurs only at the OH group, the functional group of all alcohols. The chain length is not a factor.

9.3 ISOMERISM

Compounds can have identical molecular formulas but different structures.

Ammonium cyanate and urea, the chemicals of Wöhler's important experiment, both have the molecular formula, CH_4N_2O. The atoms are just organized differently:

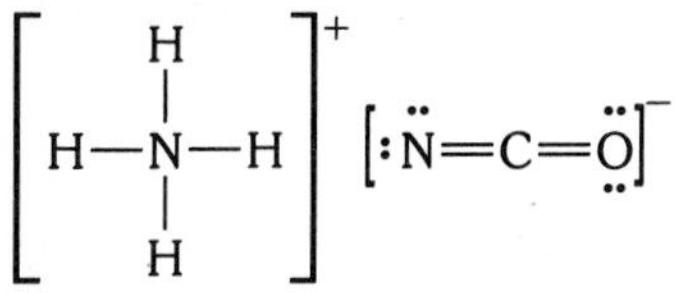

Ammonium cyanate
CH_4N_2O

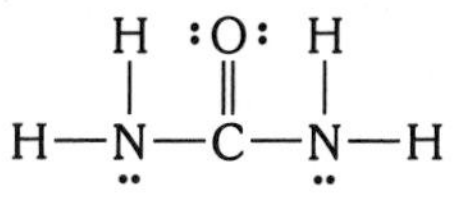

Urea
CH_4N_2O

Compounds with identical molecular formulas but different structures are called **isomers** of each other, and the existence of isomers is a phenomenon called **isomerism.** Isomerism is one of the reasons why there are so many organic compounds.

"Isomer" has Greek roots—*isos,* the same, and *meros,* parts; in other words, "the same parts" (but put together differently).

Pentane has three isomers, and all share the formula C_5H_{12}. Pay particular attention here to the differences in their *structures.* We will return to their names in Chapter 10.

Pentane
(*n*-pentane)

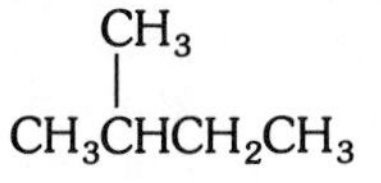

2-Methylbutane
(isopentane)

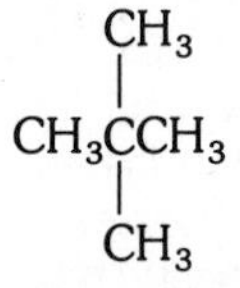

2,2-Dimethylpropane
(neopentane)

The names in parentheses are the *common* names of these compounds. The letter *n* stands for *normal,* meaning the straight-chain isomer. *Neo* signifies *new,* as in a new isomer.

Because the isomers of pentane belong to the same organic family, their chemical properties are quite similar. Often, however, isomers are in different families with widely different properties. The two isomers of C_2H_6O are examples. One isomer is ethyl alcohol and

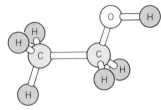

Ethyl alcohol.

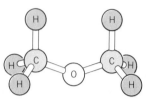

Dimethyl ether.

TABLE 9.2
Properties of Two Isomers—Ethyl Alcohol and Dimethyl Ether

Property	Ethyl Alcohol	Dimethyl Ether
Structure	CH_3CH_2OH	CH_3OCH_3
Boiling point	78.5 °C	−24 °C
Melting point	−117 °C	−138.5 °C
Density (25 °C)	0.79 g/mL (a liquid)	0.002 g/mL (a gas)
Solubility in water	Soluble in all proportions	Slightly soluble

the other is dimethyl ether. The latter is in the ether family, which has the general formula of R—O—R′. (The prime sign, ′, signifies only that the two R groups need not be identical.)

Both physically and chemically, ethyl alcohol and dimethyl ether are radically different substances, as the data in Table 9.2 show. At room temperature, the former is a liquid and the latter is a gas. Ethyl alcohol reacts with sodium. No member of the ether family does. Isomers very commonly differ this much, which explains why we almost always use structural rather than molecular formulas. Only the structure lets us see at a glance just what family a compound belongs to.

EXAMPLE 9.2   RECOGNIZING ISOMERS

Problem: Which pair of structures represents a pair of isomers?

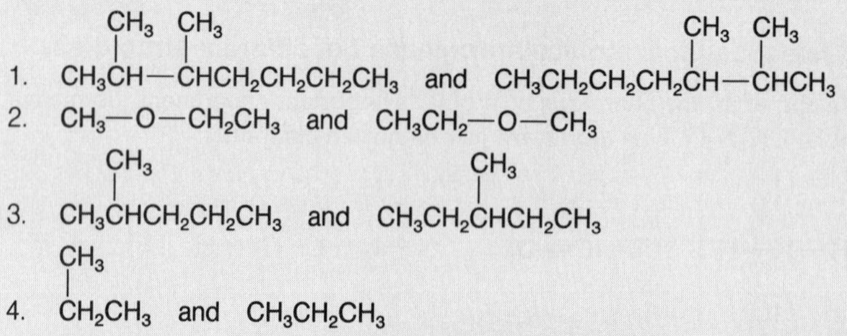

1. CH_3CH—$CHCH_2CH_2CH_2CH_3$ and $CH_3CH_2CH_2CH_2CH$—$CHCH_3$
   (with CH_3 CH_3 substituents shown above each pair)

2. CH_3—O—CH_2CH_3 and CH_3CH_2—O—CH_3

3. $CH_3CHCH_2CH_2CH_3$ and $CH_3CH_2CHCH_2CH_3$
   (with CH_3 substituents shown above)

4. CH_2CH_3 and $CH_3CH_2CH_3$
   (with CH_3 substituent shown above)

Solution: Unless you spot a difference that rules out isomerism right away, the first step is to see whether the molecular formulas are the same. If they aren't, the two structures are not isomers; and if they are, the two might be identical or they might be isomers. In this problem, the members of each pair share the same molecular formula.

Pair 1 C_9H_{20} Pair 2 C_3H_8O Pair 3 C_6H_{14} Pair 4 C_3H_8

Next, to see whether a particular pair represents isomers we try to find at least one structural difference. If we can't, the two structures are identical; they are just oriented differently on the page, or their chains are twisted into different conformations. Don't be fooled by an "east-to-west" versus a "west-to-east" kind of difference. The difference must be *internal* within the structure. (Whether you face east or west you're the same person!)

Pair 1 is an example of this east-versus-west difference in orientation. These two structures are identical. Their internal sequences are the same.

Pair 2 are likewise identical; they're just oriented differently on the page.

Pair 3 are isomers. In the first, a CH_3 group joins a five-carbon chain at the second carbon of the chain and in the second, this group is attached at the third carbon.

Pair 4 are identical. The two structures differ only in the conformations of their chains.

PRACTICE EXERCISE 5 Examine each pair to see whether the members are identical, are isomers, or are different in some other way.

(a) $H—O—CH_3$ and $CH_3—O—H$

(b) $CH_3—NH—CH_3$ and $CH_3—CH_2—NH_2$

(c) $\underset{\underset{CH_3}{|}}{CH_2CH_2CHCH_3}$ and $\underset{\underset{CH_2CH_3}{|}}{CH_3CH_2CH_2CH_2CHCH_3}$

(d) $CH_2{=}CHCH_2CH_3$ and $CH_3CH{=}CHCH_3$

(e) $CH_3CH_2\overset{\overset{O}{\|}}{C}OH$ and $HO\overset{\overset{O}{\|}}{C}CH_3$

SUMMARY

Organic and inorganic compounds Most organic compounds are molecular and the majority of inorganic compounds are ionic. Molecular and ionic compounds differ in composition, in types of bonds, and in several physical properties.

Structural features of organic molecules The ability of carbon atoms to join to each other may times in succession in straight chains and branched chains as well as into cyclic rings accounts in a large measure for the existence of several million organic compounds.

When groups within a molecule are joined by a single bond, they can rotate relative to each other about this bond. Full structural formulas of organic compounds are usually condensed by grouping the hydrogens attached to a carbon immediately by it; by letting single bonds on a horizontal line be understood; and by leaving bond angles and conformational possibilities to the informed imagination.

The families of organic compounds are organized around functional groups, which are parts of molecules at which most of the chemical reactions occur. Nonfunctional parts of molecules can sometimes be given the general symbol R—, as in R—O—H, the general symbol for all alcohols.

Isomerism Differences in the *conformations* of carbon chains do not create new compounds, but differences in the organizations of their structural parts do. Isomers are compounds with identical molecular formulas but different structures.

REVIEW EXERCISES

The answers to these review exercises are in the *Study Guide* that accompanies this book.

Organic and Inorganic Compounds

9.1 Why are the compounds of carbon generally called *organic* compounds?

9.2 Prior to 1828 scientists believed that what had to be present in order to make organic compounds from other types?

9.3 Who was the scientist who provided the first serious challenge to the idea that a special force had to be present to make organic compounds?

9.4 Describe what this scientist did that opened the way to renewed efforts to make organic compounds from inorganic compounds.

9.5 What kind of bond between atoms predominates among organic compounds?

9.6 Which of the following compounds are inorganic?
(a) CH_3OH (b) CO (c) CCl_4
(d) $NaHCO_3$ (e) K_2CO_3

9.7 Are the majority of all compounds that dissolve in water ionic or molecular? Inorganic or organic?

9.8 Explain why very few organic compounds conduct electricity either in an aqueous solution or as molten materials.

9.9 Each compound described below is either ionic or molecular. State which it most likely is, and give one reason.
(a) The compound is a colorless gas at room temperature.
(b) This compound dissolves in water. When hydrochloric acid is added, the solution fizzes and an odorless, colorless gas is released, which can extinguish a burning flame.
(c) This compound melts at 300 °C, and it burns in air.
(d) This compound melts at 675 °C, and it becomes white when heated.
(e) This compound is a liquid that does not dissolve in water.

Structural Formulas

9.10 One can write the structure of butane, lighter fluid, as follows:

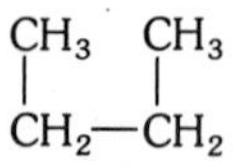

$$\begin{array}{cc} CH_3 & CH_3 \\ | & | \\ CH_2 & - CH_2 \end{array}$$

Butane

Are butane molecules properly described as straight chain or as branched chain, in the sense in which we use these terms? Explain.

9.11 Which of the following structures are possible, given the numbers of bonds that various atoms can form?
(a) $CH_2CH_2CH_3$ (b) $CH_3{=}CHCH_2CH_3$
(c) $CH_3CH{=}CH_2CH_2CH_3$

9.12 Write full (expanded) structures for each of the following *molecular* formulas. Remember how many bonds the various kinds of atoms must have. In some you will have to use double or triple bonds. (*Hint:* A trial-and-error approach will have to be used.)
(a) CH_5N (b) CH_2Br_2
(c) $CHCl_3$ (d) C_2H_6
(e) CH_2O_2 (f) CH_2O
(g) NH_3O (h) C_2H_2
(i) N_2H_4 (j) HCN
(k) C_2H_3N (l) CH_4O

9.13 Write neat condensed structures of the following:

(a)

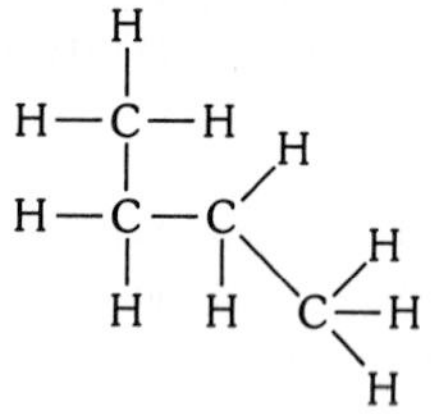

(b)

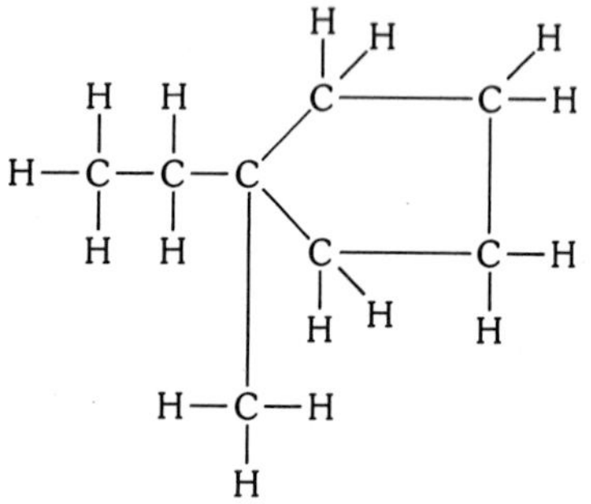

(c)

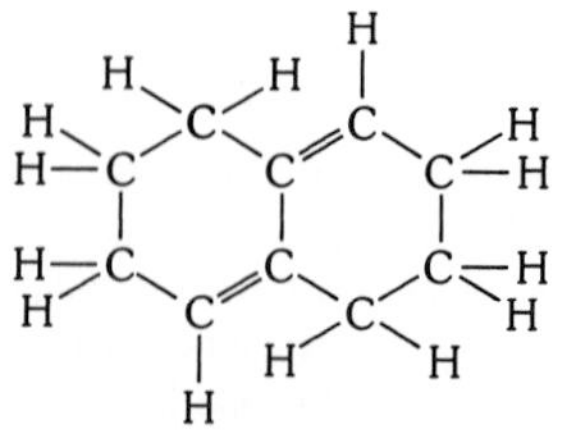

Isomers

9.14 Decide whether the members of each pair are identical, are isomers, or are unrelated.

(a) 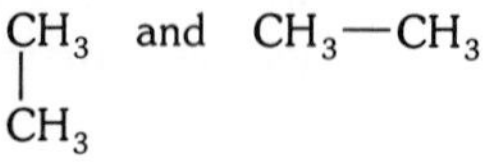 CH_3 and $CH_3{-}CH_3$
     $|$
     CH_3

(b) 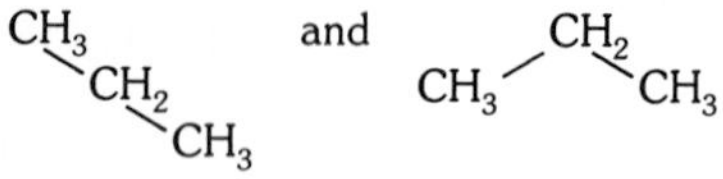 CH_3 and CH_2
       CH_2 CH_3 CH_3
         CH_3

(c) $CH_3CH_2{-}OH$ and $CH_3CH_2CH_2{-}OH$

(d) 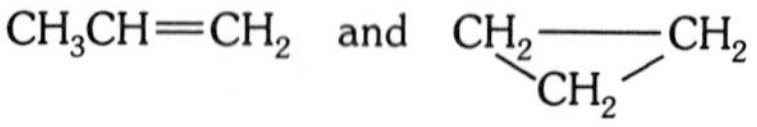 $CH_3CH{=}CH_2$ and $CH_2 {\rule{1cm}{0.4pt}} CH_2$
                          CH_2

(e) 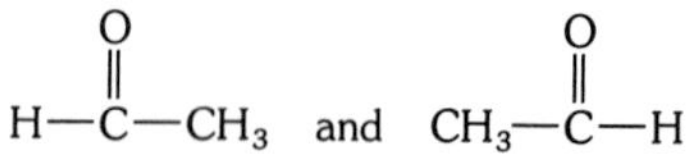 $H{-}\overset{\displaystyle O}{\overset{\|}{C}}{-}CH_3$ and $CH_3{-}\overset{\displaystyle O}{\overset{\|}{C}}{-}H$

(f) 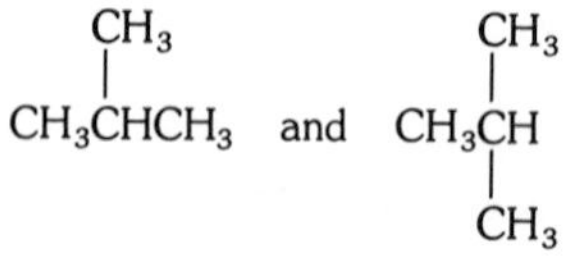 $CH_3\overset{\displaystyle CH_3}{\overset{|}{C}}HCH_3$ and $CH_3\overset{\displaystyle}{\underset{\displaystyle CH_3}{\overset{\displaystyle CH_3}{\overset{|}{C}H}}}$

(g) $CH_3CH_2CH_2{-}NH_2$ and $CH_3CH_2{-}NH{-}CH_3$

(h) 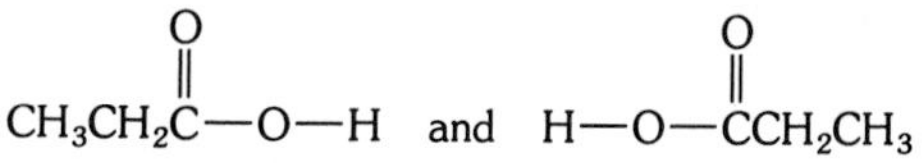 $CH_3CH_2\overset{\displaystyle O}{\overset{\|}{C}}{-}O{-}H$ and $H{-}O{-}\overset{\displaystyle O}{\overset{\|}{C}}CH_2CH_3$

(i) 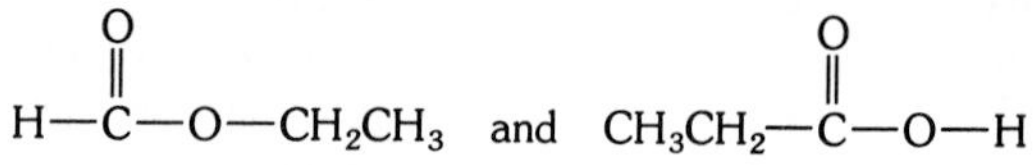 $H{-}\overset{\displaystyle O}{\overset{\|}{C}}{-}O{-}CH_2CH_3$ and $CH_3CH_2{-}\overset{\displaystyle O}{\overset{\|}{C}}{-}O{-}H$

(j) 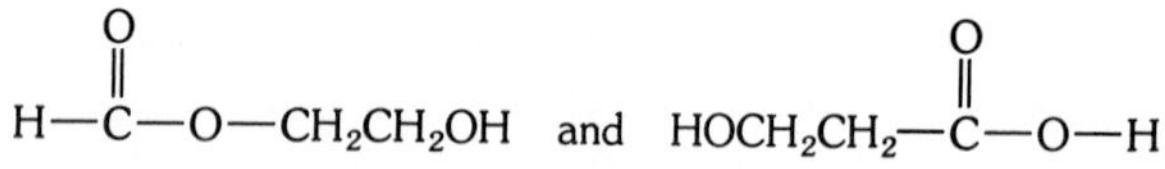 $H{-}\overset{\displaystyle O}{\overset{\|}{C}}{-}O{-}CH_2CH_2OH$ and $HOCH_2CH_2{-}\overset{\displaystyle O}{\overset{\|}{C}}{-}O{-}H$

(k) 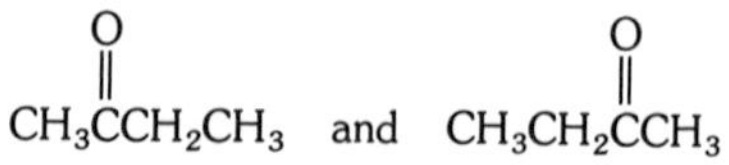 $CH_3\overset{\displaystyle O}{\overset{\|}{C}}CH_2CH_3$ and $CH_3CH_2\overset{\displaystyle O}{\overset{\|}{C}}CH_3$

(l) 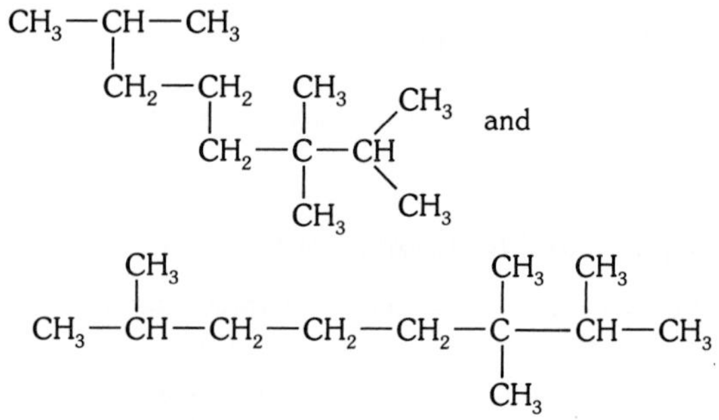 $CH_3{-}\overset{\displaystyle}{\overset{|}{C}H}{-}CH_3$
           $CH_2{-}CH_2$ CH_3 CH_3
                    $CH_2{-}\overset{\displaystyle}{\underset{\displaystyle CH_3}{\overset{|}{C}}}{-}CH$ and
                                CH_3

                  CH_3 CH_3 CH_3
$CH_3{-}\overset{|}{C}H{-}CH_2{-}CH_2{-}CH_2{-}\overset{\displaystyle}{\underset{\displaystyle CH_3}{\overset{|}{C}}}{-}\overset{|}{C}H{-}CH_3$

(m) 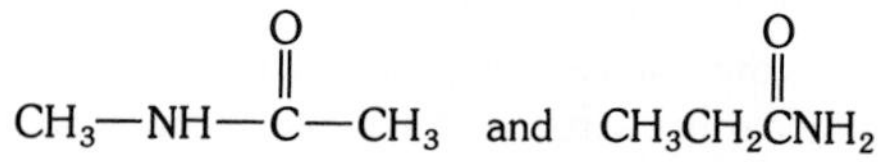 $CH_3{-}NH{-}\overset{\displaystyle O}{\overset{\|}{C}}{-}CH_3$ and $CH_3CH_2\overset{\displaystyle O}{\overset{\|}{C}}NH_2$

(n) $H{-}O{-}O{-}H$ and $H{-}O{-}H$

Families of Organic Compounds

9.15 Name the family to which each compound belongs.

(a) $CH_3CH{=}CH_2$

(b) $HOCH_2CH_2CH_3$

(c) CH_3CH_2SH

(d) $CH_3C{\equiv}CH$

(e) $CH_3CH_2\overset{\displaystyle O}{\overset{\|}{C}}{-}O{-}CH_3$

(f) $CH_3CH_2\overset{\displaystyle O}{\overset{\|}{C}}H$

(g) $CH_3CH_2CH_2\overset{\displaystyle O}{\overset{\|}{C}}OH$

(h) $CH_3\overset{\displaystyle O}{\overset{\|}{C}}CH_2CH_2CH_3$

(i) $CH_3CH_2CH_2NH_2$

(j) $CH_3{-}O{-}CH_2CH_3$

9.16 Name the families to which the compounds in Practice Exercise 9.14 belong. (A few belong to more than one family.)

Chapter 10
Hydrocarbons

Petroleum is a mixture mostly of hydrocarbons, the subjects of this chapter. At refineries such as this, petroleum is separated into useful fractions based on boiling points.

10.1 FAMILIES OF HYDROCARBONS

Hydrocarbons can be saturated or unsaturated, open chain or cyclic, and all are insoluble in water.

Hydrocarbons are organic compounds made only from carbon and hydrogen. Their molecules can have single, double, or triple bonds, as illustrated in Figure 10.1. Petroleum, discussed in Special Topic 10.1, is our chief source of hydrocarbons.

Single, Double, or Triple Bonds Define the Hydrocarbon Families. Hydrocarbons with only single bonds make up the family of **alkanes.** Table 10.1 gives the structures of the 10 simplest, open-chain alkanes. **Alkenes** are hydrocarbons with double bonds, and those with triple bonds are **alkynes.**

The ring system in benzene confers such unique properties on the molecule that compounds with the benzene ring, regardless of other functional groups, are in a class by themselves, the **aromatic compounds.** Any compound without the benzene ring (or certain rings similar to it), regardless of its functional groups, is called an **aliphatic compound.** Thus diethyl ether, once an important anesthetic, isopropyl alcohol (rubbing alcohol), and acetic acid (in vinegar) are all aliphatic.

Compounds Can Be Saturated or Unsaturated. Regardless of functional groups or rings, compounds with only single bonds are called **saturated compounds.** The alkanes are thus *saturated hydrocarbons.* Saturated compounds occur in other families, too, like diethyl ether and isopropyl alcohol. (You can see how classifications often overlap.)

When one or more multiple bonds are present, the substance is an **unsaturated compound.** Thus all alkenes and alkynes as well as all aromatic compounds are *unsaturated hydrocarbons.* The carbon–oxygen double bond in acetic acid makes it an unsaturated compound, too.

Cyclic Hydrocarbons Occur in Various Ring Sizes and Degrees of Unsaturation. Cyclic compounds can have double bonds as in cyclohexene. (*Remember:* Alkene double bonds, whether in open chains or rings, are always shown by two lines. They are not "understood.") Rings, of course, can carry attached atoms or groups of atoms called *substituents,* as the methyl group in methylcyclohexane.

The triple bond occurs rarely in nature, and we will not study it further.

$CH_3CH_2—O—CH_2CH_3$
Diethyl ether

$$CH_3—\overset{\overset{\textstyle OH}{|}}{CH}—CH_3$$
Isopropyl alcohol

$$CH_3—\overset{\overset{\textstyle O}{||}}{C}—OH$$
Acetic acid

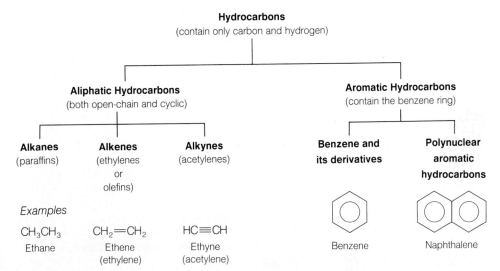

Hydrocarbons
(contain only carbon and hydrogen)

Aliphatic Hydrocarbons
(both open-chain and cyclic)

Aromatic Hydrocarbons
(contain the benzene ring)

Alkanes
(paraffins)

Alkenes
(ethylenes or olefins)

Alkynes
(acetylenes)

Benzene and its derivatives

Polynuclear aromatic hydrocarbons

Examples

CH_3CH_3
Ethane

$CH_2{=}CH_2$
Ethene (ethylene)

$HC{\equiv}CH$
Ethyne (acetylene)

Benzene

Naphthalene

FIGURE 10.1
There are several kinds of hydrocarbons. (The circles in the structures for benzene and naphthalene will be explained later in the chapter.)

TABLE 10.1
Straight-Chain Alkanes

IUPAC Name	Number of Carbon Atoms	Molecular Formula[a]	Structure	Boiling Point (°C)	Melting Point (°C)	Density (g/mL, 20 °C)
Methane	1	CH_4	CH_4	−161.5	−182.5	
Ethane	2	C_2H_6	CH_3CH_3	−88.6	−183.3	
Propane	3	C_3H_8	$CH_3CH_2CH_3$	−42.1	−189.7	
Butane	4	C_4H_{10}	$CH_3CH_2CH_2CH_3$	−0.5	−138.4	
Pentane	5	C_5H_{12}	$CH_3CH_2CH_2CH_2CH_3$	36.1	−129.7	0.626
Hexane	6	C_6H_{14}	$CH_3CH_2CH_2CH_2CH_2CH_3$	68.7	−95.3	0.659
Heptane	7	C_7H_{16}	$CH_3CH_2CH_2CH_2CH_2CH_2CH_3$	98.4	−90.6	0.684
Octane	8	C_8H_{18}	$CH_3CH_2CH_2CH_2CH_2CH_2CH_2CH_3$	125.7	−56.8	0.703
Nonane	9	C_9H_{20}	$CH_3CH_2CH_2CH_2CH_2CH_2CH_2CH_2CH_3$	150.8	−53.5	0.718
Decane	10	$C_{10}H_{22}$	$CH_3CH_2CH_2CH_2CH_2CH_2CH_2CH_2CH_2CH_3$	174.1	−29.7	0.730

[a] The molecular formulas of the open-chain alkanes fit the general formula C_nH_{2n+2}, where n = the number of carbon atoms per molecule.

More than 2 billion lb of cyclohexane are made annually in the United States, with over 90% being used to make nylon.

Cyclohexene   Methylcyclohexane

In structures of cyclic compounds, the ring system itself is usually represented simply by a polygon, a many-sided figure. A ring carbon atom is understood to be at each corner together with as many H atoms at the corner as are needed to fill out four bonds to the ring carbon. A pentagon, for example, can represent cyclopentane. This symbolism is illustrated in Table 10.2, where you can see how the numbers of atoms making up a ring can vary. (Rings of dozens of atoms are known.)

The photograph of the model of methylcyclopentane and its progressively more condensed structures further illustrate the use of a geometric figure to represent a ring.

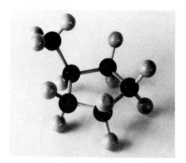

Methylcyclopentane.

Three ways to represent the structure of methylcyclopentane

Ring Atoms Can Come from Elements Other Than Carbon. Cyclic compounds with ring atoms other than carbon are called **heterocyclic compounds.** Tetrahydropyran, for example, has an oxygen atom in its ring. Many carbohydrates, like glucose, have tetrahydropyran ring systems heavily substituted by OH groups.

Petroleum (*petra,* rock; *oleum,* oil) is a complex mixture of organic compounds of which nearly all are hydrocarbons. Most, in fact, are alkanes. When this mixture is heated to its boiling point, the first vapors to leave include the smallest molecules of the lowest boiling, most volatile compounds. As these separate, the boiling point of the remaining mixture increases gradually. The vapors are cooled so they can return to the liquid state, and engineers can collect the liquids according to preset ranges of boiling points. The whole operation is called *fractional distillation,* and the separately collected product mixtures are called *fractions.* The accompanying table describes the most common fractions obtained from petroleum.

Each fraction is a mixture of hydrocarbons with an overall volatility that makes it useful in certain kinds of engines or heating devices. The highest boiling residues that cannot be distilled are used as residual fuel oil or made into asphalt or coke.

The natural gasoline fraction of crude petroleum falls far short of meeting the world's needs for this vital fuel. But organic chemists and chemical engineers have found ways to break or "crack" larger molecules in the kerosene range (and higher) into molecules small enough to work as gasoline.

Many unsaturated hydrocarbons are also obtained by cracking. They serve as cheap raw materials for many useful products, including all plastics, as well as lacquers, resins, organic solvents, and synthetic rubber.

Petroleum Fractions

Boiling Point Range (°C) (a measure of volatility)	Number of Carbons in Molecules of This Fraction	Uses
Below 20	1–4	Natural gas, heating and cooking fuel; raw material for other chemicals. (Methane is the gas used with Bunsen burners.)
20–60	5–6	Petroleum "ether" (a nonpolar solvent and cleaning fluid),
60–100	6–7	Ligroin or light naphtha (also a nonpolar solvent and cleaning fluid)
40–200	5–10	Gasoline
175–325	12–18	Kerosene, jet fuel, tractor fuel
250–400	12 and higher	Gas oil, fuel oil, diesel oil
Nonvolatile liquids	20 and higher	Refined mineral oil, lubricating oil, grease (a dispersion of soap in oil)
Nonvolatile solids	20 and higher	Paraffin wax, asphalt and tar for roads and roofing

TABLE 10.2
Some Cycloalkanes

IUPAC Name	Structure	Boiling Point (°C)	Melting Point (°C)	Density (20 °C)
Cyclopropane	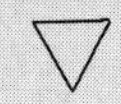 △	−33	−127	1.809 g/L (0 °C)
Cyclobutane	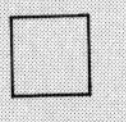 ☐	−13.1	−80	0.7038 g/L (0 °C)
Cyclopentane	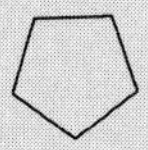 ⬠	49.3	−94.4	0.7460 g/mL
Cyclohexane	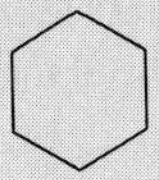 ⬡	80.7	6.47	0.7781 g/mL
Cycloheptane	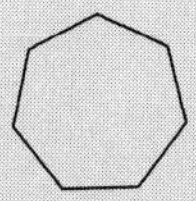 ⬡	118.5	−12	0.8098 g/mL
Cyclooctane	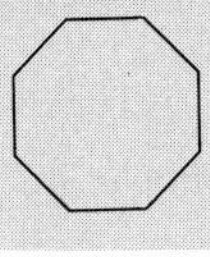 ⯃	149	14.3	0.8349 g/mL

$$
\begin{array}{c}
CH_2 \\
CH_2 \quad\quad CH_2 \\
CH_2 \quad\quad CH_2 \\
O
\end{array}
$$

Tetrahydropyran

Hydrocarbons Do Not Dissolve in Water.   Carbon–hydrogen bonds and carbon–carbon bonds of all types, single, double, or triple, are almost entirely nonpolar. Hydrocarbons, as a result, are almost completely nonpolar compounds.

One consequence is that hydrocarbons are insoluble in water. Their molecules have no polar sites that attract water molecules. Water molecules, strongly attracted to each other by hydrogen bonds, are simply unable to let hydrocarbon molecules in. But in nonpolar solvents, like ether, gasoline, or benzene, hydrocarbon molecules relatively easily slip in and among the solvent molecules, fully intermingle, and so dissolve. Gasoline, itself a mixture of hydrocarbons (mostly alkanes) and therefore nonpolar, is a good solvent for tar and grease. (If you ever use it to clean tar from surfaces, be sure to keep all flames away.)

Like Dissolves Like.   Grease and tar are relatively nonpolar materials, so their solubility in nonpolar gasoline illustrates the **like-dissolves-like** rule: Polar solvents tend to dissolve polar or ionic substances and nonpolar solvents tend to dissolve nonpolar solutes. Water molecules, for example, are polar, and so water dissolves salt (ionic) and sugar (polar molecular) but not hydrocarbons. Hydrocarbon solvents, with nonpolar molecules, dissolve nonpolar, hydrocarbon-like compounds, but not ionic or polar molecular substances. Neither salt nor sugar dissolves in gasoline.

Structures Carry Molecular "Map Signs".   Our brief survey of the physical properties of hydrocarbons has introduced the first molecular "map sign" of our study, the hydrocarbon-like feature. Substances whose molecules are entirely *or even mostly* hydrocarbon-like are insoluble in water but soluble in nonpolar solvents. When a glance at a structure tells us that it is mostly hydrocarbon-like, we can safely predict that the compound is insoluble in water. Consider, for example, the structure of lauryl alcohol,

$$CH_3CH_2CH_2CH_2CH_2CH_2CH_2CH_2CH_2CH_2CH_2CH_2OH$$
Lauryl alcohol

The long alkane-like chain dominates this molecule. The polar OH group (in this setting of a long hydrocarbon chain) cannot contribute enough polarity to the whole molecule to make lauryl alcohol soluble in water. Thus, by learning one very general fact, one molecular "map sign," we do not have to memorize a long list of separate (but similar) facts about an equally long list of separate compounds found at the molecular level of life. With the molecular "map sign" in hand — hydrocarbon-like compounds tend to be insoluble in water — we can look at the structures of hundreds of complicated compounds and confidently predict particular properties such as the likelihood of their being soluble in some solvent. Later, we will see how the abundance of hydrocarbon-like groups in the molecules of cell membranes contributes to both the stability and the flexibility of such systems.

The vapors of hydrocarbon solvents catch fire very easily, so be careful if you use these solvents for any purpose. In the right proportion in air, hydrocarbon vapors explode when ignited.

"Like" refers to likeness in molecular polarity.

PRACTICE EXERCISE 1   Which of the following is more soluble in gasoline?

$$HO-CH_2-CH-CH_2-OH \qquad CH_3-CH_2-CH-CH_2-OH$$
$$\quad\quad\quad | \qquad\qquad\qquad\qquad\qquad\qquad | $$
$$\quad\quad\quad OH \qquad\qquad\qquad\qquad\qquad\qquad CH_3$$

Glycerol   2-Methyl-1-butanol

Within a Family, Boiling Points Increase with Formula Weight. Notice in Table 10.1 how the boiling points of the straight-chain alkanes increase with chain length. Because of their small size and low polarity, the hydrocarbons with 1−4 carbons per molecule are generally gases at or near room temperature.

This correlation between size and boiling point is true of all hydrocarbon families. Hydrocarbons with 5 to about 16 carbon atoms per molecule are usually liquids at room temperature. When alkanes have about 18 or more carbon atoms per molecule, the substances are waxy solids at room temperature. Paraffin wax, for example, is a mixture of alkanes with molecules of 20 and more carbon atoms.

Most candles are made from paraffin.

Nearly All Hydrocarbons Float on Water. Still another physical property of aliphatic hydrocarbons, as illustrated by the data in Table 10.1, is that all are less dense than water. Thus such compounds float on water. When oil spills occur, some of the material floats away to damage shore habitat.

10.2 ALKANES

The formal name of a compound shows the number of carbon atoms in the parent chain, the kinds and locations of substituents on this chain, and the family to which the compound belongs.

The straight-chain alkanes in Table 10.1 form a **homologous series** because its members differ from each other by CH_2 units. Butane, for example, is the next higher **homolog** of propane.

Isomerism occurs among alkanes with four and more carbons. The two isomers of butane and the three of pentane are shown in Table 10.3, together with data showing how rapidly the number of isomers increases with carbon content.

TABLE 10.3
The Isomeric Butanes and Pentanes

Structure	IUPAC Name (Common Name)	Boiling Point (°C)
C_4H_{10} Isomers		
$CH_3CH_2CH_2CH_3$	Butane (n-butane)	0
CH_3 $\|$ CH_3CHCH_3	2-Methylpropane (isobutane)	−12
C_5H_{12} Isomers		
$CH_3CH_2CH_2CH_2CH_3$	Pentane (n-pentane)	36
CH_3 $\|$ $CH_3CHCH_2CH_3$	2-Methylbutane (isopentane)	28

CH_3
     $\|$
CH_3CCH_3
     $\|$
     CH_3
	2,2-Dimethylpropane (neopentane)	10

Numbers of Isomers of Higher Alkanes		
C_6H_{14}	5	
C_8H_{18}	18	
$C_{10}H_{22}$	75	
$C_{20}H_{42}$	366,319	
$C_{40}H_{82}$	6.25×10^{13} (estimated)	

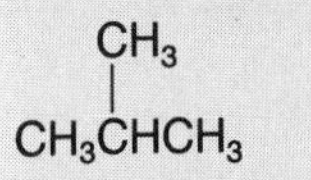

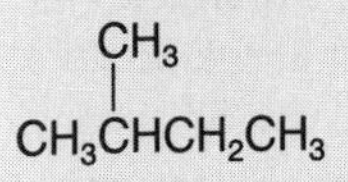

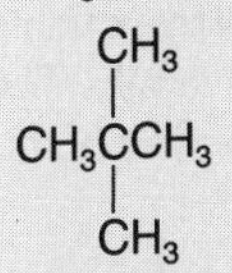

"Nomenclature" is from the Latin *nomen* (name) + *calare* (to call). Wealthy ancient Romans had slaves called *nomenclators* who were to remind their owners of the names of important people who approached them on the street.

The Names of Compounds Carry Structural Information. In chemistry, **nomenclature** refers to the rules used to name compounds. The International Union of Pure and Applied Chemistry (IUPAC) is the organization that now develops the rules of chemical nomenclature, and all chemical societies in the world belong to it. Its rules are known as the **IUPAC rules.** They are so carefully drawn that only one name could be written for each compound, and only one structure could be drawn for each name.

The IUPAC names, unfortunately, are sometimes very long and difficult to write or pronounce. It's much easier to call table sugar *sucrose* than α-D-glucopyranosyl β-D-fructofuranoside. This illustrates why shorter names, referred to as *common names,* are still widely used. We will want to learn some common names, too, and you will see that even they usually have some system to them.

IUPAC Rules for Naming Alkanes and Cycloalkanes. In the IUPAC rules, the last syllable in the name designates the family to which the compound belongs. The names of all the saturated hydrocarbons, for example, end in *-ane.* The names of hydrocarbons with double bonds end in *-ene,* and those with triple bonds end in *-yne.* The rules for the alkanes are as follows:

1. Use the ending *-ane* for all alkanes (and cycloalkanes).
2. Determine what is the longest continuous chain of carbons in the structure, and let this be the *parent chain* for naming purposes.

   For example, view the branched-chain alkane,

$$CH_3—CH_2—\overset{\overset{\displaystyle CH_3}{|}}{CH}—CH_2—CH_2—CH_3$$

as coming from

$$CH_3—CH_2—CH_2—CH_2—CH_2—CH_3$$

by replacing a hydrogen atom on the third carbon from the left with a CH_3 group.

$$CH_3—CH_2—\overset{\overset{\displaystyle CH_3}{\searrow}\overset{\displaystyle H}{\nearrow}}{CH}—CH_2—CH_2—CH_3 \longrightarrow CH_3—CH_2—\overset{\overset{\displaystyle CH_3}{|}}{CH}—CH_2—CH_2—CH_3$$

3. Attach a prefix to *-ane* that specifies the number of carbon atoms *in the parent chain.* The prefixes through C_{10} are as follows (and these must be memorized). The names in Table 10.1 illustrate their use. You can see them used for the cycloalkanes, Table 10.2, too.

We won't need to know the prefixes for the higher alkanes.

meth-	1 C	hex-	6 C
eth-	2 C	hept-	7 C
prop-	3 C	oct-	8 C
but-	4 C	non-	9 C
pent-	5 C	dec-	10 C

For example, the parent chain of our example has six carbons, so the corresponding alkane is called hexane, *hex* for six carbons and *-ane* for being in the alkane family. The branched-chain compound whose name we are devising is regarded as a derivative of this parent, hexane.

4. Assign numbers to each carbon of the parent chain starting from whichever of its ends gives the location of the first branch the lower of two possible numbers.

   For example, the correct direction for numbering our example is from left to right.

$$\underset{1}{CH_3}—\underset{2}{CH_2}—\underset{3}{\overset{\overset{\displaystyle CH_3}{|}}{CH}}—\underset{4}{CH_2}—\underset{5}{CH_2}—\underset{6}{CH_3}$$ (correct direction of numbering)

Had we numbered from right to left, the carbon holding the branch would have had a higher number.

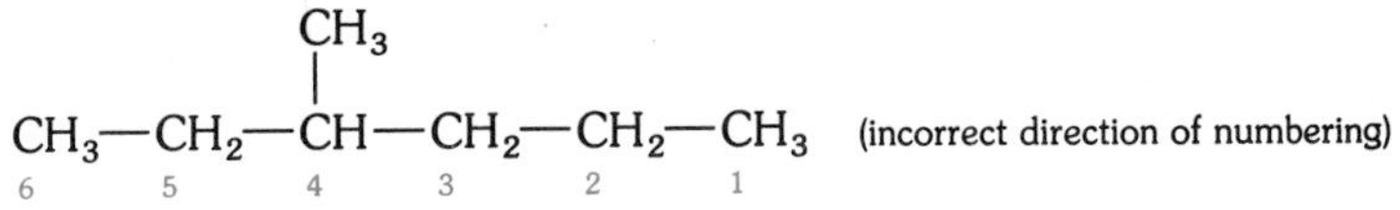

$$CH_3 - CH_2 - \underset{\underset{\displaystyle |}{\underset{\displaystyle CH_3}{}}}{CH} - CH_2 - CH_2 - CH_3 \quad \text{(incorrect direction of numbering)}$$
$${6} \quad {5} \quad {4} \quad {3} \quad {2} \quad {1}$$

5. Determine the correct name for each branch (or for any other atom or group). We must now pause and learn the names of a few such groups.

Any branch that consists only of carbon and hydrogen and has only single bonds is called an **alkyl group,** and the names of all alkyl groups end in -yl. Think of an alkyl group as an alkane minus one of its hydrogen atoms. For example,

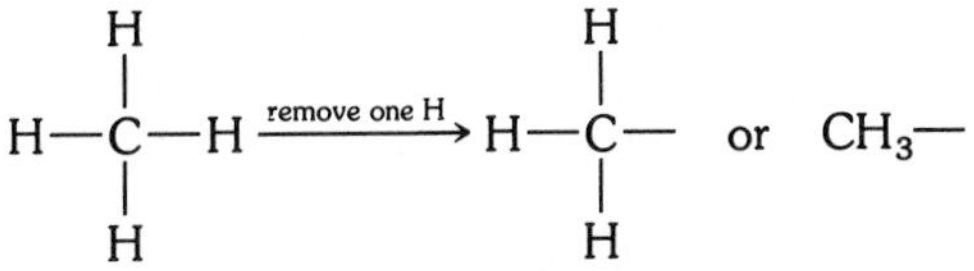

Methane   **Methyl**

Methyl.

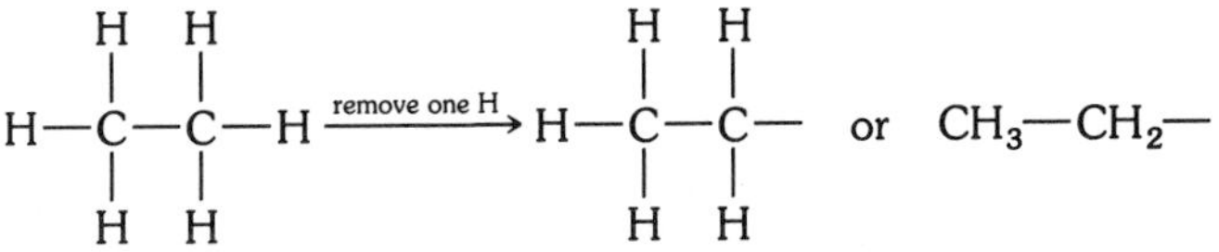

Ethane   **Ethyl**

Ethyl.

Two alkyl groups can be obtained from propane because the middle position is not equivalent to either of the end positions.

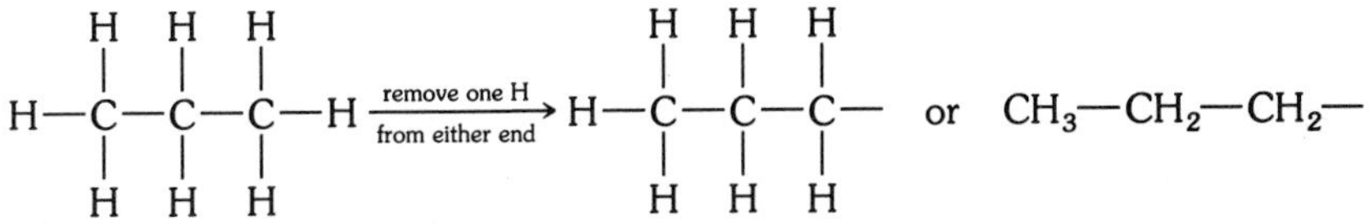

Propane   **Propyl**

Propyl.

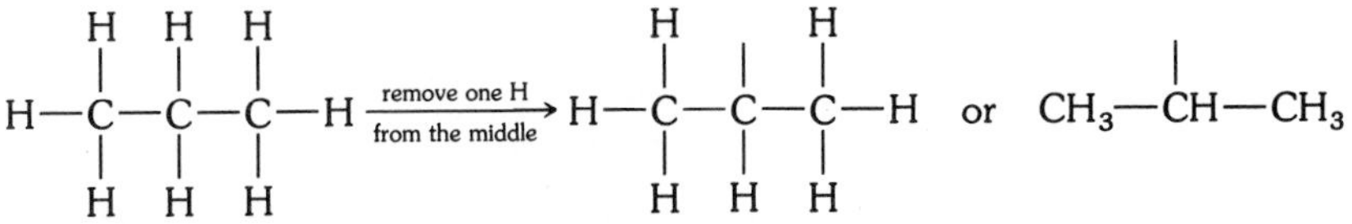

Propane   **Isopropyl**

Two alkyl groups can similarly be obtained from butane.

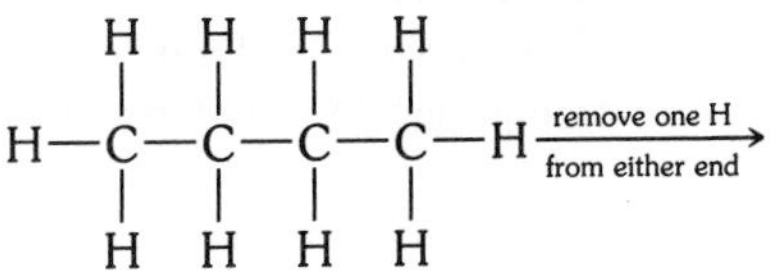

Butane

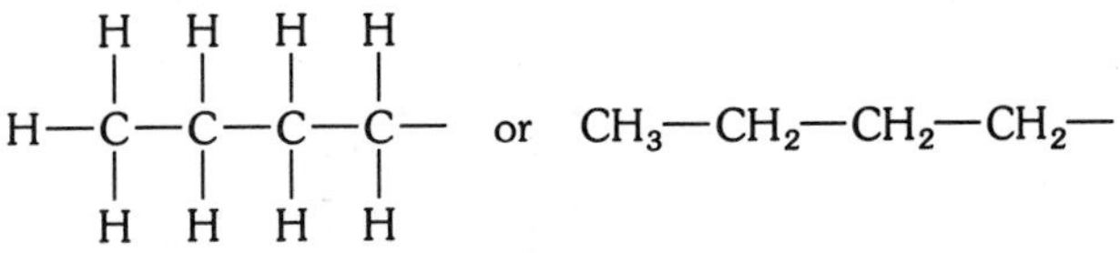

Isopropyl.

Butyl

Butyl.

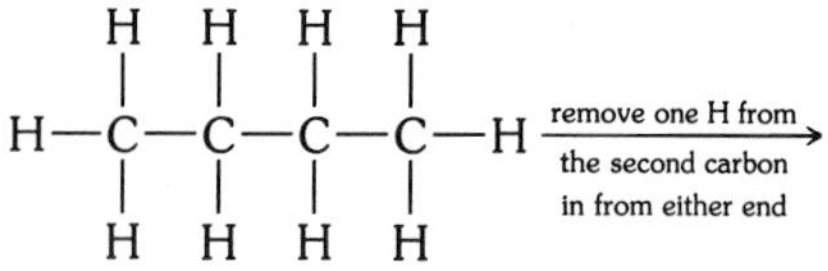

H—C—C—C—C—H remove one H from →
the second carbon
in from either end

(Butane, with H H H H on top and H H H H on bottom)

Butane

H—C—C—C—C—H   or CH_3—CH_2—CH—CH_3

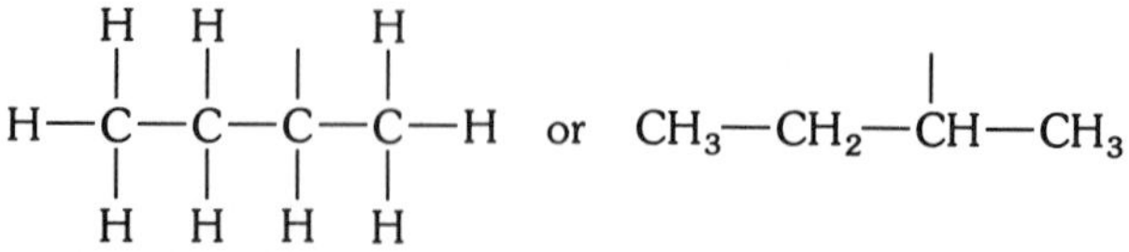

sec-Butyl
(sec = secondary)

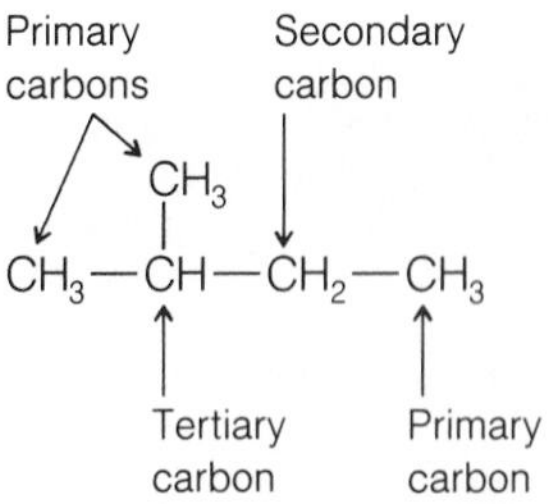

sec-Butyl.

Primary carbons   Secondary carbon

CH_3

CH_3—CH—CH_2—CH_3

Tertiary carbon   Primary carbon

This is called the *secondary* butyl group (abbreviated *sec*-butyl) because the open bonding site is at a **secondary carbon,** a carbon that is directly attached to just two other carbons. A **primary carbon** is one to which just one other carbon is directly attached. The open bonding site in the butyl group, for example, is at a primary carbon atom. A **tertiary carbon** is one that holds directly three other carbons. We will encounter a tertiary carbon in a group that we will soon study.

Butane is the first alkane to have an isomer. Its common name is isobutane, and we can derive two more alkyl groups from it.

H—C—C—C—H   remove one H from any → H—C—C—C—   or CH_3CHCH_2—
one of the CH_3 groups

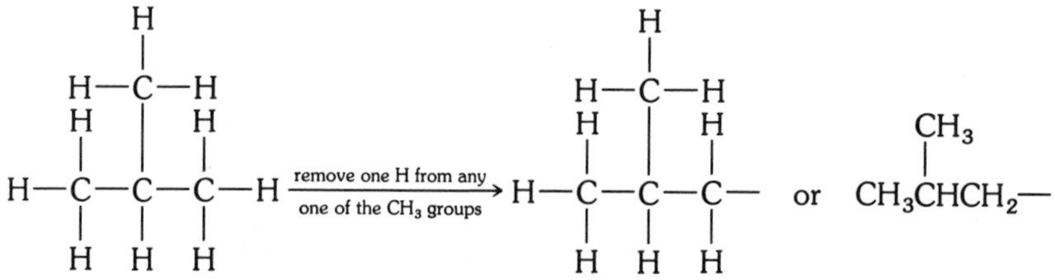

Isobutane   **Isobutyl**

(with CH_3 shown on the isobutyl "or" structure)

Isobutyl.

H—C—C—C—H   remove the lone H from → H—C—C—C—H   or CH_3CCH_3
the tertiary atom

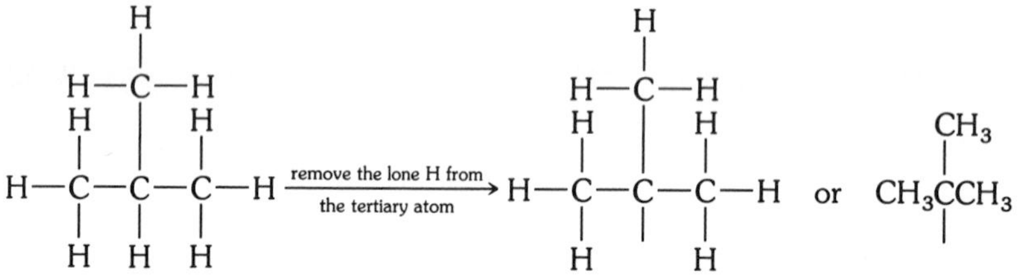

Isobutane   **t-Butyl**
(t = tertiary)

(with CH_3 shown on the t-butyl "or" structure)

t-Butyl.

Notice that the open bonding site in the *tertiary*-butyl group (abbreviated *t*-butyl) occurs at a tertiary carbon.

The names and structures of these alkyl groups must now be learned. If you have access to ball-and-stick models, make models of each of the parent alkanes and then remove hydrogen atoms to generate the open bonding sites and the alkyl groups. The *Study Guide* accompanying this book has exercises that provide drills in recognizing alkyl groups when they are positioned in different ways on the page.

The prefix *iso*- in the name of an alkyl group, such as in isopropyl or isobutyl, has a special meaning. It can be used to name any alkyl group that has the following general features.

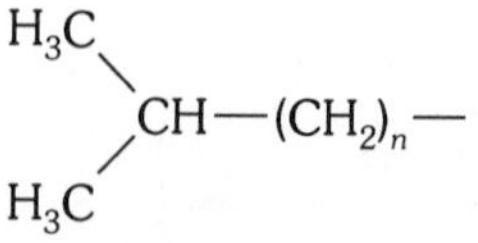

Here is another way to condense a structure. Thus $CH_3(CH_2)_3CH_3$ represents $CH_3CH_2CH_2CH_2CH_3$.

H_3C
       CH—$(CH_2)_n$—   (n = 0, 1, 2, 3, etc.)
H_3C n = 0, isopropyl
                            n = 1, isobutyl

Notice that in each of these names there is a word fragment (for example, *-prop-*, *-but-*, etc.) associated with a number of carbon atoms. Here, it specifies the *total* number of carbons in the alkyl group. Thus the isopropyl group has three carbons and the isobutyl has four.

6. Attach the name of the alkyl group or other substituent to the name of the parent as a prefix. Place the location number of the group in front of the resulting name and separate the number from the name by a hyphen.

   Returning to our original example, its name is 3-methylhexane.

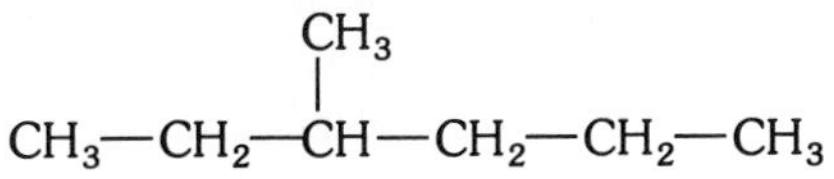

$$CH_3-CH_2-\overset{\overset{\displaystyle CH_3}{|}}{CH}-CH_2-CH_2-CH_3$$
3-Methylhexane

7. When two or more groups are attached to the parent, name each and locate each with a number. Always use *hyphens* to separate numbers from words in the IUPAC names of compounds, and arrange the names of the alkyl groups alphabetically in the final name. For example,

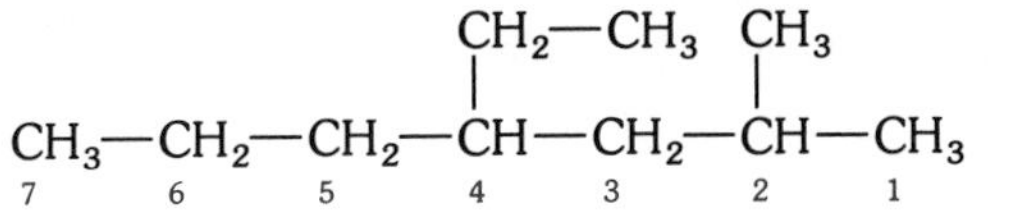

$$CH_3-CH_2-CH_2-\overset{\overset{\displaystyle CH_2-CH_3}{|}}{CH}-CH_2-\overset{\overset{\displaystyle CH_3}{|}}{CH}-CH_3$$

4-Ethyl-2-methylheptane

7 6 5 4 3 2 1

8. When two or more substituents are identical, use such prefixes as di- (for 2), tri- (for 3), tetra- (for 4), and so forth; and specify the location number of *every* group. Always separate a number from another number in a name by a *comma*. For example,

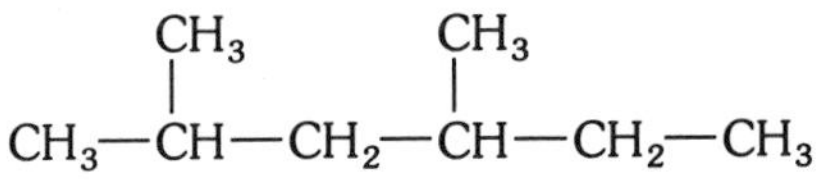

$$CH_3-\overset{\overset{\displaystyle CH_3}{|}}{CH}-CH_2-\overset{\overset{\displaystyle CH_3}{|}}{CH}-CH_2-CH_3$$

Correct name: 2,4-Dimethylhexane
Incorrect names: 2,4-Methylhexane
3,5-Dimethylhexane
2-Methyl-4-methylhexane

9. When identical groups are on the *same* carbon, repeat the number of this carbon in the name. For example,

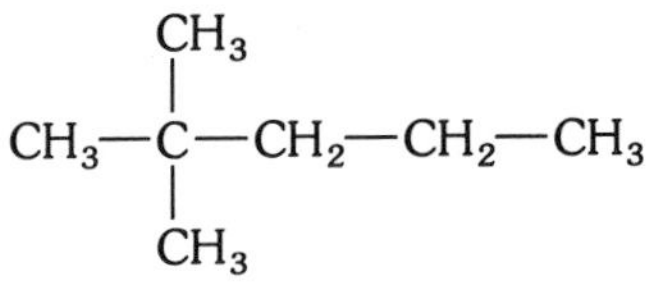

$$CH_3-\overset{\overset{\displaystyle CH_3}{|}}{\underset{\underset{\displaystyle CH_3}{|}}{C}}-CH_2-CH_2-CH_3$$

Correct name: 2,2-Dimethylpentane
Incorrect names: 2-Dimethylpentane
2,2-Methylpentane
4,4-Dimethylpentane

Another example:

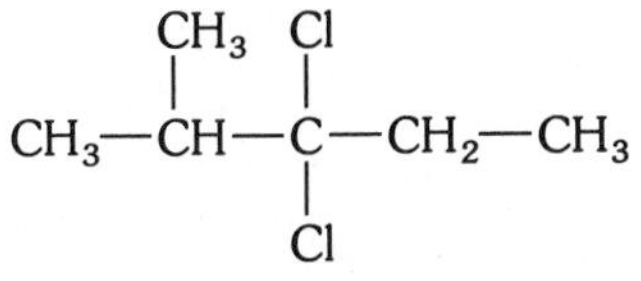

$$CH_3-\overset{\overset{\displaystyle CH_3}{|}}{CH}-\overset{\overset{\displaystyle Cl}{|}}{\underset{\underset{\displaystyle Cl}{|}}{C}}-CH_2-CH_3$$

Correct name: 3,3-Dichloro-2-methylpentane

Notice that the names of nonalkyl substituents are assembled first in the final name, so a compound such as in our previous example is viewed as a derivative of the hydrocarbon, 2-methylpentane.

10. To name a cycloalkane, place the prefix *cyclo* before the name of the straight-chain alkane that has the same number of carbon atoms as the ring. This was illustrated in Table 10.2.

11. When necessary, give numbers to the ring atoms by giving location 1 to a ring position that holds a substituent and numbering around the ring in whichever direction reaches the nearest substituent first. For example,

No number is needed when the ring has only one group. Thus,

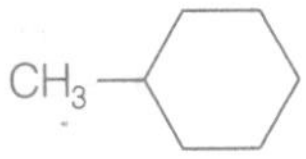

CH$_3$—⬡

is named methylcyclohexane, not 1-methylcyclohexane.

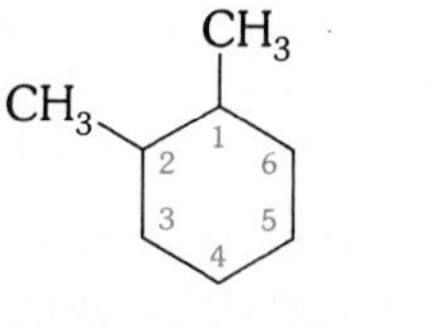

CH$_3$—⬡ (1) (2) (3) (4) (5) (6) with CH$_3$

1,2-Dimethylcyclohexane

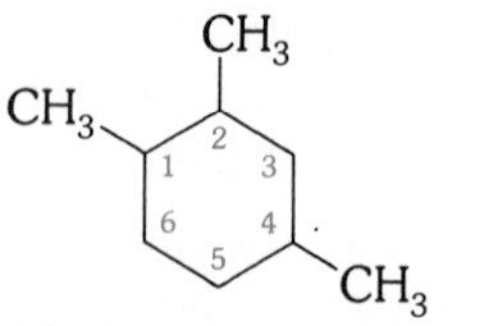

CH$_3$—⬡ (1) (2) (3) (4) (5) (6) with CH$_3$ and CH$_3$

1,2,4-Trimethylcyclohexane

These are not all of the IUPAC rules for alkanes, but they will cover all our needs. Before we give more examples of the rules for alkanes, we will introduce the names that are used for several substituents that can be attached to alkane chains:

—F	fluoro	—I	iodo
—Cl	chloro	—NO$_2$	nitro
—Br	bromo	—NH$_2$	amino

Now study the following examples of correctly named compounds. Be sure to notice that in choosing the parent chain we sometimes have to go around a corner as the chain zigzags on the page.

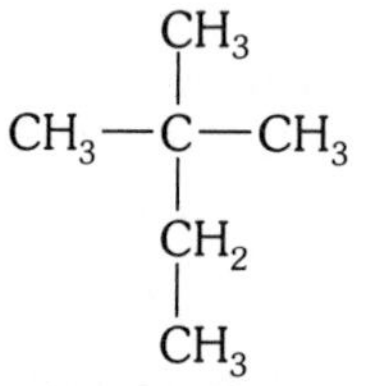

CH$_3$—C(CH$_3$)(CH$_3$)—CH$_2$—CH$_3$

2,2-Dimethylbutane
Not 2-methyl-2-ethylpropane

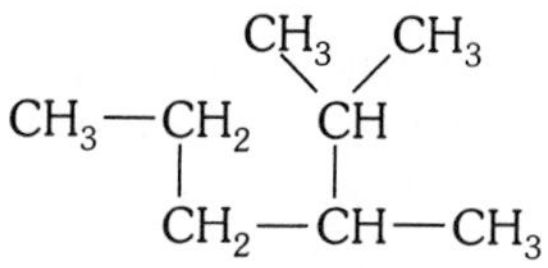

CH$_3$—CH$_2$—CH(CH$_3$)—CH$_2$—CH(CH$_3$)—CH$_3$

2,3-Dimethylhexane
Not 2-isopropylpentane

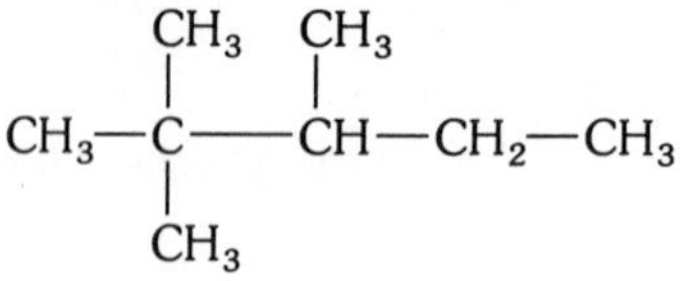

CH$_3$—C(CH$_3$)(CH$_3$)—CH(CH$_3$)—CH$_2$—CH$_3$

2,2,3-Trimethylpentane
Not 2,3-trimethylpentane
Not 2-*t*-butylbutane

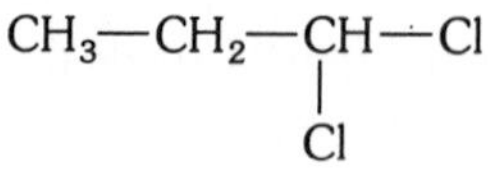

CH$_3$—CH$_2$—CH(Cl)—Cl

1,1-Dichloropropane
Not 3,3-dichloropropane
Not 3,3-chloropropane
Not 1-dichloropropane
Not 1,1-chloropropane

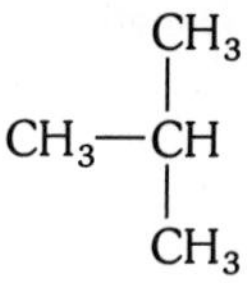

CH$_3$—CH(CH$_3$)—CH$_3$

2-Methylpropane
Not 1,1-dimethylethane
Not isobutane (which is its common name)

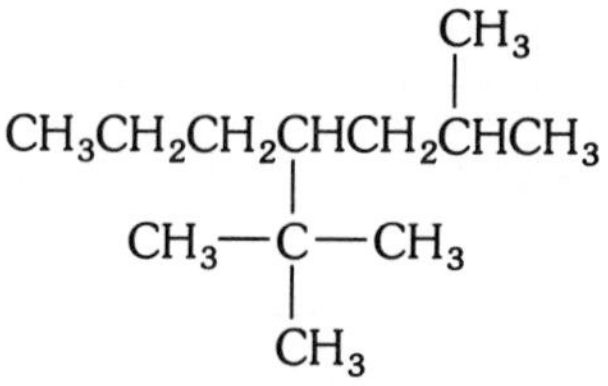

CH$_3$CH$_2$CH$_2$CHCH$_2$CHCH$_3$ with CH$_3$ and C(CH$_3$)(CH$_3$)(CH$_3$)

4-*t*-butyl-2-methylheptane
Not 4-*t*-butyl-6-methylheptane

EXAMPLE 10.1   USING THE IUPAC RULES TO NAME AN ALKANE

Problem: What is the IUPAC name for the following compound?

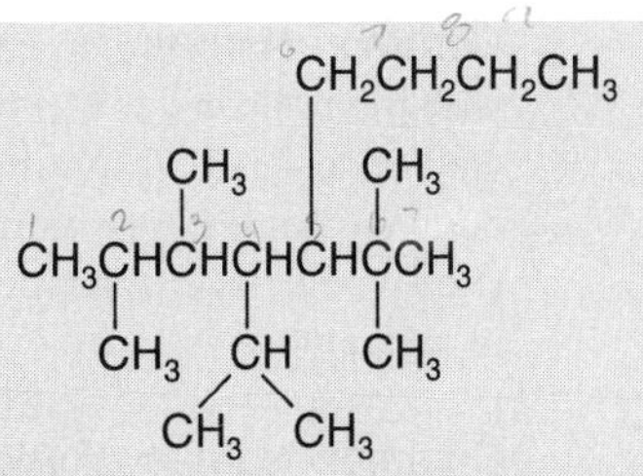

$$CH_2CH_2CH_2CH_3$$

$$CH_3 \quad\quad CH_3$$

$$CH_3CHCHCHCHCCH_3$$

$$CH_3 \quad CH \quad CH_3$$

$$CH_3 \quad CH_3$$

Solution: The compound is an alkane because it is a hydrocarbon with only single bonds, so the ending to the name is -ane. The next step is to find the longest chain even if we have to go around corners. This chain is nine carbons long, so the name of the parent alkane is *nonane*. We have to number the chain from left to right, as follows, in order to reach the first branch with the lower number.

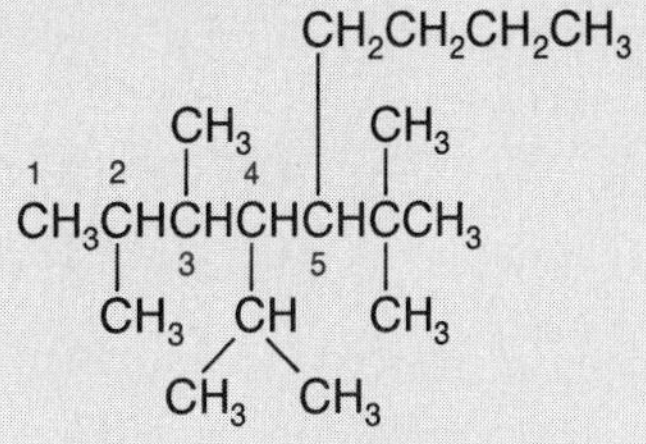

$$\overset{6}{C}H_2\overset{7}{C}H_2\overset{8}{C}H_2\overset{9}{C}H_3$$

$$CH_3 \quad\quad CH_3$$

$$\overset{1}{C}H_3\overset{2}{C}H\overset{}{C}H\overset{4}{C}H\overset{}{C}H\overset{}{C}CH_3$$

$$CH_3 \quad \overset{}{C}H \quad CH_3$$

$$CH_3 \quad CH_3$$

At carbons 2 and 3 there are the one-carbon methyl groups. At carbon 4, there is a three-carbon isopropyl group (not the propyl group, because the bonding site is at the middle carbon of the three-carbon chain). At carbon 5, there is a four-carbon *t*-butyl group. (It has to be this particular butyl group because the bonding site is at a tertiary carbon.) In alphabetizing these alkyl groups, ignore Greek numerical prefixes like di- and tri-, and ignore designations of kinds of carbons, like *sec*- or *t*-. (The *iso*- prefix, however, is alphabetized among the ''i'' words.)

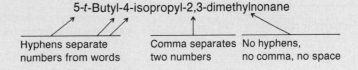

5-*t*-Butyl-4-isopropyl-2,3-dimethylnonane

Hyphens separate numbers from words   Comma separates two numbers   No hyphens, no comma, no space

PRACTICE EXERCISE 2   Write the IUPAC names of the following compounds.

(a) 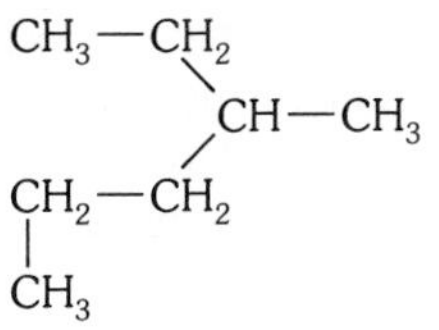
$$CH_3-CH_2$$
$$\quad\quad\quad CH-CH_3$$
$$CH_2-CH_2$$
$$|$$
$$CH_3$$

(b) 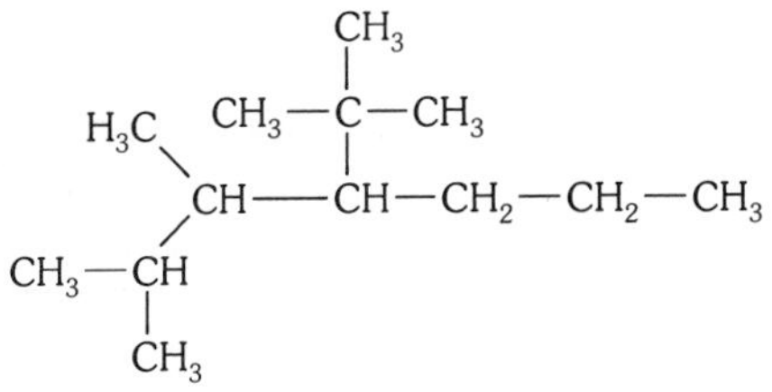
$$\quad\quad\quad\quad\quad CH_3$$
$$\quad\quad\quad\quad\quad |$$
$$H_3C \quad CH_3-C-CH_3$$
$$\quad\quad CH-CH-CH_2-CH_2-CH_3$$
$$CH_3-CH$$
$$\quad\quad |$$
$$\quad\quad CH_3$$

(c)
$$\quad\quad\quad\quad CH_3 \quad\quad CH_3 \quad\quad CH_3$$
$$\quad\quad\quad\quad | \quad\quad\quad | \quad\quad\quad |$$
$$CH_3-CH_2-CH-CH-CH-CH_2-CH-CH_3$$
$$\quad\quad\quad\quad\quad\quad |$$
$$\quad\quad\quad CH_2-CH_2-CH_2-CH_3$$

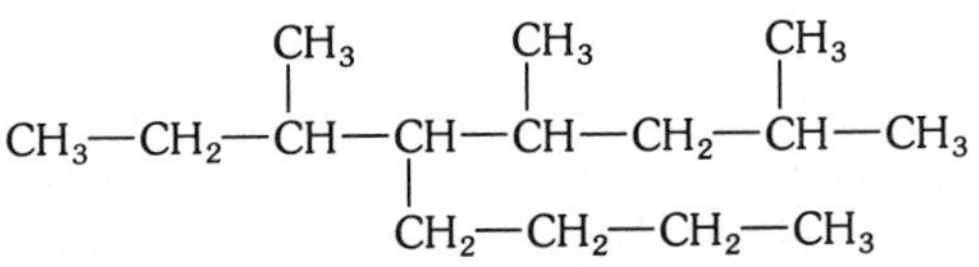

(d) 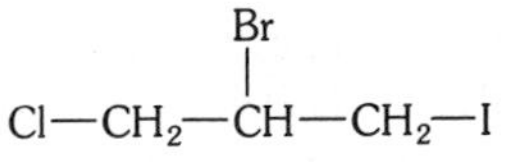
$$\quad\quad\quad\quad\quad Br$$
$$\quad\quad\quad\quad\quad |$$
$$Cl-CH_2-CH-CH_2-I$$

Common Names. In some references you might see the names of straight-chain alkanes with the prefix *n*-, as in *n*-butane, the common name of butane. It stands for *normal,* which is a way of designating that the straight-chain isomer is regarded as the *normal* isomer, as in the common names, *n*-pentane, *n*-hexane, and so forth. It is used only when isomers are possible. (You would never see *n*-propane printed as a name, for example, because there are no isomers of propane.)

Common Names of Alcohols, Amines, and Haloalkanes Employ the Names of the Alkyl Groups. The following examples of some halogen derivatives of the alkanes, called *haloalkanes,* illustrate how common names are easily constructed. The IUPAC names are given for comparison.

Structure	Common Name	IUPAC Name
CH_3Cl	Methyl chloride	Chloromethane
CH_3CH_2Br	Ethyl bromide	Bromoethane
$CH_3CH_2CH_2Br$	Propyl bromide	1-Bromopropane

CH_3CHCH_3
   |
   Cl
	Isopropyl chloride	2-Chloropropane

$CH_3CH_2CH_2CH_2Cl$	Butyl chloride	1-Chlorobutane

$CH_3CH_2CHCH_3$
     |
     Br
	sec-Butyl bromide	2-Bromobutane

CH₃—C(CH₃)(CH₃)—Br	*t*-Butyl bromide	2-Bromo-2-methylpropane
CH₃—C(CH₃)(CH₃)—Cl	*t*-Butyl chloride	2-Chloro-2-methylpropane

---

PRACTICE EXERCISE 3   Give the common names of the following compounds.

(a) $ClCH_2CH_3$ (b) $BrCH_2CH_2CH_2CH_3$ (c) CH_3CHCH_2Cl (with CH_3 branch)   (d) CH_3CCH_3 (with CH_3 and Br)

Alkanes and Cycloalkanes Have Very Few Chemical Properties. The saturated hydrocarbons are the most chemically inert of all organic compounds in our study. Alkanes do burn in air, of course. This reaction is called *combustion,* and if sufficient oxygen is present, the products are carbon dioxide and water. (In insufficient oxygen, some carbon monoxide also forms.)

No bond in alkanes is polar enough to invite attack by any of the common, ionic (or polar) inorganic acids, bases, oxidizing and reducing agents (at least at room temperature). Alkanes are utterly indifferent to concentrated sulfuric acid, sodium metal, strong alkalies, the permanganate ion, and water. This inertness, in fact, is why the alkanes are called the *paraffins,* from the Latin *parum,* little, and *affinis,* affinity. Household paraffin wax, typically used to seal homemade jellies, is a mixture of alkanes with about 20 carbons per molecule.

Mineral oil is a safe laxative (when used with care) because it is a mixture of high formula weight alkanes that undergo no chemical reactions in the intestinal tract.

Here is our second important "map sign." Alkanes and alkane-like sections of molecules are largely unaffected by reactants that attack functional groups. Alkyl groups are often called *nonfunctional groups* for this reason, as we noted in Chapter 9. You can see that we're moving ahead in our ability to read molecular structures for signs of properties.

10.3 ALKENES

The IUPAC names of alkenes end in -ene, and the double bond takes precedence over side chains in numbering the parent chain.

Table 10.4 shows the structures and some physical properties of several alkenes. As with alkanes, the first four are gases at room temperature, and all are much less dense than water. Like all hydrocarbons, alkenes are insoluble in water and soluble in nonpolar solvents.

The Parent Chain Must Include the Double Bond When Naming Alkenes.   The IUPAC rules for naming alkenes are as follows:

1. Use the ending, -ene for all alkenes.

2. As a prefix to this ending, count the number of carbon atoms in the longest sequence *that includes the double bond*. Then use the same prefix that would be used if the compound were saturated.

    For example,

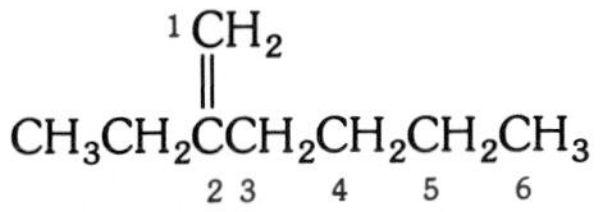

$$\overset{1}{C}H_2$$
$$\|$$
$$CH_3CH_2\underset{2\ 3}{C}\underset{4}{C}H_2\underset{5}{C}H_2\underset{6}{C}H_3$$

The parent chain has six carbons, not seven.

3. Number the parent chain from whichever end gives the lower number to the first carbon of the double bond to be reached. (This rule gives precedence to the double bond over any substituent on the parent chain.)

The location of the double bond takes precedence over any alkyl groups when numbering the chain.

TABLE 10.4
Properties of Some 1-Alkenes

IUPAC Name	Structure	Boiling Point (°C)	Melting Point (°C)	Density (g/mL at 10 °C)
Ethene	$CH_2{=}CH_2$	−104	−169	
Propene	$CH_2{=}CHCH_3$	−48	−185	
1-Butene	$CH_2{=}CHCH_2CH_3$	−6	−185	
1-Pentene	$CH_2{=}CHCH_2CH_2CH_3$	30	−165	0.641
1-Hexene	$CH_2{=}CHCH_2CH_2CH_2CH_3$	64	−140	0.673
1-Heptene	$CH_2{=}CHCH_2CH_2CH_2CH_2CH_3$	94	−119	0.697
1-Octene	$CH_2{=}CHCH_2CH_2CH_2CH_2CH_2CH_3$	121	−102	0.715
1-Nonene	$CH_2{=}CHCH_2CH_2CH_2CH_2CH_2CH_2CH_3$	147	−81	0.729
1-Decene	$CH_2{=}CHCH_2CH_2CH_2CH_2CH_2CH_2CH_2CH_3$	171	−66	0.741
Cyclopentene	 ⬠	44	−135	0.722
Cyclohexene	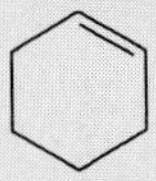 ⬡	83	−104	0.811

For example,

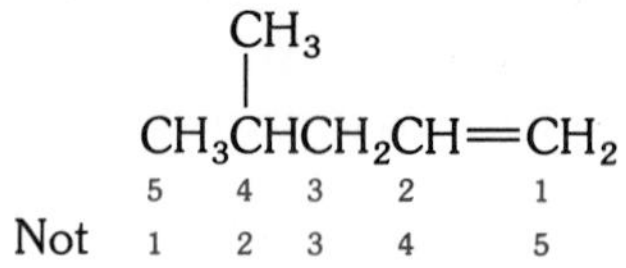

$$CH_3$$
$$|$$
$$CH_3CHCH_2CH{=}CH_2$$

The double bond is at position 1, not 4

5 4 3 2 1

Not 1 2 3 4 5

4-Methyl-1-pentene
(complete name)

Not: 2-methyl-4-pentene

4. To the name begun with rules 1 and 2, place the number that locates the first carbon of the double bond as a prefix, and separate this number from the name by a hyphen.

5. If substituents are on the parent chain or ring, complete the name obtained by rule 4 by placing the names and location numbers of the substituents as prefixes.

Remember to separate numbers from numbers by commas, but use hyphens to connect a number to a word.

EXAMPLE 10.2   NAMING AN ALKENE

Problem: Write the name of the following alkene.

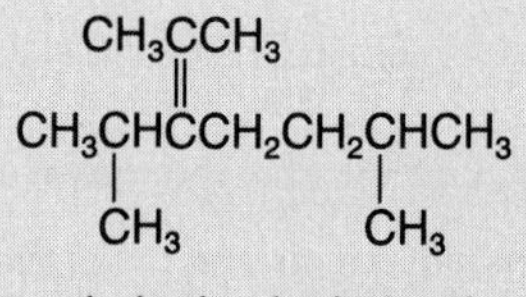

$$CH_3CCH_3$$
$$\|$$
$$CH_3CHCCH_2CH_2CHCH_3$$
$$|\qquad\qquad|$$
$$CH_3\qquad\quad CH_3$$

Solution: First, identify the longest chain *that includes the double bond,* and number it to let one of the carbons of the double bond have the lower number.

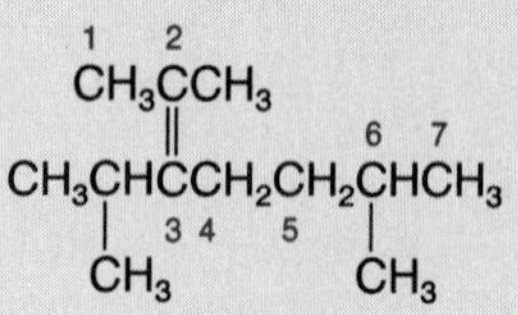

$$\overset{1\ \ 2}{CH_3CCH_3}$$
$$\|\qquad\quad \overset{6\ \ 7}{}$$
$$CH_3CHCCH_2CH_2CHCH_3$$
$$|\ \ \ _{3\ 4}\ \ _{5}\ |$$
$$CH_3\qquad\qquad CH_3$$

We can see that the parent alkene is 2-heptene. It holds two methyl groups (positions 2 and 6) and one isopropyl group (position 3). The names and location numbers are assembled as follows:

2,6-Dimethyl-3-isopropyl-2-heptene

A comma
separates
two numbers

Hyphens separate
numbers from names

PRACTICE EXERCISE 4   Write the IUPAC names of the following compounds.

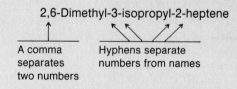

(a) 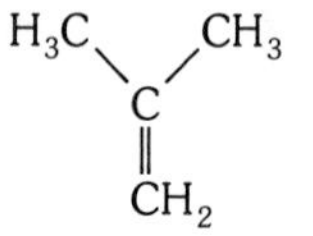 H_3C CH_3
           C
           $\|$
           CH_2

(b) 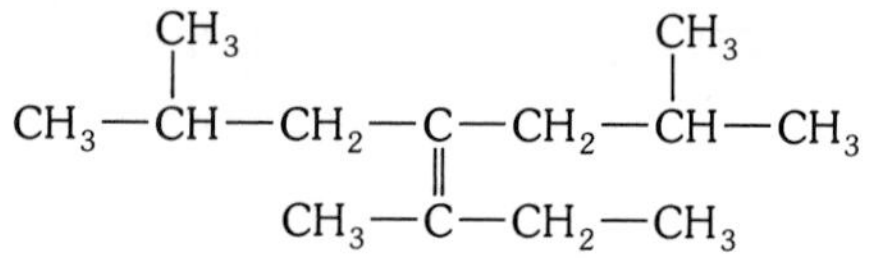 $CH_3{-}CH{-}CH_2{-}C{-}CH_2{-}CH{-}CH_3$
with CH_3 above first CH, CH_3 above last CH, and $CH_3{-}C{-}CH_2{-}CH_3$ (with $\|$) below center C

(c) $CH_3{-}CH{=}CH{-}Cl$ (d) $Br{-}CH_2{-}CH{=}CH_2$

No Free Rotation at a Double Bond Causes Geometric Isomerism among the Alkenes. All of the atoms in ethene lie in the same plane (Figure 10.2), and the lines that join them intersect at angles of 120°. If we replace a hydrogen at each end of the double bond in ethene by a methyl group, the result is 2-butene, but we can make this change in two ways. In one, the methyl groups are on the same side of the double bond, and in the other they are on opposite sides, *and we cannot twist one to make it equivalent to the other.* There is no free rotation about a double bond. The two geometric forms of 2-butene are molecules of different compounds.

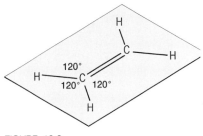

FIGURE 10.2
The geometry at a carbon-carbon double bond.

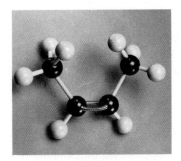

cis-2-Butene
(bp 3.7 °C)

trans-2-Butene
(bp 0.9 °C)

Substances identical both in molecular formulas and in the orders of the attachment of their atoms but with different geometries are called **geometric isomers.** The existence of such isomers is called **geometric isomerism.** When two atoms or groups are on the same side of the double bond, they are said to be *cis* to each other. When they are on opposite sides, they are *trans* to each other. (Sometimes geometric isomerism is called *cis–trans isomerism.*) These cis or trans designations can be made parts of the names of the isomers, as the examples of *cis*- and *trans*-2-butene show.

The *side* of a double bond is not the same as the *end* of a double bond.

Same side

C=C   One end

Not All Alkenes Have Cis–Trans Isomers. When the two atoms or groups *at either end* of the double bond are identical—for example, both are H or both are CH_3—this end has nothing special toward which anything at the other end can be uniquely cis or trans. In 1-butene, for example, the two H atoms on one end of the double bond are identical, so the two ways of writing 1-butene, below, represent the same substance.

1-Butene

is the same as

1-Butene

1-Butene.

If we flop one structure over, top to bottom, we get the second. But this does not reorganize the bonds to make anything different. There are no cis–trans isomers of 1-butene.

EXAMPLE 10.3   WRITING THE STRUCTURES OF CIS AND TRANS ISOMERS

Problem: Write the structures of the cis and trans isomers, if any, of the following alkene.

$$CH_3CH=CHCH_2CH_3$$

2-Pentene

Solution: First, write a carbon–carbon double bond without any attached groups. Spread the single bonds at the carbon atoms at angles of roughly 120°. Draw two of these partial structures.

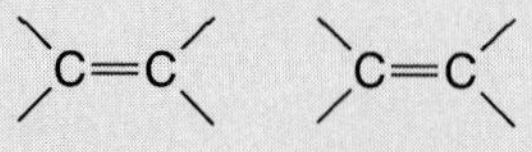

Then attach the two groups that are at one of the ends of the double bond. Attach them *identically* to make identical partial structures.

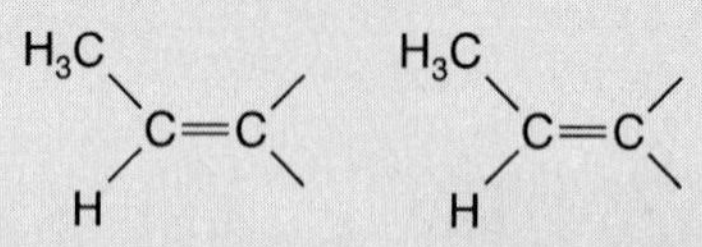

Finally, at the other end of the double bond, draw the other two groups, only this time be sure that they are switched in their relative positions.

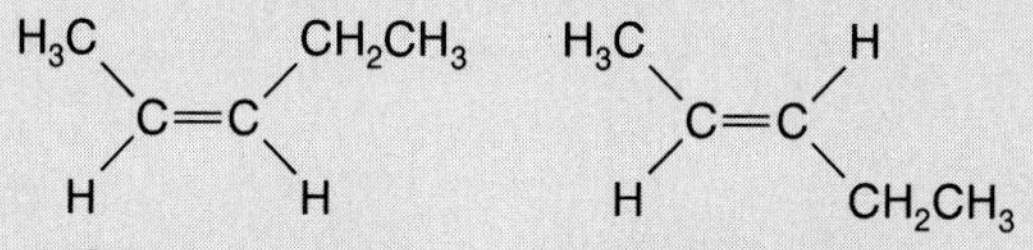

cis-2-Pentene   trans-2-Pentene

Be sure to check to see if the two structures are geometric *isomers* and not two identical structures that are merely flip-flopped on the page.

PRACTICE EXERCISE 5   Write the structures of the cis and trans isomers, if any, of the following compounds.

(a) $CH_3CH_2C=CHCH_3$ (b) $ClCH=CHCl$ (c) $CH_3C=CH_2$ (d) $ClC=CHBr$
    $|$ $|$ $|$
    CH_3 CH_3 Cl

Geometric Isomerism Also Occurs among Ring Compounds. The double bond is not the only source of restricted rotation; the ring is another. Two geometric isomers of 1,2-dimethylcyclopropane are known, for example, and neither can be twisted into the other without breaking the ring open. This costs too much energy to occur spontaneously even at quite high temperatures

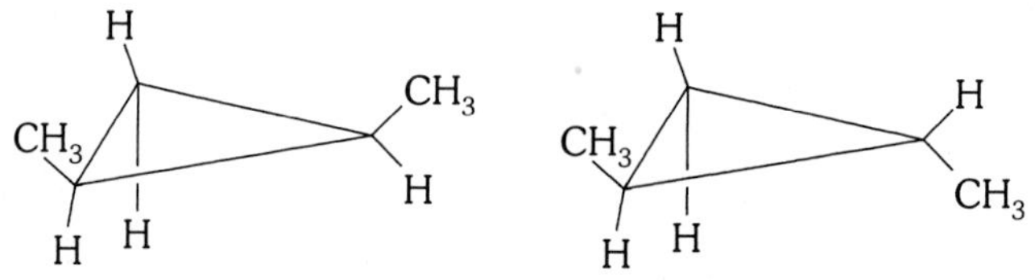

cis-1,2-Dimethylcyclopropane   trans-1,2-Dimethylcyclopropane
(b.p. 37 °C)   (b.p. 28 °C)

This kind of cis–trans difference involving ring compounds occurs often among molecules of carbohydrates.

10.4 CHEMICAL REACTIONS OF THE CARBON–CARBON DOUBLE BOND

The carbon–carbon double bond adds H₂ and H₂O.

The **addition reaction** is the most typical reaction of a carbon–carbon double bond. When it happens, the pieces of some reactant molecule become attached at opposite ends of the double bond and the double bond changes to a single bond. Thus all additions to the double bond have the following features:

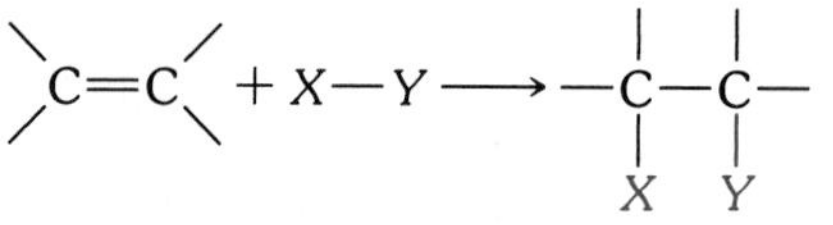

The double bond is also readily attacked by oxidizing agents such as ozone, O_3, and potassium permanganate, $KMnO_4$.

$$\text{C=C} + X—Y \longrightarrow —\overset{|}{\underset{X}{C}}—\overset{|}{\underset{Y}{C}}—$$

Chlorine and bromine readily add to an alkene group. For example,
$$CH_2=CH_2 + Br_2 \rightarrow Br—CH_2—CH_2—Br$$

Before we study examples relevant to the molecular basis of life, we should note that this reaction shows one bond of a double bond to be different from the other. One of the two breaks open, but the other holds. Time doesn't permit us to study the reasons for this difference.

Double Bonds Add Hydrogen. In the presence of a powdered metal catalyst, and under both pressure and heat, hydrogen adds to double bonds. The reaction is called the *hydrogenation* of the double bond.

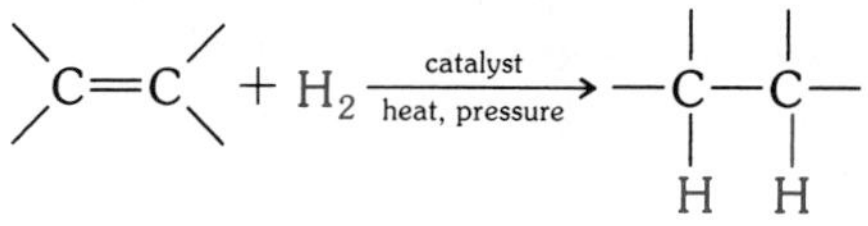

$$\text{C=C} + H_2 \xrightarrow[\text{heat, pressure}]{\text{catalyst}} —\overset{|}{\underset{H}{C}}—\overset{|}{\underset{H}{C}}—$$

Specific examples are

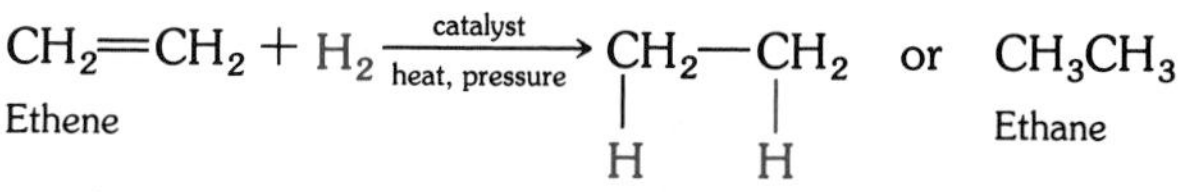

$$CH_2=CH_2 + H_2 \xrightarrow[\text{heat, pressure}]{\text{catalyst}} \underset{\text{Ethene}}{} CH_2—CH_2 \text{ or } CH_3CH_3$$

3-Methylcyclopentene + H₂ → Methylcyclopentane or (structures shown)

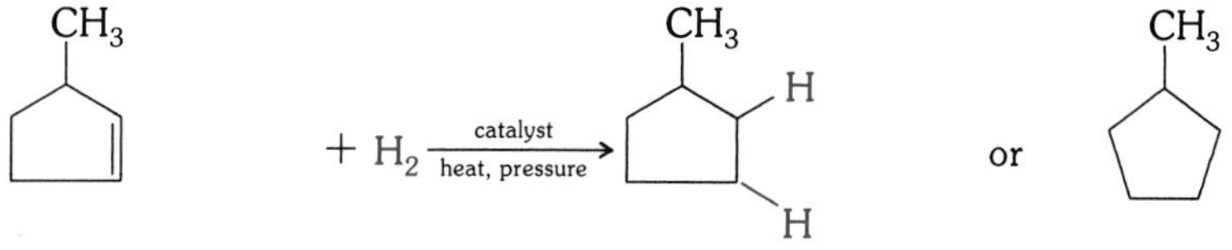

Although molecular hydrogen is not a reactant in body cells, there are substances in the body that can donate hydrogen to alkene groups, and this kind of reaction is very important at several stages in metabolism.

EXAMPLE 10.4 WRITING THE STRUCTURE OF THE PRODUCT OF THE ADDITION OF HYDROGEN TO A DOUBLE BOND

Problem: Write the structure of the product of the following reaction.

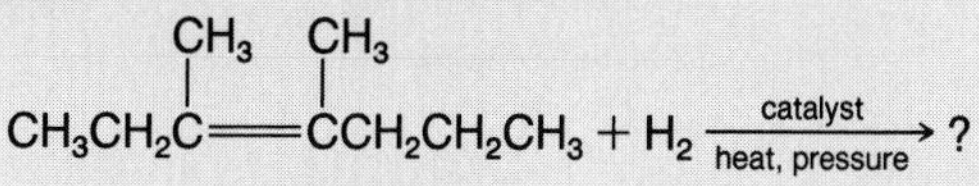

$$CH_3CH_2\overset{CH_3}{\underset{}{C}}=\overset{CH_3}{\underset{}{C}}CH_2CH_2CH_3 + H_2 \xrightarrow[\text{heat, pressure}]{\text{catalyst}} ?$$

Solution.: The only change occurs at the double bond, and the rest of the structure goes through the reaction unchanged. *This is true of all of the addition reactions we will study.* Therefore copy the structure of the alkene just as it is, except leave only a single bond where the double bond was. Then increase by one the number of hydrogens at each carbon of the original double bond. The structure of the product can be written as

$$
\begin{array}{cc}
CH_3 & CH_3 \\
| & | \\
CH_3CH_2C\!\!-\!\!-\!\!CCH_2CH_2CH_3 \\
| & | \\
H & H
\end{array}
\quad \text{or, more condensed} \quad
\begin{array}{cc}
CH_3 & CH_3 \\
| & | \\
CH_3CH_2CH\!\!-\!\!-\!\!CHCH_2CH_2CH_3
\end{array}
$$

Study Example 10.4 very carefully. It illustrates the strategy throughout all of the chapters on organic chemistry. We need a studying strategy, because there are too many specific reactions to memorize as individual "things in themselves." Yet, one of our goals is to be able to predict organic reactions. The time-tested way to handle this is to focus the memory work on the *types* of reactions and then apply this knowledge to figure out a specific example. The *types* of reactions, such as hydrogenation, not their specific illustrations, constitute the fundamental chemical properties of functional groups, the "map signs" that each functional group stands for.

We have just learned one such "map sign" for the carbon–carbon double bond. It can add hydrogen, and when it does, the double bond becomes a single bond, and each of its carbons picks up another hydrogen. This general chemical fact is what has to be learned, but learned more by working out specific examples than by straight memorizing.

PRACTICE EXERCISE 6   Write the structures of the products, if any, of the following:

(a) $CH_3CH\!\!=\!\!CH_2 + H_2 \xrightarrow[\text{heat, pressure}]{\text{catalyst}}$

(b) $CH_3CH_2CH_3 + H_2 \xrightarrow[\text{heat, pressure}]{\text{catalyst}}$

(c) $+ H_2 \xrightarrow[\text{heat, pressure}]{\text{catalyst}}$

(d) $CH_3(CH_2)_7CH\!\!=\!\!CH(CH_2)_7CO_2H + H_2 \xrightarrow[\text{heat, pressure}]{\text{catalyst}}$

Water Adds to Double Bonds.   Water adds to the carbon–carbon double bond to give an alcohol. An acid catalyst (or the appropriate enzyme) is required. Water alone or aqueous bases have no effect on alkenes whatsoever.

In the body, enzymes catalyze the addition of water to alkene double bonds.

$$
\underset{\text{Alkene}}{\begin{array}{c} \diagdown \quad \diagup \\ C\!\!=\!\!C \\ \diagup \quad \diagdown \end{array}} + H\!\!-\!\!OH \xrightarrow[\text{heat}]{H^+} \underset{\text{Alcohol}}{\begin{array}{c} | \quad | \\ -\!C\!\!-\!\!C\!\!- \\ | \quad | \\ H \quad OH \end{array}}
$$

Specific examples are

See Special Topic 10.2 for a description of how this reaction occurs.

$$CH_2\!\!=\!\!CH_2 + H\!\!-\!\!OH \xrightarrow[\substack{240\ °C \\ \text{(closed vessel)}}]{10\%\ H_2SO_4} CH_3\!\!-\!\!CH_2\!\!-\!\!OH$$
Ethene   Ethyl alcohol

$$
\underset{\text{2-Methyl propene}}{\begin{array}{c} CH_3 \\ | \\ CH_3\!\!-\!\!C\!\!=\!\!CH_2 \end{array}} + H\!\!-\!\!OH \xrightarrow[25\ °C]{10\%\ H_2SO_4} \underset{\substack{t\text{-Butyl alcohol} \\ \text{(very little forms)}}}{\begin{array}{c} CH_3 \\ | \\ CH_3\!\!-\!\!C\!\!-\!\!CH_3 \\ | \\ OH \end{array}} \quad Not \quad \underset{\text{(very little forms)}}{\begin{array}{c} CH_3 \\ | \\ CH_3\!\!-\!\!CH\!\!-\!\!CH_2\!\!-\!\!OH \end{array}}
$$

SPECIAL TOPIC 10.2   HOW THE ADDITION OF WATER TO ALKENES HAPPENS

We will use the addition of water to ethene to show how this addition reaction takes place.

Step 1. A proton, H^+, transfers from the acid catalyst to form a stronger carbon–hydrogen bond.

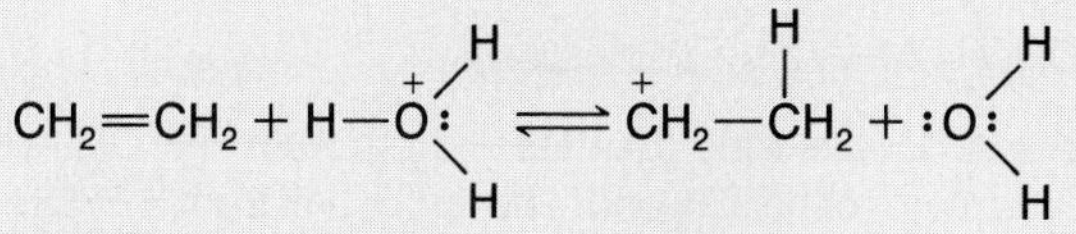

$$CH_2{=}CH_2 + H{-}\overset{+}{\underset{H}{O}}{:} \rightleftharpoons \overset{+}{C}H_2{-}CH_2 + {:}\overset{H}{\underset{H}{O}}{:}$$

| Ethene | Acid catalyst | Ethyl carbocation |

The ethyl carbocation is an example of the **carbocations** in general. These are unstable ions in which a carbon atom has a sextet of electrons, not an octet, and therefore has a positive charge. In the next step, this unstable species regains an outer octet.

Step 2. A carbon–oxygen bond forms as a water molecule attacks the ethyl carbocation.

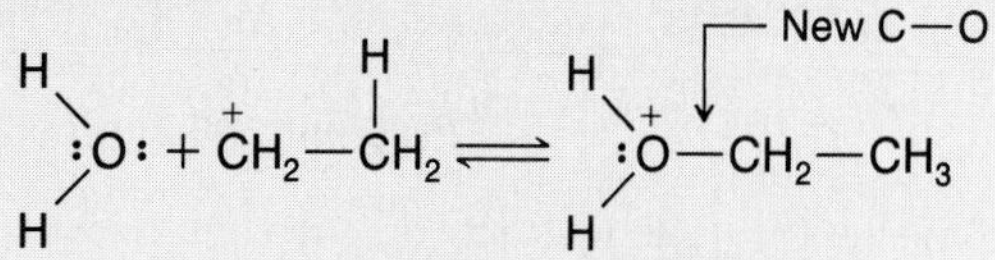

$$\underset{H}{\overset{H}{:}}\!O{:} + \overset{+}{C}H_2{-}CH_2 \rightleftharpoons \underset{H}{\overset{H}{:}}\!\overset{+}{O}{-}CH_2{-}CH_3$$

New C—O bond

Protonated form of ethyl alcohol

The positive charge is now on oxygen, but the oxygen also has an outer octet, so it's a much more stable system than the carbocation. The product is the protonated form of the alcohol. The final product emerges as a proton transfers to some acceptor (e.g., water). Thus the catalyst is recovered.

Step 3. $\quad \underset{H}{\overset{H}{:}}\!O{:} + \underset{H}{\overset{H}{:}}\!\overset{+}{O}{-}CH_2{-}CH_3 \rightleftharpoons$

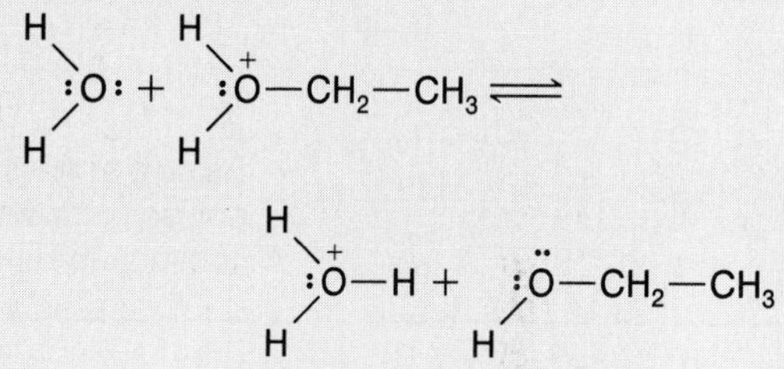

$$\underset{H}{\overset{H}{:}}\!\overset{+}{O}{-}H + {:}\underset{H}{\overset{..}{O}}{-}CH_2{-}CH_3$$

| Recovered catalyst | Ethyl alcohol |

Although water can add in two possible ways to 2-methylpropene, essentially only one route is followed, and *t*-butyl alcohol is the chief product. This behavior is general, and **Markovnikov's rule** is used to predict which product to expect.

Vladimer Markovnikov (1838–1904) was a Russian chemist.

> **Markovnikov's Rule** When an unsymmetrical reactant of the type H—G (e.g., H—OH) adds to an unsymmetrical alkene, the carbon with the greater number of hydrogens gets one more hydrogen.

The following examples illustrate Markovnikov's rule.

$$CH_3CH{=}CH_2 + H{-}OH \xrightarrow{H^+} CH_3{-}\underset{\underset{OH}{|}}{C}H{-}CH_3 \quad Not \quad CH_3{-}CH_2{-}CH_2{-}OH$$

| Propene | | Isopropyl alcohol |

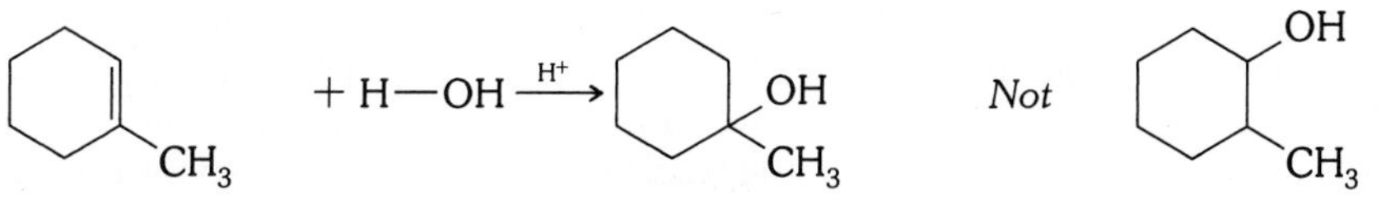

Concentrated sulfuric acid adds easily to double bonds in accordance with Markovnikov's rule. For example,

$$CH_3CH{=}CH_2 + H_2SO_4 \longrightarrow CH_3\underset{\underset{OSO_3H}{|}}{C}HCH_3$$

| 1-Methylcyclohexene | $+ H{-}OH \xrightarrow{H^+}$ | 1-Methylcyclohexanol | *Not* | |

EXAMPLE 10.5   USING MARKOVNIKOV'S RULE

Problem: What product forms in the following situation?

$$CH_3CH_2CH{=}CH_2 + H_2O \xrightarrow[\text{heat}]{H^+} ?$$

Solution: As in all the addition reactions we are studying, the carbon skelton can be copied over intact, except that a single bond is shown where the double bond was.

$$CH_3CH_2{-}CH{-}CH_2 \quad \text{(incomplete)}$$

To decide which carbon of the original double bond gets the H atom from H—OH, we use Markovnikov's rule—the H atom has to go to the CH_2 end. The —OH unit from H—OH goes to the other carbon. The structure of the product is

$$CH_3CH_2{-}\underset{\underset{OH}{|}}{CH}{-}\underset{\underset{H}{|}}{CH_2} \quad \text{or, more condensed} \quad CH_3CH_2\underset{\underset{OH}{|}}{CH}CH_3$$

sec-Butyl alcohol

Be sure at this stage to see that each carbon in the structure of the product has four bonds. If not, some mistake has been made. This is always a useful way to avoid at least some of the common mistakes made in solving a problem such as this.

PRACTICE EXERCISE 7   Write structures for the product(s), if any, that would form under the conditions shown. If no reaction occurs, write "no reaction."

(a) $CH_2{=}CHCH_2CH_3 + H_2O \xrightarrow{H^+}$

(b) $CH_2{=}\underset{\underset{CH_3}{|}}{C}{-}CH_3 + H_2O \xrightarrow{H^+}$

(c)  $CH_3{-}CH{=}\underset{\overset{|}{CH_3}}{C}{-}\bigcirc + H_2O \xrightarrow[\text{heat}]{H^+}$

10.5   THE POLYMERIZATION OF ALKENES

Hundreds to thousands of alkene molecules can join together to make one large molecule of a polymer.

Under a variety of conditions, many hundreds of ethene molecules reorganize their bonds, join together, and change into one large molecule.

$$n CH_2{=}CH_2 \xrightarrow[\text{trace of } O_2]{\text{heat, pressure}} {+}CH_2{-}CH_2{\rightarrow}_n \quad (n = \text{a large number})$$

Ethylene   Polyethylene
(ethene)   (repeating unit)

"Polymer" has Greek roots: *poly,* many, and *meros,* parts.

The product is an example of a **polymer,** a substance of very high formula weight whose molecules have repeating structural units. In polyethylene the repeating unit is $CH_2{-}CH_2$, and hundreds of these are joined together in an extremely long chain. Chain-branching reactions also occur as the polymer forms, so the final product includes both straight- and branched-chain molecules. The starting material for a polymer is called a **monomer,** and the reaction is called **polymerization.**

As we said, a variety of conditions will cause a monomer to polymerize. The catalyst, for example, can be one that generates positively charged intermediates called carbocations (Special Topic 10.2). A proton donor can convert ethylene into the ethyl carbocation:

$$G—H + CH_2=CH_2 \longrightarrow H—CH_2—CH_2^+ + G^-$$

Acid catalyst   Ethylene   Ethyl carbocation

This cation then attacks an unchanged molecule of the alkene:

$$CH_3—CH_2^+ + CH_2=CH_2 \longrightarrow CH_3—CH_2—CH_2—CH_2^+$$

But this new and longer carbocation then attacks another molecule of alkene:

$$CH_3—CH_2—CH_2—CH_2^+ + CH_2=CH_2 \longrightarrow CH_3—CH_2—CH_2—CH_2—CH_2—CH_2^+$$

You can begin to see how this works. Still another carbocation has been produced, and it can attack still another molecule of the alkene. Thus the chain grows with each step by one repeating unit, $—CH_2—CH_2—$, until it picks up some stray anion, such as an anion from the catalyst, for example, and the chain stops growing.

Polymerization is an example of a *chemical chain reaction* because the product of one step initiates the next step.

The polymerization of ethylene, like all such reactions, doesn't produce one, pure homogeneous substance. Some chains grow longer than others, and chain-branching reactions also occur, so the final polymer is a mixture of molecules. All are very large, however, and they all have the same general feature. This is why we can write the structure of a polymer simply by writing its repeating unit, as we did for polyethylene.

Like ethylene, propylene (propene) can also be polymerized, and polypropylene has taken over many uses of polyethylene in both household and hospital applications. Most indoor–outdoor carpeting is woven of polypropylene fibers.

The condensed structure of polypropylene is

$$—(CH_2—CH)_n—$$
$$\quad\quad\quad CH_3$$

Because polypropylene and polyethylene are fundamentally alkanes, they have all of the chemical inertness of this family. These polymers are, therefore, popular for making containers for food juices and medical fluids, for making refrigerator boxes and bottles, containers for chemicals, sutures, catheters, various drains, and wrappings for aneurysms.

Table 10.5 gives just a few examples of polymers, nicknamed vinyl polymers, for which the monomers are substituted alkenes. Dienes also polymerize, and natural rubber is a polymer of a diene called isoprene.

Proteins, DNA, cellulose, and starch are all polymers.

10.6 AROMATIC COMPOUNDS

The benzene ring undergoes substitution reactions instead of addition reactions despite a high degree of unsaturation.

Until well into this century, the structure of benzene was a problem. Its formula, C_6H_6, indicates that it is quite unsaturated, because the ratio of hydrogens to carbons is much lower than in hexane, C_6H_{14}, or in cyclohexane, C_6H_{12}. But benzene gives essentially none of the reactions of alkenes.

The Benzene Molecule Is Planar and All Its Hydrogens Are Equivalent. The six carbons of benzene are in a ring, and each carbon holds one hydrogen. All six hydrogens are chemically equivalent, which means that it is possible to replace any one of them by Cl, for example, to make chlorobenzene, C_6H_5Cl, and no isomeric monochlorobenzenes form. In other words, it doesn't matter which of the six hydrogens in benzene is replaced; the same chlorobenzene forms. Structure **1** is the basic organization of the atoms in benzene.

TABLE 10.5
Some Polymers of Substituted Alkenes

Polymer	Monomer	Uses
Synthetic Polymers		
Polyvinyl chloride	$CH_2{=}CHCl$	Bottles and other containers; insulation; plastic pipe
Saran	$CH_2{=}CCl_2$ and $CH_2{=}CHCl$	Packaging film; fibers; tubing
Teflon	$F_2C{=}CF_2$	Nonsticking surfaces for pots and pans
Orlon	$CH_2{=}CH{-}C{\equiv}N$	Fabrics
Polystyrene	$C_6H_5{-}CH{=}CH_2$	Foam plastics and molded items
Lucite	$CH_2{=}\overset{\overset{\textstyle CH_3}{\textstyle\vert}}{C}{-}CO_2CH_3$	Coatings; windows; molded items
Natural Polymer		
Natural rubber	$CH_2{=}\overset{\overset{\textstyle CH_3}{\textstyle\vert}}{C}{-}CH{=}CH_2$ Isoprene	Tires; hoses; boots

The older structure, **2**, is still widely used to represent benzene, although it is usually abbreviated further:

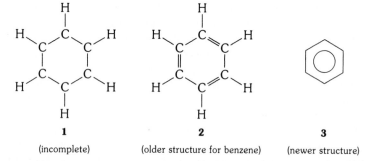

1	**2**	**3**
(incomplete)	(older structure for benzene)	(newer structure)

The trouble with **1** is that each carbon has only three bonds, not four. For decades, scientists simply made each carbon have four bonds by writing in three double bonds, as seen in structure **2**. But **2** says that benzene is a triene, a substance with three alkene groups per molecule. Trienes are known compounds, and they have the same kinds of reactions as mono-enes (e.g., ethene) and dienes — addition reactions. But benzene doesn't give addition reactions. Benzene, for example, doesn't add water like an alkene.

Because the three double bonds indicated in structure **2** are misleading in a chemical sense, scientists today often represent benzene simply by a hexagon with a circle inside, structure **3**.

The ring, often referred to as the *benzene ring,* is planar. All six carbons and all six hydrogen atoms lie in the same plane, and all of the bond angles are 120°, as seen in the scale model of Figure 10.3.

The Benzene Ring Resists Addition Reactions. The typical reactions of benzene are **substitution** reactions, not addition reactions. In a substitution, one hydrogen atom is replaced by another atom or group. Benzene reacts with chlorine or bromine, for example, but only in the presence of iron or an iron(III) chloride catalyst, to give chlorobenzene and HCl. In this equation (and others to come), C_6H_6 stands for benzene, and C_6H_5 represents the **phenyl** group, which is the benzene molecule less one H.

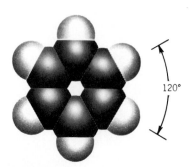

FIGURE 10.3
Scale model of a molecule of benzene.

$$C_6H_6 + Cl_2 \xrightarrow{\text{Fe or}}_{\text{FeCl}_3} C_6H_5-Cl + H-Cl$$

Benzene   Chlorobenzene

A similar reaction occurs with bromine.

Benzene also reacts, by substitution, with sulfur trioxide dissolved in concentrated sulfuric acid.

$$C_6H_6 + SO_3 \xrightarrow[\text{room temperature}]{\text{H}_2\text{SO}_4 \text{ (concd)}} C_6H_5-\overset{\displaystyle O}{\underset{\displaystyle O}{\overset{\|}{\underset{\|}{S}}}}-O-H \quad \text{or} \quad C_6H_5SO_3H$$

Benzenesulfonic acid

Benzene reacts with warm, concentrated nitric acid dissolved in concentrated sulfuric acid. We will represent nitric acid as $HO-NO_2$, instead of HNO_3, because it loses the OH group during the reaction.

$$C_6H_6 + HO-NO_2 \xrightarrow[50-55\,°C]{\text{H}_2\text{SO}_4 \text{ (concd)}} C_6H_5-NO_2 + H_2O$$

Nitrobenzene

Although we did not study the reactions, we can mention here that bromine, chlorine, concentrated sulfuric acid, and concentrated nitric acid all react vigorously with alkenes at room temperature, without extra catalysts, by addition not substitution. Clearly the benzene ring is in a class by itself.

Aromatic Compounds Are Any with Benzene-Like Rings. Any compounds with planar, highly unsaturated rings, like benzene, that *give substitution reactions at the ring instead of addition reactions* are now classified as **aromatic compounds.** This term is a holdover from the days when most of the known compounds of benzene actually had aromatic fragrances. The odor test no longer applies. A few particularly important aromatic compounds related to benzene are given in Table 10.6.

TABLE 10.6
Some Important Aromatic Compounds

Name	Structure	Uses
Toluene	CH$_3$ 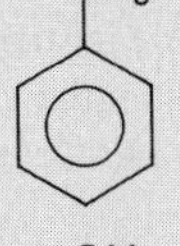	Solvent. Raw material for other aromatic compounds
Phenol	OH 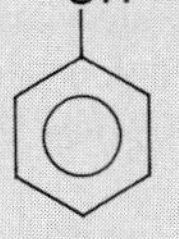	Bactericide (Lister's original antiseptic). Raw material for making aspirin
Aniline	NH$_2$ 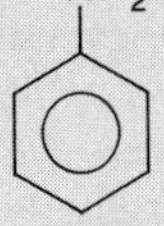	Manufacture of aniline dyes and many pharmaceuticals
Benzoic acid	CO$_2$H 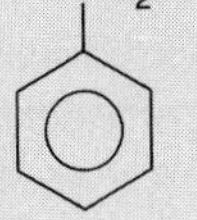	In some ointments to soften the skin. Raw material for manufacture of other aromatic compounds

Many heterocyclic compounds also have unsaturated, benzene-like rings and give substitution instead of addition reactions. They, too, are classified as aromatic. All other compounds, as we noted earlier, are **aliphatic compounds.**

Why Benzene Resists Addition Reactions. Although structure **2** for benzene isn't right in a chemical sense, it does correctly indicate that we have to account for three shared pairs of electrons, those of the three double bonds in **2.** None of these three pairs is confined between just two ring atoms, as structure **2** suggests. All three pairs move in a space that encompasses the entire ring. Thus they actually exist in more space in the benzene ring than most shared electrons. The circle in structure **3** is intended to indicate this feature.

Giving electrons more space means that their mutual repulsions are not as great, and this is what stabilizes the system. If something added just to one of the "double bonds" of structure **2,** this closed circuit, ring-encompassing space would be destroyed, because then this "double bond" would become a single bond. The powerful resistance of the benzene ring to addition reactions stems from the extra stability that three pairs of electrons find in being able to exist in a larger space than usual.

SUMMARY

Hydrocarbons The carbon frameworks of hydrocarbon molecules can be straight chains, branched chains, or rings. When only single bonds occur, the substance is said to be saturated; otherwise it is unsaturated. Regardless of how much we condense a structure, each carbon atom must always have four bonds. Most single bonds need not be shown, but carbon–carbon double and triple bonds are always indicated by two or three lines.

Being nonpolar compounds, the hydrocarbons are all insoluble in water, and many mixtures of alkanes are common, nonpolar solvents. The rule *like dissolves like* lets us predict solubilities.

Nomenclature of alkanes In the IUPAC system, a compound's family is always indicated in the name of a compound by the ending; the number of carbons in the parent chain are shown by a prefix; and the locations of side chains or groups are specified by numbers assigned to the carbons of the parent chain. Alkane-like substituents are called alkyl groups, and the names and formulas of those having from 1–4 carbon atoms must be learned. Common names are still popular, particularly when the IUPAC names are long and cumbersome to use in conversation.

Chemical properties of alkanes Alkanes and cycloalkanes are generally unreactive at room temperature toward concentrated acids and bases, toward oxidizing and reducing agents, toward even the most reactive metals, and toward water. They burn, giving off carbon dioxide and water.

Alkenes The lack of free rotation at a double bond makes geometric (cis–trans) isomers possible, but they exist only when the two groups are not identical at *either* end of the double bond. Cyclic compounds also exhibit cis–trans isomerism.

Alkenes and cycloalkenes are given IUPAC names by a set of rules very similar to those used to name their corresponding saturated forms. However, the double bond takes precedence both in selecting and in numbering the main chain (or ring). The first unsaturated carbon encountered in moving down the chain or around the ring through the double bond must have the lower number.

Addition reactions Hydrogen and water can be made to add to carbon–carbon double bonds. The addition of hydrogen gives alkanes. The addition of water, which follows Markovnikov's rule, gives alcohols.

Polymerization of alkenes The polymerization of an alkene is like an addition reaction. The alkene serves as the monomer, and one alkene molecule adds to another, and so on until a long chain with a repeating unit forms—the polymer molecule.

Aromatic properties When aromatic compounds undergo reactions at the benzene ring, substitutions rather than additions occur. In this way, the closed-circuit electron network of the ring remains unbroken.

Reactions of Unsaturated Hydrocarbons

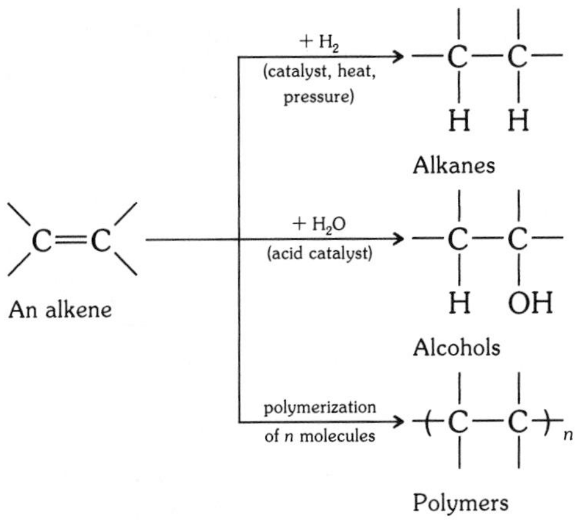

An alkene

$+ H_2$
(catalyst, heat, pressure)

Alkanes

$+ H_2O$
(acid catalyst)

Alcohols

polymerization
of n molecules

Polymers

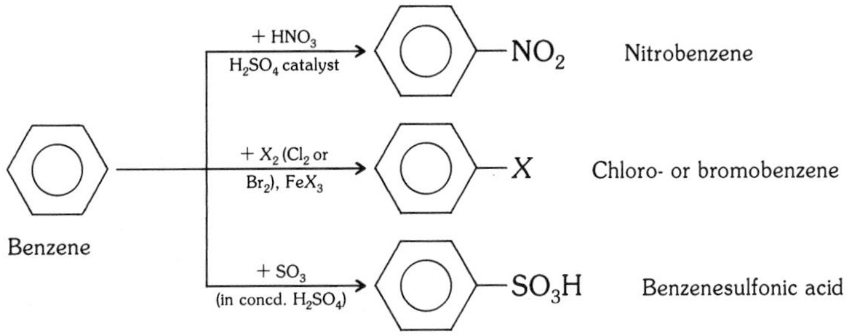

Benzene

$+ HNO_3$
H_2SO_4 catalyst

$-NO_2$ Nitrobenzene

$+ X_2$ (Cl_2 or Br_2), FeX_3

$-X$ Chloro- or bromobenzene

$+ SO_3$
(in concd. H_2SO_4)

$-SO_3H$ Benzenesulfonic acid

REVIEW EXERCISES

The answers to these review exercises are in the *Study Guide* that accompanies this book.

Types of Compounds

10.1 Which family of hydrocarbons is represented by each compound?
(a) 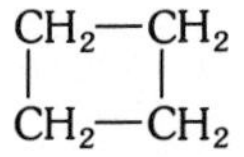 CH_2-CH_2
$\quad\;\; CH_2-CH_2$
(b) $CH_3CH{=}CHCH_2CH_2CH_2CH_2CH_2CH_3$
(c) $CH_3-C{\equiv}C-CH_3$
(d) $CH_3CH_2CH_2CH_2CH_3$

10.2 In what families of hydrocarbons are the following compounds?

(a) 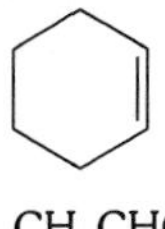   (b)

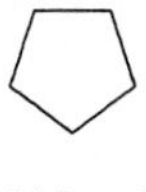

(c) $CH_3CHCH{=}CH_2$ (d) $HC{\equiv}CCH_2CH_2CH_2CH_2CH_3$
$\quad\;\;\; CH_3$

10.3 Which of the following structures represent unsaturated compounds?

(a)

Furan

(b)

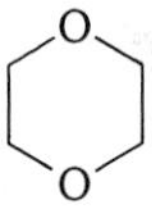

Dioxane

(c) 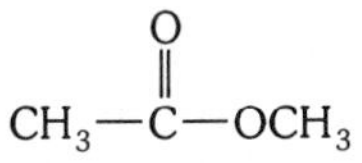 $CH_3-\overset{\displaystyle O}{\overset{\|}{C}}-OCH_3$

Methyl acetate

(d) 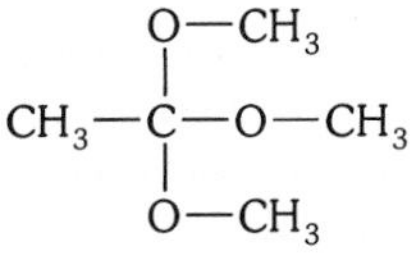 $CH_3-\overset{\displaystyle O-CH_3}{\underset{\displaystyle O-CH_3}{\overset{|}{\underset{|}{C}}}}-O-CH_3$

Methyl orthoacetate

10.4 Which compounds are saturated?

(a)

1,4-Cyclohexadiene

(b) $CH_3-C{\equiv}N$

Acetonitrile

(c)

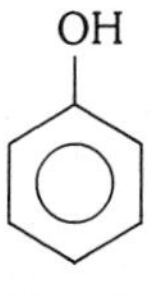

Phenol

(d)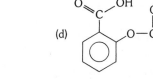

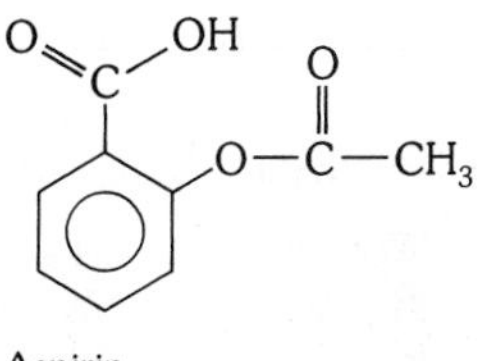

Aspirin

Structures for Cyclic Compounds

10.5 Expand the following structure of nicotinamide, one of the B vitamins.

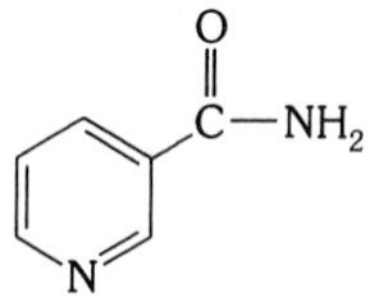

Nicotinamide

10.6 Expand the structure of thiamine, vitamin B_1. Notice that one nitrogen has four bonds and therefore carries a positive charge.

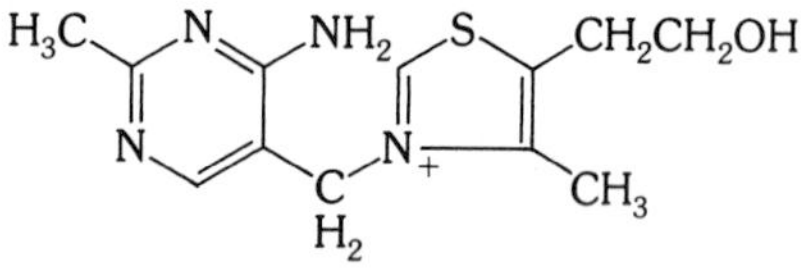

Thiamine

Physical Properties and Structure

10.7 Which compound must have the higher boiling point? Explain.

$$CH_3CH_2CH_3 \qquad CH_3CH_2CH_2CH_2CH_3$$
$$\textbf{A} \qquad\qquad\qquad \textbf{B}$$

10.8 Which compound is less soluble in a nonpolar solvent? Explain.

$$ClCH_2CH_2CH_2CH_2CH_2Cl \qquad HOCH_2CH_2CH_2CH_2CH_2OH$$
$$\textbf{A} \qquad\qquad\qquad\qquad \textbf{B}$$

10.9 Suppose that you were handed two test tubes containing colorless liquids, and you were told that one contained pentane and the other held hydrochloric acid. How could you use just water to tell which tube contained which compound without carrying out any chemical reaction?

10.10 Suppose that you were given two test tubes and were told that one held methyl alcohol, CH_3OH, and the other held hexane. How could water be used to tell these substances apart without carrying out any chemical reaction?

Nomenclature of Alkanes

10.11 There are five isomers of C_6H_{14}. Write their condensed structures and their IUPAC names.

10.12 Which of the isomers of hexane (Review Exercise 10.11) has the common name, n-hexane? Write its structure.

10.13 Which of the hexane isomers (Review Exercise 10.11) has the common name, isohexane? Write its structure.

10.14 There are nine isomers of C_7H_{16}. Write their condensed structures and their IUPAC names.

10.15 Write the condensed structures of the isomers of heptane (Review Exercise 10.14) that have the following names.
(a) n-Heptane   (b) Isoheptane

10.16 Write the IUPAC names of the following compounds.

(a)

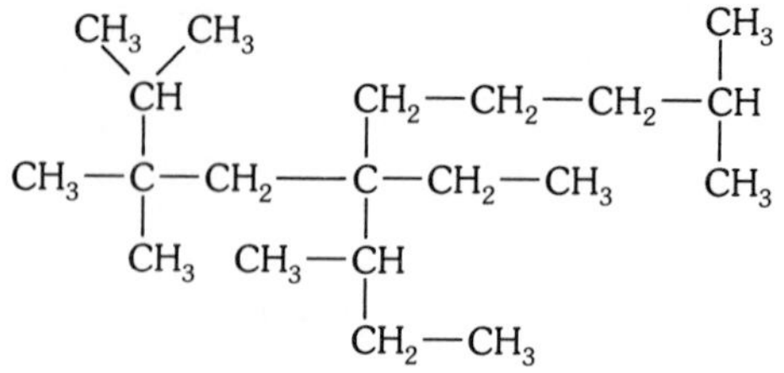

(b)

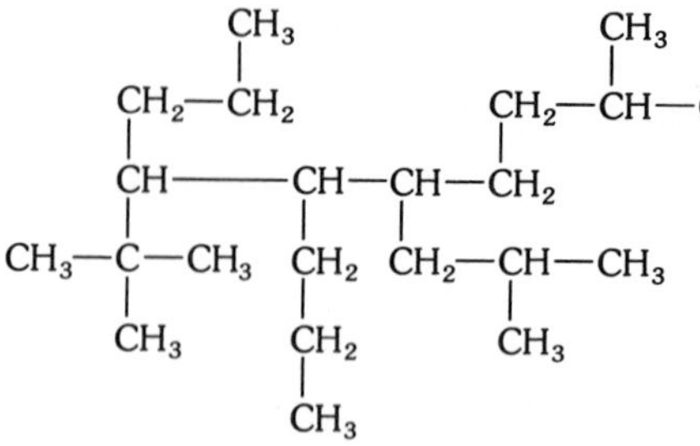

10.17 Write condensed structures for the following compounds.
(a) Isobutyl bromide   (b) Ethyl iodide
(c) Propyl chloride   (d) t-Butylcyclohexane

10.18 Write condensed structures for the following compounds.
(a) sec-Butyl chloride   (b) n-Butyl iodide
(c) Isopropyl bromide   (d) Isohexyl bromide

10.19 The following are incorrect efforts at naming certain compounds. What are the most likely condensed structures and correct IUPAC names?
(a) 1,6-Dimethylcyclohexane   (b) 2,4,5-Trimethylhexane
(c) 1-Chloro-n-butane   (d) Isopropane

10.20 The following names cannot be the correct names, but it is still possible to write structures from them. What are the correct IUPAC names and the condensed structures?
(a) 1-Chloroisobutane   (b) 2,4-Dichlorocyclopentane
(c) 2-Ethylbutane   (d) 1,3,6-Trimethylcyclohexane

Reactions of Alkanes

10.21 Write the balanced equation for the complete combustion of octane, a component of gasoline.

10.22 Gasohol is a mixture of ethyl alcohol, CH_3CH_2OH, in gasoline. Write the balanced equation for the complete combustion of ethyl alcohol. (The products are carbon dioxide and water.)

10.23 What reaction, if any, does hexane give with the following reactants?
(a) NaOH(aq)   (b) concd H_2SO_4
(c) HCl(aq)   (d) H_2O

Cis-Trans Isomerism

10.24 Which of the following pairs of structures represent identical compounds or which are isomers?

(a) CH_2=CH—$\overset{\displaystyle CH_3}{|}$ and $\overset{\displaystyle CH_3}{|}$—CH=$CH_2$

(b) [cyclohexene ring]—CH_3 and [cyclohexene ring]—CH_3

(c) [cyclopentene ring]—CH_3 and [cyclopentene ring]—CH_3

(d) CH_3—CH=$\overset{\displaystyle CH_3}{\underset{\displaystyle |}{C}}$—$CH_2$—$CH_3$ and

CH_3—CH_2—$\overset{\displaystyle CH_3}{\underset{\displaystyle |}{C}}$=CH—$CH_3$

(e) CH_3—$\overset{\displaystyle Br}{\underset{\displaystyle |}{C}}$=CH and CH_3—$\overset{\displaystyle Cl}{\underset{\displaystyle |}{C}}$=CH
$\underset{\displaystyle Cl}{}$ $\underset{\displaystyle Br}{}$

10.25 Write the structures of the cis and trans isomers, if any, of each compound.

(a) CH_3—$\overset{\displaystyle CH_3}{\underset{\displaystyle |}{C}}$=CHCH$_3$

(b) CH_3—[cyclohexane ring]—CH_3

(c) CH_3CH=CCl_2

(d) $CH_3\overset{\displaystyle}{C}$=$\overset{\displaystyle}{C}CH_3$
$\quad\;\; \underset{\displaystyle Cl}{|}\;\; \underset{\displaystyle Cl}{|}$

Nomenclature of Alkenes

10.26 Write the condensed structures of the following compounds.
(a) Propylene   (b) Isobutylene
(c) cis-2-Hexene   (d) 3-Chloro-2-pentene

10.27 Write the IUPAC names of the following compounds.
(a) $CH_3(CH_2)_7CH$=CH_2

(b) Cl—CH=CHCH$\overset{\displaystyle CH_3}{\underset{\displaystyle |}{CH}}CH_3$

(c) $CH_3CH_2CH_2\overset{\displaystyle CH_2}{\underset{\displaystyle ||}{C}}CH_2\overset{\displaystyle CH_3}{\underset{\displaystyle |}{CH}}CH_2CH_3$

(d) $CH_3\overset{\displaystyle CH_3}{\underset{\displaystyle |}{C}}CH$=CH
$\quad\;\;\; \underset{\displaystyle CH_3}{|}\;\; \underset{\displaystyle CH_3}{|}$

10.28 Write the condensed structures and the IUPAC names for all of the isomeric pentenes, C_5H_{10}. Include cis and trans isomers.

Reactions of the Carbon–Carbon Double Bond

10.29 Write equations for the reactions of 2-methylpropene with the following reactants.
(a) H_2 (Ni, heat, pressure)   (b) H_2O (H^+ catalyst)

10.30 Write equations for the reactions of 1-methylcyclopentene with the reactants listed in Review Exercise 10.29.

10.31 Write equations for the reactions of 2-methyl-2-butene with the compounds given in Review Exercise 10.29.

10.32 Write equations for the reactions of 1-methylcyclohexene with the reactants listed in Review Exercise 10.29. (Do not attempt to predict if cis or trans isomers form.)

10.33 One of the raw materials for the synthesis of nylon, adipic acid, can be made from cyclohexene by oxidation using potassium permanganate. The balanced equation for the first step in which the potassium salt of adipic acid forms is as follows:

$$3C_6H_{10} + 8KMnO_4 \longrightarrow 3K_2C_6H_8O_4$$
$$+ 8MnO_2 + 2KOH + 2H_2O$$

How many grams of potassium permanganate are needed for the oxidation of 10.0 g of cyclohexene, assuming that the reaction occurs exactly and entirely as written?

10.34 Referring to Review Exercise 10.33, how many grams of the potassium salt of adipic acid can be made if 24.0 g of $KMnO_4$ are used in accordance with the equation given?

Polymerization

10.35 Rubber cement can be made by mixing some polymerized 2-methylpropene with a solvent such as toluene. When the solvent evaporates, a very tacky and sticky residue of the polymer remains, which soon hardens and becomes the glue. Write the structure of the polymer of 2-methylpropene (called polyisobutylene) in two ways. (The structure is quite regular, like polypropylene.)
(a) One that shows four repeating units, one after the other.
(b) The condensed structure.

10.36 Safety glass is made by sealing a thin film of polyvinyl acetate between two pieces of glass. If the glass shatters, its pieces remain glued to the film and cannot fly about. Using four vinyl acetate units, write part of the structure of a molecule of polyvinyl acetate. (The structure is quite regular, like that of polypropylene.) Also write its condensed structure. The structure of vinyl acetate is as follows:

$$\overset{\displaystyle O}{\underset{\displaystyle ||}{}}$$
$$O\overset{\displaystyle}{C}CH_3$$
$$\underset{\displaystyle |}{}$$
$$CH_2=CH$$

Vinyl acetate

Aromatic Properties

10.37 Dipentene has a very pleasant, lemonlike fragrance, but it is not classified as an aromatic compound. Why?

Dipentene

10.38 Sulfanilamide, the structurally simplest of the sulfa drugs, has no odor at all, but it is still classified as an aromatic compound. Explain.

Sulfanilamide

10.39 Write equations for the reactions, if any, of benzene with the following compounds.
(a) Sulfur trioxide (in concentrated sulfuric acid)
(b) Concentrated nitric acid (in concentrated sulfuric acid)
(c) Hot sodium hydroxide solution
(d) Hydrochloric acid
(e) Chlorine (alone)

10.40 Explain why benzene strongly resists addition reactions and gives substitution reactions instead.

Predicting Reactions

10.41 Write the structures of the products to be expected in the following situations. If no reaction is to be expected, write "no reaction." To work this kind of exercise, you have to be able to do three things.
(a) *Classify* a specific organic reactant into its proper family.
(b) *Recall* the short list of chemical facts about the family. (If there is no match-up between this list and the reactants and conditions specified by a given problem, assume that there is no reaction.)
(c) *Apply* the recalled chemical fact, which might be some "map sign" associated with a functional group, to the specific situation.

Study the next two examples before continuing.

EXAMPLE 10.6 PREDICTING REACTIONS

Problem: What is the product, if any, in the following?

$$CH_3CH_2CH_2CH_3 + H_2SO_4 \longrightarrow$$

Solution: We note first that the organic reactant is an alkane, so we next turn to the list of chemical properties about all alkanes that we learned. With this family, of course, the list is very short. Except for combustion and halogenation, we have learned no reactions for alkanes,

and we assume, therefore, that there aren't any others, not even with sulfuric acid. Hence, the answer is "no reaction."

EXAMPLE 10.7 PREDICTING REACTIONS

Problem: What is the product, if any, in the following situation?

$$CH_3CH=CHCH_3 + H_2O \xrightarrow{\text{acid catalyst}} ?$$

Solution: We first note that the organic reactant is an alkene, so we review our mental "file" of reactions of the carbon–carbon double bond.

1. Alkenes add hydrogen (in the presence of a metal catalyst and under heat and pressure) to form alkanes.
2. They add water in the presence of an acid catalyst to give alcohols.
3. They polymerize.

These are the chief chemical facts about the carbon–carbon double bond that we have studied, and we see that the list includes a reaction with water in the presence of an acid catalyst. We remember that in all addition reactions the double bond changes to a single bond and the pieces of the molecule that adds end up on the carbons at this bond. We also have to remember Markovnikov's rule to tell us which pieces of the water molecule go to which carbon. However, in this specific example, Markovnikov's rule does not apply (the alkene is symmetrical).

Answer:
$$CH_3CH_2\underset{\underset{OH}{|}}{C}HCH_3$$

Now work the following parts. (Remember: The formula C_6H_6 stands for benzene and C_6H_5- is the phenyl group.)

(a) $CH_3CH_2CH=CH_2 + H_2 \xrightarrow[\text{pressure}]{\text{Ni, heat,}}$

(b) $CH_3CH=CHCH_3 + H_2O \xrightarrow{H^+}$

(c) $C_6H_6 + Cl_2 \xrightarrow{FeCl_3}$

(d) $CH_3CH_2CH_3 + (\text{concd})H_2SO_4 \rightarrow$

(e) $+ H_2O \xrightarrow{H^+}$

(f) $CH_3CH=CHCH_2CH_3 + H_2 \xrightarrow[\text{pressure}]{\text{Ni, heat,}}$

(g) $+ \text{concd. } H_2SO_4 \longrightarrow$

(h) $C_6H_5-CH=CH-C_6H_5 + H_2O \xrightarrow{H^+}$

(i) 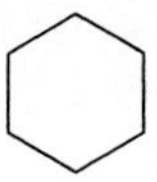 cyclohexene $+ H_2 \xrightarrow{\text{Ni, heat, pressure}}$

(j) $CH_3CH_2CH_2CH_3 + O_2 \xrightarrow{\text{complete combustion}}$
(balance the equation)

(k) $C_6H_6 + H_2O \rightarrow$

10.42 Write the structures of the products in the following situations. If no reaction is to be expected, write "no reaction".

(a) $CH_3CH=CH_2 + H_2O \xrightarrow{H^+}$

(b) $CH_2=CH-CH_2-CH=CH_2 + 2H_2 \xrightarrow{\text{Ni, heat, pressure}}$

(c) $+CH_2-CH_2+_n + H_2SO_4 \text{ (concd)} \rightarrow$

(d) $CH_3-\overset{CH_3}{\underset{}{C}}=\overset{CH_3}{\underset{}{C}}-CH_3 + H_2O \xrightarrow{H^+}$

(e) 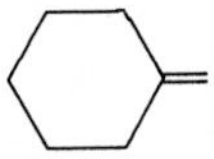 cyclohexane $=CH_2 + H_2 \xrightarrow{\text{Ni, heat, pressure}}$

(f) 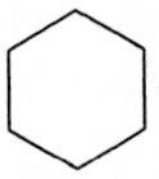 cyclohexane $+ H_2O \xrightarrow{H^+}$

(g) $C_6H_6 + Br_2 \xrightarrow{FeBr_3}$
(h) $C_6H_6 + NaOH(aq) \longrightarrow$

(i) $C_6H_5-CH_3 + O_2 \xrightarrow{\text{complete combustion}}$ (Balance the equation.)

(j) 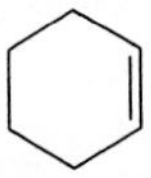 cyclohexene $+ NaOH(aq) \longrightarrow$

Petroleum (Special Topic 10.1)

10.43 What kinds of substances predominate in the mixture called petroleum?

10.44 What is the range of the carbon atoms per molecule in (a) gasoline, (b) jet fuel, and (c) paraffin wax?

10.45 What happens when engineers "crack" certain petroleum fractions?

Addition of Water to Alkenes (Special Topic 10.2)

10.46 What forms when a proton is accepted by a molecule of ethene? (Write its name and structure.)

10.47 If 2-butene accepted a proton like ethene, what would form? (Write its structure.)

10.48 What forms when a water molecule attaches itself to the ethyl carbocation? (Write its structure.) What kind of reaction changes this product into a molecule of ethyl alcohol?

10.49 What forms when a water molecule attaches itself to the carbocation that formed in Review Exercise 10.47? (Write its structure.) When this product loses a proton, what stable compound results? (Write its structure.)

Chapter 11
Alcohols, Thioalcohols, Ethers, and Amines

Sulfur analogs of alcohols make up much of the fluid that gives the skunk both respect and a bad name, especially when it's tail is up. We'll study about alcohols and thioalcohols in this chapter.

11.1 OCCURRENCE, TYPES, AND NAMES OF ALCOHOLS

Molecules of the alcohol family have an OH group attached to a saturated carbon.

The OH group of alcohols is one of the most widely occurring in nature. It is present, for example, in carbohydrates, proteins, and nucleic acids as well as in several vitamins and hormones. Rubbing alcohol, beverage alcohol, and antifreezes are all different members of the alcohol family.

Many Families of Organic Compounds Have the OH Group. In the **alcohols,** the OH group is bound to a carbon atom from which only single C—C and C—H bonds extend. Only in this situation, illustrated by several simple alcohols in Table 11.1, can the OH group be called the **alcohol group.**

In the family of the phenols the OH group is attached to a benzene ring, and in the carboxylic acids it is held by a carbon that also has a double bond to oxygen.

TABLE 11.1
Some Common Alcohols

Name[a]	Structure	Boiling Point (°C)
Methyl alcohol (methanol)	CH_3OH	65
Ethyl alcohol (ethanol)	CH_3CH_2OH	78.5
Propyl alcohol (1-propanol)	$CH_3CH_2CH_2OH$	97
Isopropyl alcohol (2-propanol)	$CH_3\underset{\underset{OH}{\mid}}{C}HCH_3$	82
Butyl alcohol (1-butanol)	$CH_3CH_2CH_2CH_2OH$	117
sec-Butyl alcohol (2-butanol)	$CH_3CH_2\underset{\underset{OH}{\mid}}{C}HCH_3$	100
Isobutyl alcohol (2-methyl-1-propanol)	$CH_3\overset{\overset{CH_3}{\mid}}{C}HCH_2OH$	108
t-Butyl alcohol (2-methyl-2-propanol)	$CH_3\overset{\overset{CH_3}{\mid}}{\underset{\underset{CH_3}{\mid}}{C}}OH$	83
Ethylene glycol (1,2-ethanediol)	$\underset{\underset{OH}{\mid}}{C}H_2-\underset{\underset{OH}{\mid}}{C}H_2$	197
Propylene glycol (1,2-propanediol)	$CH_3\underset{\underset{OH}{\mid}}{C}H-\underset{\underset{OH}{\mid}}{C}H_2$	189
Glycerol (1,2,3-propanetriol)	$\underset{\underset{OH}{\mid}}{C}H_2-\underset{\underset{OH}{\mid}}{C}H-\underset{\underset{OH}{\mid}}{C}H_2$	290

[a] The IUPAC names are in parentheses.

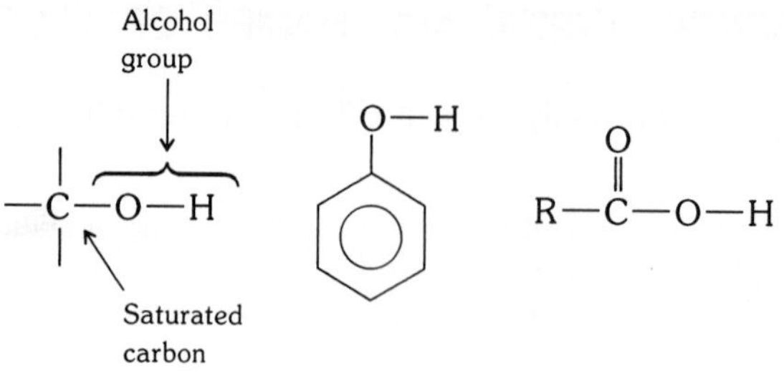

Alcohol
group

−C−O−H

Saturated
carbon

Alcohol
system

Phenol

Carboxylic
acids

$R-\overset{O}{\overset{\|}{C}}-O-H$

PRACTICE EXERCISE 1   Classify the following as alcohols, phenols, or carboxylic acids.

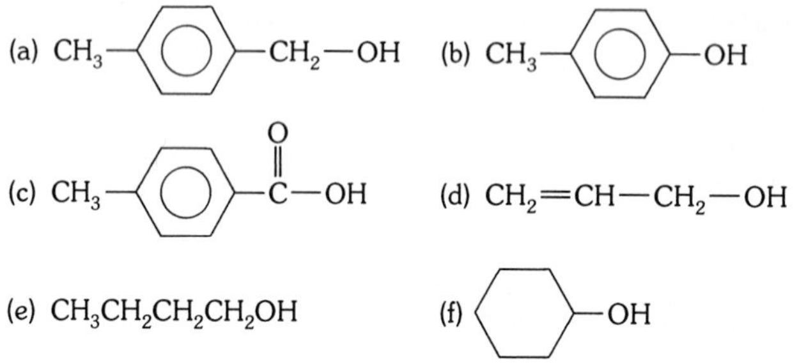

(a) CH_3—⬡—CH_2—OH   (b) CH_3—⬡—OH

(c) CH_3—⬡—$\overset{O}{\overset{\|}{C}}$—OH   (d) $CH_2{=}CH-CH_2-OH$

(e) $CH_3CH_2CH_2CH_2OH$ (f) ⬡—OH

Some Chemical Properties Vary with the Subclasses of Alcohols.   We will see in Section 11.3 that the condition of the carbon holding the OH group, whether it has zero, one, or two H atoms as well, affects how alcohols are oxidized. **Primary alcohols** (1° alcohols) are those in which the OH group is held by a 1° carbon atom. In a **secondary alcohol,** the OH group is held by a secondary (2°) carbon atom. When the OH group is on a 3° carbon atom, the alcohol is a **tertiary alcohol.**

Pronounce 1° as *primary,* 2° as *secondary,* and 3° as *tertiary.*

The R— groups in 2° and 3° alcohols do not have to be the same.

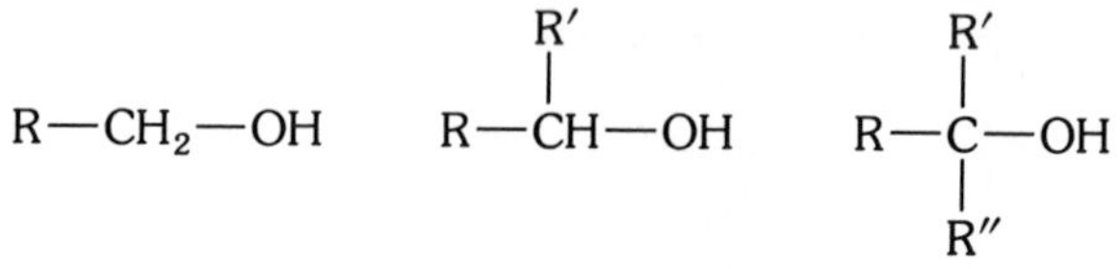

$R-CH_2-OH$ $R-\overset{R'}{\underset{}{\overset{|}{C}}H-OH$ $R-\overset{R'}{\underset{R''}{\overset{|}{\underset{|}{C}}}-OH$

Primary alcohol   Secondary alcohol   Tertiary alcohol

Monohydric alcohols, the "simple alcohols," have one OH per molecule. Some alcohols have more than one OH group. *Dihydric alcohols,* the **glycols,** have two. Ethylene glycol, an ingredient in some commercial antifreezes, is an example. *Trihydric alcohols,* like glycerol, have three alcohol groups per molecule. Many substances, particularly sugars, have several OH groups. Special Topic 11.1 describes several important, individual alcohols.

The following system, the 1,1-diol, is unstable; only rare examples are known.

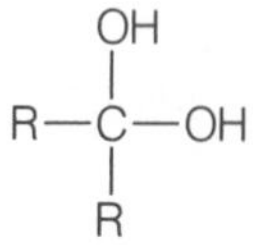

$R-\overset{OH}{\underset{R}{\overset{|}{\underset{|}{C}}}-OH$

1,1-Diols

$\underset{\underset{OH\ \ OH}{|\ \ \ \ |}}{CH_2-CH_2}$ $\underset{\underset{OH\ OH\ OH}{|\ \ \ |\ \ \ |}}{CH_2-CH-CH_2}$ $\underset{\underset{OH\ \ OH\ OH\ OH\ OH}{|\ \ \ \ |\ \ \ |\ \ \ |\ \ \ |}}{CH_2-CH-CH-CH-CH-\overset{O}{\overset{\|}{C}}-H}$

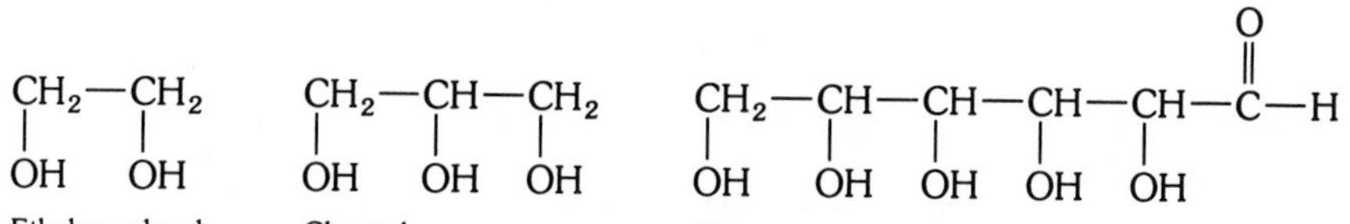

Ethylene glycol
(1,2-ethanediol)   Glycerol
(1,2,3-propanetriol)   Glucose, a sugar
(open-chain form)

In di- and polyhydroxy alcohols, the OH groups are always on different carbon atoms. Almost no stable system is known in which one carbon holds two or three OH groups.

Methyl Alcohol (Methanol, Wood Alcohol) When taken internally in enough quantity, methyl alcohol causes either blindness or death. In industry, it is used as the raw material for making formaldehyde (which is used to make polymers), as a solvent, and as a denaturant (poison) for ethyl alcohol. It is also used as the fuel in Sterno and in burners for fondue pots.

Most methyl alcohol is made by the reaction of carbon monoxide with hydrogen under high pressure and temperature:

$$2H_2 + CO \xrightarrow[\substack{temp = 350-400\ ^\circ C \\ catalyst = ZnO-Cr_2O_3}]{3000\ lb/in.^2} CH_3OH$$

Ethyl Alcohol (Ethanol, Grain Alcohol) Some ethyl alcohol is made by the fermentation of sugars, but most is synthesized by the addition of water to ethene in the presence of a catalyst. A 70% (v/v) solution of ethyl alcohol in water is used as a disinfectant.

In industry, ethyl alcohol is used as a solvent and to prepare pharmaceuticals, perfumes, lotions, and rubbing compounds. For these purposes, the ethyl alcohol is adulterated by poisons that are very hard to remove so that the alcohol cannot be sold or used as a beverage. (Nearly all countries derive revenue by taxing potable, or drinkable, alcohol.)

Isopropyl Alcohol (2-Propanol) Isopropyl alcohol is a common substitute for ethyl alcohol for giving back rubs. It is twice as toxic as ethyl alcohol, and in solutions with concentrations from 50 to 99% (v/v) it is used as a disinfectant.

Ethylene Glycol (1,2-Ethanediol), and **Propylene Glycol** (1,2-Propanediol) Ethylene and propylene glycol are the chief components in permanent-type antifreezes. Their great solubility in water and their very high boiling points make them ideal for this purpose. An aqueous solution roughly 50% (v/v) in either glycol does not freeze until about $-40\ ^\circ C\ (-40\ ^\circ F)$.

Glycerol (1,2,3-Propanetriol, Glycerin) Glycerol, a colorless, syrupy liquid with a sweet taste, is freely soluble in water and insoluble in nonpolar solvents. It is one product of the digestion of the fats and oils in our diets. Because it has three OH groups per molecule, each capable of hydrogen bonding to water molecules, glycerol can draw moisture from humid air. It is sometimes used as a food additive to help keep foods moist.

Sugars All carbohydrates consist of polyhydroxy compounds, and we will study them in a later chapter.

Phenol (Carbolic Acid) Phenol, as we have already noted, was the first antiseptic to be used by Joseph Lister (1827–1912), an English surgeon. It kills bacteria by denaturing their proteins, which destroys the abilities of these substances to function normally. (Interestingly, phenol does not denature nucleic acids, the chemicals of heredity.) Phenol, however, is a general protoplasmic poison and so is dangerous to healthy tissue. Other antiseptics have been developed since Lister's time.

PRACTICE EXERCISE 2   Classify each of the following as monohydric or dihydric. For each found to be monohydric, classify it further as 1°, 2°, or 3°. If the structure is too unstable to exist, state so.

(a) $CH_3-\underset{\underset{OH}{|}}{C}H-CH_3$ (b) (cyclohexene ring)—OH   (c) $CH_3-\underset{\underset{OH}{|}}{\overset{\overset{OH}{|}}{C}}-CH_3$

(d) (cyclohexane ring with two adjacent —OH groups)   (e) $HO-CH_2-\underset{\underset{CH_3}{|}}{\overset{\overset{CH_3}{|}}{C}}-CH_3$ (f) (benzene ring)—CH_2-OH

(g) $CH_3-\underset{\underset{CH_3}{|}}{\overset{\overset{CH_3}{|}}{C}}-OH$ (h) $CH_3-CH_2-\underset{\underset{OH}{|}}{\overset{\overset{CH_3}{|}}{C}}H$ (i) $CH_3-\underset{\underset{OH}{|}}{\overset{\overset{OH}{|}}{C}}-OH$

The Simple Alcohols Are Usually Called by Their Common Names. To write a common name, simply write the word *alcohol* after the name of the alkyl group present. For example,

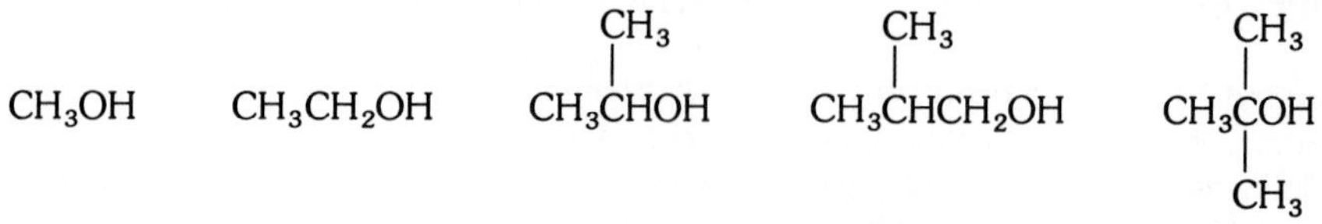

$$CH_3OH \qquad CH_3CH_2OH \qquad \overset{\overset{\displaystyle CH_3}{|}}{CH_3CHOH} \qquad \overset{\overset{\displaystyle CH_3}{|}}{CH_3CHCH_2OH} \qquad \overset{\overset{\displaystyle CH_3}{|}}{\underset{\underset{\displaystyle CH_3}{|}}{CH_3COH}}$$

These common names should be
learned.

| Methyl | Ethyl | Isopropyl | Isobutyl | t-Butyl |
| alcohol | alcohol | alcohol | alcohol | alcohol |

When the alkyl group has no common name, the IUPAC system is used, and the rules are similar to those for naming alkanes. As discussed in Appendix III, the IUPAC name is based on a *parent alcohol,* the longest chain that holds the OH group.

11.2 PHYSICAL PROPERTIES OF ALCOHOLS

Hydrogen bonding dominates the physical properties of alcohols.

Alcohol molecules are polar, and they can both donate and accept hydrogen bonds. Both boiling points and solubilities in water are affected. In this section we will establish another molecular "map sign" for relating structures to properties.

Boiling Points Increase with the Number of OH Groups per Molecule. Table 11.2 pairs alkanes with alcohols of comparable formula weight, and you can see how greatly the OH group influences boiling point. The 154 °C difference in boiling point between ethane and methyl alcohol, for example, indicates how great the attraction is between alcohol molecules. This extra attraction is from the two hydrogen bonds that can extend between molecules of monohydric alcohols (Figure 11.1), whereas no hydrogen bonds are present in alkanes.

Water molecules experience three hydrogen bonds between neighbors (Figure 11.1), so despite water's lower formula weight, it has a higher boiling point than methyl alcohol. And notice how the boiling point difference leaps when a dihydric alcohol, like ethylene glycol (Table 11.2), is compared to the alkane of comparable formula weight. Four hydrogen bonds can extend from a dihydric molecule, twice as many as occur between molecules of monohydric alcohols.

Solubilities in Water Increase with the Number of OH Groups per Molecule. Methane is insoluble in water because, as illustrated in Figure 11.2, the tightly hydrogen-bonded water molecules cannot let the nonpolar methane molecules in. Methyl alcohol, however, readily dissolves in water. Its molecules can donate and accept hydrogen bonds and so are able to slip into water's network of hydrogen bonds (see Figure 11.2).

TABLE 11.2
The Influence of the Alcohol Group on Boiling Points

Name	Structure	Formula Weight	Boiling Point (°C)	Difference in Boiling Point
Ethane	CH_3CH_3	30	−89	154
Methyl alcohol	CH_3OH	32	65	
Propane	$CH_3CH_2CH_3$	44	−42	120
Ethyl alcohol	CH_3CH_2OH	46	78	
Butane	$CH_3CH_2CH_2CH_3$	58	0	197
Ethylene glycol	$HOCH_2CH_2OH$	62	197	

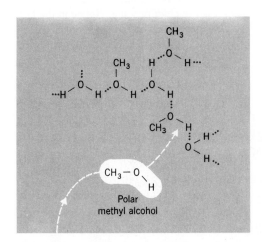

FIGURE 11.1
Hydrogen bonding in water *(a)* and in alcohols *(b)*.

As the size of a monohydric alcohol molecule increases, it becomes more and more alkane-like. In decyl alcohol, for example, the small, water-like OH group is overwhelmed by the long hydrocarbon chain:

$$CH_3CH_2CH_2CH_2CH_2CH_2CH_2CH_2CH_2CH_2OH$$

Decyl alcohol

The flexings and twistings of long chains interfere too much with water's hydrogen-bonding networks, and so water cannot let decyl alcohol molecules into solution. This alcohol and most with five or more carbons are insoluble in water. They do dissolve, however, in such nonpolar solvents as diethyl ether, benzene, and gasoline.

The more OH groups there are per molecule, the more soluble the compound is in water and the less soluble it is in nonpolar solvents. Glycerol, sugars, and even the dihydric alcohols do not dissolve in nonpolar solvents but dissolve in water.

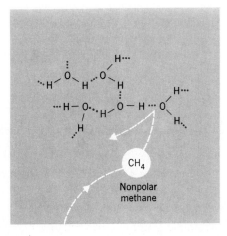

FIGURE 11.2
How a short-chain alcohol dissolves in water. *(a)* The alcohol molecule can take the place of a water molecule in the hydrogen-bonding network of water. *(b)* An alkane molecule cannot break into the hydrogen-bonding network in water, so the alkane cannot dissolve.

11.3 CHEMICAL PROPERTIES OF ALCOHOLS

The loss of water (dehydration) and the loss of hydrogen (oxidation) are two important reactions of alcohols.

Alcohols react with both inorganic and organic compounds, but we will study only the inorganic reactants in this chapter. We first mention, however, a reaction that alcohols do not have.

When an alcohol dissolves in water, it doesn't raise or lower the pH.

Alcohols Do Not Ionize in Water. The alcohols are like water, donors of neither hydroxide ions nor hydrogen ions. Alcohols are thus neither acids nor bases in the conventional sense, but are neutral compounds.

Alcohols Can Be Dehydrated to Alkenes. Heat and an acid catalyst make an alcohol split out water leaving behind a carbon–carbon double bond. The reaction is called the *dehydration of an alcohol,* and it gives an alkene. The pieces of the water molecule, one H and one OH, come from *adjacent* carbons. Specific enzymes catalyze this reaction in living systems where strongly acidic catalysts, of course, cannot occur.

In general:

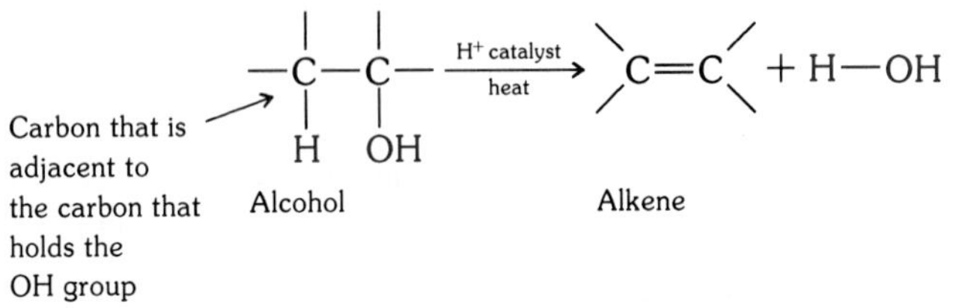

$$-\underset{H}{\overset{|}{C}}-\underset{OH}{\overset{|}{C}}- \xrightarrow[\text{heat}]{\text{H}^+ \text{ catalyst}} \diagdown C = C \diagup + \text{H—OH}$$

Carbon that is adjacent to the carbon that holds the OH group   Alcohol   Alkene

Specific examples are as follows:

$$CH_3CH_2OH \xrightarrow[\text{170–180 °C}]{\text{H}_2\text{SO}_4 \text{ (cond)}} CH_2{=}CH_2 + H_2O$$
Ethyl alcohol   Ethene

$$\underset{\text{sec-Butyl alcohol}}{CH_3CH_2\overset{\overset{\displaystyle OH}{|}}{C}HCH_3} \xrightarrow[\text{100 °C}]{\text{60\% H}_2\text{SO}_4} \underset{\substack{\text{2-Butene} \\ \text{(chief product)}}}{CH_3CH{=}CHCH_3} + \underset{\substack{\text{1-Butene} \\ \text{(by-product)}}}{CH_3CH_2CH{=}CH_2} + H_2O$$

Sometimes, as in this example, two or more alkenes seem to be possible products, because more than one adjacent carbon could supply the H atom needed to complete a molecule of H—OH. When one of the possible alkenes, however, has a double bond with more attached alkyl groups than any of the others it is the chief product. Notice in the last example that the chief product, 2-butene has two alkyl groups at the double bond (two CH_3 groups), but 1-butene has only one (the CH_3CH_2 group). We aren't concerned about this for alcohol dehydrations among body chemicals, because enzymes are very specific in directing reactions.

Enzymes are exceedingly selective in what they do and how they control reactions.

EXAMPLE 11.1   WRITING THE STRUCTURE OF THE ALKENE THAT FORMS WHEN AN ALCOHOL UNDERGOES DEHYDRATION

Problem: What is the product of the dehydration of isobutyl alcohol?

$$\underset{}{CH_3-\overset{\overset{\displaystyle CH_3}{|}}{C}H-CH_2-OH} \xrightarrow[\text{heat}]{\text{acid}} ?$$

Solution: All we have to do is rewrite the given structure, except we leave out the OH group and we omit one H from a carbon that is adjacent to the carbon holding the OH group. Doing this to isobutyl alcohol leaves:

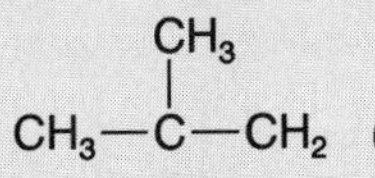

$$CH_3-\overset{\overset{\displaystyle CH_3}{|}}{C}-CH_2 \quad \text{(incomplete answer)}$$

Between the two adjacent carbons that lost the H and the OH we now have to write in another bond to make a double bond:

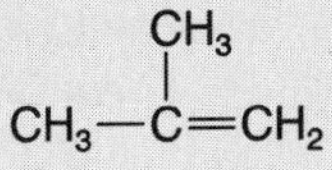

$$CH_3-\overset{\overset{\displaystyle CH_3}{|}}{C}=CH_2 \quad \text{2-Methylpropene (the answer)}$$

Each carbon now has four bonds.

PRACTICE EXERCISE 3   Write the structures of the alkenes that can be made by the dehydration of the following alcohols.

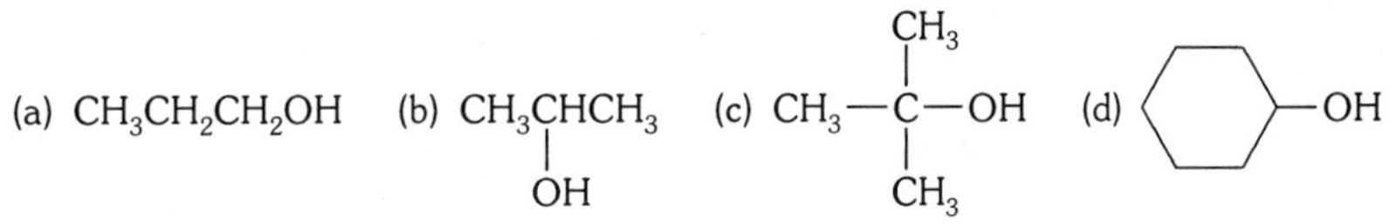

(a) $CH_3CH_2CH_2OH$ (b) $CH_3\underset{\underset{\displaystyle OH}{|}}{C}HCH_3$ (c) $CH_3-\overset{\overset{\displaystyle CH_3}{|}}{\underset{\underset{\displaystyle CH_3}{|}}{C}}-OH$ (d) hexagon—OH

Alcohols Can Be Oxidized. We learned in Section 3.1 that reactions in which electrons transfer are redox reactions, and these are common to many families of organic compounds. It is often tricky, however, to tell which atom in an organic molecule gains or loses electrons. Organic chemists, therefore, use alternative definitions of oxidation and reduction. An *oxidation* is the loss of H or the gain of O by a molecule, and a *reduction* is the loss of O or the gain of H. The oxidations of alcohols are examples of losses of H atoms.

The goal of our study of alcohol oxidations is to be able to write the structures of the products. We will not learn how to balance these reactions, so we will use unbalanced "reaction sequences," instead. We will simplify even further and use the symbol (O) for any oxidizing agent able to give the reaction. Typical oxidizing agents for lab work are potassium permanganate, $KMnO_4$, and either sodium or potassium dichromate ($Na_2Cr_2O_7$ or $K_2Cr_2O_7$). In the body, redox enzymes remove hydrogen from alcohols.

The hydrogen removed from an alcohol molecule by oxidation ends up in a water molecule. (The oxidizing agent furnishes the O atom for H_2O.) One H comes from the OH group of the alcohol, and the other H comes from the carbon that has been holding the OH group.

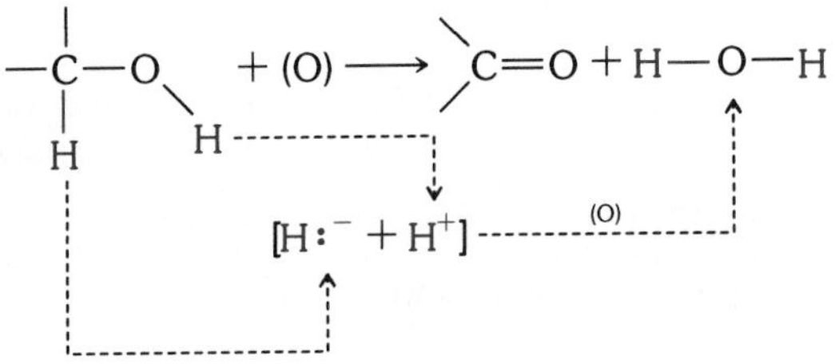

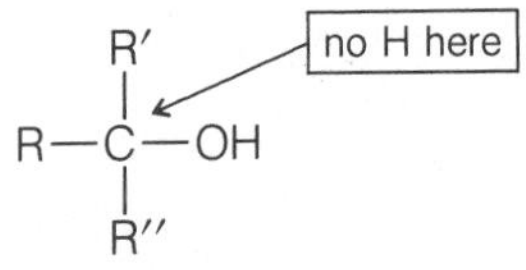

$$-\overset{|}{\underset{|}{C}}-O\overset{\diagdown}{H} + (O) \longrightarrow \overset{\diagdown}{\diagup}C=O + H-O-H$$
$$[H:^- + H^+] \cdots (O) \cdots$$

Only 1° and 2° alcohols can be oxidized this way, because 3° alcohols do not have an H atom on the carbon with the OH group.

The organic product of this oxidation has a carbon–oxygen double bond, but the actual family of this product depends on the subclass of the alcohol. The product can be an aldehyde, a carboxylic acid, or a ketone, all of which have carbon–oxygen double bonds.

$$R-\overset{\overset{\displaystyle R'}{|}}{\underset{\underset{\displaystyle R''}{|}}{C}}-OH \quad \boxed{\text{no H here}}$$

3° Alcohol system

$$R-\overset{\displaystyle O}{\overset{\|}{C}}-H$$

Aldehyde

$$R-\overset{\displaystyle O}{\overset{\|}{C}}-OH$$

Carboxylic acid

$$R-\overset{\displaystyle O}{\overset{\|}{C}}-R'$$

Ketone

Primary Alcohols Are Oxidized to Aldehydes or Carboxylic Acids. When a conventional oxidizing agent is used, like potassium permanganate, 1° alcohols are oxidized first to aldehydes and then to carboxylic acids. It is difficult to stop at the aldehyde stage, because *aldehydes oxidize more easily than alcohols.* The oxidation of a 1° alcohol, therefore, is usually done with enough oxidizing agent to take the reaction all the way to the carboxylic acid. In body cells this is not a problem, because different enzymes are needed for each step.

In general, when the right enzyme catalyzes the reaction, the aldehyde forms:

$$RCH_2OH \xrightarrow[\text{(enzyme catalyzed)}]{\text{(O)}} R-\overset{\displaystyle O}{\overset{\|}{C}}-H + H_2O$$

1° Alcohol   Aldehyde

A specific example is the oxidation of ethyl alcohol in alcoholic beverages to give an aldehyde, acetaldehyde (which causes the hangover):

$$CH_3CH_2OH \xrightarrow[\text{enzyme}]{\text{(O)}} CH_3\overset{\displaystyle O}{\overset{\|}{C}}H \qquad (+H_2O)$$

Ethyl alcohol   Acetaldehyde

KMnO$_4$ and Na$_2$Cr$_2$O$_7$ are strong oxidizing agents.

When potassium permanganate or sodium dichromate is used, then normally the intent is to carry the reaction through to the carboxylic acid, so enough oxidizing agent, as we said, is used at the start:

When Cr$_2$O$_7^-$ = (O), its orange color changes to the green Cr^{3+} ion as it oxidizes.

$$RCH_2OH \xrightarrow{\text{(O)}} (R-\overset{\displaystyle O}{\overset{\|}{C}}-H) \xrightarrow{\text{more (O)}} R-\overset{\displaystyle O}{\overset{\|}{C}}-OH$$

1° Alcohol   Aldehyde   Carboxylic acid

A specific example is

When MnO$_4^-$ = (O), its purple color gives way to a brown sludge of MnO$_2$ as it oxidizes.

$$CH_3CH_2CH_2OH \xrightarrow{\text{(O)}} CH_3CH_2\overset{\displaystyle O}{\overset{\|}{C}}OH \qquad (+H_2O)$$

Propyl alcohol   Propionic acid

EXAMPLE 11.2   WRITING THE STRUCTURE OF THE PRODUCT OF THE OXIDATION OF A PRIMARY ALCOHOL

Problem: What aldehyde and what carboxylic acid could be made by the oxidation of butyl alcohol?

Solution: First, write the structure of the given alcohol.

$$CH_3CH_2CH_2CH_2-OH$$

Then either cross out or erase the H on the OH group, and reduce by one the number of H atoms on the carbon that holds the OH group. When we do this to butyl alcohol, we have

$$CH_3CH_2CH_2CH-O \quad \text{(incomplete structure)}$$

Finally, we make the carbon–oxygen bond a double bond. Thus, the aldehyde that forms is

$$CH_3CH_2CH_2CH=O \quad \text{or} \quad CH_3CH_2CH_2\overset{\displaystyle O}{\overset{\|}{C}}-H \quad \text{(butyraldehyde)}$$

The second way that we wrote the structure of the product might make it easier to solve the second part of the problem: write the structure of the carboxylic acid that can be made from butyl alcohol. This is most easily written by inserting an oxygen atom between the carbon atom of the C=O group in the aldehyde and an H atom attached to it:

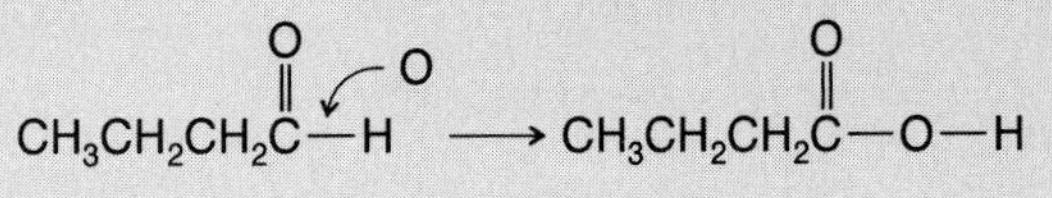

$$CH_3CH_2CH_2\overset{\overset{\displaystyle O}{\|}}{C}\!-\!H \longrightarrow CH_3CH_2CH_2\overset{\overset{\displaystyle O}{\|}}{C}\!-\!O\!-\!H \quad \text{(butyric acid)}$$

PRACTICE EXERCISE 4   Write the structures of the aldehydes and carboxylic acids that can be made by the oxidation of the following alcohols.

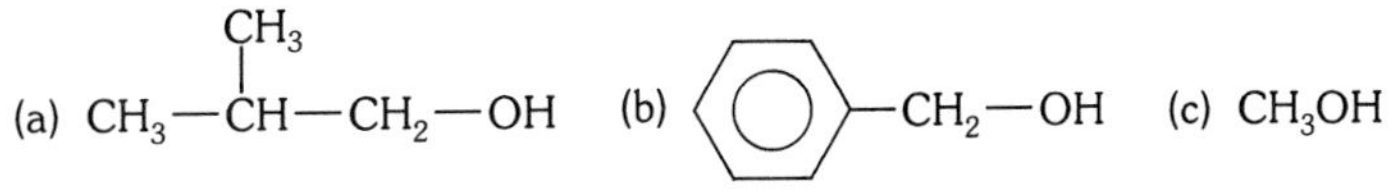

(a) $CH_3-\overset{\overset{\displaystyle CH_3}{|}}{CH}-CH_2-OH$ (b) ⬡$-CH_2-OH$ (c) CH_3OH

Secondary Alcohols are Oxidized to Ketones. Ketones strongly resist oxidation, so they are easily made by the oxidation of 2° alcohols.

In general,

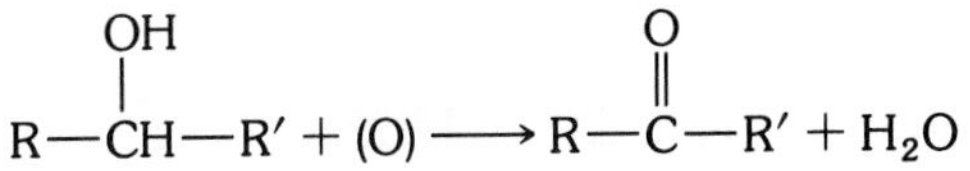

$$R-\overset{\overset{\displaystyle OH}{|}}{CH}-R' + (O) \longrightarrow R-\overset{\overset{\displaystyle O}{\|}}{C}-R' + H_2O$$

2° alcohol   Ketone

Specific examples are

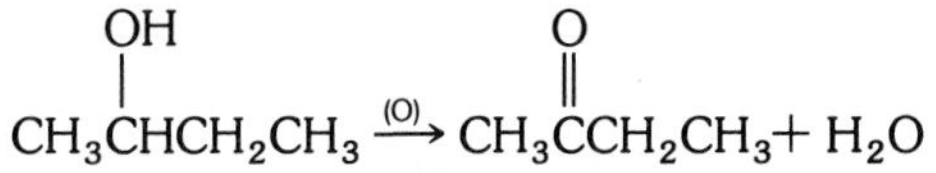

$$CH_3\overset{\overset{\displaystyle OH}{|}}{CH}CH_2CH_3 \overset{(O)}{\longrightarrow} CH_3\overset{\overset{\displaystyle O}{\|}}{C}CH_2CH_3 + H_2O$$

sec-Butyl alcohol   Butanone

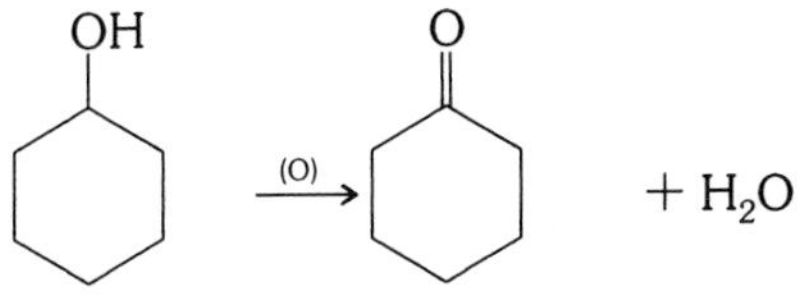

Cyclohexanol $\overset{(O)}{\longrightarrow}$ Cyclohexanone $+ H_2O$

EXAMPLE 11.3   WRITING THE STRUCTURE OF THE PRODUCT OF THE OXIDATION OF A SECONDARY ALCOHOL

Problem: What ketone forms when isopropyl alcohol is oxidized?

Solution: We work this essentially as we worked Example 11.2. We start with the structure of the given alcohol, and then we cross out or erase the H on the OH group and the H on the carbon that is holding this OH group.

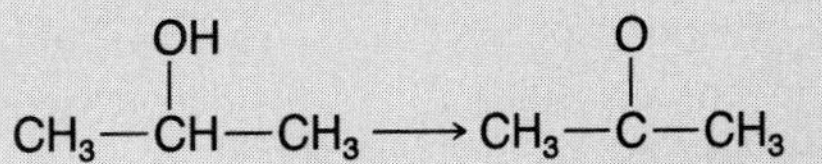

$$CH_3-\overset{\overset{\displaystyle OH}{|}}{CH}-CH_3 \longrightarrow CH_3-\overset{\overset{\displaystyle O}{|}}{C}-CH_3 \quad \text{(incomplete answer)}$$

Then we make the bond to oxygen a double bond:

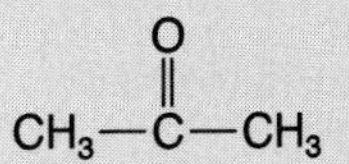

$$CH_3-\overset{\overset{\displaystyle O}{\|}}{C}-CH_3 \quad \text{Acetone (the answer)}$$

PRACTICE EXERCISE 5   Write the structures of the ketones that can be made by the oxidation of the following alcohols.

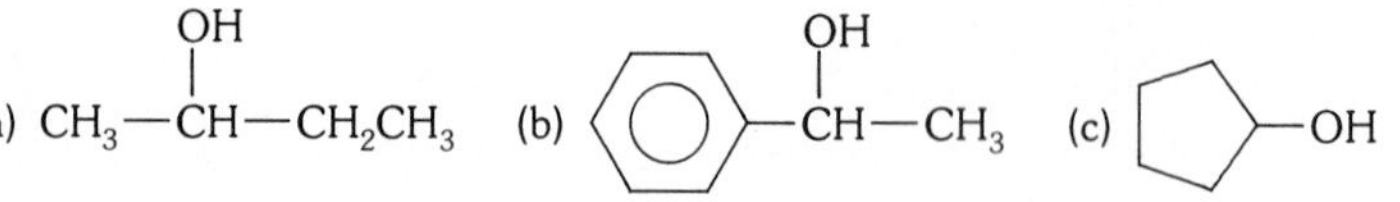

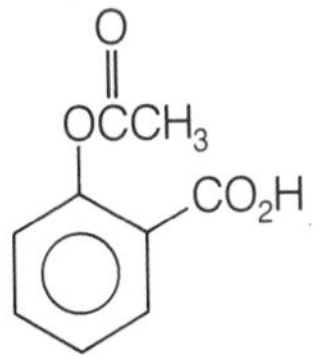

$$
\text{(a) } CH_3-\underset{\underset{OH}{|}}{CH}-CH_2CH_3 \qquad \text{(b)} \quad \bigcirc-\underset{\underset{OH}{|}}{CH}-CH_3 \qquad \text{(c)} \quad \bigcirc-OH
$$

Aspirin

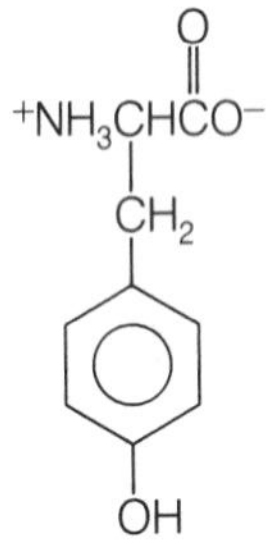

Tyrosine

Mercaptan is a contraction of *mercury capturer.* Compounds with SH groups form precipitates with mercury ions.

Lower formula weight thioalcohols are present in and are responsible for the respect usually accorded skunks.

Phenols, Unlike Alcohols, Can Neutralize Hydroxide Ions. In **phenols** the OH group is attached to a benzene ring. Many phenols are found in both nature and commerce. The simplest member of this family is called phenol, and it is a raw material for making aspirin. Tyrosine, one of the amino acids the body uses to make proteins, is also a phenol with other functional groups.

Unlike alcohols, phenols cannot be dehydrated. Although they are weak acids, phenols are strong enough to be able to neutralize sodium hydroxide. For example,

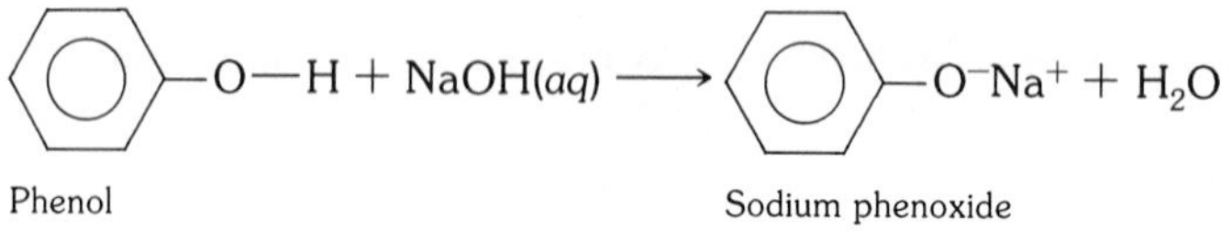

$$
\bigcirc-O-H + NaOH(aq) \longrightarrow \bigcirc-O^-Na^+ + H_2O
$$

Phenol   Sodium phenoxide

11.4   THIOALCOHOLS AND DISULFIDES

Both the SH group, an easily oxidized group, and the —S—S— system, an easily reduced group, are present in molecules of proteins.

Alcohols, R—O—H, can be viewed as alkyl derivatives of water, H—O—H. Similar derivatives of hydrogen sulfide, H—S—H, are also known, and are in the family called the **thioalcohols,** commonly called the **mercaptans.**

$$
\text{R—S—H} \qquad \text{R—S—S—R}'
$$

Thioalcohols   Disulfides
(mercaptans)

The SH group is variously called the thiol group, the mercaptan group, or the sulfhydryl group.

Table 11.3 gives the IUPAC names and structures of a few thioalcohols. Some important

TABLE 11.3
Some Common Thioalcohols

Name	Structure	Boiling Point (°C)
Methanethiol	CH_3SH	6
Ethanethiol	CH_3CH_2SH	36
1-Propanethiol	$CH_3CH_2CH_2SH$	68
1-Butanethiol	$CH_3CH_2CH_2CH_2SH$	98
Cysteine (a monomer for proteins)	$^+NH_3-\underset{\underset{CH_2-SH}{\|}}{CH}-\overset{\overset{O}{\|\|}}{C}-O^-$	(solid)

properties of proteins depend on the presence of the SH group, which is contributed to protein structure by one of the building blocks of proteins, the amino acid called cysteine (Table 11.3).

The SH Group Is Easily Oxidized to the Disulfide Group. Only one reaction of thioalcohols is important in our study of proteins. Thioalcohols are readily oxidized to **disulfides,** compounds with two sulfur atoms joined by a covalent bond, R—S—S—R'.

In general:

$$R—S—H + H—S—R + (O) \longrightarrow R—S—S—R + H_2O$$

Two molecules of   One molecule
a thioalcohol   of a disulfide

Specific example,

$$2\ CH_3SH + (O) \longrightarrow CH_3—S—S—CH_3 + H_2O$$

Methanethiol   Dimethyl disulfide

EXAMPLE 11.4   WRITING THE PRODUCT OF THE OXIDATION OF A THIOALCOHOL

Problem: What is the product of the oxidation of ethanethiol?

Solution: Because the oxidation of the SH group generates the —S—S— group, we begin simply by writing this group down.

$$—S—S—$$

Then we attach the alkyl group from the thioalcohol, one on each sulfur atom. Ethanethiol furnishes ethyl groups, so the answer is

$$CH_3CH_2—S—S—CH_2CH_3 \quad \text{(diethyl disulfide)}$$

Disulfides Are Easily Reduced to Thioalcohols. The sulfur–sulfur bond in disulfides is very easily reduced, and the products are molecules of the thioalcohols from which the disulfide could be made. We will use the symbol (H) to represent any reducing agent that can do the task, just as we used (O) for an oxidizing agent.

In general:

$$R—S—S—R + 2(H) \longrightarrow R—S—H + H—S—R$$

One molecule   Two molecules of
of disulfide   a thioalcohol

Specific example:

$$CH_3CH_2—S—S—CH_2CH_2CH_3 + 2(H) \longrightarrow$$

Ethyl propyl disulfide

$$CH_3CH_2—S—H + H—S—CH_2CH_2CH_3$$

Ethanethiol   1-Propanethiol

EXAMPLE 11.5   WRITING THE PRODUCT OF THE REDUCTION OF A DISULFIDE

Problem: What forms when the following disulfide reacts with a reducing agent?

$$CH_3—S—S—CH_2CH_3$$

Solution: The easiest approach is simply to split the disulfide molecule between the two sulfur atoms:

$$CH_3—S—S—CH_2CH_3 \dashrightarrow CH_3—S + S—CH_2CH_3 \quad \text{(incomplete)}$$

Then attach one H to each sulfur to make the —SH groups.

$$CH_3—S—H \qquad + H—S—CH_2CH_3 \qquad \text{(the answers)}$$

Methanethiol   Ethanethiol

PRACTICE EXERCISE 6   Complete the following equations by writing the structures of the products that form. If no reaction occurs (insofar as we have studied organic chemistry), write "no reaction."

(a) $CH_3—S—S—CH_3 + (H) \longrightarrow ?$ (b) $CH_3—\underset{\underset{SH}{|}}{CH}—CH_3 + (O) \longrightarrow ?$

(c) 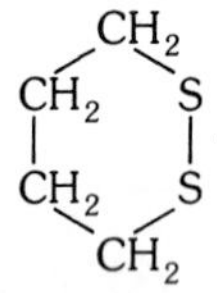 [cyclic structure with CH₂, CH₂, S, CH₂, S, CH₂] $+ (H) \longrightarrow ?$ (d) 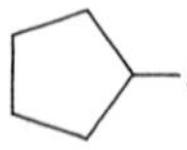 [cyclopentane]—$SH + (O) \longrightarrow ?$

11.5   ETHERS

The ethers are almost as chemically unreactive as the alkanes.

Ethers are compounds whose molecules have two organic groups joined to the same oxygen atom, R—O—R'. The two groups can be almost anything except C=O. When the C in C=O is bound to O, we have an ester, not an ether. The first three compounds given below are ethers, for example, but methyl acetate is an ester. (We will study esters in Chapter 13.)

$CH_3CH_2—O—CH_2CH_3$ $CH_3—O—$⬡   ⬡$—O—$⬡ $CH_3—\overset{\overset{O}{\|}}{C}—O—CH_3$

Diethyl ether   Methyl phenylether   Diphenyl ether   Methyl acetate (an ester, not an ether)

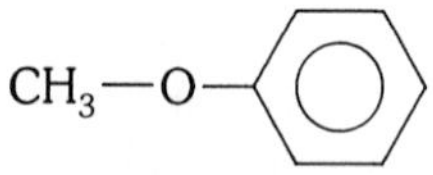

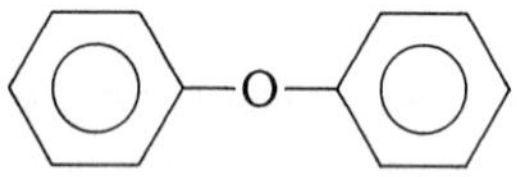

Table 11.4 gives some examples of ethers, and three are described in Special Topic 11.2, including methyl *t*-butyl ether, which is used to dissolve gallstones and so remove them without surgery.

TABLE 11.4
Some Ethers

Common Name	Structure	Boiling Point (°C)
Dimethyl ether	CH_3OCH_3	−23
Methyl ethyl ether	$CH_3OCH_2CH_3$	11
Methyl *t*-butyl ether	$CH_3OC(CH_3)_3$	55.2
Diethyl ether	$CH_3CH_2OCH_2CH_3$	34.5
Dipropyl ether	$CH_3CH_2CH_2OCH_2CH_2CH_3$	91
Methyl phenyl ether	$CH_3—O—C_6H_5$	155

SPECIAL TOPIC 11.2 SOME IMPORTANT ETHERS

Diethyl Ether (''Ether'') Diethyl ether is a colorless, volatile liquid with a pungent, somewhat irritating odor, and it was once widely used as an anesthetic. It acts as a depressant for the central nervous system and somewhat of a stimulant for the sympathetic system. It exerts an effect on nearly all tissues of the body.

Because mixtures of ether and air in the right proportions can explode, anesthesiologists avoid using ether whenever possible.

Divinyl Ether (Vinethene) Divinyl ether is another anesthetic, but it also forms an explosive mixture in air. Its anesthetic action is more rapid than that of diethyl ether.

Methyl *t*-butyl Ether The standard treatment for stones in the gallbladder or the gallbladder duct has been surgical removal of the gallbladder. Medical scientists have long sought a safe way to dissolve the stones where they are, and they have known that a chief constituent of these stones is cholesterol. Cholesterol is mostly hydrocarbon-like, so the search has been for a relatively nonpolar solvent for cholesterol that could therefore dissolve the stones without causing serious side effects and that would be at least as safe as the surgical treatment.

In 1985, medical scientists at the Mayo Clinic found that methyl *t*-butyl ether worked rapidly and well when used with four patients, and more studies are being made. This ether is less toxic than diethyl ether, it can be inserted into the gallbladder by a catheter, it remains a liquid at body temperature, and it works in just a few hours with no evidence of disagreeable side effects.

The common names of ethers are made by naming the groups attached to the oxygen and adding the word *ether,* as illustrated in Table 11.4.

Compound	Boiling Point (°C)
Pentane	36
Diethyl ether	35
1-Butanol	118

Ethers Can Accept Hydrogen Bonds. Because the ether group cannot donate hydrogen bonds, the boiling points of simple ethers are more like those of the alkanes of comparable formula weights than those of the alcohols. The oxygen of the ether group, however, can accept hydrogen bonds (Figure 11.3), so ethers are more soluble in water than alkanes. Both 1-butanol and its isomer, diethyl ether, for example, dissolve in water to the extent of about 8 g/100 mL (20°C), but pentane is completely insoluble in water.

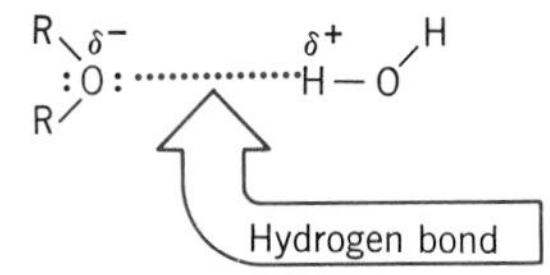

FIGURE 11.3
An ether molecule can accept a hydrogen bond from a water molecule. (The δ− end of a hydrogen bond is the acceptor end.)

Ethers Give No Reactions with Common Reactants. At room temperature, ethers do not react with strong acids, bases, oxidizing, or reducing agents. Like all organic compounds, they burn.

Ethers Can Be Made from Alcohols. We will look briefly at one method for making ethers because it relates to some of the chemistry we will study in Chapter 12.

We learned in Section 11.3 that alcohols can be dehydrated to alkenes. The precise temperature that works best has to be determined for each alcohol, because at another temperature a different kind of dehydration can occur. Water can split out *between* two alcohol molecules rather than from within one molecule, and the product is an ether.

In general,

$$\underset{\text{Two alcohol molecules}}{R{-}O{-}H + H{-}O{-}R} \xrightarrow[\text{heat}]{\text{acid catalyst,}} \underset{\text{Ether}}{R{-}O{-}R} + H_2O$$

A specific example:

$$\underset{\text{Ethyl alcohol}}{2CH_3CH_2OH} \xrightarrow[140\ °C]{H_2SO_4} \underset{\text{Diethyl ether}}{CH_3CH_2{-}O{-}CH_2CH_3} + H_2O$$

Earlier we learned that concd H_2SO_4 acts on ethyl alcohol to give ethene when the temperature is 170 °C.

The best temperature to make an ether from an alcohol is generally lower than needed to make an alkene. Our interest, however, is simply in the *possibility* of making an ether from an alcohol as well as in what particular ether can be made.

EXAMPLE 11.6   WRITING THE STRUCTURE OF AN ETHER THAT CAN FORM FROM AN ALCOHOL

Problem: If the conditions are right, butyl alcohol can be converted to an ether. What is the structure of this ether?

Solution: As usual, the structure of the starting material gives us most of the answer. We know that the ether must get its organic groups from the alcohol, so to write the structure of the ether we write one O atom with two bonds from it.

$$-O-$$

Then we attach the alkyl group of the alcohol, one to each of the bonds.

$$CH_3CH_2CH_2CH_2-O-CH_2CH_2CH_2CH_3$$

Of course, we need *two* molecules of the alcohol to make one molecule of this ether, but always remember, *we balance an equation after we have written the correct formulas for reactants and products.* Remembering that water is the other product, the balanced equation is

$$2CH_3CH_2CH_2CH_2OH \xrightarrow[\text{heat}]{\text{acid catalyst}} CH_3CH_2CH_2CH_2-O-CH_2CH_2CH_2CH_3 + H_2O$$

Butyl alcohol   Dibutyl ether

PRACTICE EXERCISE 7   Write the structures of the ethers to which the following alcohols can be converted.

(a) CH_3OH (b) $CH_3CH_2CH_2OH$ (c) 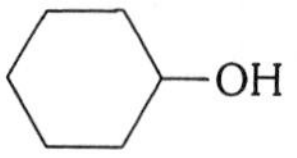 ⬡—OH

11.6 OCCURRENCE, NAMES, AND PHYSICAL PROPERTIES OF AMINES

The amino group, NH_2, has some of the properties of ammonia, including the ability to be involved in hydrogen bonding.

Both the amino group and its protonated form occur in living things in proteins, enzymes, and genes. When a carbonyl group, $C=O$, is attached to nitrogen, we have the amide system, an important functional group in proteins.

$$-NH_2 \qquad -NH_3^+ \qquad \overset{\overset{\displaystyle O}{\|}}{-C}-\overset{|}{N}-$$

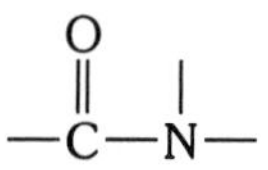

Amino group   Protonated amino group   Amide system

The Amines Are Organic Relatives of Ammonia.   The replacement of one, two, or all three hydrogen atoms of an ammonia molecule by a hydrocarbon group gives a compound classified as an **amine.** For example,

$$CH_3NH_2 \qquad CH_3NHCH_3 \qquad \overset{\overset{\displaystyle CH_3}{|}}{CH_3NCH_3} \qquad CH_3NHCH_2CH_3 \qquad ⬡—NH_2$$

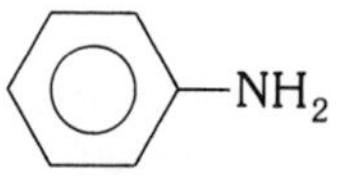

Methylamine   Dimethylamine   Trimethylamine   Methylethylamine   Aniline

These are the common, not the IUPAC names.

TABLE 11.5
Amines

Common Name	Structure	Boiling Point (°C)	Solubility in Water
Methylamine	CH₃NH₂	−8	Very soluble
Dimethylamine	CH₃NHCH₃	8	Very soluble
Trimethylamine	CH₃NCH₃ | CH₃	3	Very soluble
Ethylamine	CH₃CH₂NH₂	17	Very soluble
Diethylamine	CH₃CH₂NHCH₂CH₃	55	Very soluble
Triethylamine	CH₃CH₂NCH₂CH₃ | CH₂CH₃	89	14 g/dL
Propylamine	CH₃CH₂CH₂NH₂	49	Very soluble
Aniline	C₆H₅NH₂	184	4 g/dL

The hydrocarbon groups, as you can see, do not have to be alike, and they can be aromatic groups, as in aniline. The nitrogen atom can be a part of a ring, and such compounds, like pyridine, are classified as *heterocyclic amines*. Table 11.5 gives many other examples of amines.

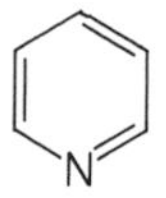

Pyridine

For a compound to be an amine, not only must its molecules have a nitrogen with three bonds, but also none of these bonds *can be directly* to a C=O group. If one is, the substance is an **amide**. Structure **1**, for example, is an amide, not an amine. Structure **2**, on the other hand, has two functional groups, a keto group and an amino group, because its N atom is not joined *directly* to C=O. One chemical difference is that amines, like ammonia, are basic and amides are not.

$$\underset{\textbf{1}}{R-\overset{\overset{O}{\parallel}}{C}-NH_2} \qquad \underset{\textbf{2}}{R-\overset{\overset{O}{\parallel}}{C}-CH_2-NH_2}$$

The common names of the simple, aliphatic amines are made by writing the names of the alkyl groups on nitrogen in front of the word *amine* (and leaving no space). Table 11.5 gives examples.

Many amines are physiologically active, and some are important drugs. Special Topic 11.3 describes a number of interesting examples.

Amines Form Weaker Hydrogen Bonds Than Alcohols. When the boiling points of compounds of about the same formula weight and chain branching are compared, amines (those of the type RNH_2) boil higher than alkanes but lower than alcohols. The size of $\delta-$ on N (in RNH_2) is less than it is on O (in ROH), because N is less electronegative than O. As a result, the size of $\delta+$ on H is less in the amine than in the alcohol. The smaller sizes of $\delta+$ and $\delta-$ in amines still permits hydrogen bonding (Figure 11.4), but it is weaker than among the alcohols. Hydrogen bonding also helps amines to be much more soluble in water than alkanes, as Figure 11.4 also explains.

Although hydrogen bonding in amines is weak, it is very important at the molecular level of life, because it stabilizes the special molecular shapes of proteins and nucleic acids.

Compound	Boiling Point (°C)
CH₃CH₃	−89
CH₃NH₂	−6
CH₃OH	65

Epinephrine and norepinephrine are two of the many hormones in our bodies. We will study the nature of hormones in a later chapter, but we can use a definition here. **Hormones** are compounds the body makes in special glands to serve as chemical messengers. In response to a stimulus somewhat unique for each hormone, such as fright, food odor, sugar ingestion, and others, the gland secretes its hormone into the circulatory system. The hormone then moves to some organ or tissue, where it activates a particular metabolic series of reactions that constitute the biochemical response to the initial stimulus. Maybe you have heard the expression, "I need to get my adrenalin flowing." Adrenalin—or epinephrine, its technical name—is made by the adrenal gland. If you ever experience a sudden fright, a trace of epinephrine immediately flows and the results include a strengthened heartbeat, a rise in blood pressure, and a release of glucose into circulation from storage—all of which get the body ready to respond to the threat.

Norepinephrine has similar effects, and because these two hormones are secreted by the adrenal gland, they are called **adrenergic agents.**

HO—, HO— (benzene ring)—CHCH$_2$NHCH$_3$ with OH
Epinephrine

HO—, HO— (benzene ring)—CHCH$_2$NH$_2$ with OH
Norepinephrine

(benzene ring)—CHCH$_2$NH$_2$ with OH
β-Phenylethanolamine

Several useful drugs mimic epinephrine and norepinephrine, and all are classified as *adrenergic drugs.* Most of them, like epinephrine and norepinephrine, are related structurally to β-phenylethanolamine (which is not a formal name, obviously). In nearly all of their uses, these drugs are prepared as dilute solutions of their amine acid salts. Several of the β-phenylethanolamine drugs are even more structurally like epinephrine and norepinephrine, because they have the structural features of 1,2-dihydroxybenzene. This compound is commonly called *catechol,* so the cate-

chol-like adrenergic drugs are called the **catecholamines.** Synthetic epinephrine (an agent in Primatene Mist), Ethylnorepinephrine, and Isoproterenol are examples.

HO—, HO— (benzene ring)—CHCHNH$_2$ with OH and CH$_2$CH$_3$
Ethylnorepinephrine
(Used against asthma in children)

HO—, HO— (benzene ring)—CHCH$_2$NHCH(CH$_3$)$_2$ with OH
Isoproterenol
(Used in treating emphysema and asthma)

Another family of physiologically active amines is the β-phenylethylamines. For example, dopamine (which is also a catecholamine) is the compound the body uses to make norepinephrine. Its synthetic form is used to treat shock associated with severe congestive heart failure.

The amphetamines are a family of β-phenylethylamines that include Dexedrine ("speed") and Methedrin ("crystal," "meth"). The amphetamines can be legally prescribed as stimulants and antidepressants, and sometimes they are prescribed for weight-control programs. However, millions of these "pep pills" or "uppers" are sold illegally, and this use of amphetamines constitutes a major drug abuse problem. The dangers of overuse include suicide, belligerence and hostility, paranoia, and hallucinations.

HO—, HO— (benzene ring)—CH$_3$CH$_2$NH$_2$
Dopamine

(benzene ring)—CH$_2$CHNH$_2$ with CH$_3$
Dexedrine

(benzene ring)—CH$_2$CHNHCH$_3$ with CH$_3$
Methedrine

In Chapter 17 we will discuss the mechanisms by which these drugs and the naturally occurring hormones work.

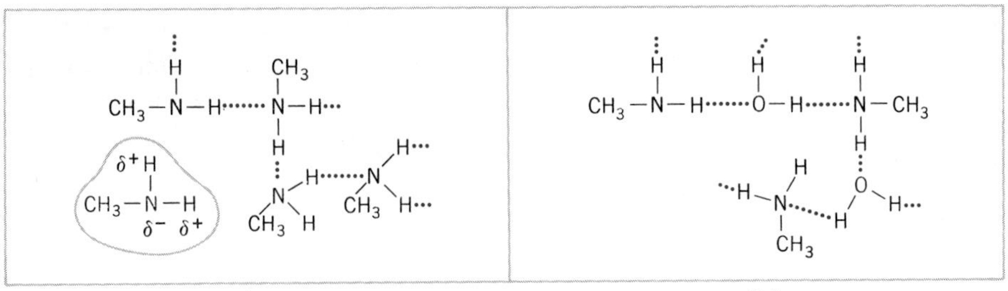

FIGURE 11.4
Hydrogen bonds in *(a)* amines and in *(b)* aqueous solutions of amines.

11.7 CHEMICAL PROPERTIES OF AMINES

The amino group is a proton acceptor, and the protonated amino group is a proton donor.

We will examine two chemical properties of amines that will be particularly important to our study of biochemicals: the basicity of amines (this section) and their conversion to amides (in Chapter 13).

Amines Can Neutralize Strong Acids. Although amines, like ammonia, are *weak* bases, they are able to establish an equilibrium in water in which a small excess of [OH$^-$] over [H$^+$] exists. The solution is therefore basic.

$$R-\overset{..}{N}H_2 + H_2O \rightleftharpoons R-NH_3^+ + OH^-$$

Amine (or
ammonia,
when R = H)

Protonated
amine (or
ammonium ion
when R = H)

Amino acids, the monomers for proteins, have the NH$_3^+$ group in acidic and neutral solutions.

Thus biochemicals with amino groups can raise the pH of a solution, a property of considerable importance at the molecular level of life, as we have often emphasized.

Compounds with amino groups can also rapidly neutralize hydronium ions.

$$R-\overset{..}{N}H_2 + H-\overset{+}{\underset{H}{O}}{:}H \longrightarrow R-\overset{H}{\underset{H}{N^+}}-H + H_2O$$

Amine (or
ammonia,
when R = H)

Hydronium
ion

Protonated
amine (or
ammonium ion
when R = H)

For example,

$$CH_3\overset{..}{N}H_2 + HCl(aq) \longrightarrow CH_3NH_3^+ Cl^- + H_2O$$

Methylamine   Hydrochloric
acid

Methylammonium
chloride

It doesn't matter whether the nitrogen atom in an amine bears one, two, or three hydrocarbon groups. The amine can still neutralize strong acids, because the reaction involves just the unshared pair of electrons on the nitrogen, not any of the bonds to the other groups. We call an amine that has accepted a proton, H$^+$, a *protonated amine*.

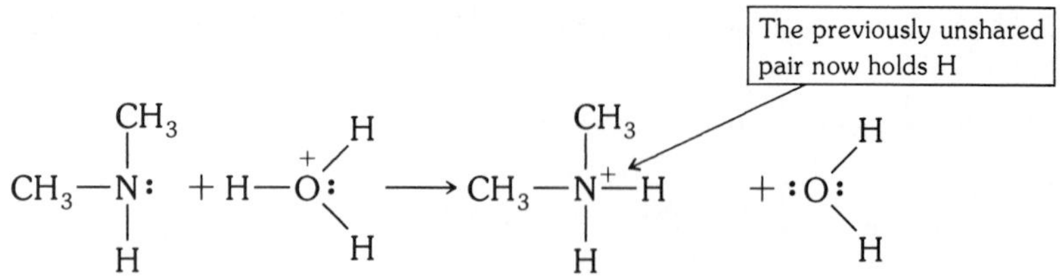

The previously unshared
pair now holds H

$$CH_3-\overset{\displaystyle CH_3}{\underset{\displaystyle H}{N}}: \ + \ H-\overset{+}{\underset{\displaystyle H}{O}}: \ \longrightarrow \ CH_3-\overset{\displaystyle CH_3}{\underset{\displaystyle H}{N}}\overset{+}{-}H \ \ +:\overset{\displaystyle H}{\underset{\displaystyle H}{O}}:$$

Dimethylamine   Hydronium   Dimethylammonium ion
                  ion   (a protonated amine)

PRACTICE EXERCISE 8   What are the structures of the positive ions that form when the following amines react completely with hydrochloric acid?

(a) aniline   (b) trimethylamine   (c) $NH_2CH_2CH_2NH_2$

Protonated Amines Are Water Soluble and Can Neutralize Hydroxide Ion.   A protonated amine and an associated anion make up an organic salt called an **amine salt.** Like all salts, they are crystalline solids at room temperature. In addition, like the salts of the ammonium ion, nearly all amine salts of strong acids are soluble in water *even when the parent amine is not.* The protonated amine in an amine salt is a fully charged *ion.* Ions are much better hydrated by water molecules than amines whose molecules have only the small, partial charges of its polar bonds.

When protonated amines neutralize the hydroxide ion they revert to their parent amines. (We will show only skeletal structures.)

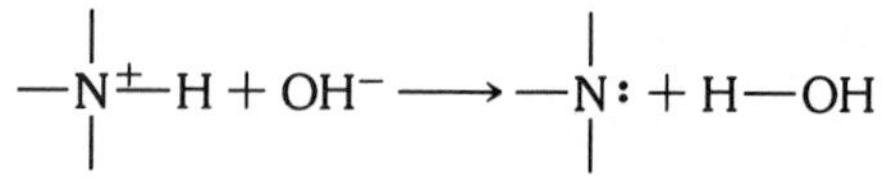

$$-\overset{|}{\underset{|}{N}}\overset{+}{-}H + OH^- \longrightarrow -\overset{|}{\underset{|}{N}}: + H-OH$$

Protonated   Amine
amine

For example,

$$CH_3NH_3{}^+ + OH^- \longrightarrow CH_3NH_2 + H_2O$$

$$CH_3CH_2\overset{+}{N}H_2CH_3 + OH^- \longrightarrow CH_3CH_2NHCH_3 + H_2O$$

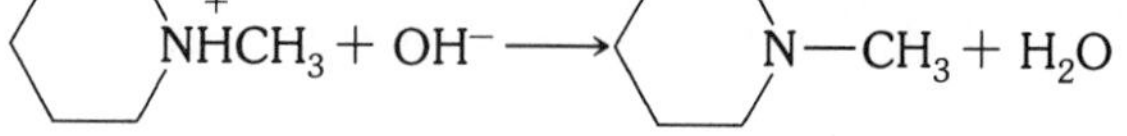

$$\bigcirc\!\!\!\overset{+}{N}HCH_3 + OH^- \longrightarrow \bigcirc\!\!\!N-CH_3 + H_2O$$

The Amino Group Is a "Solubility Switch".   An amine's solubility in water can be switched on simply by adding enough strong acid to protonate it. Triethylamine, for example, is insoluble in water, but we can dissolve it, turn on its solubility, merely by adding a strong acid, like HCl(*aq*). The amine dissolves as its protonated form is produced.

$$(CH_3CH_2)_3N: + HCl(aq) \longrightarrow (CH_3CH_2)_3\overset{+}{N}H \ Cl^-$$

Triethylamine   Triethylammonium chloride
(water insoluble)   (water soluble)

We can just as quickly and easily bring the amine back out of solution again simply by adding a strong base. Protonated amines, as we just learned, are good proton donors, and the OH$^-$ ion is a strong acceptor.

$$(CH_3CH_2)_3\overset{+}{N}H + OH^- \longrightarrow (CH_3CH_2)_3N + H_2O$$

Triethylammonium
ion
(water soluble)

Triethylamine

(water insoluble)

The amino group solubility switch:

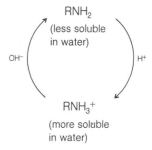

The significance of this "switching" relationship is that the solubilities of complex compounds with the amine function can be changed almost instantly simply by adjusting the pH of the medium. One important application involves medications made from alkaloids.

Alkaloids Are Complex Amines.   A number of amines obtained from the bark, roots, leaves, flowers, or fruit of various plants are useful drugs. These naturally occurring, acid-neutralizing, physiologically active amines are called **alkaloids,** and morphine, codeine, and quinine are just three typical examples (Special Topic 11.3).

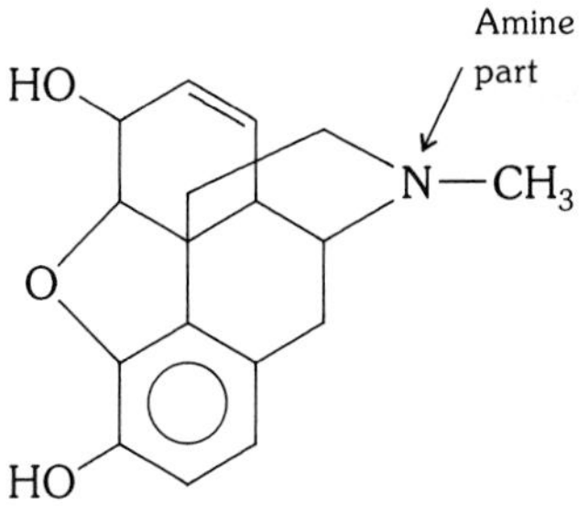

Morphine

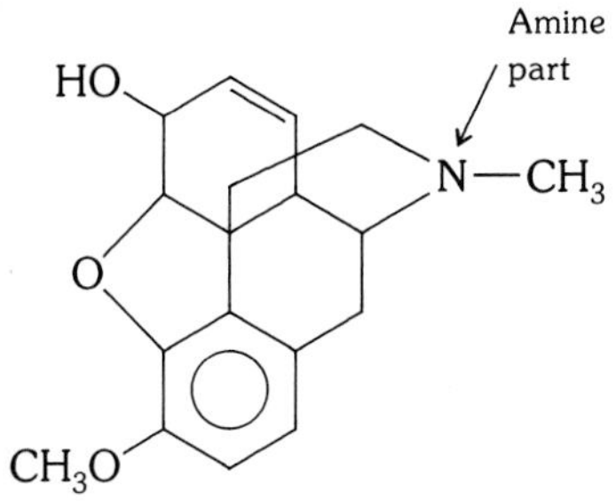

Codeine

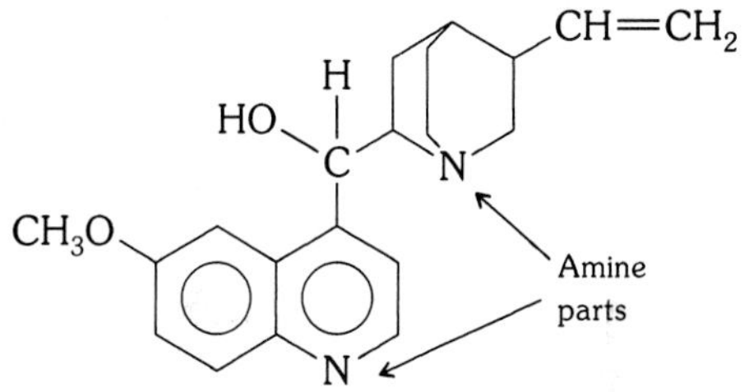

Quinine

To make it easier to administer alkaloidal drugs in the dissolved state, they often are prepared as their water-soluble amine salts. Morphine, for example, a potent sedative and painkiller, is often given as morphine sulfate, the salt of morphine and sulfuric acid. Quinine, an antimalarial drug, is available as quinine sulfate. Codeine, sometimes used in cough medicines, is often present as codeine phosphate. Special Topic 11.3 tells about a few other physiologically active amines, most of which are also prepared as their amine salts.

EXAMPLE 11.7 . WRITING THE STRUCTURE OF THE PRODUCT OF DEPROTONATING THE CATION OF AN AMINE SALT

Problem: The protonated form of amphetamine is shown below. What is the structure of the product of its reaction with hydroxide ion?

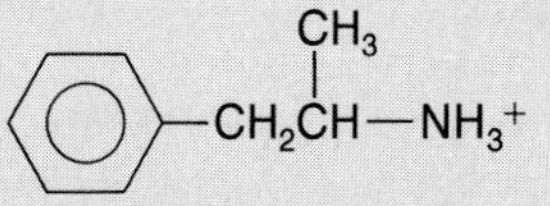

$$\text{C}_6\text{H}_5-CH_2\overset{\overset{\displaystyle CH_3}{|}}{CH}-NH_3^+$$

Solution: Because OH⁻ removes just one H⁺ from the protonated amine's cation, all we have to do is reduce the number of H atoms on the nitrogen by one and cancel the positive charge. The answer therefore is

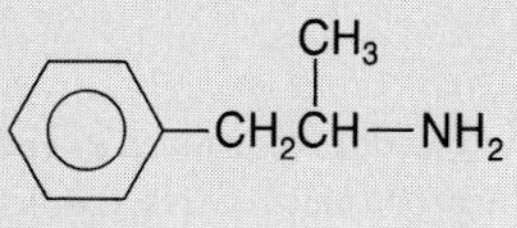

$$\text{C}_6\text{H}_5-CH_2\overset{\overset{\displaystyle CH_3}{|}}{CH}-NH_2$$

Amphetamine

PRACTICE EXERCISE 9   Write the structures of the products after the following protonated amines have reacted with OH⁻ in a 1 : 1 mole ratio.

(a)

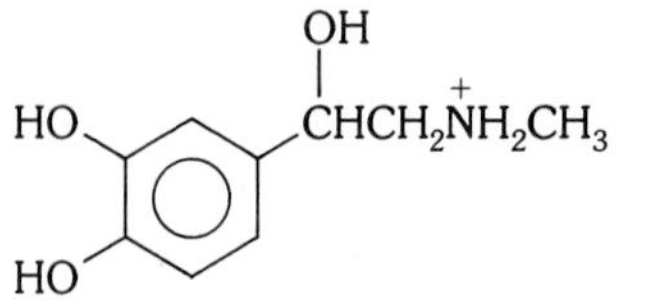

$$\text{HO} \quad \text{OH} \atop \text{HO} \quad \overset{+}{C}HCH_2\overset{+}{N}H_2CH_3$$

Epinephrine (adrenaline), a hormone given here in its protonated form. As the chloride salt in a 0.1% solution, it is injected in some cardiac failure emergencies (see also Special Topic 11.3).

(b) CH_3O, CH_3O, CH_3O — $CH_2CH_2\overset{+}{N}H_3$

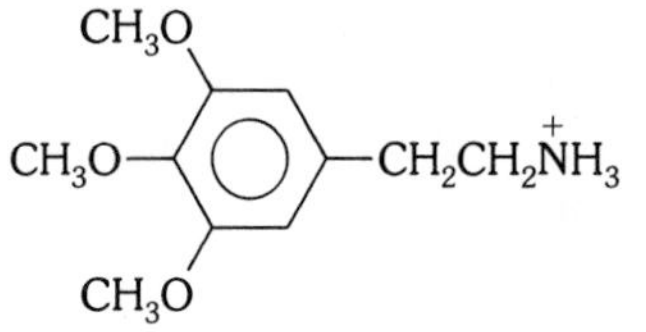

Mescaline, a mind-altering hallucinogen shown here in its protonated form. It is isolated from the mescal button, a growth on top of the peyote cactus. Indians in the southwestern United States have used it in religious ceremonies.

SUMMARY

Alcohols   The alcohol system has an OH group attached to a saturated carbon. The common names have the word *alcohol* following the name of the alkyl group.

Alcohol molecules hydrogen bond to each other and to water molecules. By the action of heat and an acid catalyst they can be dehydrated internally to give carbon–carbon double bonds or externally to give ethers. Primary alcohols can be oxidized to aldehydes and to carboxylic acids. Secondary alcohols can be oxidized to ketones. Tertiary alcohols cannot be oxidized (without breaking up the carbon chain). The OH group in alcohols does not function as either an acid or a base in the ordinary sense.

Thioalcohols   The thioalcohols or mercaptans, R—S—H, are easily oxidized to disulfides, R—S—S—R, and these are easily reduced back to the original thioalcohols.

Phenols   When the OH group is attached to a benzene ring, the system is the phenol system, and it is now acidic enough to neutralize strong bases.

Ethers   The ether system, R—O—R′, does not react at room temperature or body temperature with strong acids, bases, oxidizing agents, or reducing agents. It can accept hydrogen bonds but cannot donate them.

Amines and protonated amines   When one, two, or three of the hydrogen atoms in ammonia are replaced by an organic group (other than a carbonyl group), the result is an amine. The nitrogen atom can be part of a ring, as in heterocyclic amines. Like ammonia, the amines are weak bases, but all can form salts with strong acids. The cations in these salts are protonated amines. Amine salts are far more soluble in water than their parent amines. Protonated amines are easily deprotonated by any strong base to give back the original, and usually far less soluble amine. Thus any compound with the amine function has a "solubility switch," because its solubility in an aqueous system can be turned on by adding acid (to form the amine salt) and turned off again by adding base (to recover the amine).

Reactions studied   Without attempting to present balanced equations, or even all of the inorganic products, we can summarize the reactions studied in this chapter as follows. We have learned two "map signs" for alcohols, dehydration (either to alkenes or ethers) and oxidation. We have studied just one map sign apiece for thioalcohols, disulfides, and phenols. We've also found that ethers are as unreactive as alkanes. Finally, we studied the chemistry of the amino group as a solubility switch. Here then is a summary of these properties.

Alcohols

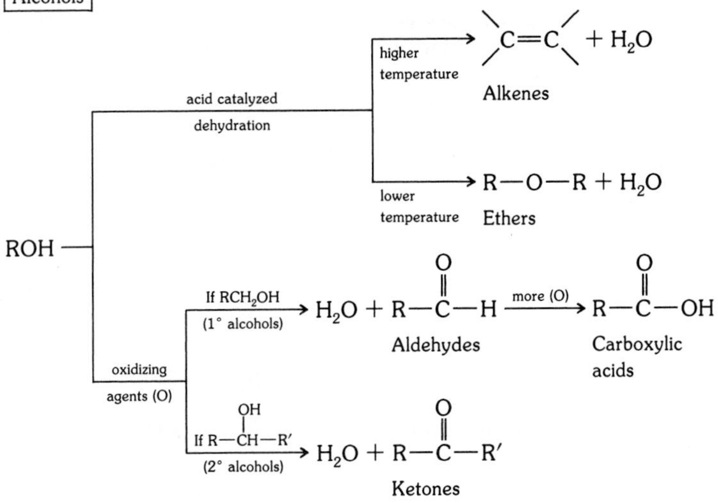

$$\underset{\text{higher temperature}}{\xrightarrow{\hspace{1cm}}} \quad \overset{}{\underset{}{C}}=\overset{}{\underset{}{C} } + H_2O$$

Alkenes

acid catalyzed
dehydration

$$\underset{\text{lower temperature}}{\xrightarrow{\hspace{1cm}}} \quad R\!-\!O\!-\!R + H_2O$$

Ethers

ROH

If RCH$_2$OH
(1° alcohols)
$$\xrightarrow{\hspace{1cm}} H_2O + R\!-\!\overset{O}{\overset{\|}{C}}\!-\!H \xrightarrow{\text{more (O)}} R\!-\!\overset{O}{\overset{\|}{C}}\!-\!OH$$

Aldehydes   Carboxylic
                      acids

oxidizing
agents (O)

If R$-$CH$-$R′ (with OH above)
(2° alcohols)
$$\xrightarrow{\hspace{1cm}} H_2O + R\!-\!\overset{O}{\overset{\|}{C}}\!-\!R'$$

Ketones

3° Alcohols are not oxidized.

Mercaptans and Disulfides

$$R\!-\!S\!-\!H \underset{\text{reduction}}{\overset{\text{oxidation}}{\rightleftharpoons}} R\!-\!S\!-\!S\!-\!R$$

Mercaptans   Disulfides

Phenols

$$\text{(benzene ring)}\!-\!OH \xrightarrow{\text{NaOH}(aq)} \text{(benzene ring)}\!-\!O^-Na^+$$

Amines

NH$_3$

RNH$_2$

R$_2$NH

R$_3$N

$$\xrightarrow{H_2O} \quad -N^+\!-\!H + OH^-$$

$$\underset{\substack{(HX = \text{a strong} \\ \text{acid})}}{\xrightarrow{HX(aq)}} \quad -N^+\!-\!H + X^-$$

Amine salts

OH$^-$

REVIEW EXERCISES

The answers to all review exercises are in the *Study Guide* that accompanies this book.

Functional Groups

11.1 Name the functional groups to which the arrows point in the structure of eugenol, an oily liquid sometimes used as a substitute for oil of cloves in formulating fragrances and perfumes.

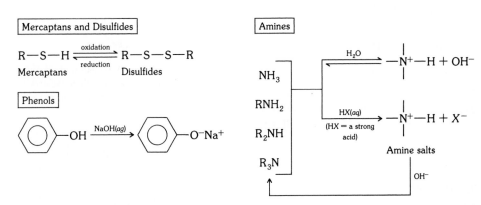

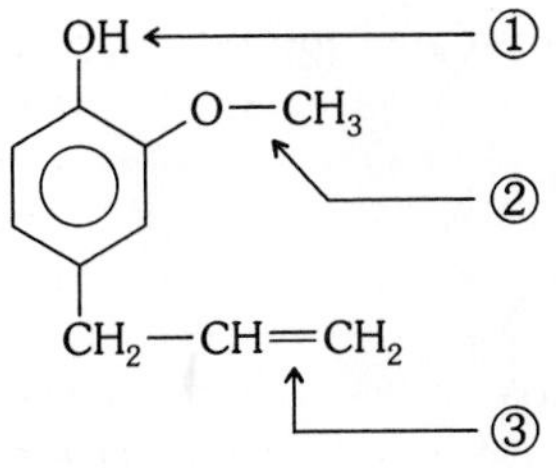

Eugenol

11.2 What are the functional groups to which the arrows point in the structure of geranial, a constituent of oil of lemon grass?

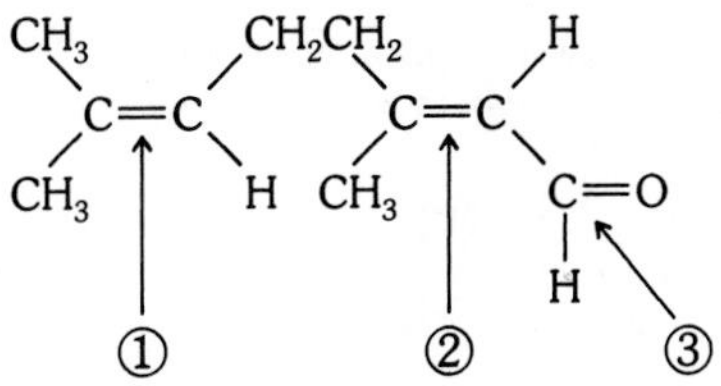

Geranial

11.3 Name the functional groups to which the arrows point in the structure of cortisone, a drug used in treating certain forms of arthritis. If a group is an alcohol, state whether it is a 1°, 2°, or 3° alcohol.

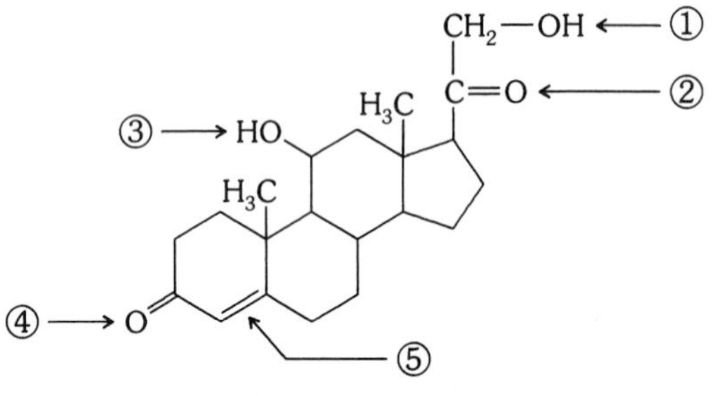

Cortisone

11.4 Give the names of the functional groups to which the arrows point in prostaglandin E₁, one of a family of compounds that are smooth muscle stimulants.

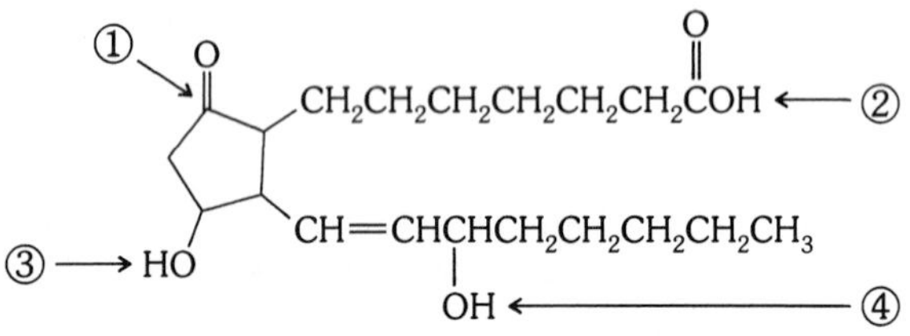

Prostaglandin E₁

11.5 The following compounds are all very active physiological agents. Name the numbered functional groups that are present in each.

(a)

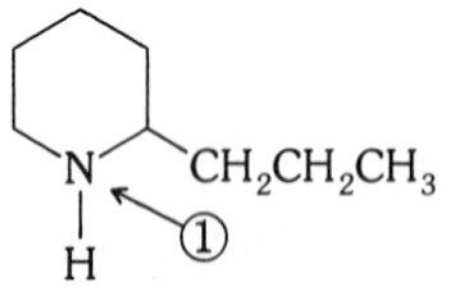

Coniine, the poison in the extract of hemlock that was involved in the death of the Greek philosopher Socrates

(b)

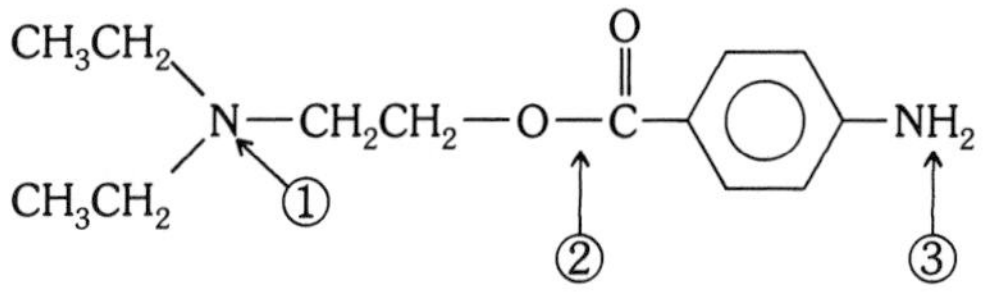

Novocaine, a local anesthetic

(c)

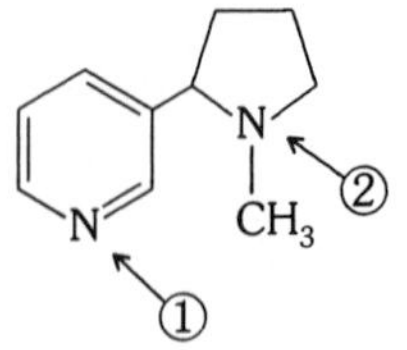

Nicotine, a poison in tobacco leaves

(d)

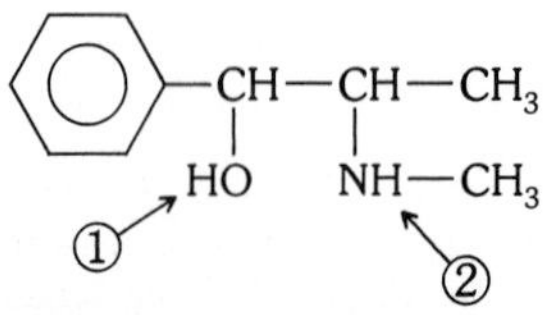

Ephedrine, a bronchodilator

11.6 Some extremely potent, physiologically active compounds are in the following list. Name the functional groups that they have.

(a)

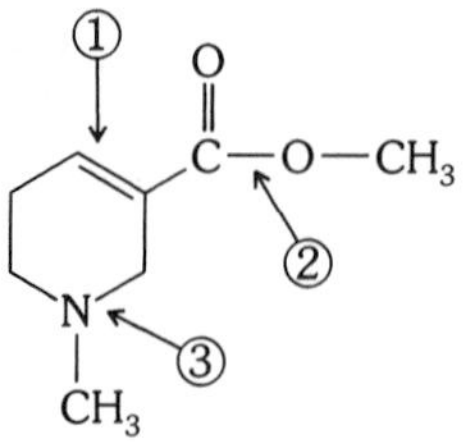

Arecoline, the most active component in the nut of the betel palm. This nut is chewed daily by millions of inhabitants of parts of Asia and the Pacific islands as a narcotic.

(b)

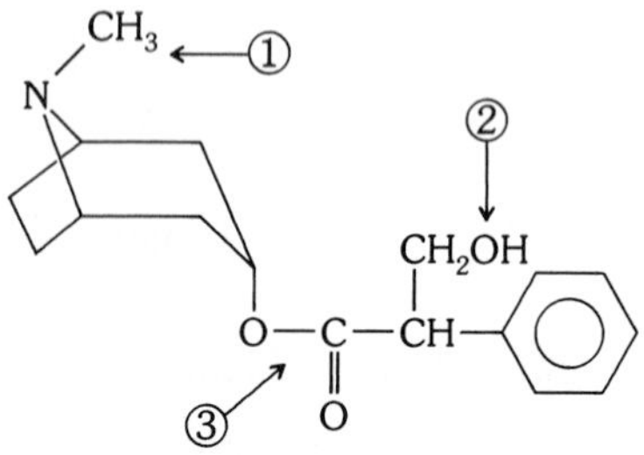

Hyoscyamine, a constituent of the seeds and leaves of henbane, and a smooth muscle relaxant. (A very similar form is called atropine, a drug used to counteract nerve poisons.)

(c)

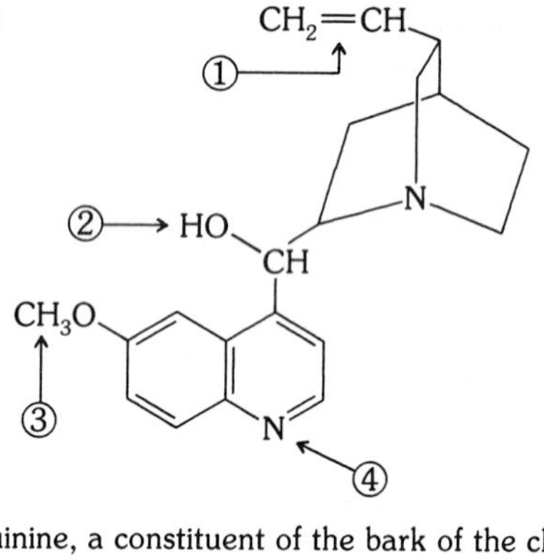

Quinine, a constituent of the bark of the chinchona tree in South America and used worldwide to treat malaria.

(d)

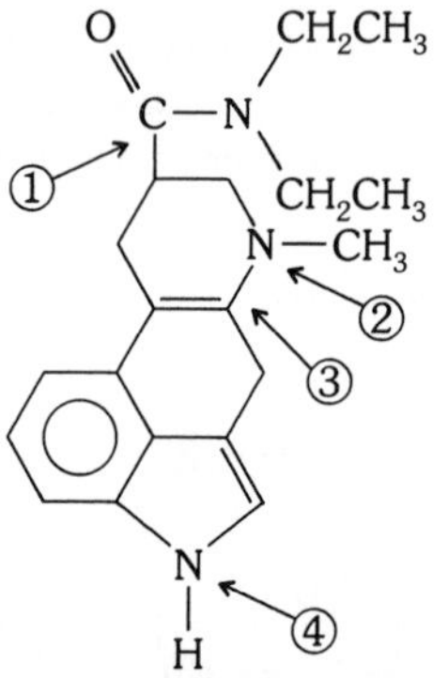

LSD (lysergic acid diethylamide), a constituent of diseased rye and a notorious hallucinogen.

Structures and Names of Alcohols

11.7 Write the structure of each compound.
   (a) Isobutyl alcohol   (b) Isopropyl alcohol
   (c) Propyl alcohol   (d) Glycerol

11.8 Write the structures of the following compounds.
   (a) Methyl alcohol   (b) t-Butyl alcohol
   (c) Ethyl alcohol   (d) Butyl alcohol

11.9 Give the common names of the following compounds.
   (a) CH_3CHCH_3 (b) $HOCH_2CH_2OH$
         |
         OH

   (c) $CH_3-\overset{\underset{|}{CH_3}}{\underset{|}{C}}-OH$ (d) $HOCH_2CHCH_2OH$
              | |
              CH_3 OH

11.10 What are the common names of the following compounds?
   (a) CH_3CH_2OH (b) $CH_3CH_2CH_2OH$
   (c) $HOCH_2CHCH_3$ (d) $HOCH_2CH_2CH_2CH_3$
              |
              CH_3

11.11 What is the structure and common name of the simplest stable dihydric alcohol?

11.12 Give the structure and the common name of the simplest stable trihydric alcohol.

Physical Properties of Alcohols

11.13 Draw a figure that illustrates a hydrogen bond between two molecules of ethyl alcohol. Use a dotted line to represent this bond, and write in the $\delta-$ and $\delta+$ symbols where they belong.

11.14 When ethyl alcohol dissolves in water, its molecules slip into the hydrogen-bond network in water. Draw a figure that illustrates this. Use dotted lines for hydrogen bonds, and place $\delta-$ and $\delta+$ symbols where they belong.

11.15 Arrange the following compounds in their order of increasing boiling points. Place the letter symbol of the compound that has the lowest boiling point on the left end of the series, and arrange the remaining letters in the correct order.

   OH
   |
   $CH_3CHCH_2CH_3$ $CH_3CH_2CH_2CH_2CH_2CH_3$
         A   B

   HO   OH
   | |
   CH_3CH-CH_2 CH_3-O-CH_3
         C   D

   ____<____<____<____
   Lowest   Highest
   bp   bp

11.16 Arrange the following compounds in their order of increasing solubility in water. Place the letter symbol of the compound with the least solubility on the left end of the series, and arrange the remaining letters in the correct order.

HO   OH   OH
| | |
$CH_3CHCHCHCH_3$ $CH_3CH_2CHCH_2CH_3$ $CH_3CH_2CH_2CH_2CH_3$
     |
     OH
   A   B   C

   ____<____<____
   Least   Most
   soluble   soluble

Chemical Properties of Alcohols

11.17 Write the structures of the alkenes that form when the following alcohols undergo acid-catalyzed dehydration.

   (a)   OH
         |
        $CH_3CHCH_2CH_3$

   (b) CH_3
         |
        $CH_3CCH_2CH_2CH_3$
         |
         OH

   (c) [cyclohexane ring with CH_3 and OH]

   (d) [benzene ring]$-CH_2CH_2OH$

   (e) [benzene ring]$-CHCH_2-$[benzene ring]
                       |
                       OH

   (f) CH_3-[cyclohexane ring]$-OH$

11.18 Write the structures of the products of the oxidation of the alcohols listed in Review Exercise 11.17. If the alcohol is a 1° alcohol, give the structures of both the aldehyde and the carboxylic acid that could be made by varying the quantities of the oxidizing agent.

11.19 Write the structure of any alcohol that could be dehydrated to give each alkene. In some cases, more than one alcohol would work.
   (a) $CH_2=CH_2$ (b) $CH_3CH=CH_2$
                             CH_3
                             |
   (c) $CH_2=C-CH_3$ (d) $CH_3CH=CHCH_3$

11.20 Write the structures of any alcohols that could be used to prepare the following compounds by an oxidation.

   (a) $CH_3CH_2\overset{O}{\overset{||}{C}}OH$ (b) $CH_3\overset{O}{\overset{||}{C}}CH_2CH_2CH_3$

   (c) [cyclohexane ring]$-\overset{O}{\overset{||}{C}}-H$ (d) $CH_3\underset{\underset{CH_3}{|}}{CH}-\overset{O}{\overset{||}{C}}OH$

11.21 Suppose you were given an unknown liquid and told that it is either t-butyl alcohol or butyl alcohol. You're also told that when a few drops of the unknown are shaken with a strong oxidizing agent, a reaction occurs. Which alcohol is it? Explain, and include a reaction sequence for the reaction that occurs.

11.22 An unknown is either pentane or 2-butanol. When it is shaken with a strong oxidizing agent, a reaction takes place. Which compound is it? Explain, including an equation.

Phenols

11.23 What is one chemical difference between phenol and cyclohexyl alcohol? Write an equation.

11.24 An unknown was either **A** or **B**.

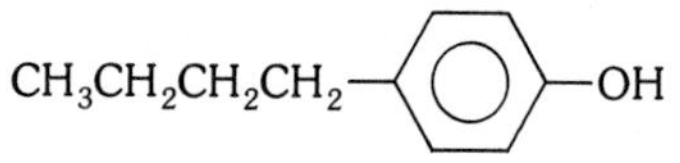

$CH_3CH_2CH_2CH_2$—⬡—OH

A

$CH_3CH_2CH_2$—⬡—CH_2OH

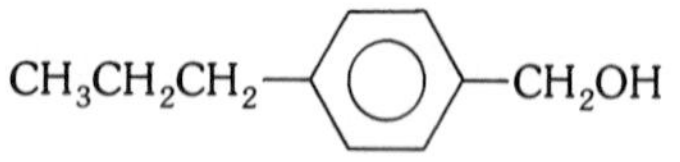

B

The unknown dissolved in aqueous sodium hydroxide, but not in water. Which compound was it? How can you tell? (Write an equation.)

11.25 An unknown was either **I** or **II**.

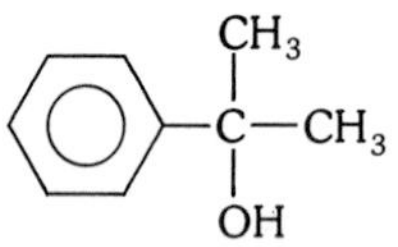

⬡—$\underset{\underset{OH}{|}}{\overset{\overset{CH_3}{|}}{C}}$—$CH_3$ HO—⬡—$\overset{\overset{CH_3}{|}}{CH}$—$CH_3$

I 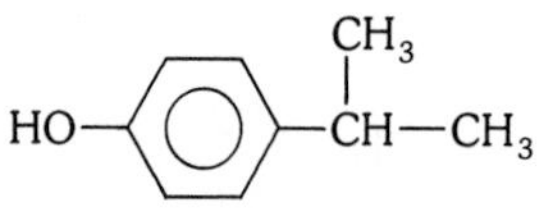   **II**

When it was shaken with aqueous KOH, no reaction occurred. Which was the unknown? How can you tell?

Thioalcohols and Disulfides

11.26 Although we did not systematically develop the rules for naming thioalcohols, the patterns in Table 11.3 make these rules fairly obvious. Write the structures of the following compounds.
(a) Isopropyl mercaptan   (b) 1-Butanethiol
(c) Dimethyl disulfide   (d) 1,2-Ethanedithiol

11.27 Complete the following reaction sequences by writing the structures of the organic products that form.
(a) $CH_3SH + (O) \longrightarrow$
(b) $CH_3\overset{\overset{CH_3}{|}}{CH}$—S—S—$\overset{\overset{CH_3}{|}}{CH}CH_3 + (H) \longrightarrow$
(c) 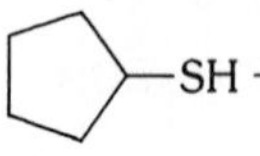 ⬠—$SH + (O) \longrightarrow$
(d) CH_3—S—S—$CH_2CH_2CH_3 + (H) \longrightarrow$

11.28 Propane, methanethiol, and ethyl alcohol have similar formula weights, but propane boils at -42 °C, methanethiol at

6 °C, and ethanol at 78 °C. What do these data suggest about hydrogen bonding in the thioalcohol family? Does it occur at all? Are the hydrogen bonds as strong as those in the alcohol family?

Ethers

11.29 Which of the following compounds have the ether function and which do not?
(a) CH_3—O—CH_2CH_2—O—CH_3
(b) CH_3—O—$\overset{\overset{O}{||}}{C}$—$CH_3$
(c) CH_3—O—$\overset{\overset{O}{||}}{C}$—O—$CH_3$
(d) 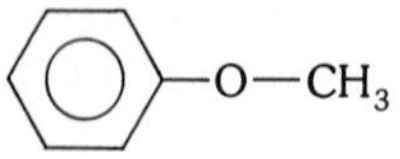 ⬡—O—CH_3
(e)
(f) CH_3—O—O—CH_3
(g) CH_2=CH—O—CH_2CH_3
(h) 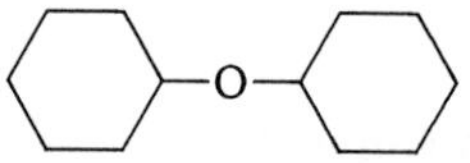 ⬡—O—⬡

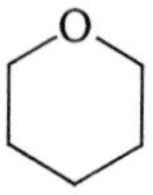

11.30 What are the structures of the ethers into which the following alcohols can be changed?
(a) $CH_3\overset{\overset{OH}{|}}{C}HCH_3$ (b) $CH_3\overset{\overset{CH_3}{|}}{C}HCH_2OH$
(c) 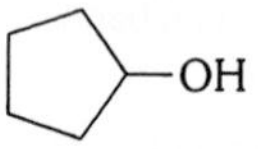 ⬠—OH   (d) 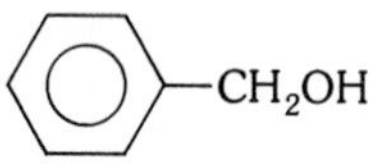 ⬡—CH_2OH

11.31 What alcohols would be needed to prepare the following ethers? Write their structures.
(a) CH_3—O—CH_3
(b) $CH_3CH_2CH_2CH_2$—O—$CH_2CH_2CH_2CH_3$

11.32 If an equimolar mixture of methyl alcohol and ethyl alcohol were heated with concentrated sulfuric acid under conditions suitable for making the ether system, what organic products would be isolated from this mixture after the reaction is over?

11.33 What happens, chemically, when the following compound is heated with aqueous sodium hydroxide?
$$CH_3CH_2—O—CH_2CH_3$$

Structures of Amines

11.34 Classify the following as aliphatic, aromatic, or heterocyclic amines or amides, and name any other functional groups, too.

(a) $CH_3-O-CH_2-\overset{\overset{O}{\|}}{C}-NH_2$

(b) $CH_3-O-\overset{\overset{O}{\|}}{C}-CH_2-NH_2$

(c) N—CH₃

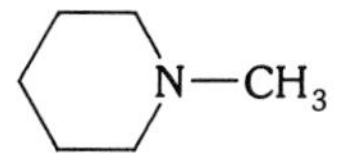

(d) $CH_3-$$-NH_2$

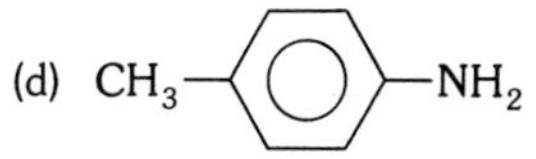

11.35 Classify each of the following as aliphatic, aromatic, or heterocyclic amines or amides. Name any other functional groups that are present.

(a)

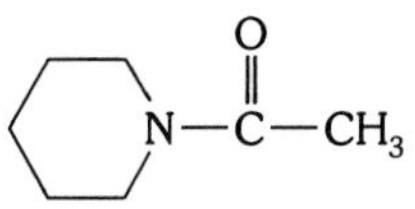

(b)

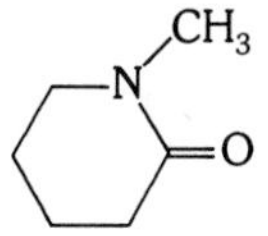

(c)

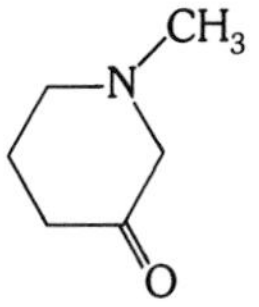

(d)

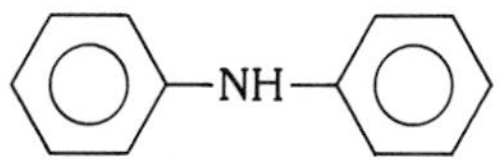

Nomenclature of Amines and Amine Salts

11.36 Give the common names of the following compounds or ions.
(a) $CH_3CH_2NHCH_2CH_3$
(b) $CH_3CH_2\overset{\overset{CH_2CH_3}{|}}{N}CHCH_3$ with CH₃
(c) $(CH_3)_3\overset{+}{N}H$
(d) $CH_3-\overset{\overset{CH_3}{|}}{\underset{\underset{CH_3}{|}}{C}}-\overset{+}{N}H_2CH_2\overset{\overset{CH_3}{|}}{C}HCH_3$

11.37 What are the common names of the following compounds?
(a) $[(CH_3)_3C]_3N$
(b) $C_6H_5NH_2$
(c) $(CH_3)_2CHNHCH_2CH_2CH_2CH_3$
(d) $(CH_3CH_2)_3\overset{+}{N}H\ Cl^-$

Chemical Properties of Amines and Amine Salts

11.38 Complete the following reaction sequences by writing the structures of the organic products. If no reaction occurs, write "no reaction."
(a) $CH_3NH_2 + NaOH(aq) \longrightarrow$
(b) $-NH_2 + HBr(aq) \longrightarrow$
(c) $NH_4Cl(aq) + NaOH(aq) \longrightarrow$

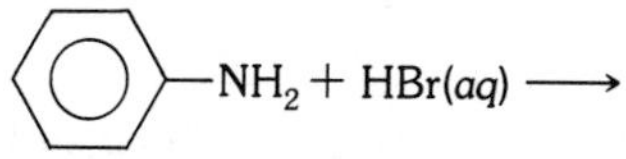

(d) $\overset{+}{N}H_2 + OH^- \longrightarrow$

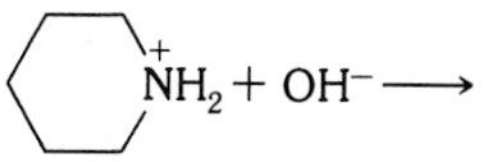

(e) $^+NH_3CH_3 + OH^- \longrightarrow$

(f) $^+NH_3-CH_2-\overset{\overset{O}{\|}}{C}-O^- + OH^- \longrightarrow$

11.39 Write the structures of the organic products that form in each situation. Assume that all reactions occur at room temperature. (Some of the named compounds are described in Review Exercises 11.5 and 11.6.) If no reaction occurs, write "no reaction."

(a) $-NH_2 + HCl(aq) \longrightarrow$

(b) $+ OH^- \longrightarrow$
Protonated form of arecoline

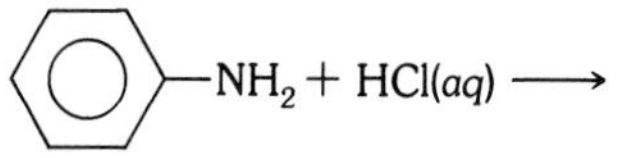

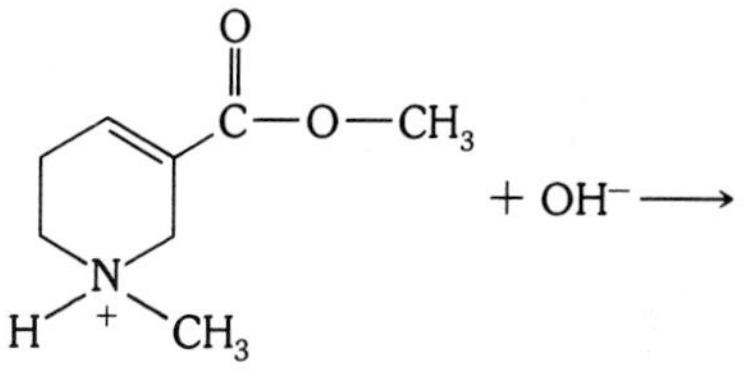

(c) $+ OH^- \longrightarrow$
Protonated form of nicotine

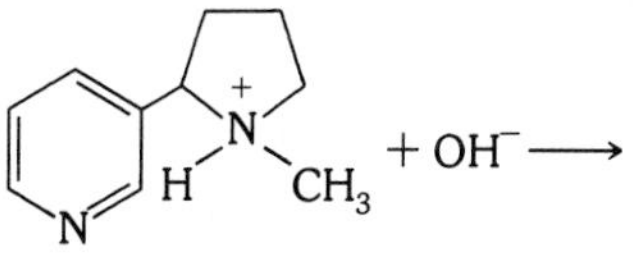

(d) $+ HCl(aq) \longrightarrow$
Ephedrine

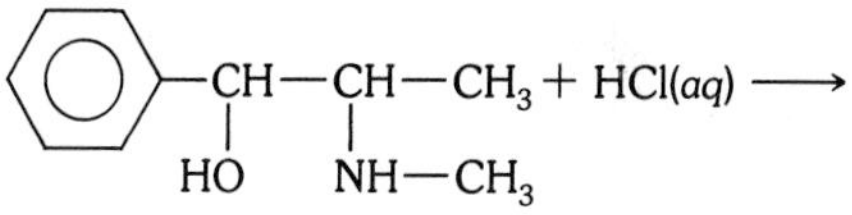

(e) $+ HCl(aq) \longrightarrow$
Hyoscamine

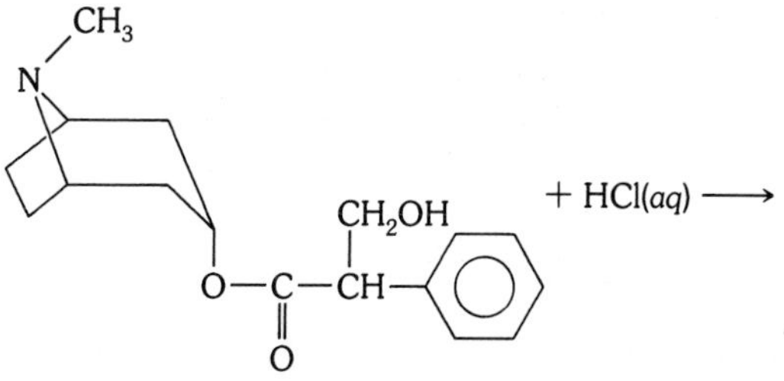

11.40 Which is the stronger base, **A** or **B**? Explain.

$CH_3CH_2\overset{\overset{O}{\|}}{C}NHCH_3$ $CH_3\overset{\overset{O}{\|}}{C}H\overset{\overset{O}{\|}}{C}CH_3$ with NH₂

A   **B**

11.41 Which is the stronger proton acceptor, **A** or **B**? Explain.

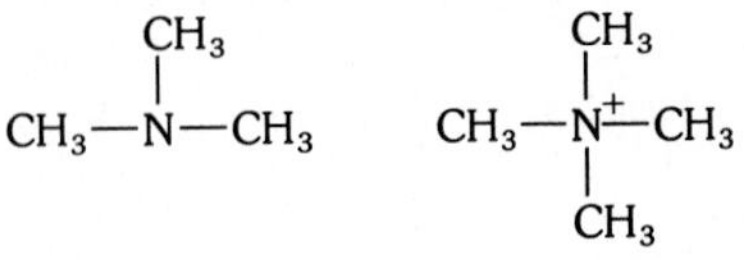

$$CH_3-N-CH_3 \quad\quad CH_3-\overset{+}{N}-CH_3$$
(with CH_3 groups on nitrogen)

A   **B**

Organic Reactions

11.42 Write the structure of the principal organic product that would be expected to form in each of the following situations. If no reaction occurs, state so.

   If the reaction is the oxidation of a primary alcohol, give the structure of the *aldehyde* only. If the conditions are sulfuric acid and heat and the reactant is an alcohol, write the structure of the *alkene* when the coefficient of the alcohol is given as "1." If it is given as "2," write the structure of the *ether*. (This violates our rule that balancing an equation is the *last* step in writing a reaction, but we need a signal to tell what kind of a reaction is meant.)

(a) $(CH_3)_2CHOH \xrightarrow{(O)}$

(b) $CH_3CH_2CH_2CH_3 + H_2SO_4 \longrightarrow$

(c) 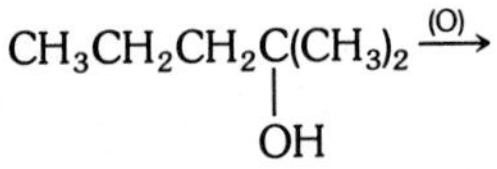 $CH_3CH_2CH_2C(CH_3)_2 \xrightarrow{(O)}$ with OH substituent

(d) $CH_3CH_2CH_2OH + NaOH\ (aq) \longrightarrow$

(e) $2CH_3OH \xrightarrow[\text{heat}]{H_2SO_4}$

(f) 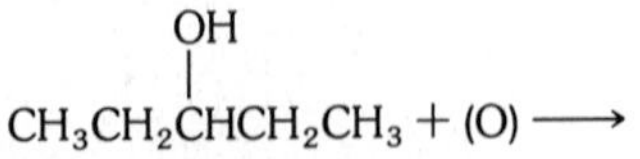 $CH_3CH_2CHCH_2CH_3 + (O) \longrightarrow$ with OH substituent

(g) $CH_3-O-CH_3 + (O) \longrightarrow$

(h) $CH_3CH=CHCH_3 + H_2O \xrightarrow[\text{heat}]{H^+}$

(i) $CH_3CH_2CH_2OH \xrightarrow[\text{heat}]{H_2SO_4}$

(j) 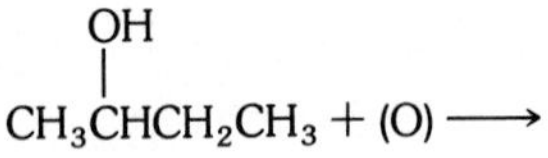 $CH_3CHCH_2CH_3 + (O) \longrightarrow$ with OH substituent

(k) 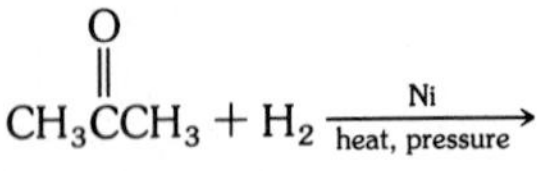 $CH_3\overset{O}{\overset{\|}{C}}CH_3 + H_2 \xrightarrow[\text{heat, pressure}]{Ni}$

(l) $CH_3(CH_2)_8CH_3 + NaOH(aq) \longrightarrow$

(m) 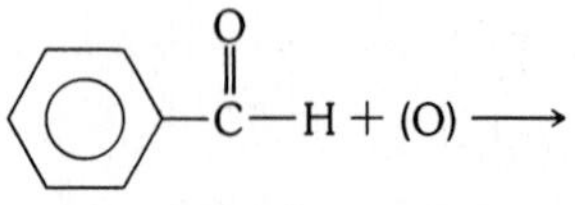 phenyl-$\overset{O}{\overset{\|}{C}}-H + (O) \longrightarrow$

(n) $CH_3CH_2CH_2NH_2 + HCl(aq) \xrightarrow{25\,°C}$

(o) $CH_3SH + (O) \longrightarrow$

11.43 Following the same directions as given in Review Exercise 11.42, write the structure of the principal organic product that would be expected to form in each situation. If no reaction occurs, write "no reaction."

(a) $CH_3CH=CH_2 + H_2 \xrightarrow[\text{pressure}]{Ni,\ heat,}$

(b) 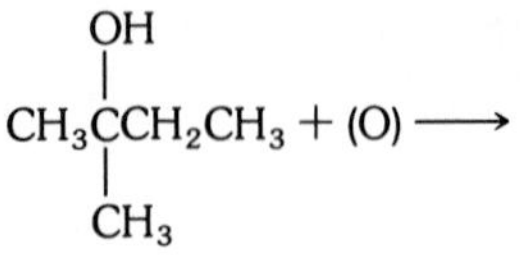 $CH_3\overset{OH}{\underset{CH_3}{C}}CH_2CH_3 + (O) \longrightarrow$ (with OH and CH_3 on same carbon)

(c) cyclopentyl$-OH \xrightarrow[\text{heat}]{H_2SO_4}$

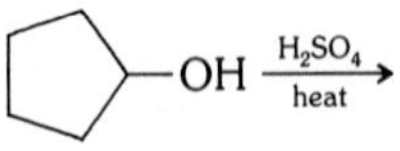

(d) $CH_3CH_2OCH_3 + HCl(aq) \longrightarrow$

(e) 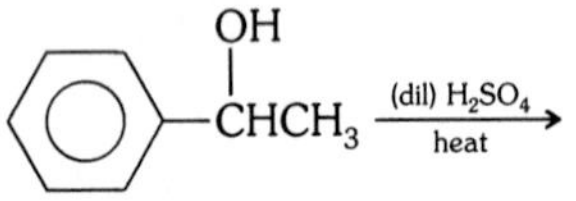 phenyl$-CHCH_3 \xrightarrow[\text{heat}]{(dil)\ H_2SO_4}$ with OH substituent

(f) 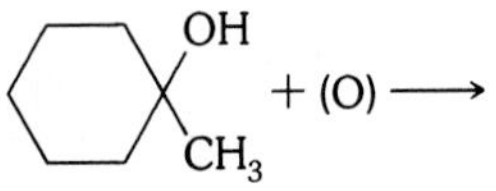 cyclohexane with OH and CH_3 $+ (O) \longrightarrow$

(g) $CH_3-O-CH_2CH=CH_2 + H_2 \xrightarrow[\text{pressure}]{Ni,\ heat,}$

(h) 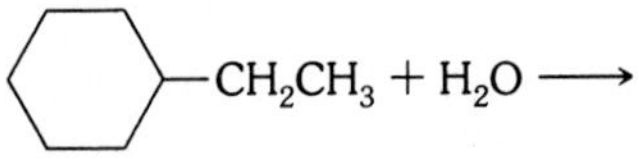 cyclohexyl$-CH_2CH_3 + H_2O \longrightarrow$

(i) 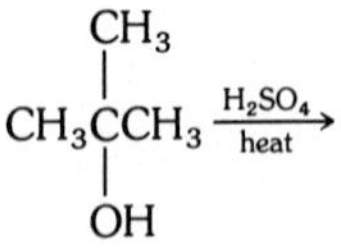 $CH_3\overset{CH_3}{\underset{OH}{C}}CH_3 \xrightarrow[\text{heat}]{H_2SO_4}$ (with CH_3 and OH)

(j) $CH_3CH_2CH_2OH + (O) \longrightarrow$

(k) 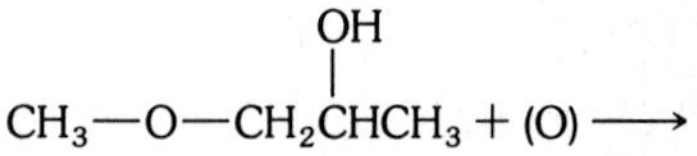 $CH_3-O-CH_2CHCH_3 + (O) \longrightarrow$ with OH substituent

(l) 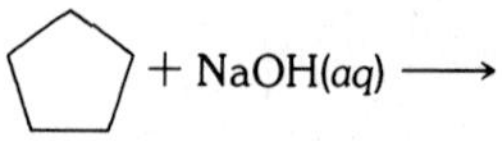 cyclopentane $+ NaOH(aq) \longrightarrow$

(m) $NH_2CH_2CH_2CH_2NH_2 + HCl(aq) \xrightarrow{25\,°C}$ (excess)

(n) 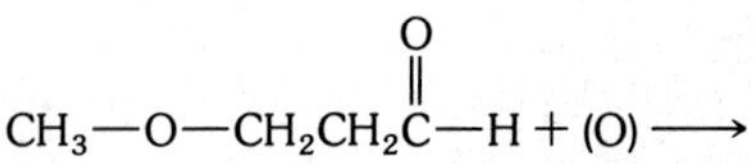 $CH_3-O-CH_2CH_2\overset{O}{\overset{\|}{C}}-H + (O) \longrightarrow$

(o) $CH_3CH_2\overset{O}{\overset{\|}{C}}CH_2CH_3 + H_2 \xrightarrow[\text{heat, pressure}]{Ni}$

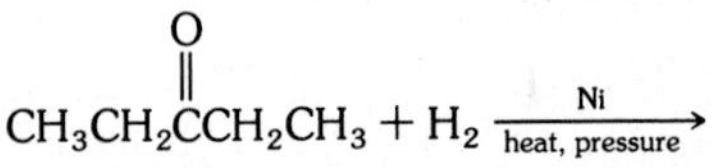

(p) $CH_3S-SCH_3 + 2(H) \longrightarrow$

Chapter 12
Aldehydes and Ketones

*Structural Features and Names of Aldehydes
    and Ketones*

The Oxidation of Aldehydes

The Reduction of Aldehydes and Ketones

*The Reactions of Aldehydes and Ketones with
    Alcohols*

*Entire buildings are reflecting surfaces in most modern cities today. One of the reactions studied
in this chapter is used to make small mirrors.*

12.1 STRUCTURAL FEATURES AND NAMES OF ALDEHYDES AND KETONES

Molecules of both aldehydes and ketones contain the carbonyl group, C=O; when this group holds at least one hydrogen atom, it is the aldehyde group.

A knowledge of some of the properties of aldehydes and ketones is essential to an understanding of carbohydrates. All simple sugars are either polyhydroxy aldehydes or polyhydroxy ketones, and many intermediates in metabolism are aldehydes or ketones. Special Topic 12.1 describes just a few of them.

The Carbonyl Group Is the Functional Group in All Aldehydes and Ketones. The carbon–oxygen double bond is called the **carbonyl group** (pronounced carbon-EEL), and it is present in aldehydes and ketones (as well as the families to be studied in Chapter 13).

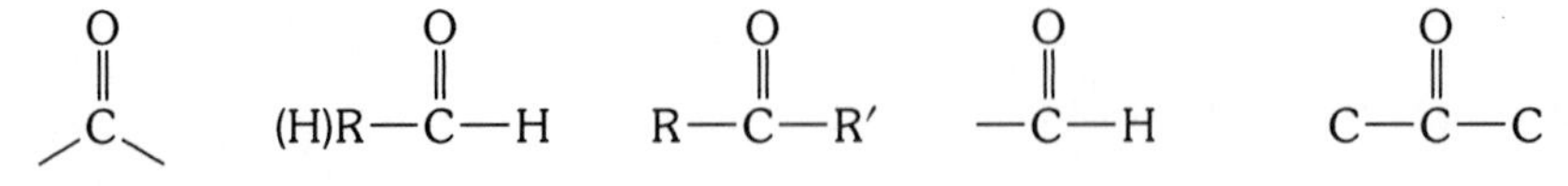

Carbonyl group	Aldehydes	Ketones	Aldehyde group	Ketone system

Notice that the carbonyl group in aldehydes must hold at least one H atom. In aldehydes this combination is called the **aldehyde group,** which is often condensed as CH=O or as CHO. To be an **aldehyde** there must be attached to the CH=O group either C or H, never N, O, or some other atom. Formaldehyde, the simplest aldehyde, $H_2C{=}O$, has a second H attached to CH=O. Other aldehydes are shown in Table 12.1.

In **ketones,** the carbonyl carbon must be flanked on both sides by C atoms only. In this setting, the carbonyl group is called the **keto group.** Several ketones are listed in Table 12.2.

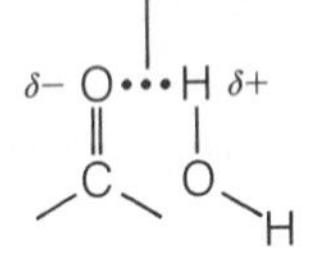

The Carbonyl Group Makes a Molecule Moderately Polar. Table 12.3 gives the boiling points of compounds of very similar formula weights but from different families. You can see that an aldehyde or ketone has a boiling point intermediate between an alkane and an alcohol. We have learned that an alkane is almost nonpolar and that an alcohol is very polar. We can justifiably infer, therefore, that aldehydes and ketones are moderately polar.

Molecules of an alcohol, of course, are able both to donate and accept hydrogen bonds. The carbonyl group cannot donate hydrogen bonds, but it can accept them, so both the aldehyde and keto groups help bring compounds into solution in water. As you can see in Tables 12.1 and 12.2, the low formula weight aldehydes and ketones are soluble in water.

Hydrogen bond

TABLE 12.1
Aldehydes

Name[a]	Structure	Boiling Point (°C)	Solubility in Water[b]
Formaldehyde (methanal)	$CH_2{=}O$	−21	Very soluble
Acetaldehyde (ethanal)	$CH_3CH{=}O$	21	Very soluble
Propionaldehyde (propanal)	$CH_3CH_2CH{=}O$	49	16 g/dL
Butyraldehyde (butanal)	$CH_3CH_2CH_2CH{=}O$	76	4 g/dL
Benzaldehyde	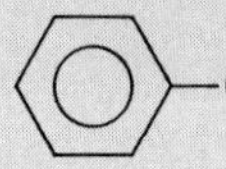 ⬡—CH=O	178	0.3 g/dL

[a] The IUPAC names are in parentheses.
[b] At 25°C.

TABLE 12.2
Ketones

Name[a]	Structure	Boiling Point (°C)	Solubility in Water[b]
Acetone (propanone)	$CH_3\overset{O}{\overset{\|}{C}}CH_3$	56	Very soluble
Methyl ethyl ketone (butanone)	$CH_3\overset{O}{\overset{\|}{C}}CH_2CH_3$	80	33 g/dL
Methyl propyl ketone (2-pentanone)	$CH_3\overset{O}{\overset{\|}{C}}CH_2CH_2CH_3$	102	6 g/dL
Diethyl ketone (3-pentanone)	$CH_3CH_2\overset{O}{\overset{\|}{C}}CH_2CH_3$	102	5 g/dL
Cyclopentanone	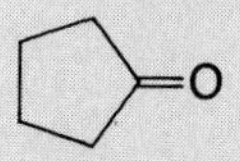 ⬠=O	129	Slightly soluble
Cyclohexanone	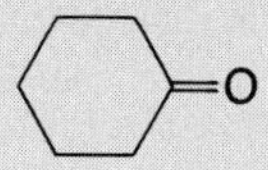 ⬡=O	156	Slightly soluble

[a] The IUPAC names are in parentheses.
[b] At 25°C.

Aldehydes Have Common Names Based on Their Corresponding Acids. What is easy about the common names of aldehydes is that they all (well, nearly all) end in -aldehyde. The prefixes to this are the same as the prefixes of the carboxylic acids, discussed below, to which the aldehydes are easily oxidized. Even exposure to air causes the following reaction to occur (slowly).

$$RCH\!=\!O \xrightarrow{\text{oxidizing agent}} RCO_2H$$

Aldehyde   Carboxylic acid

This is, therefore, a good place to study the common names of both acids and aldehydes.

The common names of the simple carboxylic acids are based on some natural sources of the acids. The one carbon acid, for example, is called formic acid because it is present in the stinging fluid of ants, and the Latin root for ants is *formica*. The prefix in formic acid is *form-*, so the one-carbon aldehyde is called *formaldehyde*. Here are the four simplest carboxylic acids and their common names together with the structures and names of their corresponding aldehydes.

$\overset{O}{\overset{\|}{H}}COH$ Formic acid $\overset{O}{\overset{\|}{H}}CH$ Formaldehyde

$CH_3\overset{O}{\overset{\|}{C}}OH$ Acetic acid $CH_3\overset{O}{\overset{\|}{C}}H$ Acetaldehyde

$R\overset{O}{\overset{\|}{C}}\!-\!OH$
Carboxylic acids

TABLE 12.3
Boiling Points and Polarities

Compound	F. Wt.	B.P. (°C)
$CH_3CH_2CH_2CH_3$	60	−0.5
$CH_3CH_2CH\!=\!O$	58	49
$CH_3\overset{O}{\overset{\|}{C}}CH_3$	58	56
$CH_3CH_2CH_2OH$	60	97

Latin *acetum*, vinegar

Formaldehyde Pure formaldehyde is a gas at room temperature, and it has a very irritating and distinctive odor. It is quite soluble in water, so it is commonly marketed as a solution called formalin (37% w/w) to which some methyl alcohol has been added. In this and more dilute forms, formaldehyde was once commonly used as a disinfectant and as a preservative for biological specimens. (Concern over formaldehyde's potential hazard to health has caused these uses to decline.) Most formaldehyde today is used to make various plastics such as Bakelite.

Acetone Acetone is valued as a solvent. It not only dissolves a wide variety of organic compounds, it also is miscible with water in all proportions. Nail polish remover is generally acetone. Should you ever use "superglue," it would be a good idea to have some acetone (nail polish remover) handy because superglue can glue your fingers together so tightly that it takes a solvent such as acetone to get them unstuck.

Acetone is a minor by-product of metabolism, but in some situations (e.g., untreated diabetes) enough is produced to give breath the odor of acetone.

Some Aldehydes and Ketones in Metabolism The aldehyde group and the keto group occur in many compounds at the molecular level of life. The following are just a few examples of substances with the aldehyde group.

Just a few of the many substances at the molecular level of life that contain the keto group are the following:

$$CH_3-\overset{\overset{\displaystyle O}{\|}}{C}-CO_2^-$$

Pyruvate ion, a product of the metabolism of glucose and fructose

$$CH_3-\overset{\overset{\displaystyle O}{\|}}{C}-CH_2-CO_2^-$$

Acetoacetate ion, a product of the metabolism of long-chain carboxylic acids, and present in blood at elevated levels in diabetes

$$HO-CH_2-\overset{\overset{\displaystyle O}{\|}}{C}-CH_2-O-PO_3^{2-}$$

Dihydroxyacetone phosphate, an intermediate in the metabolism of glucose and fructose

$$HOCH_2-\underset{\underset{\displaystyle OH}{|}}{CH}-\underset{\underset{\displaystyle OH}{|}}{CH}-\underset{\underset{\displaystyle OH}{|}}{CH}-\underset{\underset{\displaystyle OH}{|}}{CH}-\overset{\overset{\displaystyle O}{\|}}{C}-H$$

Glucose, a product of the digestion of sugars and starch

$$H-\overset{\overset{\displaystyle O}{\|}}{C}-\underset{\underset{\displaystyle OH}{|}}{CH}-CH_2-OPO_3^{2-}$$

Glyceraldehyde 3-phosphate, an intermediate in glucose metabolism

Pyridoxal, one of the vitamins (B$_6$)

Estrone, a female sex hormone

Greek *proto*, first, and *pion*, fat

$$CH_3CH_2\overset{\overset{\displaystyle O}{\|}}{C}OH$$

Propionic acid

$$CH_3CH_2\overset{\overset{\displaystyle O}{\|}}{C}H$$

Propionaldehyde

Latin *butyrum*, butter

$$CH_3CH_2CH_2\overset{\overset{\displaystyle O}{\|}}{C}OH$$

Butyric acid

$$CH_3CH_2CH_2\overset{\overset{\displaystyle O}{\|}}{C}H$$

Butyraldehyde

In the aromatic series we have the following examples:

$$\underset{\text{Benzoic acid}}{C_6H_5\overset{\displaystyle O}{\overset{\|}{C}}OH} \qquad \underset{\text{Benzaldehyde}}{C_6H_5\overset{\displaystyle O}{\overset{\|}{C}}H}$$

Common Names of Simple Ketones Include the Names of Their Hydrocarbon Groups. The following examples illustrate how common names for simple ketones can be devised. You should remember, however, that dimethyl ketone is called acetone.

Pronounce the *-one* in ket*one* as *own*.

$$\underset{\text{Ethyl methyl ketone}}{CH_3\overset{\displaystyle O}{\overset{\|}{C}}-CH_2CH_3} \qquad \underset{\text{Diethyl ketone}}{CH_3CH_2-\overset{\displaystyle O}{\overset{\|}{C}}-CH_2CH_3} \qquad \underset{\substack{\text{(Dimethyl ketone)}\\ \text{Acetone}}}{CH_3-\overset{\displaystyle O}{\overset{\|}{C}}-CH_3}$$

The name *acetone* stems from the fact that this ket*one* can be made by heating the calcium salt of *acet*ic acid.

PRACTICE EXERCISE 1   Write the structures of the following ketones.

                 (a) Ethyl isopropyl ketone   (b) Methyl phenyl ketone
                 (c) Dipropyl ketone   (d) Di-*t*-butyl ketone

---

The IUPAC rules for naming aldehydes and ketones are very similar to those for the alkanes and alcohols, and they are based on the idea of a parent (see Appendix III).

12.2 THE OXIDATION OF ALDEHYDES

The aldehyde group is easily oxidized to the carboxylic acid group, but the ketone system is difficult to oxidize.

We learned in Chapter 11 that the oxidation of a 1° alcohol to an aldehyde requires care, because aldehydes are easily oxidized to carboxylic acids. Both permanganate and dichromate ion work very well.

$$\underset{\text{1° Alcohol}}{RCH_2OH} + \underset{\substack{\text{Oxidizing}\\ \text{agent}}}{(O)} \xrightarrow{\text{relatively hard}} \underset{\text{Aldehyde}}{R-\overset{\displaystyle O}{\overset{\|}{C}}-H} + H_2O$$

$$\xrightarrow[\text{(O)}]{\text{relatively easy}} \underset{\substack{\text{Carboxylic}\\ \text{acid}}}{RCO_2H}$$

Much less care is needed to oxidize a 2° alcohol to a ketone, because ketones are considerably more difficult to oxidize than 2° alcohols.

Oxidizing agents much milder than permanganate or dichromate ion change aldehydes to carboxylic acids. We will look briefly at two mild oxidizing agents because they provide simple test tube tests for aldehydes.

Tollens' Test Produces a Silver Mirror. One very mild oxidizing agent is called **Tollens' reagent,** and it consists of an alkaline solution of the diammonia complex of the silver ion, $Ag(NH_3)_2{}^+$. It reacts with an aldehyde to oxidize it to the anion of a carboxylic acid.

Tollens' reagent deteriorates on standing, so it is freshly made before each use.

$$RCH{=}O + 2Ag(NH_3)_2{}^+ + 3OH^- \longrightarrow RCO_2{}^- + 2Ag + 2H_2O + 4NH_3$$

One product is metallic silver, and it appears as a beautiful mirror coating the inside of the test tube. (This is one technology used to make mirrors.) The appearance of a silver mirror is evidence that the silver ion has oxidized something (and so has been itself reduced to silver atoms). Neither alcohols nor other carbonyl compounds give this reaction, so the Tollens' reagent thus provides a good way to tell if an unknown carbonyl compound is an aldehyde. The test is called the **Tollens' test** (or, sometimes, the silver mirror test).

A positive Tollens' test is the formation of metallic silver in any form; as a mirror on a clean glass surface or otherwise as a grayish precipitate. Glucose, a carbohydrate that we will learn more about later, gives a positive Tollens' test.

Benedict's Test Is Another Test for an Easily Oxidized System. A basic solution of the copper(II) ion in the presence of the citrate ion is called **Benedict's reagent.** The citrate ion wraps around the Cu^{2+} ion and so keeps Cu^{2+} in solution. Otherwise, Cu^{2+} would combine with OH^- ions and precipitate. The Cu^{2+} ion, however, is still available for other reactions.

When an easily oxidized compound is added to warm Benedict's reagent, Cu^{2+} ions are reduced to Cu^+ ions. These are not protected from OH^- ions by the citrate ion, and they are changed by the base into a precipitate of copper(I) oxide, Cu_2O.

Benedict's reagent has a brilliant blue color caused by the copper(II) ion, but Cu_2O has a brick red color. In the **Benedict's test** for an easily oxidized group, the blue color disappears as the reddish precipitate forms.

Three systems, all of which occur among various carbohydrates, give positive Benedict's tests:

α-Hydroxy aldehyde α-Keto aldehyde α-Hydroxy ketone

Glucose, for example, is an α-hydroxy aldehyde, and Benedict's test is a common method for detecting glucose in urine. In certain conditions, such as diabetes, the body cannot prevent some of the excess glucose in the blood from appearing in the urine, so testing the urine for its glucose concentration has long been used in medical diagnosis.

Clinitest® tablets are a convenient solid form of Benedict's reagent. Each tablet contains all of the needed reactants. To test for glucose a few drops of urine are mixed with one tablet. As the tablet dissolves, the heat needed for the test is generated. The color that develops is compared with a color code on a chart provided with the tablets.[1] Other tests for glucose are based on enzyme-catalyzed reactions, and we will learn about them later.

12.3   THE REDUCTION OF ALDEHYDES AND KETONES

Aldehydes and ketones are reduced to alcohols when hydrogen adds to their carbonyl groups.

Reduction converts aldehydes to 1° alcohols and ketones to 2° alcohols. We will study two methods of reduction, the direct addition of hydrogen and hydride ion transfer. The overall results are the same.

Hydrogen Adds Catalytically to the Carbonyl Groups of Aldehydes and Ketones. Under heat and pressure and in the presence of a finely divided metal catalyst (like powdered nickel), aldehydes and ketones react with hydrogen as follows:

---

[1] It should be said that specialists in the control of diabetes prefer to monitor the carbohydrate status of a person with diabetes by testing the concentration of glucose in the patient's blood instead of in the urine. However, not all patients can or are willing to manage blood tests, particularly when needed frequently.

Margin notes:

The test tube must be very clean with no soapy or greasy film on its inside surface.

$Cu(OH)_2$ is extremely insoluble in water.

A carbon atom immediately adjacent to a carbonyl group is often called an alpha (α) carbon:

Alpha carbon

H_2 adds to $C=O$ under the same conditions as it adds to $C=C$.

$$R-\overset{\overset{\displaystyle O}{\|}}{C}-H + H_2 \xrightarrow[\text{pressure}]{\text{Ni, heat,}} R-CH_2-OH$$

Aldehyde   1° Alcohol

$$R-\overset{\overset{\displaystyle O}{\|}}{C}-R' + H_2 \xrightarrow[\text{pressure}]{\text{Ni, heat,}} R-\overset{\overset{\displaystyle OH}{|}}{CH}-R'$$

Ketone   2° Alcohol

Specific examples include:

$$CH_3CH_2CH_2CH{=}O + H_2 \xrightarrow[\text{pressure}]{\text{Ni, heat,}} CH_3CH_2CH_2CH_2OH$$

Butyraldehyde   Butyl alcohol

$$CH_3\overset{\overset{\displaystyle O}{\|}}{C}CH_3 + H_2 \xrightarrow[\text{pressure}]{\text{Ni, heat,}} CH_3\overset{\overset{\displaystyle OH}{|}}{CH}CH_2$$

Acetone   Isopropyl alcohol

The Hydride Ion Reduces Aldehydes and Ketones. The hydride ion, $H:^-$, a powerful reducing agent, is supplied in the body by a small number of complex organic donors, like NADH, NADPH, $FMNH_2$, and $FADH_2$. (The structures of these, as we indicated, are complex, and we will not look at them until Chapter 17. The hydride-donating structural units in them, whose symbols are NAD, NADP, FMN, and FAD, are all made from B vitamins.)

The carbonyl group is an excellent acceptor of the hydride ion. Aldehydes and ketones can accept this species as shown in the next equation, where we use NADH to illustrate one of the hydride ion donors in cells. We represent NADH as $NAD:H$, however, to emphasize that an electron pair goes with H. As you can see, the overall result, the reduction of an aldehyde or a ketone to an alcohol, is identical to the catalytic hydrogenation of the aldehyde or ketone.

The N in NADH and NADPH refers to nicotinamide, a B vitamin. The F in $FMNH_2$ or $FADH_2$ refers to riboflavin, another B vitamin.

$$NAD:H + \overset{\diagdown}{\underset{\diagup}{C}}{=}\ddot{O}\!\!:^{..} \longrightarrow NAD^+ + H-\overset{|}{\underset{|}{C}}-\ddot{O}\!\!:^- \xrightarrow{\quad H^+ \text{ (from the buffer)}\quad} H-\overset{|}{\underset{|}{C}}-\ddot{O}-H$$

Hydride   Aldehyde   Anion of   Alcohol
donor   or ketone   an alcohol

$H:^-$ is never free in water but always passes directly from donor to acceptor.

One of the many examples in cells of this kind of reduction is the conversion of the keto group in the pyruvate ion to the 2° alcohol group of the lactate ion, a step in the metabolism of glucose.

$$CH_3-\overset{\overset{\displaystyle O}{\|}}{C}-CO_2^- + NAD:H + H^+ \longrightarrow NAD^+ + CH_3-\overset{\overset{\displaystyle OH}{|}}{CH}-CO_2^-$$

Pyruvate ion   Lactate ion

EXAMPLE 12.1   WRITING THE STRUCTURE OF THE PRODUCT OF THE REDUCTION OF AN ALDEHYDE OR KETONE

Problem: What is the product of the reduction of propionaldehyde?

Solution: All of the action is at the carbonyl group. It changes to an alcohol group. Therefore all we have to do is copy over the structure of the given compound, change the double bond to a single bond, and supply the two hydrogen atoms: one to the oxygen atom of the original carbonyl group and one to the carbon atom.

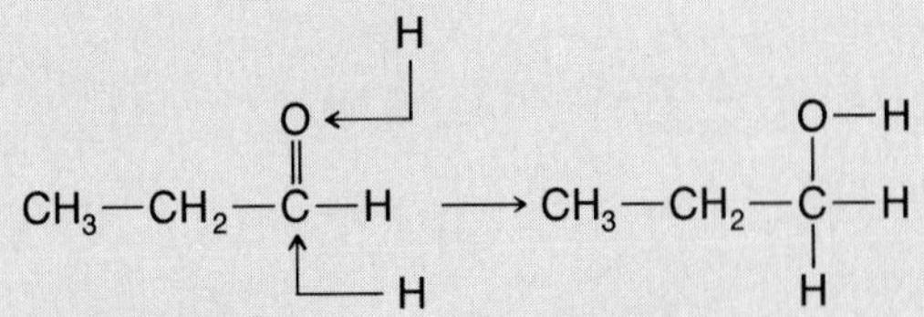

$$CH_3-CH_2-\overset{\overset{\displaystyle H}{\underset{\displaystyle \uparrow}{}}}{\underset{\underset{\displaystyle H}{\uparrow}}{\overset{\displaystyle O \leftarrow}{C}}}-H \longrightarrow CH_3-CH_2-\overset{\overset{\displaystyle O-H}{|}}{\underset{\underset{\displaystyle H}{|}}{C}}-H \quad (\text{or } CH_3CH_2CH_2OH)$$

Propionaldehyde   Propyl alcohol

PRACTICE EXERCISE 2   Write the structures of the products that form when the following aldehydes and ketones are reduced.

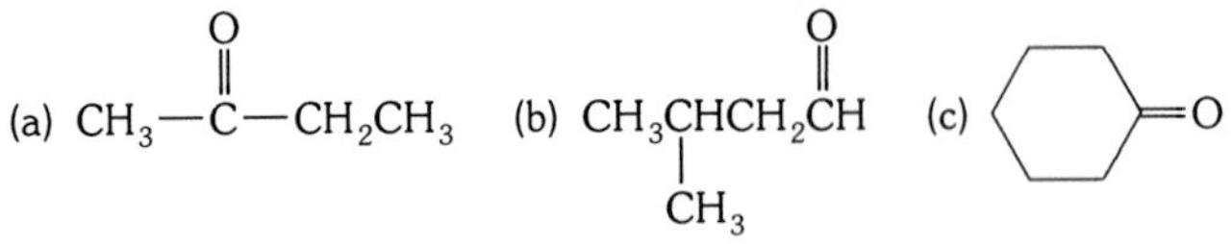

(a) $CH_3-\overset{\overset{\displaystyle O}{\|}}{C}-CH_2CH_3$ (b) $CH_3\underset{\underset{\displaystyle CH_3}{|}}{CH}CH_2\overset{\overset{\displaystyle O}{\|}}{CH}$ (c) ⬡=O

12.4   THE REACTIONS OF ALDEHYDES AND KETONES WITH ALCOHOLS

1,1-Diethers are acetals or ketals formed when aldehydes or ketones react with alcohols.

This section is background to our study of carbohydrates whose molecules have the functional groups introduced here. We will study the simplest possible examples of these groups now so that carbohydrate structures will be easier to understand later.

Hemiacetals Are Unstable Compounds Formed by the Addition of Alcohols to the Carbonyl Groups of Aldehydes.   The following equilibrium exists in a solution of an aldehyde and an alcohol.

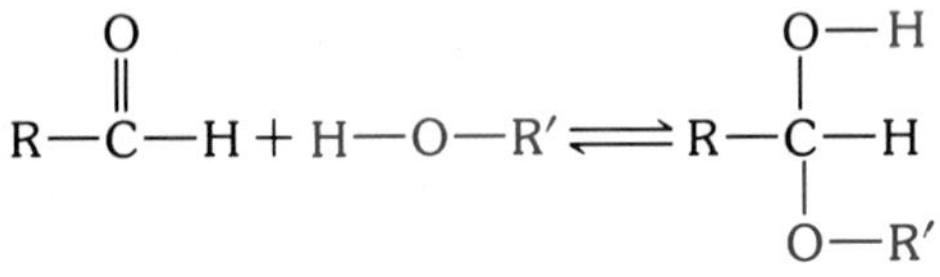

$$R-\overset{\overset{\displaystyle O}{\|}}{C}-H + H-O-R' \rightleftharpoons R-\overset{\overset{\displaystyle O-H}{|}}{\underset{\underset{\displaystyle O-R'}{|}}{C}}-H$$

Aldehyde   Alcohol   Hemiacetal

The hemiacetal system:

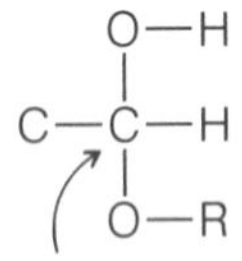

$$\overset{\overset{\displaystyle O-H}{|}}{\underset{\underset{\displaystyle O-R}{|}}{C-\overset{}{C}-H}}$$

This originally was the carbon atom of an aldehyde group

The **hemiacetal** is produced by the addition of the alcohol to the double bond of the C=O group. It always has a carbon atom that holds both an O—H and an O—R group. When these groups are this close together in the same molecule, they markedly modify each other's properties. The OH is no longer a simple alcohol and the C—O—R system is no longer an ordinary ether. The entire structural package, therefore, defines a separate family, the hemiacetals.

In the formation of a hemiacetal, the O—R unit of the alcohol molecule *always* becomes attached to the carbon of the original C=O group, and the H atom of H—O—R always goes to the oxygen of C=O group.

Except among carbohydrates, a hemiacetal is generally an unstable compound. If we try to isolate and purify one, it breaks back down during the process. The addition reaction that made the hemiacetal reverses itself, and nothing but the original aldehyde and alcohol are obtained. Thus hemiacetals generally exist only as part of an equilibrium. We're interested in them because the hemiacetal system occurs among carbohydrates and because they are intermediate structures on the way to 1,1-diethers called *acetals*. Acetals are stable enough to be isolated, and the acetal system occurs widely among carbohydrates.

EXAMPLE 12.2 WRITING THE STRUCTURE OF A HEMIACETAL GIVEN ITS PARENT ALDEHYDE AND ALCOHOL

Problem: Write the structure of the hemiacetal that is present at equilibrium in a solution of propionaldehyde and ethyl alcohol.

Solution: The best way to start is to rewrite the structure of the aldehyde, but show only one bond from carbon to oxygen:

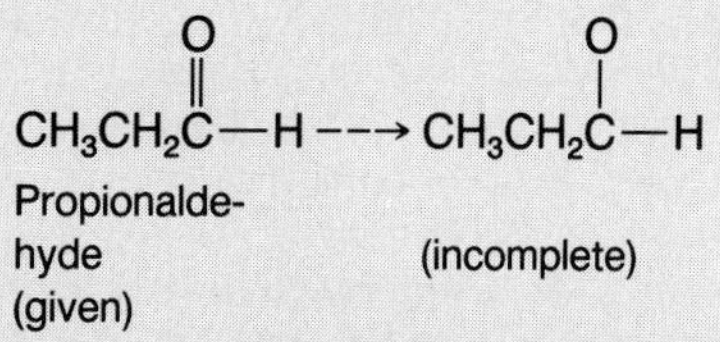

$$
\underset{\substack{\text{Propionalde-}\\\text{hyde}\\\text{(given)}}}{CH_3CH_2\overset{\overset{\displaystyle O}{\|}}{C}\!-\!H} \dashrightarrow \underset{\text{(incomplete)}}{CH_3CH_2\overset{\overset{\displaystyle O}{|}}{C}\!-\!H}
$$

Then look at the structure of the given alcohol, CH_3CH_2OH. The H on its oxygen atom ends up on the oxygen atom of the developing structure:

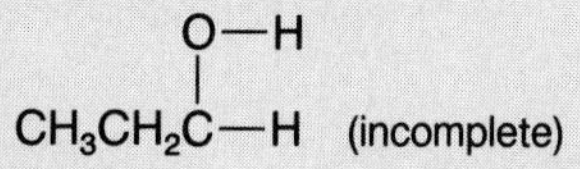

$$
\underset{\text{(incomplete)}}{CH_3CH_2\overset{\overset{\displaystyle O-H}{|}}{C}\!-\!H}
$$

Finally, the entire remainder of the alcohol molecule is attached by its oxygen atom to the carbon atom that presently has the O—H group in our developing structure:

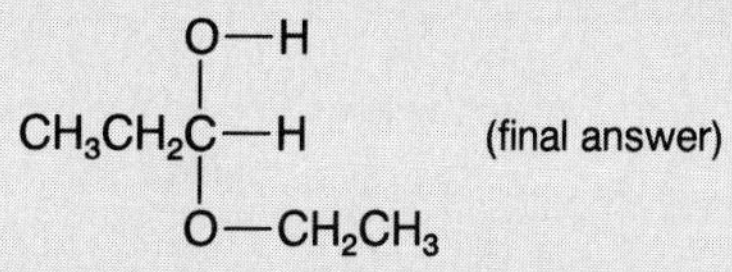

$$
\underset{\underset{\displaystyle O-CH_2CH_3}{|}}{CH_3CH_2\overset{\overset{\displaystyle O-H}{|}}{C}\!-\!H} \quad \text{(final answer)}
$$

We can leave the answer in this form, or we can redraw it to condense it a little more:

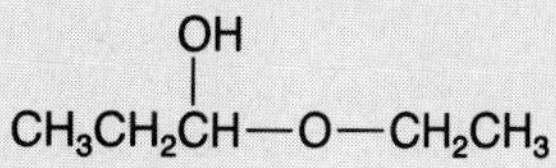

$$
CH_3CH_2\overset{\overset{\displaystyle OH}{|}}{C}H\!-\!O\!-\!CH_2CH_3
$$

PRACTICE EXERCISE 3   Write the structures of the hemiacetals that are present in the equilibria that involve the following pairs of compounds.

(a) Acetaldehyde and methyl alcohol   (b) Butyraldehyde and ethyl alcohol

(c) Benzaldehyde and propyl alcohol   (d) Formaldehyde and methyl alcohol

---

Still another skill needed for our study of carbohydrates is the ability to write the structures of the aldehyde and alcohol that are liberated by the breakdown of a hemiacetal.

EXAMPLE 12.3 WRITING THE BREAKDOWN PRODUCTS OF A HEMIACETAL

Problem: What aldehyde and alcohol form when the following hemiacetal breaks down?

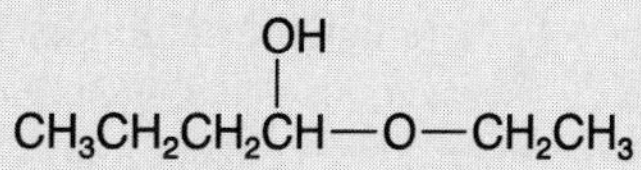

$$
CH_3CH_2CH_2\overset{\overset{\displaystyle OH}{|}}{C}H\!-\!O\!-\!CH_2CH_3
$$

Solution: This is the kind of problem where the ability to pick out the carbon atom of the original carbonyl group is especially helpful. *Remember:* To find this carbon we look for one that holds both a O—H group and a O—R system. When we find this carbon, we break its bond to the O—R unit and separate the pieces. (This was the bond that formed when the alcohol added to the aldehyde to form the hemiacetal system.) *Do not break any other bond in the given hemiacetal.*

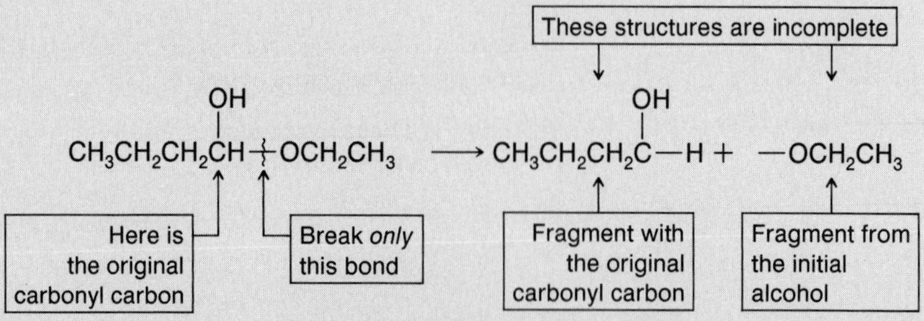

These structures are incomplete

$$CH_3CH_2CH_2CH \overset{OH}{|} \!\!+\!\! OCH_2CH_3 \longrightarrow CH_3CH_2CH_2\overset{OH}{\underset{|}{C}}\!-\!H + \quad -OCH_2CH_3$$

| Here is the original carbonyl carbon | Break *only* this bond | Fragment with the original carbonyl carbon | Fragment from the initial alcohol |

Now we have to fix the fragment with the carbon of the original C=O group into a structure with an actual C=O group. For this, we move the H atom on the fragment's oxygen over to the O atom of the other fragment. (This gives us the original alcohol molecule.) Then we write a carbon–oxygen double bond to make the carbonyl group, and we have the original aldehyde. The answer, then, is

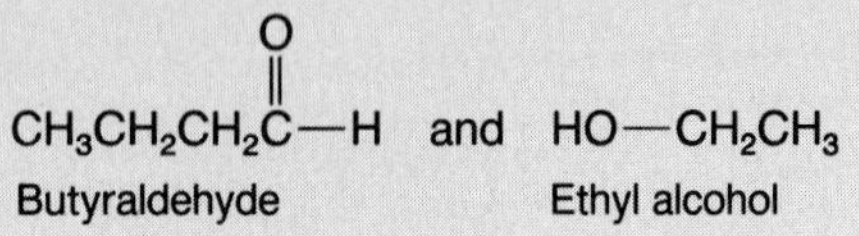

$$\underset{\text{Butyraldehyde}}{CH_3CH_2CH_2\overset{\overset{O}{\|}}{C}\!-\!H} \quad \text{and} \quad \underset{\text{Ethyl alcohol}}{HO\!-\!CH_2CH_3}$$

PRACTICE EXERCISE 4   Write the structures of the breakdown products of the following hemiacetals.

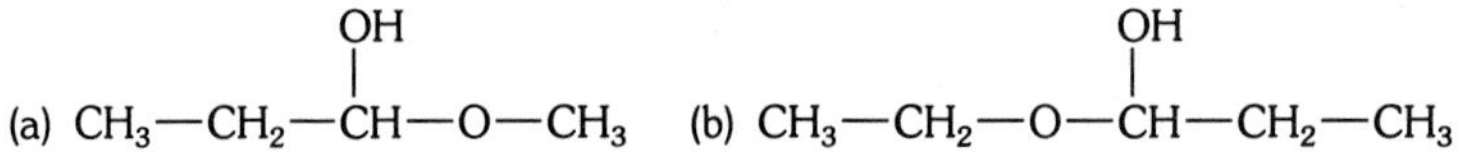

(a) $CH_3\!-\!CH_2\!-\!\overset{OH}{\underset{|}{CH}}\!-\!O\!-\!CH_3$ (b) $CH_3\!-\!CH_2\!-\!O\!-\!\overset{OH}{\underset{|}{CH}}\!-\!CH_2\!-\!CH_3$

The hemiketal system:

$$\overset{O-H}{\underset{\underset{O-R}{|}}{C\!-\!C\!-\!C}}$$

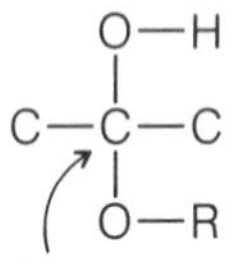

This originally was the carbon atom of a keto group

When Alcohols Add to Ketones, Hemiketals Form. Ketones behave very much like aldehydes toward alcohols. They add alcohols to form equilibria that contain a structure almost identical to a hemiacetal, only now it's called a **hemiketal.** Like the hemiacetal, the hemiketal has a carbon atom, originally the carbonyl carbon atom of the parent ketone, which holds both O—H and O—R.

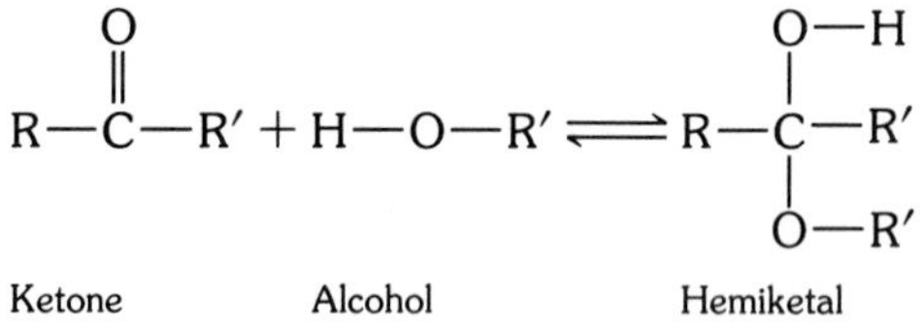

$$\underset{\text{Ketone}}{R\!-\!\overset{\overset{O}{\|}}{C}\!-\!R'} + \underset{\text{Alcohol}}{H\!-\!O\!-\!R'} \rightleftharpoons \underset{\text{Hemiketal}}{R\!-\!\overset{\overset{O-H}{|}}{\underset{\underset{O-R'}{|}}{C}}\!-\!R'}$$

Among carbohydrates fructose (levulose) has a hemiketal system.

Acetals and Ketals Are 1,1-Diether Systems Made from Hemiacetals or Hemiketals and Alcohols. Hemiacetals and hemiketals are special kinds of alcohols, and they resemble alcohols in one important property. They can be converted into ethers. They are not ordinary ethers but special kinds, 1,1-diethers, called **acetals** and **ketals.** The overall change that leads to an acetal is as follows:

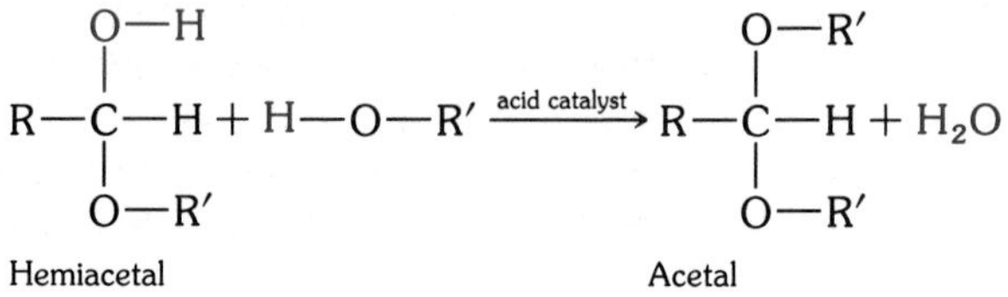

$$\underset{\text{Hemiacetal}}{R\!-\!\overset{\overset{O-H}{|}}{\underset{\underset{O-R'}{|}}{C}}\!-\!H} + H\!-\!O\!-\!R' \xrightarrow{\text{acid catalyst}} \underset{\text{Acetal}}{R\!-\!\overset{\overset{O-R'}{|}}{\underset{\underset{O-R'}{|}}{C}}\!-\!H} + H_2O$$

Hemiketals give the identical kind of reaction, but the products are called ketals. Unlike hemiacetals and hemiketals, both acetals and ketals are stable compounds that can be isolated and stored.

One difference between the formation of an acetal and an ordinary ether is that acetals form more readily. As we have already suggested, when two functional groups are very close to each other in a molecule, each modifies the properties of the other. Here, the OR group makes the nearby OH group much more reactive toward forming an ether.

EXAMPLE 12.4   FORMING AN ACETAL

Problem: What acetal can be made from acetaldehyde and methyl alcohol?

Solution: First, write the structure of the aldehyde. Then erase one of the two bonds in the carbonyl group:

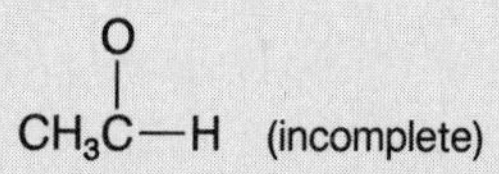

$$CH_3\overset{\overset{\textstyle O}{|}}{C}-H \quad \text{(incomplete)}$$

The structure of the acetal we seek must have two C—O bonds at the site of the original carbonyl carbon. One C—O bond is in place, so we now add another. (It will be provided by a molecule of methyl alcohol.)

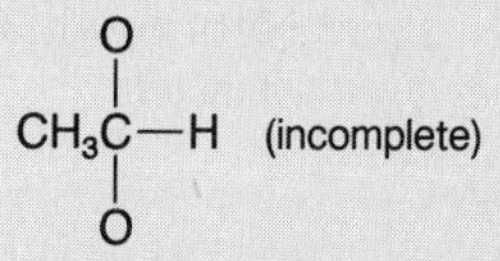

$$CH_3\overset{\overset{\textstyle O}{|}}{\underset{\underset{\textstyle O}{|}}{C}}-H \quad \text{(incomplete)}$$

Now use *two* alkyl groups from the original alcohol, methyl groups in our example. Attach one to each oxygen to get the structure of the acetal. The balanced equation for its formation is

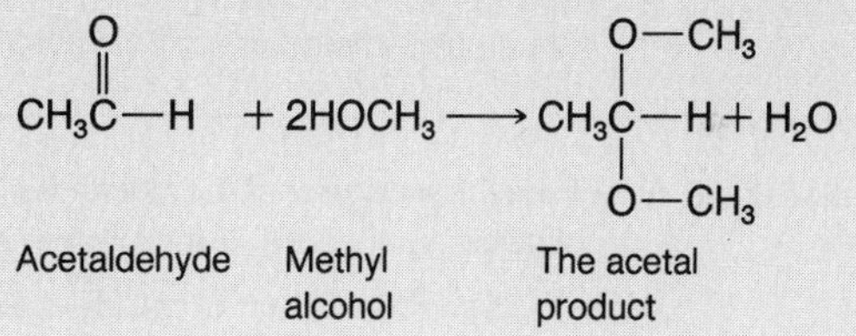

$$CH_3\overset{\overset{\textstyle O}{||}}{C}-H \;+\; 2HOCH_3 \;\longrightarrow\; CH_3\overset{\overset{\textstyle O-CH_3}{|}}{\underset{\underset{\textstyle O-CH_3}{|}}{C}}-H + H_2O$$

| Acetaldehyde | Methyl alcohol | The acetal product |

To make a ketal from a ketone and an alcohol, we'd follow the identical steps. The only difference would be an extra alkyl group in place of the aldehydic hydrogen atom.

PRACTICE EXERCISE 5   Write the structure of the acetal or ketal that could be made from each set of reactants.

(a) Propionaldehyde and methyl alcohol   (b) Acetone and propyl alcohol

---

Acids and Enzymes Catalyze the Hydrolysis of Acetals and Ketals.   Acetals and ketals are stable, as we said, but only if they are kept out of contact with aqueous acids or certain enzymes. These catalysts promote the hydrolysis of acetals or ketals to their parent alcohols and carbonyl compounds, aldehydes or ketones. We study this, the only chemical reaction of acetals and ketals we need to know, because it is the chemistry of the digestion of carbohydrates.

EXAMPLE 12.5   WRITING THE STRUCTURES OF THE PRODUCTS OF THE HYDROLYSIS OF ACETALS OR KETALS

Problem: What are the products of the following reaction?

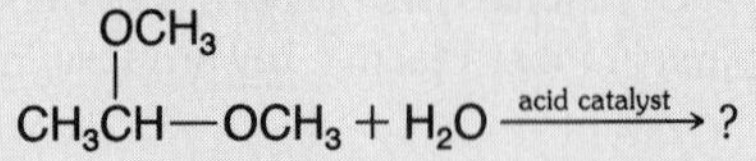

$$\underset{\displaystyle CH_3CH-OCH_3}{\overset{\displaystyle OCH_3}{|}} + H_2O \xrightarrow{\text{acid catalyst}} ?$$

Solution: First, we should double check that the given structure is of an acetal (or ketal) and isn't just something with two ordinary ether groups. So we look for a single carbon with *two* ether groups from it. Because we can find it, we've also found what was the original carbonyl carbon atom of the parent aldehyde (or ketone). Break both of its bonds to these oxygen atoms. *Do not break any other bonds.* Separate the fragments for further processing.

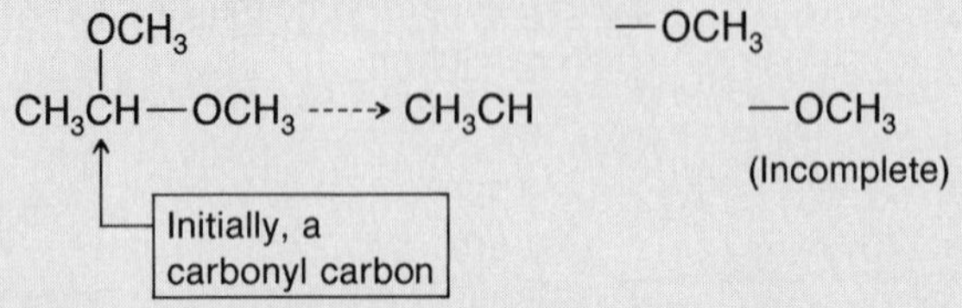

$$\underset{\displaystyle CH_3CH-OCH_3}{\overset{\displaystyle OCH_3}{|}} \dashrightarrow CH_3CH \qquad \begin{array}{l} -OCH_3 \\ -OCH_3 \\ \text{(Incomplete)} \end{array}$$

Initially, a carbonyl carbon

Next, we finish writing the carbonyl group where it belongs, at the identified carbon atom. Then we place H atoms on the oxygen atoms of the other fragments to finish writing the structures of the alcohol molecules that also form. The final products of the hydrolysis of the given acetal are

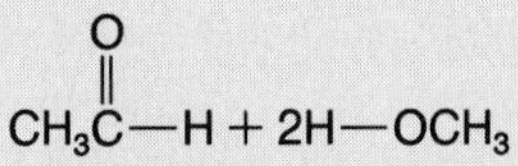

$$\underset{\displaystyle CH_3\overset{\displaystyle O}{\overset{\|}{C}}-H}{} + 2H-OCH_3$$

Notice that the oxygen atom for the new carbonyl group comes from the water molecule that hydrolyzed the acetal. The two hydrogen atoms needed to complete the structures of the two alcohol molecules also come from the water molecule.

PRACTICE EXERCISE 6   Write the structures of the aldehydes (or ketones) and the alcohols that are obtained by hydrolyzing the following compounds. If they do not hydrolyze like acetals or ketals, write "no reaction."

(a) $CH_3-O-CH_2-O-CH_3$ (b) $CH_3-O-CH_2-CH_2-O-CH_2-CH_3$

(c) 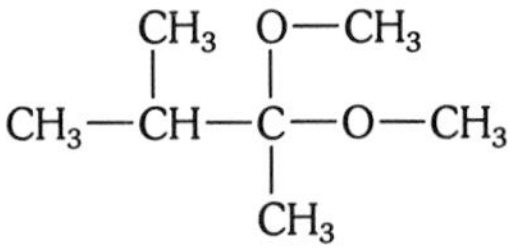 $CH_3-CH-\underset{\displaystyle CH_3}{\overset{\displaystyle CH_3}{\underset{|}{\overset{|}{C}}}}-O-CH_3$ with $O-CH_3$

SUMMARY

Physical properties of aldehydes and ketones The carbonyl group confers moderate polarity, which gives aldehydes and ketones higher boiling points and solubilities in water than hydrocarbons but lower boiling points and solubilities in water than alcohols (that have comparable formula weights).

Chemical properties of aldehydes and ketones Aldehydes are easily oxidized to carboxylic acids, but ketones resist oxidation. Aldehydes give a positive Tollen's test and ketones do not. α-Hydroxy aldehydes and ketones give the Benedict's test.

When an aldehyde or a ketone is dissolved in an alcohol, some of the alcohol adds to the carbonyl group of the aldehyde or ketone. An equilibrium forms that includes molecules of a hemiacetal (or hemiketal). The chart at the end of this summary outlines the chemical properties of the aldehydes and ketones we have studied.

Hemiacetals and hemiketals Hemiacetals and hemiketals are usually unstable compounds that exist only in an equilibrium involving the parent carbonyl compound and the parent alcohol (which generally is the solvent). Hemiacetals and hemiketals readily break back down to their parent carbonyl compounds and alcohols. When an acid catalyst is added to the equilibrium, a hemiacetal or hemiketal reacts with more alcohol to form an acetal or ketal.

Acetals and ketals Acetals and ketals are 1,1-diethers that are

stable in aqueous base or in water, but not in aqueous acid. Acids catalyze the hydrolysis of acetals and ketals, and the final products are the parent aldehydes (or ketones) and alcohols.

Aldehydes

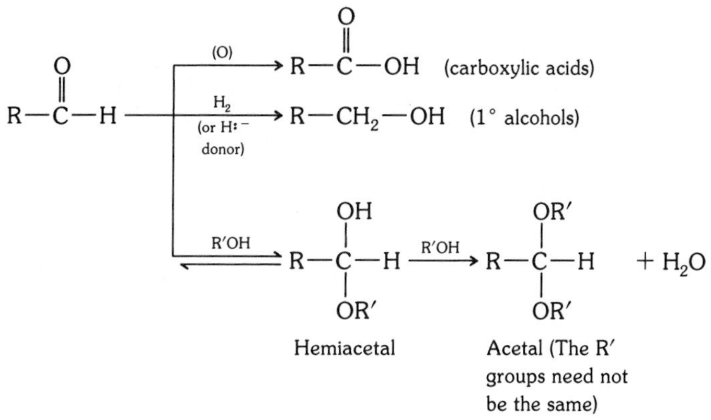

Ketones

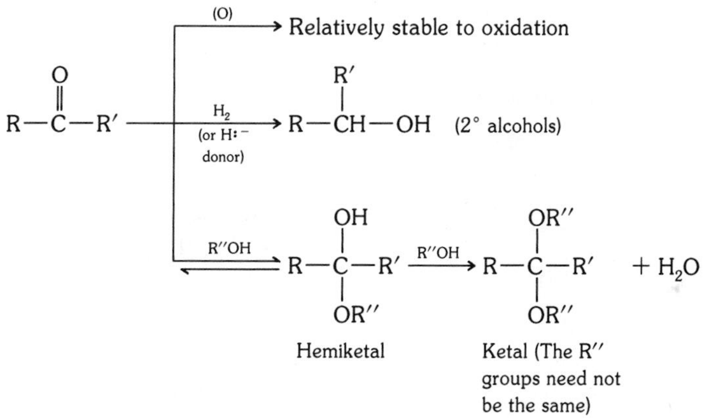

Acetals and ketals

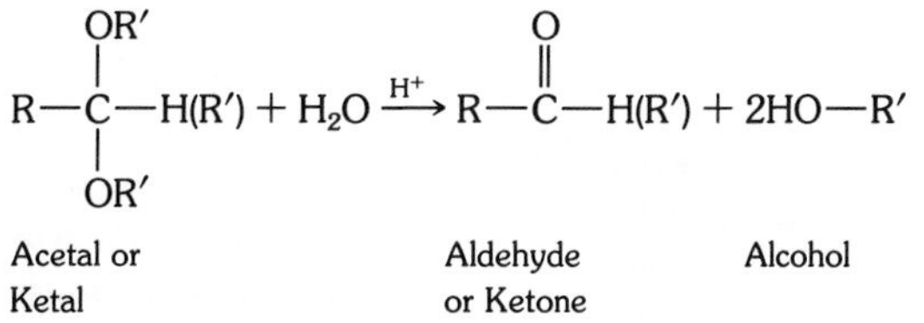

REVIEW EXERCISES

The answers to these review exercises are in the *Study Guide* that accompanies this book.

Names and Structures

12.1 What is the *structural* difference between an aldehyde and a ketone?

12.2 What are the structures of the following compounds?
(a) Butyraldehyde (b) Acetaldehyde
(c) Propionic acid (d) Formic acid

12.3 Write the structures of the following compounds.
(a) Acetic acid (b) Propionaldehyde
(c) Formaldehyde (d) Butyric acid

12.4 Give the common names of the following:

(a)

$$C_6H_5\overset{\displaystyle O}{\overset{\displaystyle \|}{C}}\text{—H}$$

(b) CH_3CHO

(c) $CH_3CH_2CH_2CO_2H$ (d) CH_3CH_2CHO

12.5 Write the common names of these compounds.

(a) HCO_2H (b) C_6H_5CHO

(c) $CH_3CH_2CO_2H$ (d) $HCHO$

12.6 If the common name of $CH_3CH_2CH_2CH_2CO_2H$ is valeric acid, then what is the most likely common name of the following compound?

$$CH_3CH_2CH_2CH_2CH{=}O$$

12.7 If the common name of the following compound:

is anisaldehyde, then what is the most likely common name of

Physical Properties of Aldehydes and Ketones

12.8 Arrange the following compounds in their order of increasing boiling points. Do this by placing the letters that identify them in the correct order, starting with the lowest-boiling compound and moving in order to the highest-boiling compound.

A

B

C

D

12.9 Arrange the following compounds in their order of increasing boiling points. Do this by placing the letters that identify them in the correct order, starting with the lowest-boiling compound on the left in the series and moving to the highest-boiling compound.

A   B   C   D

12.10 Reexamine the compounds of Review Exercise 12.8, and arrange them in their order of increasing solubility in water.

12.11 Arrange the compounds of Review Exercise 12.9 in their order of increasing solubility in water.

12.12 Draw the structure of a water molecule and an acetone molecule and align them on the page to show how the acetone molecule can accept a hydrogen bond from the water molecule. Use a dotted line to represent this hydrogen bond and place $\delta+$ and $\delta-$ symbols where they are appropriate.

12.13 Draw the structures of molecules of methyl alcohol and acetaldehyde, and align them on the page to show how a hydrogen bond (which you are to indicate by a dotted line) can exist between the two. Place $\delta+$ and $\delta-$ symbols where they are appropriate.

Oxidation of Alcohols and Aldehydes

12.14 What are the structures of the aldehydes and ketones to which the following compounds can be oxidized?

(a)
$$HO{-}\overset{\displaystyle CH_3}{\underset{\displaystyle |}{}}CHCH_2CH_3$$

(b)

(c) $C_6H_5CH_2OH$

(d)
$$CH_3\overset{\displaystyle CH_3}{\underset{\displaystyle \underset{\displaystyle CH_3}{|}}{\overset{\displaystyle |}{C}}}CH_2OH$$

12.15 Examine each of the following compounds to see if it can be oxidized to an aldehyde or to a ketone. If it can, write the structure of the aldehyde or ketone.

(a) CH_3CH_2OH

(b)
$$CH_3\overset{\displaystyle OH}{\underset{\displaystyle |}{}}CH{-}OH$$

(c)

(d)
$$CH_3\overset{\displaystyle OH}{\underset{\displaystyle |}{}}CHCH_2\overset{\displaystyle O}{\overset{\displaystyle \|}{C}}OH$$

12.16 An unknown compound, C_3H_6O, reacted with permanganate ion to give $C_3H_6O_2$, and the same unknown also gave a positive Tollens' test. Write the structures of C_3H_6O and $C_3H_6O_2$.

12.17 An unknown compound, $C_3H_6O_2$, could be oxidized easily by permanganate ion to $C_3H_4O_3$, and it gave a positive Benedict's test. Write structures for $C_3H_6O_2$ and $C_3H_4O_3$.

12.18 Which of the following compounds can be expected to give a positive Benedict's test? All are intermediates in metabolism.

(a)
$$HOCH_2\overset{\displaystyle O}{\overset{\displaystyle \|}{C}}H\underset{\displaystyle HO}{\underset{\displaystyle |}{C}}H$$

(b)
$$HOCH_2\overset{\displaystyle O}{\overset{\displaystyle \|}{C}}CH_2OH$$

(c)
$$CH_3\underset{\displaystyle \underset{\displaystyle OH}{|}}{CH}CH_2\overset{\displaystyle O}{\overset{\displaystyle \|}{C}}OH$$

(d)
$$CH_3\overset{\displaystyle O}{\overset{\displaystyle \|}{C}}CH_2\overset{\displaystyle O}{\overset{\displaystyle \|}{C}}OH$$

12.19 Which of the following compounds give a positive Benedict's test? (Most are intermediates in metabolism.)

(a) $\underset{\underset{HO}{|}}{HOCH_2CH}-\underset{\underset{OH}{|}}{CH}\overset{\overset{O}{\parallel}}{C}H$ (b) $HOCH_2CH_2\overset{\overset{O}{\parallel}}{C}CH_3$

(c) $CH_3\overset{\overset{O}{\parallel}}{C}-\overset{\overset{O}{\parallel}}{C}H$ (d) $HO\overset{\overset{O}{\parallel}}{C}-\overset{\overset{O}{\parallel}}{C}CH_3$

12.20 In what form does silver occur in the Tollens' reagent?

12.21 What is the function of the citrate ion in the Benedict's reagent?

12.22 What is the formula of the precipitate that forms in a positive Benedict's test?

12.23 Clinitest® tablets are used for what?

12.24 What is one practical commercial application of the Tollens' test?

12.25 One of the steps in the metabolism of fats and oils in the diet is the oxidation of the following compound:

$$CH_3\underset{\underset{OH}{|}}{CH}CH_2CO_2^-$$

Write the structure of the product of this oxidation.

Reduction of Aldehydes and Ketones

12.26 Consider the reaction that occurs when a hydride ion transfers from some hydride ion donor (which we can write as $M\colon H$) to acetaldehyde. (We will see several examples of this kind of reaction in our later study of metabolism.) What is the structure of the product (assuming that H^+ is also available)?

12.27 If a hydride ion donor ($M\colon H$) transfers its hydride ion to a molecule of acetone, what is the structure of the product (assuming that H^+ is also available)?

12.28 The metabolism of aspartic acid, an amino acid, occurs by a succession of steps, one of which is indicated as follows:

$$^+NH_3-\underset{\underset{CH_2CO_2^-}{|}}{CH}-CO_2^-\xrightarrow{\text{two steps}}$$

Aspartate ion

$$^+NH_3-\underset{\underset{CH_2CH=O}{|}}{CH}-CO_2^-\xrightarrow[\text{then } H^+]{NAD\colon H}\ ?$$

Write the structure of the product of the last step.

12.29 One of the steps the body uses to make long-chain carboxylic acids involves a reaction similar to the following reaction.

$$CH_3\overset{\overset{O}{\parallel}}{C}CH_2\overset{\overset{O}{\parallel}}{C}-S\text{-enzyme} + NAD\colon H\xrightarrow[\text{(then } H^+\text{)}]{}$$

$$?-CH_2-\overset{\overset{O}{\parallel}}{C}-S\text{-enzyme} + NAD^+$$
$$\mathbf{I}$$

Complete the structure of **I**.

12.30 Write the structures of the aldehydes or ketones that could be used to make the following compounds by reduction (hydrogenation).

(a) [cyclopentane ring with OH]

(b) $CH_3\underset{\underset{OH}{|}}{CH}CH_3$

(c) $C_6H_5CH_2OH$ (d) [cyclohexane ring]$-CH_2OH$

Hemiacetals and Acetals. Hemiketals and Ketals

12.31 Examine each structure and decide if it represents a hemiacetal, hemiketal, acetal, ketal, or something else.

(a) $CH_3-O-\underset{\underset{CH_3}{|}}{CH}-OH$

(b) $CH_3CH_2\underset{\underset{O-CH_3}{|}}{CH}-O-CH_3$

(c) $CH_3-O-\underset{\underset{CH_2-O-CH_3}{|}}{CH}-CH_3$

(d) $CH_3-O-\underset{\underset{O-CH_3}{|}}{C}(CH_3)_2$

12.32 Examine each structure and decide if it represents a hemiacetal, hemiketal, acetal, ketal, or something else.

(a) $CH_3-\underset{\underset{OH}{|}}{CH}CH_2-O-CH_3$

(b) $HOCH_2OCH_3$

(c) $HOCH_2\underset{\underset{OCH_3}{|}}{CH}OCH_3$

(d) $\begin{matrix} CH_2-O \\ | \qquad\qquad CH-O-CH_3 \\ CH_2-CH_2 \end{matrix}$

12.33 Either an aldehyde or a ketone could be used to make each of the following compounds by hydrogenation. Write the structure of the aldehyde or ketone suitable in each part.

(a) CH_3-O-[benzene ring]$-\underset{\underset{OH}{|}}{CH}CH_3$

(b) 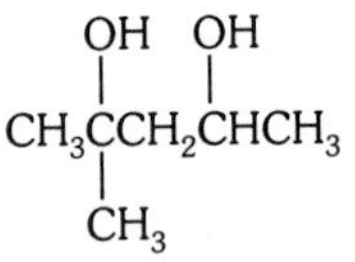
$$CH_3\overset{\overset{\displaystyle OH}{|}}{C}CH_2\overset{\overset{\displaystyle OH}{|}}{C}HCH_3$$
$$\underset{\underset{\displaystyle CH_3}{|}}{}$$

(c) $CH_3CH_2OCH_2CH_2OH$

(d) 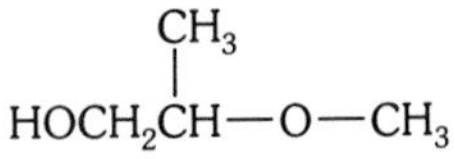 $HOCH_2\overset{\overset{\displaystyle CH_3}{|}}{C}H-O-CH_3$

12.34 Write the structures of the hemiacetals and the acetals that can form between acetaldehyde and the following alcohols.
(a) Methyl alcohol   (b) Ethyl alcohol

12.35 What are the structures of the hemiketals and the ketals that can form between acetone and the following alcohols?
(a) Methyl alcohol   (b) Ethyl alcohol

12.36 Write the structure of the hydroxyaldehyde (a compound having both the alcohol group and the aldehyde group in the same molecule) from which the following hemiacetal forms in a ring-closing reaction. (You may leave the chain of the open-chain compound coiled somewhat.)

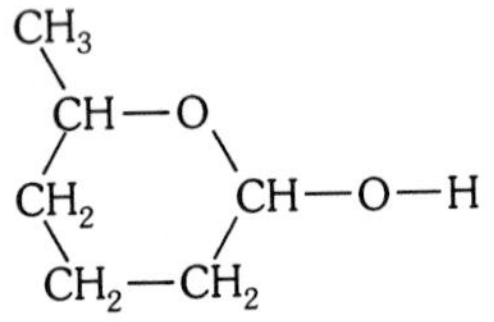

(a) Draw an arrow that points to the hemiacetal carbon.
(b) Write the structure of the open-chain form that has a free aldehyde group. (You may leave the chain coiled.)

12.37 One form in which a glucose molecule exists is given by the following structure. (*Note:* The atoms and groups that are attached to the carbon atoms of the six-membered ring must be seen as projecting *above* or *below* the ring.)

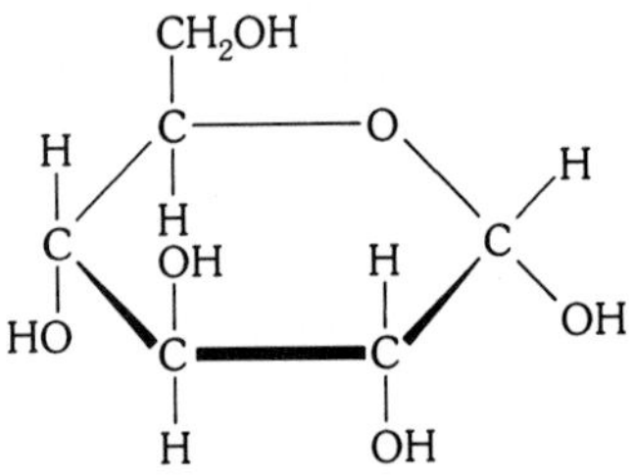

12.38 Write the structure of a hydroxy ketone (a molecule that has both the OH group and the keto group) from which the following hemiketal forms in a ring-closing reaction. (You may leave the chain of the open-chain compound somewhat coiled.)

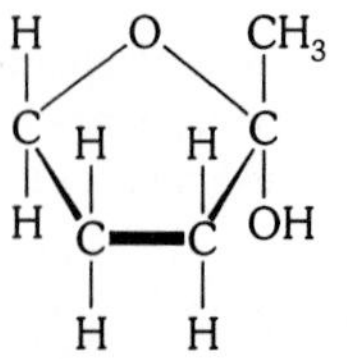

(a) Draw an arrow to the carbon of the hemiketal system that came initially from the carbon atom of a keto group.
(b) In water, fructose exists in an equilibrium with an open-chain form of the given structure. This form has a keto group in the same molecule as five OH groups. Draw the structure of this open-chain form (leaving the chain coiled somewhat as it is in the structure that was given).

12.39 Fructose occurs together with glucose in honey, and it is sweeter to the taste than table sugar. One form in which a fructose molecule can exist is given by the following structure.

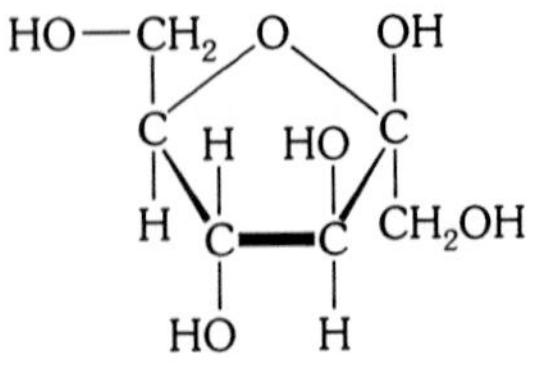

12.40 The digestion of some carbohydrates is simply their hydrolysis catalyzed by enzymes. Acids catalyze the same kind of hydrolysis of acetals and ketals. Write the structures of the products, if any, that form by the action of water and an acid catalyst on the following compounds.

(a) 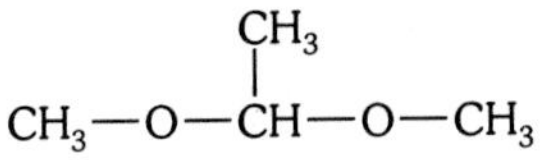 $CH_3-O-\overset{\overset{\displaystyle CH_3}{|}}{C}H-O-CH_3$

(b) $CH_3-O-\overset{\overset{\displaystyle CH_3}{|}}{C}H-CH_2-O-CH_3$

(c) $CH_3-O-\overset{\overset{\displaystyle CH_3}{|}}{\underset{\underset{\displaystyle CH_3}{|}}{C}}-O-CH_3$

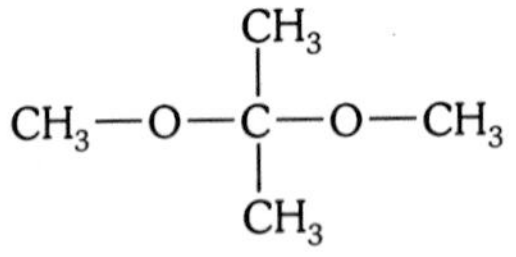

(d)

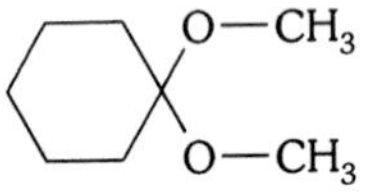

12.41 What are the structures of the products, if any, of the action of water that contains a trace of acid catalyst on the following compounds?

(a) 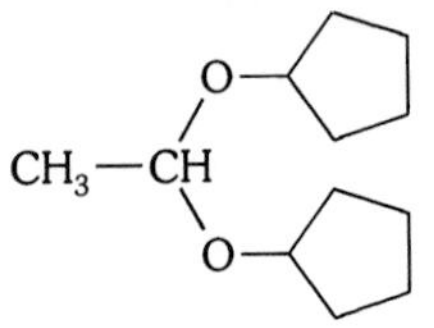   (b)

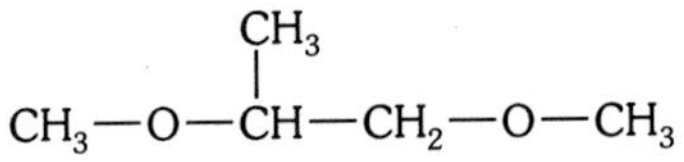

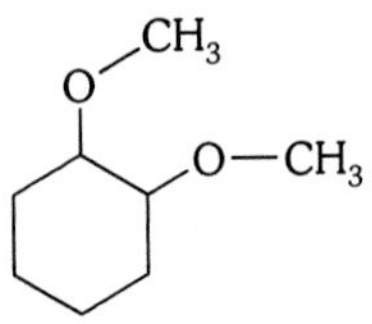

(c) 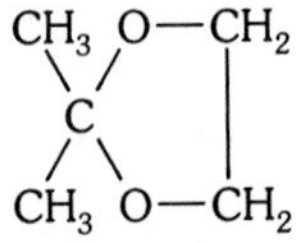   (d) 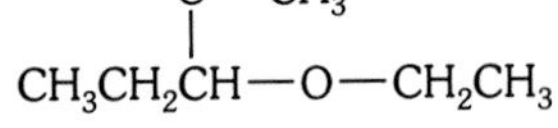 $CH_3CH_2\overset{\overset{\displaystyle O-CH_3}{|}}{C}H-O-CH_2CH_3$

12.42 Complete the following reaction sequences by writing the structures of the organic products that form. If no reaction occurs, write "no reaction." (Reviewed here are reactions of earlier chapters, too.)

(a) $CH_3CH{=}CHCH_3 + H_2 \xrightarrow[\text{heat, pressure}]{Ni}$

(b) $CH_3\overset{OH}{\underset{|}{C}}HCH_3 + (O) \rightarrow$

(c) $CH_3{-}O{-}\overset{OCH_3}{\underset{|}{C}}HCH_3 + H_2O \xrightarrow{H^+}$

(d) $CH_3\overset{O}{\overset{||}{C}}H + H_2 \xrightarrow[\text{heat, pressure}]{Ni}$

(e) $CH_3CH_2\overset{O}{\overset{||}{C}}H + (O) \rightarrow$

(f) $CH_3{-}\overset{CH_3}{\underset{\underset{CH_3}{|}}{\overset{|}{C}}}{-}OH + (O) \rightarrow$

(g) $CH_3OH + CH_3CH_2\overset{O}{\overset{||}{C}}H \rightleftharpoons$

(h) $CH_3CH_2{-}O{-}CH_2CH_3 + (O) \rightarrow$

(i) $C_6H_5\overset{CH_3}{\underset{|}{C}}H{-}O{-}CH_2CH_3 \rightleftharpoons$

(j) $CH_3\overset{O}{\overset{||}{C}}H + 2CH_3OH \xrightarrow{\text{acid catalyst}}$

12.43 Write the structures of the organic products that form in each of the following situations. If no reaction occurs, write "no reaction." (Some of the situations constitute a review of earlier chapters.)

(a) $C_6H_5\overset{O}{\overset{||}{C}}H + (O) \rightarrow$

(b) [cyclopentane ring with OH and CH_3] $+ (O) \rightarrow$

(c) $CH_3\overset{O}{\overset{||}{C}}H + CH_3CH_2CH_2OH \rightleftharpoons$

(d) $CH_3{-}O{-}CH_2CH_2{-}O{-}CH_3 + H_2O \xrightarrow{H^+}$

(e) $CH_3CH_2{-}O{-}\overset{OH}{\underset{\underset{CH_3}{|}}{\overset{|}{C}}}{-}CH_3 \rightleftharpoons$

(f) $CH_3CH_2\overset{O}{\overset{||}{C}}H + H_2 \xrightarrow[\text{heat, pressure}]{Ni}$

(g) $CH_3CH_2\overset{O{-}CH_2CH_3}{\underset{\underset{CH_3}{|}}{\overset{|}{C}}}{-}O{-}CH_2CH_3 + H_2O \xrightarrow{H^+}$

(h) $C_6H_5\overset{OH}{\underset{|}{C}}HCH_3 + (O) \rightarrow$

(i) $CH_3{-}O{-}CH_2CH{=}CH_2 + H_2 \xrightarrow[\text{heat, pressure}]{Ni}$

(j) $CH_3CH_2\overset{O}{\overset{||}{C}}H + 2CH_3OH \xrightarrow{H^+}$

12.44 Catalytic hydrogenation of compound **A**, C_3H_6O, gave **B**, C_3H_8O. When **B** was heated strongly in the presence of sulfuric acid, it changed to compound **C**, C_3H_6. The acid-catalyzed addition of water to **C** gave compound **D**, C_3H_8O; and when **D** was oxidized, it changed to **E**, C_3H_6O. Compounds **A** and **E** are isomers, and compounds **B** and **D** are isomers. Write the structures of **A**–**E**.

12.45 When compound **F**, $C_4H_{10}O$ was gently oxidized, it changed to compound **G**, C_4H_8O, but vigorous oxidation changed **F** (or **G**) to compound **H**, $C_4H_8O_2$. Action of hot sulfuric acid on **F** changed it to compound **I**, C_4H_8. The addition of water to **I** (in the presence of an acid catalyst) gave compound **J**, $C_4H_{10}O$, a compound that could not be oxidized. Compounds **F** and **J** are isomers. Write the structures of **F**–**J**.

Important Aldehydes and Ketones (Special Topic 12.1)

12.46 The odor of what compound can be detected on the breath of someone with advanced but untreated diabetes?

12.47 What aldehyde or ketone was once commonly used to preserve biological specimens?

12.48 The female hormone estrone has two functional groups. What must one of them be, and why?

12.49 Dihydroxyacetone phosphate is either compound **A** or **B**. Even if you have never seen its structure before or memorized it, you can tell which one it is. Which is it, and how can you tell?

$HOCH_2\overset{O}{\overset{||}{C}}CH_2OPO_3{}^{2-}$ $CH_3\overset{O}{\overset{||}{C}}\overset{}{\underset{\underset{OH}{|}}{C}}HOPO_3{}^{2-}$

A   **B**

12.50 When using "superglue" it's a good idea to have a supply of what compound handy? Why? How can a supply be easily purchased?

Chapter 13
Carboxylic Acids and Their Derivatives

Occurrence and Structural Features

Chemical Properties of Carboxylic Acids and Their Salts

Esters of Carboxylic Acids

Esters of Phosphoric Acid

Amides

Soaring over Yosemite Valley near Half-Dome, this hang glider trusts his life to the strength of a synthetic fabric. One topic in this chapter is how chemists have been able to convert small, simple molecules into the enormous molecules of such synthetics.

13.1 OCCURRENCE AND STRUCTURAL FEATURES

The functional groups studied in this chapter occur throughout all metabolism.

Anywhere we care to look in the living world we find compounds with the functional groups studied in this chapter. The carboxyl group or its anion, the carboxylate ion, is in all amino acids and proteins, all fatty acids (used to make fats and oils), and many of the metabolic intermediates of carbohydrates. The ester group is the chief functional group in all fats and oils. The amide group contributes every third bond to the "backbones" of all proteins.

"Carboxyl" comes from *carbonyl* + *hydroxyl*.

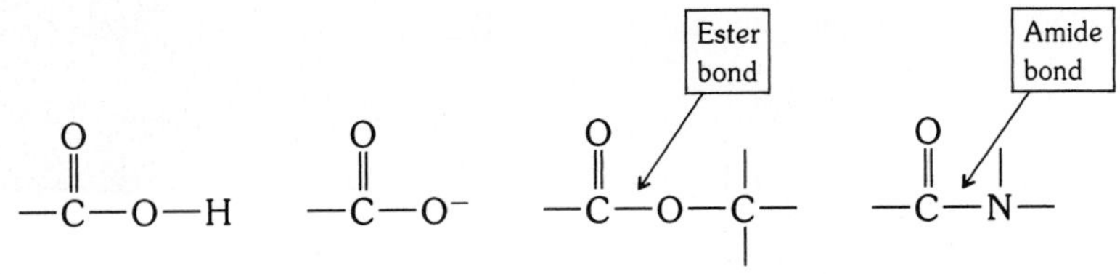

Carboxyl group | Carboxylate ion | Ester group | Amide group

Common abbreviations of the carboxyl group are CO_2H and $COOH$.

The Names of All Acid Derivatives Are Derived from the Names of the Acids.
Salts of carboxylic acids, esters, and amides are called *derivatives of carboxylic acids* because they can be made from the acids and changed back to the acids by relatively simple reactions. Since the word parts for the names of all of these derivatives are related, they should now be learned. Table 13.1 shows how these names are formed.

A number of common **carboxylic acids** are given in Table 13.2, including several obtained by the hydrolysis (digestion) of fats and oils. As we noted in Chapter 12, the common names of the acids originated in early sources of these compounds. Thus valeric acid gets its name from the Latin *valerum,* meaning to be strong. What is strong about valeric acid is its odor. All of the acids with 3–10 carbon atoms have vile odors, the odors of long unwashed athletic socks or stale locker rooms.

The simple acids with 4–20 or more carbon atoms are called the **fatty acids.**

Molecules of Carboxylic Acids Form Hydrogen Bonds between Them.
Carboxylic acids have higher boiling points than alcohols of comparable formula weights. The reason is that molecules of carboxylic acids form hydrogen-bonded pairs:

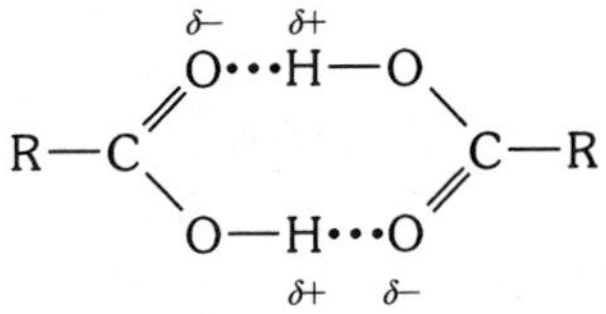

TABLE 13.1
Making Common Names for Acids and Their Derivatives

Class	Characteristic Name Ending of the Class	Characteristic Prefix in the Acid Name			
		C_1 Form-	C_2 Acet-	C_3 Propion-	C_4 Butyr-
Carboxylic acids	-ic acid	Formic acid	Acetic acid	Propionic acid	Butyric acid
Carboxylate ion	-ate (ion)	Formate ion	Acetate ion	Propionate ion	Butyrate ion
Ester	-ate	(Alkyl) formate[a]	(Alkyl) acetate	(Alkyl) propionate	(Alkyl) butyrate
Amide	-amide	Formamide	Acetamide	Propionamide	Butyramide

[a] The name of the specific alkyl group would be used in the name of a specific ester.

TABLE 13.2
Carboxylic Acids

n	Structure	Name[a]	Origin of Name[b]	Melting Point (°C)	Boiling Point (°C)	Solubility (g/100 g H_2O at 20 °C)
Straight-Chain Saturated Acids, $C_nH_{2n}O_2$						
1	HCO_2H	Formic acid (methanoic acid)	L. *formica*, ant	8	101	Soluble
2	CH_3CO_2H	Acetic acid (ethanoic acid)	L. *acetum*, vinegar	17	118	Soluble
3	$CH_3CH_2CO_2H$	Propionic acid (propanoic acid)	L. *proto*, *pion*	−21	141	Soluble
4	$CH_3(CH_2)_2CO_2H$	Butyric acid (butanoic acid)	L. *butyrum*, butter	−6	164	Soluble
5	$CH_3(CH_2)_3CO_2H$	Valeric acid (pentanoic acid)	L. *valere*, to be strong (valerian root)	−35	186	4.97
6	$CH_3(CH_2)_4CO_2H$	Caproic acid (hexanoic acid)	L. *caper*, goat	−3	205	1.08
7	$CH_3(CH_2)_5CO_2H$	Enanthic acid (heptanoic acid)	Gr. *oenanthe*, vine blossom	−9	223	0.26
8	$CH_3(CH_2)_6CO_2H$	Caprylic acid (octanoic acid)	L. *caper* goat	16	238	0.07
9	$CH_3(CH_2)_7CO_2H$	Pelargonic acid (nonanoic acid)	Pelargonium geranium	15	254	0.03
10	$CH_3(CH_2)_8CO_2H$	Capric acid (decanoic acid)	L. *caper*, goat	32	270	0.015
12	$CH_3(CH_2)_{10}CO_2H$	Lauric acid (dodecanoic acid)	Laurel	44	—	0.006
14	$CH_3(CH_2)_{12}CO_2H$	Myristic acid (tetradecanoic acid)	Myristica (nutmeg)	54	—	0.002
16	$CH_3(CH_2)_{14}CO_2H$	Palmitic acid (hexadecanoic acid)	Palm oil	63	—	0.0007
18	$CH_3(CH_2)_{16}CO_2H$	Stearic acid (octadecanoic acid)	Gr. *stear*, solid	70	—	0.0003
Miscellaneous Carboxylic Acids						
	$C_6H_5CO_2H$	Benzoic acid	Gum benzoin	122	249	0.34 (25 °C)
	$C_6H_5CH\!=\!CHCO_2H$	Cinnamic acid (trans isomer)	Cinnamon	132	—	0.04
	$CH_2\!=\!CHCO_2H$	Acrylic acid	L. *acer*, sharp	13	141	soluble
	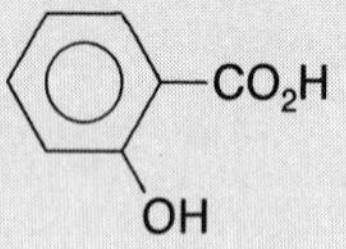	Salicylic acid	L. *salix*, willow	159	211	0.22 (25 °C)

[a] In parentheses by each common name is the IUPAC name.
[b] L. means Latin and Gr. means Greek.

This makes the *effective* formula weight of a carboxylic acid much higher than its calculated formula weight, and therefore the boiling point is higher. Thus formic acid, which has about the same formula weight as ethyl alcohol, boils at 101 °C which is 22 °C higher than ethyl alcohol.

The lower formula weight carboxylic acids ($C_1 - C_4$) are soluble in water largely because the carboxyl group has *two* oxygen atoms that can accept hydrogen bonds from water molecules. In addition, the carboxyl group has the O—H group that can donate hydrogen bonds.

13.2 CHEMICAL PROPERTIES OF CARBOXYLIC ACIDS AND THEIR SALTS

Although carboxylic acids are weak acids, they can neutralize strong bases and they react with alcohols to form esters.

Both the carboxyl group and the carboxylate ion strongly resist oxidation and reduction, and neither reacts to any appreciable extent with water. The properties that we need to know about for future use are their Brønsted acid – base relationships and the formation of esters and amides.

> In 1 *M* acetic acid, there is only about 0.5% ionization.

The Carboxylic Acids Are Weak Acids. The hydroxide ion, the carbonate ion, and the bicarbonate ion are all strong enough bases so that when any of them is supplied to a carboxylic acid (in a $1:1$ mole ratio), the acid is neutralized. This reaction is extremely important at the molecular level of life, because the carboxylic acids we normally produce by metabolism must be neutralized. Otherwise, the pH of body fluids, such as the blood, would fall too low to sustain life.

The organic product of the neutralization of a carboxylic acid is the anion of the acid, a carboxylate ion, $RCO_2{}^-$. The inorganic products vary with the base used. The neutralization occurs rapidly at room temperature. With hydroxide ion, the reaction is as follows:

$$RCO_2H + OH^- \longrightarrow RCO_2{}^- + H\text{—}OH$$

With bicarbonate ion, which is the chief base in the buffer systems of the blood, the following reaction occurs.

$$RCO_2H + HCO_3{}^- \longrightarrow RCO_2{}^- + H_2O + CO_2$$

Some specific examples are as follows:

$$CH_3CO_2H + OH^- \longrightarrow CH_3CO_2{}^- + H_2O$$

Acetic acid   Acetate ion

$$CH_3(CH_2)_{16}CO_2H + OH^- \longrightarrow CH_3(CH_2)_{16}CO_2{}^- + H_2O$$

Stearic acid   Stearate ion
(insoluble in water)   (soluble in water)

> The stearate ion is one of several organic ions in soap.

$$C_6H_5CO_2H + HCO_3{}^- \longrightarrow C_6H_5CO_2{}^- + H_2O + CO_2$$

Benzoic acid   Benzoate ion
(insoluble   (soluble in
in water)   water)

PRACTICE EXERCISE 1   Write the structures of the carboxylate anions that form when the following carboxylic acids are neutralized.

(a) $CH_3CH_2CO_2H$ (b) 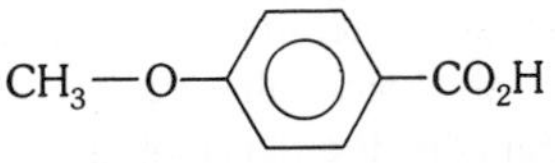 $CH_3\text{—}O\text{—}\langle\bigcirc\rangle\text{—}CO_2H$ (c) $CH_3CH\text{=}CHCO_2H$

Acetic Acid Most people experience acetic acid directly in the form of its dilute solution in water called vinegar. Because blood is slightly alkaline, acetic acid circulates as the acetate ion, and we will meet this species many times when we study metabolic pathways. The acetate ion, in fact, is one of the major intermediates in the metabolism of carbohydrates, lipids, and proteins.

Acetic acid is also an important industrial chemical, and more than 3 billion lb (23 billion mol) are manufactured each year in the United States. Acetate rayon is just one consumer product made using acetic acid.

Propionic Acid and Its Salts Propionic acid occurs naturally in Swiss cheese with a concentration that can be as high as 1%. Its sodium and calcium salts are food additives used in baked goods and processed cheese to retard the formation of molds or the growth of bacteria.

Sorbic Acid and the Sorbates   Sorbic acid, $CH_3CH=CHCH=CHCO_2H$, and its sodium or potassium salts are added in trace concentrations to a variety of foods to inhibit the growth of molds and yeasts. The sorbates often appear on ingredient lists for fruit juices, fresh fruits, wines, soft drinks, sauerkraut and other pickled products, and some meat and fish products. For food products that usually are wrapped, such as cheese and dried fruits, solutions of sorbate salts are sometimes sprayed onto the wrappers.

Sodium Benzoate Traces of sodium benzoate inhibit molds and yeasts in products that normally have pH values below 4.5 or 4.0. (The sorbates work better at slightly higher pH values, up to 6.5.) You'll see sodium benzoate on ingredient lists for beverages, syrups, jams and jellies, pickles, salted margarine, fruit salads, and pie fillings. Its concentration is low, 0.05–0.10%, and neither benzoic acid nor the benzoate ion accumulates in the body.

The Sodium and Potassium Salts of Carboxylic Acids Are Soluble in Water. The salts of carboxylic acids are true salts, assemblies of oppositely charged ions. All, therefore, are solids at room temperature. They follow the "like-dissolves-like rule" by being insoluble in nonpolar solvents, like ether or hydrocarbons. The salts with Group IA metals, but not with Group IIA metals, are soluble in water. Several salts are used as decay inhibitors in foods and bottled beverages, as described in Special Topic 13.1.

Carboxylate Ions Are Good Proton Acceptors. Because carboxylate ions are the anions of *weak* acids, they themselves must be relatively good bases. At room temperature, for example, the following neutralization of a strong acid by a carboxylate ion occurs virtually instantaneously. It is the most important reaction of the carboxylate ion we will study, because it makes the carboxylate group a neutralizer of excess acid at the molecular level of life.

$$R-CO_2^- + H-\overset{+}{O}\overset{\diagup H}{\underset{\diagdown H}{}} \longrightarrow R-CO_2H + H-OH$$

These rapid proton transfers don't require any heating either.

Specific examples:

$$C_6H_5CO_2^- + H_3O^+ \longrightarrow C_6H_5CO_2H + H_2O$$

Benzoate ion   Benzoic acid
(soluble in water)   (insoluble in water)

$$CH_3(CH_2)_{16}CO_2^- + H_3O^+ \longrightarrow CH_3(CH_2)_{16}CO_2H + H_2O$$

Stearate ion   Stearic acid
(soluble in water)   (insoluble in water)

The carboxylic acid solubility switch:

RCO₂H
(less soluble)

H⁺   OH⁻

RCO₂⁻
(more soluble)

The Carboxyl Group Is an Important Solubility "Switch". An insoluble acid, like stearic acid or benzoic acid, dissolves when the pH is high (basic) because it promptly changes to its anion, as the equations above showed. But if we next add acid to the solution to make the pH lower, the anion takes back a proton and the intact acid molecule comes out of solution. Thus by changing the pH of the medium we can switch on or off the solubility of a water-insoluble acid.

PRACTICE EXERCISE 2   Write the structures of the organic products of the reactions of the following compounds with dilute hydrochloric acid at room temperature.

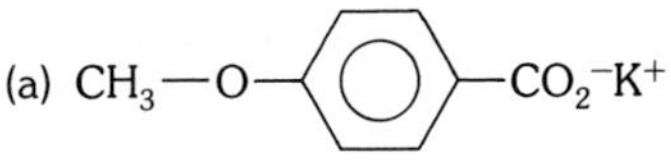

(a) CH_3—O—⟨○⟩—$CO_2^-K^+$ (b) $CH_3CH_2CO_2^-Li^+$ (c) $(CH_3CH{=}CHCO_2^-)_2Ca^{2+}$

Carboxylic Acids React with Alcohols to Form Esters.   When a solution of a carboxylic acid in an alcohol is heated, and a strong acid catalyst is present, the following species become involved in an equilibrium.

$$\underset{\substack{\text{Carboxylic}\\\text{acid}}}{R-\overset{\overset{\displaystyle O}{\|}}{C}-O-H} + \underset{\text{Alcohol}}{H-O-R'} \underset{}{\overset{H^+}{\rightleftharpoons}} \underset{\text{Ester}}{R-\overset{\overset{\displaystyle O}{\|}}{C}-O-R'} + H-OH$$

When the alcohol is in excess, Le Chatelier's principle operates and the equilibrium shifts so much to the right that this reaction is a good method for making an ester. An ester synthesis is called **esterification,** and some specific examples are as follows:

$$\underset{\text{Acetic acid}}{CH_3-\overset{\overset{\displaystyle O}{\|}}{C}-O-H} + \underset{\substack{\text{Ethyl alcohol}\\\text{(large excess)}}}{H-O-CH_2CH_3} \overset{H^+}{\longrightarrow} \underset{\text{Ethyl acetate}}{CH_3-\overset{\overset{\displaystyle O}{\|}}{C}-O-CH_2CH_3} + H_2O$$

These equations are actually forward reactions of equilibria, and the excesses of alcohols force them to the right.

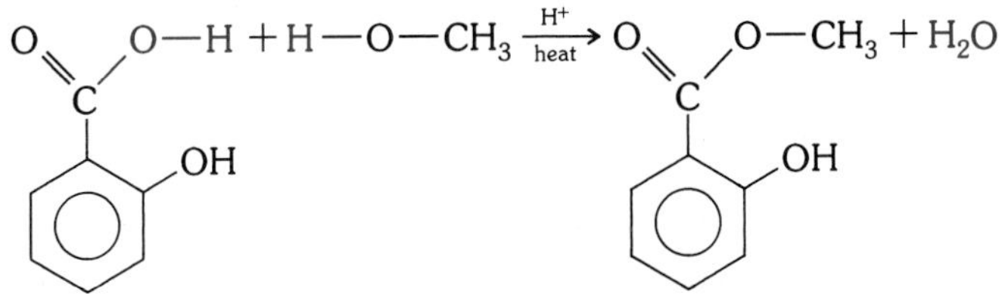

Salicylic acid + Methyl alcohol (large excess) $\overset{H^+}{\underset{\text{heat}}{\longrightarrow}}$ Methyl salicylate (oil of wintergreen) + H_2O

EXAMPLE 13.1   WRITING THE STRUCTURE OF A PRODUCT OF ESTERIFICATION

Problem: What is the ester that can be made from benzoic acid and methyl alcohol?

Solution: Write the structures of the two reactants. Sometimes it helps to let the OH groups "face" each other:

$$\underset{\text{Benzoic acid}}{C_6H_5-\overset{\overset{\displaystyle O}{\|}}{C}-O-H} + \underset{\text{Methyl alcohol}}{H-O-CH_3}$$

Although it is not important in predicting structures, always erase the OH group from the carboxyl group, not the alcohol. This will make it easier to learn to draw structures of amides later.

To obtain the pieces of the water molecule, the other product of esterification, we have to remove the OH group from the acid and the proton on oxygen from the alcohol. This leaves us with the following fragments:

$$C_6H_5-\overset{\overset{\displaystyle O}{\|}}{C} \quad \text{and} \quad -O-CH_3$$

Now all that is left to do is join these fragments.

$$C_6H_5-\overset{\overset{\displaystyle O}{\|}}{C}-O-CH_3 \quad \text{(product of the esterification)}$$
Methyl benzoate

PRACTICE EXERCISE 3   Write the structures of the esters that form by the esterification of acetic acid by the following alcohols.
(a) Methyl alcohol   (b) Propyl alcohol   (c) Isopropyl alcohol

PRACTICE EXERCISE 4   Write the structures of the esters that can be made by the esterification of ethyl alcohol by the following acids.
(a) Formic acid   (b) Propionic acid   (c) Benzoic acid

13.3 ESTERS OF CARBOXYLIC ACIDS

The ester group is split apart by water and by hydroxide ion.

Esters are compounds with the carbonyl–oxygen–carbon network, and Table 13.3 lists several common examples. Interestingly, the esters of vile-smelling acids generally have unusually pleasant fragrances, as the information in Table 13.4 shows. Special Topic 13.2 describes some important esters in more detail.

   The ester group cannot donate hydrogen bonds, because it has no HO (or HN) group, but it can *accept* them. This allows the lower formula weight esters to be relatively soluble in water.

When we digest fats and oil, ester groups are hydrolyzed.

Esters React with Water to Give the Parent Acid and Alcohol.   The hydrolysis of an ester, its reaction with water, is very slow unless an acid catalyst is used. (In the body, an enzyme acts as the catalyst.) In general,

$$R-\overset{\overset{\displaystyle O}{\|}}{C}-O-R' + H-OH \xrightarrow[\text{heat}]{H^+} R-\overset{\overset{\displaystyle O}{\|}}{C}-O-H + H-O-R'$$

Specific examples are

$$CH_3-\overset{\overset{\displaystyle O}{\|}}{C}-O-CH_2CH_3 + H_2O \xrightarrow[\text{heat}]{H^+} CH_3-\overset{\overset{\displaystyle O}{\|}}{C}-OH + HO-CH_2CH_3$$
Ethyl acetate   Acetic acid   Ethyl alcohol

$$CH_3-O-\overset{\overset{\displaystyle O}{\|}}{C}-C_6H_5 + H_2O \xrightarrow[\text{heat}]{H^+} CH_3-OH + HO-\overset{\overset{\displaystyle O}{\|}}{C}-C_6H_5$$
Methyl benzoate   Methyl   Benzoic acid
                                           alcohol

The *only* bond that breaks in the hydrolysis of an ester is the single covalent bond that joins the carbon atom of the carbonyl group to the oxygen atom. We can call it the **ester bond.** Notice also that the products are always the parents of the ester and that the names of these parents are strongly implied in the name of the ester itself. Thus methyl benzoate hydrolyzes to *methyl* alcohol and *benzo*ic acid.

TABLE 13.3
Some Esters of Carboxylic Acids

Common Name[a]	Structure	Melting Point (°C)	Boiling Point (°C)	Solubility (g/100 g H_2O, 20 °C)
Some Ethyl Esters				
Ethyl formate (ethyl methanoate)	$HCO_2C_2H_5$	−79	54	Soluble
Ethyl acetate (ethyl ethanoate)	$CH_3CO_2C_2H_5$	−82	77	7.35 (25 °C)
Ethyl propionate (ethyl propanoate)	$CH_3CH_2CO_2C_2H_5$	−73	99	1.75
Ethyl butyrate (ethyl butanoate)	$CH_3(CH_2)_2CO_2C_2H_5$	−93	120	0.51
Methyl benzoate	$C_6H_5CO_2CH_3$	−12	199	Insoluble
Methyl salicylate	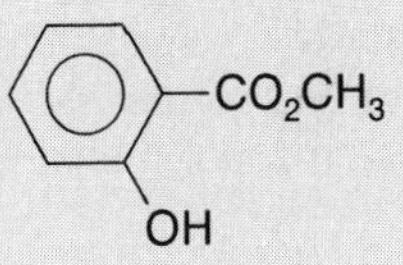	−9	223	Insoluble
Acetylsalicylic acid	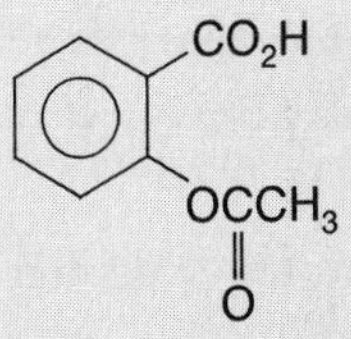	135		
Natural waxes	$CH_3(CH_2)_nCO_2(CH_2)_nCH_3$	$n = 23–33$ Carnauba wax $= 25–27$ Beeswax $= 14–15$ Spermaceti		

[a] The IUPAC names are in parentheses.

TABLE 13.4
Fragrances or Flavors of Some Esters

Name	Structure	Source or Flavor
Ethyl formate	$HCO_2CH_2CH_3$	Rum
Isobutyl formate	$HCO_2CH_2CH(CH_3)_2$	Raspberries
Pentyl acetate	$CH_3CO_2CH_2CH_2CH_2CH_2CH_3$	Bananas
Isopentyl acetate	$CH_3CO_2CH_2CH_2CH(CH_3)_2$	Pears
Octyl acetate	$CH_3CO_2(CH_2)_7CH_3$	Oranges
Ethyl butyrate	$CH_3CH_2CH_2CO_2CH_2CH_3$	Pineapples
Pentyl butyrate	$CH_3CH_2CH_2CO_2(CH_2)_4CH_3$	Apricots
Methyl salicylate	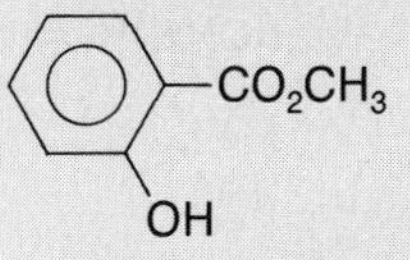	Oil of wintergreen

Esters of *p*-Hydroxybenzoic Acid, the Parabens Several alkyl esters of *p*-hydroxybenzoic acid—referred to as *parabens* on ingredient labels—are used to inhibit molds and yeasts in cosmetics, pharmaceuticals, and food.

Salicylates Certain esters and salts of salicylic acid are analgesics—pain suppressants—and antipyretics—fever reducers. The parent acid, salicylic acid, is itself too irritating to the stomach for these uses, but sodium salicylate and acetylsalicylic acid (aspirin) are commonly used. Methyl salicylate, a pleasant-smelling oil, is used in liniments, and it readily migrates through the skin.

Sodium salicylate

Acetyl salicylic acid (aspirin)

Methyl salicylate (oil of wintergreen)

Dacron Dacron, a polyester of exceptional strength, is widely used to make fabrics and film backing for recording tapes. (Actually, the name *Dacron* applies just to the fiber form of this polyester. When it is cast as a thin film, its name is *Mylar*.) Dacron fabrics have been used in surgery to repair or replace segments of blood vessels, as seen in Figure 13.1.

The formation of Dacron and many other polyfunctional polymers starts with two difunctional monomers, *aAa* and *bBb*, which are able to react with each other in such a way that a small molecule, *ab*, splits out, and the monomer fragments join end to end to make a very long polymer molecule. In principle, the polymerization can be represented as follows:

$$aAa + bBb + aAa + bBb + aAa + bBb + \cdots \text{ etc.} \longrightarrow$$
$$-A-B-A-B-A-B- \cdots \text{ etc.} + n(ab)$$
A copolymer

Because two monomers are used, the reaction is called *copolymerization*.

To make Dacron, one monomer is ethylene glycol, which has two alcohol groups. The other monomer is dimethyl terephthalate, which has two ester groups. One reaction of esters that we will not study in detail is that they can be made to react with alcohols to replace the original alkyl group in the ester by another. An alcohol molecule splits out. When ethylene glycol and dimethyl terephthalate polymerize, the *ab* molecule that splits out is methyl alcohol, and thus the terephthalate ester has become an ester of ethylene glycol instead of methyl alcohol. The copolymerization proceeds as follows:

FIGURE 13.1
The knitted tubing for this graft in an operative site is made of Dacron fibers.

etc.) $-\overset{\overset{\text{O}}{\|}}{\text{C}}-OCH_3$ + H$-OCH_2CH_2O-$H + $CH_3O-\overset{\overset{\text{O}}{\|}}{\text{C}}-$$\overset{\overset{\text{O}}{\|}}{\text{C}}-OCH_3$ + H$-OCH_2CH_2O-$H + etc.

Ethylene glycol   Dimethyl terephthalate

etc.) $-\overset{\overset{\text{O}}{\|}}{\text{C}}\left(-OCH_2CH_2O-\overset{\overset{\text{O}}{\|}}{\text{C}}-\overset{\overset{\text{O}}{\|}}{\text{C}}\right)_n O-CH_2CH_2O-$etc. + $n\,CH_3OH$

(repeating unit)
Dacron/Mylar; *n* is very large

EXAMPLE 13.2   PREDICTING THE PRODUCTS OF AN ESTER HYDROLYSIS

Problem: What are the products of the hydrolysis of the following ester?

$$CH_3-\overset{\overset{\displaystyle O}{\|}}{C}-O-CH_2CH_2CH_3$$

Solution: The crucial step is locating the ester bond, the carbonyl-to-oxygen bond. It doesn't matter in which direction this bond happens to point on the page:

$$CH_3CH_2CH_2-O-\overset{\overset{\displaystyle O}{\|}}{C}-CH_3 \quad \text{or} \quad CH_3-\overset{\overset{\displaystyle O}{\|}}{C}-O-CH_2CH_2CH_3$$

Ester bond
(carbonyl-to-oxygen bond)

The most common mistakes made by students are to break the wrong bond or to break too many bonds.

As we said, this bond is the only bond that breaks in ester hydrolysis, so break it. Erase it and separate the fragments. If the ester were written as follows:

$$CH_3CH_2CH_2-O-\overset{\overset{\displaystyle O}{\|}}{C}-CH_3 \longrightarrow CH_3CH_2CH_2-O + \overset{\overset{\displaystyle O}{\|}}{C}-CH_3 \quad \text{(incomplete)}$$

On the other hand, if the ester were written in the other direction:

$$CH_3-\overset{\overset{\displaystyle O}{\|}}{C}-O-CH_2CH_2CH_3 \longrightarrow CH_3-\overset{\overset{\displaystyle O}{\|}}{C} + O-CH_2CH_2CH_3 \quad \text{(incomplete)}$$

Both the OH and the H are supplied by H—OH, the other reactant.

Either way gives the same results. Next we attach the pieces of the water molecule so as to make the "parents" of the ester. We attach OH to the carbonyl carbon and we put H on the oxygen atom of the other fragment. The products therefore are the following, propyl alcohol and acetic acid.

$$HOCH_2CH_2CH_3 + HO-\overset{\overset{\displaystyle O}{\|}}{C}-CH_3$$

PRACTICE EXERCISE 5   Write the structures of the products of the hydrolysis of the following esters.

(a) $CH_3-O-\overset{\overset{\displaystyle O}{\|}}{C}-CH_3$ (b) $CH_3CH_2-\overset{\overset{\displaystyle O}{\|}}{C}-O-\overset{\overset{\displaystyle CH_3}{|}}{C}HCH_3$ (c) $CH_3\overset{\overset{\displaystyle CH_3}{|}}{C}H-\overset{\overset{\displaystyle O}{\|}}{C}-O-CH_2CH_2CH_3$

The Saponification of an Ester Gives the Salt of the Parent Acid and the Parent Alcohol. A strong base, like the hydroxide ion, also breaks up an ester. One product is the parent alcohol and the other is the anion of the parent carboxylic acid. This reaction of an ester with strong, aqueous alkali is called **saponification,** and it requires a full mole (not just a catalytic trace) of base for each mole of ester bonds. In general:

Latin sapo, soap, and onis, to make. Ordinary soap is made by the saponification of the ester groups in fats and oils.

$$R-\overset{\overset{\displaystyle O}{\|}}{C}-O-R' + OH^- \overset{heat}{\longrightarrow} R-\overset{\overset{\displaystyle O}{\|}}{C}-O^- + H-O-R'$$

Specific examples are as follows. (Assume that OH^- comes from NaOH or KOH.)

$$CH_3-\overset{\overset{\displaystyle O}{\|}}{C}-O-CH_2CH_3 + OH^- \xrightarrow[\text{heat}]{} CH_3-\overset{\overset{\displaystyle O}{\|}}{C}-O^- + HO-CH_2CH_3$$

Ethyl acetate   Acetate ion   Ethyl alcohol

$$CH_3-O-\overset{\overset{\displaystyle O}{\|}}{C}-C_6H_5 + OH^- \xrightarrow[\text{heat}]{} CH_3-OH + {}^-O-\overset{\overset{\displaystyle O}{\|}}{C}-C_6H_5$$

Methyl benzoate   Methyl   Benzoate ion
                                                   alcohol

EXAMPLE 13.3   WRITING THE STRUCTURES OF THE PRODUCTS OF SAPONIFICATION

Problem: What forms by the saponification of the following ester?

$$CH_3CH_2\overset{\overset{\displaystyle O}{\|}}{C}-O-CH_3$$

Solution: Ester saponification is very similar to ester hydrolysis. Therefore, first break (erase) the ester bond—*break only this bond*—and separate the fragments:

$$CH_3CH_2\overset{\overset{\displaystyle O}{\|}}{C}\mid O-CH_3 \longrightarrow CH_3CH_2\overset{\overset{\displaystyle O}{\|}}{C} + O-CH_3 \quad \text{(incomplete)}$$

Now change the fragment that has the carbonyl group into the *anion* of a carboxylic acid. Do this by attaching $-O^-$ to the carbonyl carbon atom. Then attach an H atom to the oxygen atom of the other fragment to make the alcohol molecule:

$$CH_3CH_2\overset{\overset{\displaystyle O}{\|}}{C}-O^- + H-O-CH_3 \quad \text{(the products)}$$

PRACTICE EXERCISE 6   Write the structures of the products of the saponification of the following esters.

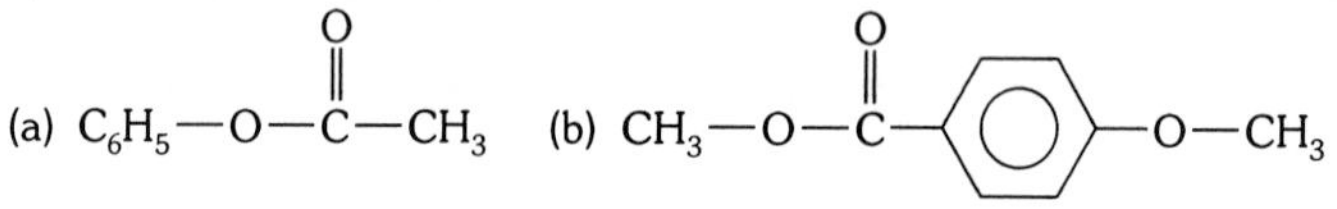

(a) $C_6H_5-O-\overset{\overset{\displaystyle O}{\|}}{C}-CH_3$ (b) $CH_3-O-\overset{\overset{\displaystyle O}{\|}}{C}-\bigcirc\!\!\!\!\!\!\!\!\!\!-O-CH_3$

13.4 ESTERS OF PHOSPHORIC ACID

Some widely distributed esters in living organisms are those of phosphoric acid, diphosphoric acid, and triphosphoric acid.

Phosphoric acid appears in several forms and anions in the body, but the three fundamental parents of them all are phosphoric acid, diphosphoric acid, and triphosphoric acid.

$$HO-\overset{\overset{\displaystyle O}{\|}}{\underset{\underset{\displaystyle OH}{|}}{P}}-OH \qquad HO-\overset{\overset{\displaystyle O}{\|}}{\underset{\underset{\displaystyle OH}{|}}{P}}-O-\overset{\overset{\displaystyle O}{\|}}{\underset{\underset{\displaystyle OH}{|}}{P}}-OH \qquad HO-\overset{\overset{\displaystyle O}{\|}}{\underset{\underset{\displaystyle OH}{|}}{P}}-O-\overset{\overset{\displaystyle O}{\|}}{\underset{\underset{\displaystyle OH}{|}}{P}}-O-\overset{\overset{\displaystyle O}{\|}}{\underset{\underset{\displaystyle OH}{|}}{P}}-OH$$

Phosphoric acid   Diphosphoric acid   Triphosphoric acid

All are polyprotic acids, but at the slightly alkaline pH values of body fluids, they cannot exist as free acids. They occur, instead, as mixtures of negative ions.

Phosphoric Acid Is the Parent Acid of the Monophosphate Esters. If you look closely at the structure of phosphoric acid, you can see that part of it resembles a carboxyl group.

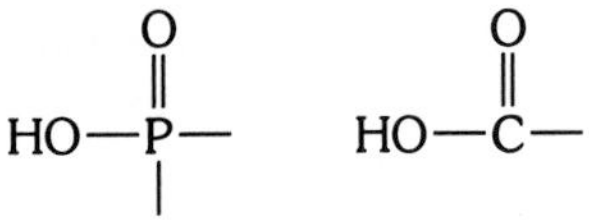

$$
\underset{\substack{\text{Part of a}\\\text{phosphoric}\\\text{acid molecule}}}{HO-\overset{\displaystyle O}{\overset{\|}{\underset{|}{P}}}-}
\qquad
\underset{\substack{\text{Part of a}\\\text{carboxylic}\\\text{acid molecule}}}{HO-\overset{\displaystyle O}{\overset{\|}{C}}-}
$$

It isn't surprising therefore that esters of phosphoric acid exist and that they are structurally similar to esters of carboxylic acids.

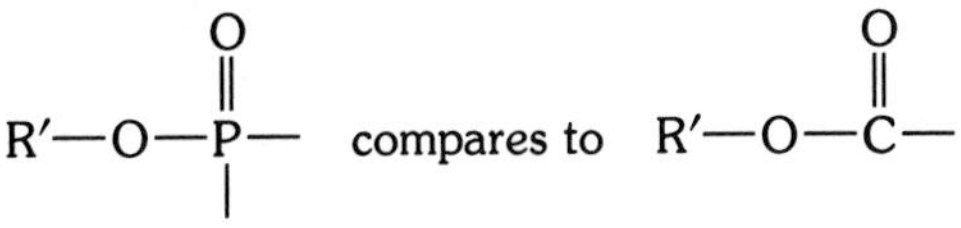

$$
\underset{\substack{\text{Part of a}\\\text{phosphate}\\\text{ester}}}{R'-O-\overset{\displaystyle O}{\overset{\|}{\underset{|}{P}}}-}
\quad\text{compares to}\quad
\underset{\substack{\text{Part of a}\\\text{carboxylate}\\\text{ester}}}{R'-O-\overset{\displaystyle O}{\overset{\|}{C}}-}
$$

Monophosphate Esters Are Also Acids. One large difference between a phosphate ester and a carboxylate ester is that a phosphate ester is still a diprotic acid. Its molecules still carry two proton-donating OH groups. Depending on the pH of the medium, therefore, a phosphate ester can exist in any one of three forms, and usually there is an equilibrium mixture of all three.

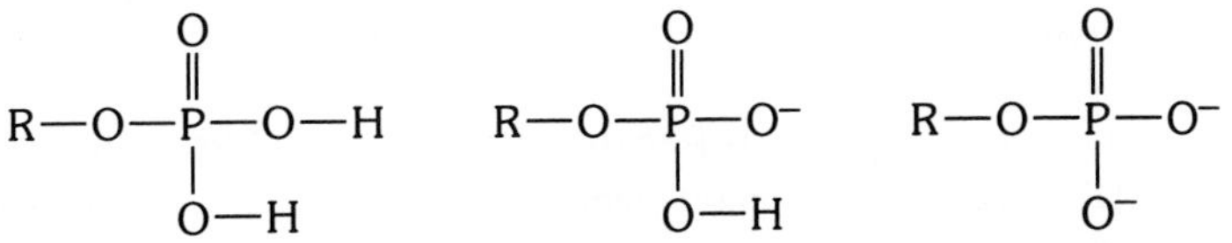

$$
\underset{\substack{\text{Phosphate ester}\\\text{(as a diprotic acid)}}}{R-O-\overset{\displaystyle O}{\overset{\|}{\underset{\underset{\displaystyle O-H}{|}}{P}}}-O-H}
\qquad
\underset{\substack{\text{Phosphate ester}\\\text{(as a singly}\\\text{ionized species)}}}{R-O-\overset{\displaystyle O}{\overset{\|}{\underset{\underset{\displaystyle O-H}{|}}{P}}}-O^{-}}
\qquad
\underset{\substack{\text{Phosphate ester}\\\text{(as a doubly}\\\text{ionized species)}}}{R-O-\overset{\displaystyle O}{\overset{\|}{\underset{\underset{\displaystyle O^{-}}{|}}{P}}}-O^{-}}
$$

—favored at low pH   —favored at pH values just below 7   —favored at pH values above 7

At the pH of most body fluids (just slightly more than 7), phosphate esters exist mostly as the double ionized species, as the dinegative ion. All forms, however, are generally soluble in water, and one reason that the body converts so many substances into their phosphate esters might be to improve their solubilities in water.

Diphosphate Esters Are Also Acids and Acid Anhydrides. A diphosphate ester actually has three functional groups: the phosphate ester group, the proton-donating OH group, and a unit called the *phosphoric anhydride system.*

$$
R-O-\overset{\displaystyle O}{\overset{\|}{\underset{\underset{\displaystyle OH}{|}}{P}}}-O-\overset{\displaystyle O}{\overset{\|}{\underset{\underset{\displaystyle OH}{|}}{P}}}-OH
$$

phosphoric anhydride system

ester group →

← proton-donating groups

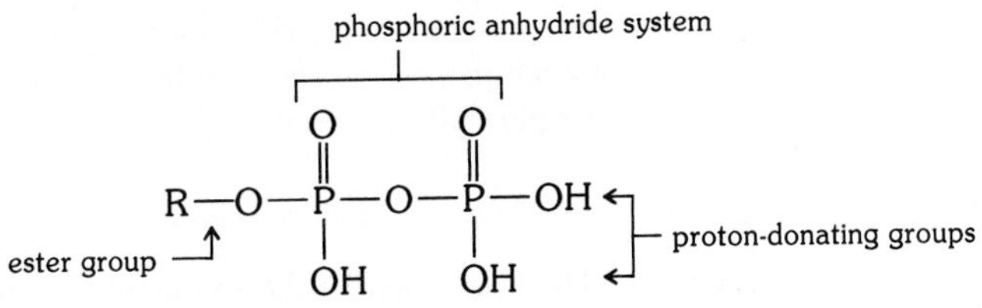

Diphosphate ester

The phosphoric anhydride unit consists of the following:

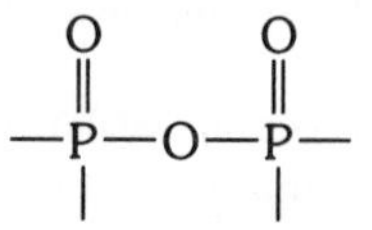

$$\begin{array}{ccc} O & & O \\ \parallel & & \parallel \\ -P-O-P- \\ | & & | \end{array}$$ The phosphoric anhydride system

Adenosine diphosphate, or ADP, is one of the many diphosphate esters in the body. We show it as a triply charged anion, because it exists largely in this fully ionized form at the pH of most body fluids.

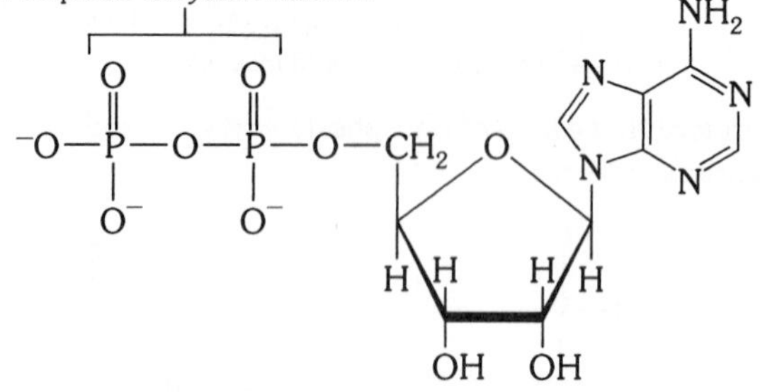

Adenosine diphosphate, ADP
(fully ionized form)

The Phosphoric Anhydride System Is a Major Source of Chemical Energy in Living Systems. ADP, as you can see, has the phosphoric anhydride system. It is one of the body's chief means of providing chemical energy, and ADP is often called an *energy-rich* compound. The strong repulsions of the electron-rich oxygens, some negatively charged, in the fully anionic form of the network account for this internal energy. These repulsions create a strong tension along the anhydride chain, which is released when the chain breaks apart in a reaction. Such reactions of phosphoric anhydride systems, therefore, are very exothermic.

The Triphosphate Esters Have Two Phosphoric Anhydride Systems. Adenosine triphosphate, or ATP, is the most common member of a small family of triphosphate esters. Because the triphosphates have two anhydride systems, they include some of the body's most energy-rich substances.

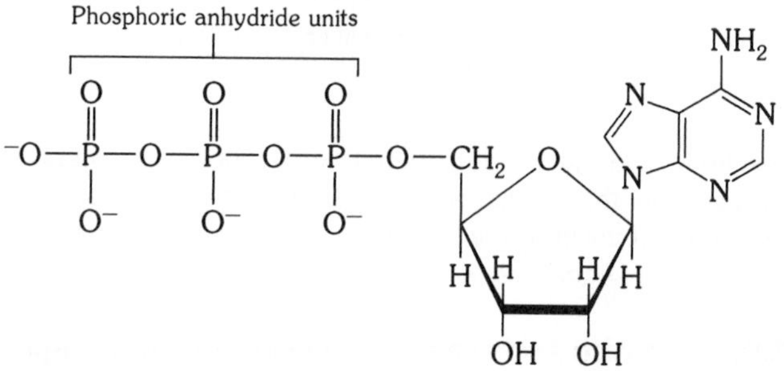

Adenosine triphosphate, ATP
(in its fully ionized form)

Living systems use triphosphates more widely to provide chemical energy than diphosphates. The contraction of a muscle, for example, can be represented as follows: We use P_i to stand for the mixture of inorganic ions, mostly $H_2PO_4^-$ and HPO_4^{2-}, produced as ATP reacts.

$$\text{Relaxed muscle} + \text{ATP} \xrightarrow{\text{enzyme}} \text{contracted muscle} + \text{ADP} + P_i$$

Although this is an extremely simplified statement, it shows that muscular work requires ATP. If the body's supply of ATP were consumed *with no way to remake it,* we'd soon lose all capacity for such work. The resynthesis of ATP from ADP and inorganic phosphate ion is one of the major uses of the chemical energy in the food we eat. ATP is synthesized only as needed. As soon as ADP and P_i appear following the use of ATP, they trigger the resynthesis of ATP. There is thus a rapid turnover of ATP in the body. The chemical energy in the food you eat drives the synthesis of roughly 40 kg (88 lb) of ATP every 24 hours. We'll study how the body does this remarkable task in Chapter 19.

13.5 AMIDES

Amides are neutral nitrogen compounds that can be hydrolyzed to carboxylic acids and ammonia (or amines).

Amides are compounds with a carbonyl–nitrogen bond, sometimes called the **amide bond.** This bond forms when amides are made, and it breaks when amides are hydrolyzed. Amides can be derived either from ammonia or from amines. Those made from ammonia are called the *simple* amides.

Amides of ammonia (simple amides) — Amides of amines — Amide group

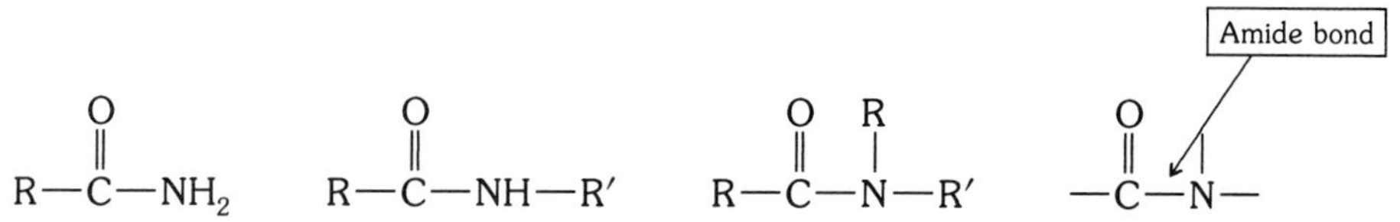

We study the amide system because all proteins are essentially polyamides whose molecules have regularly spaced amide bonds.

The amide bond is called the *peptide bond* in protein chemistry.

Hydrogen Bonding Is Strong in Amides. Table 13.5 lists several low formula weight amides. Their molecules are quite polar, and those with N—H can both donate and accept hydrogen bonds. These forces are so strong in amides that all simple amides except formamide, unlike simple esters or alcohols, are solids at room temperature. This same force also has a vital function among proteins, which, as we said, are polyamides. Special Topic 13.3 describes the important synthetic polyamide, nylon. A family of cyclic amides, the barbiturates, which have well-known physiological properties, are discussed in Special Topic 13.4.

TABLE 13.5
Amides of Carboxylic Acids

IUPAC Name	Structure	Melting Point (°C)
Formamide	$HCONH_2$	3
N-Methylformamide	$HCONHCH_3$	−5
N,N-Dimethylformamide	$HCON(CH_3)_2$	−61
Acetamide	CH_3CONH_2	82
N-Methylacetamide	$CH_3CONHCH_3$	28
N,N-Dimethylacetamide	$CH_3CON(CH_3)_2$	−20
Propionamide	$CH_3CH_2CONH_2$	79
Butyramide	$CH_3CH_2CH_2CONH_2$	115
Benzamide	$C_6H_5CONH_2$	133

The term *nylon* is a coined name that applies to any synthetic, long-chain, fiber-forming polymer with repeating amide linkages. One of the most common members of the nylon family, nylon-66, is made from 1,6-hexamethylenediamine and adipic acid:

$$NH_2CH_2CH_2CH_2CH_2CH_2CH_2NH_2$$
Hexamethylenediamine

Adipic acid

(The "66" means that each monomer has six carbon atoms.) To be useful as a fiber-forming polymer, each nylon-66 molecule should contain from 50 to 90 of each of the monomer units. Shorter molecules form weak or brittle fibers.

When molten nylon resin is being drawn into fibers, newly emerging strands are caught up on drums and stretched as they cool. Under this tension, the long polymer molecules within the fiber line up side by side, overlapping each other, to give a finished fiber of unusual strength and beauty. Part of nylon's strength comes from the innumerable hydrogen bonds that extend between the polymer molecules and that involve their many regularly spaced amide groups.

Nylon is more resistant to combustion than wool, rayon, cotton, or silk, and it is as immune to insect attack as Fiberglas. Molds and fungi do not attack nylon molecules. In medicine, nylon is used in specialized tubing, and as velour for blood contact surfaces. Nylon sutures were the first synthetic sutures and are still used.

Repeating unit in nylon-66

Amides, Unlike Amines, Are Neither Acidic Nor Basic.   Amides are neutral in an acid–base sense. They do not affect the pH of an aqueous system. The carbonyl group causes this. Because of its oxygen atom, the whole $C=O$ group is electronegative. It draws back the electron pair on the neighboring nitrogen in an amide well enough to make it unable to accept a proton (and thus act like a base). In an amine, by contrast, the lack of a neighboring $C=O$ group leaves the electron pair on N able to accept a proton and function as a base.

Amides Are Made from Amines by Acyl Group-Transfer Reactions.   A direct way to make a simple amide is to heat the ammonium salt of a carboxylic acid:

This method, however, is quite unlike the way amides are made in the living systems. Instead, special molecules serve as carriers for **acyl groups,** a unit present in all carboxylic acids and their derivatives but without independent existence.

To name an acyl group, change *-ic acid* to *-yl* in the name of the parent acid.

In the presence of the correct enzyme, an acyl group can transfer from its carrier to the amino group of another molecule and make an amide bond. Thus,

G is some organic group, but not necessarily an alkyl group. Hence, the symbol R is not used here.

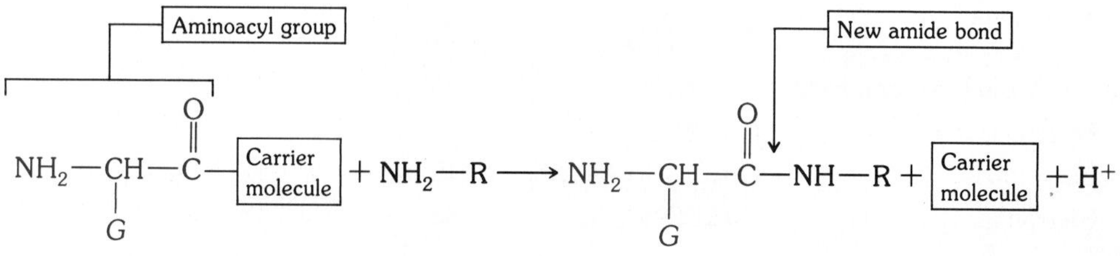

Barbiturates are amide-like compounds that are the best known, the most widely prescribed, and the most frequently abused members of a family of drugs that act as sedatives. In properly supervised doses, they mildly depress the central nervous system and cause both breathing and heartbeat to slow down. In this way they induce sleep. (In slang, they are called downers, goof balls, or barbs.)

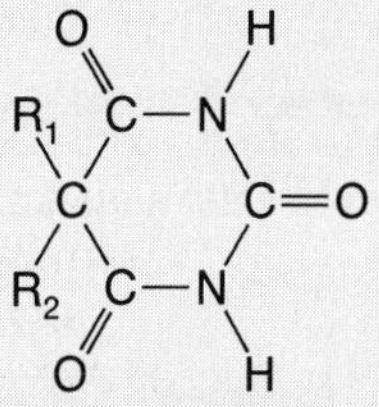

At higher doses, barbiturates give the symptoms of drunkenness, and at still higher doses breathing and heart action are so slowed that death occurs.

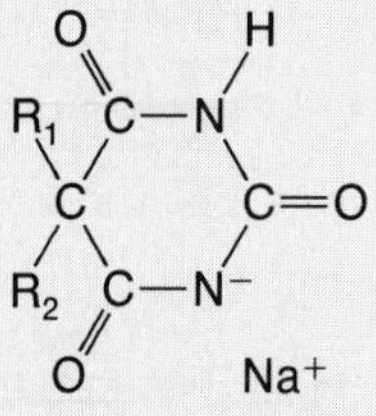

Barbiturates and alcohol are a lethal combination to many who, with a bad hangover, just want to get to sleep at once. Film star Marilyn Monroe is thought to have died in this way.

The accompanying table lists just a few of the approximately two dozen derivatives of barbituric acid that can be prescribed. Nembutal and Seconal are probably the most commonly abused.

Barbituric acids   Sodium barbiturates

Barbiturates			
$R_1 =$	$R_2 =$	Name of Acid	Name of Salt
H	H	Barbituric acid	Sodium barbiturate
CH_3CH_2-	C_6H_5-	Phenobarbital (Luminal)	Phenobarbital sodium
CH_3CH_2-	$CH_3CH_2CH_2CH-$ $\underset{CH_3}{\vert}$	Pentobarbital (Nembutal)	Pentobarbital sodium
$CH_2=CHCH_2-$	$CH_3CH_2CH_2CH-$ $\underset{CH_3}{\vert}$	Secobarbital (Seconal)	Secobarbital sodium

This aminoacyl transfer method rather than the direct method is important for our upcoming study of biochemistry. We particularly need to learn how to figure out the structure of an amide that can be made regardless of the exact nature of the acyl transfer agent.

EXAMPLE 13.4   WRITING THE STRUCTURE OF AN AMIDE THAT CAN BE MADE FROM THE ACYL GROUP OF AN ACID AND AN AMINE

Problem: What amide can be made from the following two substances, assuming that a suitable acyl group-transfer process is available?

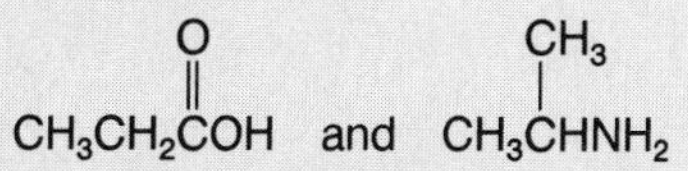

$$CH_3CH_2\overset{O}{\overset{\|}{C}}OH \quad \text{and} \quad CH_3\overset{CH_3}{\overset{\vert}{C}}HNH_2$$

Solution: The best way to proceed is to write the skeleton of the amide system and then build on it. It doesn't matter how we orient this skeleton:

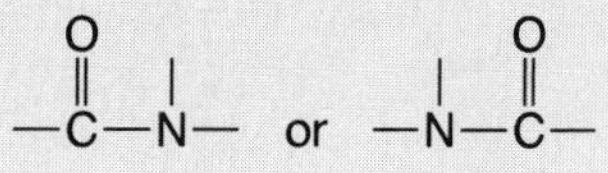

$$-\overset{O}{\overset{\|}{C}}-\overset{}{\underset{}{N}}- \quad \text{or} \quad -\overset{}{\underset{}{N}}-\overset{O}{\overset{\|}{C}}- \qquad \text{(incomplete)}$$

Then we look at the acid and see what has to be attached to the carbon atom of this skeleton, and we write it in.

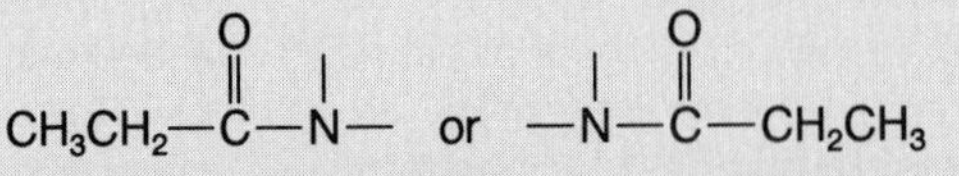

$$CH_3CH_2-\overset{\overset{\displaystyle O}{\|}}{C}-\overset{\displaystyle |}{N}- \quad or \quad -\overset{\displaystyle |}{N}-\overset{\overset{\displaystyle O}{\|}}{C}-CH_2CH_3 \qquad (incomplete)$$

Then we examine the given amine to see what organic group it carries, and we attach it to the N atom. (If there are *two* organic groups on N in the amine, we have to attach both, of course.)

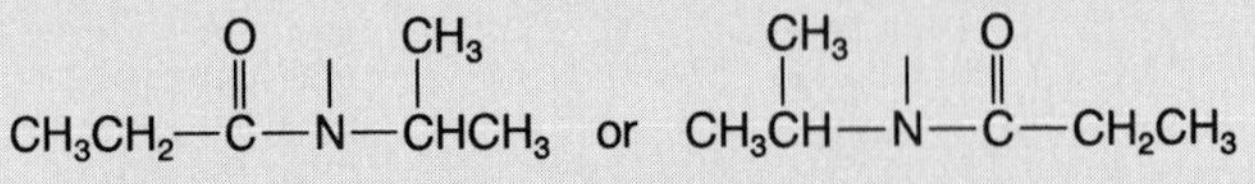

$$CH_3CH_2-\overset{\overset{\displaystyle O}{\|}}{C}-\overset{\overset{\displaystyle CH_3}{|}}{N}-CHCH_3 \quad or \quad CH_3\overset{\overset{\displaystyle CH_3}{|}}{CH}-N-\overset{\overset{\displaystyle O}{\|}}{C}-CH_2CH_3 \qquad (incomplete)$$

Finally, of the two H atoms on N in the amine, one survives, and our last step is to write it in. The final answer is

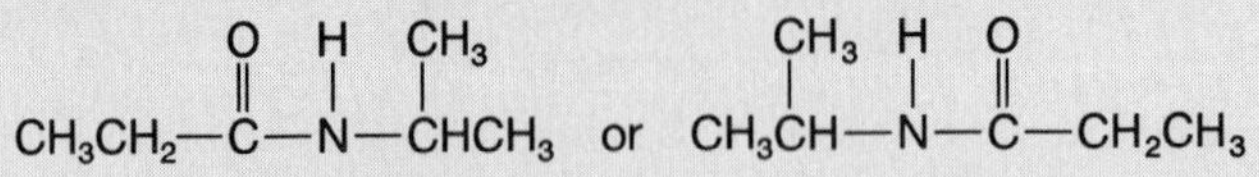

$$CH_3CH_2-\overset{\overset{\displaystyle O}{\|}}{C}-\overset{\overset{\displaystyle H}{|}}{N}-\overset{\overset{\displaystyle CH_3}{|}}{CH}CH_3 \quad or \quad CH_3\overset{\overset{\displaystyle CH_3}{|}}{CH}-\overset{\overset{\displaystyle H}{|}}{N}-\overset{\overset{\displaystyle O}{\|}}{C}-CH_2CH_3$$

These structures, of course, are identical.

PRACTICE EXERCISE 7   What amides, if any, could be made by suitable acyl group-transfer reactions from the following pairs of compounds?

(a) CH_3NH_2 and $\underset{\underset{\displaystyle CH_3}{|}}{CH_2CHCO_2}$

(b) $NH_2C_6H_5$ and CH_3CO_2H

(c) $CH_3\overset{\overset{\displaystyle O}{\|}}{C}CH_2NH_2$ and CH_3NH_2

(d) CH_3CO_2H and $\underset{\underset{\displaystyle CH_3}{|}}{CH_3NCH_3}$

Amides Can Be Hydrolyzed to Carboxylic Acids and Amines (or Ammonia). Acids, bases, and enzymes all promote the hydrolysis of amides, the reaction by which proteins are digested.

The actual products formed when an amide is hydrolyzed depend on the pH of the medium. If alkaline, the carboxylate ion and the amine (or ammonia) form. If acidic, the carboxylic acid and the protonated amine (or NH_4^+) are the products. Of course, the hydrolysis of an amide can occur without any catalyst or promoter; it is just much slower this way. To simplify, we'll write amide hydrolysis as a simple reaction with water to give the free carboxylic acid and the free amine. Here are some examples.

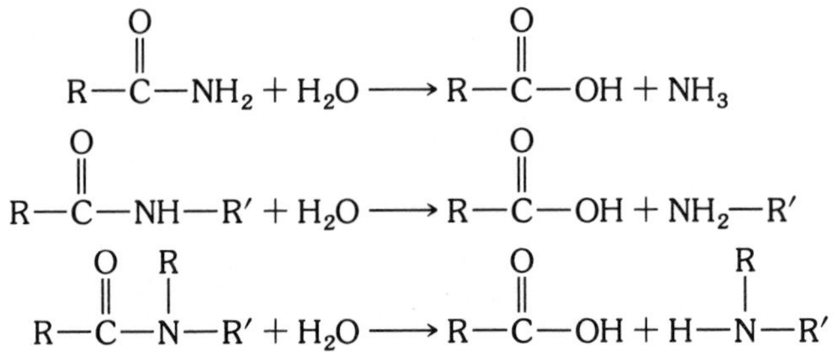

$$R-\overset{\overset{\displaystyle O}{\|}}{C}-NH_2 + H_2O \longrightarrow R-\overset{\overset{\displaystyle O}{\|}}{C}-OH + NH_3$$

$$R-\overset{\overset{\displaystyle O}{\|}}{C}-NH-R' + H_2O \longrightarrow R-\overset{\overset{\displaystyle O}{\|}}{C}-OH + NH_2-R'$$

$$R-\overset{\overset{\displaystyle O}{\|}}{C}-\overset{\overset{\displaystyle R}{|}}{N}-R' + H_2O \longrightarrow R-\overset{\overset{\displaystyle O}{\|}}{C}-OH + H-\overset{\overset{\displaystyle R}{|}}{N}-R'$$

EXAMPLE 13.5 WRITING THE PRODUCTS OF THE HYDROLYSIS OF AN AMIDE

Problem: Acetophenetidin (phenacetin) has long been used in some brands of headache remedies. (APC tablets, for example, consist of aspirin, phenacetin, and caffeine.)

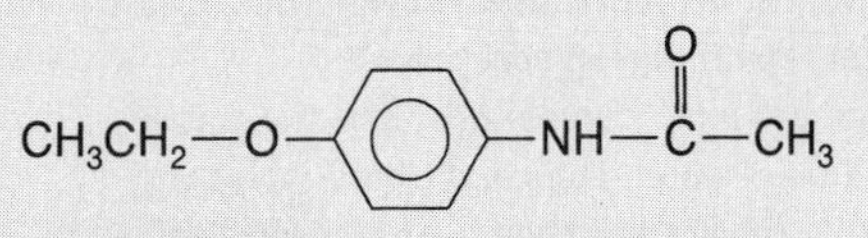

$$CH_3CH_2-O-\bigcirc-NH-\overset{O}{\underset{\|}{C}}-CH_3$$

Acetophenetidin

If this compound is an amide, what are the products of its hydrolysis?

Solution: Acetophenetidin does have the amide bond—carbonyl to nitrogen—so it can be hydrolyzed. (The functional group on the left side of this structure is that of an ether, and ethers do not react with water.)

Because the amide bond breaks when an amide is hydrolyzed, simply erase this bond from the structure and separate the parts. *Do not break any other bond.*

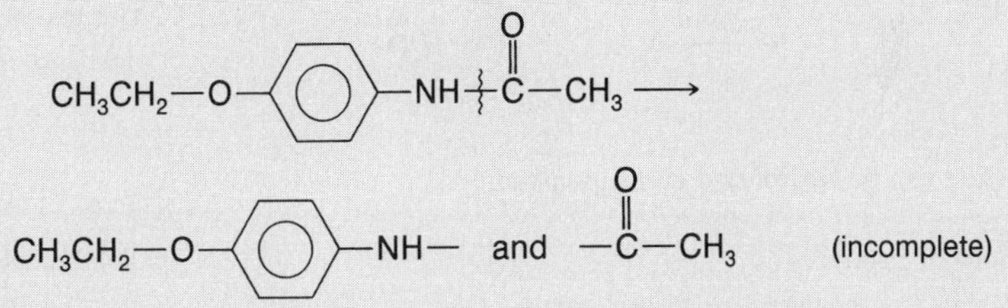

$$CH_3CH_2-O-\bigcirc-NH\!\!\mid\!\!\overset{O}{\underset{\|}{C}}-CH_3 \longrightarrow$$

$$CH_3CH_2-O-\bigcirc-NH- \text{ and } -\overset{O}{\underset{\|}{C}}-CH_3 \quad \text{(incomplete)}$$

We know that the hydrolysis uses HO—H to give a carboxylic acid and an amine, so we put a HO group on the carbonyl group in the appropriate fragment and we put H on the nitrogen of the other fragment. The products, therefore, are

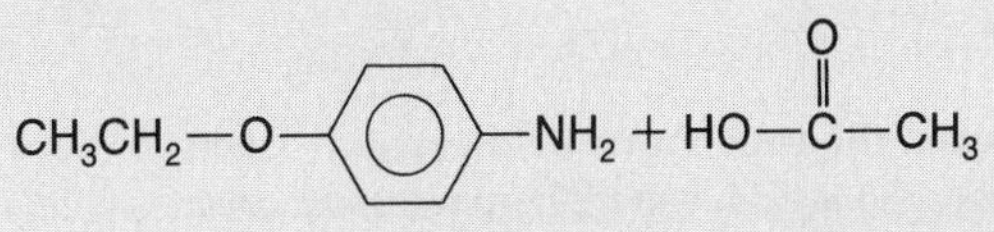

$$CH_3CH_2-O-\bigcirc-NH_2 + HO-\overset{O}{\underset{\|}{C}}-CH_3$$

PRACTICE EXERCISE 8 For any compounds in the following list that are amides, write the products of their hydrolysis.

(a) $C_6H_5-\overset{O}{\underset{\|}{C}}-NH-CH_3$

(b) $C_6H_5-\overset{O}{\underset{\|}{C}}-CH_2-NH_2$

(c) $C_6H_5-NH-\overset{O}{\underset{\|}{C}}-CH_3$

(d) $CH_3-\overset{O}{\underset{\|}{C}}-NH-CH_2CH_2-NH-\overset{O}{\underset{\|}{C}}-CH_3$

(use an excess of water)

SUMMARY

Acids and their salts The carboxyl group, CO_2H, is a polar group that confers moderate water solubility to a molecule without preventing its solubility in nonpolar solvents. This group is resistant to oxidation and reduction. Carboxylic acids are strong proton donors toward hydroxide ions. Toward water, carboxylic acids are weak acids. Carboxylate anions, therefore, are good proton acceptors.

Salts of carboxylic acids are ionic compounds, and the potassium and sodium salts are soluble in water. Hence, the carboxyl group is one of nature's important "solubility switches." An insoluble acid becomes soluble in base, but it is thrown out of solution again by the addition of acid.

The derivatives of acids—salts, esters, and amides—can be made from the acids and are converted back to the acids. We can organize the reactions we have studied for the carboxylic acids as follows:

$$RCO_2H \begin{cases} \xrightarrow[\ce{<=>}]{H_2O} RCO_2^- + H_3O^+ \\ \xrightarrow{OH^-} RCO_2^- + H_2O \\ \xrightarrow{R'OH,\ H^+} RCO_2{-}R' + H_2O \end{cases}$$

Esters Esters can be hydrolyzed and saponified.

Reactions of Esters:

$$R{-}\overset{\overset{\displaystyle O}{\|}}{C}{-}O{-}R' \begin{cases} \xrightarrow{H_2O\ (excess),\ H^+} R{-}\overset{\overset{\displaystyle O}{\|}}{C}{-}OH + HOR' \\ \xrightarrow[\text{saponification}]{OH^-} R{-}\overset{\overset{\displaystyle O}{\|}}{C}{-}O^- + HOR' \end{cases}$$

Esters and anhydrides of the phosphoric acid system Esters of phosphoric acid, diphosphoric acid, and triphosphoric acid occur in living systems largely as anions, because these esters are also polyprotic acids. In addition, those of di- and triphosphoric acid are phosphoric anhydrides, which are energy-rich compounds.

Amides The carbonyl–nitrogen bond, the amide bond, can be formed by letting an amine or ammonia react with anything that can transfer an acyl group. Amides are neither basic nor acidic, but are neutral. They can be hydrolyzed to their parent acids and amines. We can summarize the reactions involving amides as follows:

$$\left. \begin{array}{l} NH_3 \\ RNH_2 \\ R_2NH \end{array} \right\} \xrightarrow{R{-}\overset{\overset{\displaystyle O}{\|}}{C}{-}\boxed{Carrier}} R{-}\overset{\overset{\displaystyle O}{\|}}{C}{-}\overset{\overset{\displaystyle H(R)}{|}}{N}{-}H(R) + \boxed{Carrier}$$

$$R{-}\overset{\overset{\displaystyle O}{\|}}{C}{-}\overset{\overset{\displaystyle H(R)}{|}}{N}{-}H(R) + H_2O \longrightarrow R{-}\overset{\overset{\displaystyle O}{\|}}{C}{-}OH + H{-}\overset{\overset{\displaystyle H(R)}{|}}{N}{-}H(R)$$

Brief survey of what functional groups are attacked by specific reactants

1. Groups affected by oxidizing agents, (O)
   (a) Alcohol groups (1° and 2°, not 3°) are converted to carbonyl groups.

   $$-\overset{\overset{\displaystyle |}{}}{\underset{\underset{\displaystyle H}{|}}{C}}{-}O{-}H + (O) \longrightarrow \overset{\diagdown}{\underset{\diagup}{}}C{=}O + H_2O$$

   1° alcohols are oxidized to aldehydes or carboxylic acids.
   2° alcohols are oxidized to ketones.
   (b) Thioalcohols are oxidized to disulfides.

   $$2R{-}S{-}H + (O) \longrightarrow R{-}S{-}S{-}R + H_2O$$

   (c) Aldehydes are oxidized to carboxylic acids.

   $$RCHO + (O) \longrightarrow RCO_2H$$

2. Groups that can be reduced by hydrogen or hydrogen donors
   (a) Carbon–carbon double bonds become saturated.

   $$\overset{\diagdown}{\underset{\diagup}{}}C{=}C\overset{\diagup}{\underset{\diagdown}{}} + H_2 \xrightarrow[\text{pressure}]{Ni,\ heat,} H{-}\overset{\overset{\displaystyle |}{}}{\underset{\underset{\displaystyle |}{}}{C}}{-}\overset{\overset{\displaystyle |}{}}{\underset{\underset{\displaystyle |}{}}{C}}{-}H$$

   (b) Disulfides are reduced to thioalcohols.

   $$R{-}S{-}S{-}R + 2(H) \longrightarrow 2RSH$$

   (c) Aldehydes are reduced to 1° alcohols.

   $$RCHO + 2(H) \longrightarrow RCH_2OH$$

   (d) Ketones are reduced to 2° alcohols.

   $$R{-}\overset{\overset{\displaystyle O}{\|}}{C}{-}R + 2(H) \longrightarrow R{-}\overset{\overset{\displaystyle OH}{|}}{C}H{-}R$$

3. Groups split apart by water (hydrolysis)
   (a) Acetals (and ketals) are hydrolyzed to aldehydes (or ketones) and alcohols.

   $$R{-}\overset{\overset{\displaystyle OR'}{|}}{\underset{\underset{\displaystyle OR'}{|}}{C}}{-}H(R) + H_2O \xrightarrow[\text{enzyme}]{H^+\ or} R{-}\overset{\overset{\displaystyle O}{\|}}{C}{-}H(R) + 2HOR'$$

   (b) Esters are hydrolyzed to acids and alcohols.

   $$RCO_2R' + H_2O \xrightarrow{H^+} RCO_2H + HOR'$$

   (c) Amides are hydrolyzed to acids and amines (or ammonia).

   $$R{-}\overset{\overset{\displaystyle O}{\|}}{C}{-}\overset{\overset{\displaystyle H(R)}{|}}{N}{-}H(R) + H_2O \longrightarrow RCO_2H + H{-}\overset{\overset{\displaystyle H(R)}{|}}{N}{-}H(R)$$

4. Groups that neutralize acids at room temperature
   (a) Amines

   $$RNH_2 + H^+ \longrightarrow RNH_3^+$$

(b) Carboxylate ions

$$RCO_2^- + H^+ \longrightarrow RCO_2H$$

5. Groups that neutralize bases at room temperature
   (a) Protonated amines

$$RNH_3^+ + OH^- \longrightarrow RNH_2 + H_2O$$

   (b) Carboxylic acids

$$RCO_2H + OH^- \longrightarrow RCO_2^- + H_2O$$

6. Miscellaneous reactions involving water
   (a) Water adds to alkene double bonds (under Markovnikov's rule) to give alcohols.
   (b) Water is a product in the following reactions:
       (1) Dehydration of an alcohol
       (2) Formation of acetals or ketals
       (3) Esterification
       (4) Formation of amides from ammonium salts of acids
       (5) The oxidations of:
           1° and 2° alcohols
           Thioalcohols

REVIEW EXERCISES

The answers to these review exercises are in the *Study Guide* that accompanies this book.

Structures and Names of Carboxylic Acids and Their Salts

13.1 What is the structure of the carboxyl group, and in what way does it differ from the functional group in an alcohol? In a ketone?

13.2 Fatty acids are carboxylic acids obtained from what substances?

13.3 What is the common name of the acid in vinegar? In sour milk?

13.4 Write the structures of the following substances.
   (a) Butyric acid   (b) Acetic acid
   (c) Formic acid   (d) Benzoic acid

13.5 Write the structures of the following:
   (a) Sodium acetate   (b) Propionic acid
   (c) Sodium benzoate   (d) Potassium butyrate

Physical Properties of Carboxylic Acids

13.6 Draw a figure that shows how two acetic acid molecules can pair in a hydrogen-bonded form.

13.7 The hydrogen-bond system in formic acid includes an array of molecules, one after the other, each carbonyl oxygen of one molecule attracted to the HO group of the next molecule in line. Represent this linear array of hydrogen-bonded molecules of formic acid by a drawing.

13.8 Arrange the following compounds in their order of increasing solubility in water. Do this by arranging their identifying letters in a row in the correct order, placing the letter of the least soluble on the left.

CH_3CO_2H	$CH_3(CH_2)_5CO_2H$	$CH_3(CH_2)_6CH_3$
A	B	C

13.9 Arrange the following compounds in their order of increasing boiling points by arranging their identifying letters in a row in the correct order. Place the letter of the lowest-boiling compound on the left.

HO_2CCO_2H	CH_3CO_2H	CH_3OH	$CH_3CH_2CH_3$
A	B	C	D

Carboxylic Acids as Weak Acids

13.10 Arrange the following compounds in their order of increasing acidity by writing their identifying letters in a row in the correct sequence. (Place the letter of the least acidic compound on the left.)

CH_3CO_2H	HNO_3	CH_3CH_2OH	C_6H_5OH
A	B	C	D

13.11 Give the order of increasing acidity of the following compounds. Arrange their identifying letters in the order that corresponds to their acidity, with the letter of the least acidic compound on the left.

$C_6H_5CH_2$—OH CH_3——⟨O⟩——OH

  A   B

$C_6H_5CO_2H$ H_2SO_4

  C   D

13.12 Write the net ionic equation for the complete reaction, if any, of sodium hydroxide with the following compounds at room temperature.

(a) $CH_3CH_2CO_2H$ (b) HO_2C——⟨O⟩——CH_3

(c) CH_3CH_2OH

13.13 What are the net ionic equations for the reactions of the following compounds with aqueous potassium hydroxide at room temperature?

(a) ⟨benzene ring with CO_2H and CO_2H groups⟩

(b) $HOCH_2\overset{O}{\overset{\|}{C}}CH_2CH_2CO_2H$

(c) C_6H_5—OH

Salts of Carboxylic Acids

13.14 Which compound, **A** or **B,** is more soluble in water? Explain.

$$CH_3(CH_2)_8CO_2H \qquad CH_3(CH_2)_8CO_2Na$$
$$\textbf{A} \qquad\qquad\qquad \textbf{B}$$

13.15 Which compound, **A** or **B,** is more soluble in ether? Explain.

CH$_3$—⟨benzene ring⟩—CO$_2$H   CH$_3$—⟨benzene ring⟩—CO$_2$K

A   **B**

13.16 Suppose you added 0.1 mol of hydrochloric acid to an aqueous solution that contained 0.1 mol of the compound given in each of the following parts. If any reaction occurs rapidly at room temperature, write its net ionic equation.

(a) CH$_3$—⟨benzene ring⟩—CO$_2^-$ (b) CH$_3$CO$_2$H

(c) $^-$O$_2$CCH$_2$CH$_2$CO$_2^-$

13.17 Suppose that you have each of the following compounds in a solution in water. What reaction, if any, would occur rapidly at room temperature if an equimolar quantity of hydrochloric acid were added? Write net ionic equations.
(a) CH$_3$CH$_2$CO$_2$H   (b) CH$_3$CH$_2$CH$_2$CO$_2^-$
(c) HOCH$_2$CH$_2$CH$_2$CO$_2^-$

Esterification

13.18 What are the structures of the products of the esterification by methyl alcohol of each compound?
(a) Formic acid

(b) CH$_3$CH$_2$$\overset{\overset{\displaystyle CH_3}{|}}{C}HCO_2$H

(c) Cl—⟨benzene ring⟩—CO$_2$H

(d) Oxalic acid, HO$\overset{\overset{\displaystyle O}{||}}{C}$—$\overset{\overset{\displaystyle O}{||}}{C}$OH (Show the esterification of both of the carboxyl groups.)

13.19 When propionic acid is esterified by each of the following compounds, what are the structures of the esters that form?
(a) Ethyl alcohol
(b) Isobutyl alcohol
(c) Phenol
(d) HOCH$_2$CH$_2$OH (ethylene glycol). Show the esterification of both alcohol groups.

Structures and Physical Properties of Esters

13.20 Write the structures of the following compounds.
(a) Methyl formate   (b) Ethyl benzoate

13.21 What are the structures of the following compounds?
(a) Isopropyl propionate   (b) Isobutyl butyrate

13.22 Arrange the following compounds in their order of increasing boiling points. Do this by placing their identifying letters in a row, starting with the lowest-boiling compound on the left.

CH$_3$CH$_2$CH$_2$CH$_2$CO$_2$H   CH$_3$CH$_2$OCH$_3$
A   **B**

CH$_3$CH$_2$CO$_2$CH$_3$ CH$_3$CH$_2$CO$_2$CH$_2$CH$_3$
C   **D**

13.23 Arrange the following compounds in their order of increasing solubilities in water by placing their identifying letters in the correct sequence, beginning with the least soluble on the left.

CH$_3$(CH$_2$)$_4$CO$_2$Na   CH$_3$(CH$_2$)$_2$CO$_2$H
A   **B**

CH$_3$(CH$_2$)$_4$CO$_2$CH$_3$ CH$_3$(CH$_2$)$_6$OCH$_3$
C   **D**

Reactions of Esters

13.24 Write the equation for the acid-catalyzed hydrolysis of each compound. If no reaction occurs, write "no reaction."

(a) CH$_3$$\overset{\overset{\displaystyle CH_3}{|}}{C}$H—O—$\overset{\overset{\displaystyle O}{||}}{C}CH_3$ (b) CH$_3$$\overset{\overset{\displaystyle O}{||}}{\underset{\underset{\displaystyle CH_3}{|}}{C}}$HC—O—CH$_3$

(c) CH$_3$—O—CH$_2$CO$_2$H (d) HOCH$_2$$\overset{\overset{\displaystyle O}{||}}{C}$—O—CH$_3$

13.25 What are the equations for the acid-catalyzed hydrolyses of the following compounds? If no reaction occurs, write "no reaction."

(a) CH$_3$CH$_2$—O$\overset{\overset{\displaystyle O}{||}}{C}$—C$_6H_5$

(b) CH$_3$CH$_2$$\overset{\overset{\displaystyle O}{||}}{C}$—O—C$_6H_5$

(c) CH$_3$CH$_2$—O—⟨benzene ring⟩—$\overset{\overset{\displaystyle O}{||}}{C}CH_3$

(d) CH$_3$—O—$\overset{\overset{\displaystyle O}{||}}{C}CH_2CH_2$$\overset{\overset{\displaystyle O}{||}}{C}$—O—CH$_3$

13.26 The digestion of fats and oils involves the complete hydrolysis of molecules such as the following. What are the structures of its hydrolysis products?

CH$_3$(CH$_2$)$_{10}$$\overset{\overset{\displaystyle O}{||}}{C}$—O—CH$_2$—CH—CH$_2$—O—$\overset{\overset{\displaystyle O}{||}}{C}$(CH$_2$)$_{16}CH_3$

$\qquad\qquad\qquad\qquad\qquad$ O—$\overset{\underset{\displaystyle ||}{\underset{\displaystyle O}{}}}{C}$(CH$_2$)$_8CH_3$

13.27 Cyclic esters are known compounds. What is the structure of the product of the hydrolysis of the following compound?

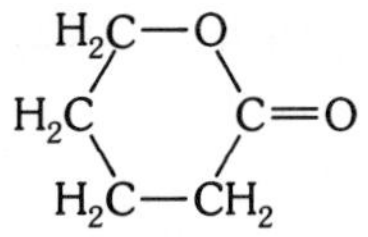

$$\begin{array}{c} H_2C\!-\!O \\ H_2C \qquad C\!=\!O \\ H_2C\!-\!CH_2 \end{array}$$

13.28 What are the structures of the products of the saponification of the compounds in Review Exercise 13.24?

13.29 What forms, if anything, when the compounds of Review Exercise 13.25 are subjected to saponification by aqueous KOH? Write their structures.

13.30 What are the products of the saponification of the compound given in Review Exercise 13.26? (Assume that aqueous NaOH is used.)

13.31 Write the structure of the organic ion that forms when the compound of Review Exercise 13.27 is saponified.

Phosphate Esters and Anhydrides

13.32 Write the structures of the following compounds.
(a) Monoethyl phosphate
(b) Monomethyl diphosphate
(c) Monopropyl triphosphate

13.33 State one apparent advantage to the body of its converting many compounds into phosphate esters.

13.34 What part of the structure of ATP is particularly responsible for its being described as an *energy-rich* compound? Explain.

Synthesis of Amides

13.35 Examine the following acyl group-transfer reaction.

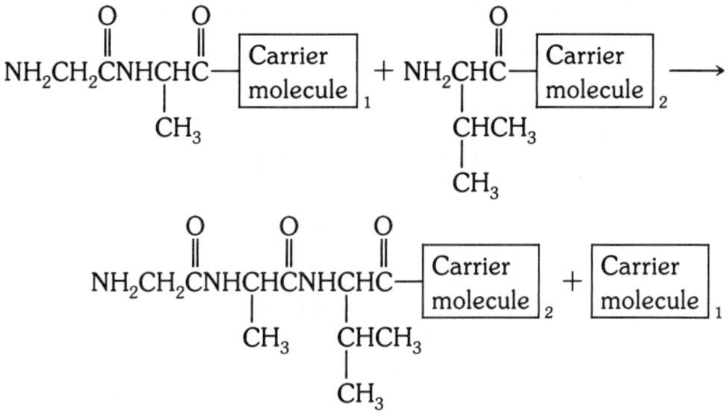

$$\underset{\underset{CH_3}{|}}{\overset{O}{\underset{\|}{NH_2CH_2CNHCHC}}}\!-\!\boxed{\begin{array}{c} Carrier \\ molecule \end{array}}_1 + \underset{\underset{\underset{CH_3}{|}}{CHCH_3}}{\overset{O}{\underset{\|}{NH_2CHC}}}\!-\!\boxed{\begin{array}{c} Carrier \\ molecule \end{array}}_2 \longrightarrow$$

$$\underset{\underset{CH_3}{|}}{\overset{O}{\underset{\|}{NH_2CH_2CNHCHCNHCHC}}}\!-\!\boxed{\begin{array}{c} Carrier \\ molecule \end{array}}_2 + \boxed{\begin{array}{c} Carrier \\ molecule \end{array}}_1$$

(a) Which specific acyl group transferred? (Write its structure.)
(b) How many amide bonds are in the product?

13.36 Write the structures of the amides that can be made from the following pairs of compounds.
(a) Acetic acid and ammonia
(b) Propionic acid and methylamine
(c) Benzoic acid and dimethylamine
(d) Formic acid and ethylamine

13.37 What are the structures of the amides that can be made from the following pairs of compounds?
(a) Butyric acid and methylamine
(b) Acetic acid and dimethylamine

(c) Formic acid and ammonia
(d) Propionic acid and ammonia

Reactions of Amides

13.38 What are the products of the hydrolysis of the following compounds? (If no hydrolysis occurs, state so.)

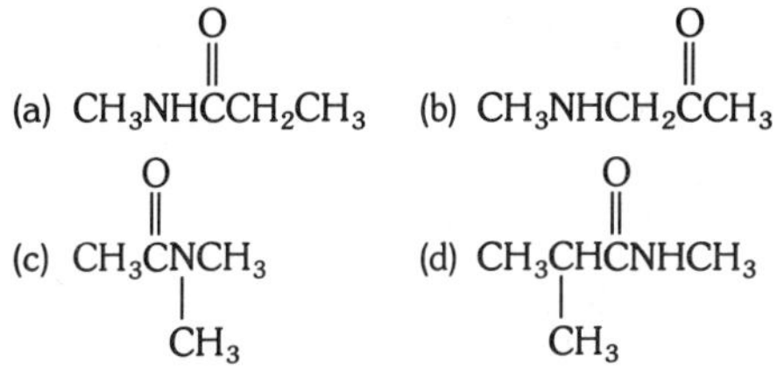

(a) $CH_3NHCCH_2CH_3$ (b) $CH_3NHCH_2CCH_3$

(c) CH_3CNCH_3 (d) $CH_3CHCNHCH_3$

13.39 Write the structures of the products of the hydrolysis of the following compounds. If no reaction occurs, state so. If more than one bond is subject to hydrolysis, be sure to hydrolyze all of them.

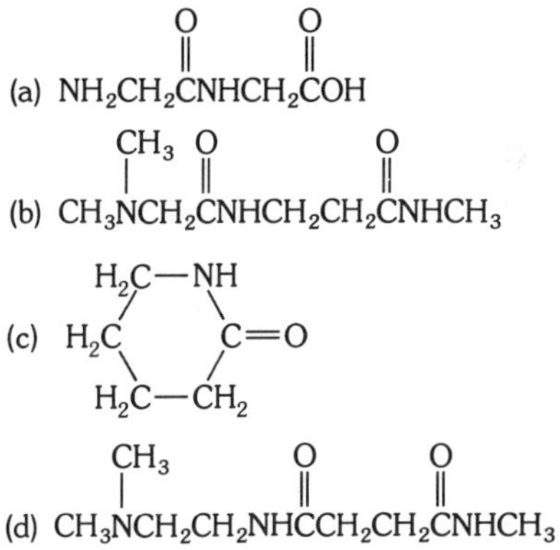

(a) $NH_2CH_2CNHCH_2COH$

(b) $CH_3NCH_2CNHCH_2CH_2CNHCH_3$

(c) $\begin{array}{c} H_2C\!-\!NH \\ H_2C \qquad C\!=\!O \\ H_2C\!-\!CH_2 \end{array}$

(d) $CH_3NCH_2CH_2NHCCH_2CH_2CNHCH_3$

Review of Organic Reactions

13.40 What are all of the functional groups we have studied that can be changed by each of the following reactants? Write the equations for the reactions, using general symbols such as ROH or RCO_2H, and so forth, to illustrate these reactions, and name the organic families to which the reactants and products belong.
(a) Water, with either an acid or an enzyme catalyst.
(b) Hydrogen (or a hydride ion donor) and any needed catalysts and special conditions.
(c) An oxidizing agent represented by (O), such as $Cr_2O_7{}^{2-}$ or $MnO_4{}^-$, but not oxygen used in combustion.

13.41 We have described three functional groups that typify those involved in the chemistry of the digestion of carbohydrates, fats and oil, and proteins. What are the names of these groups and to which type of food does each belong?

13.42 Complete the following reaction sequences by writing the structures of the organic products. If no reaction occurs, state so. (These constitute a review of this and earlier chapters on organic chemistry.)

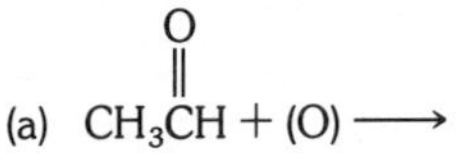

(a) $CH_3CH + (O) \longrightarrow$

(b) $CH_3CH_2CH_3 + H_2SO_4 \longrightarrow$

(c) $CH_3OH + CH_3CO_2H \xrightarrow[\text{heat}]{H^+}$

(d) $CH_3CH_2\overset{\overset{\displaystyle OH}{|}}{C}HCH_3 \xrightarrow[\text{heat}]{(O)}$

(e) $CH_3\overset{\overset{\displaystyle OH}{|}}{C}HCH_3 \xrightarrow[\text{heat}]{H_2SO_4}$

(f) $CH_3O\overset{\overset{\displaystyle O}{||}}{C}\overset{\underset{\displaystyle CH_3}{|}}{C}HCH_3 + H_2O \xrightarrow{H^+}$

(g) $CH_3CH_2CH{=}CH_2 + H_2O \xrightarrow{H^+}$

(h) $CH_3\overset{\overset{\displaystyle O-CH_3}{|}}{C}H-O-CH_3 + H_2O \xrightarrow{H^+}$

(i) $CH_3\overset{\overset{\displaystyle O}{||}}{\underset{\underset{\displaystyle CH_3}{|}}{C}}HC-O-CH_3 + NaOH(aq) \longrightarrow$

(j) $CH_3-O-CH_2CH_2\overset{\overset{\displaystyle O}{||}}{C}CH_3 + H_2O \longrightarrow$

(k) $CH_3CH_2CO_2H + NaOH(aq) \longrightarrow$

(l) $CH_3\overset{\overset{\displaystyle O}{||}}{C}CH_3 + H_2 \xrightarrow[\text{heat, pressure}]{Ni}$

(m) $CH_3O\overset{\overset{\displaystyle O}{||}}{C}CH_3 + H_2O \xrightarrow[\text{heat}]{H^+}$

(n) $CH_3(CH_2)_8CH_3 + NaOH(aq) \longrightarrow$

(o) $CH_3CH_2\overset{\overset{\displaystyle OCH_2CH_3}{|}}{C}HOCH_2CH_3 + H_2O \xrightarrow[\text{heat}]{H^+}$

(p) $\text{C}_6\text{H}_5{-}\overset{\overset{\displaystyle O}{||}}{C}{-}H + (O) \rightarrow$

(q) $CH_3CH_2\overset{\overset{\displaystyle O}{||}}{C}-O-CH_2CH_3 + NaOH(aq) \xrightarrow[\text{heat}]{}$

(r) $CH_3-\overset{\overset{\displaystyle O}{||}}{C}-H + 2CH_3OH \xrightarrow[\text{heat}]{H^+}$

(s) $CH_3\overset{\overset{\displaystyle O}{||}}{C}CH_3 + (O) \longrightarrow$

(t) $CH_3-O-\overset{\overset{\displaystyle OCH_3}{|}}{C}HCH_2CH_2CH_3 + NaOH(aq) \longrightarrow$

(u) $CH_3CH_2OH + H_2O \longrightarrow$

(v) $CH_3CH_2CO_2H + CH_3OH \xrightarrow[\text{heat}]{H^+}$

(w) $CH_3CH_2CH_2NH_2 + HCl(aq) \xrightarrow[25\,°C]{}$

(x) $H\overset{\overset{\displaystyle O}{||}}{C}NH_2 + H_2O \xrightarrow[\text{heat}]{}$

(y) $CH_3SH + (O) \longrightarrow$

(z) $CH_3NH\overset{\overset{\displaystyle O}{||}}{C}CH_2CH_2\overset{\overset{\displaystyle O}{||}}{C}NHCH_3 + H_2O \xrightarrow{\text{enzyme}}$
    (excess)

13.43 Write the structures of the organic products, if any, that form in the following situations. If no reaction occurs, state so. Some of these constitute a review of the reactions of earlier chapters.

(a) $CH_3CH_2CO_2^- + HCl(aq) \longrightarrow$

(b) $CH_3CH_2CH_2CH_3 + (O) \longrightarrow$

(c) $CH_3CH_2-O-CH_2CH_2\overset{\overset{\displaystyle O}{||}}{C}CH_3 + NaOH \longrightarrow$

(d) $CH_3CH_2CH_2-O-\overset{\overset{\displaystyle O}{||}}{C}-CH_2CH_2-O-\overset{\overset{\displaystyle O}{||}}{C}-CH_3$
    $+ H_2O \xrightarrow{H^+}$
    (excess)

(e) $C_6H_5-\overset{\overset{\displaystyle O-CH_2CH_3}{|}}{C}H-OCH_2CH_3 + H_2O \xrightarrow{H^+}$

(f) $CH_3\overset{\overset{\displaystyle OH}{|}}{\underset{\underset{\displaystyle CH_3}{|}}{C}}CH_2CH_3 \xrightarrow{(O)}$

(g) $CH_3CH_2CO_2H + CH_3\overset{\overset{\displaystyle CH_3}{|}}{C}HOH \xrightarrow[\text{heat}]{H^+}$

(h) $CH_3(CH_2)_9CH_3 + NaOH \longrightarrow$

(i) $CH_3-O-\overset{\overset{\displaystyle O}{||}}{C}CH_2CH_2-O-\overset{\overset{\displaystyle O}{||}}{C}CH_2CH_3$
    $+ NaOH(aq) \longrightarrow$
    (excess)

(j) $CH_3(CH_2)_6CO_2H + NaOH(aq) \longrightarrow$

(k) $^-O_2CCH_2CH_2CH_2CO_2^- + HCl(aq) \longrightarrow$
    (excess)

(l) $CH_3OH + HO_2CCH_2CH_2CH_2CO_2H \xrightarrow[\text{heat}]{H^+}$
    (excess)

(m) $C_6H_5-\overset{\overset{\displaystyle O}{||}}{C}-O-CH_2CH_3 + NaOH(aq) \xrightarrow[\text{heat}]{}$

(n) $C_6H_5-\overset{\overset{\displaystyle O}{||}}{C}-H + 2CH_3OH \xrightarrow[\text{heat}]{H^+}$

(o) $CH_3(CH_2)_3CH_3 + (O) \longrightarrow$

(p) $CH_3\overset{\overset{\displaystyle O}{\|}}{C}CH_2CH_3 + (O) \longrightarrow$

(q) $HO\overset{\overset{\displaystyle O}{\|}}{C}CH_2CH_2CH_3 + NaOH(aq) \xrightarrow[25\ °C]{}$

(r) $CH_3CH_2-\overset{\overset{\displaystyle OCH_2CH_3}{|}}{\underset{\underset{\displaystyle CH_3}{|}}{C}}-O-CH_2CH_3 + H_2O \xrightarrow{H^+}$

(s) ⬠ $+ NaOH(aq) \rightarrow$

(t) $CH_3-O-\overset{\overset{\displaystyle O}{\|}}{C}CH_2CH_2\overset{\overset{\displaystyle O}{\|}}{C}-O-CH_3 + H_2O \xrightarrow{enzyme}$
(excess)

(u) $NH_2CH_2CH_2CH_2NH_2 + HCl(aq) \xrightarrow[25\ °C]{}$
(excess)

(v) $CH_3-OCH_2CH_2\overset{\overset{\displaystyle O}{\|}}{C}-H + (O) \longrightarrow$

(w) $CH_3CH_2OH + CH_3(CH_2)_4CO_2H \xrightarrow[heat]{H^+}$

(x) $CH_3CH_2-O-\overset{\overset{\displaystyle CH_3}{|}}{CH}-O-CH_2CH_3 + H_2O \xrightarrow[heat]{H^+}$

(y) $CH_3CH_2\overset{\overset{\displaystyle O}{\|}}{C}CH_2CH_3 + H_2 \xrightarrow[heat,\ pressure]{Ni}$

(z) $CH_3S-SCH_3 + 2(H) \longrightarrow$

Important Acids and Salts (Special Topic 13.1)

13.44 Which simple carboxylate ion is an intermediate in metabolism?

13.45 The carboxylic acids or salts used as food additives are most often used to retard the growth of what?

Important Esters (Special Topic 13.2)

13.46 The parabens are esters of what acid? What are they used for?

13.47 What is the purpose of an analgesic? An antipyretic?

13.48 What advantage does aspirin have over its parent acid, salicylic acid, as an analgesic and antipyretic?

13.49 What is the common name of the methyl ester of salicylic acid?

13.50 Dacron is a *copolymer*. What does this mean?

13.51 How are Mylar and Dacron alike? How are they different?

13.52 What alcohol is used to make Dacron?

Nylon (Special Topic 13.3)

13.53 What is the functional group in nylon?

13.54 The strength of a nylon fiber comes in part from what weak bond?

13.55 Newly forming nylon fibers are stretched as they are produced. How does this improve the fiber's strength?

13.56 What does 66 refer to in "nylon-66?"

Barbiturates (Special Topic 13.4)

13.57 What functional group is present in the barbiturates?

13.58 Mild, well-regulated doses of barbiturates do what to the central nervous system? What functions slow down?

13.59 A combination of a barbiturate with what other drug has been known to be lethal at doses under which neither is lethal when taken individually?

Chapter 14
Carbohydrates

What an incredible amount of chemical work goes on in the world's rice paddies. The energy in sunlight fashions simple molecules into the complex forms found in carbohydrates, the subject of this chapter, and other substances.

14.1 BIOCHEMISTRY—AN OVERVIEW

Building materials, energy, and information are basic essentials for life.

Biochemistry is the systematic study of the chemicals of living systems, their organization, and the principles of their participation in the processes of life.

Life Has a Molecular Basis. The molecules of living systems are lifeless, yet life has a molecular basis. When they are isolated from their cells or when they are studied in living environments, the chemicals at the foundation of life obey all of the known laws of chemistry and physics. In isolation, however, not one single compound of a cell has life. Thus the cell's intricate organization is as important to life as the chemicals themselves. The cell is the smallest unit of matter that lives and can make a new cell exactly like itself. A change in its environment—a change in temperature, pH, or pressure, or the appearance of hostile substances—can quickly render a cell nothing more than a lifeless bag of chemicals.

An ancient writer said that we once were dust and to dust we will return. Compared with dust and ashes scattered by the four winds, an organism is a finely ordered, high-energy system. The marvelous interlude between dust and dust that we call life endures only as long as the organism can maintain this dynamic organization of parts and defend it against the many natural tendencies to disintegration and decay.

Our Vital Needs Are Materials, Energy, and Information. Whether it be a plant or an animal, all life has three absolutely vital needs: materials, energy, and information. Our purpose in the remainder of this text is to study the molecular basis of meeting these three needs and how one organism, the human body, uses the substances that satisfy these needs. As we look in a general way at life at the subcellular and molecular levels, we will also learn of several chemical dangers constantly faced and almost always met by a living system.

We will begin in this and the next three chapters to study the organic materials of life, starting with the three great classes of foodstuffs: carbohydrates, lipids, and proteins. We use their molecules to build and run ourselves and to try to stay in some state of repair. Proteins are particularly important in both the structures and functions of cells, whether of plants or animals. Plants rely heavily on carbohydrates for cell walls, and animals obtain considerable energy from carbohydrates made by plants. Lipids (fats and oils) serve many purposes and are rich in energy. Because of their central catalytic role in regulating chemical events in cells, we will study a special family of proteins, the enzymes. These are a part of the cell's information networks.

All of the materials taken in the diet have to be processed by the digestive system and distributed by the bloodstream and the lymph if cells are to obtain what they need in forms suitable for that species. We must, therefore, study the fluid networks that process materials, distribute them, and carry wastes away, with a special emphasis on the molecular basis of using oxygen and releasing carbon dioxide.

Energy for Living Also Has a Molecular Basis. How can gulps of air, a peanut butter sandwich, and a glass of milk be used by the body to make or repair, say, muscle cells? How do these materials translate into such a whirlwind of energy as an active little child? Or an athlete? The molecular basis of energy for life forms another broad topic of our study. And as we study biochemical energetics and its enzymes and metabolic pathways, we'll have numerous occasions to peer deeply into the molecular basis of some disorders and diseases.

Some Chemicals Store Information for Living. Enzymes are actually a part of the molecular basis of information, and some kind and form of information are essential to life. Without plans or blueprints, materials and energy could combine to produce only rubble and rubbish. Monkeys swinging hammers would only reduce a stack of lumber to splinters. Carpenters, using no more raw energy, can make the same lumber into a building, because they have both a plan and experience.

Cornstarch, potato starch, table sugar, and cotton are all carbohydrates.

Meat is rich in protein.

Butter, lard, margarine, and corn oil are all lipids.

Individual genes are sections of molecules of a polymer called DNA, which is one kind of nucleic acid.

Enzymes, however, do not originate the plans of a living system. They only help to carry them out. The blueprint for an individual member of a species is encoded in the molecular structures of a family of substances called nucleic acids. Nucleic acids direct the synthesis of a cell's enzymes, and each individual has a unique set.

Our systems are probably nowhere more vulnerable to atomic radiations or to chemicals that mimic them than at the cells' nucleic acids. We will close our study of the molecular basis of life with a study of radiations in health and disease, a topic that is better appreciated *after* the importance of nucleic acids and enzymes has made a deep impression.

We have an exciting trip ahead of us. In the preceding chapters we have slowly and carefully built a solid foundation of chemical principles. It's been like a mountain climbing trip where the route for a large part of the trek is through country with few grand vistas but still with a beauty of its own. Now we're moving to elevations where the vistas begin to open. It's like a new beginning, and we start it with a study of the first of the three chief classes of food materials, the carbohydrates.

14.2   MONOSACCHARIDES

The structure of glucose is the key to the structures of most carbohydrates.

Carbohydrates are polyhydroxy aldehydes, polyhydroxy ketones, or substances that yield these by simple hydrolysis (by reaction with water).

The oxidized and reduced forms of polyhydroxy aldehydes and ketones, as well as certain amino derivatives, are also in the family of carbohydrates.

The Monosaccharides Are the Monomers for the Hydrolyzable Carbohydrates. Carbohydrates that do not react with water are called **monosaccharides** or sometimes **simple sugars.** Virtually all have names ending in *-ose*, and they have the general formula $(CH_2O)_n$. Subclasses based on total carbon content and functional group are useful and are devised as follows:

1.   The number of carbons in one molecule, or the value of n in $(CH_2O)_n$.

If the number of carbons is	The monosaccharide is a
3	triose
4	tetrose
5	pentose
6	hexose, etc.

2.   The nature of the carbonyl group. If an aldehyde group is present, the monosaccharide is an **aldose.** If a keto group is present, the monosaccharide is a **ketose.**

Combinations of these terms are often used. **Glucose,** for example, is a hexose with an aldehyde group, so it is an **aldohexose. Fructose,** a hexose with a keto group, is a **ketohexose.**

Oligosaccharide molecules yield 3–10 monosaccharide molecules when they are hydrolyzed.

Disaccharides and Polysaccharides Are Hydrolyzable Carbohydrates. The molecules of **disaccharides,** such as the nutritionally important **sucrose, maltose,** and **lactose,** can be hydrolyzed to two monosaccharide molecules. Starch and cellulose are called **polysaccharides,** because each of their molecules can be hydrolyzed to hundreds of monosaccharide molecules, those of glucose.

Most Mono- and Disaccharides Are Reducing Sugars. Carbohydrates that give positive tests with Tollens' or Benedict's reagents are called **reducing sugars** because something in the reagent, Ag^+ or Cu^{2+}, is reduced in a positive test (to Ag or Cu_2O). All monosaccharides and nearly all disaccharides — sucrose is an exception — are reducing sugars, but polysaccharides are not.

Glucose Is the Centerpiece of Carbohydrate Chemistry. The widespread distribution of cellulose and starch in the enormous numbers and kinds of plants very likely make glucose the most abundant organic species on earth. Cellulose, a polymer of glucose, is present in the walls of virtually all plant cells, and starch, another glucose polymer, is the way plants store chemical energy. Special Topic 14.1 describes in general terms how glucose units are made in plants by photosynthesis.

Glucose is also by far the most common carbohydrate in blood and is often called **blood sugar.** The structure and properties of glucose, as you can see, are thus central to carbohydrate chemistry.

Glucose is $C_6H_{12}O_6$ and many of its properties can be understood in terms of its being 2,3,4,5,6-pentahydroxyhexanal.

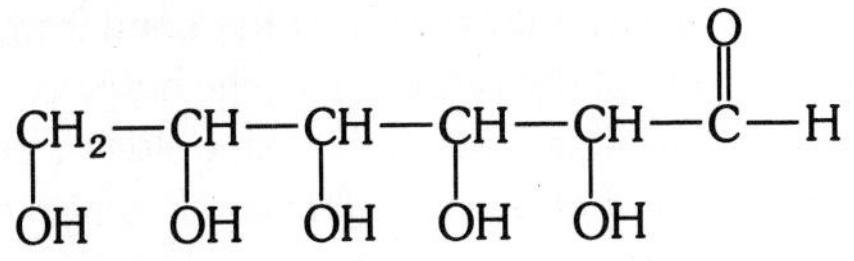

Glucose is also called corn sugar, because it can be made by the hydrolysis of cornstarch.

$$\underset{\underset{OH}{|}}{CH_2}-\underset{\underset{OH}{|}}{CH}-\underset{\underset{OH}{|}}{CH}-\underset{\underset{OH}{|}}{CH}-\underset{\underset{OH}{|}}{CH}-\overset{\overset{O}{\|}}{C}-H$$

Basic structure of glucose

This structure, however, turns out to be just one of three forms. The other two are ring structures, as we'll study next.

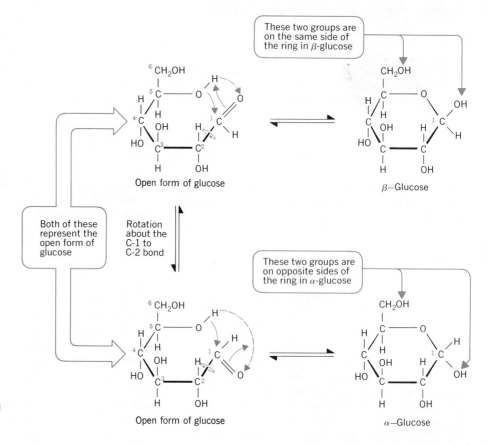

These two groups are on the same side of the ring in β-glucose

Open form of glucose

β—Glucose

Both of these represent the open form of glucose

Rotation about the C-1 to C-2 bond

These two groups are on opposite sides of the ring in α-glucose

Open form of glucose

α—Glucose

FIGURE 14.1
The α and β forms of glucose arise from the same intermediate, the open-chain form. Depending on how the aldehyde group, CH=O, is pointing when the ring closes, one ring form or the other results.

The H at C-2 and C-5 and the OH at C-3 (Figure 14.1) do not stick *inside* the ring. They stick *above or below the plane* of the ring.

Glucose Exists Almost Entirely in Cyclic Forms with Easily Opened Rings. Figure 14.1 gives the structures of the three forms of glucose, including the open form we just looked at. (It's shown in Figure 14.1 in a coiled configuration to make it easier to see its relationship to the cyclic forms). In an aqueous solution, all three forms are in dynamic equilibrium:

$$\alpha\text{-Glucose} \rightleftharpoons \text{open-chain form} \rightleftharpoons \beta\text{-glucose}$$

The key to this equilibrium is the instability of the hemiacetal system, studied in Section 12.4 where we learned that its breakdown gives a parent aldehyde and alcohol group.

The ordinary form of crystalline glucose is the cyclic α-form. Its hemiacetal system is at ring carbon-1, where there is both an ether linkage and an OH group. Notice for future reference that this OH group is on the opposite side of the ring from the CH_2OH group. Because the hemiacetal system is an integral part of the ring of α-glucose, the ring is not as stable as ordinary six-membered rings. The spontaneous breaking up and reforming of the hemiacetal group inevitably causes the ring to open and close. The breakup opens the ring and the hemiacetal's parent aldehyde group emerges. Its parent alcohol group also appears, but it is down the chain *on the same molecule*. What is now important to the glucose equilibrium is that groups are free to rotate about single bonds in open-chain systems.

When a rotation of one-half a turn takes place at the bond from carbon-1 to 2 before the ring snaps shut again, the OH group of the reformed cyclic hemiacetal acquires an orientation opposite to what it had in α-glucose. It is now on the same side of the ring as the C-6 CH_2OH group. We thus have another form of glucose, β-glucose. It is also a cyclic hemiacetal and is therefore somewhat unstable. It can open up again, experience chain rotations, and then close, either as the same β form or, after a rotation, as the α form. These ring openings and closings are what take place in the dynamic equilibrium of aqueous glucose.

The two cyclic forms of glucose differ only in the orientation of the OH group at C-1, upward in β-glucose and downward in α-glucose. The projections of all of the other OH groups

1. First write a six-membered ring with an oxygen in the upper right-hand corner.

2. Next "anchor" the —CH$_2$OH on the carbon to the left of the oxygen. (Let all the —H's attached to ring carbons be "understood.")

   or condense to

3. Continue in a *counterclockwise* way around the ring, placing the —OH groups first down, then up, then down.

   or condense to

4. Finally, at the last site on the trip, how the last —OH is positioned depends on whether the α or the β form is to be written. The α is "down," the β "up."

   β-Glucose   or condense to   β-Glucose   α-Glucose

If this detail is immaterial, or if the equilibrium mixture is intended, the structure may be written as:

   or condense to

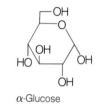
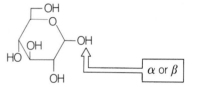

α or β

FIGURE 14.2
How to draw the cyclic forms of glucose in a highly condensed way.

do not change. At equilibrium the proportions are 36% α-glucose, 0.02% open-chain glucose, and 64% β-glucose. Figure 14.2 shows how to write these structures easily and quickly.

The open-chain form of glucose occurs only in solution, but it is the form attacked by Tollens' or Benedict's reagents. As the open-form molecules react and are thus removed from the equilibrium, they are replaced by a steady shifting of the equilibrium from closed forms to the open form. This is why glucose, despite the fact that it exists almost entirely in one cyclic form or the other, still gives the chemical properties of a pentahydroxy aldehyde (and why it's still all right to define a monosaccharide as a polyhydroxy aldehyde — or ketone — rather than as a cyclic hemiacetal).

This is another example of Le Chatelier's principle in action.

Galactose, an Isomer of Glucose, Also Exists in Cyclic Hemiacetal Forms. Galactose, an aldohexose and an isomer of glucose, occurs in nature mostly as structural units in larger molecules such as the disaccharide lactose. Galactose differs from glucose only in the orientation of the C-4 OH group. Like glucose, it is a reducing sugar, and it exists in solution in three forms, alpha, beta, and open.

Lactose is milk sugar.

α-Galactose   Open form of galactose   β-Galactose

Fructose Is an Important Ketohexose.   Fructose is found together with glucose and sucrose in honey and in fruit juices. It, too, can exist in more than one form, including cyclic hemiketals. We'll show only one.

Fructose — open forms   β-Fructose

Because its molecules have the α-hydroxyketone system, fructose is a reducing sugar. We obtain fructose when we digest table sugar, and the mono- and diphosphoric acid esters of fructose are important compounds in metabolism.

The important monosaccharides are glucose, galactose, and fructose. All are reducing sugars. All occur in key foodstuffs, and all are metabolized by the body.

Deoxycarbohydrates Have Fewer OH Groups.   A deoxycarbohydrate is one lacking an OH group where normally this group is expected. 2-Deoxyribose, for example, is the same as ribose except that there is no OH group at C-2, just two H's instead. Ribose and 2-deoxyribose are building blocks of nucleic acids, ribose for the ribonucleic acids (or RNAs) and deoxyribose for deoxyribonucleic acid (or DNA). Each of these aldopentoses can exist in three forms, two cyclic hemiacetals and an open form. We'll show only one of the closed forms of each.

β-Ribose   β-2-Deoxyribose

14.3   OPTICAL ISOMERISM AMONG THE CARBOHYDRATES[1]

Carbohydrate molecules possess a unique handedness that adapts them to the enzymes that catalyze their reactions.

Galactose is one of these isomers of glucose.

Glucose is one of 16 isomers, not counting those that differ only in the easily convertible α and β forms. (Each of the 16 isomers exists in these cyclic forms.) We refer to a different kind of isomerism, one new to our study, to a kind in which molecules are as closely related as a left hand is to the right hand.

Molecules Can Be as Nearly Alike as an Object and Its Mirror Image and Still Be Different.   One of the 16 isomers of glucose has molecules that are exact mirror images to those of naturally occurring glucose, and yet this isomer is useless to the life of a cell. Figure 14.3 shows two structures with a mirror between them. To the left of the mirror is naturally occurring α-glucose. If you put this molecular model in front of a real mirror, then carefully

[1] Where time is short, this section can be omitted without causing problems in later chapters.

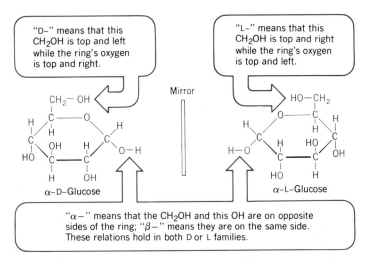

FIGURE 14.3
The D and L forms of α-glucose.

made a model of what you saw in the mirror as the image, you would have the structure shown on the right. It's as if you put your left hand in front of a mirror and made a model of what you see as an image. The model would be of your *right* hand.

The structures in Figure 14.3 are so closely alike that we want to name them both α-glucose, so to designate which is which, we call the structure of natural glucose α-D-glucose and that of its mirror image α-L-glucose. Thus we say that one glucose is in the D family of carbohydrates, and the other is in the L family.

The essential feature that places a carbohydrate in the **D family** is having the CH_2OH group *above* the plane of the ring when the ring's O atom is to its *right* and in the *upper-right-hand corner* (see Figure 14.3). Members of the **L-family** also have the CH_2OH group topside, but over on the right side of the ring with the ring's oxygen atom to its left. Thus, if you look back at the cyclic forms of galactose, you will see that they are also in the D family.

The other isomers of glucose differ in the ways in which the OH groups project, up or down, around the ring. The 16 isomers occur as 8 pairs, each pair consisting of molecules related as an object to its mirror image.

Two Molecules Are Truly Identical Only if They Superimpose. If, as we have said, α-D- and α-L-glucose are *isomers,* then they are truly *different* substances. One striking difference, as we noted, is the uselessness of α-L-glucose as a nutrient to living things. These two isomers are no more identical than a left-hand and a right-hand glove. The right-hand glove is relatively useless to the left hand. It doesn't fit properly. (This is an advance hint that the ability of an enzyme to fit to a particular molecule is important.)

If the two gloves were identical, we should be able to *superimpose* them. We should be able to perform an imaginary blending and merging of one with its mirror image, and do this so successfully that they would coincide identically and simultaneously in every part.

Try this with our left and right hands. They're related as an object to its own mirror image (ignoring wrinkles, fingerprints, scars, and jewelry). Line up your hands palms facing and imagine one blending into and through the other. It may seem that everything is merging identically — thumb to thumb, little finger to little finger, and so forth — but the palms are coming out on opposite sides. If you turn one hand over to get the palms to come out right, then the fingers won't superimpose. Your left and right hands, although as closely related as an object to its mirror image, do not superimpose. This is how structurally close D and L isomers are.

The test of superimposability is the single most definitive test for judging if two molecular models are of the same compound.

Molecules That Do Not Superimpose with Their Mirror Images Are Chiral. A large number of substances have molecules that are different in the way that two gloves differ.

Chiral is from the Greek *cheir*, meaning ''hand.''

These molecules, like gloves, also have *handedness*. The technical term is **chirality.** A molecule with chirality does not superimpose with its mirror image, and is called a **chiral molecule.** Any molecule, like one of H_2, H_2O, or CH_4, that does superimpose with its mirror image lacks chirality or handedness and is called an *achiral molecule.* Achiral molecules are always symmetrical in some way.

Isomers whose molecules are related as object to mirror image that cannot be superimposed are called **enantiomers.** α-D-Glucose and α-L-glucose are thus related as enantiomers.

Enantiomers Have Optical Activity. Using your hands as a pair of enantiomers, think for a moment about how similar they are. They have identical ''formula weights'' and ''molecular formulas.'' All ''bond angles'' are identical, like the angle between the thumb and the little finger. All ''bond distances'' are also identical, like the distance between the tips of the thumb and little finger in each hand. If your fingertips were electronegative atoms joined to a heavy central atom (the palm area), the ''molecules'' of your two hands would *have* to have identical polarities.

The result is that we would expect enantiomers to have identical physical properties: boiling points, melting points, solubilities, and densities, for example. And this is precisely observed among pairs of enantiomers. There is only one physical difference between two enantiomers. (Luckily; otherwise, how would we tell them apart?)

The physical difference concerns the behavior of two enantiomers to plane-polarized light, the kind of light that gets filtered out by Polaroid sunglasses. If plane-polarized light is sent through a solution of α-D-glucose, the solute molecules make the plane of the plane-polarized light rotate to the right a certain number of degrees. You would have to twist your head the same way if, wearing Polaroid sunglasses, you wanted to filter out the polarized light *emerging from the solution.* (In the lab, a special instrument called a polarimeter is used to measure the number of degress of twist.) Any compound that is able to rotate the plane of plane-polarized light in the way we just described is said to have **optical activity.** All enantiomers are optically active.

The number of degrees of rotation depends on the concentration and the distance through the solution, *but the direction, right or left, is a unique property of each enantiomer.* α-D-Glucose twists the plane of plane-polarized light to the right. In the same container at the identical concentration, α-L-glucose twists the plane the identical number of degrees in the opposite direction, to the left. This one property, the *direction,* is the physical difference between enantiomers. We say that α-D-glucose is *dextrorotatory* after a Latin root, *dexter,* meaning to the right. α-L-Glucose is *levorotatory,* where *levo* signifies to the left. Because of their effects on polarized light, isomers related as enantiomers have long been called **optical isomers,** and this kind of isomerism is called **optical isomerism.**

The designations D or L are no more than the names of two families. There is no automatic association between the names D and L and the properties of dextro- or levorotatory. Thus, naturally occurring fructose is in the D family, but is strongly levorotatory (and is nicknamed *levulose*). Naturally occurring glucose is also in the D family, but is strongly dextrorotatory (and is sometimes called *dextrose*). Signs of rotation are sometimes placed within the name of an enantomer. You may see a bottle with (+)-glucose listed as an ingredient. The (+) sign means dextrorotatory, just as the (−) sign, meaning levorotatory, might appear on a label of fructose.

Toward Achiral Substances, Enantiomers Have Identical Chemical Properties. When either D- or L-glucose is mixed with a reactant whose molecules or ions are achiral, like those of H_2 or $Ag(NH_3)_2^+$ (Tollens' reagent), the enantiomers have identical reactions. This shouldn't be too surprising; by analogy, either of your chiral hands ''reacts'' identically toward achiral objects, like broom handles and water glasses. The tremendous ''trifle'' in nature occurs when reactants are themselves chiral. Then enantiomers display dramatic differences in reactivity.

Molecules of All Enzymes Are Chiral. Virtually all reactions of living systems require enzyme catalysts, and all of nature's enzymes are chiral and in the same optical family. Those of the other family do not occur naturally.

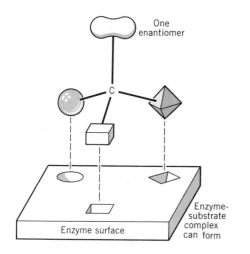 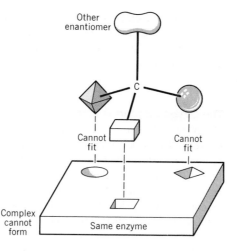

FIGURE 14.4
Since an enzyme is chiral, it can accept substrate molecules of only one pair of enantiomers. To illustrate this difference, we have used simple geometric forms. Notice that the forms labeled enantiomers are in an object to mirror image relationship, but they can't be superimposed. On the left, the enzyme can accept as a substrate the molecule of one enantiomer. On the right, the same enzyme can't accept a molecule of the other enantiomer, because the shapes don't match.

To generalize, we will use the term **substrate** for any compound undergoing an enzyme-catalyzed reaction. The "moment of truth" occurs when a substrate molecule approaches the surface of an enzyme molecule. Unless the former can physically fit to the surface of the latter, no catalysis occurs. Figure 14.4 uses simple models of enantiomers to show how one can fit to a chiral surface but the other cannot. This is why L-glucose is useless in cells. Its molecules are like left hands rummaging in a box of right-hand gloves. Nothing fits. Thus subtle differences in molecular geometries are as much matters of life and death as gross structural features.

The fit of molecules of substrate and enzyme is as dependent on complementary shapes as the fit of a key to its lock.

14.4 DISACCHARIDES

The three nutritionally important disaccharides are maltose, lactose, and sucrose.

The relationships of the disaccharides that we will study to the monosaccharides just discussed are seen in the following "word equations."

$$\text{Maltose} + \text{H}_2\text{O} \longrightarrow \text{glucose} + \text{glucose}$$
$$\text{Lactose} + \text{H}_2\text{O} \longrightarrow \text{glucose} + \text{galactose}$$
$$\text{Sucrose} + \text{H}_2\text{O} \longrightarrow \text{glucose} + \text{fructose}$$

You can see why glucose is of such central interest to carbohydrate chemists.

Maltose Is Made from Two Glucose Units. Maltose or malt sugar is not found widely in nature, although it is present in germinating grain. It occurs in corn syrup, which is made from cornstarch, and it forms from the partial hydrolysis of starch. The two glucose units in maltose are linked by an acetal bridge from carbon-1 of one unit to carbon-4 of the other.

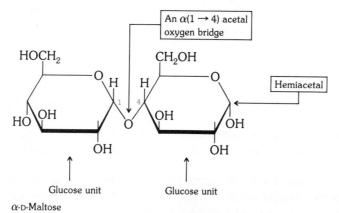

α-D-Maltose

If the OH group on the far right projected upward instead of downward, the structure would be that of β-D-maltose.

Because acetals can be hydrolyzed, maltose reacts with water. (Acids or the enzyme maltase catalyze the reaction.) The oxygen bridge breaks and two molecules of glucose form. The geometry of the bridge bears on this reaction, as we'll next see, so we need a way to describe it.

The Oxygen Bridge in Maltose Is $\alpha(1 \rightarrow 4)$.   In the structure of maltose, the bridge from C-1 of one glucose unit to C-4 of the other points downward at C-1, opposite from the reference CH_2OH group. Therefore this bridge is called an *alpha* bridge, and is given the symbol $\alpha(1 \rightarrow 4)$. Had the bridge pointed in the beta direction it would be designated as $\beta(1 \rightarrow 4)$, but the resulting molecule would not have been maltose. Another disaccharide, cellobiose, has a $\beta(1 \rightarrow 4)$ bridge between two glucose units. This is not a trifle; we can digest maltose but not cellobiose. Humans have the enzyme maltase to catalyzes the digestion of maltose, but have no enzyme to hydrolyze cellobiose.

Cellobiose is obtained by the partial hydrolysis of cellulose.

Maltose Also Has a Hemiacetal Group.   The glucose unit on the C-4 side of the $\alpha(1 \rightarrow 4)$ bridge has a hemiacetal group so this part of the maltose molecule can open and close. Thus maltose, like glucose, can exist in three forms, its own α, β, and open forms. In all three, the $\alpha(1 \rightarrow 4)$ bridge holds. The ring opening and closing action occurs only at the hemiacetal part. The availability of the open, aldehyde form makes maltose a reducing sugar. It gives positive Tollens' and Benedict's tests.

Lactose Is Made from a Galactose and a Glucose Unit.   Lactose or milk sugar occurs in the milk of mammals, 4 – 6% in cow's milk and 5 – 8% in human milk. It is obtained commercially as a by-product in the manufacture of cheese. Like maltose, lactose has an acetal oxygen bridge. From C-1 of its galactose unit there is a $\beta(1 \rightarrow 4)$ bridge to C-4 of a glucose unit. The glucose unit therefore still has a free hemiacetal system, so lactose is a reducing sugar, and it exists in its own two ring forms plus the aldehyde form in which one ring is open.

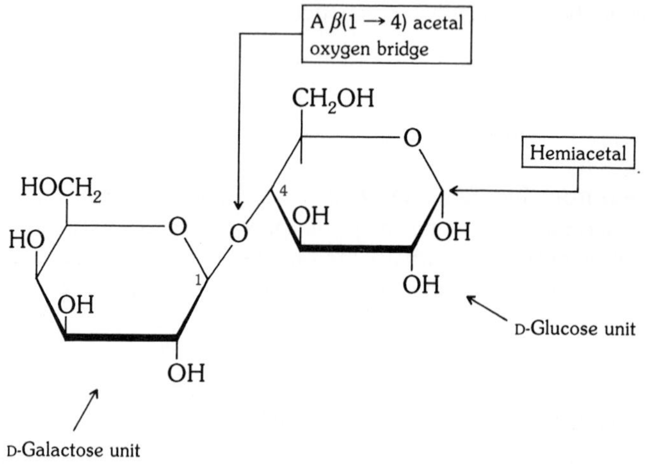

α-D-Lactose

Sucrose Is Made from a Glucose and a Fructose Unit.   Sucrose, our familiar table sugar, is obtained from sugar cane or from sugar beets. Structurally, as you can see below, it has no hemiacetal or hemiketal group, so neither ring in sucrose can open and close spontaneously in water. Hence, sucrose cannot give positive tests with Tollens' or Benedict's reagents. It's our only common nonreducing sugar.

Beet sugar and cane sugar are identical compounds — sucrose.

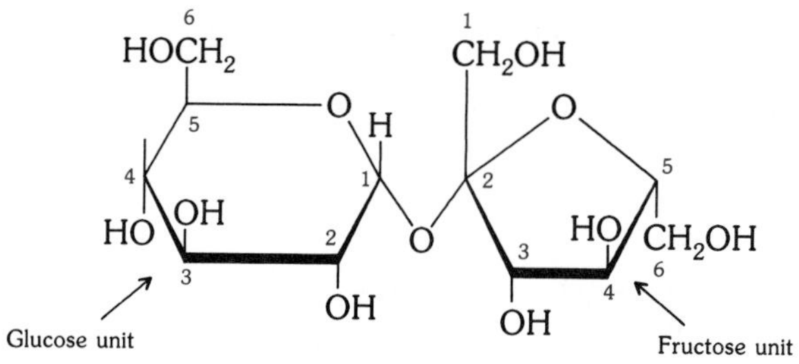

Glucose unit

Fructose unit

D-Sucrose

The 50:50 mixture of glucose and fructose that forms when sucrose is hydrolyzed is called *invert sugar,* and it makes up the bulk of the carbohydrate in honey.

14.5 POLYSACCHARIDES

Polysaccharides are polymers of monosaccharides held together by oxygen bridges similar to those in disaccharides.

Much of the glucose a plant makes by photosynthesis (Special Topic 14.1) goes to make cellulose and other substances that it needs to build its cell walls and its rigid fibers. Cotton is about 98% cellulose. Some glucose is also stored for the food needs of the plant. Free glucose, however, is too soluble in water for storage, so most is converted to a much less soluble form, starch. This polymer of glucose is particularly abundant in plant seeds. The plant uses its glucose units as a nutrient much as we do when we harvest the seeds for our own food needs.

Starch Is a Polymer of Glucose Units Linked by Acetal Oxygen Bridges. The basic structure of starch is shown in Figure 14.5. **Starch** is actually a mixture of two kinds of polymers of α-glucose. One, **amylose,** is linear, and has $\alpha(1 \rightarrow 4)$ oxygen bridges. Its long molecules are coiled with hydrogen bonds between OH groups stabilizing each coil.

The other component of starch, **amylopectin,** is branched, and has both $\alpha(1 \rightarrow 4)$ and $\alpha(1 \rightarrow 6)$ bridges. The later link the C-1 ends of long amylose molecules to C-6 positions on other long amylose chains, as also seen in Figure 14.5. There are hundreds of such links per molecule, so amylopectin is heavily branched. This branching makes coiling impossible, so its OH groups are more exposed. Amylopectin is thus more soluble in cold water than amylose, but neither dissolves well. The "solution" is actually a colloidal dispersion, because it gives the Tyndall effect.

Natural starches are about 10–20% amylose and 80–90% amylopectin. Neither is a reducing sugar and neither gives a positive Tollens' or Benedict's test. One unique test that starch does give is called the **iodine test** for starch, and it can detect extremely minute traces of starch[2] in water. When a drop of iodine reagent is added to starch, an intensely purple color develops as the iodine molecules become trapped within the vast network of starch molecules.

The acetal bridges in starch are easily hydrolyzed in the presence of acids or the enzyme amylase, which humans have. Its complete digestion gives us glucose. Its partial hydrolysis produces smaller polymer molecules that make up a substance known as *dextrin,* which has been used to manufacture mucilage and paste.

So-called *soluble* starch is partially hydrolyzed starch, and its smaller molecules more easily dissolve in water.

[2] The starch–iodine reagent is made by dissolving iodine, I_2, in aqueous sodium iodide, NaI. Iodine by itself is very insoluble in water, but iodine molecules combine with iodide ions to form the triiodide ion, I_3^-. Molecular iodine is readily available from this ion if some reactant is able to react with it.

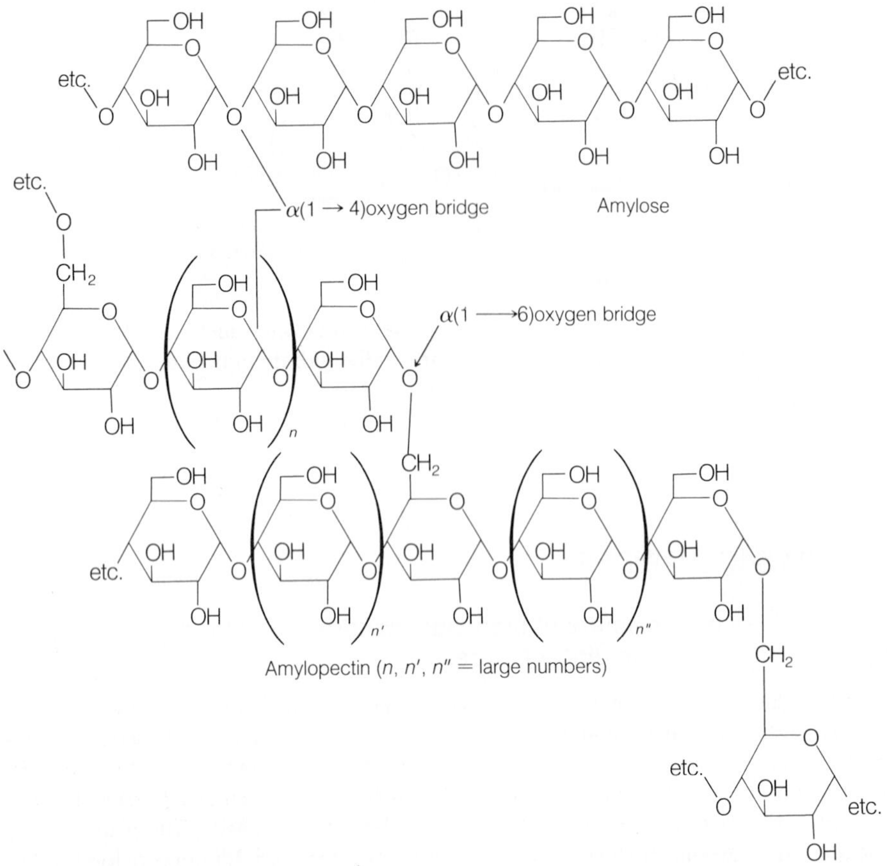

FIGURE 14.5
The glucose polymers in starch, amylose and amylopectin. Depending on the origin of the starch, the formula weight varies from 50,000 to several million. (A formula weight of 1 million corresponds to about 6000 glucose units per polymer molecule.)

$\alpha(1 \rightarrow 4)$oxygen bridge   Amylose

$\alpha(1 \rightarrow 6)$oxygen bridge

Amylopectin (n, n', n'' = large numbers)

Glycogen is sometimes called animal starch.

Glycogen Is the Animal Form of Amylopectin. Molecules of **glycogen** consist of glucose units ranging in number from 1700 to 600,000 and are essentially like those of amylopectin, but more branched.

Glycogen is the way that we store glucose, and it is found chiefly in the liver and in muscle tissue. A network of enzymes handles its formation when glucose is in plentiful supply after a meal rich in carbohydrates. Another set of enzymes handles the debranching and hydrolysis of glycogen when glucose is needed between meals or in other periods of fasting.

Cellulose Is a Linear Polymer of Glucose Joined by $\beta(1 \rightarrow 4)$ Bridges. Unlike starch or glycogen, cellulose is a polymer of the beta form of glucose. Its molecules (Figure 14.6) are unbranched, so they bear a resemblance to amylose, but all of the bridges are $\beta(1 \rightarrow 4)$ instead of $\alpha(1 \rightarrow 4)$. In cotton, the typical molecule has from 2000 to 9000 glucose units.

Humans have no enzyme that can catalyze the hydrolysis of the beta bridge in cellulose, so none of the huge supply of cellulose in the world's plants, or the cellobiose that could be made from it, is nutritionally useful to us. Many bacteria have this enzyme, however, and some strains dwell in the stomachs of cattle and other animals. Bacterial action converts cellulose in hay and other animal feed into other molecules that the larger animals then use.

Just two of the units in cellulose constitute cellobiose.

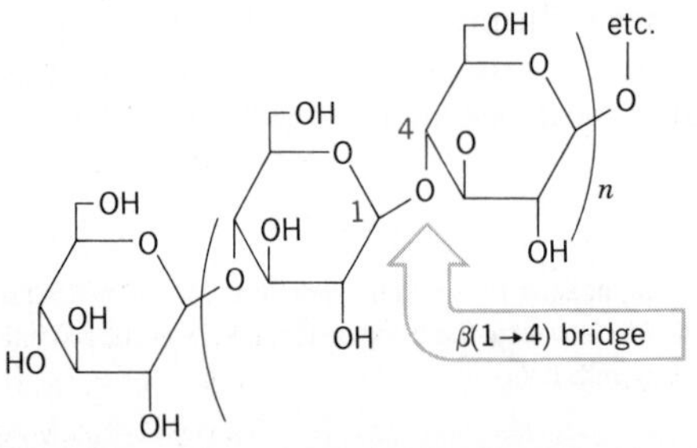

FIGURE 14.6
Cellulose, a linear polymer of β-D-glucose. The value of n varies from 1000 to 13,000 (or 2000–26,000 glucose units) in different varieties of cotton. The strength of a cotton fiber comes in part from the thousands of hydrogen bonds that can exist between parallel and overlapping cellulose molecules.

$\beta(1 \rightarrow 4)$ bridge

SUMMARY

Carbohydrates Carbohydrates are polyhydroxy aldehydes or ketones or substances that can be hydrolyzed to these compounds. Those that can't be hydrolyzed are the monosaccharides, which in pure forms exist as cyclic hemiacetals or cyclic hemiketals and are reducing sugars.

Monosaccharides The three nutritionally important monosaccharides are glucose, galactose, and fructose. Glucose is the chief carbohydrate in blood. In water, three forms are present. Two are cyclic hemiacetals designated as the α and β forms, and the third is an open-chain polyhydroxy aldehyde. Galactose, which differs from glucose only in the orientation of the OH at C-4, is obtained (together with glucose) from the hydrolysis of lactose. Fructose is a reducing ketohexose. Deoxysugars such as 2-deoxyribose, lack an OH group at one available position.

Disaccharides The disaccharides break up into two monosaccharides when they react with water. Maltose is made of two glucose units joined by an $\alpha(1 \rightarrow 4)$ acetal bridge. In a molecule of lactose (milk sugar), a galactose unit joins a glucose unit by a $\beta(1 \rightarrow 4)$ oxygen bridge. In sucrose (cane or beet sugar), there is an oxygen bridge from C-1 of a glucose unit to C-2 of a fructose unit. Both maltose and lactose retain hemiacetal systems, so both are reducing sugars and exist in solution in three interconvertible forms, like the monosaccharides. They also exist in α and β forms. Sucrose is a nonreducing disaccharide. The digestion of these disaccharides is by the following reactions:

$$\text{Maltose} + H_2O \longrightarrow \text{glucose} + \text{glucose}$$
$$\text{Lactose} + H_2O \longrightarrow \text{glucose} + \text{galactose}$$
$$\text{Sucrose} + H_2O \longrightarrow \text{glucose} + \text{fructose}$$

Polysaccharides Three important polysaccharides of glucose are starch (a plant product), glycogen (an animal product), and cellulose (a plant fiber). In molecules of each, $(1 \rightarrow 4)$ acetal bridges occur. They're alpha bridges in starch and glycogen and beta bridges in cellulose. In the amylopectin portion of starch as well as in glycogen, numerous $\alpha(1 \rightarrow 6)$ bridges also occur. No polysaccharide gives a positive test with Tollens' or Benedict's reagents. Starch gives a positive iodine test. As starch is hydrolyzed its molecules successively break down to dextrins, maltose, and finally, glucose. Humans have enzymes that catalyze the hydrolysis of $\alpha(1 \rightarrow 4)$ bridges but not $\beta(1 \rightarrow 4)$ bridges.

Optical activity Molecules of carbohydrates possess chirality (handedness), and each carbohydrate is optically active. Molecules of carbohydrates exist as mirror-image pairs called enantiomers, and enantiomer molecules do not superimpose. Enantiomers have identical physical properties except in the direction they rotate the plane of plane-polarized light. They also have identical chemical properties toward all achiral reactants. Toward chiral substances, including enzymes, enantiomers react differently, and the reactions of carbohydrates in living cells are catalyzed only by enzymes to which their molecules can physically fit.

All nutritionally important carbohydrates belong to the D family, which describes a relationship of parts within their molecules. Their enantiomers are in the L family.

REVIEW EXERCISES

The answers to these review exercises are in the *Study Guide* that accompanies this book.

Biochemistry

14.1 Substances in the diet must provide raw materials for what three essentials for life?

14.2 What are the three broad classes of nutrients in foods?

14.3 What two kinds of compounds are most involved in the molecular basis of information? Which one carries the genetic "blueprints"?

14.4 What is as important to the life of a cell as the chemicals that make it up or that it receives?

Carbohydrate Terminology

14.5 Examine the following structures and identify by letter(s) which structure(s) fit(s) each of the labels. If a particular label is not illustrated by any structure, state so.

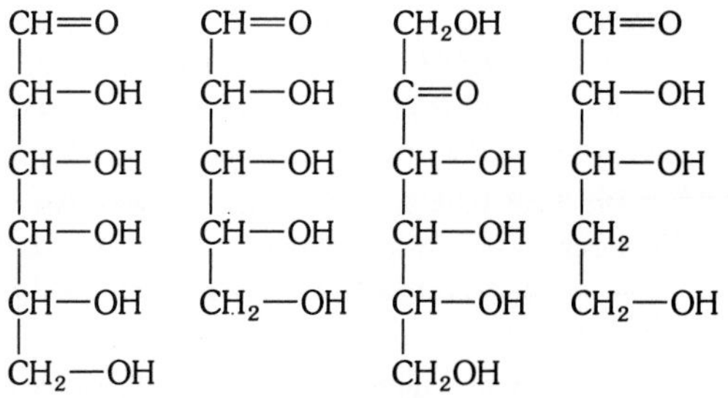

A	B	C	D
CH=O	CH=O	CH₂OH	CH=O
CH—OH	CH—OH	C=O	CH—OH
CH—OH	CH—OH	CH—OH	CH—OH
CH—OH	CH—OH	CH—OH	CH₂
CH—OH	CH₂—OH	CH—OH	CH₂—OH
CH₂—OH		CH—OH	
		CH₂OH	

(a) Ketose(s)   (b) Deoxy sugar(s)

(c) Aldohexose(s)   (d) Aldopentose(s)

14.6 Write the structure (open-chain form) that illustrates
(a) Any aldopentose   (b) Any ketotetrose

14.7 A sample of 0.0001 mol of a carbohydrate reacted with water in the presence of a catalyst to give 1 mol of glucose. Classify this carbohydrate as a mono-, di-, or polysaccharide.

14.8 An unknown carbohydrate failed to give a positive Benedict's test. Classify it as a reducing or a nonreducing carbohydrate.

Monosaccharides

14.9 What is the name of the most abundant carbohydrate in (a) blood and in (b) corn syrup?

14.10 What is the name of the ketose present in honey?

14.11 What is the structure of the open form of the following cyclic hemiacetal? (Write the open form with its chain coiled in the same way it is coiled in the closed form.)

14.12 Examine the following structure. If you judge that it is either a cyclic hemiketal or a cyclic hemiacetal, write the structure of the open-chain form (coiled in like manner as the chain of the ring).

14.13 Mannose, an isomer of glucose, is identical with glucose except that in the cyclic structures the OH at C-2 in mannose projects on the same side of the ring as the CH₂OH group. Write the structures of the three forms of mannose that are in equilibrium in a solution in water. Identify which corresponds to α-mannose and which to β-mannose.

14.14 Allose is identical with glucose except that in its cyclic forms the OH group at C-3 projects on the opposite side of the ring from the CH₂OH group. Write the structures of the three forms of allose that are in equilibrium in an aqueous solution. Which structures are α- and β-allose?

14.15 If less than 0.05% of all galactose molecules are in their open-chain form at equilibrium in water, how can galactose give a strong, positive Tollens' test, a test good for the aldehyde group?

14.16 At equilibrium in an aqueous solution, there are three forms of glucose, with one being just a trace of the open form. Suppose that in some enzyme-catalyzed process the β form is removed from this equilibrium. What becomes of the other forms?

14.17 Study the cyclic form of fructose on page 292 again. If its designation as β-fructose signifies a particular relationship between the CH₂OH group at C-5 and the OH group at C-2, what must be the cyclic formula of α-fructose?

14.18 Is the following structure that of α-fructose, β-fructose, or something else? How can you tell?

14.19 Write the cyclic structure of α-3-deoxyribose and draw an arrow that points to its hemiacetal carbon.

14.20 Could 4-deoxyribose exist as a cyclic hemiacetal with a five-membered ring (one of whose atoms is O)? Explain.

Optical Activity

14.21 When we say that α-D-glucose has *chirality*, what does this mean?

14.22 Would β-L-glucose give a positive Tollens' test? Explain.

14.23 Among the cyclic forms of the aldohexoses, what specifically is true about those in the D family? In the L family?

14.24 Suppose a 1 M solution of an aldohexose in a given container causes the plane of plane-polarized light to rotate 26° to the right.
(a) Is this compound dextrorotatory or levorotatory?
(b) What will a 1 M solution of the enantiomer of this compound do to plane-polarized light in the identical container?

14.25 In what *structural* way are α-D-fructose and α-L-fructose related? (You need not write actual structures to answer this.)

14.26 Why do D-glucose and L-glucose respond so differently to enzymes in the body?

Disaccharides

14.27 What are the names of the three nutritionally important disaccharides?

14.28 What is invert sugar?

14.29 Why isn't sucrose a reducing sugar?

14.30 Examine the following structure and answer the questions about it.

(a) Does it have a hemiacetal system? Where? (Draw an arrow to it or circle it.)

(b) Does it have an acetal system? Where? (Circle it.)

(c) Does this substance give a positive Benedict's test? Explain.

(d) In what specific structural way does it differ from maltose?

(e) Write the cyclic structures and names of the products of the acid-catalyzed hydrolysis of this compound. (This compound, incidentally, is cellobiose.)

14.31 Trehalose is a disaccharide found in young mushrooms and yeast, and it is the chief carbohydrate in the hemolymph of certain insects. On the basis of its structural features, respond to the following:

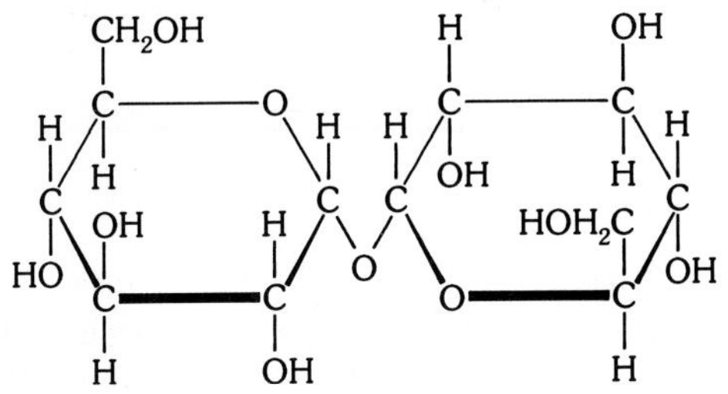

(a) Is trehalose a reducing sugar? Explain.

(b) Identify, by name only, the products of the hydrolysis of trehalose.

14.32 Maltose has a hemiacetal system. Write the structure of maltose in which this group has changed to the open form.

14.33 Write the open form of lactose.

Polysaccharides

14.34 Name the polysaccharides that give only glucose when they are completely hydrolyzed.

14.35 What is the main structural difference between amylose and cellulose?

14.36 How are amylose and amylopectin alike structurally?

14.37 How are amylose and amylopectin different structurally?

14.38 Why can't humans digest cellulose?

14.39 What is the iodine test? Describe the reagent, state what it is used to test for, and describe what is seen in a positive test.

14.40 How do amylopectin and glycogen compare structurally?

14.41 How does the body use glycogen?

Photosynthesis (Special Topic 14.1)

14.42 Where did the chemical energy in glucose or starch originate?

14.43 In terms of reactants and products, what is accomplished by photosynthesis?

14.44 What compound in leaves is able to absorb solar energy and channel it into photosynthesis?

14.45 What species of plants handle most of the photosynthesis on this planet?

14.46 What events in nature make photosynthesis just part of a huge cycle?

Chapter 15
Lipids

To face an Antarctic winter, these chinstrap penguins have stored considerable chemical energy in the form of fat. In this chapter we learn about the fats and oils of nutritional importance.

15.1 WHAT LIPIDS ARE

The edible fats and oils consist of hydrocarbon-like molecules that are esters of long-chain fatty acids and glycerol.

When undecomposed plant or animal material is crushed and ground with a nonpolar solvent such as ether, whatever dissolves is classified as a **lipid.** This operation catches a large variety of relatively nonpolar substances, and all are lipids. Thus this broad family is defined not by structure but by the technique used to isolate its members, solvent extraction. Among the many substances that won't dissolve in nonpolar solvents are carbohydrates, proteins, other very polar organic substances, inorganic salts, and water.

Extraction is the process of shaking or stirring a mixture with a solvent that dissolves just part of the mixture.

Lipids Are Broadly Subdivided According to the Presence of Saponifiable Groups.
One of the major classes of lipids, the **saponifiable lipids,** consists of compounds with one or more groups that can be hydrolyzed or saponified. In nearly all examples, these are ester groups. A number of families are in this class, including the waxes, the neutral fats, the phospholipids, and the glycolipids. The **nonsaponifiable lipids** lack groups that can be hydrolyzed or saponified. These include cholesterol and many sex hormones. The chart in Figure 15.1 outlines the many kinds of lipids.

Butterfat and salad oils are neutral fats.

Cholesterol and several sex hormones are steroids.

Plant Waxes Are Simple Esters with Long Hydrocarbon Chains.   The waxes, the simplest of the saponifiable lipids, occur as protective coatings on fruit and leaves as well as on fur, feathers, and skin. Nearly all **waxes** are esters of long-chain monohydric alcohols and long-chain monocarboxylic acids, both of which contain an *even* number of carbons. As many as 26–34 carbon atoms can be incorporated in *each* of the alcohol and acid units, which makes the waxes almost totally hydrocarbon-like.

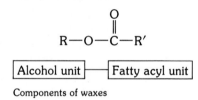

An acyl group has the general structure:

Lanolin is used to make cosmetic skin lotions.

Any particular wax, like beeswax, consists of a mixture of similar compounds that share the kind of structure shown above. In molecules of lanolin (wool fat), however, the alcohol

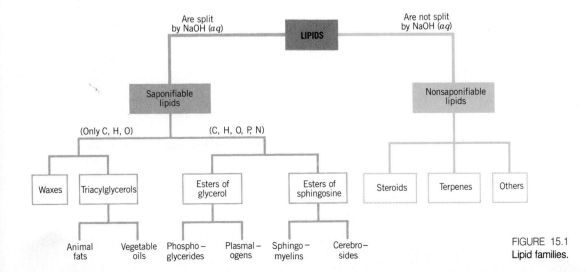

FIGURE 15.1
Lipid families.

portion is contributed by steroid alcohols, which have large ring systems that we'll study in Section 15.4. Waxes exist in sebum, a secretion of human skin that helps to keep the skin supple.

PRACTICE EXERCISE 1   One particular ester in beeswax can be hydrolyzed to give a straight-chain primary alcohol with 26 carbons and a straight-chain carboxylic acid with 28 carbons. Write the structure of this ester.

In the older literature, the triacyl-glycerols are called the **triglycer-ides.**

The Triacylglycerols Are Triesters of Glycerol. The molecules of the most abundant lipids are the **triacylglycerols,** which are esters of glycerol and long-chain monocarboxylic acids, the **fatty acids.** Unlike more complex lipids, triacylglycerol molecules have no ionic sites, so sometimes they are called the *neutral fats.* Their general structure is

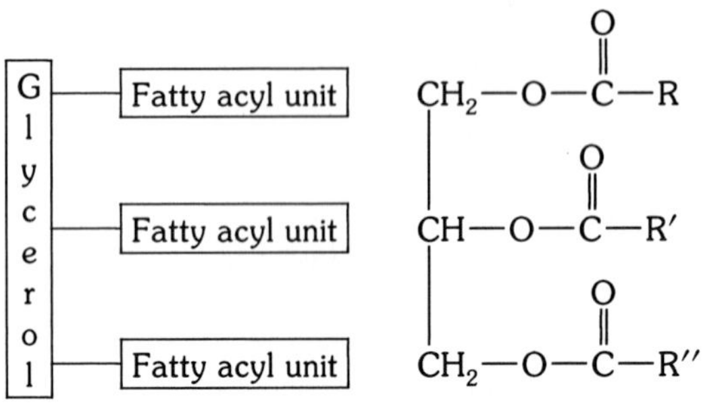

$$
\begin{array}{l}
\text{CH}_2\text{—O—}\overset{\displaystyle O}{\overset{\|}{\text{C}}}\text{—R} \\[1em]
\text{CH—O—}\overset{\displaystyle O}{\overset{\|}{\text{C}}}\text{—R}' \\[1em]
\text{CH}_2\text{—O—}\overset{\displaystyle O}{\overset{\|}{\text{C}}}\text{—R}''
\end{array}
$$

Glycerol — Fatty acyl unit / Fatty acyl unit / Fatty acyl unit

The triacylglycerols include lard (hog fat), tallow (beef fat), butterfat—all animal fats—as well as such plant oils as olive oil, cottonseed oil, corn oil, peanut oil, soybean oil, coconut oil, and linseed oil. At room temperature, animal fats are solids and vegetable oils are liquids.

The three acyl units in a typical triacylglycerol molecule are from three different fatty acids. Fats and oils are thus mixtures of different molecules that share common structural features. Although we can't draw, for example, the structure of cottonseed oil, we can describe a typical molecule, like that of structure **1.**

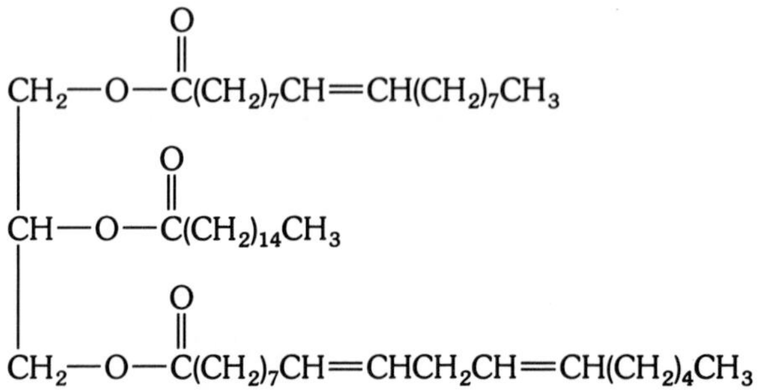

$$
\begin{array}{l}
\text{CH}_2\text{—O—}\overset{\displaystyle O}{\overset{\|}{\text{C}}}\text{(CH}_2)_7\text{CH}=\text{CH(CH}_2)_7\text{CH}_3 \\[1em]
\text{CH—O—}\overset{\displaystyle O}{\overset{\|}{\text{C}}}\text{(CH}_2)_{14}\text{CH}_3 \\[1em]
\text{CH}_2\text{—O—}\overset{\displaystyle O}{\overset{\|}{\text{C}}}\text{(CH}_2)_7\text{CH}=\text{CHCH}_2\text{CH}=\text{CH(CH}_2)_4\text{CH}_3
\end{array}
$$

1

In a particular fat or oil, certain fatty acids predominate, others either are absent or are present in trace amounts, and virtually all of the molecules are triacylglycerols. Data on the fatty acid compositions of several fats and oils are listed in Table 15.1.

The Fatty Acids Are Mostly Long-Chain, Unbranched Monocarboxylic Acids. The fatty acids obtained from the lipids of most plants and animals share the following features.

1. They are usually monocarboxylic acids, RCO_2H
2. The R group is usually an unbranched chain.
3. The number of carbon atoms is almost always even.
4. The R group can be saturated, or it can have one or more double bonds, which are cis. There are CH_2 units between double bonds.

TABLE 15.1
Fatty Acids Obtained from Neutral Fats and Oils

| Type | Fat or Oil | Average Composition of Fatty Acids (%) | | | | | |
		Myristic Acid	Palmitic Acid	Stearic Acid	Oleic Acid	Linoleic Acid	Others
Animal Fats	Butter	8–15	25–29	9–12	18–33	2–4	a
	Lard	1–2	25–30	12–18	48–60	6–12	b
	Beef tallow	2–5	24–34	15–30	35–45	1–3	b
Vegetable Oils	Olive	0–1	5–15	1–4	67–84	8–12	
	Peanut		7–12	2–6	30–60	20–38	
	Corn	1–2	7–11	3–4	25–35	50–60	
	Cottonseed	1–2	18–25	1–2	17–38	45–55	
	Soybean	1–2	6–10	2–4	20–30	50–58	c
	Linseed		4–7	2–4	14–30	14–25	d
Marine Oils	Whale	5–10	10–20	2–5	33–40	—	e
	Fish	6–8	10–25	1–3	—	—	e

[a] Also, 3–4% butyric acid, 1–2% caprylic acid, 2–3% capric acid, 2–5% lauric acid.
[b] Also, linolenic acid, 1%.
[c] Also, linolenic acid, 5–10%.
[d] Also, linolenic acid, 45–60%.
[e] Large percentages of other highly unsaturated fatty acids.

The most abundant saturated fatty acids are palmitic acid, $CH_3(CH_2)_{14}CO_2H$, and stearic acid, $CH_3(CH_2)_{16}CO_2H$, which have 16 and 18 carbons, respectively. Refer back to Table 13.2 for the other saturated fatty acids used to make triacylglycerols. They are the acids below acetic acid with even numbers of carbons, like butyric, caproic, caprylic, and capric acids. Their acyl units, however, are present in only relatively small amounts in triacylglycerols.

The unsaturated fatty acids most commonly obtained from triacylglycerols are listed in Table 15.2 and include oleic, linoleic, and linolenic acids, all with C_{18} skeletons. Oleic acid is the most abundant and most widely distributed fatty acid in nature. The double bonds in the unsaturated fatty acids of Table 15.2 are cis.

The vegetable oils rely more on the unsaturated fatty acids, like oleic and linoleic acid, for acyl groups than the animal fats, so they have more double bonds per molecule (see Table 15.1). The vegetable oils are therefore described as *polyunsaturated*. The saturated fatty acyl units of palmitic and stearic acids are far more common in animal fats.

TABLE 15.2
Common Unsaturated Fatty Acids

Name	Number of Double Bonds	Total Number of Carbons	Structure	Melting Point (°C)
Palmitoleic acid	1	16	$CH_3(CH_2)_5CH{=}CH(CH_2)_7CO_2H$	32
Oleic acid	1	18	$CH_3(CH_2)_7CH{=}CH(CH_2)_7CO_2H$	4
Linoleic acid	2	18	$CH_3(CH_2)_4CH{=}CHCH_2CH{=}CH(CH_2)_7CO_2H$	−5
Linolenic acid	3	18	$CH_3CH_2CH{=}CHCH_2CH{=}CHCH_2CH{=}CH(CH_2)_7CO_2H$	−11
Arachidonic acid	4	20	$CH_3(CH_2)_4CH{=}CHCH_2CH{=}CHCH_2CH{=}CHCH_2CH{=}CH(CH_2)_3CO_2H$	−50

The prostaglandins were discovered in the mid-1930s by a Swedish scientist, Ulf von Euler (Nobel prize, 1970), but they didn't arouse much interest in medical circles until the late 1960s, largely through the work of Sune Bergstrom. It became apparent that these compounds, which occur widely in the body, affect a large number of processes. Their general name comes from an organ, the prostate gland, from which they were first obtained. About 20 are known, and they occur in four major subclasses designated as PGA, PGB, PGE, and PGF. (A subscript is generally placed after the third letter to designate the number of alkene double bonds that occur outside of the five-membered ring.) The structures of some typical examples are shown here.

Prostaglandins are made from C_{20} fatty acids such as arachidonic acid. By coiling a molecule of this acid, as shown, you can see how its structure needs only a ring closure (suggested by the dashed arrow) and three more oxygen atoms to become PGF_2. The oxygen atoms are all provided by molecular oxygen itself.

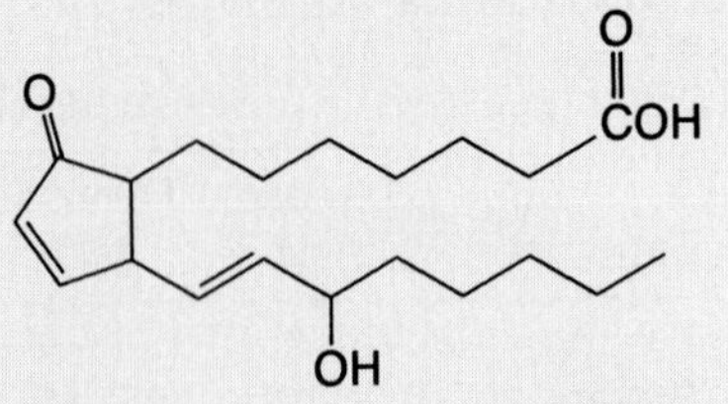

Prostaglandins as Chemical Messengers The prostaglandins are like hormones in many ways, except that they do not act globally, that is, over the entire body. They do their work within the cells where they are made or in nearby cells, so they are sometimes called *local hormones.* This is perhaps why the prostaglandins have such varied functions; they occur and express their roles in such varied tissues. They work together with hormones to modify the chemical messages that hormones bring to cells. In some cells, the prostaglandins inhibit enzymes and in others they activate them. In some organs, the prostaglandins help to regulate the flow of blood within them. In others, they affect the transmission of nerve impulses.

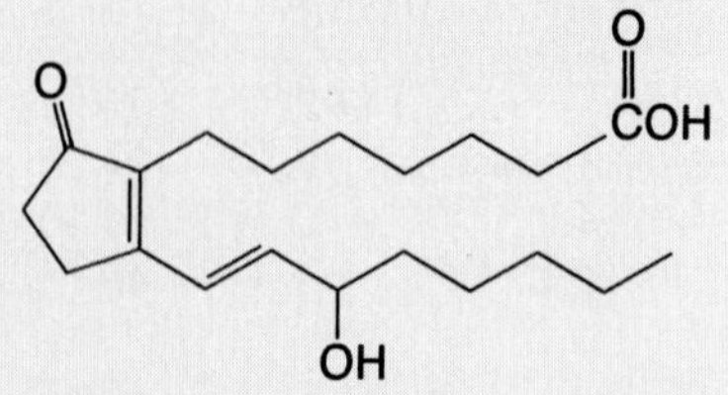

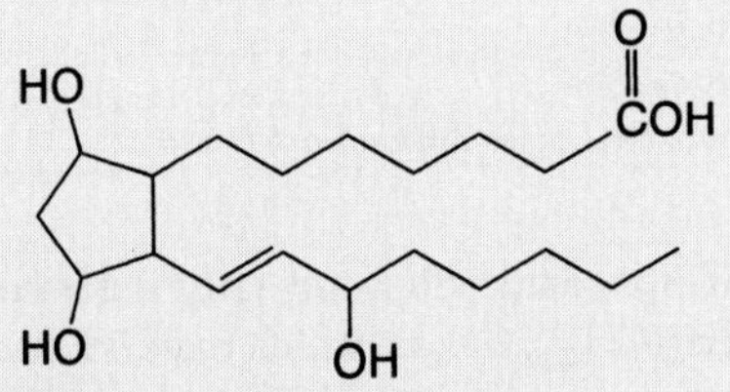

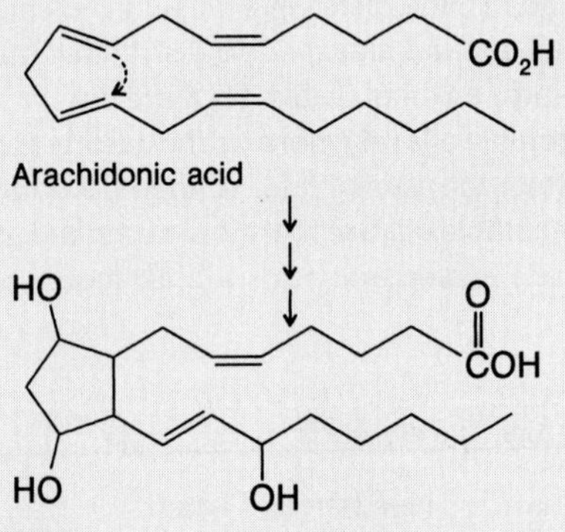

Some prostaglandins enhance inflammation in a tissue, and it is interesting that aspirin, an inflammation reducer, does exactly the opposite. This effect is caused by aspirin's ability to inhibit the work of an enzyme needed in the synthesis of prostaglandins.

Prostaglandins as Pharmaceuticals In experiments that use prostaglandins as pharmaceuticals, they have been found to have an astonishing variety of effects. One prostaglandin induces labor at the end of a pregnancy. Another stops the flow of gastric juice while the body heals an ulcer. Other possible uses are to treat high blood pressure, rheumatoid arthritis, asthma, nasal congestion, and certain viral diseases.

PGA₁

PGB₁

PGF₁

Arachidonic acid

PGF₂

The presence of cis alkene groups, which give kinks to the side chains, affects the melting points of the fatty acids, as you can see in Table 15.2. As more alkene groups are added, the melting points decrease because the structural kinks at double bonds inhibit the kind of close packing in crystals required for stronger, higher melting systems. This is why the animal fats, with fewer alkene groups, are likely to be solids at room temperature and the vegetable oils are normally liquids.

THE OMEGA-3 FATTY ACIDS AND HEART DISEASE

Omega-3 refers to the location of a double bond third in from the far end of a long-chain fatty acid, particularly those with 18, 20, and 22 carbons. Just as ω is the last letter in the Greek alphabet, so the omega position in a fatty acid is the one farthest from the carboxyl group. Arachidonic acid (Table 15.2) is an omega-6 C_{20} fatty acid because it has a double bond at the sixth carbon from the omega end. Linolenic acid (Table 15.2) is an omega-3 fatty acid. Two other omega-3 acids are considered by some to be important in metabolism:

$$CH_3CH_2CH{=}CH(CH_2CH{=}CH)_4CH_2CH_2CH_2CO_2H$$
ω-3-Eicosapentaenoic acid
$$CH_3CH_2CH{=}CH(CH_2CH{=}CH)_5CH_2CH_2CO_2H$$
ω-3-Docosahexaenoic acid

The basis of the interest in these is that Eskimos have low incidences of heart disease despite relatively high cholesterol levels in their diets (from fish oils and fish liver). Some scientists believe that the high level of the omega-3 fatty acids in marine oils provides protection against disease.

PRACTICE EXERCISE 2   To visualize how a cis double bond introduces a kink into a molecule, write the structure of oleic acid in a way that correctly shows the cis geometry of the alkene group. (Without the double bond, as in stearic acid, the entire side chain can stretch out into a perfect zigzag conformation, which makes it easy for two side chains to nestle very close to each other.)

The properties of the fatty acids are those to be expected of compounds with carboxyl groups, double bonds (where present), and long hydrocarbon chains. Thus they are insoluble in water and soluble in nonpolar solvents. They are neutralized by bases and form salts. They can be esterified. Those that have alkene groups react with hydrogen in the presence of a catalyst and pressure.

The *prostaglandins* are an unusual family of fatty acids with 20 carbons, five-membered rings, and a wide variety of effects in the body (see Special Topic 15.1).

In the late 1980s, one small group of fatty acids, the omega-3 fatty acids, appeared in scientific debates about the value of fish or marine oils in the diet. Special Topic 15.2 describes them further.

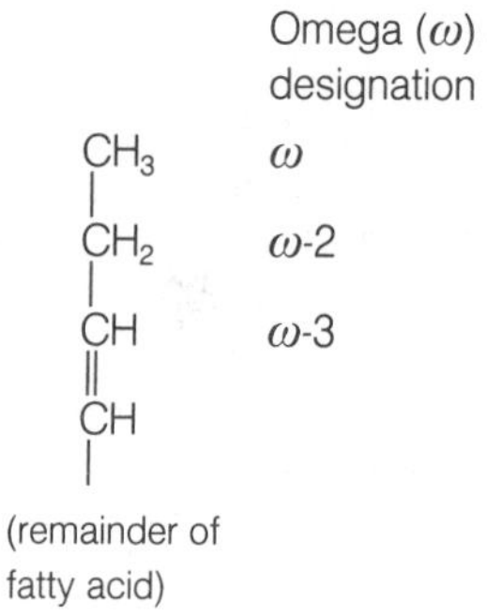

	Omega (ω) designation
CH_3	ω
CH_2	ω-2
CH	ω-3
CH	
(remainder of fatty acid)	

15.2 CHEMICAL PROPERTIES OF TRIACYLGLYCEROLS

Triacylglycerols can be hydrolyzed (digested), saponified, and hydrogenated.

During Digestion, Triacylglycerols Are Hydrolyzed.   Enzymes in the digestive tracts of humans and animals catalyze the hydrolysis of the ester links in triacylglycerols. In general,

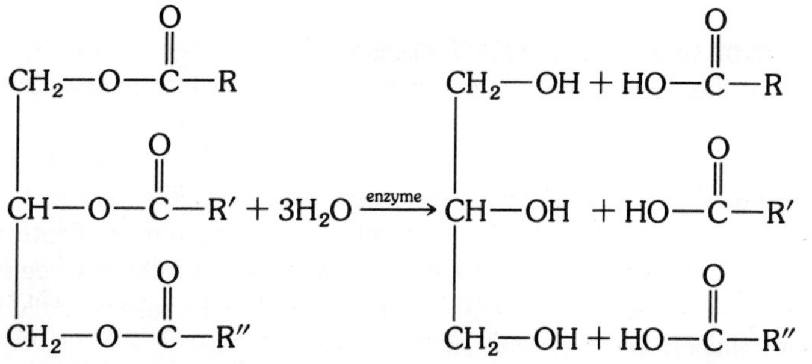

Triacylglycerol   Glycerol   Fatty acids

This hydrolysis of triacylglycerols is what happens when we digest fats and oils.

A specific example is

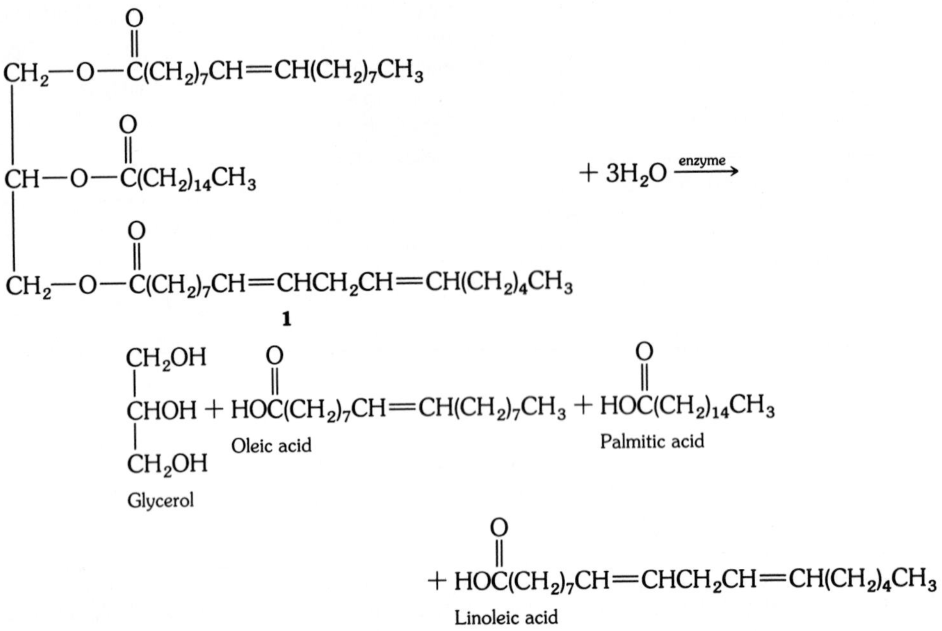

$$
\begin{array}{l}
CH_2\!-\!O\!-\!\overset{\displaystyle O}{\overset{\|}{C}}(CH_2)_7CH\!=\!CH(CH_2)_7CH_3 \\[2mm]
CH\!-\!O\!-\!\overset{\displaystyle O}{\overset{\|}{C}}(CH_2)_{14}CH_3 \\[2mm]
CH_2\!-\!O\!-\!\overset{\displaystyle O}{\overset{\|}{C}}(CH_2)_7CH\!=\!CHCH_2CH\!=\!CH(CH_2)_4CH_3
\end{array}
\qquad + 3H_2O \xrightarrow{\;enzyme\;}
$$

1

$$
\begin{array}{l}
CH_2OH \\
CHOH + HO\overset{\displaystyle O}{\overset{\|}{C}}(CH_2)_7CH\!=\!CH(CH_2)_7CH_3 + HO\overset{\displaystyle O}{\overset{\|}{C}}(CH_2)_{14}CH_3 \\
CH_2OH
\end{array}
$$

Glycerol   Oleic acid   Palmitic acid

$$
+ HO\overset{\displaystyle O}{\overset{\|}{C}}(CH_2)_7CH\!=\!CHCH_2CH\!=\!CH(CH_2)_4CH_3
$$

Linoleic acid

Soaps Are Made by the Saponification of Triacylglycerols. The saponification of the ester links in triacylglycerols by the action of a strong base (e.g., NaOH or KOH) gives glycerol and a mixture of the salts of fatty acids. These salts are soaps, and how they exert their detergent action is described in Special Topic 15.3. In general,

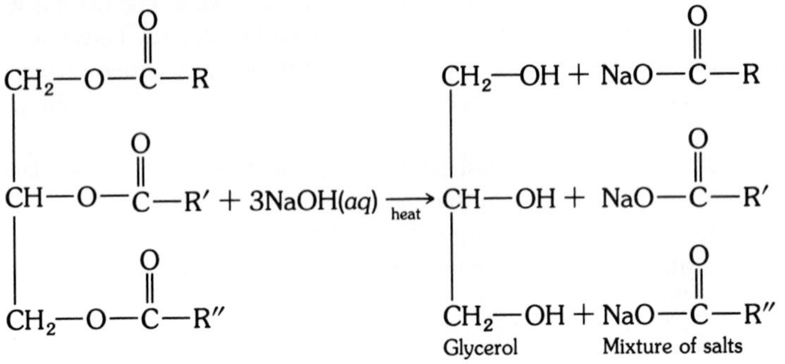

$$
\begin{array}{l}
CH_2\!-\!O\!-\!\overset{\displaystyle O}{\overset{\|}{C}}\!-\!R \\[2mm]
CH\!-\!O\!-\!\overset{\displaystyle O}{\overset{\|}{C}}\!-\!R' + 3NaOH(aq) \xrightarrow[heat]{} \\[2mm]
CH_2\!-\!O\!-\!\overset{\displaystyle O}{\overset{\|}{C}}\!-\!R''
\end{array}
\qquad
\begin{array}{l}
CH_2\!-\!OH + NaO\!-\!\overset{\displaystyle O}{\overset{\|}{C}}\!-\!R \\[2mm]
CH\!-\!OH + NaO\!-\!\overset{\displaystyle O}{\overset{\|}{C}}\!-\!R' \\[2mm]
CH_2\!-\!OH + NaO\!-\!\overset{\displaystyle O}{\overset{\|}{C}}\!-\!R''
\end{array}
$$

Glycerol   Mixture of salts

PRACTICE EXERCISE 3   Write a balanced equation for the saponification of **1** with sodium hydroxide.

The Hydrogenation of Vegetable Oils Gives Solid Shortenings. When hydrogen is made to add to some of the double bonds in vegetable oils, the oils become like animal fats, both physically and structurally. One very practical consequence is that they change from being liquids to solids at room temperature. Many people prefer solid, lardlike shortening for cooking, instead of liquid oils. Therefore the manufacturers of such products as Crisco and Spry take inexpensive, readily available vegetable oils, like corn oil and cottonseed oil, and partially hydrogenate them. Hydrogen is made to add catalytically to some (not all) of the alkene groups in the molecules of the oils. Unlike natural lard, these partially hydrogenated vegetable shortenings have no cholesterol.

Hydrogenated vegetable oils are chemically identical to animal fats.

Soap Water is a very poor cleansing agent because it can't penetrate greasy substances, the "glues" that bind soil to skin and fabrics. When just a little soap is present, however, water cleans very well, especially warm water. Soap is a simple chemical, a mixture of the sodium or potassium salts of the long-chain fatty acids obtained by the saponification of fats or oils.

Detergents Soap is just one kind of detergent. All detergents are surface-active agents that lower the surface tension of water. All consist of ions or molecules that have long hydrocarbon portions plus ionic or very polar sections at one end. The accompanying structures illustrate these features and show the varieties of detergents that are available.

Although soap is manufactured, it is not called a synthetic detergent. This term is limited to detergents that are not soap, that is, not the salts of naturally occurring fatty acids obtained by the saponification of lipids. Most synthetic detergents are salts of sulfonic acids, but others have different kinds of ionic or polar sites. The great advantage of synthetic detergents is that they work in hard water and are not precipitated by the hardness ions: Mg^{2+}, Ca^{2+}, and the two ions of iron. The anions of the fatty acids present in soap form messy precipitates with these ions. The anions of synthetic detergents do not have this property.

The accompanying figure shows how detergents work. In part *a* we see the hydrocarbon tails of the deter gent work their way into the hydrocarbon environment of the grease layer. ("Like dissolves like" is the principle at work here.) The ionic heads stay in the water phase, and the grease layer becomes pincushioned with electrically charged sites. In part *b* we see the grease layer breaking up, aided with some agitation or scrubbing. Part *c* shows a magnified view of grease globules studded with ionic groups and, being like charged, these globules repel each other. They also tend to dissolve in water, so they are ready to be washed down the drain.

$$CH_3(CH_2)_{14}CO_2^-Na^+$$

Soap—an anionic detergent

$$CH_3(CH_2)_{13}OSO_3^-Na^+$$

A sodium alkyl sulfate—an anionic detergent

$$CH_3(CH_2)_8-\bigcirc-SO_3^-Na^+$$

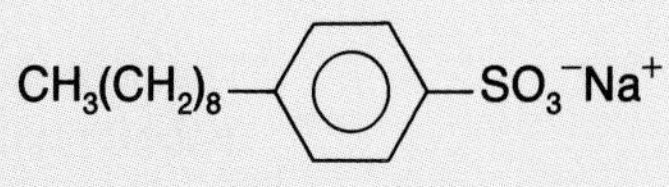

A sodium alkylbenzenesulfonate — an anionic detergent

$$CH_3(CH_2)_{11}\overset{+}{N}(CH_3)_3Cl^-$$

A trimethylalkylammonium ion—a cationic detergent

$$CH_3(CH_2)_8(OCH_2CH_2O)_nH$$

A nonionic detergent

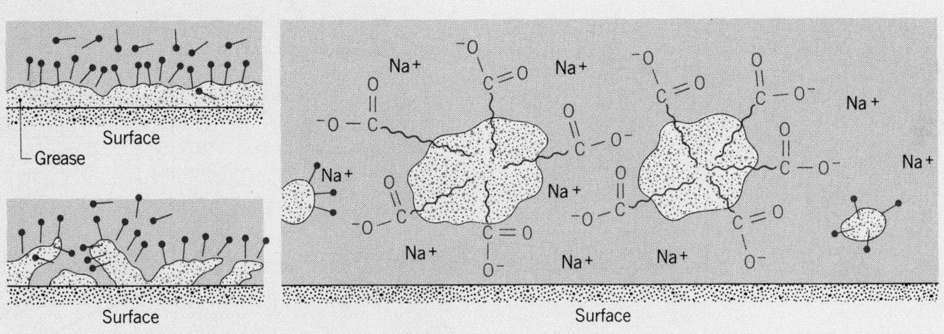

How detergents work.

PRACTICE EXERCISE 4   Write the balanced equation for the complete hydrogenation of the alkene links in structure **1**.

The chief lipid material in margarine is produced from vegetable oils in the same way. The hydrogenation is done with special care so that the final product can melt on the tongue, a property that makes butterfat so pleasant. (If all the alkene groups in a vegetable oil were hydrogenated, instead of just some of them, the product would be like beef or mutton fat, relatively hard materials that would not melt on the tongue.)

The popular brands of peanut butter whose peanut oils do not separate are made by the partial hydrogenation of the oil in real peanut butter. The lipid present becomes a solid at room temperature (and therefore it can't separate).

15.3 PHOSPHOLIPIDS

Phospholipid molecules have very polar or ionic sites in addition to long hydrocarbon chains.

Phospholipids are esters either of glycerol or sphingosine, which is a long-chain, dihydric amino alcohol with one double bond.

$$CH_3(CH_2)_{12}CH{=}CHCH{-}CH{-}CH_2{-}OH$$
$$\underset{OH}{|} \quad \underset{NH_2}{|}$$

Sphingosine

The phospholipids all have very polar but small molecular parts that are extremely important in the formation of cell membranes. We will survey their structures largely to demonstrate how they are both polar and hydrocarbon-like.

Phosphoglycerides Have Phosphate Units Plus Two Acyl Units. Molecules of **phosphoglycerides** have two ester bonds from glycerol to fatty acids plus one ester bond to phosphoric acid. The phosphoric acid unit, in turn, is joined by a phosphate ester link to a small alcohol molecule. Without this link, the compound is called phosphatidic acid.

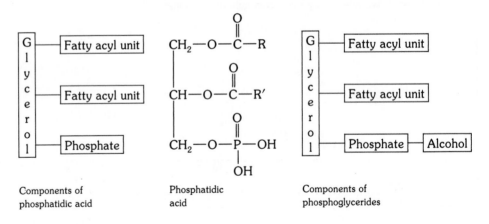

Components of   Phosphatidic   Components of
phosphatidic acid   acid   phosphoglycerides

Lecithin is from the Greek *lekitos*, egg yolk — a rich source of this phospholipid.

The three principal phosphoglycerides are esters between phosphatidic acid and either choline, ethanolamine, or serine, forming, respectively, phosphatidylcholine (lecithin), **2**, phosphatidylethanolamine (cephalin), **3**, and phosphatidylserine, **4**.

$$HOCH_2CH_2\overset{+}{N}(CH_3)_3 \qquad HOCH_2CH_2NH_2 \qquad \underset{\underset{NH_3^+}{|}}{HOCH_2CHCO_2^-}$$

Choline   Ethanolamine   Serine
(a cation)   (an amino acid)

As the structures of **2**, **3**, and **4** show, one part of each phosphoglyceride molecule is very polar because it carries full electrical charges. The remainder is nonpolar and hydrocarbon-like.

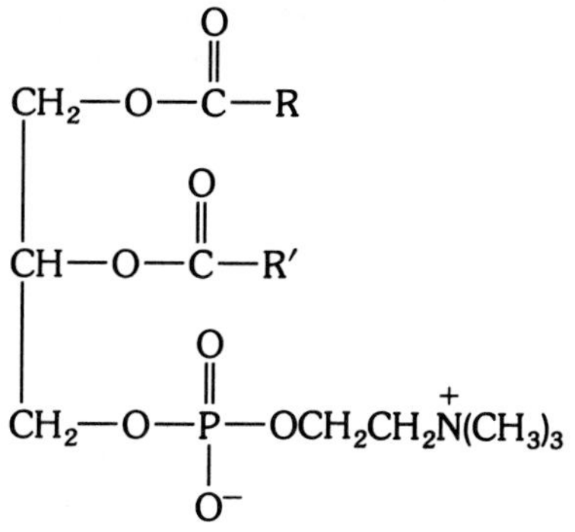

$$CH_2-O-\overset{\overset{\displaystyle O}{\|}}{C}-R$$

$$CH-O-\overset{\overset{\displaystyle O}{\|}}{C}-R'$$

$$CH_2-O-\overset{\underset{\displaystyle O^-}{|}}{\overset{\overset{\displaystyle O}{\|}}{P}}-OCH_2CH_2\overset{+}{N}(CH_3)_3$$

2
Phosphatidylcholine (lecithin)

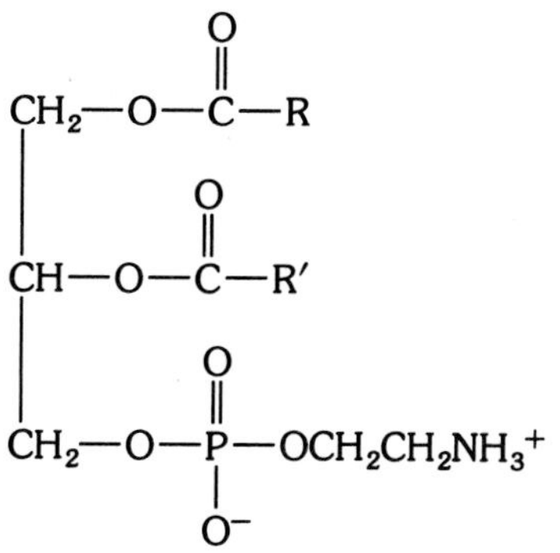

$$CH_2-O-\overset{\overset{\displaystyle O}{\|}}{C}-R$$

$$CH-O-\overset{\overset{\displaystyle O}{\|}}{C}-R'$$

$$CH_2-O-\overset{\underset{\displaystyle O^-}{|}}{\overset{\overset{\displaystyle O}{\|}}{P}}-OCH_2CH_2NH_3^+$$

3
Phosphatidylethanolamine (cephalin)

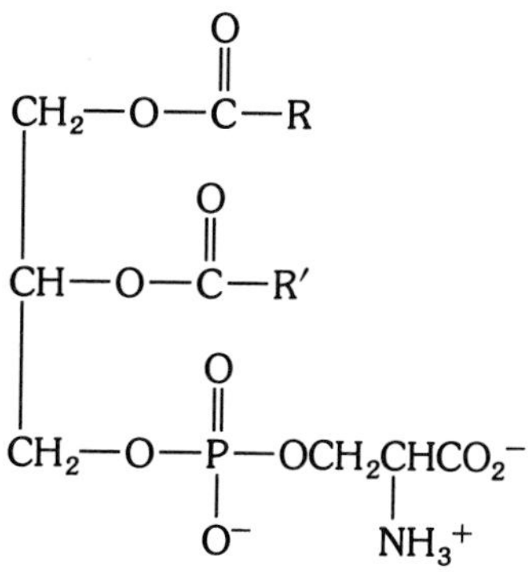

$$CH_2-O-\overset{\overset{\displaystyle O}{\|}}{C}-R$$

$$CH-O-\overset{\overset{\displaystyle O}{\|}}{C}-R'$$

$$CH_2-O-\overset{\underset{\displaystyle O^-}{|}}{\overset{\overset{\displaystyle O}{\|}}{P}}-OCH_2\underset{\underset{\displaystyle NH_3^+}{|}}{CH}CO_2^-$$

4
Phosphatidylserine

When pure, lecithin is a clear, waxy solid that is very hygroscopic. In air, it is quickly attacked by oxygen, which makes it turn brown in a few minutes. Lecithin is a powerful emulsifying agent for triacylglycerols, and this is why egg yolks, which contain it, are used to make the emulsions found in mayonnaise, ice cream, custards, and cake batter.

Cephalin is from the Greek *kephale,* head. Cephalin is found in brain tissue.

The Plasmalogens Have Both Ether and Ester Groups. The **plasmalogens** make up another family of glycerol-based phospholipids, and they occur widely in the membranes of nerve cells and muscle cells. They differ from the other phosphoglycerides by the presence of an unsaturated ether group instead of an acyl group at one end of the glycerol unit.

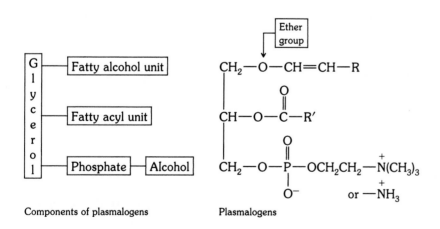

Components of plasmalogens

Plasmalogens

The Sphingolipids Are Based on Sphingosine, Not Glycerol. The two types of sphingosine-based lipids or **sphingolipids** are the sphingomyelins and the cerebrosides, and they are also important constituents of cell membranes. The sphingomyelins are phosphate diesters of sphingosine. Their acyl units occur as acylamido parts, and they come from unusual fatty acids that are not found in neutral fats.

The cerebrosides are not actually phospholipids. Instead they are **glycolipids,** lipids with a sugar (i.e., glycose) unit and not a phosphate ester system. The sugar unit, with its many OH groups, provides a strongly polar site, and it is usually a D-galactose or a D-glucose unit, or an amino derivative of these.

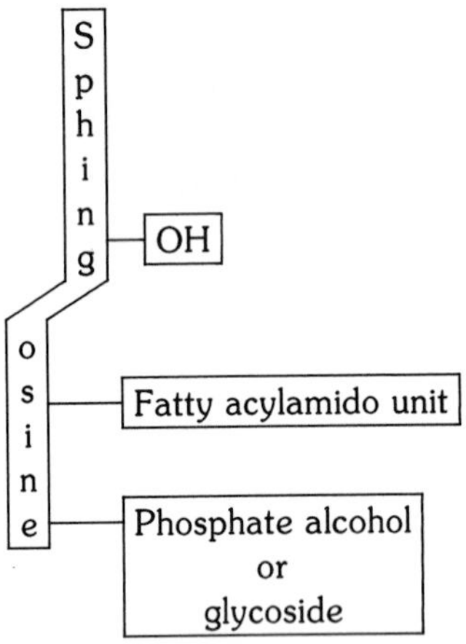

Component of sphingolipids

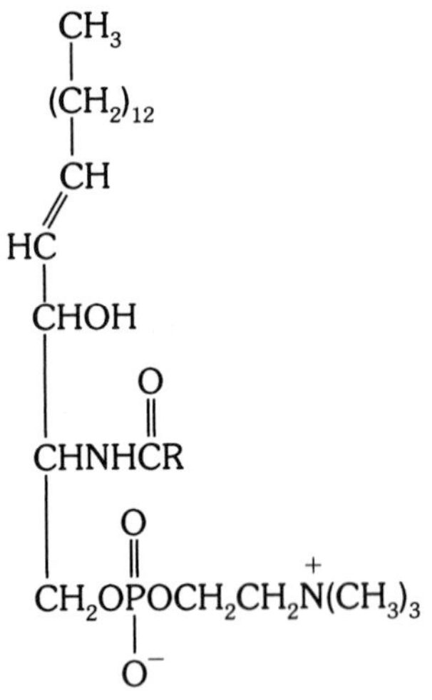

Sphingomyelins

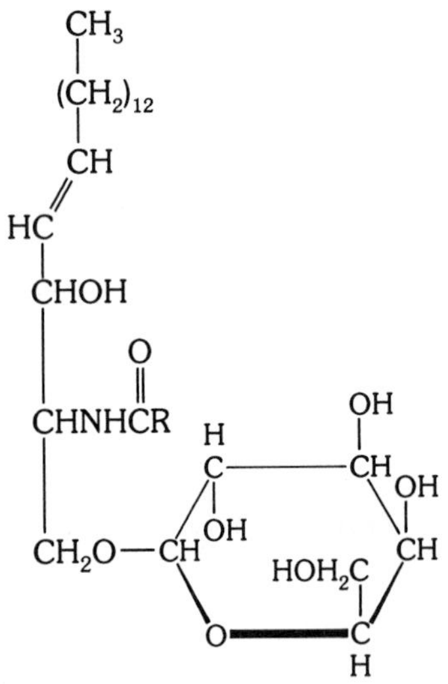

Cerebrosides

15.4 STEROIDS

Cholesterol and other steroids are nonsaponifiable lipids.

Steroids are high formula weight compounds whose molecules include the characteristic four-ring feature called the steroid nucleus. It consists of three six-membered rings and one five-membered ring, as seen in structure **5**. Several steroids are very active, physiologically.

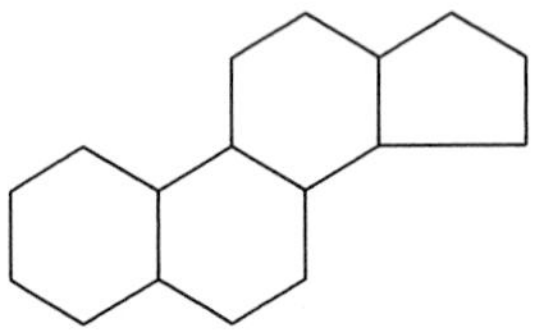

Steroid nucleus

5

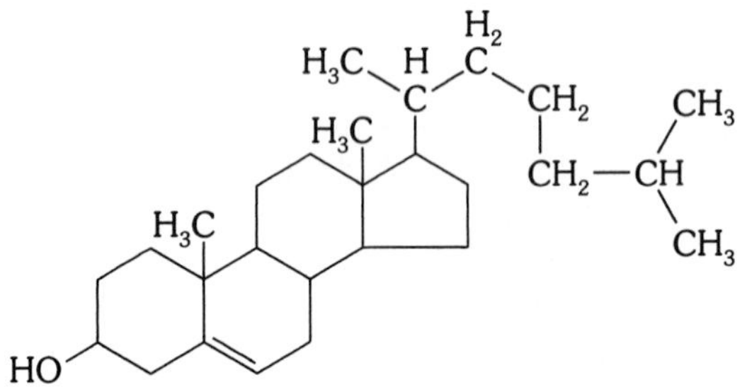

Cholesterol
(Greek: *chole,* bile; *stereos,* solid; -ol, alcohol)

Table 15.3 lists several steroids and their functions, and you can see the large range of properties they have in the body.

Steroid alcohols are called sterols.

Cholesterol Molecules Are Built into Cell Membranes. Cholesterol is an unsaturated steroid alcohol that, together with cholesterol esters, makes up a significant part of the membranes of cells and is the chief constituent in gallstones. It's the body's raw material for making the bile salts and the steroid hormones, including the sex hormones listed in Table 15.3.

Cholesterol enters the body via the diet, but about 800 mg/day is normally synthesized in the liver from acetate units. The relationship between cholesterol and the risk of heart disease will be discussed when we study the metabolism of lipids in Chapter 20.

TABLE 15.3
Important Steroids

Vitamin D₃ Precursor
Irradiation of this derivative of cholesterol by ultraviolet light opens one of
the rings to produce vitamin D₃. Meat
products are sources of this compound.

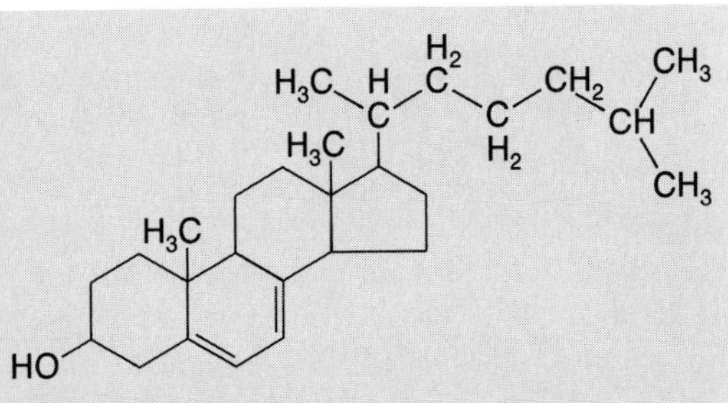

7-Dehydrocholesterol

Ultraviolet light

**Vitamin D₃ is an antirachitic factor. Its
absence leads to rickets, an infant and
childhood disease characterized by
faulty deposition of calcium phosphate
and poor bone growth.**

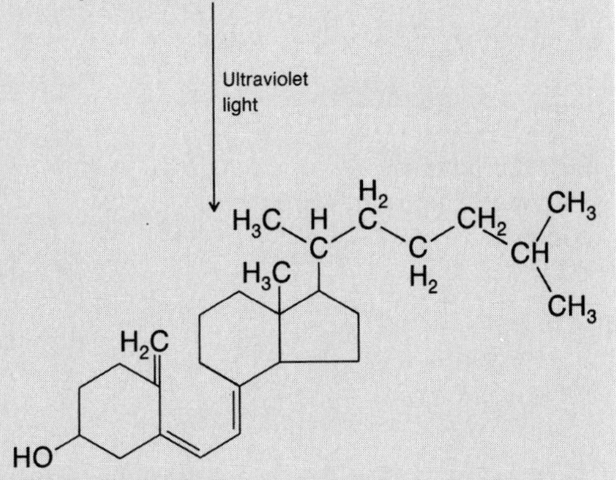

Vitamin D₃

Bile Acid
Cholic acid is found in bile in the form
of its sodium salt. This and closely
related salts are the bile salts; they are
powerful surface active agents that aid
in the digestion of lipids and in the
absorption of vitamins A, D, E, and K
from the intestinal tract.

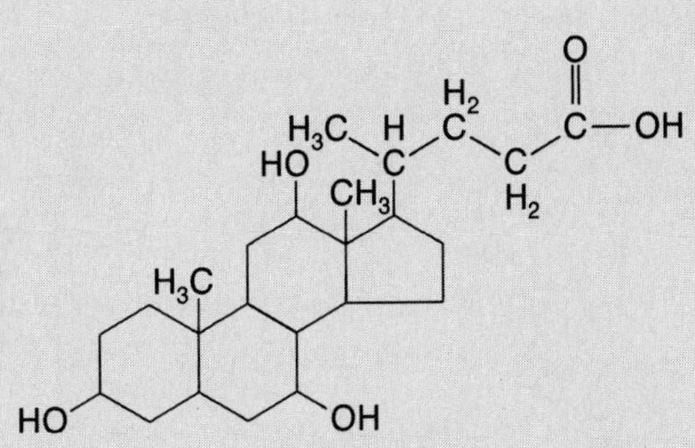

Cholic acid

Adrenocortical Hormone
Cortisol is one of the 28 hormones
secreted by the cortex of the adrenal
gland. Cortisone, very similar to
cortisol, is another such hormone.
When cortisone is used to treat
arthritis, the body changes much of it
to cortisol by reducing a keto group to
the 2° alcohol group that you see in
the structure of cortisol.

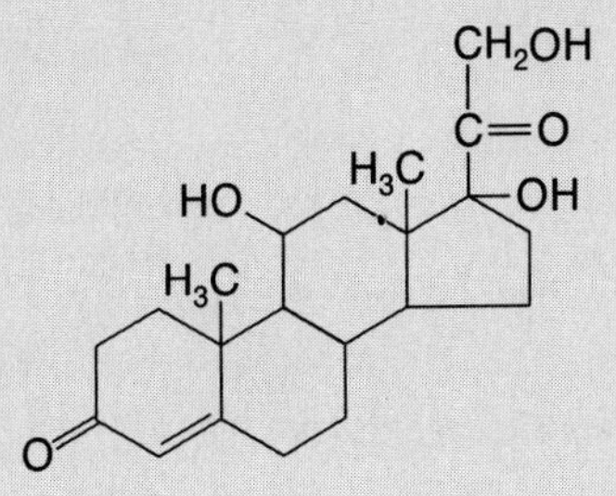

Cortisol

TABLE 15.3   (continued)

Cardiac Aglycone
Digitoxigenin is found in many poisonous plants, notably digitalis, as a
complex glycoside. In small doses it
stimulates the vagus mechanism and
increases heart tone. In larger doses it
acts as a potent poison.

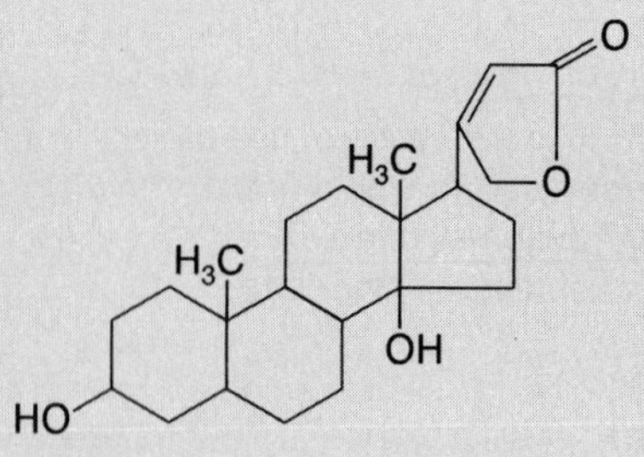

Digitoxigenin

Sex Hormones
Estradiol is a human estrogenic
hormone.

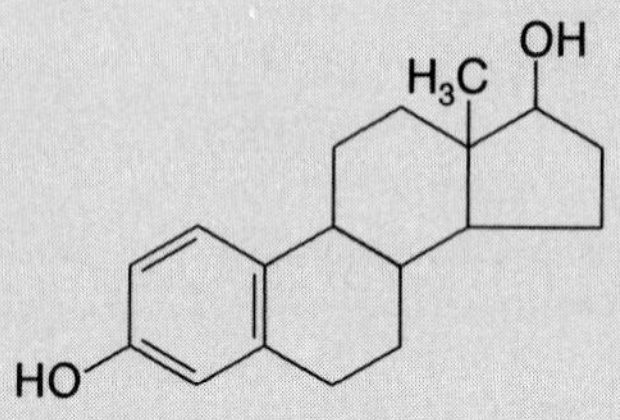

Estradiol

Progesterone, a human pregnancy
hormone, is secreted by the *corpus lu-
teum*.

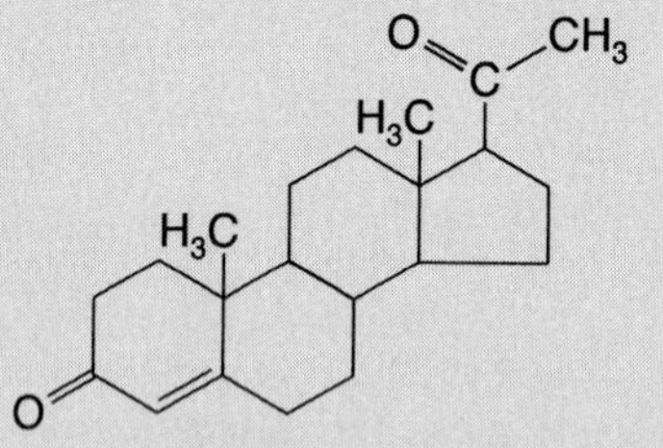

Progesterone

Testosterone, a male sex hormone,
regulates the development of reproductive organs and secondary sex
characteristics.

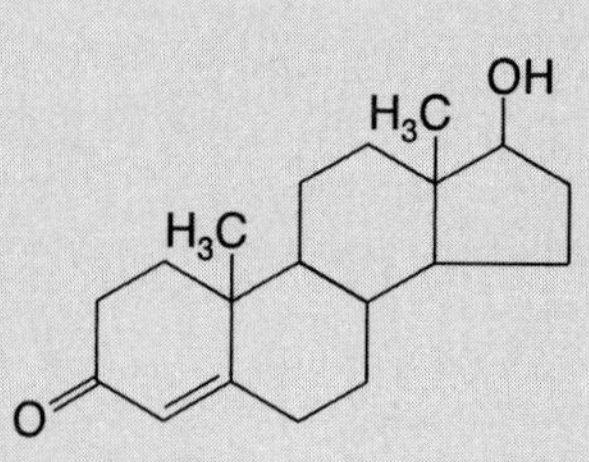

Testosterone

Androsterone is another male sex
hormone.

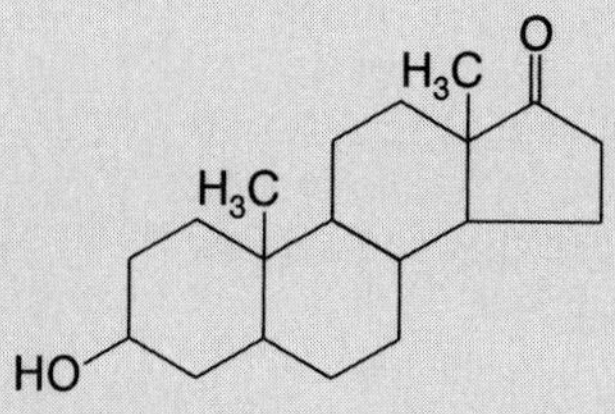

Androsterone

TABLE 15.3 (continued)

Synthetic Hormones in Fertility Control

Most oral contraceptive pills contain one or two synthetic, hormone-like compounds. (Synthetics must be used because the real hormones are broken down in the body.)

Synthetic estrogens

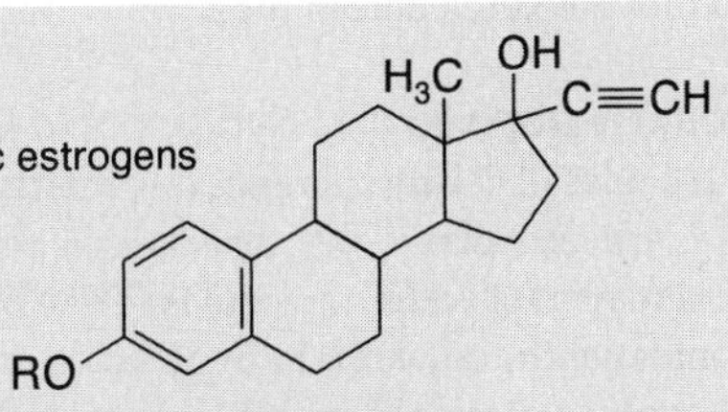

If R = H, ethynylestradiol
R = CH$_3$, mestranol

The most widely used pills have a combination of an estrogen (20–100 μg for mestranol and 20–50 μg for ethinyl estradiol) plus a progestin (0.35–2.5 mg depending on the compound).

Synthetic progestins

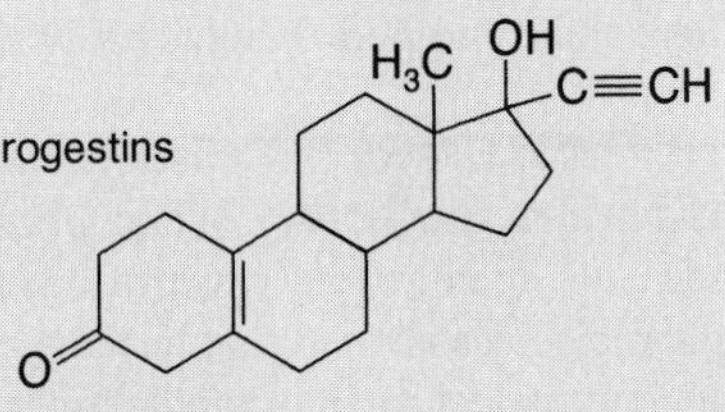

Norethynodrel

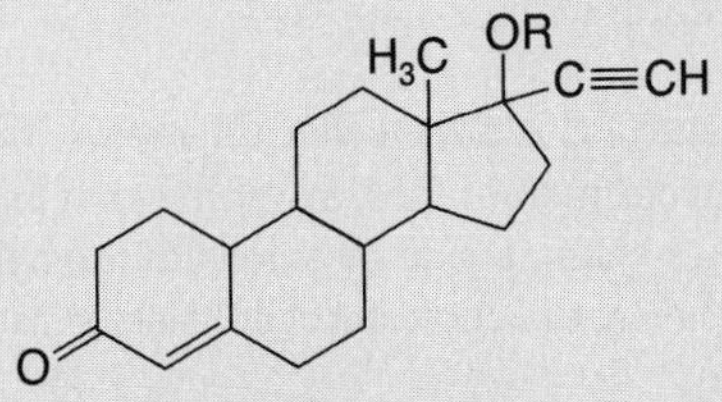

If R = H, norethindrone

R = C—CH$_3$, norethindrone acetate

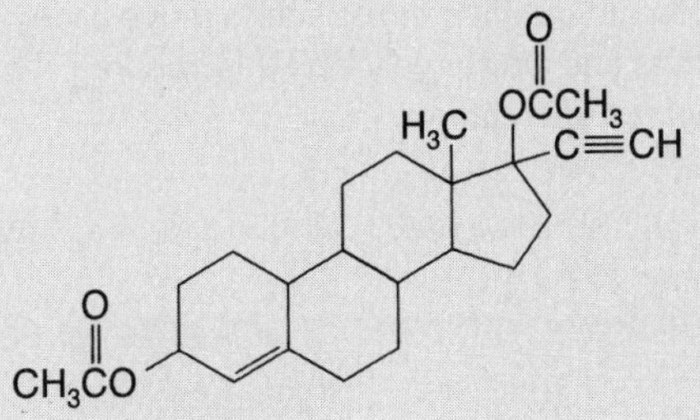

Ethynodiol diacetate

15.5 CELL MEMBRANES

Cell membranes consist of a lipid bilayer that includes molecules of proteins and cholesterol.

Cell membranes are made of both lipids and proteins. The principal lipids are the phospholipids, the glycolipids, and cholesterol. Some proteins provide closable molecular channels through which ions and molecules pass across the membrane. Other proteins act as "pumps" to move solutes across a cell membrane.

Both Hydrophilic and Hydrophobic Groups Are Necessary for Cell Membranes.   The molecules in cell membranes have parts that are either very polar or are fully ionic and others that are nonpolar. The polar or ionic sites are called **hydrophilic groups,** because they are able to attract water molecules. They become positioned in membranes where they can be in contact with the water in body fluids. In the phospholipids, these hydrophilic groups are the phosphate diester units, which have ionic sites. In a glycolipid the sugar unit with its many OH groups is the hydrophilic group.

The nonpolar, hydrocarbon sections of membrane lipids are called **hydrophobic groups** because they are water avoiding. They become positioned in the membrane away from water as much as possible. Substances like the phospholipids or glycolipids, with both hydrophilic and hydrophobic groups, are called **amphipathic compounds.**

In the Lipid Bilayer of Cell Membranes, Hydrophobic Groups Intermingle between the Membrane Surfaces.   When phospholipids or glycolipids are mixed with water, their molecules spontaneously form a **lipid bilayer,** a sheetlike array that consists of two layers of lipid molecules aligned side by side, as illustrated in Figure 15.2. The hydrophobic "tails" of the lipid molecules intermingle in the center of the bilayer away from water molecules. In a sense, these "tails" dissolve in each other, following the "like-dissolves-like" rule. The hydrophilic "heads" stick out into the aqueous phase in contact with water. These water-avoiding and water-attracting properties are the major "forces," not covalent bonds, that stabilize the membrane.

Cholesterol Stabilizes Membranes.   Molecules of cholesterol and cholesterol esters, also found in cell membranes, are somewhat long and flat. In the lipid bilayer they occur with their long axes lined up side by side with the hydrocarbon chains of the other lipids. In contrast to these chains, the cholesterol units are more rigid, so they keep the membrane from being too fluidlike.

The Lipid Bilayer is Self-Sealing.   If a pin were stuck through a cell membrane and then pulled out, the lipid layer would close back spontaneously. This flexibility is allowed because, as we said, no covalent bonds hold neighboring lipid molecules to each other. Only the net forces of attraction that we imply when we use the terms *hydrophobic* and *hydrophilic* are at work. Yet, the bilayer is strong enough to hold a cell together, and it is flexible enough to let things in and out. Water molecules move back and forth freely through the bilayer, but other molecules and ions largely lack this freedom. Their migrations depend on the protein components of the membrane.

Hydrophilic — from the Greek *hydor*, water, and *philos*, loving. *Hydrophobic* — from the Greek *phobikos*, hating

polar head

two polar tails

Phospholipid symbol

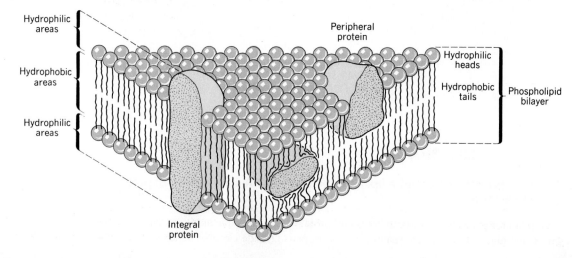

Hydrophilic areas

Hydrophobic areas

Hydrophilic areas

Peripheral protein

Hydrophilic heads

Hydrophobic tails

Phospholipid bilayer

Integral protein

FIGURE 15.2
Cell membrane.

Membrane Proteins Help to Maintain Concentration Gradients. If the cell membrane were an ordinary dialyzing membrane, any kind of small molecule or ion could move freely back and forth. The health of cells, however, demands that only some things be let in and that others be let out. This means that between one side and the other of the membrane there have to be many concentration gradients. A **gradient** is the existence of an unevenness in the value of some physical property throughout a system. A concentration gradient exists in a solution, for example, when one region of the solution has a higher concentration of solute than another.

As the data in the table in the margin show, both sodium ions and potassium ions have quite different concentrations in the fluids on the inside of a cell as compared to the fluids on the outside. Thus between the inside and the outside of a cell there is a considerable concentration gradient for both of these ions. *This gradient must be maintained at all costs* against nature's spontaneous tendency to remove concentration gradients.

Here is where some of the proteins in cell membranes carry out a vital function. One kind of membrane protein can move sodium ions against their gradient. When too many sodium ions leak to the inside of a cell, they are "pumped" back out by a special molecular machinery called the sodium–potassium pump. The same pump can move potassium ions back inside a cell. This movement of any solute against its concentration gradient is an example of **active transport,** and other reactions in cells supply the chemical energy that lets it work.

Ion	Concentration (mmol/L)	
	Plasma	Cells
Na^+	135–145	10
K^+	3.5–5.0	125

Some Proteins in Cell Membranes Are Receptors for Hormones and Neurotransmitters. A **receptor molecule** is one whose unique shape enables it to fit only to the molecule of a compound that it is supposed to receive, its substrate. It is thus able to "recognize" the molecules of just one compound from among the hundreds whose molecules bump against it. This is roughly how specific hormones are able to find only the cells that use them and bypass all others. The hormone molecule stops only where it is able to fit to a receptor. This mechanism also helps neurotransmitter molecules to move quickly from one nerve cell to the next across the very narrow gap between them. Once a receptor molecule in a cell membrane accepts its unique substrate, further biochemical changes occur. An enzyme or a gene in the cell might be activated, for example.

Neurotransmitters are organic molecules that help carry nerve signals from the end of one nerve cell to the beginning of the next.

SUMMARY

Lipids Lipids are ether-extractable substances in animals and plants, and they include saponifiable esters and nonsaponifiable compounds. The esters are generally of glycerol or sphingosine with their acyl portions contributed by long-chain carboxylic acids (fatty acids). Because molecules of all lipids are mostly hydrocarbon-like, lipids are soluble in nonpolar solvents but not in water. The fatty acids obtained from lipids by hydrolysis generally have long chains of even numbers of carbons, seldom are branched, and often have one or more alkene groups. The alkene groups are generally cis.

Molecules of the waxy coatings on leaves and fruit, or in beeswax or sebum, are simple esters between long-chain monohydric alcohols and fatty acids.

Triacylglycerols Molecules of neutral fats, those without electrically charged sites or sites that are similarly polar, are esters of glycerol and a variety of fatty acids, both saturated and unsaturated. Vegetable oils have more double bonds per molecule than animal fats. The triacylglycerols can be hydrogenated, hydrolyzed (digested), and saponified.

Phosphoglycerides Molecules of the phosphoglycerides are esters both of glycerol and of phosphoric acid. A second ester bond from the phosphate unit goes to a small alcohol molecule that can also have a positively charged group. Thus this part of a phosphoglyceride is strongly hydrophilic.

Spingomyelins Sphingomyelins are esters of sphingosine, a dihydric amino alcohol. They also have a strongly hydrophilic phosphate system.

Glycolipids Also sphingosine based, the glycolipids use a monosaccharide instead of the phosphate-to-small alcohol unit to provide the hydrophilic section. Otherwise, they resemble the sphingomyelins.

Steroids Steroids are nonsaponifiable lipids with the steroid nucleus of four fused rings (three being C_6 rings and one a C_5 ring). Several steroids are sex hormones, and oral fertility control drugs mimic their structure and functions. Cholesterol, the raw material used by the body to make other steroids, is also manufactured by the body.

Membranes A double layer of phospholipids or glycolipids plus cholesterol and proteins make up the lipid bilayer part of a cell membrane. The hydrophobic tails of these amphipathic lipids inter-

mingle within the bilayer, away from the aqueous phase. The hydrophilic heads are in contact with the aqueous medium.

The proteins on or in the bilayer serve as conduits, receptors, or are parts of pumps, such as the sodium–potassium pump, that work to maintain important concentration gradients, or as channels for small ions or molecules.

REVIEW EXERCISES

The answers to these review exercises are in the *Study Guide* that accompanies this book.

Lipids in General

15.1 Crude oil is soluble in ether, yet it isn't classified as a lipid. Explain.

15.2 Cholesterol has no ester group, yet we classify it as a lipid. Why?

15.3 Ethyl acetate has an ester group, but it isn't classified as a lipid. Explain.

15.4 What are the criteria for deciding if a substance is a lipid?

Waxes

15.5 One component of beeswax has the formula $C_{34}H_{68}O_2$. When it is hydrolyzed, it gives $C_{16}H_{32}O_2$ and $C_{18}H_{38}O$. Write the most likely structure of this compound.

15.6 When all the waxes from the leaves of a certain shrub are separated, one has the formula of $C_{60}H_{120}O_2$. Its structure is **A**, **B**, or **C**. Which is it most likely to be? Explain why the others can be ruled out.

$$CH_3(CH_2)_{56}CO_2CH_2CH_3 \qquad CH_3(CH_2)_{29}CO_2(CH_2)_{28}CH_3$$
$$\textbf{A} \qquad\qquad\qquad \textbf{B}$$
$$CH_3(CH_2)_{28}CO_2(CH_2)_{29}CH_3$$
$$\textbf{C}$$

Fatty Acids

15.7 What are the structures and the names of the two most abundant saturated fatty acids?

15.8 Write the structures and names of the unsaturated fatty acids that have 18 carbons each and that have no more than three double bonds. Show the correct geometry at each double bond.

15.9 Write the equations for the reactions of palmitic acid with (a) NaOH(aq) and (b) CH_3OH (when heated in the presence of an acid catalyst).

15.10 What are the equations for the reactions of oleic acid with (a) Br_2, (b) KOH(aq), (c) H_2 (in the presence of a catalyst and under pressure), and (d) CH_3CH_2OH (heated in the presence of an acid catalyst)?

15.11 Which of the following acids, **A** or **B**, is more likely to be obtained by the hydrolysis of a lipid? Explain.

$$\begin{array}{c} CH_3 \\ | \\ CH_3CH(CH_2)_{11}CO_2H \end{array} \qquad CH_3(CH_2)_{12}CO_2H$$
$$\textbf{A} \qquad\qquad\qquad \textbf{B}$$

15.12 State what kinds of chemicals the prostaglandins are without writing structures.

Triacylglycerols

15.13 Write the structure of a triacylglycerol that involves linolenic acid, oleic acid, and myristic acid, besides glycerol.

15.14 What is the structure of a triacylglycerol made from glycerol, stearic acid, oleic acid, and palmitic acid?

15.15 Write the structures of all the products that would form from the complete digestion of the following lipid.

$$\begin{array}{l} \quad\quad\quad\quad O \\ \quad\quad\quad\quad \| \\ CH_2{-}O{-}C(CH_2)_7CH{=}CH(CH_2)_7CH_3 \\ | \quad\quad\quad O \\ | \quad\quad\quad \| \\ CH{-}O{-}C(CH_2)_{12}CH_3 \\ | \quad\quad\quad O \\ | \quad\quad\quad \| \\ CH_2{-}O{-}C(CH_2)_7CH{=}CH(CH_2)_7CH_3 \end{array}$$

15.16 Write the structures of the products that are produced by the saponification of the triacylglycerol whose structure was given in Review Exercise 15.15.

15.17 The hydrolysis of a lipid produced glycerol, lauric acid, linoleic acid, and oleic acid in equimolar amounts. Write a structure that is consistent with these results. Is there more than one structure that can be written? Explain.

15.18 The hydrolysis of 1 mol of a lipid gave 1 mol each of glycerol and oleic acid and 2 mol of lauric acid. Write a structure consistent with this information.

15.19 What is the structural difference between the triacylglycerols of the animal fats and the vegetable oils?

15.20 Products such as corn oil are advertised as being "polyunsaturated." What does this mean in terms of the structures of the molecules that are present? And corn oil is "more polyunsaturated" than what?

15.21 What chemical reaction is used in the manufacture of oleomargarine?

15.22 Lard and butter are chemically almost the same substances, so what is it about butter that makes it so much more desirable as a spread for bread than, say, lard or tallow?

Phospholipids

15.23 What are the names of the two chief kinds of phospholipids?

15.24 In structural terms, how do the phosphoglycerides and plasmalogens differ?

15.25 How are the sphingomyelins and cerebrosides different structurally?

15.26 What structural unit provides the most polar group in a molecule of a glycolipid? (Name it.)

15.27 Phospholipids are not classified as neutral fats. Explain.

15.28 Phospholipids are particularly common in what part of a cell?

15.29 What are the names of the two types of sphingosine-based lipids?

15.30 Are the sugar units that are incorporated into the cerebrosides bound by glycosidic links or by ordinary ether links? How can one tell? Which kind of link is more easily hydrolyzed (assuming an acid catalyst)?

15.31 The complete hydrolysis of 1 mol of a phospholipid gave 1 mol each of the following compounds: glycerol, linolenic acid, oleic acid, phosphoric acid, and the cation, $HOCH_2CH_2\overset{+}{N}(CH_3)_3$.
   (a) Write a structure of this phospholipid that is consistent with the information given.
   (b) Is the substance a phosphoglyceride or a sphingolipid? Explain?
   (c) Is it an example of a lecithin or a cephalin? Explain?

15.32 When 1 mol of a certain phospholipid was hydrolyzed, there was obtained 1 mol each of lauric acid, oleic acid, phosphoric acid, glycerol, and $HOCH_2CH_2NH_2$.
   (a) What is a possible structure for this phospholipid?
   (b) Is it a sphingolipid or a phosphoglyceride? Explain.
   (c) Is it a cephalin or a lecithin? Explain.

Steroids

15.33 What is the name of the steroid that occurs as a detergent in our bodies?

15.34 What is the name of a vitamin that is made in our bodies from a dietary steroid by the action of sunlight on the skin?

15.35 Give the names of three steroidal sex hormones.

15.36 What is the name of a steroid that is part of the cell membranes in many tissues?

Cell Membranes

15.37 Describe in your own words what is meant by the *lipid bilayer* structure of cell membranes.

15.38 How do the hydrophobic parts of phospholipid molecules avoid water in a lipid bilayer?

15.39 Besides lipids, what kinds of substances are present in a cell membrane?

15.40 What kinds of forces are at work in holding a cell membrane together?

15.41 Immediately after you add a teaspoon of sugar to a hot cup of coffee, can a concentration gradient for sugar be present? Explain what this is. What happens to it, given enough time, even without stirring the mixture? What causes this change to occur? What are the chances of the original gradient ever restoring itself spontaneously?

15.42 Which has the higher level of sodium ion, plasma or cell fluid?

15.43 Does cell fluid or plasma have the higher concentration of potassium ion?

15.44 In which fluid, plasma or cell fluid, would the level of sodium ion increase if the sodium ion gradient could not be maintained?

15.45 What is meant by *active transport* in a cell membrane?

15.46 What does the sodium–potassium pump in a cell membrane do?

15.47 Name the functions that the proteins of a cell membrane can serve.

The Prostaglandins (Special Topic 15.1)

15.48 Name the fatty acid that is used to make the prostaglandins.

15.49 What effect is aspirin believed to have on prostaglandins, and how is this related to aspirin's medicinal value?

The Omega-3 Fatty Acids and Heart Disease (Special Topic 15.2)

15.50 What is it about the structure of linolenic acid that lets us call it an omega-3 acid?

15.51 What source of the omega-3 acids is relatively rich in the C_{20} and C_{22} acids?

15.52 Why have the omega-3 acids aroused the interests of people in nutrition and in medicine?

Detergent Action (Special Topic 15.3)

15.53 Which is the more general term, soap or detergent? Explain.

15.54 What kind of chemical is soap?

15.55 For household laundry work, which product is generally preferred, a synthetic detergent or soap? Why?

15.56 Why are soap and sodium alkyl sulfates called *anionic* detergents?

15.57 Explain in your own words how a detergent can loosen oils and greases from fabrics.

Chapter 16
Proteins

Proteins make up the chief parts of muscles, which Nadia Comaneci conditioned to perfection on her way to Olympic gold in balance beam competition.

16.1 AMINO ACIDS. THE BUILDING BLOCKS OF PROTEINS

Living things select from among the molecules of about 20 α-amino acids to make the polypeptides in proteins.

Proteins, found in all cells and in virtually all parts of cells, constitute about half of the body's dry weight. They give strength and elasticity to skin. As muscles and tendons, they function as the cables that enable us to move the levers of our bones. They reinforce our teeth and bones much as thick steel rods reinforce concrete. The molecules of antibodies, of hemoglobin, and of the various kinds of albumins in our blood serve as protectors and as the long-distance haulers of substances, such as oxygen or lipids, that otherwise do not dissolve well in blood. Other proteins form parts of the communications network of our nervous system. Some proteins are enzymes, hormones, and gene regulators that direct and control all forms of repair, construction, and energy conversion in the body. No other class of compounds is involved in such a variety of functions, all essential to life. They deserve the name *protein*, taken from the Greek *proteios,* "of the first rank."

Polypeptides Are Made from α-Amino Acids. The dominant structural units of **proteins** are high formula weight polymers called **polypeptides.** Metal ions and small organic molecules or ions are often present as well. The relationship of these parts to whole proteins is shown in Figure 16.1. Many proteins, however, are made entirely of polypeptides.

The monomer units for polypeptides are α-**amino acids,** which have the general structure given by **1**. About 20 amino acids are used and most of them are used several times in the same polypeptide. Hundreds of **amino acid residues** are individually derived from one or another of the various α-amino acids. They are joined together in a single polypeptide molecule, but before we can study how polypeptides are put together we must learn more about amino acids.

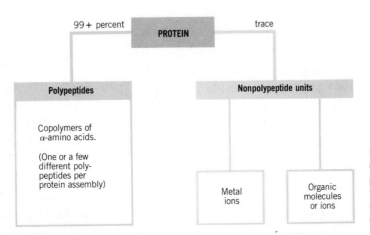

FIGURE 16.1
Components of proteins. Some proteins consist exclusively of polypeptide molecules, but most also have nonpolypeptide units such as small organic molecules or metal ions, or both.

TABLE 16.1
Amino Acids $^+NH_3-CH-CO_2^-$
                              |
                              G

Type	Side Chain, G	Name	Symbol	pI
Side chain is nonpolar	—H	Glycine	Gly	5.97
	—CH₃	Alanine	Ala	6.00
	—CH(CH₃)₂	Valine	Val	5.96
	—CH₂CH(CH₃)₂	Leucine	Leu	5.98

	—CHCH₂CH₃
	    |
	   CH₃
		Isoleucine	Ile	6.02

	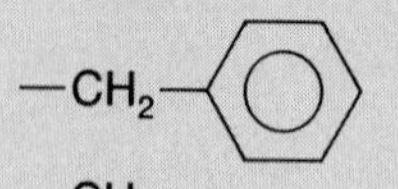 —CH₂— (phenyl ring)	Phenylalanine	Phe	5.48
	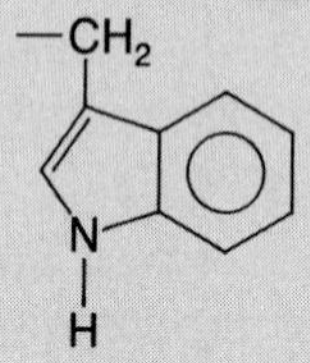 —CH₂ (indole ring)	Tryptophan	Trp	5.89
	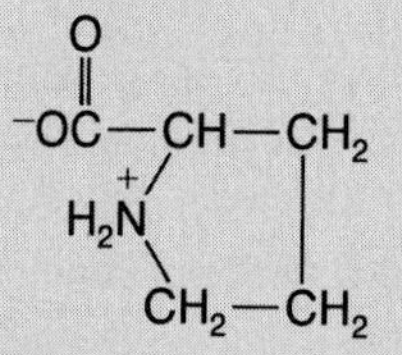 (complete structure, proline ring)	Proline	Pro	6.30
Side chain has a hydroxyl group	—CH₂OH	Serine	Ser	5.68

	—CHOH
	   |
	  CH₃
		Threonine	Thr	5.64

	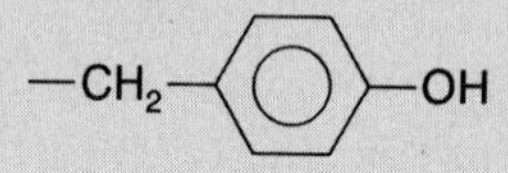 —CH₂— (ring) —OH	Tyrosine	Tyr	5.66
Side chain has a carboxyl group (or an amide)	—CH₂CO₂H	Aspartic acid	Asp	2.77
	—CH₂CH₂CO₂H	Glutamic acid	Glu	3.22
	—CH₂CONH₂	Asparagine	Asn	5.41
	—CH₂CH₂CONH₂	Glutamine	Gln	5.65
Side chain has a basic amino group	—CH₂CH₂CH₂CH₂NH₂	Lysine	Lys	9.74
	—CH₂CH₂CH₂NH—C—NH₂ (with NH)	Arginine	Arg	10.76
	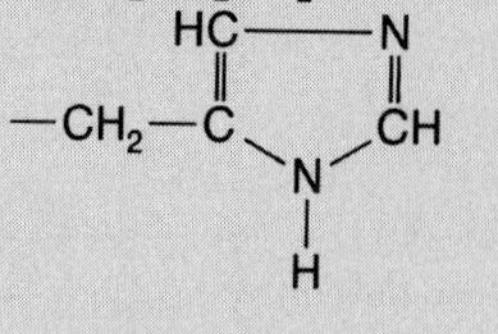 —CH₂— (imidazole ring)	Histidine	His	7.59
Side chain contains sulfur	—CH₂SH	Cysteine	Cys	5.07
	—CH₂CH₂SCH₃	Methionine	Met	5.74

The same set of 20 amino acids, Table 16.1, is used by all species of plants and animals. In rare instances, a few others are present in certain polypeptides, and some of the 20 listed in Table 16.1 occur in slightly modified forms.

Hereafter, when we say "amino acid," we'll mean α-amino acid.

The α-Amino Acids Are Based on a Dipolar Form of α-Amino Acetic Acid but Have Different Side chains at the α Position. As you can see in Table 16.1, the amino acids differ in the groups, G, called **side chains**, located at the α position in **1**.

In the solid state, amino acids exist entirely in the form shown by **1**, which is called a **dipolar ion.** It is a neutral particle but has a positive and a negative charge on different sites. (We can call them molecules because they are electrically neutral.) Because they are exceedingly polar, amino acids, like salts, have melting points that are considerably higher than those of most molecular compounds, and they are insoluble in nonpolar solvents, but soluble in water.

Structure **1** is actually an internally neutralized molecule. We can imagine that it started out with a regular amino group, NH_2, and an ordinary carboxyl group, CO_2H. But then the amino group, a proton acceptor, took a proton from the carboxyl group, a proton donor, to give **1**. Of course, **1**, has its own (weaker) proton-donating group, NH_3^+, and its own (also weaker) proton-accepting group, CO_2^-, so these dipolar ions can neutralize acids or bases of sufficient strength, such as H_3O^+ and OH^-. (In fact, amino acids can serve as buffers.)

For most amino acids to exist as dipolar ions, **1**, in water, the pH has to be about $6-7$. If we make the pH much lower (more acidic) or much higher (more basic), the form of the amino acid changes. If we add enough strong acid like HCl(aq), for example, to a solution of an amino acid at a pH initially about $6-7$, and lower the pH to about 1, the CO_2^- groups take on protons. They change to structure **2**. Now they are cations that can migrate to a cathode.

These shifts of H+ ions are illustrations of Le Chatelier's principle at work.

$$^+NH_3-CH-CO_2^-$$

$$| \atop G$$

1

$$^+NH_3-CH-CO_2H \qquad\qquad NH_2-CH-CO_2^-$$

$$| \atop G \qquad\qquad\qquad\qquad | \atop G$$

2   **3**

If we add enough strong base to raise the pH to about 11, then most of the amino acid molecules transfer protons from their NH_3^+ groups to OH^- ions, and change to structure **3**. This is an anion, so it can migrate to an anode.

A pH Exists for Each Amino Acid, Its Isoelectric Point, at Which No Net Migration in an Electric Field Occurs. In an aqueous solution of an amino acid, a dynamic equilibrium exists between **1**, **2**, and **3**. If now a current is passed between electrodes dipping into such a solution, cations of form **2** migrate to the cathode. Anions of form **3** migrate to the anode. Neutral molecules, **1**, do not migrate. A molecule with equal numbers of positive and negative charges, like **1**, is said to be an **isoelectric molecule,** and it cannot migrate in an electric field.

Because the equilibrium is *dynamic,* a migrating cation could flip a proton to some acceptor, thereby become **1** and isoelectric (neutral), and stop dead. In another instant, it could shed another proton, become **3**, and so turn around and head for the anode. Similarly, an anion on its way to the anode might pick up a proton, become neutral, and also stop dead. Then it might take another proton, become **2**, and turn itself around. In the meantime, an isoelectric molecule, **1**, might either donate or accept a proton, become electrically charged, and start its own migration. (It rather reminds one of amusement park bumper cars that move in every direction.) The question is, what overall net migration occurs and how is this net effect influenced by the pH of the solution?

Although much coming and going occurs in an amino acid solution, the net molar concentrations of the species stay the same at equilibrium. If either **2** or **3** is in any molar excess, because of the pH, then some *net* migration will occur toward one electrode or the other. If the net molar concentration of **2**, for example, is greater than that of **3**, some statistical net movement to the cathode will occur.

Remember, however, that these equilibria can be shifted by adding acid or base. By carefully adjusting the pH, in fact, we can so finely tune the concentrations at equilibrium that no net migration occurs. At the right pH, the rates of proton exchange are such that each unit that is not **1** spends an equal amount of time as **2** and as **3**. (And the concentrations of **2** and **3** are very low.) In this way any net migration toward one electrode is blocked.

The pH at which no net migration of an amino acid can occur in an electric field is called the **isoelectric point** of the amino acid, and its symbol is **pI**. Table 16.1 includes a column of these pI values. Now let's see what this has to do with proteins.

Proteins, Like Amino Acids, Have Isoelectric Points. As we will soon see, all proteins have NH_3^+ or CO_2^- groups or can acquire them by a change in pH. Whole protein molecules, therefore, can also be isoelectric at the right pH. Each protein thus has its own isoelectric point. But now think of what can happen if the pH is changed. The entire electrical condition of a huge protein molecule can be made either cationic or anionic almost instantly, at room temperature, by adding strong acid or base, by changing the pH of the medium.

Such changes in the electrical charge of a protein have serious consequences at the molecular level of life. Being electrically charged can dramatically affect chemical reactions, for example, or greatly alter protein solubility. If they are to serve their biological purposes, some proteins must not be allowed to go into solution and others must not be permitted to precipitate. We'll return to this concept in this and later chapters, but the discussion focuses our attention again on how important it is that an organism control the pH values of its fluids.

We will next survey the types of side chains in amino acids and how they affect the properties of polypeptides and proteins. These properties include how a polypeptide molecule will spontaneously fold and twist into its distinctive and absolutely vital final shape. You should memorize the structures of a minimum of five amino acids that illustrate the types we are about to study; glycine, alanine, cysteine, lysine, and glutamic acid are suggested. How to use Table 16.1 to write their structures is described in the following worked example.

EXAMPLE 16.1   WRITING THE STRUCTURE OF AN AMINO ACID

Problem: What is the structure of cysteine?

Solution: Doing this kind of problem depends on two things: (1) On remembering what is common to all amino acids

$$^+NH_3 - CH - CO_2^-$$
$$|$$

and then (2) either looking up or remembering the side chain for the particular amino acid. For cysteine, this is CH_2SH, so simply attach this group to the α carbon. Cysteine is

$$^+NH_3 - CH - CO_2^-$$
$$|$$
$$CH_2SH$$

The memorization of amino acid structures involves learning the structures of the side chains and fixing them in the mind to the name of the amino acid.

PRACTICE EXERCISE 1   Write the structures of the dipolar ionic forms of glycine, alanine, lysine, and glutamic acid.

Several Amino Acids Have Hydrophobic Side Chains. The first amino acids in Table 16.1, including alanine, have essentially nonpolar, hydrophobic side chains. When a long polypeptide molecule folds into its distinctive shape, these hydrophobic groups tend to be folded next to each other rather than next to highly polar groups or to water molecules in the solution.

Some Amino Acids Have Hydrophilic OH Groups on Their Side Chains. The second set of amino acids in Table 16.1 consists of those whose side chains carry alcohol or phenol OH groups, which are polar and hydrophilic. They can donate and accept hydrogen bonds. As a long polypeptide chain folds into its final shape, these side chains tend to stick out into the surrounding aqueous phase to which they are attracted.

Two Amino Acids Have Carboxyl Groups on Their Side Chains. The side chains of aspartic and glutamic acid carry proton-donating CO_2H groups. Because body fluids are generally slightly basic, the protons available from them have been neutralized, so these groups actually occur mostly as CO_2^- groups. The solution has to be made quite acidic to prevent this. This forces protons back onto the side-chain CO_2^- groups, and this is why the pI values of aspartic and glutamic acid are low.

Aspartic acid and glutamic acid often occur as asparagine and glutamine in which their side-chain CO_2H groups have become amide groups, $CONH_2$, instead. These are also polar, hydrophilic groups, but they are not electrically charged. They are neither proton donors nor proton acceptors, so the pI values of asparagine and glutamine are higher than those of aspartic or glutamic acids.

PRACTICE EXERCISE 2   Write the structure of aspartic acid (in the manner of **1**) with the side-chain carboxyl (a) in its carboxylate form, and (b) in its amide form.

Lysine, Arginine, and Histidine Have Basic Groups on Their Side Chains. The extra NH_2 group on lysine makes its side chain basic and hydrophilic. A solution of lysine has to be made basic to prevent this group from existing in its protonated form, NH_3^+. This is why the pI value of lysine, 9.47, is relatively high. Arginine and histidine have similarly basic side chains.

PRACTICE EXERCISE 3   Write the structure of arginine in the manner of **1**, but with its side-chain amino group in its protonated form. (Put the extra proton on the $=NH$ unit, not the $-NH_2$ unit of the side chain.)

PRACTICE EXERCISE 4   Classify the side chain of the following amino acid as hydrophilic or hydrophobic. Does this side chain have an acidic, basic, or neutral group?

$$^+NH_3CHCO_2^-$$
$$|$$
$$CH_2CH_2CONH_2$$

Cysteine and Methionine Have Sulfur-Containing Side Chains. The side chain in cysteine has an SH group. As we studied in Section 14.4, molecules with this group are easily oxidized to disulfide systems, and disulfides are easily reduced to SH groups:

$$2RSH \xrightleftharpoons[(H)]{(O)} R\text{---}S\text{---}S\text{---}R + H_2O$$

Cysteine and its oxidized form, cystine, are interconvertible by oxidation and reduction, a property of far reaching importance in some proteins.

$$\underset{\underset{\text{Cysteine}}{}}{\overset{\overset{O}{\|}}{{}^-OCCHCH_2}\text{---}S\text{---}H} \xrightarrow{\overset{(O)}{\text{oxidation}}} \overset{\overset{O}{\|}}{{}^-OCCHCH_2}\text{---}S + H_2O$$

$$\underset{\underset{\text{Cysteine}}{NH_3^+}}{\overset{\overset{O}{\|}}{{}^-OCCHCH_2}\text{---}S\text{---}H} \xleftarrow[\text{2(H)}]{\text{reduction}} \underset{\underset{\text{Cystine}}{NH_3^+}}{\overset{\overset{O}{\|}}{{}^-OCCHCH_2}\text{---}S}$$

The **disulfide link** contributed by cystine is especially prevalent in the proteins that have a protective function, such as those in hair, fingernails, and the shells of certain crustaceans.

All Amino Acids Except Glycine Have Molecules that are Chiral[1].

All the amino acids except glycine are optically active and can exist as enantiomers. For each possible pair of enantiomers, however, nature supplies just one of the two (with a few rare exceptions). All the naturally occurring amino acids, moreover, belong to the same optical family, the L family. What this means is illustrated in Figure 16.2. It also means that all the proteins in our bodies, including all enzymes, are made from L-amino acids and are all chiral.

The mirror-image molecules of the L-amino acids are the D-amino acids, which can be synthesized in the laboratory.

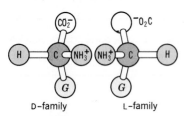

d—family   L—family

FIGURE 16.2
The two possible optically active families of α-amino acids. The absolute configuration on the right, which is in the L family, represents virtually all the naturally occurring α-amino acids.

16.2   PRIMARY STRUCTURES OF PROTEINS

The backbones of all polypeptides of all plants and animals have a repeating series of N—C—C(=O) units.

Protein structures are more complicated by far than those of carbohydrates or lipids, and every aspect of their structures is vital at the molecular level of life. We'll begin, therefore, with a broad overlook at the levels of protein structure.

Protein Structure Involves Four Features.   There are four levels of complexity in the structures of most proteins. Disarray at any level almost always renders the protein biologically useless. The first and most fundamental level, the **primary structure** of a protein, concerns only the sequence of amino acid residues joined by carbonyl-nitrogen covalent bonds called *peptide bonds,* in the polypeptide(s) of the protein. Peptide bonds, as we will see, are nothing more than amide bonds.

The next level, the **secondary structure,** also concerns just individual polypeptides. It entails noncovalent forces, particularly the hydrogen bond, and it consists of the particular way in which a long polypeptide strand has coiled or in which strands have intertwined or lined up side to side.

The **tertiary structure** of a polypeptide concerns the further bending, kinking, or twisting of secondary structures. If you've ever played with a coiled door spring, you know that the coil (secondary structure) can be bent and twisted (tertiary structure). Noncovalent forces, generally hydrogen bonds, stabilize these shapes.

[1] This small subsection as well as Review Exercise 16.10 should be omitted if Section 14.3 was not studied.

Finally, we deal with the intact protein. Its **quaternary structure** is the way in which individual polypeptides (each with all previous levels of structure), and any other molecules or ions, come together in one grand whole.

In this section, we'll study the primary structural features of polypeptides, whose central feature is the peptide bond.

The Peptide Bond Joins Aminoacyl Units Together in a Polypeptide. The **peptide bond** is the chief covalent bond that forms when amino acids are put together in a cell to make a polypeptide. It's nothing more than an amide system, carbonyl to nitrogen. To illustrate it and to show how polypeptides acquire their primary structure, we will begin by simply putting just two amino acids together.

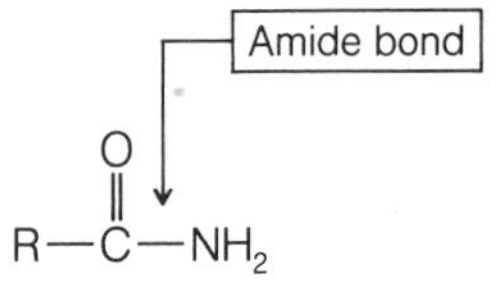

Suppose that glycine acts at its carboxyl end and alanine acts as its amino end such that, by a series of steps (not given in detail but indicated by a dashed arrow, ------>), a molecule of water splits out and a carbonyl-to-nitrogen bond is created.

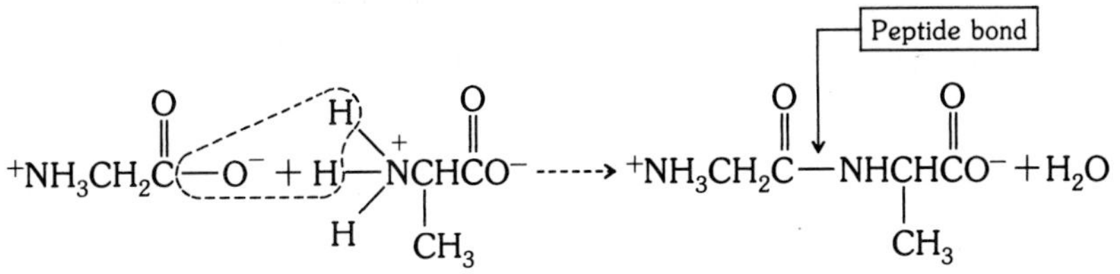

Simple amides are

O
‖
R—C—NH₂
(Amide bond)

$$^+NH_3CH_2C(-O^- \; + \; H-NCHCO^- \; ------\rightarrow \; ^+NH_3CH_2C-NHCHCO^- + H_2O$$

Glycine (Gly)   Alanine (Ala)   A dipeptide (Gly·Ala)

4

How a cell causes a peptide bond to form is a major topic under the chemistry of heredity.

Of course, there is no reason why we could not picture the roles reversed so that alanine acts at its carboxyl end and glycine at its amino end. This results in a different dipeptide (but an isomer of the first).

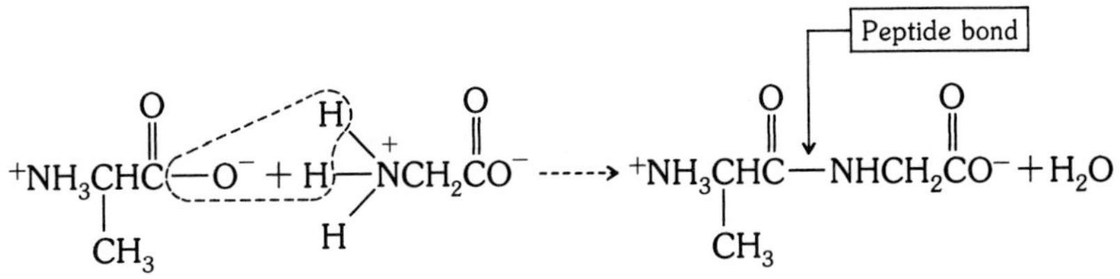

$$^+NH_3CHC(-O^- \; + \; H-NCH_2CO^- \; ------\rightarrow \; ^+NH_3CHC-NHCH_2CO^- + H_2O$$

Alanine (Ala)   Glycine (Gly)   Another dipeptide (Ala·Gly)

5

The product of the union of any two aminoacyl units by a peptide bond is called a **dipeptide,** and all dipeptides have the following features:

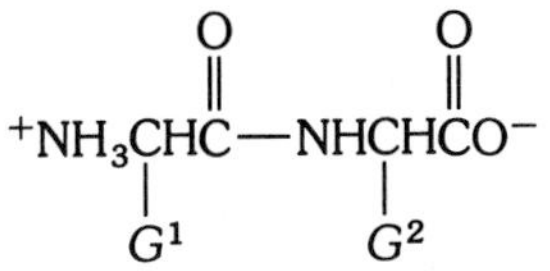

$$^+NH_3CHC-NHCHCO^-$$

G¹   G²

Dipeptide

Structures **4** and **5** differ only in the sequence in which the side chains, H and CH₃, occur on α carbons. This is fundamentally how polypeptides also differ, in their sequences of side chains.

EXAMPLE 16.2   WRITING THE STRUCTURE OF A DIPEPTIDE

Problem: What are the two possible dipeptides that can be put together from alanine and cysteine?

Solution: Both must have the same backbone, so we write two of these first (and we follow the convention that such backbones are always written in the N to C (left to right) direction):

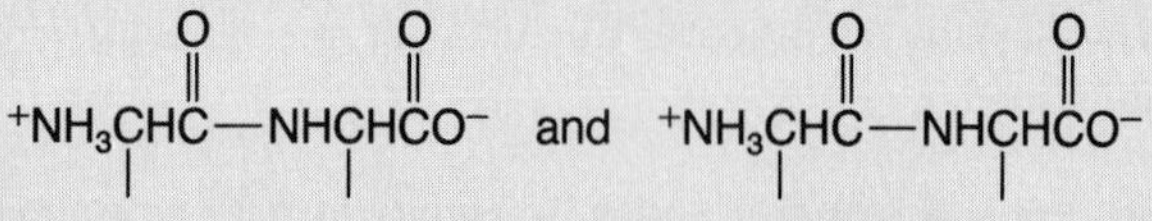

$$\text{}^{+}NH_3CHC\overset{O}{\overset{\|}{}}-NHCHC\overset{O}{\overset{\|}{}}O^- \quad \text{and} \quad \text{}^{+}NH_3CHC\overset{O}{\overset{\|}{}}-NHCHC\overset{O}{\overset{\|}{}}O^-$$

Then, either from memory or by the use of Table 16.1, we recall the two side chains, CH_3 for alanine and CH_2SH for cysteine. We simply attach these in their two possible orders to make the finished structures:

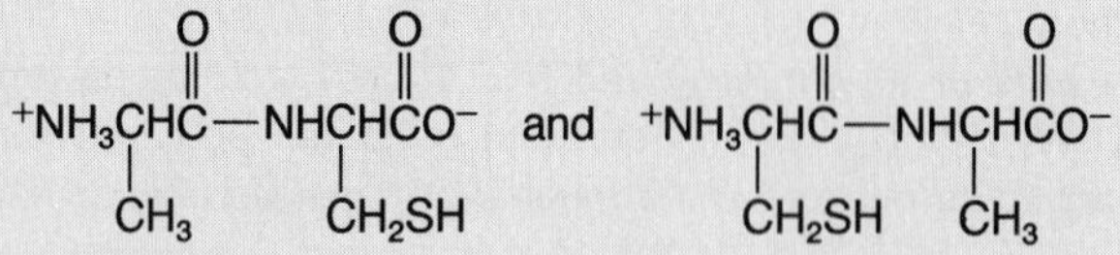

$$\text{}^{+}NH_3CHC\overset{O}{\overset{\|}{}}-NHCHCO^- \quad \text{and} \quad \text{}^{+}NH_3CHC\overset{O}{\overset{\|}{}}-NHCHCO^- \qquad \text{(answers)}$$
$$\qquad CH_3 \qquad CH_2SH \qquad\qquad CH_2SH \quad CH_3$$

It would be worthwhile at this time simply to memorize the easy repeating sequence in a dipeptide, because it carries forward to higher peptides:

$$\text{nitrogen} - \text{carbon} - \text{carbonyl} - \text{nitrogen} - \text{carbon} - \text{carbonyl}$$

PRACTICE EXERCISE 5   Write the structures of the two dipeptides that can be made from alanine and glutamic acid.

Three-Letter Symbols for Aminoacyl Units Simplify the Writing of Polypeptide Structures.

One-letter symbols have also been agreed to by biochemists.

Each amino acid has been assigned a three-letter symbol, given in Table 16.1. To use them in writing a polypeptide structure, we have to follow certain rules. The convention is that a series of three-letter symbols, each separated by a raised dot, represents a polypeptide structure, provided that the first symbol (reading left to right) is the free amino end, $^{+}NH_3$, and the last symbol has the free carboxylate end, CO_2^-. The structure of the dipeptide **4**, for example, can be rewritten as Gly · Ala, and its isomer **5** as Ala · Gly. In both, the backbones are identical.

Dipeptides still have $^{+}NH_3$ and CO_2^- groups, so a third amino acid can react at either end. In general,

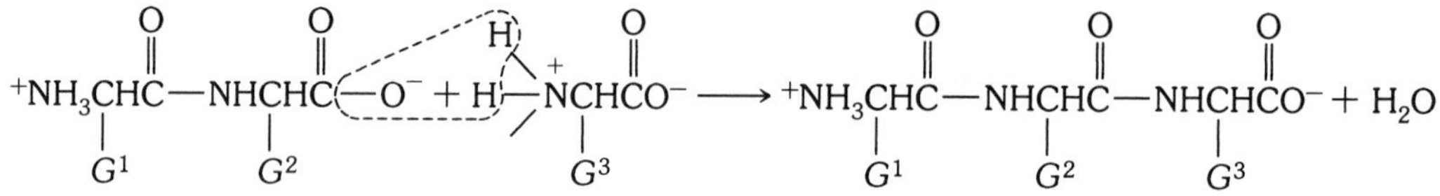

$$^{+}NH_3CHC\overset{O}{\overset{\|}{}}-NHCHC(-O^- + H-NCHCO^- \longrightarrow \text{}^{+}NH_3CHC\overset{O}{\overset{\|}{}}-NHCHC-NHCHCO^- + H_2O$$
$$\quad G^1 \qquad G^2 \qquad\qquad G^3 \qquad\qquad G^1 \qquad G^2 \qquad G^3$$

A tripeptide

6

A specific example is

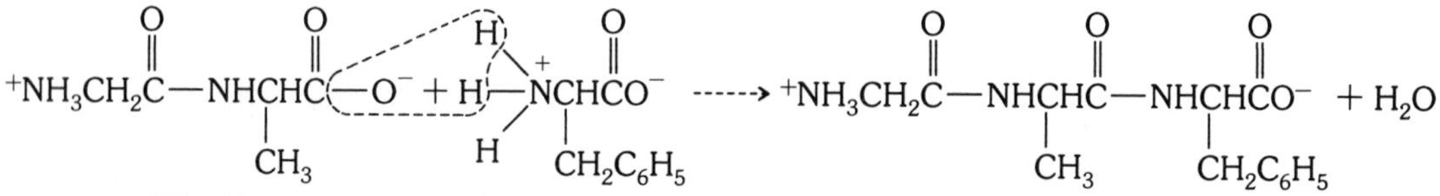

$$^{+}NH_3CH_2C\overset{O}{\overset{\|}{}}-NHCHC(-O^- + H-NCHCO^- \dashrightarrow \text{}^{+}NH_3CH_2C-NHCHC-NHCHCO^- + H_2O$$
$$\qquad CH_3 \qquad\quad H \quad CH_2C_6H_5 \qquad\qquad\qquad CH_3 \qquad CH_2C_6H_5$$

Gly · Ala   Phe   Gly · Ala · Phe
               Phenylalanine   (a tripeptide)

This tripeptide, Gly·Ala·Phe, is only one of six possible tripeptides that involve these three different amino acids. The set of all possible sequences for a tripeptide made from glycine, alanine, and phenylalanine is as follows:

Gly·Ala·Phe   Ala·Gly·Phe   Phe·Gly·Ala
Gly·Phe·Ala   Ala·Phe·Gly   Phe·Ala·Gly

Each of these tripeptides still has groups at each end, $^+NH_3$ and CO_2^-, that can interact with still another amino acid to make a tetrapeptide. And this product would still have the end groups from which the chain could be extended still further. You can see how a repetition of this many hundreds of times can produce a long polymer, a polypeptide.

The Sequence of Side Chains on the Repeating N—C—C(=O) Backbone Is the Primary Structure of All Polypeptides.

All polypeptides have the following skeleton in common. They differ in length (n) and in the kinds and sequences of side chains.

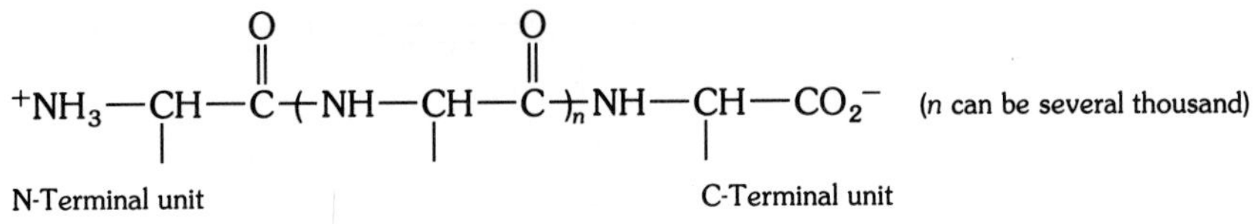

$$^+NH_3—CH—\overset{\overset{O}{\|}}{C}(\!-\!NH—CH—\overset{\overset{O}{\|}}{C}\!-\!)_n NH—CH—CO_2^- \quad (n \text{ can be several thousand})$$

N-Terminal unit   C-Terminal unit

(Notice, for later referen̲   ̲nit for the
residues with the free α-N   ̲tide bond,
as we said, is the chief cov   ̲ypeptides.
The disulfide bond is the ̲   ̲the amino
acyl units.

As the number of am   ̲number of
possible polypeptides rise   ̲corporated,
each used only once, ther̲   ̲polypeptide
with only 20 amino acid

Disulfide Bonds Can Gi   ̲ther. If the
SH group on the side ch   ̲molecules,
then mild oxidation is al   ̲as shown in
Figure 16.3. This cross-   ̲eptide molecule, in which case a clo

The symbol for cystine in a protein structure is

Cys
|
Cys

—CH₂—S—H

Polypeptide backbone

(a)

S—H H—S

(b)

S—S

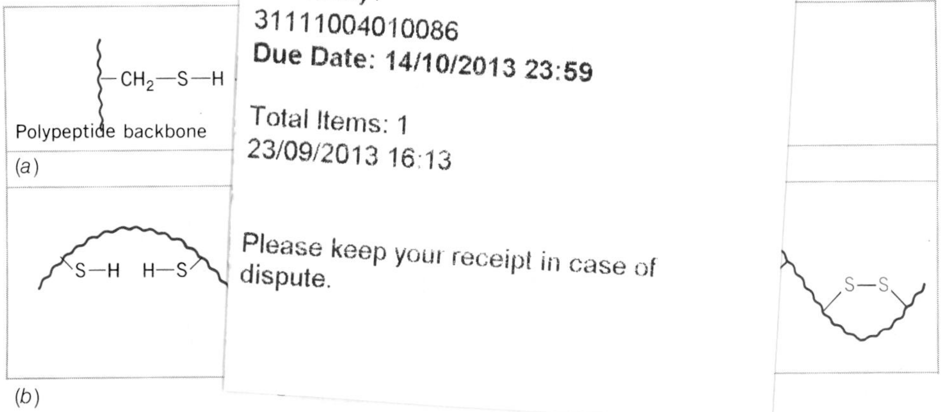

FIGURE 16.3
The disulfide link in polypeptides. (a)
Two neighboring strands are joined. (b)
Loops within the same strand can form.

Some polypeptides feature both kinds of S—S cross-linking, and one example is the hormone insulin (Figure 16.4), which is considered to be a relatively simple polypeptide. You can find three disulfide bonds, one that creates a loop, and two that hold the two insulin subunits together.

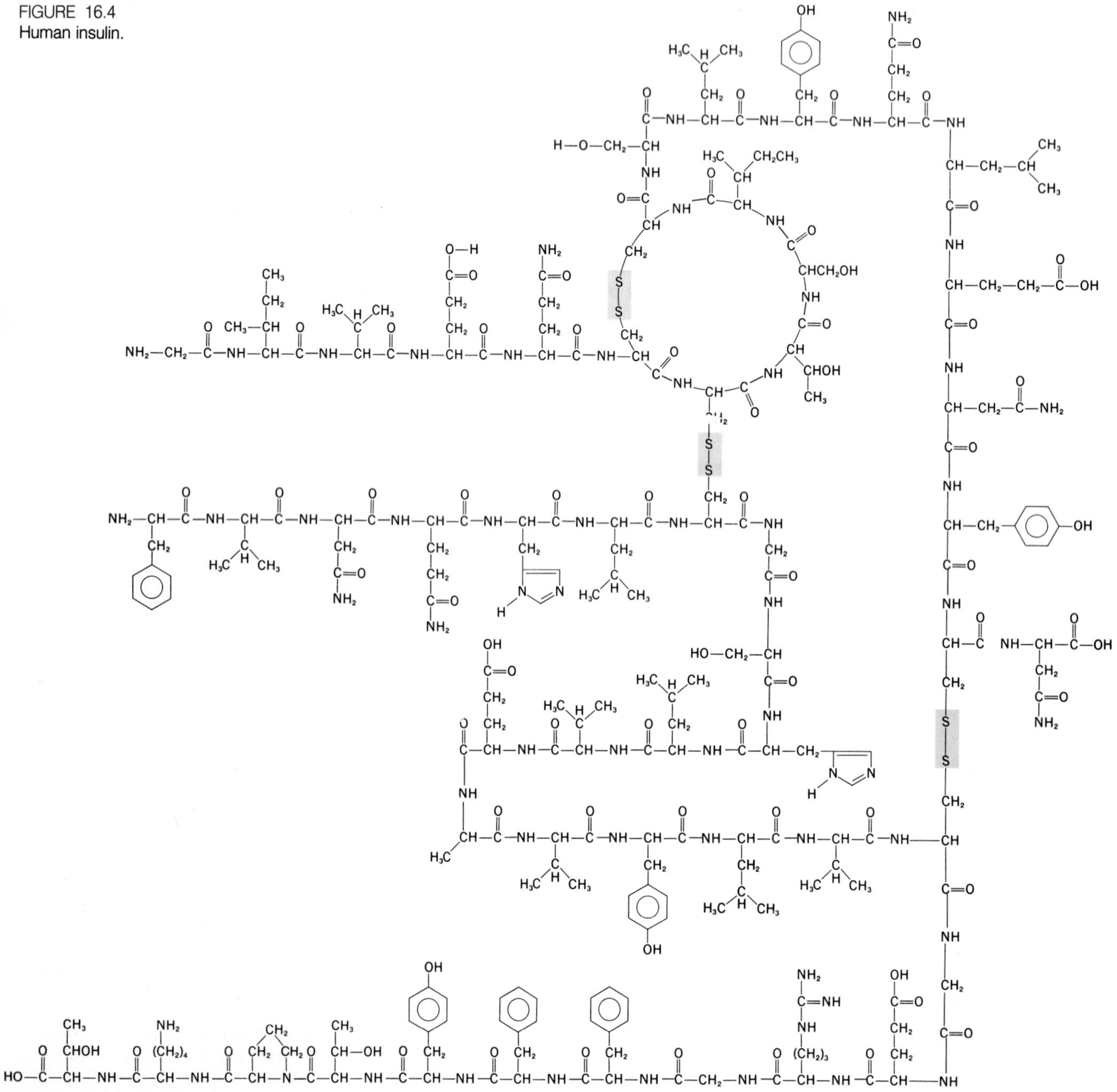

FIGURE 16.4
Human insulin.

16.3 SECONDARY STRUCTURES OF PROTEINS

The α helix, the β-pleated sheet, and the triple helix are three kinds of secondary protein structures.

Once a cell puts together a polypeptide, noncovalent forces of attraction between parts of the molecule make the molecule twist into a particular shape. The hydrogen bond is the chief noncovalent force, and we'll see in this section how it can stabilize shapes of polypeptides. Bear in mind that the proteins can become biologically useless if any aspect of their shapes is altered.

The α Helix Is a Major Secondary Structure of Polypeptides. The α helix is a coiled configuration of a polypeptide strand (see Figure 16.5.) Linus Pauling and R. B. Corey discovered it from a study of X-ray data. In the α **helix,** the polypeptide backbone coils as a right-handed screw with all its side chains sticking to the outside.

Linus Pauling won the 1954 Nobel prize in chemistry for this work.

Hydrogen Bonds Stabilize α Helices. Hydrogen bonds extend from the oxygen atoms of carbonyl groups to hydrogen atoms of NH groups farther along the backbone. Individually, a single hydrogen bond is a weak force of attraction, but when there are hundreds of them up and down a coiled polypeptide, they add up much as the individual "forces" that hold a zipper strongly shut. Generally, only segments of polypeptides, not entire lengths, are in an α-helix configuration.

$$-N-H \cdots O=C$$

Hydrogen bond

The designation α was picked only because this secondary structure was the first to be identified.

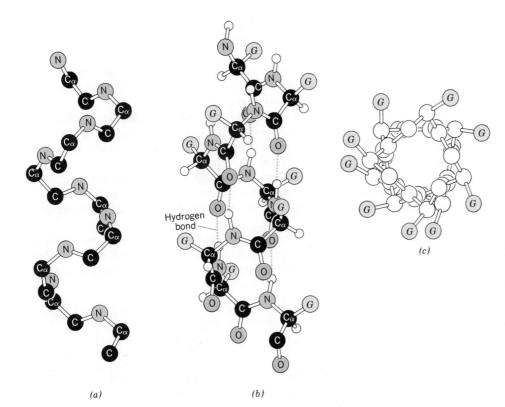

Hydrogen bond

(a) (b) (c)

FIGURE 16.5
The α helix. (a) The polypeptide backbone atoms on a helical thread. (b) Showing the H atoms and the side-chain groups, G, on the α-carbons. (c) Expanded view down the long axis of an α-helix showing how the side chains project to the outside. (Adapted by permission from L. Stryer, *Biochemistry*, 3rd ed., p. 26. W. H. Freeman and Company, New York, 1988.)

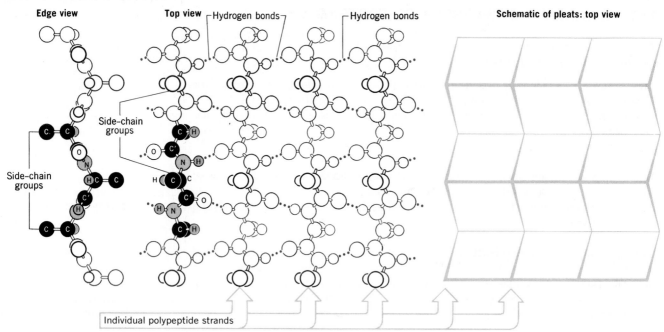

Edge view   **Top view**   Hydrogen bonds   Hydrogen bonds   **Schematic of pleats: top view**

Side–chain groups

Side–chain groups

Individual polypeptide strands

FIGURE 16.6
The β-pleated sheet.

The β-Pleated Sheet Is a Side-by-Side Array of Polypeptide Molecules. Pauling and Corey also discovered that molecules in some proteins line up side by side, to form a sheetlike array that is somewhat pleated (see Figure 16.6). This is another kind of secondary structure in which hydrogen bonds hold things together. The pleated sheet is the dominant feature in fibroin, the protein in silk. In other proteins, these sheets seldom contribute much to the overall structure.

The Polypeptides in Collagen Are in a Triple Helix. Another important secondary structure is found in a small family of proteins called the collagens. These are the proteins that give strength to bone, teeth, cartilage, tendons, and skin.

The polypeptide units in collagen are called tropocollagen, and each tropocollagen molecule consists of three polypeptide chains. Each chain has about 1000 amino acid residues, which are twisted together to form a **triple helix.**

Some of the aminoacyl units are hydroxylated derivatives of lysine and proline made with the help of vitamin C *after* the initial polypeptide is made. Thus vitamin C is essential to the formation of strong bones. In some of the types of collagen, sugar molecules are incorporated as glycosides of side-chain OH groups.

The individual strands in tropocollagen helices are in a very open helix that is not stabilized by hydrogen bonds. These open helices, however, wrap around each other into a right-handed helical cable within which hydrogen bonds are at work. In addition, some disulfide cross-links are present.

A microfiber or *fibril* of collagen forms when individual tropocollagen cables overlap lengthwise, as seen in Figure 16.7. The mineral deposits in bones and teeth become tied into the protein at the gaps between the heads of tropocollagen molecules and the tails of others.

A collagen fibril only 1 mm in diameter can hold a mass as large as 10 kg (22 lb).

FIGURE 16.7
Collagen. *(a)* The tropocollagen cable is made of three polypeptide molecules, each in an α helix (not pictured), wrapped together. *(b)* Individual tropocollagen cables, shown here by wavy lines, line up side by side with overlapping to give collagen fibrils. The circles represent gaps into which minerals can deposit in bones and teeth. The colored lines with solid color dots represent cross-links.

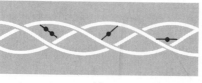

(a)

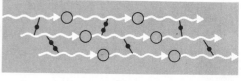

(b)

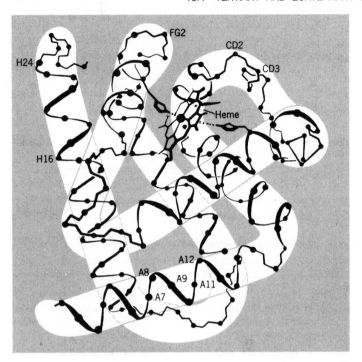

FIGURE 16.8
Myoglobin. The tubelike forms outline the segments that are in an α helix. The flat structure is the heme unit. The black dots identify (α-carbons. The side-chain groups at FG2, H16, H24, CD2, and CD3 are hydrophilic and are somewhat exposed to the aqueous medium. Hydrophobic groups that are somewhat tucked inside are at A7, A8, A9, A11, and A12, to note a few. (Reproduced by permission from R. E. Dickerson, "X-Ray Analysis of Proteins," in H. Neurath (ed.), *The Proteins,* Academic Press, New York; copyright © 1964, all rights reserved.)

16.4 TERTIARY AND QUATERNARY STRUCTURES OF PROTEINS

Tertiary structures are the results of the folding, bending, and twisting of secondary structures.

Once primary and secondary structures are in place, the final shaping of a protein occurs. All these activities happen spontaneously in cells, sometimes in a matter of seconds after the polypeptide molecule has been made and sometimes it takes several minutes. The "rules" followed by the polypeptide to give these shapes are still not known with any confidence. It's one of the remaining frontiers of protein research.

Those who manufacture proteins by genetic engineering, which we'll describe in a later chapter, have sometimes been surprised and disappointed that their products have been biologically useless lumps of unshaped molecules. Even though the correct primary structure was manufactured, the exact conditions in the solution needed for the higher levels to appear spontaneously could not be duplicated.

Tertiary Protein Structure Involves the Folding and Kinking of Secondary Structure.
When α helices take shape, their side chains tend to project outward where, in an aqueous medium, they would be in contact with water molecules. Even in water-soluble proteins, however, as many as 40% of the side chains are hydrophobic. Because such groups can't break up the hydrogen-bonding networks among the water molecules, the entire α helix undergoes further twisting and folding until the hydrophobic groups, as much as possible, are tucked to the inside, away from the water, and the hydrophilic groups stay exposed to the water. Thus the final shape of the polypeptide, its **tertiary structure,** emerges in response to simple molecular forces set up by the water-avoiding and the water-attracting properties of the side chains.

The tertiary structure of myoglobin, the oxygen-holding protein in muscle tissue, is given in Figure 16.8. It consists of just one polypeptide plus a nonprotein, heme (Figure 16.9). About 75% of the myoglobin molecule is in an α helix that is further folded as the figure shows. Virtually all its hydrophobic groups are folded inside and its hydrophilic groups are on the outside.

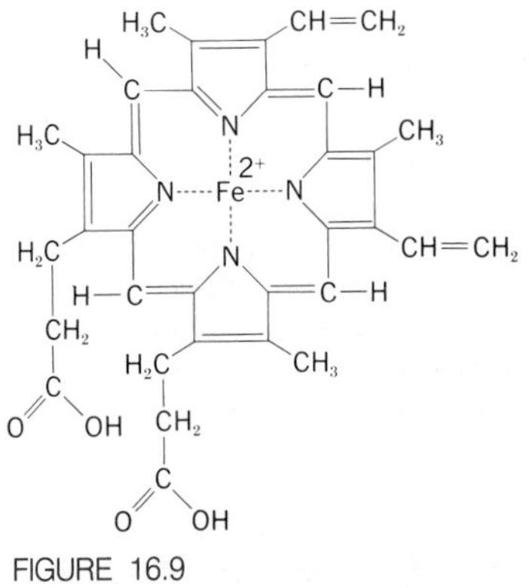

FIGURE 16.9
The heme molecule with its Fe^{2+} ion.

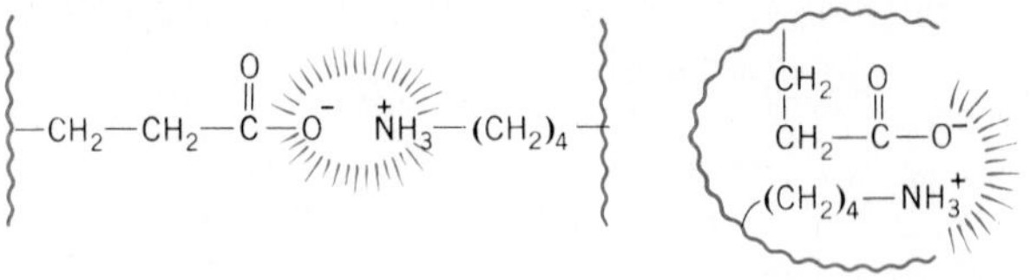

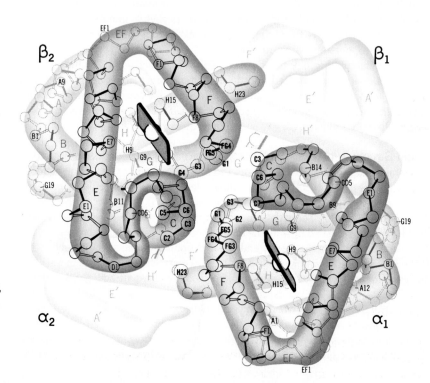

FIGURE 16.10

The salt bridge. This attraction between full and unlike charges can *(a)* hold one polypeptide to another or *(b)* stabilize a loop or coil within the same molecule.

"Prosthetic" is from the Greek *prosthesis,* an addition.

Polypeptides Incorporate Any Prosthetic Groups into Their Tertiary Structures. A nonprotein, organic compound that associates with a polypeptide, like heme in myoglobin, is called a **prosthetic group.** It is often the focus of the protein's biological purpose. Heme, for example, is the actual oxygen holder in myoglobin. It serves the same function in **hemoglobin (Hb),** the oxygen carrier in blood.

Ionic Bonds Also Stabilize Tertiary Structures. Another force that can stabilize a tertiary structure is the attraction between a full positive and a full negative charge, each occurring on a particular side chain. At the pH of body fluids, the side chains of both aspartic acid and glutamic acid carry CO_2^- groups. The side chains of lysine and arginine carry NH_3^+ groups. These oppositely charged groups naturally attract each other, like the attraction of oppositely charged ions in an ionic crystal. The attraction is called a **salt bridge** (see Figure 16.10).

When Polypeptides Group Together, the Quaternary Structures of Some Proteins Take Final Form. Proteins, like myoglobin, have finished shapes at the tertiary level. They are made up of single polypeptide molecules, sometimes with prosthetic groups.

Many proteins, however, are aggregations of two or more polypeptides, and these aggregations constitute **quaternary structures.** One molecule of the enzyme phosphorylase, for example, consists of two tightly aggregated molecules of the same polypeptide. If the two become separated, the enzyme can no longer function. Individual molecules of polypeptides that make up an intact protein molecule are called the protein's *subunits.*

FIGURE 16.11

Hemoglobin. Four polypeptide chains, each with one heme molecule represented here by the colored, flat plates that contain spheres (Fe^{2+} ions), are nestled together. (From R. E. Dickerson and I. Geis, *The Structure and Action of Proteins,* W. A. Benjamin, Inc., Menlo Park, CA, copyright © 1969. All rights reserved. Used by permission.)

The decisive importance of the primary structure to all other structural features of a polypeptide or its associated protein is illustrated by the grim story of sickle-cell anemia. This inherited disease is widespread among those whose roots are in central and western Africa.

In its mild form, where only one parent carries the genetic trait, the symptoms of sickle-cell anemia are seldom noticed except when the environment has a low partial pressure of oxygen, as at high altitudes. In the severe form, when both parents carry the trait, the infant usually dies by the age of 2 unless treatment is begun early. The problem is an impairment in blood circulation traceable to the altered shape of hemoglobin in sickle-cell anemia, particularly after the hemoglobin has delivered oxygen and is on its way back to the heart and lungs for more.

The fault at the molecular level lies in a β subunit of hemoglobin. One of the amino acid residues should be glutamic acid but is valine, instead. Thus instead of a side-chain CO_2^- group, which is electrically charged and hydrophilic, there is an isopropyl side chain, which is neutral and hydrophobic. Normal hemoglobin, symbolized as HHb, and sickle-cell hemoglobin, HbS, therefore have different patterns of electrical charges. Both have about the same solubility in well-oxygenated blood, but HbS without its oxygen load tends to precipitate inside red cells. This distorts the cells into a telltale sickle shape. The distorted cells are harder to pump, and they sometimes clump together and plug capillaries. Sometimes they split open. Any of these events places a greater strain on the heart. The error in one side chain seems miniscule, but it is far from small in human terms.

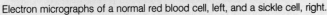
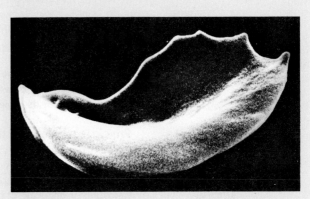

Electron micrographs of a normal red blood cell, left, and a sickle cell, right.

Hemoglobin has four subunits, two of one kind (designated α subunits) and two of another (called the β subunits). Each of the four subunits carries a heme molecule (see Figure 16.11). Salt bridges and hydrogen bonds hold the subunits together. These forces do not work unless each subunit has the appropriate primary, secondary, and tertiary structural features. If even one amino acid residue is wrong, the results can be very serious, as in the example of sickle-cell anemia, described in Special Topic 16.1.

16.5 COMMON PROPERTIES OF PROTEINS

Even small changes in the pH of a solution can affect a protein's solubility and its physiological properties.

Although proteins come in many diverse biological types, they generally have similar chemical properties because they have similar functional groups.

Protein Digestion Is Hydrolysis. The digestion of a protein is nothing more than the hydrolysis of its peptide bonds (amide linkages). The end product is a mixture of amino acids. We can illustrate this by an equation for the digestion of a pentapeptide.

$$^+NH_3CH_2\overset{\overset{O}{\|}}{C}-NH\overset{}{C}H\overset{\overset{O}{\|}}{C}-NH\overset{}{C}H\overset{\overset{O}{\|}}{C}-NH\overset{}{C}H\overset{\overset{O}{\|}}{C}-NHCHCO_2^- + 4H_2O \xrightarrow[\text{enzyme}]{H^+ \text{ or}}$$

$$\underset{CH_3}{|} \quad \underset{\underset{\underset{CO_2H}{|}}{\overset{CH_2}{|}}{CH_2}}{|} \quad \underset{\overset{CH_2}{|}}{C_6H_5} \quad \underset{\overset{(CH_2)_4}{|}}{NH_2}$$

$$^+NH_3CH_2CO_2^- + {}^+NH_3CHCO_2^- + {}^+NH_3CHCO_2^- + {}^+NH_3CHCO_2^- + {}^+NH_3CHCO_2^-$$

$$\underset{CH_3}{|} \quad \underset{\underset{\underset{CO_2H}{|}}{\overset{CH_2}{|}}{CH_2}}{|} \quad \underset{\overset{CH_2}{|}}{C_6H_5} \quad \underset{\overset{(CH_2)_4}{|}}{NH_2}$$

Glycine   Alanine   Glutamic   Phenylalanine   Lysine
                              acid

In three-letter symbolism, this equation is

$$Gly \cdot Ala \cdot Glu \cdot Phe \cdot Lys + 4H_2O \longrightarrow Gly + Ala + Glu + Phe + Lys$$

The simplicity of this equation shows why scientists like the three-letter symbols.

Since all animals use the same amino acids to make proteins, and since they can obtain the same set from plants (or each other!), a rather remarkable kinship exists throughout the living kingdom at its molecular level.

Protein Denaturation Is the Loss of Protein Shape. It isn't necessary to hydrolyze peptide bonds to denature a protein, that is, to destroy its ability to perform its biological function. All that has to happen is some disruption of secondary or higher structural features. **Denaturation** is the disorganization of the overall molecular shape of a protein. It can occur as an unfolding or uncoiling of helices, or as the separation of subunits (see Figure 16.12).

Usually, denaturation is accompanied by a major loss in solubility. When egg white is whipped or is heated, for example, as when you cook an egg, the albumin molecules unfold and become entangled among themselves. The system now is insoluble in water and it no longer lets light through. The heat of a steam sterilizer similarly denatures bacterial proteins and so kills the bacteria.

Table 16.2 has a list of several reagents or physical forces that cause denaturation, together with explanations of how they work. How effectively a given denaturing agent is depends on the kind of protein. The proteins of hair and skin and of fur or feathers quite strongly resist denaturation because they are rich in disulfide links.

Protein Solubility Depends Greatly on pH. Because some side chains as well as the end groups of polypeptides bear electrical charges, the entire molecule bears a net charge. Because these groups are either proton donors or proton acceptors, the net charge is easily changed by changing the pH. The CO_2^- groups, for example, become electrically neutral CO_2H groups when they pick up protons as a strong acid is added.

Suppose that the net charge on a polypeptide is 1—, caused by one extra CO_2^- group. When acid is added, we might imagine the following change, where the elongated shape is the polypeptide system.

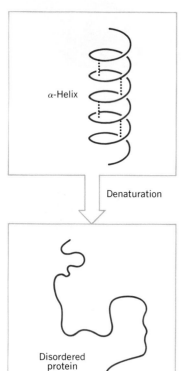

α-Helix

Denaturation

Disordered
protein
strand

FIGURE 16.12
A protein is denatured when it loses secondary, tertiary, or quaternary structure.

$$NH_3^+ \quad CO_2^- \quad CO_2^- \quad + H_3O^+ \longrightarrow \quad NH_3^+ \quad CO_2^- \quad CO_2H \quad + H_2O$$

Polypeptide, net   Added   Polypeptide, net
charge = 1—   acid   charge = 0

TABLE 16.2
Denaturing Agents for Proteins

Denaturing Agent	How the Agent May Operate
Heat	Disrupts hydrogen bonds by making molecules vibrate too violently. Produces coagulation, as in the frying of an egg
Microwave radiation	Causes violent vibrations of molecules that disrupt hydrogen bonds
Ultraviolet radiation	Probably operates much like the action of heat (e.g., sunburning)
Violent whipping or shaking	Causes molecules in globular shapes to extend to longer lengths and then entangle (e.g., beating egg white into meringue)
Detergents	Probably affect hydrogen bonds and salt bridges
Organic solvents (e.g., ethanol, 2-propanol, acetone)	May interfere with hydrogen bonds because these solvents can also form hydrogen bonds. Quickly denature proteins in bacteria, killing them (e.g., the disinfectant action of 70% ethanol)
Strong acids and bases	Disrupt hydrogen bonds and salt bridges. Prolonged action leads to actual hydrolysis of peptide bonds
Salts of heavy metals (e.g., salts of Hg^{2+}, Ag^+, Pb^{2+})	Cations combine with SH groups and form precipitates. (These salts are all poisons.)
Solutions of urea	Disrupt hydrogen bonds. (Urea, being amide-like, can form hydrogen bonds of its own.)

Now the polypeptide is isoelectric and neutral. On the other hand, a polypeptide might have a net charge of 1+, caused by one extra NH_3^+ group. The addition of OH^- can cause the following change, which makes the polypeptide isoelectric and neutral.

$$\text{Polypeptide, net charge} = 1+ \quad (NH_3^+ \quad CO_2^- \quad NH_3^+) \; + \; OH^- \; \longrightarrow \; \text{Polypeptide, net charge} = 0 \quad (NH_3^+ \quad CO_2^- \quad NH_2) \; + \; H_2O$$

Like each amino acid, each protein has a characteristic pH, its isoelectric point, at which its net charge is zero and at which it cannot migrate in an electric field. One major significance of this is that polypeptide molecules that are neutral can aggregate and clump together to become particles of enormous size that simply drop entirely out of solution (see Figure 16.13). A protein is least soluble in water when the pH equals the protein's isoelectric point. Therefore, whenever a protein must be in solution to work, as is true for many enzymes, the pH of the medium must be kept away from the protein's isoelectric point. Buffers in body fluids have the task of ensuring this.

An example of the effect of pH on solubility is given by casein, milk protein, whose pI value is 4.7. As milk turns sour, its pH drops from its normal value of 6.3–6.6 to 4.7, and more and more casein molecules become isoelectric, clump together, and separate as curds. As long as the pH of milk is something *other than* the pI for casein, the protein remains colloidally dispersed.

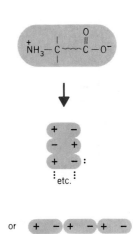

FIGURE 16.13
Several isoelectric protein molecules (top) can aggregate into very large clusters that no longer dissolve in water.

16.6   CLASSES OF PROTEINS

Three criteria for classifying proteins are their solubility in aqueous systems, their compositions, and their biological functions.

We began this chapter with hints about the wide diversities of the kinds and uses of proteins. Now that we know about their structures, we can better understand how so many types of proteins with so many functions are possible. The following three major classifications of proteins and their several examples give substance to the chapter's introduction.

Proteins Can Be Classified According to Solubility.   When proteins are classified according to their solubilities, there are two major families, the **fibrous proteins** and the **globular proteins.**

Fibrous proteins are insoluble in water and include the following special types.

When meat is cooked, some of its collagen changes to gelatin, which makes the meat easier to digest.

1. **Collagens.** Found in bone, teeth, tendons, skin, and soft connective tissue. When such tissue is boiled with water, the portion of its collagen that dissolves is called gelatin.

2. **Elastins.** Seen in many places where collagen is found, but particularly in ligaments, the walls of blood vessels, and the necks of grazing animals. Elastin is rich in hydrophobic side chains. Cross-links between elastin strands are important to its recovery after stretching.

Elastin is not changed to gelatin by hot water.

3. **Keratins.** Found in hair, wool, animal hooves, nails, and porcupine quills. The keratins are rich in disulfide links.

4. **Myosins.** The proteins in contractile muscle.

5. **Fibrin.** The protein of a blood clot. During clotting, it forms from its precursor, fibrinogen, by an exceedingly complex series of reactions.

Globular proteins, which are soluble in water or in water that contains certain salts, include the following:

1. **Albumins.** Found in egg white and in blood. In the blood the albumins serve many functions. Some are buffers. Some carry water-insoluble molecules of lipids or fatty acids. Some carry metal ions that could not be dissolved in water that is even slightly alkaline, such as Cu^{2+} ions.

2. **Globulins.** Including the gamma globulins that are part of the body's defense mechanism against diseases. The globulins need the presence of dissolved salts to be soluble in water.

Proteins Can Be Classified According to Composition.   The nature of the prosthetic group in these proteins provides another way of classifying proteins.

1. Glycoproteins are proteins with sugar units. Gamma globulin is an example.

2. Hemoproteins are proteins with heme units such as hemoglobin, myoglobin, and certain cytochromes (enzymes that help cells use oxygen).

3. Lipoproteins are proteins that carry lipid molecules, including cholesterol.

4. Metalloproteins are proteins that incorporate a metal ion, such as many enzymes do.

5. Nucleoproteins are proteins bound to nucleic acids, such as ribosomes and some viruses.

6. Phosphoproteins are proteins with a phosphate ester to a side chain —OH group, such as in serine. Milk casein is an example.

Proteins Can Be Classified According to Biological Function.   Perhaps no other system more clearly dramatizes the importance of proteins than classifying them by their biological function.

1. **Enzymes.** The body catalysts.

2. **Contractile muscle.** Proteins that are stationary filaments (myosin) and moving filaments (actin).

3. **Hormones.** Growth hormone, insulin, and many others.

4. **Neurotransmitters.** Enkephalins, endorphins, and many others.

5. **Storage proteins.** Proteins that store nutrients that the organism will need such as seed proteins in grains, casein in milk, ovalbumin in egg white, and ferritin, the iron-storing protein in human spleen.

6. **Transport proteins.** Proteins that carry things from one place to another. Hemoglobin and the serum albumins are examples already mentioned. Ceruloplasmin is a copper-carrying protein.

7. **Structural proteins.** Proteins that hold a body structure together, such as collagen, elastin, keratin, and proteins in cell membranes.

8. **Protective proteins.** Proteins that help the body to defend itself. Examples are the antibodies and fibrinogen.

9. **Toxins.** Poisonous proteins. Examples are snake venom, diphtheria toxin, and *clostridium botulinus* toxin (a toxic substance that causes some types of food poisoning).

SUMMARY

Amino acids About 20 α-amino acids supply the amino acid residues that make up a polypeptide. In the solid state or in water at a pH of roughly 6–7, most amino acids exist as dipolar ions. Their isoelectric points, the pH values of solutions in which they are isoelectric, are in this pH range of 6–7. Those with CO_2H groups on side chains have lower pI values. Those with basic side chains have higher pI values. Several amino acids have hydrophobic side chains, but the side chains in others are strongly hydrophilic. The SH group of cysteine opens the possibility of disulfide cross-links between polypeptide units.

Polypeptides Amino acid residues are held together by peptide (amide) bonds, so the repeating unit in polypeptides is —NH—CH—CO—. Each aminoacyl unit has its own side chain. This repeating system with a unique sequence of side chains constitutes the primary structure of a polypeptide.

Once this is fashioned, the polypeptide coils and folds into higher features, secondary and tertiary, that are stabilized largely by hydrogen bonds and the water-avoiding or water-attracting properties of the side chains. The most prominent secondary structures are the α helix, the β-pleated sheet, and the collagen triple helix.

Proteins Many proteins consist just of one kind of polypeptide. Many others have nonprotein, organic (prosthetic) groups, or metal ions. And still other proteins, those with quaternary structure, involve two or more polypeptides whose molecules aggregate in definite ways, stabilized sometimes by salt bridges. Thus the terms *proteins* and *polypeptide* are not synonyms, although for some specific proteins they turn out to be.

Because of their higher levels of structure, proteins can be denatured by agents that do nothing to peptide bonds. The acidic and basic side chains of polypeptides affect protein solubility, and when a protein is in a medium whose pH equals the protein's isoelectric point, the substance is least soluble. The amide bonds (peptide bonds) of proteins are hydrolyzed during digestion.

REVIEW EXERCISES

The answers to the following review exercises are in the *Study Guide* that accompanies this book.

Amino Acids

16.1 What structure will nearly all the molecules of glycine have at a pH of about 1?

16.2 In what structure will most of the molecules of alanine be at a pH of about 12?

16.3 Pure alanine does not melt, but at 290 °C it begins to char. However, the ethyl ester of alanine, which has a free NH_2 group, has a low melting point, 87 °C. Write the structure of this ethyl ester, and explain this large difference in melting point.

16.4 The ethyl ester of alanine (Review Exercise 16.3) is a much stronger base—more like ammonia—than alanine. Explain this.

16.5 Which of the following amino acids has the more hydrophilic side chain? Explain.

$$^+NH_3CHCO_2^-$$ with CH attached to CH_3 and CH_3 (A)

$$^+NH_3CHCO_2^-$$ with $CH_2CH_2CH_2NHCNH_2$ and NH_2^+ (B)

16.6 Which of the following amino acids has the more hydrophobic side chain? Explain.

$$^+NH_3CHCO_2^-$$

$$CH_2$$

A

$$^+NH_3CHCO_2^-$$

$$CH_2$$

$$OH$$

B

16.7 One of the possible forms for lysine is

$$^+NH_3CHCO_2H$$

$$CH_2CH_2CH_2CH_2NH_3^+$$

(a) Is lysine most likely to be in this form at pH 1 or at pH 11? Explain.

(b) Would lysine in this form migrate to the anode, to the cathode, or not migrate at all in an electric field?

16.8 Aspartic acid can exist in the following form.

$$NH_2CHCO_2^-$$

$$CH_2CO_2^-$$

(a) Would this form predominate at pH 2 or pH 10? Explain.

(b) To which electrode, the anode or the cathode — or to neither — would aspartic acid in this form migrate in an electric field?

16.9 What net charge do glutamic acid molecules have at pH 1?

16.10 In what optical family are all amino acids (except glycine)?

Polypeptides

16.11 Write the conventional, condensed structure of the dipeptides that can be made from lysine and glycine.

16.12 What are the condensed structures of the dipeptides that can be made from cysteine and glutamic acid? (Do not use the three-letter symbols.)

16.13 Using three-letter symbols, write the structures of all of the tripeptides that can be made from lysine, glutamic acid, and alanine.

16.14 Write the structures in three-letter symbols of all of the tripeptides that can be made from cysteine, glycine, and alanine.

16.15 What is the conventional structure of Val·Ile·Phe?

16.16 The artificial sweetener in aspartame (the chief ingredient in NutraSweet) is the methyl ester of Asp·Phe. When the peptide bond in aspartame is hydrolyzed (and only this bond), one of the products cannot exist as an anion in water, even when the pH is 12. Write a conventional, condensed structure for aspartame.

16.17 Write the conventional structure for Ala·Val·Phe·Gly·Leu.

16.18 Write the conventional structure for Asp·Thr·Lys·Glu·Tyr.

16.19 Compare the side chains in the pentapeptide of Review Exercise 16.17 (call it A) with those in:

Lys·Glu·Asp·Thr·Ser   (which we can call B)

(a) Which of the two, A or B, is the more hydrocarbon-like?

(b) Which is probably more soluble in water? Explain.

16.20 Compare the side chains in the pentapeptide of Review Exercise 16.18, which we'll label C, with those in Phe·Leu·Gly·Ala·Val, which we can label D. Which of the two would tend to be less soluble in water? Explain.

16.21 If the tripeptide Gly·Cys·Ala were subjected to mild oxidizing conditions, what would form? Write the structure of the product using three-letter symbols.

16.22 Write the structure(s) of the product(s) of the mild reduction of the following polypeptide.

Higher Levels of Protein Structure

16.23 What kind of force makes it possible for a polypeptide to be stabilized in the shape of an α helix?

16.24 The side-by-side alignment of polypeptides in a β-pleated sheet is maintained through the agency of what kind of force?

16.25 Give a brief description of the way in which polypeptide strands are organized in collagen.

16.26 What function does vitamin C perform in the formation of strong bones?

16.27 What factors affect the bending and folding of α helices in the presence of an aqueous medium? Are enzymes required?

16.28 What is meant by a salt bridge? Draw structures to illustrate your answer.

16.29 In what way does hemoglobin represent a protein with quaternary structure (in general terms only)?

16.30 How do myoglobin and hemoglobin compare (in general terms only)?

(a) Structurally, the quaternary level.

(b) Where they are found in the body?

(c) In terms of their prosthetic group(s).

(d) In terms of their functions in the body.

Properties of Proteins

16.31 What products form when the following polypeptide is completely digested?

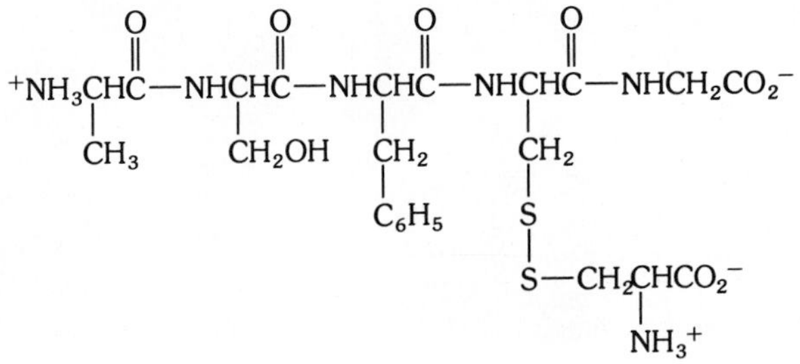

16.32 The *partial* digestion of a polypeptide gave several products that were not amino acids. Some were dipeptides, for example. Which of the following compounds are possible products, and which could not possibly have formed? Explain.

(a) 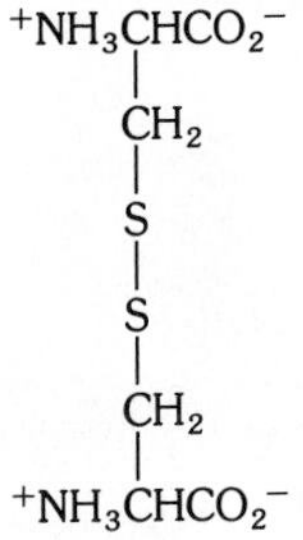

(b) 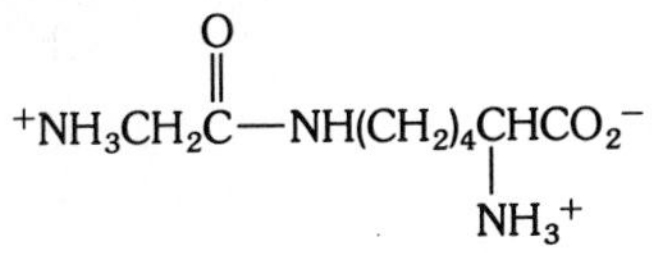

(c) 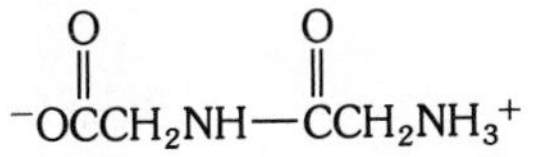

(d) 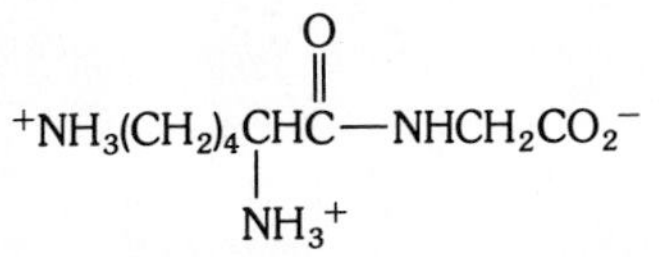

16.33 Explain why a protein is least soluble in an aqueous medium that has a pH equal to the protein's pI value.

16.34 What is the difference between the digestion and the denaturation of a protein?

Types of Proteins

16.35 What experimental criterion distinguishes between fibrous and globular proteins?

16.36 What is the relationship between collagen and gelatin?

16.37 How are collagen and elastin alike? How are they different?

16.38 What experimental criterion distinguishes between the albumins and the globulins?

16.39 What is fibrin and how is it related to fibrinogen?

16.40 What general name can be given to a protein that carries a lipid molecule?

Sickle-Cell Anemia (Special Topic 16.1)

16.41 What is the primary *structural* fault in the hemoglobin of sickle-cell anemia?

16.42 What happens in blood cells in sickle-cell anemia that causes their shapes to become distorted?

16.43 What problems are caused by the distorted shapes of the red cells?

Chapter 17
Enzymes, Hormones and Neurotransmitters

Hormones guided the remarkable transformation of a pupa into this monarch butterfly. In this chapter we'll learn the general principles of how hormones work.

17.1 ENZYMES

The catalytic abilities of enzymes often depend on cofactors that are made from B vitamins.

Nearly all of the body's thousands of catalysts, its enzymes, are proteins. A few, like ribonuclease P, have enzymic activity caused solely by a nucleic acid component, but we will not study them. We begin our study with some general properties of enzymes.

Enzymes Are Unusually Specific in the Substrates They Accept. Enzyme activity occurs when molecules of the enzyme and those of a reactant, the **substrate,** physically fit together. Because the molecules of each substrate have unique shapes, *each reaction needs its own specific enzyme.*

Few enzymes are so specific that they catalyze just one particular reaction of one compound. Most enzymes possess *relative specificity.* Enzymes that catalyze the hydrolysis of esters, for example, usually handle a variety of esters, not just one, but they usually do not catalyze the hydrolysis of other functional groups.

Even small changes in molecular shape prevent enzyme action.

Enzymes Display Remarkable Rate Enhancements. An enzyme increases the *rate* of a reaction by providing a reaction pathway, a mechanism, with a lower energy of activation than otherwise. Even small reductions of energy barriers give spectacular rate increases. The enzyme carbonic anhydrase (CA), for example, catalyzes the interconversion of carbon dioxide and water with bicarbonate ion and hydrogen ion.

$$CO_2 + H_2O \xrightleftharpoons{\text{carbonic anhydrase}} HCO_3^- + H^+$$

Carbonic anhydrase is an equilibration catalyst, and each molecule is able to handle 600,000 molecules of CO_2 in the forward reaction each second! This is 10 million times faster than the uncatalyzed reaction and the fastest known rate for any handled by an enzyme.

Enzymes Catalyze Both the Forward and Reverse Reactions of Equilibria. It is important to realize that an equilibration catalyst, like carbonic anhydrase, speeds up the *restoration of the equilibrium.* When some stress upsets it, the catalyst increases the rate of whichever reaction, forward or reverse, restores equilibrium. Whether the equilibrium shifts to the right or the left doesn't depend on the catalyst at all. It depends on such factors as the relative concentrations of reactants and products; on whether other reactions feed substances into the equilibrium or continuously remove its products; and on the temperature.

All an equilibration catalyst does is speed up whatever shift in an equilibrium is mandated by other conditions. In active tissue, for example, where metabolism generates CO_2, the above equilibrium is forced to the right. The stress here is an increase in the CO_2 concentration, so the equilibrium shifts to reduce this stress, to lower the concentration of CO_2. This, of course, puts CO_2 into the chemical form in which it must be, largely as HCO_3^-, to be carried in the blood to the lungs. When such blood arrives at the lungs, the equilibrium must shift back again because now the stress changes. By exhaling air, CO_2 is being removed from the blood. Now the equilibrium, aided again by carbonic anhydrase, shifts back to reduce the stress (loss of CO_2) by making more CO_2.

The equilibrium *must* shift to the right to make H_2CO_3 when the supply of CO_2 is high—a consequence of Le Chatelier's principle.

Most Enzymes Consist of Polypeptides Plus Cofactors. The molecules of most enzymes include a nonpolypeptide component called a **cofactor,** and where the cofactor is missing there is no enzymic activity. The wholly polypeptide part of such an enzyme is called the **apoenzyme.**

Some cofactors are simply metal ions like the Zn^{2+} ion in carbonic anhydrase or the Fe^{2+} ion in the cytochromes, enzymes used in biological oxidations. Table 17.1 gives other examples.

TABLE 17.1
Trace Elements of Nutritional Importance to Humans

Element	Uses
Chromium	Needed for glucose metabolism and the proper action of insulin
Manganese	Required for normal nerve function, the development of sound bones, and for reproduction
Iron	Essential to hemoglobin and myoglobin and several enzymes
Cobalt	Part of the vitamin B_{12} molecule
Copper	A cofactor in respiratory enzymes and necessary for strong collagen, elastin, and the myelin sheaths of nerve cells
Zinc	A cofactor in carbonic anhydrase and other enzymes
Selenium	Exact function is unclear
Molybdenum	Used in the metabolism of nucleic acids
Iodine	Used to make thyroxine, a hormone, not an enzyme
Fluorine	Needed (as F^-) to make strong teeth, but not needed to make any enzyme
Others	Nickel, silicon, tin, and vanadium are possibly trace elements in humans, but deficiency diseases are not known

In other enzymes, the cofactor is an organic species and called a **coenzyme.** Some enzymes have both a coenzyme and a metal ion cofactor.

B Vitamins Are Used to Make Coenzymes. The **vitamins** are nonprotein organic compounds that must be in the diet because the body cannot make them, at least not at rates rapid enough to do much good. They are needed in trace amounts and are not carbohydrates, lipids, proteins, or amino acids. To qualify as a vitamin, the compound (or set of closely related compounds each having the same nutritional value) must also be a nutrient whose absence in the diet causes a disease, a vitamin-deficiency disease. These include xerophthalmia (an eye disease), rickets, beriberi, pellagra, and scurvy. A table of the vitamins important to humans is in Appendix IV together with good dietary sources, recommended daily intakes, and the dangers of deficiencies.

Thiamine is vitamin B_1, and its absence in the diet causes beri beri.

Thiamine diphosphate, structure **1**, is a coenzyme made as a diphosphate ester of one of the B-vitamins, thiamine. (The shaded part of the structure shows the thiamine unit.)

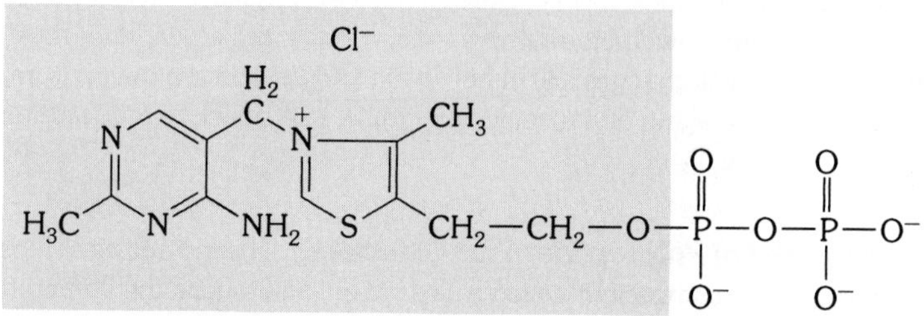

Thiamine diphosphate

1

Lack of nicotinamide in the diet causes pellagra, a problem wherever corn (maize) is the major food.

Nicotinamide, another B vitamin, is part of the structure of nicotinamide adenine dinucleotide, **2a**, usually represented simply as NAD^+ (or sometimes, just NAD). Notice that the bottom half of the NAD^+ molecule is from adenosine monophosphate, AMP.

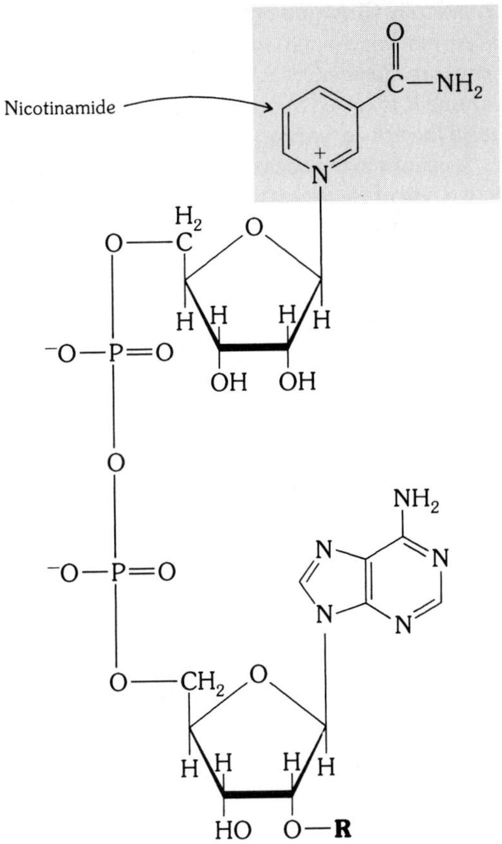

Nicotinamide

a NAD$^+$ **R**=H
b NADP$^+$ **R**=OPO$_3$$^{2-}$

2

Riboflavin

FMN

3

Still another important coenzyme, nicotinamide adenine dinucleotide phosphate, **2b**, is a phosphate ester of NAD$^+$. It's usually referred to as NADP$^+$ (or, sometimes, just NADP). Both NAD$^+$ and NADP$^+$ are coenzymes in enzymes used for biological oxidation–reduction reactions.

The P in NADP$^+$ refers to the extra phosphate ester unit.

Equations for enzyme-catalyzed reactions are usually written with the symbol of the coenzyme as an actual reactant or as a product. The following equation, for example, is the initial step in the body's metabolism of ethyl alcohol, and the enzyme has the cofactor NAD$^+$, **2a**.

$$CH_3CH_2OH + NAD^+ \longrightarrow CH_3CH{=}O + NAD{:}H + H^+$$

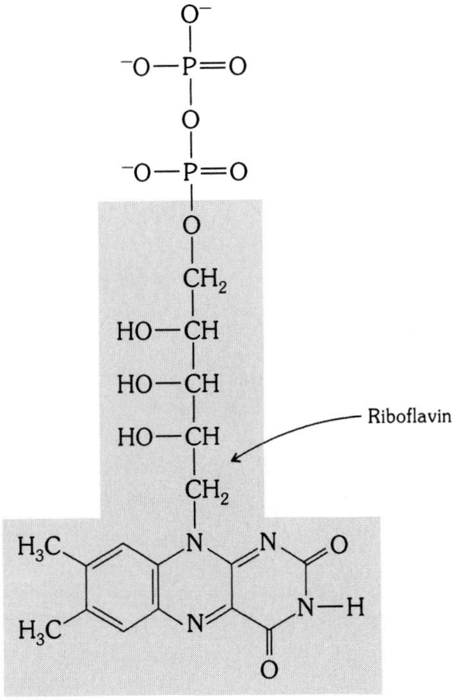

Ethyl alcohol   Acetaldehyde   Reduced form   Hydrogen
                                 of NAD$^+$ ion (buffered)

The NAD$^+$ accepts a hydride ion, H$:^-$, from ethyl alcohol and so is reduced to NAD$:$H (or NADH), the reduced form of NAD$^+$. A proton is also lost from the alcohol and is neutralized by a buffer in the surrounding solution.

Biochemists have a particularly useful way to represent a reaction like this.

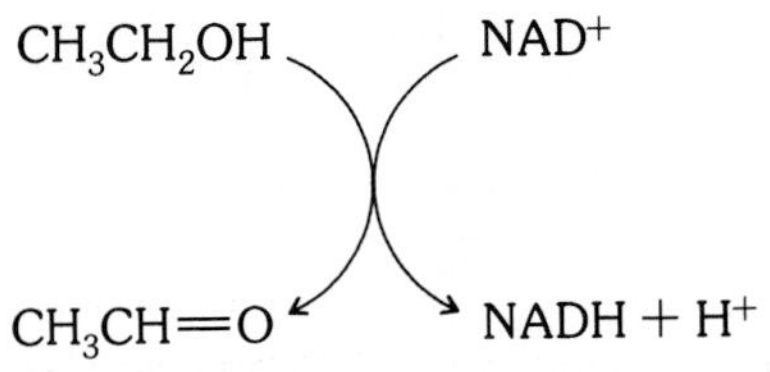

The advantages of such a display will become apparent as we continue.

Enzymes Are Recovered from Their Chemically Changed Forms.   A catalyst, of course, isn't truly a catalyst unless it undergoes no *permanent* change, something our example seems to cause. In the body, however, most enzymes are members of a series of enzymes in a chain of reactions. Thus a reaction altering one enzyme is followed at once by a reaction restoring it. For example, the NAD^+ reduced to NADH in the above reaction is recovered in the next step in which still another B vitamin, riboflavin, functions in the coenzyme. The coenzyme is FMN (flavin mononucleotide), **3**, and it can accept a pair of electrons from NAD : H by the following reaction. (The H^+ comes out of the buffer system.)

Riboflavin is vitamin B_2, and its absence in the diet causes a breakdown of tissues around the mouth, nose, and tongue.

$$NADH + FMN + H^+ \longrightarrow NAD^+ + FMNH_2$$

We can display this and the previous reaction together as follows:

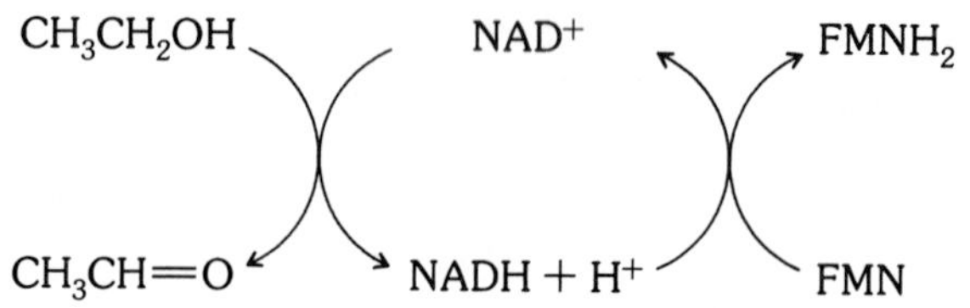

Thus the NAD^+-containing enzyme is restored, but now the FMN-containing enzyme has its coenzyme in its reduced form, $FMNH_2$. We won't continue this discussion now except to say that $FMNH_2$ passes on its load of hydrogen and electrons in yet another step and so is reoxidized to FMN.

The main points of our discussion are that B vitamins are key parts of coenzymes, and that the catalytic activities of the associated enzymes involve the molecules of these vitamins directly.

Flavin adenine dinucleotide, or FAD, is a near relative of FMN that also contains riboflavin. Its reduced form is $FADH_2$, and it is also used in biological oxidations.

Enzymes Are Named after Their Substrates or Reaction Types.   Nearly all enzymes have names that end in *-ase*. The prefix is either from the name of the substrate or from the kind of reaction. An *esterase,* for example, aids the hydrolysis of esters. A *lipase* works on the hydrolysis of lipids. A *peptidase* catalyzes the hydrolysis of peptide bonds. Similarly, an *oxidase* is an enzyme that catalyzes an oxidation, and a *reductase* handles a reduction. An *oxidoreductase* handles a redox equilibrium. A *transferase* catalyzes the transfer of a group from one molecule to another, and a *kinase* is a special transferase that handles phosphate groups.

Whenever we see *-ase* as a suffix in the name of any substance or type of reaction, the word is the name of an enzyme.

Not all enzyme names are chemically informative, like the above. The peptidases *trypsin* and *pepsin,* for example, have old names that neither end in *-ase* nor disclose their substrates. To minimize the confusion, if all enzymes had such names, an International Enzyme Commission has developed a precise system of classifying and naming enzymes, but our needs will not include their work.

Enzymes Can Occur as a Set of Isoenzymes.   Identical reactions are often catalyzed by any one of a small set of **isoenzymes,** enzymes with the same cofactors but with slightly different apoenzymes. The enzyme creatine kinase or CK is an example of one such set. Any member of the set catalyzes the transfer of a phosphate group in the following equilibrium:

CK is the newer symbol for an older one, CPK (for the older name, creatine phosphokinase).

ATP is adenosine triphosphate, which we described together with ADP in Section 13.4.

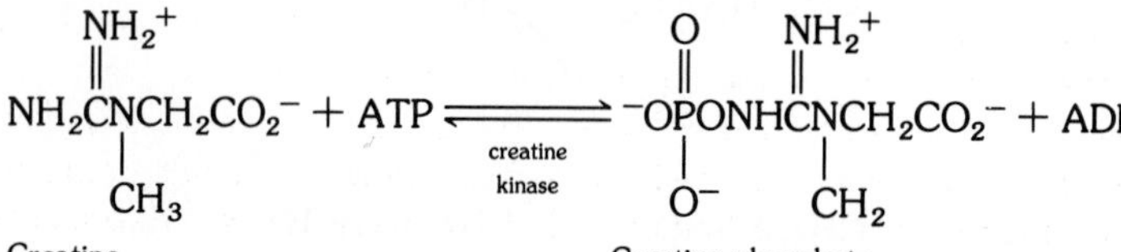

Creatine   Creatine phosphate

Various pairings of two different polypeptide chains, called *M* and *B*, are involved in the apoenzymes of the three CK coenzymes: CK(*MM*), CK(*BB*), and CK(*MB*). CK(*MM*) predominates in skeletal muscle tissue (hence the use of *M* as one subunit's symbol). CK(*BB*) predominates in brain tissue. The hybrid creatine kinase, CK(*MB*), is found almost exclusively in heart muscle where it accounts for 15–20% of total creatine kinase activity, the rest being caused by CK(*MM*). As we will see later in this chapter, the isoenzymes of creatine kinase are important in the clinical diagnosis of a heart attack.

Some references call isoenzymes by the name **isozymes.**

17.2 THE ENZYME–SUBSTRATE COMPLEX

Enzymes have one or more active sites as well as substrate-binding sites that function when enzyme–substrate complexes forms.

The Catalytic Work of an Enzyme Is Handled by Its Active Site. The active site of an enzyme consists of *catalytic groups,* organic groups that do the catalytic work. As we indicated in Section 17.1, these groups are sometimes provided by the coenzyme. Figure 17.1 gives the general idea of how a coenzyme might join to an apoenzyme to make the complete enzyme.

The substrate-binding sites are found in other parts of an enzyme molecule. These are functional groups whose shapes and polarities let them attract substrate molecules, guide them to the active site, and hold them long enough for the reaction to occur.

A Substrate Fits to Its Enzyme As a Key Fits to Its Lock. The combination of molecules of the substrate with its enzyme is called the enzyme–substrate complex. It forms and breaks down as part of the following series of chemical equilibria.

$$E + S \rightleftharpoons E-S \rightleftharpoons E-S^* \rightleftharpoons E-P \rightleftharpoons E + P$$

Enzyme Substrate Enzyme–substrate complex Substrate-activated E—S complex Enzyme–product complex Enzyme (recovered) Product

We can now see why enzymes are so specific. Only when the molecular shapes of enzyme and substrate match properly can the two molecules fit together. The specificity of the substrate–enzyme fit is so much like that of a key for just one tumbler lock that this model of how enzymes work is called the lock-and-key theory.

Figure 17.2, part (*a*), shows how partial charges and complementary shapes might bring the substrate molecule directly over the active site. The figure supposes that the reaction is one in which a substrate bond breaks. No doubt some twisting and bond stretching in the substrate occurs as forces of attraction pull it in. The effect of this is to activate the substrate for the reaction as it is pulled close to the catalytic group.

A bond-breaking reaction usually makes new functional groups, so the product molecules, still held to the enzyme, could now have polarities quite unlike those of the reactant, as seen in Figure 17.2b. Where reactant and enzyme attracted each other, product and enzyme could well now repel each other. The product fragments, therefore, would quickly leave the active site, as seen in part (*c*) of the figure. This clears the enzyme for reuse.

In some reactions, according to the induced-fit theory, the substrate molecule causes a change in the configuration of the enzyme (see Figure 17.3). Such behavior might clear solvent cages away from key functional groups to expose them for easier participation at lower energy costs. By whatever mechanism the enzyme–substrate complex forms, an enzyme's rate enhancement comes from its ability to hold and activate its substrate by means of forces of attraction situated in geometrically well-matched configurations.

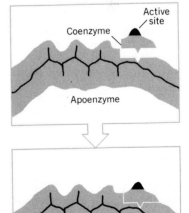

FIGURE 17.1

A coenzyme, which contributes the active site, joins to an apoenzyme in the formation of many kinds of complete enzymes.

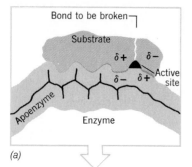

(a)

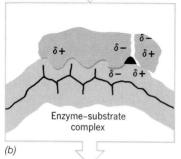

(b)

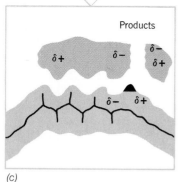

(c)

FIGURE 17.2
The lock-and-key model for enzyme action. (a) The enzyme and its substrate fit together to form an enzyme–substrate complex. (b) A reaction, such as the breaking of a chemical bond, occurs. (c) The product molecules separate from the enzyme.

FIGURE 17.3
Induced fir theory. (a) A molecule of an enzyme, hexokinase, has a gap into which a molecule of its substrate, glucose, can fit. (b) The entry of the glucose molecule induces a change in the shape of the enzyme molecule, which now surrounds the substrate entirely. (Courtesy of T. A. Steitz, Department of Molecular Biophysics and Biochemistry, Yale University, New Haven, CT.)

17.3 THE REGULATION OF ENZYMES

Enzymes are switched on and off by inhibitors, antimetabolites, genes, poisons, hormones, and neurotransmitters.

A cell cannot be doing everything at once. Some of its possible reactions have to be shut down while others occur. One way to keep a reaction switched off is to prevent its enzyme from forming, as we will learn when we study how genes can be controlled (Chapter 21). Hormones and neurotransmitters, to be studied in Section 17.5, also regulate enzymes. Still other regulatory mechanisms operate, which we'll study here.

Enzymes Can Be Activated by the Chemical Removal of an Inhibitor Molecule. Some enzymes are kept in an inactive form by the work of an **inhibitor,** a substance that combines with the enzyme to cover the active site or to change the enzyme's shape.

The inhibitor can be a *product* of the reaction that the enzyme catalyzes, or a product of a reaction farther down a series. The action of such an inhibitor is called **feedback inhibition** because, as its level increases, its molecules more and more react with ("feed back" to) and inhibit the molecules of an enzyme needed to make them. The amino acid isoleucine, for example, is made from another amino acid, threonine, by a series of steps, each with its own enzyme. Molecules of isoleucine are able to combine reversibly with the enzyme for the first step, E_1, and inhibit it.

$$\underset{\text{Threonine}}{\overset{\overset{\displaystyle NH_3^+}{|}}{CH_3CHCHCO_2^-}} \xrightarrow{E_1} \xrightarrow{E_2} \xrightarrow{E_3} \xrightarrow{E_4} \xrightarrow{E_5} \underset{\text{Isoleucine}}{\overset{\overset{\displaystyle NH_3^+}{|}}{CH_3CH_2CHCHCO_2^-}}$$

When sufficient quantities of isoleucine have been made, their further synthesis stops, because all of E_1 is inactivated.

The beautiful feature of feedback inhibition is that the system making a product shuts down automatically when enough is made. Then, as the cell consumes this product, even those of its molecules serving as inhibitors are consumed. Now the enzyme needed to make more product is released from inhibition. The inhibition is thus reversible. This phenomenon, very common in nature, is sometimes called **homeostasis,** the response of an organism to a stimulus that starts a series of events that restore the system to the original state.

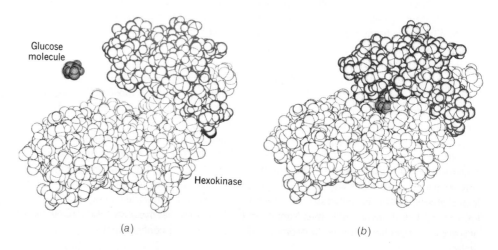

Glucose molecule

Hexokinase

(a)

(b)

An inhibitor doesn't have to be a product of the enzyme's own work. It can be something else the cell makes, or it could be a medication. What its molecules must do is bind, without reacting, somewhere on the enzyme, and they must be removable as needed.

Substrate Molecules Can Activate Enzymes.   Some enzymes have more than one active site. When no substrate is around, the enzyme adopts a configuration in which the shapes of both active sites are poorly matched to substrate molecules. When substrate concentration builds up, however, their molecules, beating against the enzyme, manage to force entry onto one active site. As this successfully happens, the whole enzyme alters in shape, and the other active site becomes activated. Now *both* active sites are well matched to the substrate, and they continue to work as long as substrate is present. Here, then, is another example of how the activity of an enzyme is related to the concentration of a substrate.

Nerve Signals Can Cause Nonsubstrate Molecules to Activate Enzymes.   Nonsubstrate molecules that activate enzymes are called **effectors,** and calmodulin and troponin are two important examples. Both are proteins with calmodulin present in most cells and troponin largely in muscle cells, like heart muscle. They are able to activate enzymes needed for chemical work initiated by nerve signals, like muscle contraction.

For calmodulin or troponin to be effectors, they must first combine with calcium ion. The problem is that this ion is normally almost entirely absent inside a cell. It cannot occur in solution inside the cell at a concentration higher than about 10^{-7} M, because cellular fluid has a level of phosphate ion of about 10^{-3} M. This is sufficiently high to make all but the 10^{-7} M concentration of Ca^{2+} insoluble as $Ca_3(PO_4)_2$. So the calcium ion has to stay largely outside the cell until the instant it is needed.

A nerve signal establishes this need, and it opens a *calcium channel* through the cell membrane. The Ca^{2+} ions can now move inside, and they bind to calmodulin or troponin before they can precipitate as calcium phosphate. This binding activates these effectors and they are now able to activate other enzymes. Once these other enzymes do their work, and functionally this is completed when the nerve signal is over, the calcium channels close. Then an active transport mechanism pipes the calcium ions to the outside, and the effector is inactivated. This is one mechanism whereby a nerve signal launches specific chemical activity in a cell, and how the activity is shut off.

To summarize this, a nerve signal sends calcium ion inside a cell where it binds to and thereby activates calmodulin or troponin. These effectors then activate other enzymes. When the nerve signal stops, calcium ion is pumped out of the cell, the effectors are deactivated, and the enzymes they activate become inactive.

Some Enzymes Are Initially in Inactive Forms Called Zymogens.   Several enzymes are first made as **zymogens,** polypeptides with molecular portions folded over active sites. When the active enzyme is needed, an activator (another enzyme) clips off these portions to expose the active sites. Circulating in the blood, for example, is a zymogen called plasminogen. In Section 17.4 we'll see how its conversion to the enzyme, plasmin, is important following the formation of a blood clot. Several digestive enzymes also are made as zymogens, as we'll learn more about in the next chapter.

Antimetabolites Inhibit Bacterial Enzymes.   Antibiotics are members of a broad family of compounds called **antimetabolites,** substances that inhibit or prevent the normal metabolism of a disease-causing bacterial system. Some antimetabolites work by inhibiting an enzyme that the bacterium needs for its own growth. Both the sulfa drugs and penicillin work this way (see Special Topics 17.1 and 17.2).

An antimetabolite is called an *antibiotic* when it is the product of the growth of a fungus or a natural strain of bacteria.

The sulfa drugs work by interfering with an enzyme in several disease-causing bacteria. These, unlike humans, require *para*-aminobenzoic acid to make folic acid. (This is a vitamin for humans, but we must obtain folic acid intact in the diet.) The structures of the sulfa drugs and *para*-amino benzoic acid are similar, as seen below, and affected bacteria make the mistake of using sulfa drugs to make folic acid when the drug molecules are abundantly present. Once it is made (in a slightly altered form, now), it won't work in the bacterial enzyme system that depends on true folic acid. Consequently, a metabolic process in the bacteria shuts down and the organism dies.

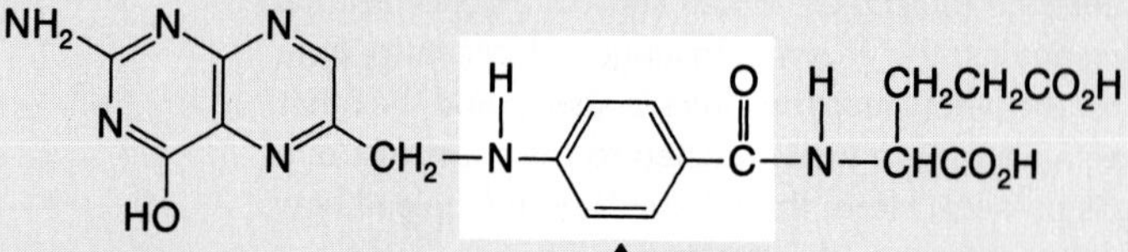

Folic acid

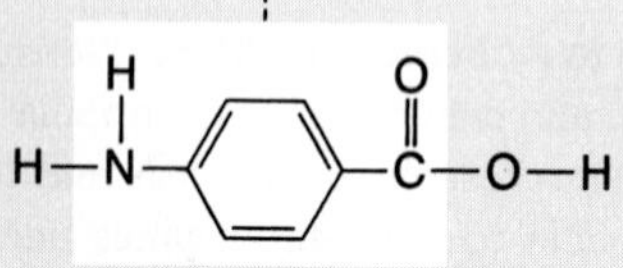

para-Aminobenzoic acid

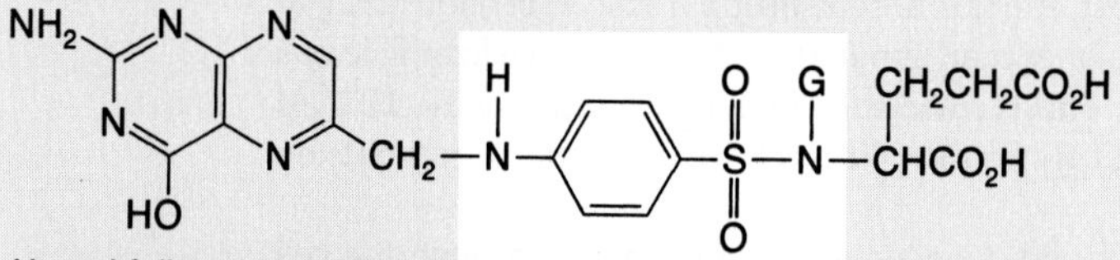

Altered folic acid in which a portion of a sulfa drug molecule has been incorporated

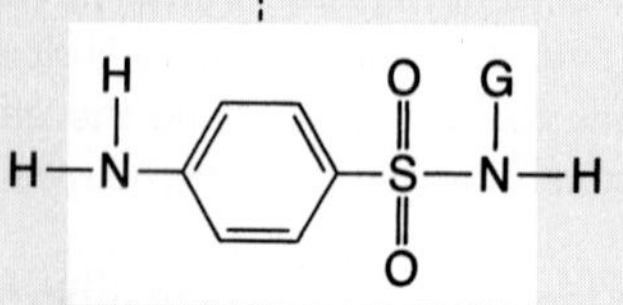

Sulfa drug (general structure)

If G = H   sulfanilamide

G = 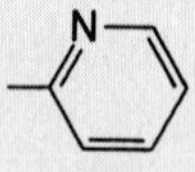   sulfapyridine

G = 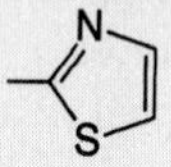   sulfathiazole

Examples of sulfa drugs

Poisons Can Permanently Inhibit Key Enzymes. The most dangerous **poisons** are effective even at very low concentrations because they are powerful inhibitors of enzymes. The cyanide ion is a very dangerous poison, for example, and it forms a strong complex with copper ions. These ions are cofactors found in an enzyme that is needed by cells to use oxygen, but the enzymes cannot work when their copper ions are combined with the cyanide ion. Thus cyanide ion is able to shut down cellular respiration, and death follows promptly.

Enzymes that have SH groups are denatured and deactivated by such heavy metal ions as Hg^{2+}, Pb^{2+}, Cu^{2+}, and Ag^+, all poisons. Nerve gas poisons and their weaker cousins, the organophosphate insecticides, inactivate enzymes of the nervous system.

The wall of a bacterial cell is a molecular structure outside of the cell membrane itself. It's essentially one huge, enveloping molecule, a gigantic polymer made from carbohydrate derivatives and amino acids. As the bacterium nears the end of making this wall, there is a final cross-link that must be made. Penicillin prevents this reaction by inhibiting its enzyme, so the organism never quite gets itself put together and eventually dies.

There is no human enzyme that reacts with penicillin, so this drug can do no harm by itself to the human body. Some people, however, are allergic to it. For them, penicillin can cause severe illness, even death.

Interestingly, the substrate that the bacterium uses in its last wall-building step incorporates D-alanine, not L-alanine, the optical isomer that humans use.

(R varies with the different types of penicillins)

Penicillin

17.4   ENZYMES IN MEDICINE

The specificity of the enzyme for its substrate and the slight differences in properties of isoenzymes provide several unusually sensitive methods of medical diagnosis.

Enzymes that normally work only inside cells are not found in the blood, except at extremely low concentrations. When cells are diseased or injured, however, their enzymes sometimes spill into the bloodstream. By detecting such enzymes in blood and measuring their levels, much can be learned about the disease or injury.

Enzyme Assays of Blood Use Substrates as Chemical "Tweezers". Despite the enormous complexity of blood as a mixture, analyses for trace components, like enzymes, are relatively easy to carry out. The substrate for the enzyme is used as a chemical "tweezers" to tweak (find) just its own enzyme and nothing else. If none of the enzyme is present, nothing happens to the substrate reagent. Otherwise, the extent of the reaction with the substrate is a measure of the concentration of the enzyme.

Viral Hepatitis Is Detected by the Presence of GPT and GOT in Blood. Heart, muscle, kidney, and liver tissue all contain the enzyme glutamate : pyruvate aminotransferase, mercifully abbreviated GPT. The liver, however, has about three times as much GPT as any other tissue, so the appearance of GPT in the blood generally indicates liver damage or a virus infection of the liver, such as viral hepatitis.

The level of another enzyme, glutamate : oxaloacetate aminotransferase or GOT, also increases in viral hepatitis, but the GPT level goes much higher than the GOT level. The ratio of GPT to GOT in the serum of someone with viral hepatitis is typically 1.6, compared with a level of 0.7 – 0.8 in healthy individuals. Notice that we speak here of *ratios*, not absolute amounts, which are normally *very* low.

Heart Attacks Cause Increased Levels of Three Enzymes in Blood. A myocardial infarction is the withering of a portion of the heart muscle following some blockage of the blood vessels supplying this muscle with oxygen and nutrients. The blockage can be caused by deposits, by hardening, or by a blood clot. If the patient survives, the withered muscle becomes scar tissue.

The popular term for this set of events is a *heart attack.*

When a patient appears with the symptoms traditionally associated with a heart attack, much is at stake in making a correct diagnosis and ruling a myocardial infarction either in or out. There are decisions that have to be based on the likelihood of living or dying and on the need to be in an intensive coronary care unit. The diagnosis can now be made with an exceptionally high reliability by the analysis of the serum for several enzymes and isoenzymes.

Molecules of proteins carry electrical charges, and both the number of charges and their signs depend on the pH of the medium. A procedure called *electrophoresis* is used to separate a mixture of proteins, and it is based on such charges and their pH dependence.

First, a surface is prepared from a porous but mechanically sturdy solid, like filter paper, or a gel made from a polymer such as polyacrylamide. This surface is wetted with a solution that contains ions and whose pH has been adjusted to some predetermined value. Electrical poles or plates are attached at opposite ends of this wetted surface, as seen in the accompanying figure, and electricity can flow because ions are present to carry it.

Next, a solution of a mixture of proteins is deposited as a very narrow band across the middle of the strip. The molecules of proteins — they could be isoenzymes — begin to travel because they are also electrically charged. However, when they have different *kinds* of charge (positive or negative), they are bound to travel in opposite directions. And when they have different *quantities* of charge and possibly even different masses, they travel at different rates. Thus, slowly the different proteins move out from the band where they were deposited, and they become increasingly separated, each into its own narrow zone or band. These bands cannot be seen directly, but there are many techniques, including color tests, for locating them and measuring the relative amounts of the individual proteins or isoenzymes in them.

Despite being *catalytically* the same, isoenzymes can be separated by electrophoresis. For example, the *MM*, *BB*, and *MB* isoenzymes of creatine kinase can be separated, and an increase in the CK(*MB*) band in blood serum helps to diagnose a myocardial infarction. The five isoenzymes of lactate dehydrogenase, LD, can also be separated and their relative amounts determined using electrophoresis.

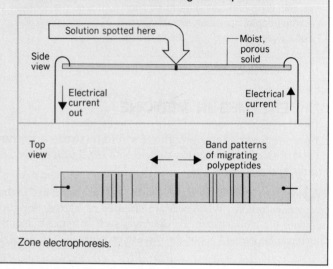

Zone electrophoresis.

When a myocardial infarction occurs, the levels of three enzymes that are normally confined to heart muscle cells begin to rise in blood serum (see Figure 17.4). These enzymes are CK (page 346), GOT (described just above), and LD, which stands for lactate dehydrogenase. LD catalyzes the oxidation of lactate to pyruvate. A clinical report from an actual patient who had a myocardial infarction is shown in Table 17.2. You can see how sharply the levels of these three enzymes rose between day I and day II.

The line of data labeled "CK(*MB*)" and the data in the columns headed by "LD ISOENZYME" provided the clinching evidence for the infarction. As we learned on page 347, the CK enzyme occurs as three isoenzymes, CK(*MM*), CK(*BB*), and CK(*MB*). The technique that uses

The older (and still often used) symbol for LD is LDH.

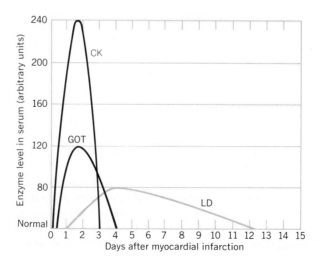

FIGURE 17.4
The concentrations of three enzymes in blood serum increase after a myocardial infarction. CK is creatine kinase; GOT is glutamate:oxaloacetate aminotransferase; and LD is lactate dehydrogenase.

TABLE 17.2
Clinical Report: Myocardial Infarction

DAY 1	DAY II	DAY III
DATE *5-25*	DATE *5-26*	DATE *5-27*
CK *51*	CK *552*	CK *399*
CK (MB) *Negative*	CK (MB) *Moderate Positive*	CK (MB) *Weak Positive*
GOT *19*	GOT *91*	GOT *117*
LD *88*	LD *151*	LD *247*
LD ISOENZYME	LD ISOENZYME	LD ISOENZYME
% of Total LD activity	% of Total LD activity	% of Total LD activity
LD_1 *28.3* %	LD_1 *33.8* %	LD_1 *37.9* %
LD_2 *32.9* %	LD_2 *32.3* %	LD_2 *32.8* %
LD_3 *19.7* %	LD_3 *16.3* %	LD_3 *15.7* %
LD_4 *11.6* %	LD_4 *9.1* %	LD_4 *7.6* %
LD_5 *7.5* %	LD_5 *8.5* %	LD_5 *6.0* %
CB	*CB*	*B.L.*

	Normal Range[a]		LD Isoenzymes Normals	
CK	Male	5–75 mU/ml	LD_1	14–29%
	Female	5–55 mU/ml	LD_2	29–40%
			LD_3	18–28%
GOT		5–20 mU/ml	LD_4	7–17%
			LD_5	3–16%
LD		30–110 mU/ml		

INTERPRETATION: The presence of a CK(MB) and/or an LD_1 greater than LD_2 strongly suggest myocardial infarction.

[a] One International Unit (U) of activity is the reaction under standard conditions of one micromole per minute of a particular substrate used in the test. (Data courtesy of Dr. Gary Hemphill, Clinical Laboratories, Metropolitan Medical Center, Minneapolis, MN.)

a chemical substrate to determine the serum CK level cannot tell these isoenzymes apart, because all three catalyze the reaction with the substrate.

To be sure that the rise in serum CK level is caused by injury to the *heart* tissue, the clinical chemist has to analyze specifically for the CK(*MB*) isoenzyme common to heart muscle. A technique called *electrophoresis* is used and is described in Special Topic 17.3. Electrophoresis separates the individual isoenzymes, and then each is analyzed separately. As the chart shows, the patient's serum CK(*MB*) level did rise.

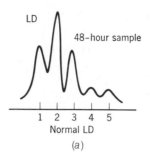
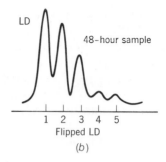

FIGURE 17.5
The lactate dehydrogenase isoen-zymes. (a) The normal pattern of the relative concentrations of the five isoenzymes. (b) The pattern after a myocardial infarction. Notice the reversal in relative concentration be-tween LD_1 and LD_2. This is the LD flip.

The numbered subscripts of the LD isoenzymes are simply the order in which their molecules sep-arate during electrophoresis.

As additional confirmation of an infarction, the serum LD fraction is further separated by electrophoresis into the five LD isoenzymes, and each is individually analyzed. The *relative* concentrations of these five differ distinctively between a healthy person (Figure 17.5a) and one who has suffered an infarction (Figure 17.5b). Of particular importance are the relative concentrations of the LD_1 and the LD_2 isoenzymes. Normally the LD_1 level is less than that of the LD_2, but following a myocardial infarction what is called an "$LD_1 - LD_2$ flip" occurs. The relative concentrations of LD_1 and LD_2 become reversed and the level of LD_1 rises higher than that of LD_2. When both the CK(*MB*) band and the $LD_1 - LD_2$ flip occur, the diagnosis of a mycardial infarction is essentially 100% certain.

The Determination of Glucose.   The detection of glucose in urine is important to people with diabetes, because in poorly managed diabetes this sugar is present in urine. One commercially available test, the Clinistix® test, uses porous paper test strips impregnated with two enzymes and an aromatic compound that can be oxidized to a dye by hydrogen peroxide. If glucose is in the drop of urine used to moisten a Clinistix® test strip, one of the enzymes, glucose oxidase, catalyzes a reaction that uses glucose to make hydrogen peroxide. The other enzyme, a peroxidase, then catalyzes the oxidation of the aromatic compound by hydrogen peroxide to give the dye. If the color appears in a few seconds, the urine contains glucose. Other enzyme-based tests for glucose in blood are also available.

Diastix is another enzyme-based test strip for urinary glucose.

A Natural Blood Clot Dissolving Enzyme Can Be Activated Medically Using Other Enzymes.   The formation of a blood clot involves a multistep process that changes fibrino-gen, a serum-soluble, circulating protein, into fibrin, a serum-insoluble protein. The fibrin molecules form the brush-heaplike structure of a blood clot, and while the clot forms, a trace amount of another protein, plasminogen, becomes absorbed onto it. Plasminogen is a zymo-gen, as we learned in the previous section, and when it is absorbed into a clot it becomes more susceptible to activation than when it is simply in circulation. In time, something else circulat-ing in the blood, *tissue plasminogen activator* or tPA, acts on plasminogen to complete its activation. The tPA converts plasminogen to an active enzyme plasmin. Plasmin is a catalyst for dissolving the fibrin of blood clots.

About 150,000 people a year die from clots in their lungs.

Therapy for a myocardial infarction is intended to open blocked capillaries in the heart muscle as rapidly as possible before the oxygen starvation of the surrounding tissue spreads the damage too widely. Obviously, the sooner this therapy is applied following an infarction, the better the chances that long-term heart muscle damage will be light. Three enzymes are commercially available for introduction into the blood stream to bring about the dissolution of clots: streptokinase, tissue plasminogen activator or tPA made by genetic engineering (Chap-ter 21), and a modified streptokinase called APSAC (for acylated plasminogen-streptokinase activator complex). A report in 1990 of the first large-scale direct comparison of tPA and streptokinase found that neither drug was better than the other in improving the survival of heart attack victims. The chief difference at that time was cost; one dose of tPA was over $2000 whereas one of streptokinase was about $200. (APSAC is somewhat less expensive than tPA but more than streptokinase.)

This therapy is called *thrombolytic therapy* because it lyses (breaks down) thrombi (blood clots).

Aspirin given with streptokinase increases its effectiveness.

Calcium Channel Blockers Are Used to Reduce the Risk of a Heart Attack.   We learned earlier how nerve signals, calcium ions, and certain effectors initiate muscular work. Calcium ions, let into heart muscle cells by calcium channels (opened by nerve signals), are thus involved in the contraction of heart muscle during each heart beat. Some medications called calcium channel blockers are able to block a percentage of these channels. This reduces the vigor of the heart beat, which is sought for patients at risk of heart attack. A reduced flux of calcium ions means fewer activations of the effectors of the enzymes involved in the heart beat.

17.5   HORMONES AND NEUROTRANSMITTERS

Hormones carry chemical signals from endocrine glands to target cells, and neurotransmitters bring chemical signals from one nerve cell to the next.

Like any complex organism, the body is made of many highly specialized parts. Information of some sort must flow among these parts, therefore, so that everything is well coordinated. This flow is handled by chemical messengers and electrical signals.

Target Cell Receptors Recognize Chemical Messengers.   In the broadest terms, the messengers are molecules or ions that bind to molecules called **receptors** at membranes of cells that are the messenger's particular **target cells.** A complex forms between molecules of the messenger and the receptor that then induces some kind of change at the target cell.

The receptor molecules are proteins.

The change itself is the actual "message." It might be the speeding up or slowing down of some reaction, or it might be a change in the permeability of a cell membrane to the flow of an ion or a molecule. The change induced by the messenger might even be the activation of a gene. In nerve cells, called neurons, the change could be the excitation of a neighboring neuron causing it to send on a signal, or the change might be the opposite, the inhibition of a signal.

Broadly speaking, there are two kinds of chemical messengers, *primary* and *secondary.* The former bring messages and the latter pass them on. The chief primary chemical messengers are hormones and neurotransmitters. They differ not so much in *how* they work as in *at what distance* and *where* they work (see Figure 17.6).

Hormones are chemicals made in specialized organs, called the *endocrine glands,* and they travel in the blood to target cells that might be as much as 15–20 cm away. Hormones vary widely in structure, from the simple epinephrine (Special Topic 11.3) to steroids (Table 15.3) to polypeptides, like insulin (Figure 16.4).

Greek *hormon,* arousing.

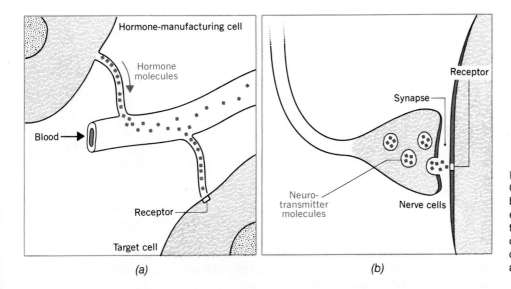

(a)   *(b)*

FIGURE 17.6
Chemical communication in the human body. (*a*) A hormone travels from an endocrine gland, where it is made, through the bloodstream to its target cell. (*b*) A neurotransmitter travels from one nerve cell to the next nerve cell across the synapse.

TABLE 17.3
The Principal Human Endocrine Glands and Their Hormones

Gland or Tissue	Hormone	Major Function of Hormone
Thyroid	Thyroxine	Stimulates rate of oxidative metabolism and regulates general growth and development
	Thyrocalcitonin	Lowers level of Ca^{2+} in blood
Parathyroid	Parathormone	Regulates the levels of calcium and phosphate ions in blood
Pancreas, β cells	Insulin	Decreases blood glucose level
α cells	Glucagon	Elevates blood glucose level
Adrenal medulla	Epinephrine	Elevates blood glucose level and heart beat
Adrenal cortex	Cortisone and related hormones	Control carbohydrate, protein, mineral, salt, and water metabolism
Anterior pituitary	Thyrotropic hormone	Stimulates thyroid gland functions
	Adenocorticotropic hormone	Stimulates development and secretion of adrenal cortex
	Growth hormone	Stimulates body weight and rate of growth of skeleton
	Gonadotropic hormones	Stimulate gonads
	Prolactin	Stimulates lactation
Posterior pituitary	Oxytocin	Causes contraction of some smooth muscles
	Vasopressin	Inhibits secretion of water from the body by way of the urine
Ovary (follicle)	Estrogens	Influence development of sex organs and female characteristics
Ovary (corpus luteum)	Progesterone	Influences menstrual cycle; prepares uterus for pregnancy; maintains pregnancy
Uterus (placenta)	Estrogens and progesterone	Function in maintenance of pregnancy
Testes	Testosterone	Responsible for development and maintenance of sex organs and secondary male characteristics
Digestive system	Several gastrointestinal hormones	Integration of digestive processes

Human growth hormone, now available by genetic engineering, was found (1990) to cause significant improvements in muscle mass, bone density, internal organs, and skin thickness in men aged 61–81 with growth hormone deficiency but otherwise in good health.

A hormone is released into circulation when its endocrine gland receives a unique signal. This might be something conveyed by one of our senses, such as an odor, or it might be a stress, or a variation in the concentration of a particular substance in a body fluid. The hormone insulin, for example, is released when the concentration of glucose in blood increases.

Some hormones, such as sex hormones, activate genes to direct the synthesis of enzymes. Other hormones, like epinephrine, activate enzymes. Others, like insulin and growth hormone, alter the permeabilities of target cell membranes to enable the migration of chemicals.

Table 17.3 provides a partial list of the body's hormones. Except for certain hormones involved in processes we will study in later chapters, we leave the study of the body's overall responses to hormones to other courses.

The prostaglandins, described in Special Topic 15.1, are now classified as hormones. They are called *local hormones* because they work where they are made and do not travel like regular hormones.

Neurotransmitters are chemicals made in neurons and sent over very short distances to neighboring neurons, and we'll return to them later.

The Formation and Splitting of cyclic-AMP Is a Major Mechanism for Passing a Message into a Cell.

The mechanisms of the action of hormones and neurotransmitters, in many situations, are very similar. To illustrate, we will study one such mechanism, one in which a cyclic nucleotide conveys the message across a cell membrane.

Cyclic refers to the *extra* ring of the phosphate diester system.

Cyclic nucleotides, particularly 3′,5′-cyclic AMP, are important secondary chemical messengers. How cyclic AMP works is sketched in Figure 17.7. At the top we see a hormone — it could just as well be a neurotransmitter — that can combine with a receptor molecule on the cell surface. A "lock-and-key" mechanism ensures that the hormone bypasses all cells without a matching receptor.

The complex formed between hormone and receptor activates an enzyme, adenylate cyclase, also an integral part of the cell membrane. This activation is the specific chemical change caused by the hormone. But the process thus launched continues, because the hormone's message has yet to get inside the cell. The enzyme, using a part of its molecule sticking on the inside of the cell, promptly catalyzes the conversion of ATP into cyclic AMP and diphosphate ion, PP_i.

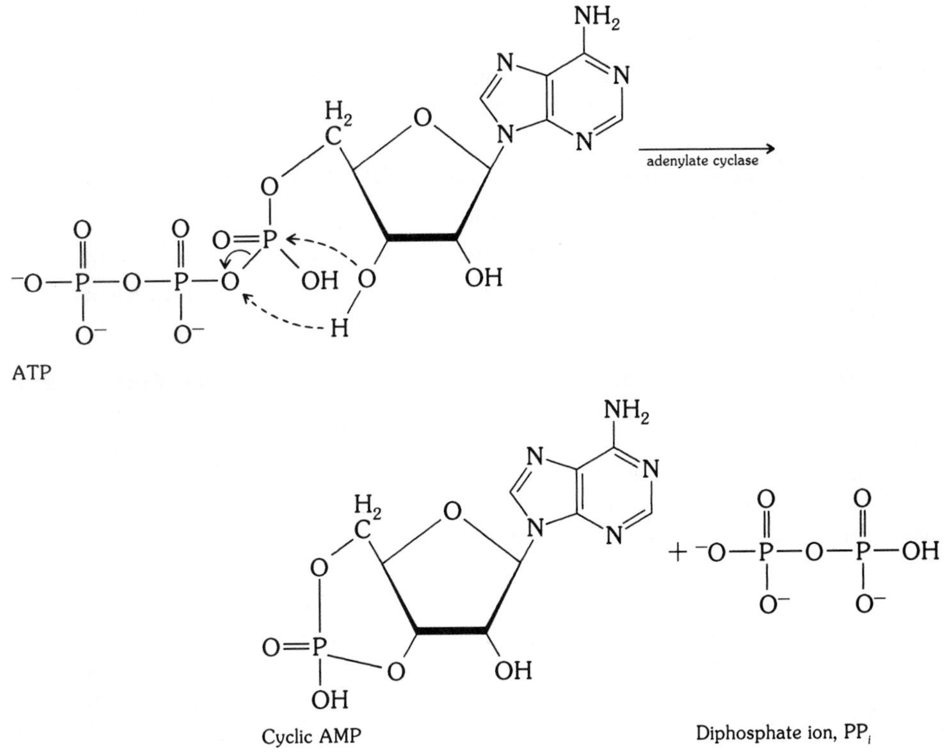

ATP

Cyclic AMP

Diphosphate ion, PP_i

The newly formed cyclic AMP, functioning as the secondary messenger, now activates an enzyme, which, in turn, catalyzes a reaction. This last event is what the message of the primary messenger was all about. Finally, another enzyme catalyzes the hydrolysis of cyclic AMP to AMP, and this shuts off the cycle. Energy-producing reactions in the cell will now remake ATP from the AMP. Let's summarize the steps.

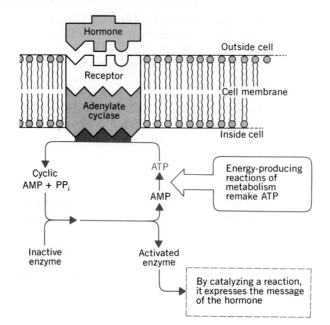

FIGURE 17.7
The activation of the enzyme, adenylate cyclase, by a hormone (or a neurotransmitter). The hormone–receptor complex activates this enzyme, which then catalyzes the formation of cyclic AMP. This, in turn, activates an enzyme inside the cell.

1. A signal releases the hormone or neurotransmitter.
2. It travels to its target cell — next door for a neurotransmitter but some farther distance away for a hormone.
3. The primary messenger molecule finds its target cell by a lock and key mechanism, and binds to a receptor. The resulting complex activates adenylate cyclase, which is part of the cell membrane.
4. Adenylate cyclase catalyzes the conversion of ATP to cyclic AMP.
5. Cyclic AMP activates an enzyme inside the cell.
6. The enzyme catalyzes a reaction, one that corresponds to the primary message.
7. Cyclic AMP is hydrolyzed to AMP, which is reconverted to ATP, and the system returns to the pre-excited state.

Neurotransmitters Move Rapidly Across the Synaptic Gap and Bind to Receptors. A partial list of neurotransmitters is given in Table 17.4. Some are nothing more than simple amino acids. Others are β-phenylethylamines or catecholamines (Special Topic 11.3, page 234), and many are polypeptides.

Each nerve cell has a fiber-like part called an *axon* that reaches to the face of the next neuron or to one of its filament-like extensions called *dendrites* (Figure 17.8). A nerve impulse

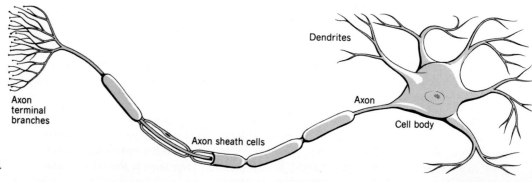

FIGURE 17.8
One kind of neuron or nerve cell.

TABLE 17.4
Neurotransmitters

Monoamines

Acetylcholine

$$(CH_3)_3\overset{+}{N}CH_2CH_2O\overset{\displaystyle O}{\overset{\|}{C}}CH_3$$

Dopamine

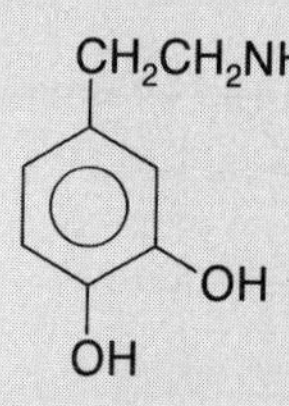

$CH_2CH_2NH_2$ ... OH, OH

Norepinephrine

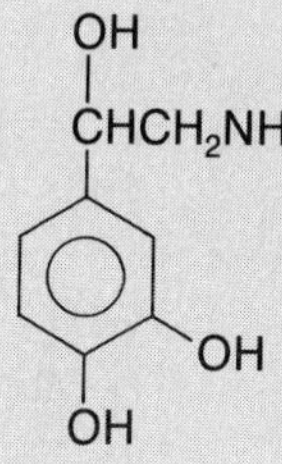

OH, $CHCH_2NH_2$ ... OH, OH

Serotonin

HO ... $CH_2CH_2NH_2$ ... N—H

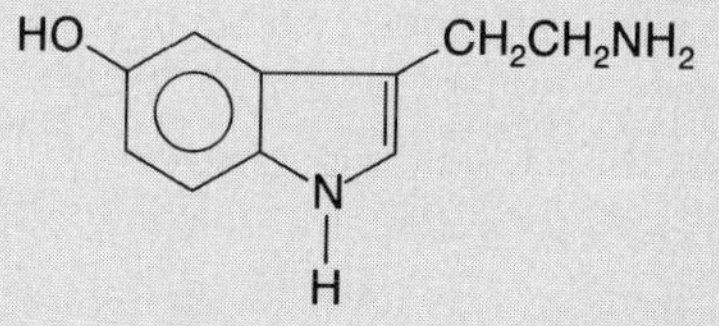

Amino Acids

Glycine $\qquad$ $^+NH_3CH_2CO_2^-$

γ-Aminobutyric acid $\qquad$ $^+NH_3CH_2CH_2CH_2CO_2^-$

Glutamic acid $\qquad$ $^+NH_3CHCO_2^-$
$\qquad\qquad\qquad\quad$ $CH_2CH_2CO_2H$

Neuropeptides

Met-Enkephalin $\qquad$ Tyr · Gly · Gly · Phe · Met

Leu-Enkephalin $\qquad$ Tyr · Gly · Gly · Phe · Leu

β-Endorphin $\qquad$ Tyr · Gly · Gly · Phe · Met · Thr · Ser · Glu · Lys · Ser

$\qquad\qquad\qquad$ Gln · Thr · Pro · Leu · Val · Thr · Leu · Phe · Lys · Asn

$\qquad\qquad\qquad$ Ala · Ile · Val · Lys · Asn · Ala · His · Lys · Lys · Gly · Gln

Substance P $\qquad$ Arg · Pro · Lys · Pro · Gln · Gln · Phe · Phe · Gly · Leu · Met—NH$_2$

Angiotensin II $\qquad$ Asp · Arg · Val · Tyr · Ile · His · Pro · Phe—NH$_2$

Somatostatin $\qquad$ Ala · Gly · Cys · Lys · Asn · Phe
$\qquad\qquad\qquad\qquad\qquad\qquad\qquad\qquad\qquad$ Phe
$\qquad\qquad\qquad\qquad$ Cys $\qquad\qquad\qquad\qquad\qquad$ Try
$\qquad\qquad\qquad\qquad$ Ser · Thr · Phe · Thr

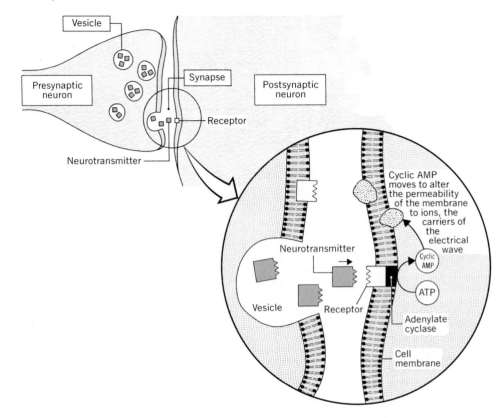

FIGURE 17.9
Neurotransmission. Neurotransmitter molecules, released from vesicles of the presynaptic neuron, travel across the synapse. At the postsynaptic neuron, they find their receptors, and the cyclic AMP system similar to that shown in Figure 17.7 becomes activated.

The traveling wave of electrical charge moves rapidly, but still not as rapidly as electricity moves in electrical wires.

consists of a traveling wave of electrical charge that sweeps down the axon as small ions migrate at different rates between the inside and the outside of the neuron. The problem is how to get this impulse launched into the next neuron so that it can continue along the length of the nerve fiber. This is solved by a *chemical* communication from one neuron to the next.

Between the terminal of an axon and the start of the next neuron, there is a very narrow, fluid-filled gap called the *synapse.* Neurotransmitters move across the synapse when the electrical wave causes them to be released from their tiny storage sacs or *vesicles* near the ends of axons. When neurotransmitter molecules lock to their receptors on the other side of the synapse, a nucleotide cyclase enzyme is activated, such as adenylate cyclase (see Figure 17.9).

Now the formation of cyclic AMP is catalyzed, and newly formed cyclic AMP initiates whatever change is programmed by the chemicals in the target neuron. An enzyme then deactivates adenylate cyclase by catalyzing the release of the neurotransmitter molecule. If it were a hormone, it would be swept away in the bloodstream, but it's not. It's still a neurotransmitter in the synapse, so unless the system wants it to act again, it must be removed or deactivated.

The ANS nerves handle the signals that run the organs that have to work autonomously (without conscious effort), such as the heart and the lungs.

The Neurotransmitter Acetylcholine Is Deactivated by Hydrolysis.   One method used to remove the neurotransmitter is to break it up by a chemical reaction. Acetylcholine, for example, a neurotransmitter in the autonomic nervous system or ANS, is catalytically hydrolyzed to choline and acetic acid. The enzyme is choline acetyltransferase.

$$\underset{\text{Acetycholine}}{(CH_3)_3\overset{+}{N}CH_2CH_2O\overset{\overset{\displaystyle O}{\|}}{C}CH_3} + H_2O \underset{\underset{\text{acetyltransferase}}{\text{choline}}}{\rightleftharpoons} \underset{\text{Choline}}{(CH_3)_3\overset{+}{N}CH_2CH_2OH} + \underset{\text{Acetic acid}}{HO\overset{\overset{\displaystyle O}{\|}}{C}CH_3}$$

Within two milliseconds (2×10^{-3} s) of the appearance of acetylcholine molecules in the synapse, they are all broken down — but not before they have served their purpose. The synapse is now cleared for a fresh release of acetylcholine from the presynaptic neuron if the signal for its release continues. If the signal does not come, then the action is shut down.

The Botulinus Toxin Prevents the Synthesis of Acetylcholine. The botulinus toxin, the extremely powerful toxic agent made by the food poisoning *botulinus bacillus*, works by preventing the *synthesis* of acetylcholine. Without it, the cholinergic nerves of the ANS can't work.

Nerve Gas Poisons Deactivate Choline Acetyltransferase. The nerve gas poisons, which include some of the more powerful insecticides, work by deactivating the enzyme needed for the hydrolysis of acetylcholine. In the absence of this enzyme, acetyltransferase, the signal cannot be turned off. It continues unabated until the heart fails.

An antidote for nerve gas poisoning, atropine, works by blocking the receptor protein for acetylcholine, so despite the continuous presence of this neurotransmitter, it isn't able to complete the signal-sending work. This tones the system down, and other processes slowly restore the system to normal.

Other blockers of the receptor protein for acetylcholine are some local anesthetics such as nupercaine, procaine, and tetracaine. Drugs that block the action of a neurotransmitter are called **antagonists** to the neurotransmitter. The neurotransmitter itself is sometimes referred to as an **agonist**.

Some Neurotransmitters Are Pulled Back from the Synaptic Gap. Norepinephrine is both a hormone and a neurotransmitter. The adrenal medulla secretes it into the bloodstream in emergencies when it must be made available to all of the nerve tissues that use it. When it works as a neurotransmitter, it is deactivated by being reabsorbed by the neuron that released it, where it is then degraded. (Some is also deactivated right within the synapse.)

Drugs that Inactivate the Monoamine Oxidases Are Used to Treat Depression. One place where norepinephrine works is in the brainstem where mood regulation is centered. Its degradation is catalyzed by enzymes called the **monoamine oxidases** or **MAO.** If these enzymes are themselves made inactive, then an excess of norepinephrine builds up, which can spill back into the synapse and send signals on. Such a deactivation of the MAOs is sometimes desired, particularly when the level of norepinephrine is low for any reason, and signal-sending activity dies down too much. Some of the antidepressant drugs, such as iproniazid, for example, work by inhibiting the monoamine oxidases.

The ANS nerves that use acetylcholine are called the *cholinergic nerves*.

The nerves that use norepinephrine are called the *adrenergic nerves* (after an earlier name for norepinephrine, noradrenaline).

Iproniazid

Amitriptyline (Elavil)

Imipramine (Tofranil)

Amitriptyline (Elavil) and imipramine (Tofranil) are antidepressants that work by inhibiting the reabsorption of norepinephrine by the presynaptic neuron. Without this reabsorption and subsequent degradation of norepinephrine by the monoamine oxidases, its level and its signalsending work stays high.

Dopamine Excesses Occur in Schizophrenia. Dopamine, like norepinephrine, is also a monoamine neurotransmitter. It occurs in neurons of the midbrain that are involved with

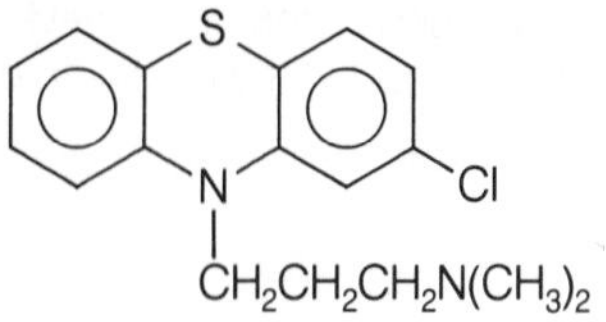

Chlorpromazine

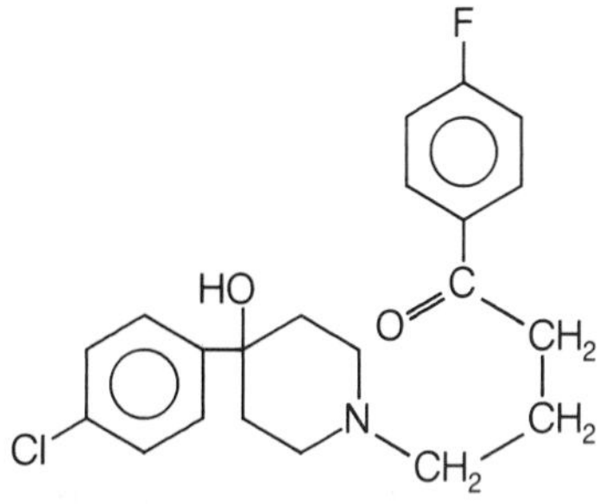

Haloperidol

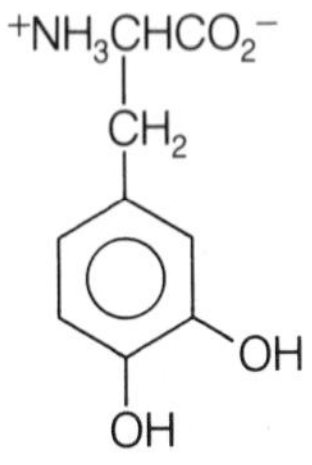

L-DOPA

feelings of pleasure and arousal as well as with the control of certain movements. In schizophrenia, the neurons that use dopamine are overstimulated, because either the releasing mechanism or the receptor mechanism is overactive.

Drugs commonly used to treat schizophrenia, such as chlorpromazine (e.g., Thorazine) and haloperidol (Haldol), bind to dopamine receptors and thus inhibit its signal-sending work.

Amphetamine Abuses Cause Schizophrenia-Like Symptoms. Stimulants such as the amphetamines (cf. Special Topic 11.3, page 234) work by triggering the release of dopamine into the arousal and pleasure centers of the brain. The effect on the brain is called a "high." But it is easy to abuse amphetamines. When this occurs, there is an overstimulation closely resembling that of schizophrenia, with delusions of persecution, hallucinations, and other disturbances of the thought processes.

Dopamine-Releasing Neurons Have Degenerated in Parkinson's Disease. When the dopamine-using neurons in the brain have degenerated, as in Parkinson's disease, an extra supply of dopamine itself is then needed to compensate. This is why a compound called L-DOPA (*levorotatory dihydroxyphenylalanine*) is used. The neurons that still work can use it to make extra dopamine.

GABA Inhibits Nerve Signals. The normal function of some neurotransmitters is to *inhibit* signals instead of to initiate them. Gamma-aminobutyric acid (GABA) is an example, and as many as one-third of the synapses in the brain have GABA available.

The inhibiting work of GABA can be made even greater by mild tranquilizers such as diazepam (e.g., Valium) and chlordiazepoxide hydrochloride (Librium), as well as by ethanol. The augmented inhibition of signals reduces anxiety, affects judgment, and induces sleep. Of course, you've probably heard of the widespread abuse of Valium and Librium, to say nothing of alcohol.

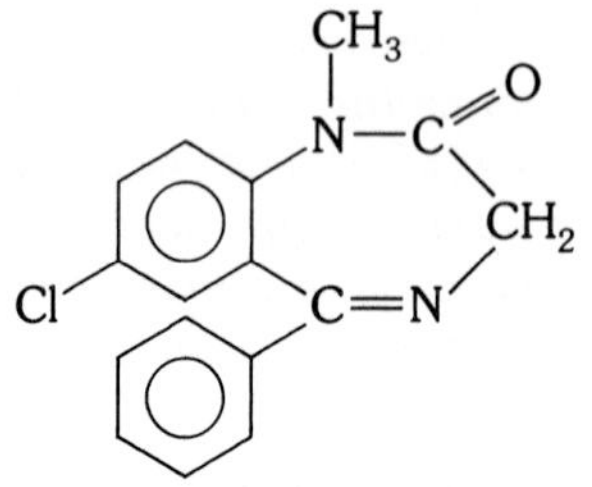

Diazepam (Valium)

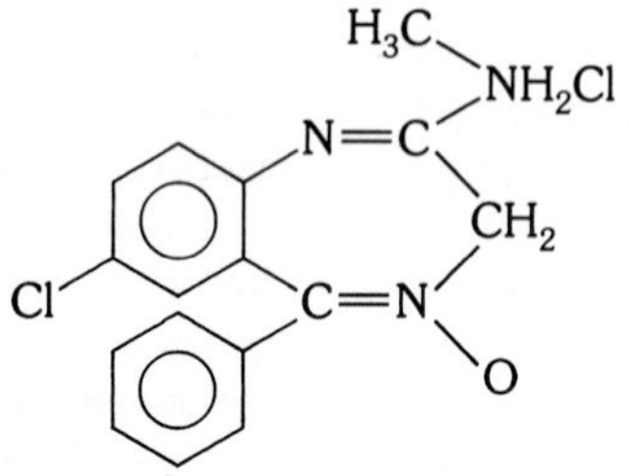

Chlordiazepoxide
hydrochloride (Librium)

Greek, *chorea,* dance.

GABA Is Deficient in Huntington's Chorea. The victims of Huntington's chorea, a hereditary neurological disorder, suffer from speech disturbances, irregular movements, and a steady mental deterioration, all related to a deficiency of GABA. Unhappily, GABA can't be administered in this disease, because it can't move out of circulation and into the regions of the brain where it works.

En- or *end-,* within; *kephale,* brain; *-orph-,* from morphine.

Several Polypeptides Act as Painkilling Neurotransmitters. As we said, some neurotransmitters are relatively small polypeptides. One type includes the *enkephalins;* another consists of the *endorphins.* As indicated in Table 17.4, they are powerful pain inhibitors. One of them, dynorphin, is the most potent painkiller yet discovered, being 200 times stronger than morphine, an opium alkaloid that is widely used to relieve severe pain. Sites in the brain that strongly bind molecules of morphine also bind those of the enkephalins, so these natural painkillers are now often referred to as the body's natural opiates.

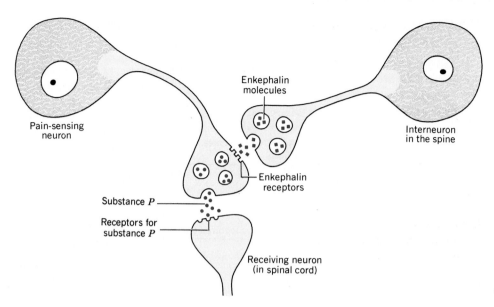

FIGURE 17.10
The inhibiton of the relase of Substance P by enkephalin helps to reduce the intensity of the pain signal to the brain.

The Enkephalins Inhibit the Release of Substance P. Substance P is a pain-signaling polypeptide neurotransmitter. According to one theory (see Figure 17.10), when a pain-transmitting neuron is activated, it releases substance P into the synapse. However, butting against such neurons are other neurons that can release enkephalin. And these, when released, inhibit the work of substance P. In this way, the intensity of the pain signal can be toned down. This might explain the delay of pain that sometimes occurs during an emergency when the brain and the body must continue to function to escape the emergency.

Substance P appears to be involved in the link between the nervous system and the body's immune system. It is known that some forms of arthritis flare-up under stress, and stress deeply involves the nervous system. But arthritis is generally regarded as chiefly a disease of the immune system. In tests on rats with arthritis, flare-ups could be induced by injections of substance P. What interests scientists about this is the possibility of controlling the severity of arthritis by somehow diminishing substance P levels in the affected joints.

One of the interesting developments in connection with the endorphins is that acupuncture — a procedure developed in China to alleviate pain — might work by stimulating the production and release of endorphins.

Many Neurotransmitters Exert More Than One Effect. Several neurotransmitters can be received by more than one kind of receptor. For example, at least three types of receptors for the opiates have been identified. Such receptor multiplicity may explain how some neurotransmitters have multiple effects. The opiates thus not only reduce pain, they induce sleep, affect gut motility, and modulate the immune system depending on which receptors are involved and in which tissue.

What we have done in this section is look at some *molecular* connections between conditions of the nervous system and particular chemical substances. This whole field is one of the most rapidly moving areas of scientific investigation today, and during the next several years we may expect to see a number of dramatic advances both in our understanding of what is happening and in the strategies of treating mental diseases.

SUMMARY

Enzymes Enzymes are the catalysts in cells. Some consist wholly of one or more polypeptides and others include, besides polypeptides, a cofactor — an organic coenzyme, a metal ion, or both. Some coenzymes are phosphate esters of B vitamins, and in these examples the vitamin unit usually furnishes the enzyme's active site. Because they are mostly polypeptide in nature, enzymes are vulnerable to all of the conditions that denature proteins. The name of an enzyme, which almost always ends in -ase, usually discloses either the identity of its substrate or the kind of reaction it catalyzes.

Some enzymes occur as small families called isoenzymes in which the polypeptide components vary slightly from tissue to tissue in the body.

An enzyme is very specific both in the kind of reaction it catalyzes and in its substrate. Enzymes make possible reactions rates that are substantially higher than the rates of uncatalyzed reactions. Equilibration enzymes, like carbonic anhydrase, favor neither the forward nor the reverse reaction but help to reestablish equilibrium quickly when it is upset.

Theories of enzyme action When an enzyme–substrate complex forms, the active site is brought up to the part of the substrate that is to react. Binding sites on the enzyme guide the substrate molecule in, and sometimes a change in the conformation of the enzyme is induced by the substrate. The recognition of the enzyme by the substrate occurs as a lock-and-key model that involves complementary shapes and electrical charges.

Regulation of enzymes Some enzymes are activated by genes. Others, like plasmin, are made as zymogens (e.g., plasminogen) that become enzymes when an activating reaction removes a molecular piece that is covering the active-site. Nonsubstrate molecules called effectors are able to activate some enzymes. Certain effectors must themselves first be activated by Ca^{2+} ions in a process involving nerve signals that open calcium channels in cell membranes. Some enzymes, situated in the membranes of cells (or small bodies within cells), are activated by the interaction between a hormone or a neurotransmitter and its receptor protein.

Enzymes can be reversibly inhibited by competitive feedback that involves a product of the enzyme's action, or by a similar action by something that isn't a product. Some of the most dangerous poisons bind to active sites and irreversibly block the work of an enzyme, or they carry enzymes out of solution by a denaturant action.

Many antibiotics and other antimetabolites work by inhibiting enzymes in bacteria.

Medical uses of enzymes The serum levels of many enzymes rise when the tissues or organs that hold these enzymes are injured or diseased. By monitoring these serum levels, and by looking for certain isoenzymes, many diseases can be diagnosed—for example, viral hepatitis and myocardial infarctions. Enzymes are also used in analytical systems that measure concentrations of substrates, such as in tests for glucose.

Certain enzymes, like tPA, can be administered to a heart attack victim to activate the clot-dissolving enzyme, plasmin. Then, to reduce the likelihood of another heart attack, calcium channel blockers can be given to reduce the vigor of the heart beat by lowering the amount of enzyme involved with it.

Hormones Endocrine glands secrete hormones, and these primary chemical messengers travel to their target cells in the blood where they activate a gene, or an enzyme, or affect the permeability of a cell membrane. They recognize their own target cells by a lock-and-key type of recognition that involves their receptor proteins.

Neurotransmitters In response to an electrical signal, vesicles in an axon release a neurotransmitter that moves across the synapse. Its molecules bind to a receptor protein on the next neuron, and then the pattern is much like that of hormones. One common response to the formation of a neurotransmitter–receptor protein complex is the activation of adenylate cyclase, which triggers the formation of cyclic AMP. In turn, cyclic AMP sets off other events, such as the activation of an enzyme that catalyzes a reaction that is ultimately what the "signal" of the neurotransmitter was all about.

Neurotransmitters include amino acids, monoamines, and polypeptides. Some neurotransmitters *activate* some response in the next neuron, whereas others *deactivate* some activity. A number of medications work by interfering with neurotransmitters.

REVIEW EXERCISES

The answers to these review exercises are in the *Study Guide* which accompanies this book.

Nature of Enzymes

17.1 What is meant by the term *enzyme* with respect to its (a) function and (b) composition (in general terms only)?

17.2 To what does the term *specificity* refer in enzyme chemistry?

17.3 Define and distinguish among the following terms.
(a) apoenzyme (b) cofactor (c) coenzyme

Coenzymes

17.4 What B vitamin is involved in the NAD^+/NADH system?

17.5 The active part of either FAD or FMN is furnished by which vitamin?

17.6 Complete and balance the following equation.

$$CH_3\overset{\overset{\displaystyle OH}{|}}{C}HCH_3 + NAD^+ \longrightarrow$$

$$CH_3\overset{\overset{\displaystyle O}{\|}}{C}CH_3 + \underline{\quad} + \underline{\quad}$$

17.7 Complete and balance the following equation.

$$\underline{\quad} + NADH + FMN \longrightarrow NAD^+ + \underline{\quad}$$

Kinds of Enzymes

17.8 What is most likely the substrate for each of the following enzymes?
(a) Sucrase (b) Glucosidase
(c) Protease (d) Esterase

17.9 What *kind* of reaction does each of the following enzymes catalyze?
(a) An oxidase   (b) Transmethylase
(c) Hydrolase   (d) Oxidoreductase

17.10 What is the difference between lactose and lactase?

17.11 What is the difference between a hydrolase and hydrolysis?

17.12 What are isoenzymes (in general terms)?

17.13 What are the three isoenzymes of creatine kinase? Give their symbols and where they are principally found.

Theory of How Enzymes Work

17.14 What name is given to that part of an enzyme where the catalytic work is carried out?

17.15 In general terms, how are the specificities and rate enhancements of enzymes explained?

17.16 What is the induced-fit theory?

Enzyme Activation and Inhibition

17.17 Feedback inhibition of an enzyme works in what way?

17.18 Why is feedback inhibition an example of a homeostatic mechanism?

17.19 How can a substrate molecule serve as the activator of an enzyme with more than one catalytic site?

17.20 Plasminogen is a zymogen.
(a) What does this mean?
(b) What is the name of the enzyme made from it, and what does this enzyme do?

17.21 Troponin is an *effector*.
(a) What does this mean?
(b) What activates troponin?
(c) How is the presence of this troponin–activator controlled?
(d) How is activated troponin deactivated?

17.22 How do the following poisons work?
(a) CN^-
(b) Hg^{2+}
(c) Nerve gases or organophosphate insecticides

17.23 What are antimetabolites, and how are they related to antibiotics?

17.24 The following overall change is accomplished by a series of steps, each with its own enzyme.

$$^{2-}O_3POCH_2\overset{\overset{\displaystyle O}{\displaystyle \|}}{C}HCOPO_3{}^{2-} \longrightarrow \longrightarrow \longrightarrow {}^{2-}O_3POCH_2\underset{\underset{\displaystyle OPO_3{}^{2-}}{\displaystyle |}}{C}HCO_2{}^-$$

$$\underset{OH}{|}$$

1,3-Diphosphoglycerate   2,3-Diphosphoglycerate
(1,3-DPG)   (2,3-DPG)

One of the enzymes in this series is inhibited by 2,3-DPG. What kind of control is exerted by 2,3-DPG on this series? (Name it.)

Enzymes in Medicine

17.25 If an enzyme such as CK or LD is normally absent from blood, how can a *serum* analysis for either tell anything? (Answer in general terms.)

17.26 What is the significance of the CK(*MB*) band in trying to find out if a person has had a heart attack and not just some painful injury in the chest region?

17.27 What is the LD flip, and how is it used in diagnosis?

17.28 What happens, chemically, in a positive Clinistix® test for glucose? (Give the answer in terms of the steps in the series of changes that can be described by words, not equations.)

17.29 A clot consists of what protein and how does it occur in blood before a clot forms?

17.30 What causes the conversion of plasminogen to plasmin? What can be done clinically to make this happen in order to reduce the risk of a heart attack?

17.31 How do calcium channel blockers reduce the risk of a heart attack?

Hormones and Neurotransmitters

17.32 In what general ways do hormones and neurotransmitters resemble each other?

17.33 What function does adenylate cyclase have in the work of at least some hormones?

17.34 How is cyclic AMP involved in the work of some hormones and neurotransmitters?

17.35 After cyclic AMP has caused the activation of an enzyme inside a cell, what happens to the cyclic AMP that stops its action until more is made?

17.36 What are the names of the sites of the synthesis of (a) hormones and (b) neurotransmitters?

17.37 What does the lock-and-key concept have to do with the work of hormones and neurotransmitters?

17.38 In general terms, name three ways by which hormones work, and give an example of a hormone for each.

17.39 What happens to acetylcholine after it has worked as a neurotransmitter? What is the name of the enzyme that catalyzes this change? In chemical terms, what specifically does a nerve gas poison do?

17.40 How does atropine counter nerve gas poisoning?

Neurotransmitters and Medicine

17.41 How does a local anesthetic such as procaine affect the functioning of acetylcholine as a neurotransmitter?

17.42 What, in general terms, are the monoamine oxidases, and in what way are they important?

17.43 What does iproniazid do chemically in the neuron-signaling that is carried out by norepinephrine?

17.44 In general terms, how do antidepressants such as amitriptyline or imipramine work?

17.45 Which neurotransmitter is also a hormone, and what is the significance of this dual character to the body?

17.46 The overactivity of which neurotransmitter is thought to be one biochemical problem in schizophrenia?

17.47 How do the schizophrenia-control drugs chlorpromazine and haloperidol work?

17.48 How can the amphetamines, when abused, give schizophrenia-like symptoms?

17.49 How does L-DOPA work in treating Parkinson's disease?

17.50 Which common neurotransmitter in the brain is a signal inhibitor? How do such tranquilizers as Valium and Librium affect it?

17.51 Why is enkaphalin called one of the body's own opiates? How does it appear to work?

Sulfa Drugs (Special Topic 17.1)

17.52 In what way does a sulfa drug molecule interfere with a bacterial enzyme?

Penicillin (Special Topic 17.2)

17.53 In what way does penicillin interfere with a bacterium?

Electrophoresis (Special Topic 17.3

17.54 What is the overall result of electrophoresis?

17.55 Describe in general terms how electrophoresis works.

Chapter 18
Extracellular Fluids of the Body

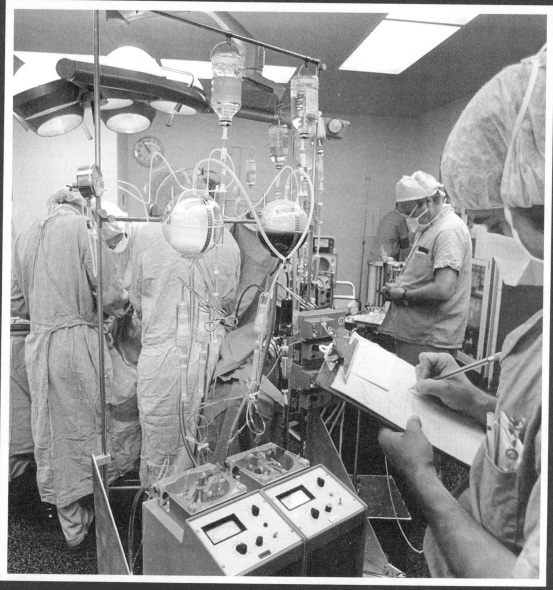

During a major operation, the acid/base balance of the blood must be carefully monitored. This balance is the chief topic of this chapter.

18.1   DIGESTIVE JUICES

The end products of the complete digestion of the nutritionally important carbohydrates, lipids, and proteins are monosaccharides, fatty acids, glycerol, and amino acids.

Life engages two environments, the outside environment we commonly think of, and the internal environment, which we usually take for granted. When healthy, our bodies have nearly perfect control over their internal environment, and so we are able to handle large changes outside fairly well; for example, large temperature fluctuations, chilling winds, stifling humidity, and a fluctuating atmospheric tide of dust and pollutants.

The fluids of the internal environment make up about 20% of the mass of the body.

Synovial fluids are the viscous lubricants of joints.

Cells Exist in Contact with Interstitial Fluids and Blood.   The **internal environment** consists of all the **extracellular fluids** or those that aren't actually inside cells. About three-quarters consists of the **interstitial fluid,** the fluid in the spaces or interstices *between* cells. The blood makes up nearly all the rest. The lymph, cerebrospinal fluid, digestive juices, and synovial fluids are also extracellular fluids.

The chemistry occurring inside cells, in the *intracellular fluid,* has been and will continue to be a major topic of our study. Here we will focus on two of the extracellular fluids, the digestive juices and the blood.

The Digestive Tract Is a Convoluted Tube Running through the Body with Access to Several Solutions of Hydrolytic Enzymes.   The principal parts of the digestive tract are given in Figure 18.1. The **digestive juices** are dilute solutions of electrolytes and hydrolytic enzymes (or their zymogens) either in the cells lining the intestinal tract or in solutions that enter the tract from various organs.

Saliva Provides α-Amylase, a Starch-Splitting Enzyme.   The flow of **saliva** is stimulated by the sight, smell, taste, and even the thought of food. Besides water (99.5%), saliva includes a food lubricant called **mucin** (a glycoprotein) and an enzyme, **α-amylase**. This enzyme catalyzes the partial hydrolysis of starch to dextrins and maltose, and it works best at the pH of saliva, 5.8–7.1. Proteins and lipids pass through the mouth essentially unchanged.

The dextrins are the partial breakdown products of starch.

Gastric Juice Starts the Digestion of Proteins with Pepsin.   When food arrives in the stomach, the cells of the gastric glands are stimulated by hormones to release the fluids that together make up **gastric juice.** One kind of gastric gland secretes mucin, which coats the stomach to protect it against its own digestive enzymes and its acid. Mucin is continuously produced and only slowly digested. If for any reason its protection of the stomach is hindered, part of the stomach itself could be digested, and this would lead to an ulcer.

The pH of gastric juice is normally in the range 0.9–2.0.

An enzyme that catalyzes the digestion of proteins is called a protease.

Another gastric gland secretes hydrochloric acid at a concentration of roughly 0.1 mol/L, about a million times more acidic than blood. The acid coagulates proteins and activates a protease. Protein coagulation retains the protein in the stomach longer for exposure to the protease.

Another gastric gland secretes the zymogen, **pepsinogen.** Pepsinogen is changed into **pepsin,** a protease, by the action of hydrochloric acid and traces of pepsin. The optimum pH of pepsin is in the range of 1–1.5, which is found in the stomach fluid. Pepsin catalyzes the only important digestive work in the stomach, the hydrolysis of some of the peptide bonds of proteins to make shorter polypeptides.

Adult gastric juice also has a lipase, but it does not start its work until it arrives in the higher pH medium of the upper intestinal tract.

The gastric juice of infants is less acidic than the adult's. To compensate for the protein-coagulating work normally done by the acid, infant gastric juice contains rennin, a powerful protein coagulator. Because the pH of an infant's gastric juice is higher than that of the adult, its lipase gets an early start on lipid digestion.

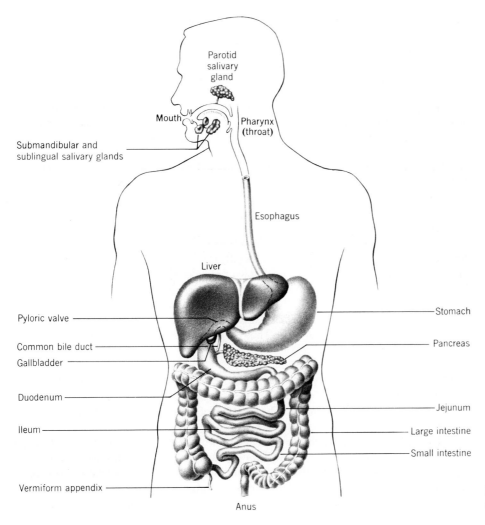

Parotid
salivary
gland

Mouth

Pharynx
(throat)

Submandibular and
sublingual salivary glands

Esophagus

Liver

Pyloric valve

Common bile duct

Gallbladder

Duodenum

Ileum

Vermiform appendix

Anus

Stomach

Pancreas

Jejunum

Large intestine

Small intestine

FIGURE 18.1
Organs of the digestive tract.

The churning and digesting activities in the stomach produce a liquid mixture called **chyme.** This is released in portions through the pyloric valve into the duodenum, the first 12 in. of the upper intestinal tract.

Pancreatic Juice Furnishes Several Zymogens and Enzymes. As soon as chyme appears in the duodenum, hormones are released that circulate to the pancreas and induce this organ to release two juices. One is almost entirely dilute sodium bicarbonate, which neutralizes the acid in chyme. The other is the one usually called **pancreatic juice.** It carries enzymes or zymogens that become involved in the digestion of practically everything in food. It contributes an **α-amylase** similar to that present in saliva, a **lipase, nucleases,** and zymogens for protein-digesting enzymes.

The nucleases include ribonuclease (RNAase) and deoxyribonuclease (DNAase).

The conversion of the proteolytic zymogens to active enzymes begins with a "master switch" enzyme called **enteropeptidase.** It is released from cells that line the duodenum when chyme arrives, and it then catalyzes the formation of trypsin from its zymogen, trypsinogen.

$$\text{Trypsinogen} \xrightarrow{\text{enteropeptidase}} \text{trypsin}$$

Enteropeptidase used to be called enterokinase.

Trypsin then catalyzes the change of the other zymogens into their active enzymes.

These proteases must exist as zymogens first or they will catalyze the self-digestion of the pancreas, which does happen in acute pancreatitis.

$$\text{Procarboxypeptidase} \xrightarrow{\text{trypsin}} \text{carboxypeptidase}$$
$$\text{Chymotrypsinogen} \xrightarrow{\text{trypsin}} \text{chymotrypsin}$$
$$\text{Proelastase} \xrightarrow{\text{trypsin}} \text{elastase}$$

Trypsin, chymotrypsin, and elastase catalyze the hydrolysis of large polypeptides to smaller ones. **Carboxypeptidase,** working in from C-terminal ends of small polypeptides, carries the action further to amino acids and di- or tripeptides.

Bile Salts Are Powerful Surfactants Necessary to Manage Dietary Lipids and Fat-Soluble Vitamins.   In order to digest most lipids, the lipase in pancreatic juice needs the help of the powerful detergents in bile, called the **bile salts.** These help to emulsify water-insoluble fatty materials and so greatly increase the exposure of lipids to water and lipase. Triacylglycerols are hydrolyzed to fatty acids, glycerol, and some monoacylglycerols.

The structure of a typical bile salt was given in Table 15.3, page 313.

Bile is a juice that enters the duodenum from the gallbladder. Its secretion is stimulated by a hormone released when chyme contains fatty material. Bile is also an avenue of excretion because it carries cholesterol and breakdown products of hemoglobin. These and further breakdown products constitute the bile pigments, which give color to feces.

The bile salts also assist in the absorption of the fat-soluble vitamins (A, D, E, and K) from the digestive tract into the blood. This work reabsorbs some bile pigments, some of which eventually leave the body via the urine. Thus the bile pigments are responsible for the color of both feces and urine.

Cells of the Intestines Carry Several Digestive Enzymes.   The term **intestinal juice** embraces not only a secretion but also the enzyme-rich fluids found inside certain kinds of cells that line the duodenum and jejunum. The secretion of some of these cells delivers an amylase and enteropeptidase, which we just described.

These intestinal cells last only about 2 days before they self-digest. They are constantly being replaced.

The other enzymes supplied by this region work within their cells on digestible compounds already being absorbed. An **aminopeptidase,** working inward from the N-terminal ends of small polypeptides, digests them to amino acids. The enzymes **sucrase, lactase,** and **maltase,** handle the digestion of disaccharides: sucrose to fructose and glucose; lactose to galactose and glucose; and maltose to glucose. An intestinal lipase and enzymes for the hydrolysis of nucleic acids are also present.

As fatty acids, glycerol, and monoacylglycerols migrate through the cells of the duodenal lining, much is reconstituted into triacylglycerols, which are taken up by the lymph system rather than the blood.

Some Vitamins and Essential Amino Acids Are Made in the Large Intestine.   No digestive functions are performed in the large intestine. Microorganisms in residence there, however, make vitamins K and B, plus some essential amino acids. These are absorbed by the body, but their contributions to overall nutrition in humans is not large.

Water and sodium chloride are reabsorbed from the large intestine, and undigested matter (including fiber), and some water make up the feces.

18.2   BLOOD AND THE ABSORPTION OF NUTRIENTS BY CELLS

The balance between the blood's pumping pressure and its colloidal osmotic pressure tips at capillary loops.

The nervous system with its neurotransmitters is the other line of communication.

The circulatory system, Figure 18.2, is one of our two main lines of chemical communication between the external and internal environments. All of the veins and arteries together are called the *vascular compartment.* The *cardiovascular compartment* includes this plus the heart.

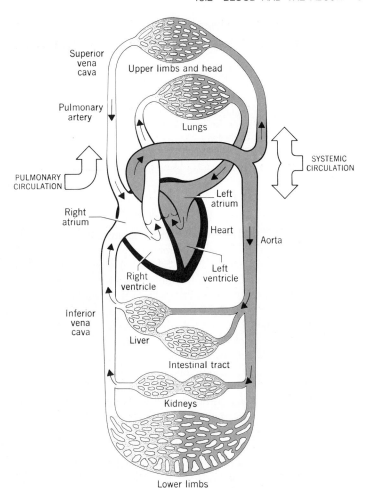

FIGURE 18.2
Human circulatory system. When oxygen-depleted venous blood (light areas) returns to the heart, it is pumped into the capillary beds of the alveoli in the lungs to reload oxygen and get rid of carbon dioxide. Then the freshly oxygenated blood (dark areas) is distributed by the arteries throughout the body, including the heart muscle.

The Blood Moves Nutrients, Oxygen, Messengers, Wastes, and Disease Fighters throughout the Body. Blood in the pulmonary branches moves through the lungs where waste carbon dioxide in the blood is exchanged for oxygen from freshly inhaled air. The oxygenated blood then moves to the rest of the system via the systemic branches.

At the intestinal tract, the blood picks up the products of digestion. Most of these are immediately monitored at the liver, and many alien chemicals are modified so they can be eliminated. In the kidneys, the blood is purified of nitrogen wastes, particularly urea, and it replenishes some of its buffer supplies. The pH of blood and its electrolyte balance depend hugely on the chemical work of the kidneys.

At endocrine glands, the blood picks up and circulates hormones whose secretions are often in response to something present in the blood.

White cells in blood give protection against bacteria; red cells or *erythrocytes* carry oxygen and a waste product, bicarbonate ion; and the platelets in blood are needed for blood clotting and other purposes. The blood also carries several zymogens needed for the blood clotting mechanism.

About 8% of the body's mass is blood. In the adult, the blood volume is 5–6 L.

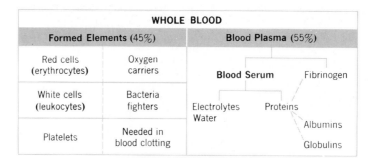

FIGURE 18.3
Major components of blood.

The Proteins in Blood Are Vital to Its Colloidal Osmotic Pressure. The principal types of substances in whole blood are summarized in Figure 18.3. One kilogram of blood plasma contains 80 g of proteins: albumins (54–58%), globulins (40–44%), and fibrinogen (3–5%). *Albumins* help carry hydrophobic molecules such as fatty acids, other lipids, and steroid hormones, and they contribute 75–80% of the osmotic effect of the blood. Some globulins carry ions (e.g., Fe^{2+} and Cu^{2+}) that otherwise could not be soluble in a fluid with a pH slightly greater than 7. The *γ-globulins* help to protect the body against infectious disease. *Fibrinogen* is converted to an insoluble form, *fibrin,* when a blood clot forms.

About one-quarter of the plasma proteins are replaced each day.

Figure 18.4 shows the quantities of various components of the major body fluids. The **electrolytes** of blood, the inorganic ions, dominate. The sodium ion is the chief cation in both blood and interstitial fluid, and the potassium ion is the major cation inside cells. A sodium–potassium pump, a special protein complex that uses energy, maintains these gradients. Both ions are needed to maintain osmotic pressure relationships, and both are a part of the regulatory system for acid–base balance.

The protein content of blood plasma is a major difference between this fluid and interstitial fluid. This is the principal reason why blood has a higher osmotic pressure than interstitial fluid.[1] The *total* osmotic pressure of blood is caused by all of the dissolved and colloidally dispersed solutes: electrolytes, organic compounds and ions, and proteins. However, the small ions and molecules can dialyze back and forth between the blood and the interstitial compartment. The large protein molecules can't do this, so it is their presence that gives to blood the higher effective osmotic pressure. This contribution to the blood's osmotic pressure made by colloidally dispersed substances is called the **colloidal osmotic pressure** of blood.

The osmolarity of plasma is about 290 mOsm/L.

As a consequence of the higher osmotic pressure of blood, water tends to flow into the blood from the interstitial compartment. Of course, this can't be allowed to happen everywhere and continually or the interstitial spaces and then the cells would eventually become too dehydrated to maintain life.

Fluids that Leave the Blood Must Return in Equal Volume. The blood vessels undergo extensive branching until the narrowest tubes called the capillaries are reached. Blood enters a capillary loop (Figure 18.5) as arterial blood, but it leaves on the other side of the loop as venous blood. During the switch, fluids and nutrients leave the blood and move into the interstitial fluids and then into the tissue cells themselves. *In the same volume* the fluids must return to the blood, but now they must carry the wastes of metabolism.

The lymph system makes antibodies and it has white cells that help defend the body against infectious diseases.

The rate of this diffusion of fluids throughout the body is sizable, about 25–30 liters/per second. Some fluids return to circulation by way of the lymph ducts, which are thin-walled, closed-end capillaries that bed in soft tissue.

---

[1] As a reminder and a useful memory aid, high solute concentration means high osmotic pressure; and solvent flows in osmosis or dialysis from a region where the solute is dilute to a region where it is concentrated. The "goal" of this flow is to even out the concentrations everywhere.

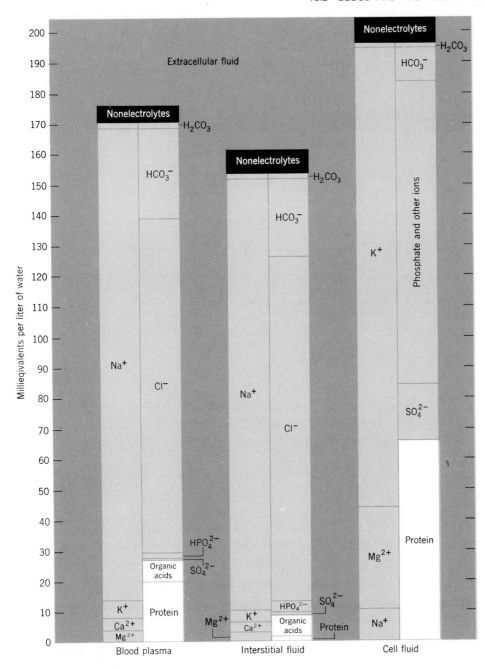

FIGURE 18.4
Electrolyte composition of body fluids. (Adapted by permission from J. L. Gamble, *Chemical Anatomy, Physiology and Pathology of Extracellular Fluids,* 6th ed. Harvard University Press, Cambridge, MA, 1954.)

Blood Pressure Overcomes Osmotic Pressure on the Arterial Side of a Capillary Loop. On the arterial side of a capillary loop, the blood pressure is high enough to overcome the natural tendency of fluids to move *into* the blood. Water and dissolved solutes are instead forced out of the blood and into the surrounding tissue where exchanges of chemicals occur.

Osmotic Pressure Overcomes Blood Pressure on the Venous Side of a Capillary Loop. As blood emerges from the thin constriction of a capillary into the venous side, its pressure drops. Now it is too low to prevent the natural diffusion of fluids back into the

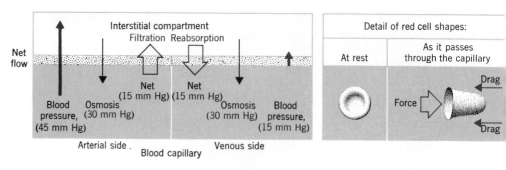

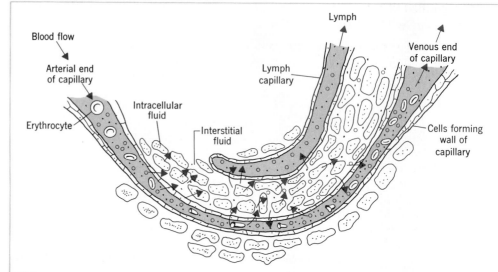

FIGURE 18.5
The exchange of nutrients and wastes at capillaries. As indicated at the top, on the arterial side of a capillary loop the blood pressure counteracts the pressure from dialysis and osmosis, and fluids are forced to leave the bloodstream. On the venous side of the loop, the blood pressure has decreased below that of dialysis and osmosis, so fluids flow back into the bloodstream. On the top right and the bottom is shown how a normal red cell distorts as it squeezes through a capillary loop. Red cells in sickle-cell anemia do not pass through as smoothly. The bottom drawing also shows how some fluids enter the lymph system.

bloodstream. By this time, of course, the fluids are carrying waste products. These relationships are illustrated in Figure 18.5, which shows how the colloidal osmotic pressure contributed by the macromolecules in blood, particularly the albumins, make the difference in determining the direction of diffusion.

Blood Loses Albumins in the Shock Syndrome. When the capillaries become more permeable to blood proteins, as they do in such trauma as sudden severe injuries, major surgery, and extensive burns, the proteins migrate out of the blood. Unfortunately, this protein loss also means the loss of the colloidal osmotic pressure that helps fluids to return from the tissue areas to the bloodstream. As a result, the total volume of circulating blood drops quickly, and this drastically reduces the blood's ability to carry oxygen and to remove carbon dioxide. The drop in blood volume and the resulting loss of oxygen supply to the brain sends the victim into *traumatic shock*.

The prompt restoration of blood volume is mandatory in the treatment of shock.

Blood Also Loses Proteins in Kidney Disease and Starvation. Sometimes the proteins in the blood are lost at malfunctioning kidneys. The effect, although gradual, is a slow but unremitting drop in the blood's colloidal osmotic pressure. Fluids accumulate in the interstitial regions. Because this takes place more slowly and water continues to be ingested, there is no sudden drop in blood volume as in shock. The victim appears puffy and waterlogged, a condition called *edema*.

Greek, *oidema,* swelling.

Edema can also appear at one stage of starvation, when the body has metabolized its circulating proteins to make up for the absence of dietary proteins.

Any obstruction in the veins can also cause edema, as in varicose veins and certain forms of cancer. Now it is the venous blood pressure that rises, creating a back pressure that reduces the rate at which fluids can return to circulation from the tissue areas. The localized swelling that results from a blow is a temporary form of edema caused by injuries to the capillaries.

18.3 THE CHEMISTRY OF THE EXCHANGE OF RESPIRATORY GASES

The binding of oxygen to hemoglobin is cooperative, and it is affected by the pH, PCO_2, and the PO_2 of the blood.

The carrier of oxygen in blood is **hemoglobin,** a complex protein found inside red blood cells (erythrocytes). It consists of four subunits, each with one molecule of heme, the actual oxygen-binding unit. When hemoglobin is oxygen-free, it is sometimes called *deoxyhemoglobin,* and when it carries oxygen it is *oxyhemoglobin.*

Each red cell carries about 2.8×10^8 molecules of hemoglobin.

The First Oxygen Molecule to Bind to Hemoglobin Activates the Binding of Three More Oxygen Molecules. A fully oxygenated hemoglobin molecule carries four oxygen molecules. For maximum efficiency in moving oxygen out of the lungs, all hemoglobin molecules should leave fully loaded. None should leave the lungs partially oxygenated.

What ensures full oxygenation are changes in the shapes of the hemoglobin subunits caused by the binding of the first oxygen molecule. This is very similar to the way a substrate can activate an enzyme with more than one catalytic site (Section 17.3). The first molecule to force its way in induces configurational changes to make all other sites far more receptive. Such changes to hemoglobin's remaining three subunits, caused by the binding of oxygen to the first subunit, make it easier for the next three oxygen molecules to bind. They flood into the partially oxygenated molecule more easily than into an empty molecule. Thus the oxygenation is *cooperative.* Structural changes permitted by hemoglobin cooperate with its own oxygenation.

Carbon monoxide binds 150–200 times more strongly to hemoglobin than oxygen and thus prevents the oxygenation of hemoglobin and causes internal suffocation.

The Relatively High PO_2 in the Lungs Aids the Oxygenation of Hemoglobin. The partial pressure of oxygen (PO_2) is higher in the lungs than anywhere else in the body, being 100 mm Hg in freshly inhaled air in alveoli and only about 40 mm Hg in oxygen-depleted tissues. Oxygen, therefore, naturally migrates from the lungs into the blood stream. It's as if the higher partial pressure *pushes* oxygen into the blood.

About 20% of a smoker's hemoglobin is more or less permanently tied up by carbon monoxide.

Deoxyhemoglobin Pulls Oxygen into the Lungs. To simplify the discussion, we will now represent deoxyhemoglobin as HHb, where the first H is a potential hydrogen ion and Hb is the rest of the molecule. We'll now also ignore the fact that one HHb unit can bind *four* oxygen molecules. With this in mind, we can represent the oxygenation of hemoglobin as the *forward* reaction in the following equilibrium, where oxyhemoglobin is represented as the anion, HbO_2^-

$$\text{HHb} + O_2 \rightleftharpoons HbO_2^- + H^+ \qquad (18.1)$$

Hemoglobin   Oxyhemoglobin

Thus in an oxygen-rich region, the lungs, the equilibrium shifts to the right, because this is the way to absorb the stress of a high PO_2.

The Oxygenation of Hemoglobin Generates Acid. Notice in Equilibrium 18.1 that HHb is evidently a weak acid *that becomes stronger as it oxygenates;* the product of the forward reaction emerges not as a molecule, but as two ions, HbO_2^- and H^+. *The presence of H^+ in Equilibrium 18.1 means that a change in pH shifts the equilibrium one way or another,* a fact of great importance at the molecular level of life, as we will see.

The Neutralization of the Acid Aids the Elimination of CO_2.

As soon as the H^+ ion forms, and while the red cell is still in the lungs, the H^+ ion is neutralized by HCO_3^- in the blood. We will show only the forward reaction of what is actually an equilibrium, because that is how the equilibrium shifts when both H^+ and HCO_3^- are at relatively high levels.

As we learned in the last chapter, carbonic anhydrase is the body's fastest-working enzyme, and it has to be fast because the red cells are always on the move.

$$H^+ + HCO_3^- \xrightarrow{\text{carbonic anhydrase}} CO_2 + H_2O \qquad (18.2)$$

In the red cell   In the red cell   In the red cell

This switch from the appearance of H^+ as a product to its disappearance as a reactant is called the **isohydric shift**. The bicarbonate ion that participates in this shift is the way that most waste CO_2 comes to the lungs. This waste is thus released for removal by exhaling.

It's an altogether beautiful and remarkable example of coordinated chemical activity at the molecular level of life. *The uptake of O_2 by hemoglobin simultaneously produces the H^+ ion needed for the formation and release of waste CO_2 from bicarbonate ion.* Let's now see how this waste is picked up at the cells that have produced it and how it is carried to the lungs; and let's also see how this works cooperatively with the *release* of oxygen at cells needing it.

Circulating Oxyhemoglobin Can Give Up Oxygen Only Where Cells Have Made Waste Carbon Dioxide.

When a tissue has done some chemical work and has used up some oxygen, it has also made some waste carbon dioxide. This waste launches the reversal of the events we described above.

Waste CO_2 diffuses from the tissue into the blood partly because of the higher PCO_2 (50 mm Hg) in active tissue versus its value in blood (40 mm Hg). Once in the blood, the CO_2 migrates inside a red cell where it encounters carbonic anhydrase. Equation 18.2 therefore promptly reverses.

CO_2 molecules diffuse in body fluids 30 times more easily than O_2 molecules, so the partial pressure gradient for CO_2 need not be as steep as that for O_2.

$$H_2O + CO_2 \xrightarrow{\text{carbonic anhydrase}} HCO_3^- + H^+ \qquad (18.2—\text{reversed})$$

From the working tissue   In the red cell at working tissue

Thus the uptake of CO_2 generates hydrogen ions inside the red cell, exactly what is needed to get oxygen released. If you'll look back to Equilibrium 18.1, you will see that an increase in the level of H^+ can only shift Equilibrium 18.1 backward:

$$H^+ + HbO_2^- \longrightarrow HHb + O_2 \qquad (18.1—\text{reverse})$$

Just made in the red cell at working tissue   In the red cell, just arrived   In the red cell   Will diffuse into the tissue

This reaction, another isohydric shift, not only neutralizes the acid generated by the waste CO_2, it also makes oxyhemoglobin give up its oxygen. Notice the cooperation. Working tissue needing oxygen has made CO_2 and, hence, it has indirectly made the H^+ required to release oxygen from newly arrived HbO_2^- and replenish its own oxygen supply.

No conditioning at a low altitude can get the cardiovascular system ready for a low PO_2 at a high altitude.

The changes in shapes of the hemoglobin subunits now operate in reverse, and all oxygen molecules smoothly leave. It's all or nothing again, and the efficiency of the unloading of oxygen is so high that if one O_2 molecule leaves, the other three follow essentially at once. Partially deoxygenated hemoglobin units do not slip through and go back to the lungs.

To summarize the chemical reactions we have just studied, we can write the following equations. The cancel lines show how we can arrive at the overall net results.

Oxygenation:

$$HHb + O_2 \longrightarrow HbO_2^- + \cancel{H^+}$$

In red cell   From air   Goes in red cell to tissues   In red cell

$$\text{H}^+ + HCO_3^- \xrightarrow{\text{CA}} CO_2 + H_2O$$

CA is carbonic anhydrase.

Just made	In red cell (but from tissues)	In red cell (but will be exhaled)		

$$HHb + O_2 + HCO_3^- \longrightarrow HbO_2^- + CO_2 + H_2O \qquad (18.3)$$

Net effect of oxygenating hemoglobin:

In red cell	From air	In red cell (but from tissues)	Goes in red cell to tissues	Leaves the lungs in exhaled air

Deoxygenation:

$$CO_2 + H_2O \xrightarrow{\text{CA}} HCO_3^- + \text{H}^+$$

Waste from tissues		In red cell but still by tissue	Will help release oxygen

$$\text{H}^+ + HbO_2^- \longrightarrow HHb + O_2$$

Just made	In red cell; by tissue	In red cell; will return to lungs	Goes into tissue needing it

$$CO_2 + H_2O + HbO_2^- \longrightarrow HHb + HCO_3^- + O_2 \qquad (18.4)$$

Net effect of deoxygenating oxyhemoglobin:

Waste from tissue	In red cell; by tissues	In red cell; will go to lungs	Goes in blood to lungs	Goes into tissue

Some CO_2 Is Carried to the Lungs on Hemoglobin. There is another chemical reaction involving waste carbon dioxide that we must now mention. Not all of the waste CO_2 winds up as HCO_3^-. Some reacts with the hemoglobin that has just been freed by deoxygenation:

$$CO_2 + HHb \longrightarrow HbCO_2^- + H^+ \qquad (18.5)$$

Carbaminohemoglobin

Waste from tissues	Just released by deoxygenation	In red cells	Will react with HbO_2^-

This is actually the forward reaction of an equilibrium. The product, $HbCO_2^-$, is called **carbaminohemoglobin,** and it is one form in which some of the waste CO_2 travels in the blood back to the lungs. (Most of this waste, however, travels as HCO_3^-.)

Notice in Equation 18.5 that this reaction of CO_2 also generates acid so the CO_2 that reacts this way also aids in the release of O_2 from oxyhemoglobin. The H^+ produced in Equation 18.5 takes its own isohydric shift and helps expel O_2 from HbO_2^- (Equation 18.1 — reverse). When the cell reaches the lungs where CO_2 can be exhaled and where H^+ is generated by oxygenation, Equation 18.5 runs in reverse to get rid of the CO_2 bound to $HbCO_2^-$.

Bicarbonate Ion Travels Largely Outside the Red Cell. As we said, most of the waste CO_2 is carried as HCO_3^-. As it forms inside red cells, it migrates into the serum. The cause of this migration is the migration of chloride ion into the red cell. The Cl^- ion can form a weak bond to hemoglobin, and this attracts it in. To keep everything electrically neutral, for every chloride ion that moves into the red cell a like-charged ion must leave, and HCO_3^- does. This switch is called the **chloride shift.**

At the lungs the shift reverses because HCO_3^- ions cannot help but be drawn back into the red cells. That is where carbonic anhydrase is, and CO_2 is now leaving the lungs. In accordance with Le Chatelier's principle, the equilibrium shifts as much as possible to replace it. This shift is Equation 18.2 in reverse, so the red cell needs HCO_3^- ion.

$$HCO_3^- \;+\; H^+ \xrightarrow{\text{carbonic anhydrase}} CO_2 \;+\; H_2O \qquad (18.2\text{—reverse})$$

Moving into red cell (with Cl^- moving out)

Leaves the lungs (drawing HCO_3^- in)

Myoglobin Binds Oxygen More Strongly Than Does Hemoglobin. Myoglobin is a heme-containing protein in red muscle tissue such as heart muscle. Its function is to bind and store oxygen for the needs of such tissue. Unlike hemoglobin, myoglobin has only one polypeptide unit and only one heme unit per molecule. But it binds oxygen more strongly than hemoglobin, so *myoglobin is able to take oxygen from oxyhemoglobin:*

HMb = myoglobin

$$HbO_2^- + HMb \longrightarrow HHb + MbO_2^-$$

This ability is vital to heart muscle which, as much as the brain, must have an assuredly continuous supply of oxygen. When oxymyoglobin, MbO_2^- gives up its oxygen for the cell's needs, it can at once get a fresh supply from the circulating blood. Not only does this cell now have CO_2 and H^+ available to deoxygenate HbO_2^-, it also has the superior oxygen affinity of its own myoglobin to draw more O_2 into the cell.

Fetal Hemoglobin Binds Oxygen Better Than Adult Hemoglobin. The hemoglobin in a fetus is slightly different from that of an adult, and it also binds oxygen more strongly than does adult hemoglobin. This ensures that the fetus can successfully pull oxygen from the mother's oxyhemoglobin and satisfy its own needs. Soon after birth, fetal hemoglobin changes to adult hemoglobin.

18.4   ACID–BASE BALANCE OF THE BLOOD

The proper treatment of acidosis or of alkalosis depends on knowing if the underlying cause is a metabolic or a respiratory disorder.

Acidosis is sometimes called *acidemia* and alkalosis is called *alkelemia.*

Acid–base balance exists when the pH of blood is in the range of 7.35–7.45. A decrease in pH, acidosis, or an increase, alkalosis, is serious and requires prompt attention, because all the equilibria that involve H^+ in the oxygenation or the deoxygenation of blood are sensitive to pH. If the pH falls below 6.8 or rises above 7.8, life is not possible.

Disturbances in Either Metabolism or Respiration Can Upset the Blood's Acid–Base Balance. In general, acidosis results from either the retention of acid or the loss of base by the body, and these can be induced by disturbances in either metabolism or respiration. Similarly, alkalosis results either from the loss of acid or from the retention of base, and some disorder in either metabolism or respiration can be the underlying cause.

A malfunction in respiration can be caused by any kind of injury to the *respiratory centers.* These are units in the brain that sense changes in the pH and PCO_2 of the blood and instruct the lungs to breathe either more rapidly or more slowly. Another cause of a malfunction of respiration is any kind of injury or disease of the lungs.

Normal values (arterial blood):

$PCO_2 = 35\text{–}45$ mm Hg
$[HCO_3^-] = 19\text{–}24$ meq/L
(1.0 meq HCO_3^- = 61 mg HCO_3^-)

What we'll do in this section is study four situations, metabolic and respiratory acidosis as well as metabolic and respiratory alkalosis. We will learn how the values of pH, PCO_2, and serum $[HCO_3^-]$ change in each situation. (Normal values are given in the margin.)

Metabolic Acidosis Receives a Respiratory Compensation, Hyperventilation. In **metabolic acidosis,** the lungs and the respiratory centers are working, and the problem is metabolic. Acids are being produced faster than they are neutralized, or they are being exported too slowly.

Excessive loss of base, such as from severe diarrhea, can also result in metabolic acidosis. (In diarrhea, the alkaline fluids of the duodenum leave the body, and as base migrates to replace them, there will be a depletion of base somewhere else, such as in the blood, at least for a period of time.)

As the pH of the blood falls and the molar concentration of H^+ rises, there are parallel *but momentary* increases in the values of PCO_2 and $[HCO_3^-]$. The value of PCO_2 starts to increase because the carbonate buffer, working hard to neutralize the extra H^+, manufactures CO_2.

$$H^+ \quad + HCO_3^- \longrightarrow H_2O + CO_2$$

From
acidosis

The kidneys work harder during this situation to try to keep up the supply of HCO_3^-.

The chief compensation for metabolic acidosis, however, involves the respiratory system. The respiratory centers, which are sensitive to changes in PCO_2, instruct the lungs to blow the CO_2 out of the body. The lungs, in other words, hyperventilate. As the equation given just above shows, the loss of each molecule of CO_2 means a net neutralization of one H^+ ion.

Hyperventilation, however, is overdone. So much CO_2 is blown out that PCO_2 actually decreases. Thus as the blood pH decreases, so too do the values of PCO_2 (from hyperventilation) and $[HCO_3^-]$ (from the reaction with H^+). We can summarize the range of a number of clinical situations that involve metabolic acidosis as follows:

Clinical Situations of Metabolic Acidosis

Lab results:	pH↓ (7.20); PCO_2↓ (30 mm Hg); $[HCO_3^-]$↓↓ (12 meq/L)
Typical patient:	An adult male comes to the clinic with a severe infection. He does not know that he has diabetes.
Range of causes:	Diabetes mellitus; severe diarrhea (with loss of HCO_3^-); kidney failure (to export H^+ or to make HCO_3^-); prolonged starvation; severe infection; aspirin overdose, alcohol poisoning.
Symptoms:	Hyperventilation (because the respiratory centers have told the lungs to remove excess CO_2 from the blood); increased urine output (to remove H^+ from the blood); thirst (to replace water lost as urine); drowsiness; headache; restlessness; disorientation.
Treatment:	If the kidneys function, use isotonic HCO_3^- intravenously to restore HCO_3^- level, thereby neutralizing H^+ and raising PCO_2. In addition, restore water. In diabetes, use insulin therapy. If the kidneys do not function, hemodialysis must be tried.

(↓) means a decrease from a normal and (↑) means an increase. Some typical values are in parentheses. Note that some changes do not necessarily bring values outside the normal ranges.

Respiratory Acidosis Is Compensated by a Metabolic Response. In **respiratory acidosis,** either the respiratory centers or the lungs have failed, and the lungs are hypoventilating *because they cannot help it.* The blood now cannot help but retain CO_2. But the retention of CO_2 means the retention of H^+, because the carbonic anhydrase-catalyzed equilibrium favors H^+ when CO_2 is high (Le Chatelier's principle).

$$CO_2 + H_2O \xrightleftharpoons{\text{carbonic anhydrase}} H^+ + HCO_3^-$$

Clinical Situations of Respiratory Acidosis

Lab results:	pH↓ (7.21); PCO_2↑ (70 mm Hg); $[HCO_3^-]$↑ (27 meq/L)
Typical patient:	Chain smoker with emphysema or anyone with chronic obstructive pulmonary disease.
Range of causes:	Emphysema, severe pneumonia, asthma, anterior poliomyelitis, or any cause of shallow breathing such as an overdose of narcotics, barbiturates, or general anesthesia; severe head injury.

Symptoms:	Shallow breathing (which is involuntary).
Treatment:	Underlying problem must be treated; possibly intravenous sodium bicarbonate; possibly hemodialysis.

The body responds metabolically as best it can to respiratory acidosis by making more HCO_3^- in the kidneys to neutralize the acid and by exporting H^+ via the urine.

Metabolic Alkalosis Also Receives a Respiratory Compensation, Hypoventilation.

In **metabolic alkalosis,** the system has lost acid; or it has retained base (HCO_3^-), or it has been given an overdose of base (e.g., antacids). Metabolic alkalosis can also be caused by a kidney-associated decrease in the serum levels of K^+ or Cl^-. The loss of these ions means the retention of Na^+ and HCO_3^- ions, because these work in tandem and oppositely. The loss of acid could be from prolonged vomiting, which removes the gastric acid. This is followed by an effort to borrow serum H^+ to replace it, and the pH of the blood increases. Improperly operated nasogastric suction can also remove too much gastric acid.

Whatever the cause, the respiratory centers sense an increase in the level of base in the blood (as the level of acid drops), and they instruct the lungs to retain the most readily available neutralizer of base it has, namely, CO_2, which removes OH^- as follows:

$$CO_2 + OH^- \longrightarrow HCO_3^-$$

To help retain CO_2 so that it can neutralize base, the lungs hypoventilate. Thus metabolic alkalosis leads to hypoventilation.

Notice carefully that hypoventilation alone cannot be used to tell whether the patient has metabolic *alkalosis* or respiratory *acidosis*. Either condition means hypoventilation. But one condition, respiratory acidosis, could be treated by intravenous sodium bicarbonate, a base. This would aggravate metabolic alkalosis.

You can see that the lab data on pH, PCO_2, and $[HCO_3^-]$ must be obtained to determine which kind of condition is actually present. Otherwise, the treatment used could be just the opposite of what should be done. People working in emergency care situations get the requisite lab data rapidly, and they must be able to interpret the data on the spot.

Clinical Situations of Metabolic Alkalosis

Lab results:	pH↑ (7.53); PCO_2↑ (56 mm Hg); $[HCO_3^-]$↑ (45 meq/L)
Typical patient:	Postsurgery patient with persistent vomiting.
Range of causes:	Prolonged loss of stomach contents (vomiting or nasogastric suction); overdose of bicarbonate or of medications for stomach ulcers; severe exercise, or stress, or kidney disease (with loss of K^+ and Cl^-); overuse of a diuretic.
Symptoms:	Hypoventilation (to retain CO_2); numbness, headache, tingling; possibly convulsions.
Treatment:	Isotonic ammonium chloride (a mild acid), intraveneously with great care. Replace K^+ loss.

Respiratory Alkalosis Is Compensated Metabolically by a Reduced Bicarbonate Level.

In **respiratory alkalosis,** the body has lost acid usually by some involuntary hyperventilation—hysterics, prolonged crying, or overbreathing at high altitudes—or by the mismanagement of a respirator. The respiratory centers have lost control, and the body expels CO_2 too rapidly. The loss of CO_2 means the loss of a base neutralizer from the blood.

An overdose of "bicarb" ($NaHCO_3$) can result from a too aggressive use of this home remedy for "heartburn."

Compensation by hypoventilation is obviously limited by the fundamental need of the body for some oxygen.

Ammonium ion acts as a neutralizer as follows:

$$NH_4^+ + OH^- \longrightarrow NH_3 + H_2O$$

Hence, the level of base rises; the pH rises. To compensate, the kidneys excrete base, HCO_3^-, so the serum level of HCO_3^- falls.

Extreme respiratory alkalosis can occur to mountain climbers, like climbers of Mount Everest (8848 m, 29,030 ft). At its summit, the barometric pressure is 253 mm Hg and the PO_2 of the air is only 43 mm Hg (as compared to 149 mm Hg at sea level). Hyperventilation brings their arterial PCO_2 down to only 7.5 mm Hg (compared to a normal of 40 mm Hg) and the blood pH is above 7.7!

Tissue that gets too little O_2 is in a state of **hypoxia**. If it gets none at all, it is in a state of **anoxia**.

Clinical Situations of Respiratory Alkalosis

Lab results:	pH↑ (7.56); PCO_2↓ (23 mm Hg); $[HCO_3^-]$↓ (20 meq/L)
Typical patient:	Someone nearing surgery and experiencing anxiety.
Range of causes:	Prolonged crying; rapid breathing at high altitudes; hysterics; fever; disease of the central nervous system; improper management of a respirator.
Symptoms:	Hyperventilation (that can't be helped). Convulsions may occur.
Treatment:	Rebreathe one's own exhaled air (by breathing into a sack, like Lucy in the "Peanuts" cartoon); administer carbon dioxide; treat underlying causes.

Take careful notice that hyperventilation alone cannot be used to tell what the condition is. Either metabolic *acidosis* or respiratory *alkalosis* is accompanied by hyperventilation, but the treatments are opposite in nature.

Combinations of Primary Acid–Base Disorders Are Possible. We have just surveyed the four *primary acid–base disorders*. Combinations of these are often seen, and health care professionals have to be alert to the ways in which the lab data vary in such combinations. Someone with diabetes, for example, might also suffer from an obstructive pulmonary disease. Diabetes causes metabolic acidosis and a *decrease* in $[HCO_3^-]$. The pulmonary disease causes respiratory acidosis with an *increase* in $[HCO_3^-]$. In combination, then, the lab data on bicarbonate level will not be in the expected pattern for either. We will not carry the study of such complications further. We mention them only to let you know that they exist. There are standard ways to recognize them.[2]

[2] See, for example, H. Valtin and F. J. Gennari, *Acid–Base Disorders, Basic Concepts and Clinical Management*, 1987. Little, Brown and Company, Boston.

18.5   ACID–BASE BALANCE AND SOME CHEMISTRY OF KIDNEY FUNCTION

Both filtration and chemical reactions in the kidneys help to regulate the acid–base balance of the blood.

Diuresis is the formation of urine in the kidneys, and it is an integral part of the body's control of its levels of electrolytes and buffers in blood.

The net urine production is 0.6–2.5 L/day.

Urea Is the Chief Nitrogen Waste Exported in the Urine.   Huge quantities of fluids diffuse from the blood each day in the kidneys. Solutes, but not colloidal particles (e.g., protein molecules), also leave the blood. Then active transport processes in kidney cells pull virtually all of the nonwaste solutes back into the blood. Most of the wastes are left in the urine that is being made. Control over these active transport processes by the kidneys means that the kidneys can adjust the concentrations of several solutes in the blood, including buffers and other electrolytes.

Urea is the chief nitrogen waste (30 g/day), but creatinine (1–2 g/day), uric acid (0.7 g/day), and ammonia (0.5 g/day) are also excreted with the urine. In acidosis, the kidneys export acid, and in alkalosis they put base into the urine. If the kidneys are injured or diseased and cannot function, wastes build up in the blood, which leads to a condition known as *uremic poisoning.*

Ur-, of the urine; *-emia,* of the blood. *Uremia* means substances of the urine present in the blood.

The Hormone Vasopressin Helps Control Water Loss.   A nonapeptide hormone, **vasopressin,** instructs the kidneys to retain or to excrete water and thus helps to regulate the overall concentrations of substances in blood. The hypophysis, where vasopressin is made, releases it when the osmotic pressure of blood rises by as little as 2%. At the kidneys, vasopressin promotes the reabsorption of water, and therefore it is often called the antidiuretic hormone (ADH).

In *diabetes insipidus,* vasopressin secretion is blocked and unchecked diuresis can make from 5 to 12 L of urine a day.

A higher-than-normal osmotic pressure (hypertonicity) means a higher concentration of solutes and colloids in blood. The released vasopressin therefore helps the blood to retain water and thus keep the blood from becoming even more concentrated. In the meantime, the thirst mechanism is stimulated to bring in water to dilute the blood.

Conversely, if the osmotic pressure of blood decreases (becomes hypotonic) by as little as 2%, the hypophysis retains vasopressin. None reaches the kidneys, so the water that has diffused from the bloodstream does not return as much. Remember that a low osmotic pressure means a low concentration of solutes, so the absence of vasopressin at the kidneys when the blood is hypotonic lets urine form. This reduces the amount of water in the blood and thereby raises the concentrations of its dissolved matter. You can see that with the help of vasopressin a normal individual can vary the intake of water widely and yet preserve a stable, overall concentration of substances in blood.

The Hormone Aldosterone Helps the Blood Retain Sodium Ion.   The adrenal cortex makes **aldosterone,** a hormone that works to stabilize the sodium ion level of the blood. This steroid hormone is secreted if the blood's sodium ion level decreases. When aldosterone arrives at the kidneys, it initiates reactions that return sodium ions from the urine being made to the blood. Of course, to keep things isotonic in the blood, the return of sodium ions also requires the return of water.

Conversely, if the sodium ion level of the blood increases, then aldosterone is not secreted, and sodium ions that have diffused from the blood are permitted to stay in the urine being made. This takes more water, too, so more urine forms.

Urine taken after several hours of fasting normally has a pH of 5.5–6.5.

The Kidneys Make HCO_3^- for the Blood's Buffer System.   We have seen that breathing is the body's most direct means of controlling acid as it removes or retains CO_2. The kidneys are the body's means of controlling base, as they make or remove HCO_3^-.

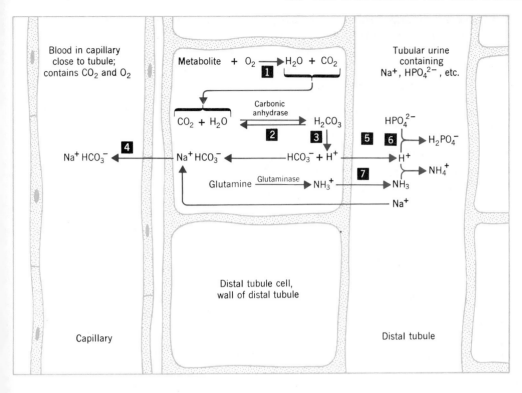

FIGURE 18.6
Acidification of the urine. The numbers refer to the text discussion.

The kidneys also adjust the blood's levels of HPO_4^{2-} and $H_2PO_4^-$, the anions of the phosphate buffer. Moreover, when acidosis develops, the kidneys can put H^+ ions into the urine. Some neutralization of these ions by HPO_4^{2-} and by NH_3 takes place, but the urine becomes definitely more acidic as acidosis continues, as we've mentioned before.

In severe acidosis, the pH of urine can go as low as 4.

Figure 18.6 shows the various reactions that take place in the kidneys, particularly during acidosis. (The numbers in the following boxes refer to this figure.) The breakdown of metabolites $\boxed{1}$ makes carbon dioxide, which enters the equilibrium whose formation is catalyzed by carbonic anhydrase $\boxed{2}$. The ionization of carbonic acid $\boxed{3}$ makes both bicarbonate ion and hydrogen ion. The bicarbonate ion goes into the bloodstream $\boxed{4}$ but the hydrogen ion is put into the tubule, $\boxed{5}$ where urine is accumulating. This urine already contains sodium ions and monohydrogen phosphate ions, but to make step $\boxed{5}$ possible, *some positive ion has to go with the HCO_3^-*. Otherwise, there would be no net electrical balance. The kidneys have the ability to select Na^+ to go with HCO_3^- at $\boxed{4}$. The kidneys can make Na^+ travel one way and H^+ the other. Newly arrived H^+ can be buffered by HPO_4^{2-} in the developing urine $\boxed{6}$. Moreover, the kidneys have an ability not generally found in other tissues to synthesize ammonia and use it to neutralize H^+ $\boxed{7}$. Thus the ammonium ion also appears in the urine.

The Kidneys Excrete Organic Anions. When acidosis has a metabolic origin (e.g., diabetes), the level of organic anions in the blood increases. These are the negative ions of the organic acids whose accelerated production causes the acidosis. The kidneys let these anions stay in the urine, but there is a limit to how concentrated the urine can become in total solutes. Hence, the more the kidneys let solutes stay in the urine, the more must they also let water stay to keep the urine dilute. As a result, the patient with metabolic acidosis can experience a general dehydration as the system borrows water from other fluids to make urine. Usually, the thirst mechanism brings in replacement water, and the patient drinks copious amounts.

The Kidneys Can Export HCO$_3^-$.   In alkalosis, the kidneys can put bicarbonate ion into the urine, and it no longer uses HPO$_4^{2-}$ to neutralize H$^+$. Both actions raise the pH of the urine, and in severe alkalosis it can go over pH 8.

SUMMARY

Digestion α-Amylase in saliva begins the digestion of starch. Pepsin in gastric juice starts the digestion of proteins. In the duodenum, trypsinogen (from the pancreas) is activated by enteropeptidase (from the intestinal juice) and becomes trypsin, which helps to digest proteins. It also activates chymotrypsin (from chymotrypsinogen), carboxypeptidase (from procarboxypeptidase), and elastin (from proelastin). These also help to digest proteins. The pancreas supplies an important lipase, which, with the help of the bile salts, catalyzes the digestion of saponifiable lipids. The bile salts also aid in the absorption of the fat soluble vitamins, A, D, E, and K.

Intestinal juice supplies enzymes for the digestion of disaccharides, nucleic acids, small polypeptides, and lipids.

The end products of the digestion of proteins are amino acids; of carbohydrates, glucose, fructose, and galactose; and of the triacylglycerols, fatty acids and glycerol. Complex lipids are also hydrolyzed, and nucleic acids yield phosphate, pentoses, and heterocyclic amines.

Blood   Proteins in blood give it a colloidal osmotic pressure that assists in the exchange of nutrients at capillary loops. Albumins are carriers for hydrophobic molecules and serum-soluble metallic ions. γ-Globulins help defend the body against bacterial infections. Fibrinogen is the precursor of fibrin, the protein of a blood clot.

Among the electrolytes, anions of carbonic and phosphoric acid are involved in buffers, and all ions are involved in regulating the osmotic pressure of the blood. The chief cation in blood is Na$^+$, and the chief cation inside cells is K$^+$.

The blood transports oxygen and products of digestion to all tissues. It carries nitrogen wastes to the kidneys. It unloads cholesterol and heme breakdown products at the gallbladder. And it transports hormones to their target cells. Lymph, another fluid, helps to return some substances to the blood from tissues.

Sudden failure to retain the protein in blood leads to an equally sudden loss in blood volume and a condition of shock. Slower losses of protein, as in kidney disease or starvation, lead to edema.

Respiration   The relatively high PO_2 in the lungs helps to force O$_2$ into HHb. This creates HbO$_2^-$ and H$^+$. In an isohydric shift, the H$^+$ is neutralized by HCO$_3^-$, which is returning from working tissues that make CO$_2$, which leaves during exhaling. Some of the H$^+$ also converts HbCO$_2^-$ to HHb and CO$_2$.

In deoxygenating HbO$_2^-$ at cells that need oxygen, the influx of CO$_2$ makes HCO$_3^-$ and H$^+$. The H$^+$ then moves (isohydric shift) to HbO$_2^-$ and breaks it down to HHb and O$_2$. Both oxygenation and deoxygenation of blood is done with the cooperative flexibility of the shapes of the subunits of hemoglobin.

In red muscle tissue, myoglobin's superior ability to bind oxygen ensures that such tissue obtains oxygen from the deoxygenation of oxyhemoglobin. Fetal hemoglobin also has a superior oxygen-binding ability.

Acid–base balance   The body uses the bicarbonate ion of the carbonate buffer to inhibit acidosis by irreversibly removing H$^+$ when the lungs release CO$_2$. The HCO$_3^-$ ion is replaced by the kidneys, which can also put excess H$^+$ into the urine. The carbonic acid, H$_2$CO$_3$, in the carbonate buffer works to control alkalosis by neutralizing OH$^-$. Metabolic acidosis, with hyperventilation, and metabolic alkalosis, with hypoventilation, arise from dysfunctions in metabolism. Respiratory acidosis, with hypoventilation, and metabolic alkalosis, with hyperventilation, occur when the respiratory centers or the lungs are not working.

Diuresis   The kidneys, with the help of hormones and changes in blood pressure, blood osmotic pressure, and concentrations of ions, monitor and control the concentrations of solutes in blood. Vasopressin tells the kidneys to keep water in the bloodstream. Aldosterone tells the kidneys to keep sodium ion (and therefore water also) in the blood stream. In acidosis, the kidneys transfer H$^+$ to the urine and replace some of the HCO$_3^-$ lost from the blood. In alkalosis the kidneys put some HCO$_3^-$ into urine.

REVIEW EXERCISES

The answers to these review exercises are in the *Study Guide* which accompanies this book.

Digestion

18.1 What are the names of the two chief extracellular fluids?

18.2 Name the fluids that have digestive enzymes or digestive zymogens.

18.3 What enzymes or zymogens are there, if any, in each of the following?

   (a) Saliva   (b) Gastric juice

   (c) Pancreatic juice   (d) Bile

   (e) Intestinal juice

18.4 Name the enzymes and the digestive juices that supply them (or their zymogens) that catalyze the digestion of each of the following.

   (a) Large polypeptides   (b) Triacylglycerols

   (c) Amylose   (d) Sucrose

   (e) Di- and tripeptides   (f) Nucleic acids

18.5 What are the end products of the complete digestion of each of the following?

(a) Proteins (b) Carbohydrates
(c) Triacylglycerols

18.6 What functional groups are hydrolyzed when each of the substances in Review Exercise 18.5 is digested? (Refer back to earlier chapters if necessary.)

18.7 In what way does enteropeptidase function as a "master switch" in digestion?

18.8 What would happen if the pancreatic zymogens were activated within the pancreas?

18.9 What services do the bile salts render in digestion?

18.10 What does mucin do (a) for food in the mouth, and (b) for the stomach?

18.11 What is the catalyst for each of the following reactions?
(a) Pepsinogen → pepsin
(b) Trypsinogen → trypsin
(c) Chymotrypsinogen → chymotrypsin
(d) Procarboxypeptidase → carboxypeptidase
(e) Proelastase → elastase

18.12 Rennin does what for an infant?

18.13 Why is gastric lipase unimportant to digestive processes in the adult stomach but useful in the infant stomach?

18.14 In terms of where they work, what is different about intestinal juice compared to pancreatic juice?

18.15 What secretion neutralizes chyme, and why is this work important?

18.16 What happens to the molecules of glycerol and fatty acids that form from digestion?

18.17 In a patient with a severe obstruction of the bile duct the feces appear clay colored. Explain why the color is light.

18.18 When the gallbladder is surgically removed, lipids of low formula weight are the only kinds that can be easily digested. Explain.

Substances in Blood

18.19 In terms of their general composition, what is the greatest difference between blood plasma and interstitial fluid?

18.20 What is the largest contributor to the net osmotic pressure of the blood as compared to the interstitial fluid?

18.21 What is fibrinogen? Fibrin?

18.22 What services are performed by albumins in blood?

18.23 What does γ-globulin do?

18.24 In what two different regions are Na^+ and K^+ ions mostly found?

Exchange of Nutrients at Capillary Loops

18.25 What two opposing forces are at work on the arterial side of a capillary loop? What is the net result of these forces, and what does the net force do?

18.26 On the venous side of a capillary loop there are two opposing forces. What are they, what is the net result, and what does this cause?

18.27 Explain how a sudden change in the permeability of the capillaries can lead to shock.

18.28 Explain how each of the following conditions leads to edema.
(a) Kidney disease (b) Starvation
(c) A mechanical blow

Chemistry of the Exchange of Respiratory Gases

18.29 What are the respiratory gases?

18.30 What compound is the chief carrier of oxygen to actively metabolizing tissues?

18.31 The binding of oxygen to hemoglobin is described as cooperative. What does this mean, in general terms?

18.32 Write the equilibrium expression for the oxygenation of hemoglobin. In what direction does this equilibrium shift when:
(a) The pH decreases?
(b) The PO_2 decreases?
(c) The red cell is in the lungs?
(d) The red cell is in a capillary loop of an actively metabolizing tissue?
(e) CO_2 comes into the red cell?
(f) HCO_3^- ions flood into the red cell?

18.33 Using chemical equations, describe the isohydric shift when a red cell is (a) in actively metabolizing tissues and (b) in the lungs.

18.34 In what two ways does the oxygenation of hemoglobin in red cells in alveoli help to release CO_2?

18.35 In what way does waste CO_2 at active tissues help to release oxygen from the red cell?

18.36 In what way does extra H^+ at active tissue help release oxygen from the red cell?

18.37 Where is carbonic anhydrase found in the blood, and what function does it have in the management of the respiratory gases in (a) an alveolus and (b) actively metabolizing tissues?

18.38 What are the two main forms in which waste CO_2 moves to the lungs?

18.39 What is the chloride shift?

18.40 In what way is the superior oxygen-binding ability of myoglobin over hemoglobin important?

18.41 Aquatic diving animals are known to have much larger concentrations of myoglobin in their red muscle tissue than humans. How is this important to their lives?

18.42 Fetal hemoglobin has a higher oxygen-binding ability than adult hemoglobin. Why is this important to the fetus?

Acid – Base Balance of the Blood

18.43 Construct a table using arrows (↑) or (↓) and typical lab data that summarize the changes observed in respiratory and metabolic acidosis and alkalosis. The column headings should be as follows:

Condition	pH	PCO_2	$[HCO_3^-]$

18.44 With respect to the *directions* of the changes in the values of pH, PCO_2, and $[HCO_3^-]$ in both respiratory acidosis and metabolic acidosis, in what way are the two types of acidosis the same? In what way are they different?

18.45 Hyperventilation is observed in what two conditions that relate to the acid–base balance of the blood? In one, giving carbon dioxide is sometimes used, and in the other, giving isotonic HCO_3^- can be a form of treatment. Which treatment goes with which condition and why?

18.46 In what two conditions that relate to the acid–base balance of the blood is hypoventilation observed? Isotonic ammonium chloride or isotonic sodium bicarbonate are possible treatments. Which treatment does with which condition, and how do they work?

18.47 In which condition relating to acid–base balance does hyperventilation have a beneficial effect? Explain.

18.48 Hyperventilation is part of the *cause* of the problem in which condition relating to the acid–base balance of the blood?

18.49 Hypoventilation is the body's way of helping itself in which condition that relates to the acid–base balance of the blood?

18.50 In which condition that concerns the acid–base balance of the blood is hypoventilation part of the *problem* rather than the cure?

18.51 How can a general dehydration develop in metabolic acidosis?

18.52 Which condition, metabolic or respiratory acidosis or alkalosis, results from each of the following situations?
(a) Hysterics
(b) Overdose of bicarbonate
(c) Emphysema
(d) Narcotic overdose
(e) Diabetes
(f) Overbreathing at a high altitude
(g) Severe diarrhea
(h) Prolonged vomiting
(i) Cardiopulmonary disease
(j) Barbiturate overdose

18.53 Referring to Review Exercise 18.52, which is happening in each situation, hyperventilation or hypoventilation?

18.54 Why does hyperventilation in hysterics cause alkalosis?

18.55 Explain how emphysema leads to acidosis.

18.56 Prolonged vomiting leads to alkalosis. Explain.

18.57 Uncontrolled diarrhea can cause acidosis. Explain.

Blood Chemistry and the Kidneys

18.58 If the osmotic pressure of the blood has increased, what, in general terms, has changed to cause this?

18.59 How does the body respond to an increase in the osmotic pressure of the blood?

18.60 If the sodium ion level of the blood falls, how does the body respond?

18.61 Alcohol in the blood suppresses the secretion of vasopressin. How does this affect diuresis?

18.62 In what ways do the kidneys help to reduce acidosis?

Chapter 19
Molecular Basis of Energy for Living

Reserve energy is vital in a race, and we'll study here how the body manages the storage and mobilization of chemical energy.

19.1   OVERVIEW OF BIOCHEMICAL ENERGETICS

High-energy phosphates, such as ATP, are the body's means of trapping the energy of the oxidation of the products of digestion.

We cannot use solar energy directly, like plants. We cannot use steam energy, like a locomotive. We need *chemical* energy for living. We obtain it from food, and we use it to make high-energy molecules. Then their energy drives the chemical processes behind muscular work, signal sending, and chemical manufacture in tissue. Our principal source of chemical energy is the **catabolism** (i.e., the breaking down) of carbohydrates and fatty acids, although we can also use proteins for energy.

Catabolism is from the Greek, *cata-*, down; *ballein,* to throw or cast. Its opposite is *anabolism,* reactions that make larger molecules from smaller ones.

The Resynthesis of ATP from ADP and P_i Is the Centerpiece of Biochemical Energetics.   In Section 13.4 we first learned about triphosphate esters and why their phosphoric anhydride units make them so energy rich. From the lowest to the highest forms of life, one of them, **adenosine triphosphate** or **ATP,** is the principal carrier of energy within living systems. The synthesis of ATP is the chief means such systems use to trap the energy available by the oxidation of metabolites. Then ATP's energy is released in chemical reactions that operate muscles and nerves, and that make other compounds.

P_i is mostly $H_2PO_4^-$ and HPO_4^{2-}.

ATP breaks down to ADP and P_i as these changes occur. Once this happens, ATP must be remade or no more work by the organism is possible. One of the major purposes of catabolism is to transfer chemical energy from carbohydrates and lipids in such a way that ATP is resynthesized from adenosine diphosphate, ADP, and inorganic phosphate ion, P_i. We'll now take an overview of all of the pathways for ATP synthesis in humans.

The Appearance of ADP and P_i Initiate the Respiratory Chain.   The resynthesis of ATP from ADP and P_i is under feedback control, so that when the supply of ATP is high, no more is made. Only as ATP is used is first one mechanism for its resynthesis and then another thrown into action.

Figure 19.1 is a broad outline of the metabolic pathways to ATP. Think of the last one, the **respiratory chain,** as being at the bottom of a tub, nearest the drain through which ATP will leave for some use. As with water leaving a tub, the first water to move is nearest the plug. So also with biochemical energetics; when ATP has to be made, the respiratory chain goes into action. It's the body's major source of ATP, and depends on oxygen and organic suppliers of hydride, H:⁻.

When the level of ATP drops and the levels of ADP + P_i rise, the rate of breathing is also accelerated.

The Citric Acid Cycle Is a Major Supplier of Hydride Ion to the Respiratory Chain.   When the respiratory chain starts, other cycles of metabolism stir. The citric acid cycle, next in from the bottom in Figure 19.1, commonly is thrown into action. The purpose of the **citric acid cycle** is to supply hydride ion to the respiratory chain. Several intermediates of this cycle are donors of H:⁻.

The citric acid cycle cannot operate without its own fuel, which is **acetyl coenzyme A,** often written as acetyl CoA.

$$CH_3\overset{\displaystyle O}{\overset{\displaystyle \|}{C}}-S-CoA$$

Acetyl coenzyme A

Acetyl Coenzyme A Can Be Made by the Catabolism of Carbohydrates.   All three food groups can be sources of energy, but carbohydrates and fatty acids are most often used in a well-nourished person.

Carbohydrates supply acetyl coenzyme A either from starch or from glucose by a pathway called **glycolysis.** Its reactions break glucose to the pyruvate ion and then a short pathway converts this to acetyl CoA.

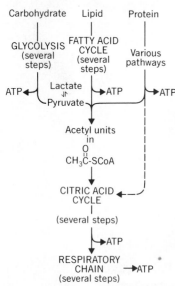

FIGURE 19.1
The major pathways for making ATP

Glycolysis also produces some ATP independently of the respiratory chain. This is vital when a cell is temporarily starved for oxygen and the respiratory chain cannot operate fast enough. In this circumstance, glycolysis makes some ATP until the oxygen supply is reestablished.

The full sequence of oxygen-consuming reactions from glucose through glycolysis, acetyl CoA, and the respiratory chain is called the **aerobic sequence** of glucose catabolism. Glycolysis alone, when run without oxygen, is called the **anaerobic sequence** of glucose catabolism. Athletes "go anaerobic" during particularly strenuous efforts, like a 100-m dash. Such a pace couldn't be sustained for a long-distance run, because of the limits to which tissues can tolerate the absence of oxygen and the wastes from anaerobic glycolysis. Rapid breathing and less strenuous efforts (or rest) sooner or later get the system back on the aerobic track.

Aerobic signifies the use of air. *Anaerobic,* stemming from "not air," means in the absence of the use of oxygen.

Acetyl CoA Is Also Made by the Catabolism of Fatty Acids. A series of reactions called the **fatty acid cycle** breaks fatty acids, two carbons at a time, into acetyl CoA, which is fed into the respiratory chain. Some of the intermediates in this cycle can also feed hydride ion directly to the respiratory chain. A great deal of ATP production can come from the catabolism of fatty acids, and long, sustained athletic efforts draw on these substances.

Acetyl CoA Can Also Be Made by the Catabolism of Amino Acids. Most amino acids can be catabolized to acetyl units or to intermediates of the citric acid cycle itself. The bodies of undernourished or starving people, who obtain insufficient dietary carbohydrate and fat, draw on the proteins of their own tissues to provide ATP. In doing this, they not only waste away, they also upset osmotic pressure relationships internally. The osmolarity of the blood decreases as its albumins are slowly used up, which leads to water retention in the interstitial and intracellular compartments and general edema.

With this as an overview of biochemical energetics, let's now look at the respiratory chain in more detail.

19.2 THE RESPIRATORY CHAIN

The flow of electrons to oxygen in the respiratory chain creates a proton gradient in mitochondria that drives the synthesis of ATP.

The term *respiration* refers to more than just breathing. It includes the chemical reactions that use oxygen in cells. Oxygen is reduced to water, and the following equation is the most basic statement we can write for what happens:

$$(\colon) \quad + 2H^+ + \cdot \ddot{O} \cdot \longrightarrow H-\ddot{O}-H + energy$$

| Pair of electrons | Pair of protons | Atom of oxygen | Molecule of water |

The electrons and protons come mostly from intermediates in the catabolism of sugars and fats. In the following discussion, remember that when any molecule loses electrons it is oxidized and the acceptor is reduced.

The gain of e^- or of e^- carriers such as $H\colon^-$ is *reduction;* the loss of e^- or of $H\colon^-$ is *oxidation.*

The Respiratory Enzymes Are Agents of Electron Transfers. The long series of oxidation–reduction reactions that make up the respiratory chain are catalyzed by the **respiratory enzymes.** The flow of electrons from the initial donor is irreversibly down an energy hill all the way to oxygen.

The cell's principal site of ATP synthesis is the *inner* membrane of a mitochondrion, Figure 19.2. Some tissues have thousands of these tiny organelles in the cytoplasm of a single cell, and each inner membrane has innumerable clusters of respiratory enzymes. One very important property of this membrane is that it is impermeable to protons, H^+, except at tiny channels that lead to an enzyme for the synthesis of ATP.

A cell in the flight muscle of a wasp has about a million mitochondria.

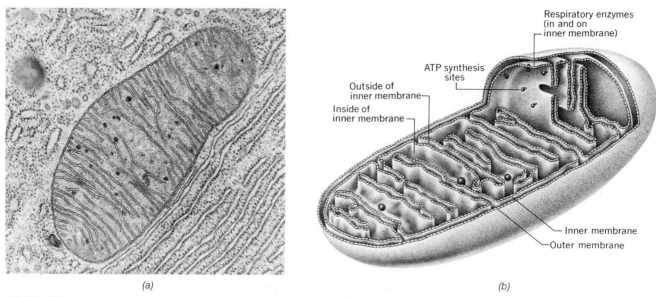

FIGURE 19.2

A mitochondrion. (a) Electron micrograph ($\times 53,000$) of a mitochondrion in a pancreas cell of a bat. (b) Perspective showing the interior. The respiratory enzymes are incorporated into the inner membrane. On the inside of this membrane are enzymes that catalyze the synthesis of ATP. (Micrograph courtesy of Dr. Keith R. Porter.)

Think of MH_2 as $M\!:\!H$, which becomes M when $H\!:^-$ and H^+ leave.

The whole layout of the respiratory chain is shown in Figure 19.3. The first respiratory enzyme carries the coenzyme NAD^+ (Section 17.1). NAD^+ accepts $H\!:^-$ from a donor molecule, frequently an intermediate in the citric acid cycle. We will let MH_2 stand for any metabolite that can donate $H\!:^-$. Then we can write the equation (next page) for the first step in the respiratory chain.

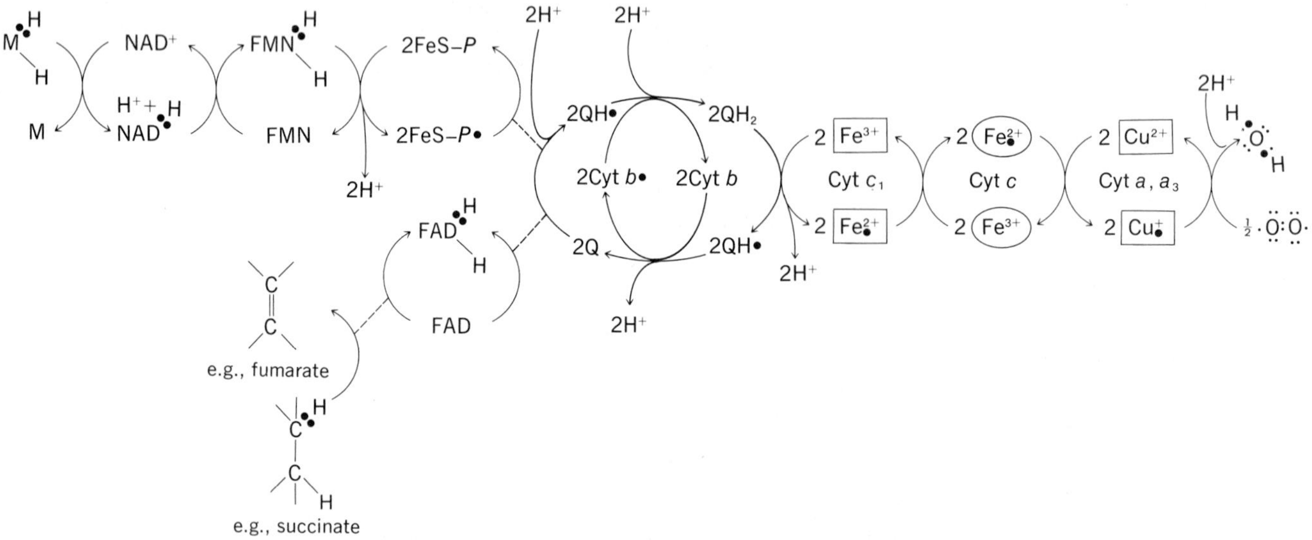

FIGURE 19.3

The respiratory chain. The boldface dots highlight the electrons that are passed along in the process from a donor to oxygen. Notice the three pairs of H^+ ions that disappear from the top side, which is inside the inner membrane. These cause configurational changes in certain membrane proteins that force H^+ ions to the outside of the inner membrane (see also Figure 19.4).

$$MH_2 + NAD^+ \longrightarrow M + NADH + H^+$$

The electron flow has now started toward oxygen. The electron pair, in the bond holding H, has moved from MH_2 to NADH. The proton is buffered.

The next enzyme in the chain is FMN with the coenzyme riboflavin (see also Section 17.1). The hydride in NADH now transfers to FMN, so we can write the following equation. (The proton is supplied by the buffer.)

$$NADH + H^+ + FMN \longrightarrow FMNH_2 + NAD^+$$

This restores the NAD^+ enzyme, and moves the electron pair one more step down the respiratory chain.

Think of NADH as NAD:H, in which the dots in boldface are those that move along the respiratory chain from MH_2.

Think of $FMNH_2$ as FMN:H, which becomes FMN when it loses H:$^-$ and H$^+$.

As Electrons Flow in the Respiratory Chain, Protons Move Across the Inner Mitochondrial Membrane.

What happens next is the transfer of just the pair of electrons. The hydrogen nucleus of H:$^-$ leaves the carrier as H$^+$, which promptly forces a change in the configuration of a membrane bound protein. This change causes protons to move across the inner mitochondrial membrane and to be injected into the fluid on the outside of the inner membrane. This step begins the most critical work of the respiratory chain, the buildup of an H$^+$ gradient across the inner mitochondrial membrane.

The species that accepts electrons as H$^+$ forms is called the *iron–sulfur protein,* which we'll symbolize by FeS—P. The iron in FeS—P occurs as Fe^{3+}. By accepting one electron, it is reduced to the Fe^{2+} state, so we need two units of FeS—P to handle the pair of electrons now about to leave $FMNH_2$. This step can be written as

$$FMNH_2 + 2FeS—P \longrightarrow FMN + 2FeS—P\cdot + 2H^+$$

We have used FeS—$P\cdot$ to represent the reduced form of the iron–sulfur protein in which Fe^{2+} occurs.

Visualize these enzymes and their reactions as occurring in the structure of the inner membrane itself.

The upper-left-hand corner of Figure 19.3 restates these three steps with the symbolism we learned in Chapter 17. It shows how enzymes are restored as well, and how electrons move down the chain.

At this stage, two electrons have moved from MH_2 to two molecules of the iron–sulfur protein, and hydrogen ions have been added to the fluid on the *outside* of the inner mitochondrial membrane. This region now has a higher hydrogen ion concentration than is on the inside of the inner membrane. The flow of electrons in the respiratory chain, in other words, is setting up a proton gradient. Protons, remember, cannot diffuse back again just anywhere along this inner membrane.

The Chemiosmotic Theory Connects the Proton Gradient to ATP Synthesis.

What we are developing here is the **chemiosmotic theory**, first proposed in 1961 by Peter Mitchell of England (Nobel prize, 1981). The basic principles of this theory are as follows:

Principles of the Chemiosmotic Theory

1. The synthesis of ATP occurs at an enzyme, now called **ATP synthase,** located on the *inside* of the inner mitochondrial membrane.
2. ATP production is driven by a flow of protons that occurs from the outside to the inside of the inner membrane.
3. The flow of protons is through special ports in the membrane and down a concentration gradient of protons across the inner membrane.
4. The energy to create the proton gradient is provided by the flow of electrons within the respiratory chain of enzymes, which are integral parts in the inner membrane.
5. The inner membrane is a closed envelope except for special ports for the flow of protons and special systems that convey needed solutes into or out of the innermost mitochondrial compartments.

Getting back to the respiratory chain, its entire function is to transport electrons from a metabolite, MH_2, to oxygen while setting up the H^+ gradient. It is quite complicated, as you no doubt have noticed, and from this point on it becomes even more so. We won't go into the finer details. Figure 19.3 does show them, including a branch in the chain that involves an FAD−enzyme, which also feeds electrons into the main chain.

Cyto-, cell; *-chrome*, pigment. The cytochromes are colored substances.

As you can notice in Figure 19.3, the rest of the chain involves further transfers of electrons through a series of enzymes. These include several called the cytochromes until finally one is reached, cytochrome oxidase (Cyt a, a_3 in Figure 19.3) that catalyzes the conversion of oxygen to water. Interestingly, this enzyme involves the Cu^{2+}/Cu^+ pair of ions, and this is why copper ion is an essential trace element in nutrition. It's also the metal ion system affected by cyanide ion, CN^-, which poisons cytochrome oxidase, shuts down the respiratory chain, and causes rapid death.

Protons Migrating through Special Portals Activate the Release of ATP Made and Held by ATP Synthase.   Figure 19.4 places the events of the respiratory chain into the context of the inner membrane. It also shows one of the portals through which protons can move back into the innermost compartment of the mitochondrion. These protons are a result of the operation of the respiratory chain, which we can summarize, starting with MH_2 and NAD^+, by the following equation.

The pH of the fluid on the outside of the inner membrane is 1.4 units less (hence, more acidic) than the fluid in the innermost compartment of a mitochondrion.

$$MH_2 + nH^+ \quad + \tfrac{1}{2}O_2 \xrightarrow{\text{respiratory chain}} M + H_2O + nH^+$$

From inside the inner membrane   Now on the outside of the inner membrane

The value of n averages between 9 and 12.

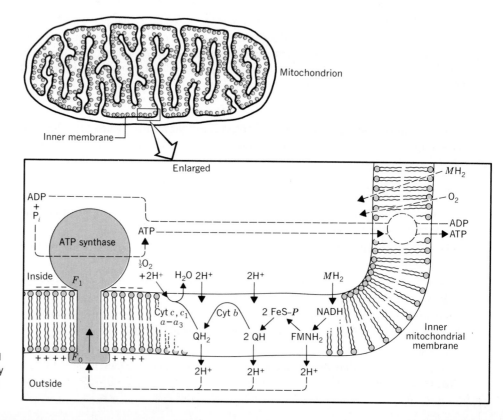

Mitochondrion

Inner membrane

FIGURE 19.4
Cutaway of a portion of the inner mitochondrial membrane that shows one respiratory chain and one proton conduit, labeled as F_0—F_1. ATP synthase is located in F_1. As indicated on the right. ADP can move inside only if ATP moves outside; the movements of these two are coupled.

Embedded in the inner mitochondrial membrane are complexes of proteins that form a tube through it. The tube is a conduit for protons, the only conduit for them. It terminates on the inside of the inner membrane with ATP synthase. As we said, this is the enzyme that catalyzes the formation of ATP from ADP and P_i.

When ATP synthase works, however, the newly formed ATP remains stuck to it inhibiting its further action. This is an example of the inhibition of an enzyme by the *product* of its own reaction. The flow of protons through the portal, however, interacts with the ATP synthase causing it to release its ATP. The proton flow, remember, is made possible by the work of the respiratory chain, so it is in this indirect way that ATP is made by the oxidation of metabolites via the respiratory chain. Thus this entire process for making ATP is often called **oxidative phosphorylation.** Each molecule of MH_2 entering the chain at NAD^+ generates about three ATPs. Each metabolite entering at FAD leads to about two ATPs.

Several Antibiotics and Drugs Inhibit Oxidative Phosphorylation. Amytal, one of the barbiturates, and rotenone, a powerful insecticide, both block the respiratory chain between NAD^+ and cytochrome b. As we've noted already, cyanide ion deactivates cytochrome oxidase. The antibiotic antimycin A stops the chain between cytochromes b and c.

Now that we've taken a careful look at the "plug" end of the energy mobilization "tub," the respiratory chain and oxidative phosphorylation, we will move back one major step to the supplier of electrons for the respiratory chain, the citric acid cycle.

19.3 THE CITRIC ACID CYCLE

Acetyl CoA is used to make the citrate ion that is then broken down bit by bit to CO_2 and units of $(H:^- + H^+)$, which funnel into the respiratory chain.

Figure 19.5 gives the steps of the citric acid cycle, a series of reactions that break down acetyl groups, which are fed to it from acetyl CoA.[1] The two carbon atoms of the acetyl group end up in molecules of CO_2, and the hydrogen atoms are fed into the respiratory chain. These reactions occur in the innermost compartment of a mitochondrion, inside the inner membrane.

The Citric Acid Cycle Dismantles Acetyl Groups. To launch the citric acid cycle, an acetyl group transfers from acetyl CoA to oxaloacetate ion to give the citrate ion. Now begins a series of reactions by which the citrate ion is degraded until another oxaloacetate ion is recovered. The numbers of the following steps match those in Figure 19.5.

1. Citrate is dehydrated to give the double bond of *cis*-aconitate. (This is the dehydration of an alcohol.)

2. Water adds to the double bond of *cis*-aconitate to give an isomer of citrate called isocitrate. Thus the net effect of steps 1 and 2 is to switch the alcohol group in citrate to a different carbon atom. However, this changes the alcohol from tertiary to secondary, from one that cannot be oxidized to one that can.

3. The secondary alcohol group in isocitrate is dehydrogenated (oxidized) to give oxalosuccinate. NAD^+ accepts the hydrogen, so ATP can now be made as $NAD:H$ passes its electron pair down the respiratory chain.

4. Oxalosuccinate loses a carboxyl group — it decarboxylates — to give α-ketoglutarate.

5. α-Ketoglutarate now undergoes a very complicated series of reactions, all catalyzed by one team of enzymes that includes coenzyme A. The results are the loss of a carboxyl group and another dehydrogenation. The hydrogen is accepted by NAD^+, so more ATP

Because chemical reactions create the gradients, we have the *chemi-* part of the term *chemiosmotic.*

This migration through a semipermeable membrane explains the *-osmotic* part of the term *chemiosmotic.*

Rotenone is a naturally occurring insecticide.

Hans Krebs won a share of the 1953 Nobel prize in medicine and physiology for his work on the citric acid cycle.

At physiological pH, the acids in the cycle exist largely as their anions.

---

[1] You should be aware that the ctiric acid cycle goes by two other names as well: the *tricarboxylic acid cycle* and the *Krebs' cycle.* You might encounter any of these names in other references.

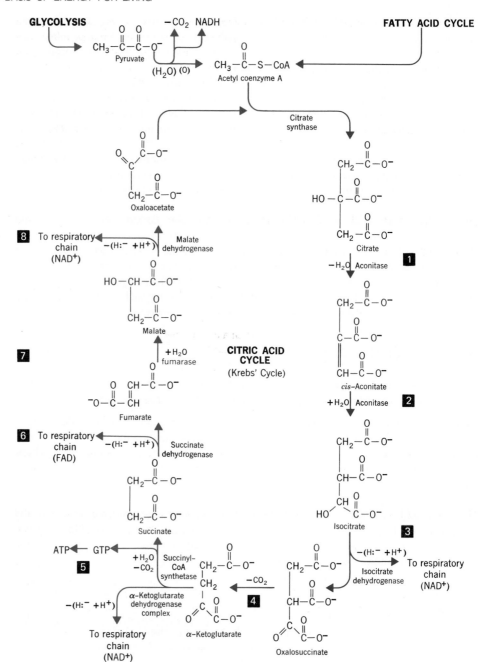

FIGURE 19.5
The citric acid cycle. The boxed
numbers refer to the text discussion.
The names of the enzymes for each
step are given by the arrows.

can now be made by the respiratory chain. The product, not shown in Figure 19.5, is the
coenzyme A derivative of succinic acid. It is converted to the succinate ion, which is in
Figure 19.5, in a reaction that also generates guanosine triphosphate, GTP. This is
another high-energy triphosphate, which is similar to ATP. GTP is able to phosphorylate
ADP to make one ATP.

6. Succinate donates hydrogen to FAD (not to NAD^+), and the fumarate ion forms. More
ATP can be made via FAD's involvement in the respiratory chain.

7. Fumarate adds water to its double bond, and malate forms.

8. The secondary alcohol in malate gives up hydrogen to NAD^+, and more ATP can be made. The cycle is now closed. We have remade one molecule of the carrier, oxalo-acetate. It can accept another acetyl group from acetyl coenzyme A. If the demand for ATP is still high enough, meaning that considerable ADP and P_i remains, another turn of the cycle can now occur.

A maximum of 12 molecules of ATP are generated from the degradation of each acetyl group by the citric acid cycle. Three are the result of step 3, four at step 5, two at step 6, and three at step 8.

We'll next see how glycolysis can lead to acetyl CoA as well as how it makes some ATP independently of the respiratory chain.

19.4 ENERGY FROM CARBOHYDRATES

The end product of glycolysis is pyruvate ion when oxygen is available, but lactate ion under anaerobic conditions.

Glycolysis, as we noted earlier, is a series of reactions that catabolize glucose while a small but important amount of ATP is made.

Greek, *glykos*, sugar or sweet; *-lysis*, dissolution.

Anaerobic Glycolysis Ends in Lactate. When a cell receives oxygen at a slower rate than necessary, glycolysis ends with the lactate ion, and this kind of glycolysis is referred to as *anaerobic glycolysis* or the **anaerobic sequence.** The equation for this, overall, is

$$C_6H_{12}O_6 + 2ADP + 2P_i \xrightarrow[\text{glycolysis}]{\text{anaerobic}} 2CH_3\overset{\overset{\displaystyle OH}{|}}{C}HCO_2^- + 2H^+ + 2ATP$$

$$\text{Glucose} \qquad\qquad\qquad \text{Lactate}$$

Except during extensive exercise, glycolysis is operated with sufficient oxygen, and it is aerobic.

Glycolysis Begins with Either Glycogen or Glucose. Figure 19.6 outlines the steps to pyruvate (or lactate) from glucose or glycogen. Those leading to fructose 1,6-diphosphate are actually up an energy hill, because they consume ATP. But this is like pushing a sled or bike up the short backside of a long hill. It is an investment in energy more than repaid by the long, downhill slide that follows. When glycolysis starts with glycogen instead of glucose, this initial investment in ATP is slightly smaller.

Some of the steps involve chemistry that we couldn't take the time to study earlier, so we won't go through glycolysis step by step. One step, however, is particularly important to an understanding of the metabolic difference between the aerobic and anaerobic alternatives. This is the oxidation of glyceraldehyde 3-phosphate to 1,3-diphosphoglycerate. It's roughly in the middle of the sequence in Figure 19.6.

$$^{2-}O_3OCH_2\overset{\overset{\displaystyle OH}{|}}{C}HCH{=}O + P_i + NAD^+ \longrightarrow {}^{2-}O_3OCH_2\overset{\overset{\displaystyle OH}{|}}{C}H{-}\overset{\overset{\displaystyle O}{\|}}{C}OPO_3^{2-} + NADH + H^+$$

Glyceraldehyde
3-phosphate

1,3-Diphosphoglycerate

This step requires NAD^+, so NADH is another product. When oxygen is available, the NADH is reoxidized to NAD^+. But when the cell lacks sufficient oxygen, there must be another way to restore the enzyme to NAD^+. Otherwise, even glycolysis would shut down, and the cell would be entirely without a way to regenerate ATP.

No reaction in the body is truly complete until its enzyme is fully restored.

The restoration of the reduced NAD^+ – enzyme, when oxygen isn't available, is done by transferring the $H:^-$ in NADH to the keto group of pyruvate. This group can accept $H:^-$ (along with H^+ from the buffer system), and change to the alcohol group of lactate. (See the

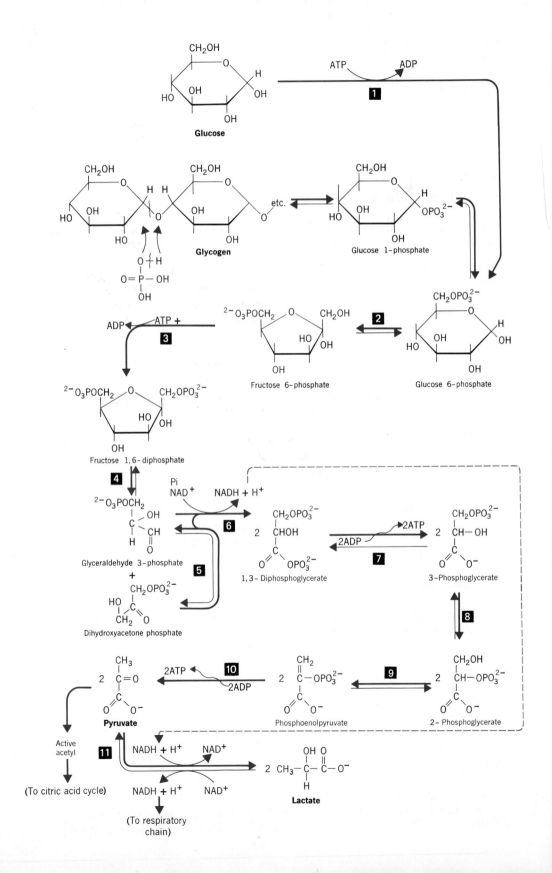

FIGURE 19.6
Glycolysis.

long dashed line in Figure 19.6.) This is why lactate is the end product of anaerobic glycolysis. Lactate serves to store H:⁻ until the cell once again becomes aerobic.

Anaerobic Glycolysis Allows Some ATP Synthesis During an Oxygen Debt. Anaerobic glycolysis provides a vital backup for the organism when some tissue uses oxygen faster than it is brought in, perhaps because of severe exercise. During such periods, the system is running an **oxygen debt,** because oxygen is insufficiently available to run the respiratory chain rapidly enough. But anaerobic glycolysis can still make some ATP.

Of course, there are limits. The longer the cell operates anaerobically, the more lactate accumulates. Eventually the system has to slow down to let respiration bring back oxygen to metabolize lactate.

By describing the production of lactate rather than lactic acid we have obscured the generation of acid during glycolysis. This acid, however, is promptly neutralized by the buffer so that the lactate ion is the product.

Of course, extensive physical exercise that forces considerable tissue to operate anaerobically can overtax the buffer. A form of metabolic acidosis called *lactic acid acidosis* is the result. Hyperventilation is initiated to blow out carbon dioxide and thus remove acid.

Galactose and Fructose Also Use the Glycolytic Pathway. We have studied glycolysis as if glucose is its only starting material. Glycolysis also accommodates galactose and fructose. Near the top of Figure 19.6, we see fructose entering the scheme. And galactose is changed by its own reactions to an early intermediate in glycolysis. Thus glycolysis is central to the catabolism of all dietary monosaccharides.

The Pentose Phosphate Pathway of Glucose Catabolism Makes an Essential Reducing Agent, NADPH. The biosyntheses of some substances in the body, such as the fatty acids, require a reducing agent. Fatty acids are almost entirely alkane-like, and alkanes are the most reduced types of organic compounds. The reducing agent used in the biosynthesis of fatty acids is NADPH, the reduced form of $NADP^+$.

$NADP^+$ is a phosphate derivative of NAD^+.

The body's principal route to NADPH is the **pentose phosphate pathway** of glucose catabolism. This complicated series of reactions (which we'll not study in detail) is very active in adipose tissue, where fatty acid synthesis occurs.

The balanced equation for the complete oxidation of one glucose molecule via the pentose phosphate pathway is as follows:

$$\text{Glucose 6-phosphate} + 12NADP^+ \xrightarrow{\text{pentose phosphate pathway}} 6CO_2 + 12NADPH + 12H^+ + P_i$$

19.5 ENERGY FROM FATTY ACIDS

Acetyl groups produced by the oxidation of fatty acids are fed into the citric acid cycle and respiratory chain.

The degradation of fatty acids takes place inside mitochondria by a repeating series of steps known as the **fatty acid cycle,** or as **beta oxidation.** Figure 19.7 outlines its chief steps.

The Fatty Acid Cycle Sends H:⁻ Directly into the Respiratory Chain and Makes Acetyl CoA for the Citric Acid Cycle. To enter the cycle, a fatty acid has to be joined to coenzyme A. It costs one ATP to do this, but now the fatty acyl unit is activated. (The ATP itself breaks down to AMP and PP_i, so the actual cost in high-energy phosphate is *two* high-energy phosphate bonds. Breaking ATP to AMP is the equivalent of breaking two molecules of ATP to two of ADP.)

PP_i

AMP is adenosine monophosphate.

The repeating sequence of the fatty acid cycle consists of four steps. Each turn of the cycle degrades the fatty acyl unit by two carbons, and produces one molecule of $FADH_2$, one of NADH, and one of acetyl coenzyme A. The now shortened fatty acyl unit is carried again

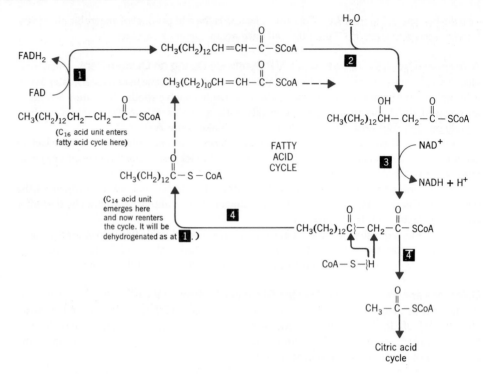

FIGURE 19.7
The fatty acid cycle (β oxidation). The numbers refer to the numbered steps discussed in the text.

through the four steps, and the process is repeated until no more two-carbon acetyl units can be made.

The $FADH_2$ and the NADH fuel the respiratory chain. The acetyl groups pass into the citric acid cycle, or they enter the general pool of acetyl coenzyme A that the body draws from to make other substances (e.g., cholesterol). Let's now look at the four steps in greater detail. The numbers that follow refer to Figure 19.7.

Although the whole series isn't exactly a cycle of a true variety, like the citric acid cycle, it's still referred to as the fatty acid *cycle*.

1. The first step is dehydrogenation. FAD accepts ($H:^- + H^+$) from the α and the β carbons of the fatty acyl unit.

$$CH_3(CH_2)_{12}\overset{\beta}{C}H_2-\overset{\alpha}{C}H_2-\overset{\overset{\displaystyle O}{\|}}{C}-SCoA + FAD \xrightarrow{\boxed{1}}$$
Palmityl coenzyme A

$$CH_3(CH_2)_{12}CH=CH-\overset{\overset{\displaystyle O}{\|}}{C}-SCoA + FADH_2 \longrightarrow (H:^- + H^+) \longrightarrow$$
An α,β-unsaturated
acyl derivative of
coenzyme A

FAD

To
respiratory
chain

2. The second step is hydration. Water adds to the alkene double bond and a secondary alcohol group forms.

$$CH_3(CH_2)_{12}CH=CH-\overset{\overset{\displaystyle O}{\|}}{C}-SCoA + H_2O \xrightarrow{\boxed{2}} CH_3(CH_2)_{12}\overset{\overset{\displaystyle OH}{|}}{C}H-CH_2-\overset{\overset{\displaystyle O}{\|}}{C}-SCoA$$
β-Hydroxyacyl derivative
of coenzyme A

3. The third step is another dehydrogenation, a loss of $(H{:}^- + H^+)$. This oxidizes the secondary alcohol to a keto group. Notice that these steps end in the oxidation of the β position of the original fatty acyl group to a keto group. This is why the fatty acid cycle is sometimes called *beta oxidation.*

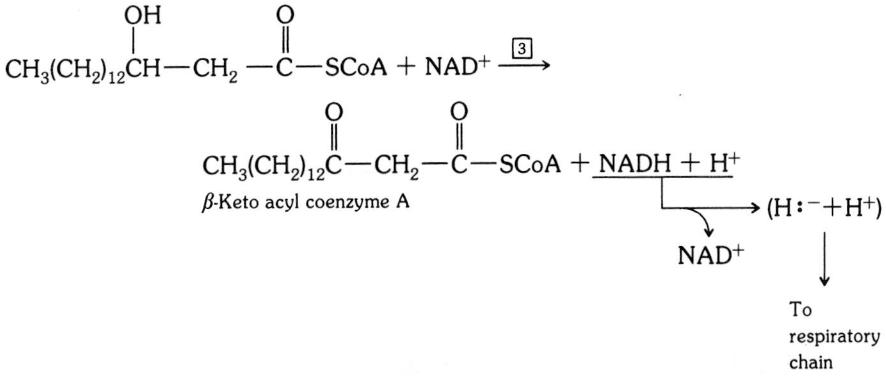

$$CH_3(CH_2)_{12}\underset{\underset{OH}{|}}{CH}-CH_2-\underset{\underset{O}{\|}}{C}-SCoA + NAD^+ \xrightarrow{\boxed{3}}$$

$$CH_3(CH_2)_{12}\underset{\underset{O}{\|}}{C}-CH_2-\underset{\underset{O}{\|}}{C}-SCoA + \underline{NADH + H^+}$$

β-Keto acyl coenzyme A

$\longrightarrow (H{:}^- + H^+)$

NAD$^+$

To respiratory chain

4. The fourth step breaks the bond between the α and the β carbons. This has been weakened by the stepwise oxidation of the β carbon, and now this bond breaks to release one unit of acetyl Coenzyme A.

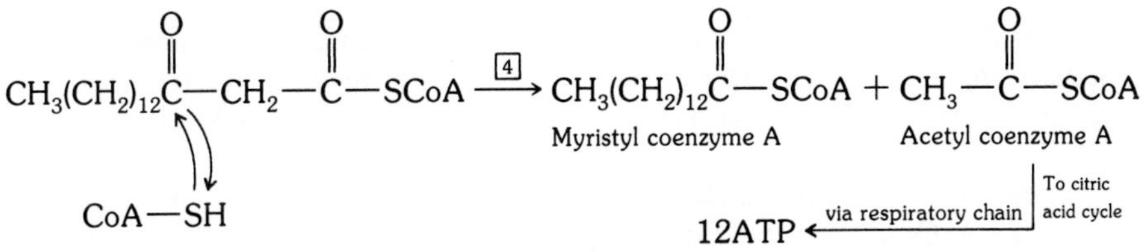

$$CH_3(CH_2)_{12}\underset{\underset{O}{\|}}{C}-CH_2-\underset{\underset{O}{\|}}{C}-SCoA \xrightarrow{\boxed{4}} CH_3(CH_2)_{12}\underset{\underset{O}{\|}}{C}-SCoA + CH_3-\underset{\underset{O}{\|}}{C}-SCoA$$

CoA—SH

Myristyl coenzyme A   Acetyl coenzyme A

To citric acid cycle

$12ATP \xleftarrow{\text{via respiratory chain}}$

The remaining acyl unit, the original now shortened by two carbons, goes through the cycle of steps again: dehydrogenation, hydration, dehydrogenation, and cleavage. After seven such cycles, one molecule of palmityl coenzyme A is broken into eight molecules of acetyl coenzyme A.

Franz Knoop directed much of the research on the fatty acid cycle, so this pathway is sometimes called *Knoop oxidation.*

A Maximum of 131 Molecules of ATP Are Made by the Fatty Acid Cycle. Table 19.1 shows how the maximum yield of ATP from the oxidation of one unit of palmityl CoA adds up to 131 ATPs. The net from palmitic acid is 2 ATP molecules fewer, or 129 ATP, because the activation of the palmityl unit, joining it to CoA, requires this initial investment, as we mentioned earlier.

TABLE 19.1
Maximum Yield of ATP from Palmityl CoA

Seven Turns of the Cycle Produce	ATP from Each Energy-Rich Intermediate	Total ATP Produced
7FADH$_2$	2	14
7NADH	3	21
8CH$_3\overset{O}{\overset{\|}{C}}$—SCoA	12	96
Deduct two high-energy phosphate bonds for activating the acyl unit		131 ATP
		−2
Net ATP yield per palmityl unit		129 ATP

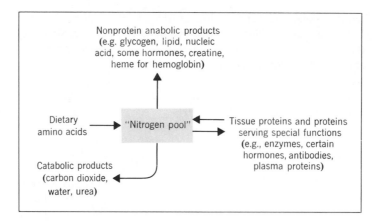

FIGURE 19.8
The nitrogen pool.

19.6 ENERGY FROM AMINO ACIDS

When not needed for making proteins, amino acids are catabolized for energy or are converted into glucose or fatty acids.

The polypeptides in enzymes have a particularly rapid turnover.

Amino Acids Enter the Body's Nitrogen Pool. Amino acids, the end products of protein digestion, are rapidly taken up by circulation and enter the general **nitrogen pool,** the name given to the whole collection of nitrogen compounds wherever they are found (see Figure 19.8).

Some amino acids are used to make replacements for worn out enzymes, hormones, and tissue proteins. Others are used to make nonprotein nitrogen compounds such as heme, some neurotransmitters, and nucleic acids. Any left over are catabolized. Their chemical energy is used to make high-energy phosphates. Their atoms end up in carbon dioxide, water, and urea, the products of the complete catabolism of amino acids. The catabolism of hemoglobin is described in Special Topic 19.1.

Amino Acids Are Sometimes Used to Make Glucose. Liver cells can use intermediates in the catabolism of many amino acids to make glucose. We'll say more about it in Chapter 20, but this synthesis of glucose from smaller molecules is called **gluconeogenesis** (*gluco-,* glucose; *neo-,* new; *-genesis,* creation), and it is indispensable to the brain. The brain's normal source of energy is the aerobic sequence starting with glucose taken from circulation. As this supply of glucose decreases between meals, during fasting, or in starvation, the liver makes glucose from molecular bits and pieces of amino acids and fatty acids.

When new glucose isn't needed and the supply of amino acids is still higher than needed to make proteins, then the molecular rummage from the partial catabolism of amino acids goes into making fatty acids. The body's synthesis of fatty acids is called **lipigenesis,** and we'll learn how it happens in Chapter 20.

The major point we make here is that the chemical energy in amino acids can be used directly to make high-energy phosphates or it can be put into the chemical energy of new glucose or fatty acids, which the body later and elsewhere taps. Excess amino acids are not excreted. You can gain weight on an all protein diet, and place severe strains on the liver and kidneys as they work overtime to convert amino groups to urea and export this waste.

We won't study in detail how each amino acid is catabolized, because each requires its own particular scheme, usually quite complicated. Three general types of reactions occur, however, which we will describe here: oxidative deamination, transamination, and direct deamination. We'll study these and how they apply to certain selected amino acids.

Oxidative Deamination Removes the Amino Group. **Oxidative deamination** is the replacement of the amino group of an amino acid by a keto group. The amino group leaves as

$$NH_2 - \overset{\overset{\textstyle O}{\|}}{C} - NH_2$$
Urea

Erythrocytes have life spans of about 120 days, and then they split open. The heme in hemoglobin is degraded to pigments, called bile pigments, which are eliminated via the feces and, to some extent, in the urine. These pigments all have chains of four rings that were present in heme.

Carbon skeleton of the bile pigments

The rings have varying numbers of double bonds according to the state of oxidation of the pigment.

The first breakdown product of hemoglobin consists of a slightly broken molecule, called verdohemoglobin. This splits into globin (the polypeptides of hemoglobin), iron(II) ion, and a greenish pigment called *biliverdin* [Latin *bilis* (bile) + *virdus* (green)]. Globin enters the nitrogen pool. Iron is conserved in a storage protein called ferritin and is reused.

Biliverdin is changed in the liver to a reddish-orange pigment called *bilirubin* [Latin *bilis* (bile) + *rubin* (red)]. This is removed from circulation by the liver, which transfers it to the bile. In this fluid bilirubin finally enters the intestinal tract. It is the principal bile pigment in humans.

The pathway from hemoglobin to bilirubin after the rupture of an erythrocyte as well as the fate of bilirubin are shown in the accompanying figure.

Once in the intestinal tract, bacterial enzymes convert bilirubin to a colorless substance called mesobilirubinogen. This is further processed to form a substance known as *bilinogen*, which usually goes by other names reflecting differences in destination rather than structure. Thus bilinogen leaving the body in feces is called *stercobilinogen* (Latin, *stercus*, dung). But some bilinogen is reabsorbed via the bloodstream, comes to the liver, and finally leaves the body in the urine. Now it is called *urobilinogen.* Some bilinogen is reoxidized to give *bilin,* a brownish pigment. Depending on its destination, bilin is called *stercobilin* or *urobilin.*

Jaundice (French, *jaune,* yellow) is a condition symptomatic of a malfunction in heme catabolism. The bile pigments accumulate in the plasma in concentrations high enough to impart a yellowish coloration to the skin. Jaundice can result from one of three kinds of malfunctions—too rapid breakup of red cells, liver infection or cirrhosis, and obstruction of the bile duct that carries bile from the gallbladder.

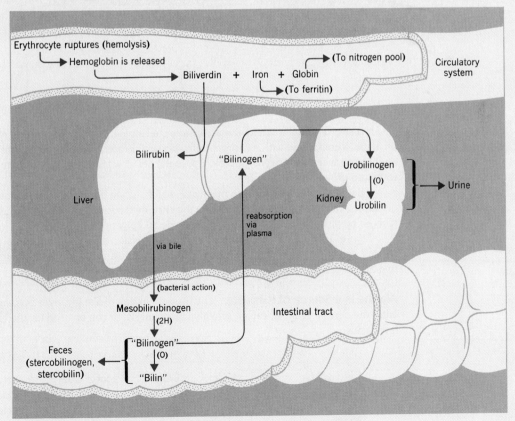

The formation and the elimination of the products of the catabolism of hemoglobin.

an ammonium ion. This reaction is largely used as a step in one of the processes that shuttles amino groups to urea.

Two coupled steps are involved. In the display that follows, the step on the left is called **transamination,** because an amino group is transferred to an acceptor. In the next step, the amino group's nitrogen is detached and changed to the ammonium ion. This enters a complex cycle of reactions called the *urea cycle* (which we will not study in detail) and is changed to urea.

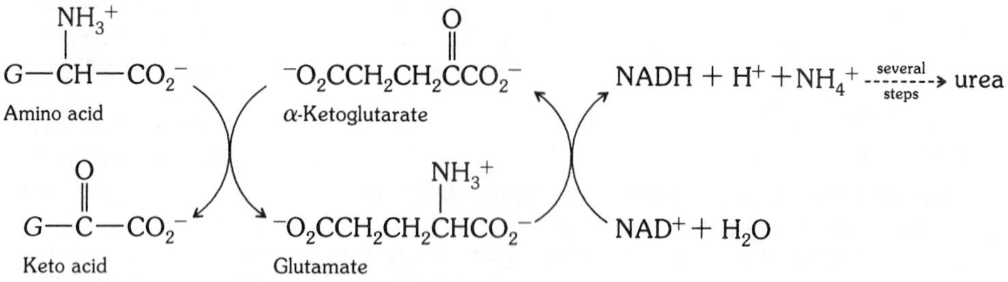

Notice that the α-ketoglutarate—glutamate pair provides a switching mechanism to convey nitrogen from amino acids to urea, in the direction of catabolism.

Direct Deamination Can Occur Without an Oxidation. Two amino acids, including serine, experience the simultaneous loss both of water and ammonia in a reaction that removes an amino group without consuming an oxidizing agent (such as NAD^+). This can take place in serine because the carbon adjacent to the amino group's carbon atom holds an alcohol group that can be dehydrated. Here's how **direct deamination** happens:

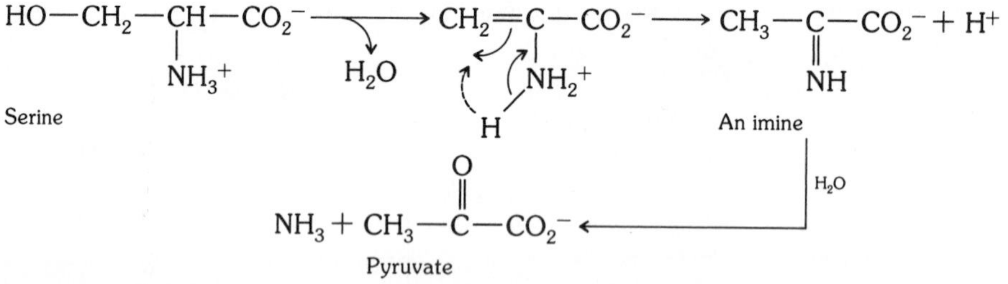

> Imine groups hydrolyze easily because they can add water and then split out ammonia.

The first step is dehydration to give an unsaturated amine. This spontaneously rearranges into an imine, a compound with a carbon–nitrogen double bond. Water adds to this double bond, and the product spontaneously breaks up, so the net effect is the hydrolysis of the imine group to a keto group and ammonia.

Thus serine breaks down to pyruvate, which, as we learned in connection with aerobic glycolysis, can send an acetyl group into the citric acid cycle. It can also be used to make new glucose by gluconeogenesis or new fatty acids by lipigenesis.

Figure 19.9 gives an overview of the catabolism of several individual amino acids, placing them into the context of major metabolic pathways. We will study the catabolism of two amino acids in more detail.

Alanine Is a Source of Pyruvate. The transamination of alanine gives pyruvate, which can enter the citric acid cycle, can go into gluconeogenesis, or can be used for the biosynthesis of fatty acids.

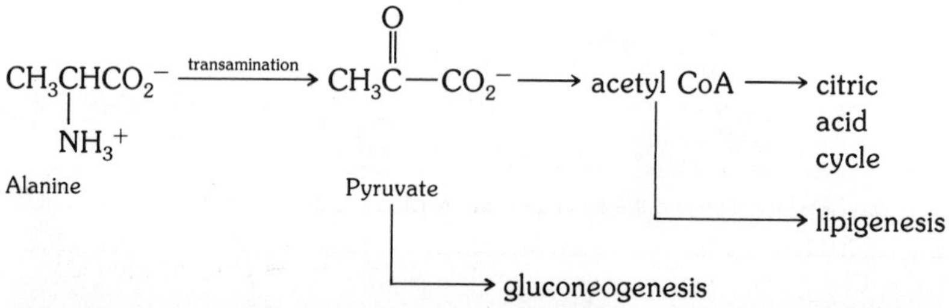

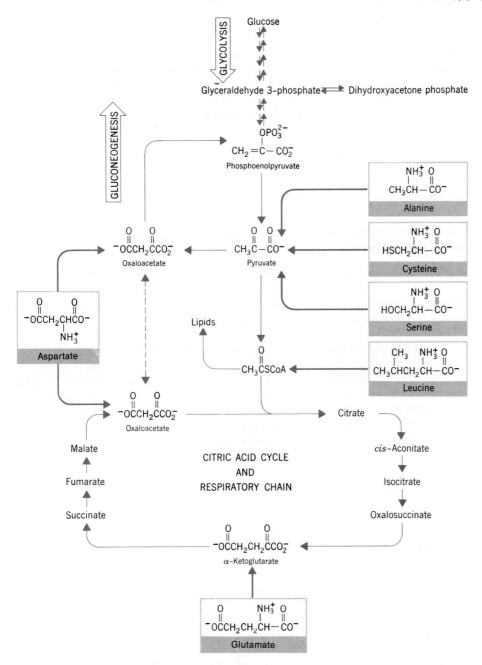

FIGURE 19.9
The catabolism of some amino acids.

Aspartic Acid Is a Source of Oxaloacetic Acid. The transamination of aspartic acid gives oxaloacetate, an intermediate in both gluconeogenesis and the citric acid cycle.

$$^-O_2CCH_2CHCO_2^- \longrightarrow {}^-O_2CCH_2\overset{\overset{\displaystyle O}{\|}}{C}CO_2^- \longrightarrow \text{citric acid cycle}$$
$$\underset{NH_3{}^+}{|}$$

Aspartic acid   Oxaloacetate

                       $\longrightarrow$ gluconeogenesis

Notice in Figure 19.9 that oxaloacetate occurs in three places, as an intermediate in the citric acid cycle, as the product of the oxidative deamination of aspartate, and as a raw material for making new glucose. For a *net gain* of glucose molecules via gluconeogenesis, the oxaloacetate in the citric acid cycle cannot be counted as available to make new glucose. Only oxaloacetate made from amino acids can give a net gain of glucose this way. This is why gluconeogenesis under conditions of starvation necessarily breaks down body proteins. It needs some of their amino acids to make new glucose via oxaloacetate.

19.7   ACIDOSIS AND ENERGY PROBLEMS

An acceleration of the fatty acid cycle tips some equilibria in a direction that leads to ketoacidosis.

Cells of certain tissues have to engage in gluconeogenesis in two serious conditions, starvation and uncontrolled diabetes mellitus. In starvation, the blood sugar level drops because of nutritional deficiencies, so the body (principally the liver) tries to compensate by making glucose. The consequences are fatal unless the underlying causes are treated.

The Level of Acetyl CoA Increases When Gluconeogenesis Is Accelerated. If you look back to Figure 19.9, you will see that gluconeogensis consumes oxaloacetate, the carrier of acetyl units in the citric acid cycle (Figure 19.5). When oxaloacetate is diverted from the citric acid cycle, acetyl coenzyme A cannot put its acetyl group into the cycle. Yet acetyl coenzyme A continues to be made by the fatty acid cycle, so acetyl CoA levels build up.

As the level of acetyl CoA increases in the liver, it creates a stress on the following equilibrium, so it shifts to the right to use up acetyl CoA and make acetoacetyl CoA.

The stress on this equilibrium is an increase in the level of acetyl CoA, so by Le Chatelier's principle the equilibrium shifts to the right to absorb this stress.

$$2\ \underset{\text{Acetyl CoA}}{CH_3\overset{O}{\overset{\|}{C}}-SCoA} \rightleftharpoons \underset{\text{Acetoacetyl CoA}}{CH_3\overset{O}{\overset{\|}{C}}CH_2\overset{O}{\overset{\|}{C}}-SCoA} + CoASH$$

As the level of acetoacetyl CoA builds up, a two-step process causes its overall hydrolysis of acetoacetate.

$$\underset{}{CH_3\overset{O}{\overset{\|}{C}}CH_2\overset{O}{\overset{\|}{C}}SCOA} + H_2O \xrightarrow{\text{two steps}} \underset{\text{Acetoacetate}}{CH_3\overset{O}{\overset{\|}{C}}CH_2\overset{O}{\overset{\|}{C}}O^-} + CoASH$$

The net effect, starting from acetyl CoA, is the following:

$$2\ \text{Acetyl CoA} + H_2O \longrightarrow \underset{\text{Acetoacetate}}{CH_3\overset{O}{\overset{\|}{C}}CH_2CO_2^-} + 2\ CoASH + H^+$$

Notice the hydrogen ion. It makes the situation very dangerous, and an increased synthesis of "new" glucose was the cause.

Accelerated Acetoacetate Production Leads to Acidosis. The acid produced by the formation of acetoacetate must be neutralized by the buffer. Under an increasingly rapid production of acetoacetate and hydrogen ion, the blood buffer slowly loses ground. A condition of acidosis sets in. It is *metabolic* acidosis, because the cause lies in a disorder of metabolism. Because the chief species responsible for this acidosis has a keto group, the condition is often called **ketoacidosis.**

Blood Levels of the Ketone Bodies Increase in Starvation and Diabetes.   The *acetoacetate ion* is called one of the **ketone bodies**. The two others are *acetone* and the *β-hydroxybutyrate ion*. Both are produced from the acetoacetate ion. Acetone arises from acetoacetate by the loss of the carboxyl group:

$$H_2O + CH_3\overset{O}{\underset{\|}{C}}CH_2\overset{O}{\underset{\|}{C}}O^- \longrightarrow CH_3\overset{O}{\underset{\|}{C}}CH_3 + HCO_3^-$$

Acetoacetate   Acetone

β-Hydroxybutyrate is called a ketone body despite its not having a keto group.

β-Hydroxybutyrate is produced when the keto group of acetoacetate is reduced by NADH:

$$CH_3\overset{O}{\underset{\|}{C}}CH_2CO_2^- + NADH + H^+ \longrightarrow CH_3\overset{OH}{\underset{|}{C}}HCH_2CO_2^- + NAD^+$$

Acetoacetate   β-Hydroxybutyrate

The ketone bodies enter general circulation. Because acetone is volatile, most of it leaves the body via the lungs, and individuals with severe ketoacidosis have "acetone breath," the noticeable odor of acetone on the breath.

Acetoacetate and β-hydroxybutyrate can be used in skeletal muscles to make ATP. Heart muscle uses these two for energy in preference to glucose. Even the brain, given time, can adapt to using these ions for energy when the blood sugar level drops in starvation or prolonged fasting. The ketone bodies are thus not in themselves abnormal constituents of blood. Only when they are produced at a rate faster than the blood buffer can handle them are they a problem.

The vapor pressure of acetone at body temperature is about 400 mm Hg making its loss by evaporation from the blood in the lungs easy.

The Conditions of Ketonemia, Ketonuria, and "Acetone Breath" Collectively Constitute Ketosis.   Normally, the levels of acetoacetate and β-hydroxybutyrate in the blood are, respectively, 2 and 4 μmol/dL. In prolonged, undetected, and untreated diabetes, these values can increase as much as 200-fold. The condition of excessive levels of ketone bodies in the blood is called **ketonemia**.

As ketonemia becomes more and more advanced, the ketone bodies begin to appear in the urine, a condition called **ketonuria**. When there is a combination of ketonemia, ketonuria, and acetone breath, the overall state is called **ketosis**. The individual is now described as *ketotic*. As ketosis becomes more severe, the associated ketoacidosis worsens and the pH of the blood continues its fatal descent.

1 μmol = 1 micromole = 10^{-6} mol

The Urinary Removal of Organic Anions Means the Loss of Base from the Blood.   For the anions of the ketone body to remain in the urine, the kidneys have to balance their negative charges by positive ions to keep everything electrically neutral. The Na^+ ions, the most abundant cations, are used. One Na^+ ion has to leave with each acetoacetate ion, for example. This loss of Na^+ is often referred to as the "loss of base" from the blood, although Na^+ is not a base. But the loss of one Na^+ stems from the appearance of one acetoacetate ion *plus one H^+ ion* that the blood had to neutralize. Thus each Na^+ that leaves the body corresponds to the loss of one HCO_3^- ion, the true base, consumed in neutralizing one H^+. Hence, the loss of Na^+ is taken as an indicator of the loss of this true base.

Another way to understand the urinary loss of Na^+ as the loss of base from the blood is that a Na^+ ion has to accompany a bicarbonate ion when it goes from the kidneys into the blood. The kidneys manufacture HCO_3^- ions normally in order to replenish the blood buffer system. The greater the number of Na^+ ions that have to be left in the *urine* in order to clear ketone bodies from the blood, the less the amount of true base, HCO_3^-, that can be put into the *blood*.

Condition	$[HCO_3^-]_{blood}$ (mmol/L)
Normal	22–30
Mild acidosis	16–20
Moderate acidosis	10–16
Severe acidosis	<10

Diuresis Must Accelerate to Handle Ketosis.   The solutes that are leaving the body in the urine cannot, of course, be allowed to make the urine too concentrated. Otherwise, osmotic pressure balances are upset. Increasing quantities of water must therefore be excreted. To

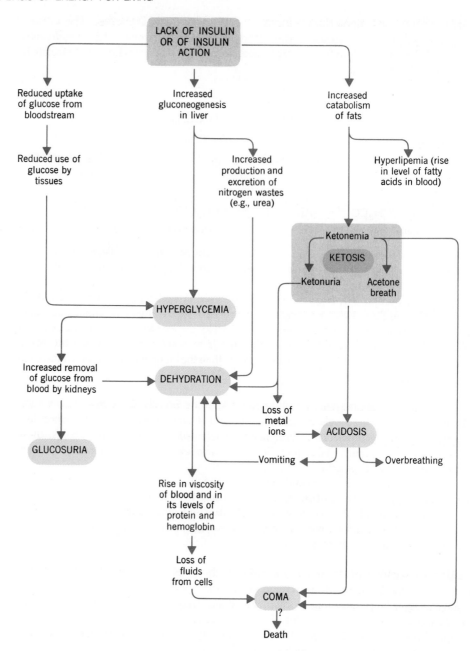

FIGURE 19.10
The principal sequence of events in
untreated diabetes.

Polyuria is the technical name for
the overproduction of urine.

satisfy this need, the individual has a powerful thirst. Other wastes, such as urea, are also
being produced at higher than normal rates, because amino acids are being sacrificed in
gluconeogenesis. These wastes add to the demand for water to make urine.

Internal Water Shortages in Ketosis Spell Dehydration of Critical Tissues. If, during a
state of ketosis, insufficient water is drunk, then water is simply taken from extracellular fluids.
The blood volume therefore tends to drop, and the blood becomes more concentrated. It also
thickens and becomes more viscous, which makes the delivery of blood more difficult.

Because the brain has the highest priority for blood flow, some of this flow is diverted
from the kidneys to try to ensure that the brain gets what it needs. This only worsens the
situation in the kidneys, and they have an increasingly difficult time clearing wastes. As the

water shortage worsens, some water is borrowed from the intracellular supply. This, in addition to a combination of other developments, leads to coma and eventually death.

Figure 19.10 outlines the succession of events in untreated type I diabetes. It is nothing short of remarkable how the absence of one chemical, insulin, can release such a vast train of biochemical events. But at the molecular level of life, this is the kind of story that occurs very often.

SUMMARY

High-energy compounds Triphosphate esters such as adenosine triphosphate, ATP, are the chief storehouse of chemical energy in living systems. When they react and make some of their energy available to cellular processes, they break down to organic disphosphate (e.g., ADP) and inorganic phosphate ion (P_i). (Sometimes they break down to monophosphates, like AMP.) The chief task of biochemical energetics is to remake high-energy compounds using the chemical energy in foods.

Respiratory chain A series of electron-transfer enzymes called the respiratory enzymes occur together as groups called respiratory assemblies in the inner membranes of mitochondria. These enzymes process metabolites (MH_2), which often are obtained by the operation of the citric acid cycle. Either NAD^+ or FAD can accept ($H:^- + H^+$) from a metabolite. Their reduced forms, NADH or $FADH_2$, then pass on the electrons until cytochrome oxidase uses them (together with H^+) to reduce oxygen to water.

Oxidative phosphorylation In the chemiosmotic process, the flow of electrons in the respiratory chain releases protons that create changes in the configurations of proteins of the mitochondrial membrane. Such changes force protons across the membrane, and this creates a proton gradient. The H^+ ions flow back through portals in this membrane that end at a molecule of ATP synthase still holding a molecule of ATP made from ADP and P_i. With the arrival of the protons, a configurational change occurs in ATP synthase to expel the ATP and expose the enzyme to more ATP synthesis. Various drugs and antibiotics can block the respiratory chain.

Citric acid cycle This cycle provides hydride ion to the respiratory chain. Acetyl groups from acetyl coenzyme A are joined to a four-carbon carrier, oxaloacetate, to make citrate. This six-carbon salt of a tricarboxylic acid then is degraded bit by bit as ($H:^- + H^+$) is fed to the respiratory chain. Each acetyl unit leads to the synthesis of a maximum of 12 ATP molecules.

Glycolysis Glycolysis uses glucose or glycogen to make pyruvate, which can be converted to acetyl CoA, the fuel of the citric acid cycle. Glycolysis also generates some ATP. (Galactose and fructose enter glycolysis, too.) The glycolysis pathway has one enzyme that under aerobic conditions sends hydride directly to the respiratory chain.

Under anaerobic conditions, glycolysis ends at the lactate ion, not the pyruvate ion. Under a low oxygen supply, the keto group in pyruvate is reduced to lactate so as to store $H:^-$ and thereby regenerate one of the enzymes. This enables glycolysis to be a source of ATP when insufficient oxygen is available.

The sequence from glucose through acetyl CoA, the citric acid cycle and the respiratory chain is called the aerobic sequence of glucose catabolism.

Pentose phosphate pathway The body's need for NADPH to make fatty acids is met by catabolizing glucose through the pentose phosphate pathway.

The catabolism of fatty acids Fatty acyl groups, after being pinned to coenzyme A, are catabolized by the fatty acid cycle. By a succession of four steps: dehydrogenation, hydration of a double bond, oxidation of the resulting alcohol, and cleavage of the bond from the α to the β carbon—one turn of the cycle removes one two-carbon acetyl group. The cycle then repeats as the shortened fatty acyl group continues to be degraded. Each turn of the cycle produces one $FADH_2$ and one NADH, which pass $H:^-$ to the respiratory chain for the synthesis of ATP. Each turn also sends one acetyl group into the citric acid cycle which, via the respiratory chain, leads to several more ATPs. The net ATP production is 129 ATPs per palmityl residue.

Amino acid distribution The nitrogen pool receives amino acids from the diet, from the breakdown of proteins in body fluids or tissues, and from any synthesis of nonessential amino acids that occurs. Amino acids are used to build and repair tissue, replace proteins of body fluids, make nonprotein nitrogen compounds, provide chemical energy if needed, and supply molecular parts for gluconeogenesis or lipigenesis.

Amino acid metabolism By reactions of transamination and oxidative deamination α-amino acids shuffle amino groups between themselves and intermediates of the citric acid cycle. Deaminated amino acids eventually become acetyl coenzyme A, acetoacetyl CoA, pyruvate, or an intermediate in the citric acid cycle. The skeletons of most amino acids can be used to make glucose, fatty acids, or the ketone bodies. Their nitrogen atoms become part of urea.

Ketoacidosis Acetoacetate, β-hydroxybutyrate, and acetone build up in the blood—ketonemia—in starvation or in diabetes. The first two are normal sources of energy in some tissues. When made faster than metabolized, however, they cause a loss of bicarbonate ion from the blood buffer, which leads to ketoacidosis.

The kidneys try to leave the anionic ketone bodies in the urine, but this takes Na^+ (for electrical neutrality), and water (for osmotic pressure balances). The loss of Na^+ in the urine is called a loss of "base," because its loss means less is available to accompany replacement HCO_3^- needed by the blood.

Under developing ketoacidosis, the kidneys have extra nitrogen wastes and, in diabetes, extra glucose to be exported in the urine. For these, more water is needed, and unless it is brought in by the thirst

mechanism, it has to be sought from within. But the brain has first call on blood flowage, so the kidneys suffer more. Eventually, if these events continue unchecked, the victim goes into a coma and dies.

REVIEW EXERCISES

The answers to these review exercises are in the *Study Guide* that accompanies this this book.

High-Energy Phosphates

19.1 Complete the following structure of ATP.

$$\text{Adenosine}-\text{O}-\overset{\overset{\textstyle O}{\|}}{\underset{\underset{\textstyle O^-}{|}}{P}}-$$

19.2 Write the structure of ADP and of AMP in the manner started by Review Exercise 19.1.

19.3 At physiological pH, what does the term *inorganic phosphate* stand for? (Give formulas and names.)

19.4 What is the overall central purpose of biochemical energetics?

Overview of Metabolic Pathways

19.5 In the general area of biochemical energetics, what is the purpose of each of the following pathways?
(a) Respiratory chain   (b) Anaerobic glycolysis
(c) Citric acid cycle   (d) Fatty acid cycle

19.6 What prompts the respiratory chain to go into operation?

19.7 In general terms, the intermediates that send electrons down the respiratory chain come from what metabolic pathway that consumes acetyl groups?

19.8 Arrange the following sets of terms in sequence in the order in which they occur or take place. Place the identifying letter of the first sequence of a set to occur on the left of the row of numbers.
(a) Citric acid cycle   pyruvate   respiratory chain
        1   2   3
    glycolysis   acetyl CoA
        4   5
(b) Respiratory chain   fatty acid cycle
        1   2
    citric acid cycle   acetyl CoA
        3   4

19.9 The *aerobic sequence* begins with what metabolic pathway and ends with which pathway?

Respiratory Chain

19.10 What is missing in the following basic expression for what must happen in the respiratory chain?

$$\tfrac{1}{2}O_2 + 2H^+ \longrightarrow H_2O$$

19.11 What general name is given to the set of enzymes involved in electron transport?

19.12 Write the following display in the normal form of a chemical equation.

(a) Which specific species is oxidized? (Write its structure.)
(b) Which species is reduced?

19.13 Write the following equation in the form of a display like that shown in Review Exercise 19.12.

$$\underset{\substack{|\\OH}}{CH_3CHCO_2^-} + NAD^+ \longrightarrow \underset{\substack{\|\\O}}{CH_3CCO_2^-} + NADH + H^+$$

19.14 Arrange the following in the order in which they receive and pass on electrons.

FMN   Cyt a NAD$^+$ FeSP

19.15 What does cytochrome oxidase do?

19.16 What is FAD, and where is it involved in the respiratory chain?

Oxidative Phosphorylation and the Chemiosmotic Theory

19.17 Across which cellular membrane does the respiratory chain establish a gradient of H$^+$ ions? On which side of this membrane is the value of the pH lower?

19.18 According to the chemiosmotic theory, the flow of what particles most directly leads to the production of ATP?

19.19 If the inner mitochondrial membrane is broken, the respiratory chain can still operate, but the phosphorylation of ADP that normally results stops. Explain why in general terms.

19.20 Complete and balance the following equation. (Use $\tfrac{1}{2}$ as the coefficient of oxygen as shown.)

$$MH_2 + H^+ \qquad + \tfrac{1}{2}O_2 \longrightarrow$$
From inside the inner membrane

19.21 Briefly describe the theory presented in this chapter that explains how a flow of a pair of protons across the inner mitochondrial membrane initiates the synthesis of ATP from ADP and P$_i$. Be sure to mention how ATP synthase is involved.

Citric Acid Cycle

19.22 What makes the citric acid cycle start up?

19.23 What chemical unit is degraded by the citric acid cycle? Give its name and structure.

19.24 How many times is a secondary alcohol group oxidized in the citric acid cycle?

19.25 Water adds to a carbon–carbon double bond how many times in one turn of the citric acid cycle?

19.26 The conversion of pyruvate to an acetyl unit is both an oxidation and a decarboxylation.

(a) If *only* decarboxylation occurred, what would form from pyruvate? Write the structure of the other product in:

$$CH_3-\overset{\overset{O}{\|}}{C}-\overset{\overset{O}{\|}}{C}-O^- + H^+$$
$$\longrightarrow \underline{\hspace{1cm}} + O{=}C{=}O$$

(b) If this product is oxidized, what is the name and the structure of the product of such oxidation?

19.27 One cofactor in the enzyme assembly that catalyzes the oxidative decarboxylation of pyruvate requires thiamine, one of the B vitamins. Therefore in beriberi, the deficiency disease for this vitamin, the level of what substance can be expected to rise in blood serum (and for which an analysis can be made as part of the diagnosis of beriberi)?

19.28 The enzyme for the conversion of isocitrate to oxalosuccinate (Figure 19.5) is stimulated by one of these two substances, ATP or ADP. Which one is the likelier activator? Explain.

Catabolism of Glucose

19.29 Fill in the missing substances and balance the following incomplete equation:

$$\underset{\text{Glucose}}{C_6H_{12}O_6} + 2ADP + \underline{\hspace{0.6cm}} \longrightarrow \underset{\text{Lactate}}{2C_3H_5O_3^-} + 2H^+ + \underline{\hspace{0.6cm}}$$

19.30 What particular significance does glycolysis have when a tissue is running an oxygen debt?

19.31 What happens to pyruvate under (a) aerobic conditions and (b) anaerobic conditions?

19.32 What happens to lactate when an oxygen debt is repaid?

19.33 The pentose phosphate pathway uses $NADP^+$, not NAD^+. What forms from $NADP^+$, and how does the body use it (in general terms)?

Catabolism of Fatty Acids

19.34 How are long-chain fatty acids activated for entry into the fatty acid cycle?

19.35 Complete the following equations for one turn of the fatty acid cycle by which a six carbon fatty acyl group is catabolized.

(a) $CH_3CH_2CH_2CH_2CH_2\overset{\overset{O}{\|}}{C}-SCoA + FAD \rightarrow$
$$\underline{\hspace{2cm}} + \underline{\hspace{1.5cm}}$$

(b) $\underline{\hspace{2.5cm}} + H_2O \rightarrow \underline{\hspace{2cm}}$
(c) $\underline{\hspace{2.5cm}} + NAD^+ \rightarrow$
$$\underline{\hspace{3cm}} + NADH + H^+$$
(d) $\underline{\hspace{2.5cm}} + CoA{-}SH \rightarrow$
$$\underline{\hspace{2cm}} + \underline{\hspace{1.5cm}}$$

19.36 Write the equations for the four steps in the fatty acid cycle as it operates on butyryl CoA. How many more turns of the cycle are possible after this one?

19.37 How is the FAD–enzyme recovered from its reduced form, $FADH_2$, when the fatty acid cycle operates?

19.38 How is the reduced form of the NAD^+–enzyme used in the fatty acid cycle restored to its oxidized form?

19.39 Why is the fatty acid cycle sometimes called beta oxidation?

19.40 Myristic acid, $CH_3(CH_2)_{12}CO_2H$, can be catabolized by the fatty acid cycle just like palmitic acid.

(a) How many units of acetyl CoA can be made from it?

(b) In producing this much acetyl CoA, how many times does $FADH_2$ form and then deliver its hydrogen to the respiratory chain?

(c) Referring again to part (a), how many times does NADH form as acetyl CoA is produced, and then deliver its hydrogen to the respiratory chain?

(d) Complete the following table by supplying the missing numbers of molecules that are involved in the catabolism of myristic acid to acetyl CoA.

Intermediate	Maximum Number of ATP from Each	Total Number of ATP Possible from Each as Acetyl CoA Forms
__ FADH$_2$	____	____
__ NADH	____	____
__ CH$_3\overset{\overset{O}{\|}}{C}$—SCoA	____	====
	Sum = _____	

Deduct _____ high-energy phosphate bonds for activating the myristyl group   − _____

Net ATP produced for each myristyl group as it changes to acetyl CoA   _____

19.41 What specific function does the citric acid cycle have in the use of fatty acids for energy?

Nitrogen Pool

19.42 What is the nitrogen pool?

19.43 What are four ways in which amino acids are used in the body?

19.44 When the body retains more nitrogen than it excretes in all forms, the system is said to be on a *positive nitrogen balance*. Would this state characterize infancy or old age?

19.45 What happens to amino acids that are obtained in the diet but aren't needed to make any nitrogenous compounds?

19.46 In the conditions of starvation or diabetes, what can the amino acids be used for?

The Catabolism of Amino Acids

19.47 Write the structure of the keto acid that forms when phenyl-alanine undergoes transamination with α-ketoglutarate.

19.48 When valine and α-ketoglutarate undergo transamination, what new keto acid forms? Write its structure.

19.49 By means of two successive equations, one a transamination and the other an oxidative deamination, write the reactions that illustrate how the amino group of alanine can be removed as NH_4^+.

19.50 Write the structure of the keto acid that forms by the direct deamination of threonine.

Ketoacidosis

19.51 Which nutrient becomes increasingly important as a source of energy (ATP) if the net effect of either starvation or diabetes is the reduced availability of glucose as a source of ATP?

19.52 The catabolism of the nutrient of Review Exercise 19.51 produces what intermediate that is further catabolized by the citric acid cycle?

19.53 Two molecules of acetyl CoA can combine to give the coenzyme A derivative of what keto acid? Give the structure of this keto acid.

19.54 Give the names and structures of the ketone bodies. (For those that occur as anions at body pH, give the structures as anions.)

19.55 What is ketonemia?

19.56 What is ketonuria?

19.57 What is meant by acetone breath?

19.58 Ketosis consists of what collection of conditions?

19.59 What is ketoacidosis? What form of acidosis is it, metabolic or respiratory?

19.60 The formation of which particular compound most lowers the supply of HCO_3^- in ketoacidosis?

19.61 What are the reasons for the increase in the volume of urine that is excreted in someone with untreated type I diabetes?

19.62 If the ketone bodies (other than acetone) can normally be used by heart and skeletal muscle, what makes them dangerous in starvation or in diabetes?

19.63 Why does the rate of urea production increase in untreated, type I diabetes?

19.64 When a physician refers to the loss of Na^+ as the loss of *base*, what is actually meant?

Catabolism of Heme (Special Topic 19.1)

19.65 Arrange the names of the following substances in the order in which they appear during the catabolism of heme.

Biliverdin	Heme	Hemoglobin	Bilirubin
1	2	3	4

Mesobilirubinogen	Bilin	Bilinogen
5	6	7

19.66 Briefly describe what jaundice does to the body.

19.67 Describe how the following kinds of jaundice arise.
(a) The jaundice of hemolysis.
(b) The jaundice of hepatic diseases.

19.68 Why should an obstruction of the bile duct cause jaundice?

Chapter 20
Metabolism and Molecule Building

To make something useful out of a pile of simple things, as at this Amish barn raising, requires young energy and old knowledge. So it is in making complex organisms out of simple molecules. We'll learn in this chapter how the body supplies metabolic energy.

20.1 METABOLIC INTERRELATIONSHIPS, AN OVERVIEW

Acetyl CoA and pyruvate stand at major metabolic crossroads.

We will begin this chapter with a broad overview of metabolism, using Figure 20.1. This will place the pathways of catabolism, studied in Chapter 19, into the context of those of molecule building, anabolism, in this chapter.

We start at the top of the figure with the products of digestion, chiefly the monosaccharides, amino acids, fatty acids, and glycerol. The amino acids enter the nitrogen pool. The monosaccharides make up what is called **blood sugar,** but this is almost entirely glucose. Some goes quickly to replenish glycogen reserves.

The fatty acids might be put into storage as fat in tissue called *adipose tissue,* which centers mostly around the middle of the body and cushions internal organs.

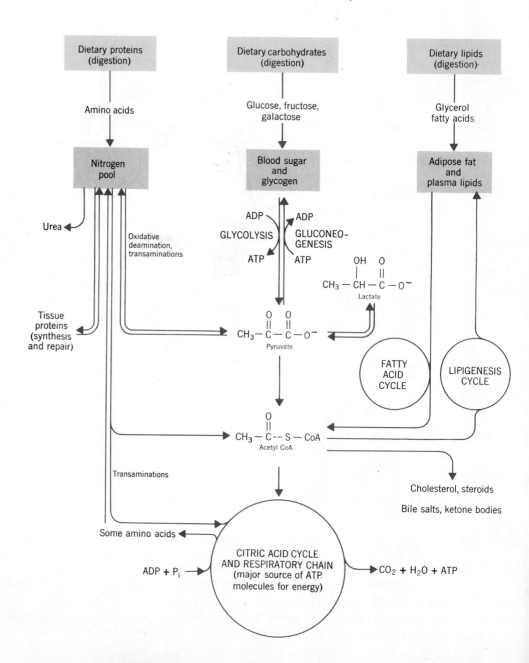

FIGURE 20.1
Interrelationships of major metabolic pathways.

The many ways in which amino acids can be used are next on the left in Figure 20.1, and you can see how they can become involved in all of the other pathways shown. Glycolysis makes some ATP and pyruvate, which leads to more ATP. Gluconeogenesis would appear to be the reverse of glycolysis, but it isn't quite that, as we will see.

Toward the right in the figure we see the fatty acid cycle feeding acetyl CoA toward the citric acid cycle and the respiratory chain (which, for simplicity, haven't been separated in the figure). And we note that acetyl CoA is the raw material for making fatty acids by lipigenesis. It's also the starting material for cholesterol, steroid hormones, bile salts, and the ketone bodies.

Everything is tied to everything else, and Figure 20.1 displays this claim. Notice particularly how many pathways converge either on acetyl CoA or pyruvate. No matter how complex compounds become in the body, many of them can be used to make others.

We will continue our study with the metabolism of glycogen, the chief storage form for glucose units in the body.

20.2 GLYCOGEN METABOLISM

Much of the body's control over the blood sugar level is handled by its regulation of the synthesis and breakdown of glycogen.

The digestion of the starch and disaccharides in the diet gives glucose, fructose, and galactose, but their catabolic pathways all converge very quickly to that of glucose itself. Galactose, for example, is changed by a few steps in the liver to glucose 1-phosphate. Fructose is changed to a compound that occurs early in glycolysis. So our emphasis will be on glucose metabolism, and we'll start with the glucose in circulation.

Wide Variations in the Blood Sugar Level Signal Something Wrong.   The concentration of monosaccharides in whole blood, expressed in milligrams per deciliter (mg/dL), is called the **blood sugar level.** This is very close to the glucose level, because glucose is overwhelmingly the major monosaccharide. When determined after several hours of fasting, the blood sugar level, called the **normal fasting level,** is $65-95$ mg/dL ($70-110$ mg/dL in plasma). In a condition of **hypoglycemia,** the blood sugar level is *below* normal, and in **hyperglycemia** it is *above* the normal fasting level.

mg/dL = milligrams per deciliter, where 1 dL = 100 mL

-*glyc*-, sugar
-*emia*, in blood
hypo-, under, below
hyper-, above, over
renal, of the kidneys
-*uria*, in urine

When the blood sugar level become too hyperglycemic (becomes high), the kidneys are unable to put back into the blood all the glucose that left it in a glomerulus. Glucose then appears in the urine, a condition called **glucosuria.** The blood sugar level above which this happens is called the **renal threshold** for glucose, and it is in the range of about $140-160$ mg/dL, sometimes higher.

Hypoglycemia Can Make You Faint.   Your brain relys almost entirely on glucose for its chemical energy, so if hypoglycemia develops rapidly, you can experience dizziness and may faint. People with diabetes who unknowingly take too much insulin experience this in a severe form known as *insulin shock*. The brain consumes about 120 g/day of glucose, and a quick onset of hypoglycemia starves the brain cells. They do have the ability to switch over to other nutrients, but brain cells cannot do this very rapidly.

Persistent Hyperglycemia Indicates Diabetes.   Whenever hyperglycemia develops and tends to persist, something is wrong with the mechanisms for withdrawing glucose from circulation. Diabetes is a common cause of hyperglycemia, but there are other possible causes.

In an individual with a sustained hyperglycemia, some glucose combines with hemoglobin to give glycohemoglobin (or glycosylated hemoglobin). The measurement of the level of this substance has become the best way to monitor the average blood sugar level of a diabetic, better than any direct measurements of blood glucose. Its level doesn't fluctuate as widely as the blood sugar level, and it doesn't have to be determined as frequently.

Excess Blood Glucose Normally Is Withdrawn from Circulation.

When there is more than enough glucose in circulation to meet energy needs, the body does not eliminate the excess but conserves its chemical energy. There are two ways to do this. One is to convert glucose to fat, and we'll study how this is done later in this chapter. The other is to synthesize glycogen, which we'll discuss here. (We studied glycogen in Section 14.5.)

Liver and muscle cells can convert glucose to glycogen by a series of steps called **glycogenesis** ("glycogen creation"). The liver holds 70–110 g of glycogen, and the muscles, taken as a whole, contain 170–250 g. When muscles need glucose, they take it back out of glycogen. When the blood needs glucose because the blood sugar level has dropped too much, the liver hydrolyzes as much of its glycogen reserves as needed and then puts the glucose into circulation. The overall series of reactions in either tissue that hydrolyzes glycogen is called **glycogenolysis** (lysis or hydrolysis of glycogen). This process is controlled by several hormones.

Epinephrine Stimulates Glycogenolysis.

When muscular work is begun, the adrenal medulla secretes the hormone **epinephrine.** In muscle tissue, and to some extent in the liver, epinephrine activates glycogenolysis by the long series steps, all occurring extremely rapidly. One epinephrine molecule can trigger the mobilization of thousands of glucose units, which then are ready to supply energy that the body needs.

The end product of glycogenolysis isn't actually glucose but glucose 1-phosphate. Cells that can do glycogenolysis also have an enzyme called phosphoglucomutase, which catalyzes the conversion of glucose 1-phosphate to its isomer, glucose 6-phosphate:

> An estimated 30,000 molecules of glucose are released from glycogen for each molecule of epinephrine that initiates glycogenolysis.

Glucose 1-phosphate   Glucose 6-phosphate

Glucose Is Trapped in the Muscle Cell When It Is in the Form of Glucose 6-Phosphate.

Glucose 6-phosphate, rather than glucose, is the form in which a glucose unit must be to enter a pathway that produces ATP. *It is also in a form that cannot migrate out of muscle cells.* It won't be lost from tissue needing it during exercise. Thus glycogenolysis in muscle tissue is an important supplier of energy. When the supply of muscle glycogen is low, muscle cells can take glucose from circulation, *trap it as glucose 6-phosphate,* and then convert this to glycogen.

Glucagon Activates Liver Glycogenolysis and Thus Affects the Blood Sugar Level.

> The glucagon molecule has 29 amino acid residues.

The α cells of the pancreas make a polypeptide hormone, **glucagon,** which helps to maintain a normal blood sugar level. When the blood sugar level drops, these cells release glucagon. Its target tissue is the liver, where it is an excellent activator of glycogenolysis.

Unlike epinephrine, glucagon *inhibits* glycolysis, so this action helps to keep the supply of glucose up. Glucagon, also unlike epinephrine, does not cause an increase in blood pressure or pulse rate, and it is longer acting than epinephrine.

Liver Cells Can Release Glucose to Circulation.

Glucose units released from liver glycogen as glucose 6-phosphate are converted to glucose by an enzyme that the liver has, but not the muscles, glucose 6-phosphatase. It catalyzes the hydrolysis of glucose 6-phosphate to glucose and inorganic phosphate.

$$\text{Glucose 6-P} + H_2O \xrightarrow{\text{glucose 6-phosphatase}} \text{glucose} + P_i$$

Glucose can now leave the liver and so help raise the blood sugar level. During periods of fasting, therefore, the overall process in the liver from glucose 1-phosphate to glucose 6-phosphate to glucose is a major supplier of glucose for the blood. Glucagon, which triggers this, is thus an important regulator of the blood sugar level.

The brain depends on the liver during fasting to maintain its favorite source of chemical energy, circulating glucose. (We are beginning to see in chemical terms how vital the liver is to the performance of other organs.) When circulating glucose is taken up by a brain cell, it is promptly trapped in the cell by being converted to glucose 6-phosphate.

Several hereditary diseases involve the glucose — glycogen interconversion, and some are discussed in Special Topic 20.1.

The letter P is often used to represent the whole phosphate group in the structures of phos-phate ester intermediates in metabolism.

Human Growth Hormone Stimulates the Release of Glucagon.

Growth requires energy, so the action of glucagon that helps to supply a source of energy, glucose, aids in the work of the human growth hormone. In some situations, such as a disfiguring condition known as acromegaly, there is an excessive secretion of human growth hormone that promotes too high a level of glucose in the blood. This is undesirable because a prolonged state of hyperglycemia from any cause can lead to diabetes and to some of the blood-capillary related complications of diabetes.

Acromegaly is sometimes called *giantism* because certain bone structures and visceral organs be-come enlarged.

Insulin Strongly Lowers the Blood Sugar Level.

The β cells of the pancreas make and release **insulin,** a polypeptide hormone. Its release is stimulated by an increase in the blood sugar level, such as normally occurs after a carbohydrate-rich meal. As insulin moves into action, it finds its receptors at the cell membranes of muscle and adipose tissue. The insulin — receptor complexes somehow make it possible for glucose molecules to move easily into the cells, and this, of course, lowers the blood sugar level.

Not all cells depend on insulin to take up glucose. Brain cells, for example, and cells in the kidneys, the intestinal tract, red blood cells, and in the lenses of the eyes take up glucose directly. If too much insulin gets into circulation, as by an error in insulin therapy, the individual's blood sugar level falls too low, which leads to insulin shock, as we've already mentioned.

If the receptors are continuously overloaded with work in someone who consumes a great deal of sugar, they can wear out, and the individual becomes diabetic.

Adipose tissue is fatty tissue that surrounds internal organs.

Life-saving first aid for someone in insulin shock is sugared fruit juice or candy to counter the hypoglyce-mia.

Somatostatin Inhibits Glucagon and Slows the Release of Insulin.

The hypothalamus, a specific region in the brain, makes **somatostatin,** another hormone that participates in the regulation of the blood sugar level. When the β cells of the pancreas secrete insulin, which helps to *lower* the blood sugar level, the α cells should not at the same time release glucagon, which helps to *raise* this level. Somatostatin acts at the pancreas to inhibit the release of glucagon as well as to slow down the release of insulin. It thus helps to prevent a wild swing in the blood sugar level that insulin alone might cause.

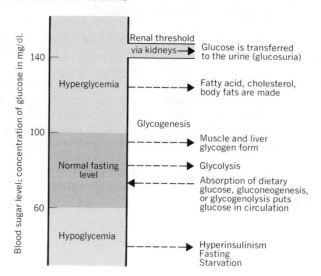

FIGURE 20.2
Factors that affect the blood sugar level.

20.3   GLUCOSE TOLERANCE

The ability of the body to tolerate swings in the blood sugar level is essential to health.

Your **glucose tolerance** is the ability of your body to manage its blood sugar level within the normal range. We'll take an overview here of the many factors that contribute to glucose tolerance.

When glucose enters the bloodstream from the intestinal tract, some stays in circulation and some is removed by various tissues. Muscle cells, for example, trap glucose and use it either to make ATP or to replenish glycogen reserves. Liver cells can similarly trap glucose, but the liver, unlike muscle tissue, is able to release glucose back into the bloodstream when the blood sugar level must be raised.

Glycolysis can happen to glucose and so lower the blood sugar level. The end product, as we learned in Chapter 19, is either pyruvate or lactate, depending on the oxygen supply. When extensive anaerobic glycolysis occurs, however, the lactate level in blood increases.

Neo, new; *neogenesis*, new creation; *gluconeogenesis*, the synthesis of new glucose.

Lactate still has useful chemical energy, and much is conserved. For roughly every six lactates, liver cells are able to convert five to glucose. This synthesis of glucose from smaller molecules is called **gluconeogenesis**, and it requires triphosphate energy. The sixth lactate is catabolized to make it. Gluconeogenesis, therefore, is a process for increasing the blood sugar level. (We'll learn more about it at the end of this section.)

We can see from all these processes that many factors affect the blood sugar level. Some tend to raise it and some do the opposite. Figure 20.2 summarizes them in a different kind of display.

The Glucose Tolerance Test Measures Glucose Tolerance.   In the **glucose tolerance test,** the individual is given a drink that contains glucose, generally 75 g for an adult and 1.75 g/kg of body weight for children. Then the blood sugar level is checked at regular intervals of time.

Figure 20.3 gives typical plots of this level versus time. The lower curve is that of a person with normal glucose tolerance, and the upper curve is of one whose glucose tolerance is typical of a person with diabetes. In both, the blood sugar level rises sharply at first. The

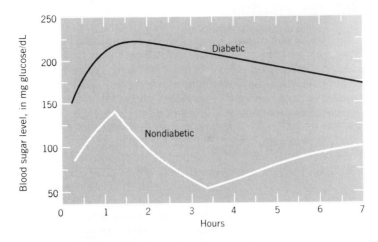

FIGURE 20.3
Glucose tolerance curves.

healthy person, however, soon manages the high level and brings it back down with the help of a normal flow of insulin and somatostatin. In the diabetic, the level comes down only very slowly and remains essentially in the hyperglycemic range throughout.

Notice that in the normal individual, the blood sugar level can sometimes drop to a mildly hypoglycemic level. Such hypoglycemia is possible also in someone who has eaten a carbohydrate-rich breakfast. With glucose pouring into the bloodstream, the release of a bit more insulin than needed can occur. This leads to the overwithdrawal of glucose from circulation, and midmorning brings dizziness and sometimes even fainting (even falling asleep in class). The prevention isn't more sugared doughnuts but a balanced breakfast.

Unhappily, most people with midmorning sag go into another round of sugared coffee and sugared rolls. The glucose gives a short lift, but then an oversupply of insulin restores the mild hypoglycemia of the sag.

Glucose Tolerance Is Poor in Diabetes. The subject of glucose tolerance is nowhere more discussed than in connection with diabetes. **Diabetes** is defined clinically as a disease in which the blood sugar level persists in being much higher than warranted by the dietary and nutritional status of the individual. Invariably, a person with untreated diabetes has glucosuria, and the discovery of this condition often triggers the clinical investigations that are necessary to rule diabetes in or out.

As discussed in Special Topic 20.2, there are two broad kinds of diabetes, type I and type II. Type I diabetics are unable to manufacture insulin (at least not enough), and they need daily insulin therapy to manage their blood sugar levels. Maintaining a relatively even and normal blood sugar level is the best single strategy that such diabetics have for the prevention of some of the vascular and neural problems that can complicate their health later. Most type II diabetics are able to manage their blood sugar levels by a good diet, weight control, exercise, and sometimes the use of oral medications.

Gluconeogenesis Is Not Quite the Exact Reverse of Glycolysis. The overall scheme of **gluconeogenesis,** by which glucose is made from smaller molecules, is given in Figure 20.4. Excess lactate as well as several amino acids can be used as a starting materials. GTP in the figure stands for guanosine triphosphate, another high-energy phosphate like ATP. Thus triphosphate energy is needed, as we'd expect, to take small molecules back up an energy hill to complex, energy-rich carbohydrates.

Three steps in glycolysis cannot be directly reversed, those involving the curved arrows in heavy print in Figure 20.4. However, the liver and the kidneys have special enzymes that create bypasses. They are integral parts of the enzyme team for gluconeogenesis. The other steps in gluconeogenesis are run as reverse shifts in equilibria that occur in the opposite direction in glycolysis.

The name for this disorder is from the Greek *diabetes,* to pass through a siphon, and *mellitus,* honey sweet— meaning to pass urine that contains sugar. We'll call it diabetes for short. In severe, untreated diabetes, the victim's body wastes away despite efforts to satisfy a powerful thirst and hunger. To the ancients, it seemed as if the body were dissolving from within.

In the United States there are nearly 6.5 million known cases of diabetes, and almost as many others are believed to have this disease. It ranks third behind heart disease and cancer as the cause of death.

Fewer than 10% of all cases of diabetes are of the severe, insulin-dependent variety in which the β cells of the pancreas are unable to make and secrete insulin. This is **type I diabetes,** and insulin therapy is essential. It is also called **insulin-dependent diabetes mellitus** or **IDDM.** In the past, because most victims contracted it before they were 20, it was called *juvenile-onset diabetes.*

The rest of all those with diabetes have a form called **type II diabetes, or noninsulin-dependent diabetes mellitus, NIDDM.** Most victims are able to manage their blood sugar levels by diet and exercise alone, without insulin injections. Their problem is actually not a lack of insulin but with a breakdown in the machinery for taking advantage of it. Most get type II diabetes when they are over 40, so it has been called *adult-onset diabetes.*

Type I Diabetes Develops in Six Stages D. S. Eisenbarth, a diabetes specialist, divides the onset of type I diabetes into six stages. We'll review them as background for illustrations of equilibrium chemistry and factors that shift equilibria.

The first stage is simply a genetic condition. Both people and experimental animals who develop type I diabetes have at least one defective gene, almost certainly more, that contributes to their susceptibility to this disease. At least one such gene is among the major histocompatibility genes that code for the antigens of tissue transplantation. (An *antigen* is an alien chemical that the immune system's *antibodies* destroy.) This genetic problem can now be recognized by various assays before the individual has any other symptoms of diabetes.

The second stage is a triggering incident. It can be a viral infection. The mumps virus, for example, causes diabetes in some. Usually, the onset of virus-caused type I diabetes occurs slowly over a few years.

The third stage is the appearance in the blood of certain antibodies. A virus, for example, can alter substances on the membranes of the pancreatic β cells so that the body's immune system sees them as foreign antigens and so makes antibodies against them. In this sense, type I diabetes might be caused by antibodies against its own pancreatic β cells, and so be an autoimmune disease. This means that the body's immune system fails to recognize the proteins of its own body and sets out to destroy them and, therefore, itself. The presence of such antibodies can also be detected before diabetes happens.

The fourth stage in the onset of type I diabetes is the gradual loss of the ability to secrete insulin. How well the pancreas secretes insulin in response to a dose of glucose given to someone in a fasting state can be measured, and this ability shows a decline before the traditional symptoms of diabetes appear.

The fifth stage is diabetes and persistent hyperglycemia. Most of the pancreatic β cells have disappeared.

In the sixth stage the destruction of β cells is complete.

Immune-Suppressant Therapy Works If the Problem Is Caught in Its Early Stages Because the immune system becomes involved with an autoimmune reaction during the onset of type I diabetes, then one type of treatment might be to use drugs, like cyclosporin, that suppress the immune system. Cyclosporin is used to suppress the rejection of transplanted organs, like kidney transplants. When used in the early stages of the onset of type I diabetes, cyclosporin prevents insulin dependence in a significant fraction of individuals tested.

Insulin Receptors Are a Problem in Type II Diabetes The onset of type II diabetes is much slower than that for type I, and obesity is a factor in most cases. The number of insulin-receptor proteins on the surfaces of target cells declines in obesity. Obesity usually involves a diet that is far richer in sugars and other carbohydrates than normal, and some scientists believe these substances evoke such a continuous presence of insulin that the receptor proteins literally wear out faster than they can be replaced. When the weight is reduced, particularly in connection with physical exercise, the relative numbers of receptor proteins rebound.

A mutation of the insulin receptor protein has been found to occur in some with non-insulin-dependent diabetes. This renders the receptor less effective and the victims have a high insulin resistance.

SPECIAL TOPIC 20.2   (CONTINUED)

Glycosylation of Proteins May Cause the Long-Term Complications of Diabetes The immediate complications are an elevated blood sugar level, metabolic acidosis, and eventual death from coma and uremic poisoning. Insulin therapy corrects these immediate problems, but it deals less well with the longer term complications.

The continuous presence of a high level of blood glucose shifts certain chemical equilibria in favor of glycosylated compounds. The aldehyde group of the open form of glucose, for example, can react with amino groups to form products called *Schiff bases*.

$$-CH=O + H_2N- \rightleftharpoons -CH=N- + H_2O$$

| Aldehyde | Amino | A Schiff |
| group | group | base |

Hemoglobin, for example, gives this reaction, and a high level of glucose shifts this equilibrium to the right. The level of glycosylated hemoglobin thus increases. (As we have mentioned, the measurement of the level of this compound is now regarded as the best way to monitor how well a person is managing his blood sugar level.)

Any material to which glucose has access and which has NH_2 groups—all proteins and genes, for example—can be glycosylated. Since the reaction is reversible, when the glucose level is brought down and kept within a normal range, the Schiff base level also declines.

The problem with the Schiff bases in the long term is that they undergo molecular rearrangements that give more permanent products, called Amadori compounds, in which the C=N double bond has migrated to C=C positions. After a time, the formation of the Amadori compounds is not reversible.

One such complication that probably is caused by these reactions occurs in the basement membrane of blood capillaries. They thicken as diabetes progresses; the condition is called *microangiopathy*. (The basement membrane is the protein support structure that encases the single layer of cells of a capillary.) Microangiopathy is believed to lead to the other complications, most of which involve the vascular system or the neural networks: kidney problems, gangrene of the lower limbs, and blindness.

Diabetes is the leading cause of new cases of blindness in the United States, and it is the second most common cause of blindness, overall. (During an eye examination ophthalmologists can detect the development of microangiopathy in the retina of the eye before other symptoms of diabetes are recognized.)

Blindness from Diabetes May Also Reflect the Reduction of Glucose to Sorbitol Glucose is reduced by the enzyme aldose reductase to sorbitol. It's a minor reaction in cells of the lens of the eye, but an abundance of glucose shifts equilibria in favor of too much sorbitol. Sorbitol, unlike glucose, tends to be trapped in lens cells, and as the sorbital concentration rises so does the osmotic pressure in the fluid. This draws water into the lens cells, which generates pressure and leads to cataracts.

The same kinds of swelling might occur in peripheral nerve cells, too, and cause them to deteriorate and no longer be able to assist in motor nerve functions. The osmotic swelling might also be the cause of poorer circulation into peripheral capillaries, which eventually opens the way for gangrene or for a breakdown of the filtration mechanisms in the kidneys.

Diet Control Is Mandatory The nature of the diet—both quality and quantity—does appear to be a decisive factor in the onset of type II diabetes. In countries with very low levels of refined sugars in the national diets, the incidence of type II diabetes is rare.

The best single treatment of diabetes is any effort that keeps strict control on the blood sugar level. What must at all costs be avoided are the episodes of upward surges followed by precipitous declines as either the diet or the insulin treatments are not well managed.

β-Cell Transplants May Bring a Cure to Type I Diabetes In the early 1980s, scientists discovered that β cells, stripped of neighboring cells, can be transplanted. They need not even be inserted into the receiver's pancreas, and they start to make insulin in a few weeks. Apparently the cells *adjacent to* the β cells are responsible for inducing the body's immune-centered rejection process.

Human β cells work best, of course, but those from pigs and cows also appear to be usable provided that they are encapsulated in very small spheres. These spheres have microscopic holes large enough to let insulin molecules escape but not large enough to let antibodies inside. These techniques have cured type I diabetes in experimental animals.

When diabetics have reached the stage where enough kidney damage has occurred to warrant a kidney transplant, both a kidney and a pancreas are sometimes transplanted in the same operation. The success rate of this procedure for the management of diabetes has been encouraging.

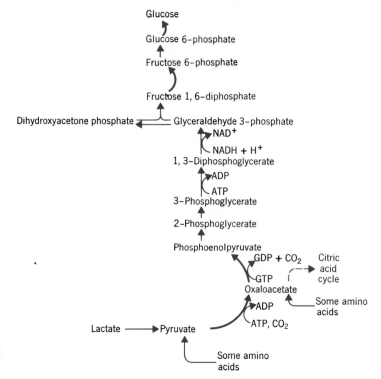

FIGURE 20.4
Gluconeogenesis. The straight arrows signify steps that are the reverse of corresponding steps in glycolysis. The heavy, curved arrows denote steps that are unique to gluconeogenesis.

20.4 ABSORPTION, DISTRIBUTION, AND SYNTHESIS OF LIPIDS

Several lipoprotein complexes in the blood transport triacylglycerols, fatty acids, cholesterol, and other lipids.

The complete digestion of triacylglycerols produces glycerol and a mixture of long-chain fatty acids. These, together with some monoacylglycerols (from incomplete digestion), leave the digestive tract. As they migrate across the intestinal barrier they are extensively reconstituted so that what enters circulation consists mostly of triacylglycerols.

Lipids are insoluble in water, and they are carried in blood by proteins with which they form **lipoprotein complexes**. Defects in this system can cause heart disease, as discussed in Special Topic 20.3, page 424, a survey of recent developments in the relationship of dietary fat, cholesterol and heart attacks.

> Remember that cholesterol is a nonsaponifiable lipid so it can't be hydrolyzed by digestion.

Adipose Tissue Is the Chief Lipid Storage Depot. There are two kinds of adipose tissue, brown and white. Both are associated with internal organs, and they cushion the organs against bumps and shocks. They also insulate them from swings in temperature. White adipose tissue stores energy as triacylglycerols chiefly on behalf of the energy budgets of other tissues. Brown adipose tissue appears to serve simply to generate heat, by the catabolism of fatty acids, to keep the temperature of the body's core steady.

> Because lipids are water insoluble, they attract the least amount of associated water in storage.

A 70-kg adult male has about 12 kg of triacylglycerol in storage. If he had to exist on no food, just water and a vitamin–mineral supplement, and if he needed 2500 kcal/day, this fat would supply his caloric needs for 43 days. Of course, during this time the body proteins would also be wasting away to serve gluconeogenesis, and metabolic acidosis would be a problem of growing urgency.

> These data are for information; they're certainly not recommendations!

Fatty Acids Can Be Made from Acetyl CoA. Acetyl CoA stands at a major metabolic crossroads, as we saw in Figure 20.1. It can be made from any monosaccharide in the diet, from virtually all amino acids, and from fatty acids. Once made, it can be shunted into the citric acid cycle where its chemical energy can be used to make ATP; or its acetyl group can be made into other compounds that the body might need, such as fatty acids. We'll look in some detail at **lipigenesis,** the synthesis of fatty acids. It provides an excellent example of a multienzyme complex in action.

Whenever acetyl Coenzyme A molecules are made within mitochondria but aren't needed for the citric acid cycle and respiratory chain, they are exported to the cytosol. The enzymes for lipigenesis are found there, not within the mitochondria, which illustrates the general rule that the body segregates catabolism from anabolism.

The cytosol is the fluid outside of cellular organelles such as mitochondria and nuclei.

As might be expected, because lipigenesis is in the direction of climbing an energy hill, the cell has to invest some energy of ATP to make fatty acids from smaller molecules. The first payment occurs in the first step in which the bicarbonate ion reacts with acetyl CoA.

$$CH_3-\overset{\displaystyle O}{\overset{\displaystyle \|}{C}}-SCoA \; + HCO_3^- + ATP \longrightarrow$$

Acetyl CoA

$$^-O-\overset{\displaystyle O}{\overset{\displaystyle \|}{C}}-CH_2-\overset{\displaystyle O}{\overset{\displaystyle \|}{C}}-SCoA + 2H^+ + ADP + P_i$$

Malonyl CoA

Because the symbols ATP, ADP, and P_i are not given with their electrical charges, we cannot provide an electrical balance to equations such as this.

This activates the acetyl system for lipigenesis.

The enzyme that now takes over is actually a huge complex of seven enzymes called *fatty acid synthase* (see Figure 20.5). In the center of this complex is a molecular unit long enough to serve as a swinging arm carrier. It's called the **acyl carrier protein,** or ACP, and like the boom of a construction crane, this arm swings from site to site in the synthase. Thus the arm brings what it carries over first one enzyme and then another, and at each stop a reaction is catalyzed that contributes to chain lengthening. Let's see how it works.

Enzymes

1. Malonyl transferase
2. 3-Ketoacyl-ACP synthase
3. 3-Ketoacyl-ACP reductase
4. 3-Hydroxyacyl-ACP dehydratase
5. Enoyl-ACP reductase
6. Acetyl transferase
7. Acyl carrier protein

FIGURE 20.5
The lipigenesis cycle. At the top, an acetyl group is activated and joined as a malonyl unit to an arm of the acyl carrier protein, ACP. Another acetyl group transfers from acetyl CoA to site E. In a second transfer, this acetyl group and is then joined to the malonyl unit as CO_2 splits back out. This gives a β-ketoacyl system whose keto group is reduced to CH_2 by the next series of steps. One turn of the cycle adds a CH_2CH_2 unit to the growing acyl chain.

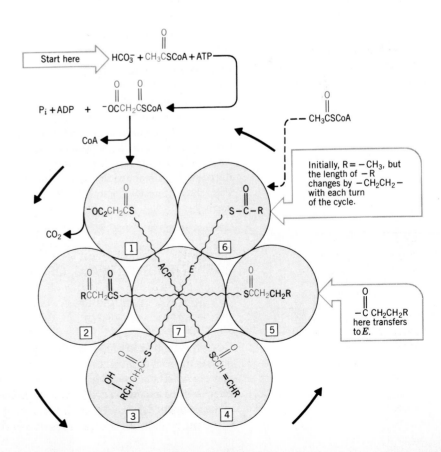

The malonyl unit in malonyl CoA, which we just made, transfers to the swinging arm of ACP. In the meantime, a similar reaction occurs to another molecule of acetyl CoA at a different site on the synthase, called simply E.

Next, the acetyl group on E transfers to the malonyl group on ACP. As it does, carbon dioxide, the initial activator, is ejected; it has served its activating purpose. A four-carbon derivative of ACP, acetoacetyl ACP, forms. The E unit is vacated.

$$CH_3\overset{O}{\underset{\|}{C}}-S-E + {}^-O\overset{O}{\underset{\|}{C}}CH_2\overset{O}{\underset{\|}{C}}-S-ACP \longrightarrow CH_3\overset{O}{\underset{\|}{C}}CH_2\overset{O}{\underset{\|}{C}}-S-ACP + CO_2 + E$$
Acetoacetyl ACP

Acetoacetyl ACP, now on the swinging boom, is moved over first one enzyme and then another until its keto group is reduced to CH_2. First, the keto group is reduced to a 2° alcohol. Then this is dehydrated to introduce a double bond. And the double bond is next reduced to the saturated system in butyryl ACP, and we have reduced the keto group to CH_2. Notice that NADPH, the reducing agent made by the pentose phosphate pathway of glucose catabolism, is used here, not NADH.

The butyryl group is now transferred to the vacant E unit of the synthase, the unit that initially held an acetyl group. This ends one complete turn of the cycle. To recapitulate, we have gone from two two-carbon acetyl units to one four-carbon butyryl unit. The steps now repeat so that the C_4 unit is elongated to a six-carbon unit.

In the next turn, this six-carbon acyl group will be elongated to an eight-carbon group. And the process will repeat until the chain is as long as the system requires. Overall, the net equation for the synthesis of the palmitate ion from acetyl CoA is as follows, and you can see how much high-energy phosphate (ATP) and how much reducing agent (NADPH) is required per mole of palmitate ion.

$$8CH_3\overset{O}{\underset{\|}{C}}SCoA + 7ATP + 14NADPH \longrightarrow$$
$$CH_3(CH_2)_{14}CO_2^- + 7ADP + 7P_i + 8CoASH + 14NADP^+ + 6H_2O$$

Cholesterol Is Also Made from Acetyl CoA. Every membrane of every cell in the body requires cholesterol, and cholesterol is the raw material for making steroid hormones and bile salts. If insufficient cholesterol comes in from the diet, the liver has enzyme systems that catalyze its synthesis from acetyl CoA. Over two dozen steps are involved.

One of the enzymes is deactivated by cholesterol, so it illustrates feedback control of a biochemical process. If the system functions normally, whenever the supply of cholesterol is sufficiently high, its further synthesis remains switched off. But when the supply falls enough, the cholesterol deactivating the enzyme is removed for use elsewhere, and cholesterol synthesis restarts. Special Topic 20.3 tells what can happen when the system isn't functioning normally.

Steroid nucleus

Cholesterol

Essential Amino Acids for Adults*

Isoleucine
Leucine
Lysine
Methionine
Phenylalanine
Threonine
Tryptophan
Valine

* These plus histidine are believed to be essential to infants.

20.5 THE SYNTHESIS OF AMINO ACIDS

The body can manufacture a number of amino acids from intermediates in the catabolism of nonprotein substances.

Certain Amino Acids Must Be Provided by the Diet. We do not need all of the 20 amino acids in the diet, because we can make roughly half of them. Those that we *must* obtain by the diet are called the **essential amino acids,** named in the margin. *Essential* in this context refers *only* to a *dietary* need. In a larger context, the body must have all 20 amino acids to make polypeptides.

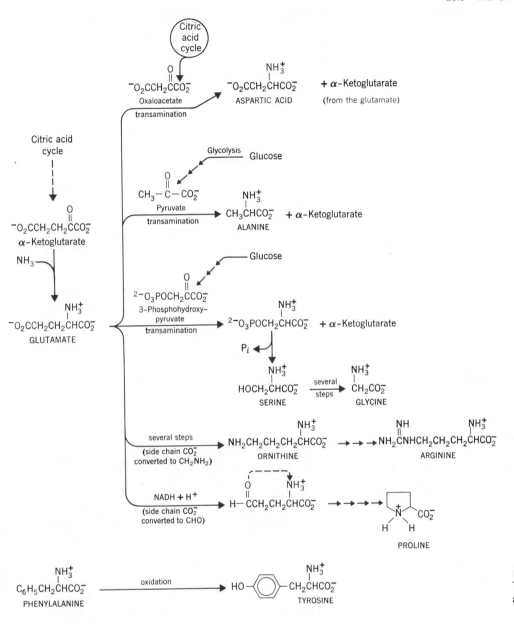

FIGURE 20.6
The biosynthesis of some nonessential amino acids.

A number of food materials, particularly vegetables and grains, are deficient in one or more essential amino acids. Rice, for example, is low in lysine. Beans are low in valine. Corn carries too little tryptophan and lysine. Soybeans are low in methionine and cysteine. So is cassava, a root widely used in many third-world countries. If any one of these is the sole food in the diet, severe malnutrition results.

Kwashiorkor, from a Ghan dialect meaning the condition the first-born enters when the second-born arrives, is one of the world's widespread protein-deficiency diseases. While the first-born receives mother's milk, it gets perfectly balanced protein. There is none finer, although cow's milk and egg white are nearly as good. When the second-born arrives, the first goes onto a diet mostly of corn or cassava gruel with little or no milk, eggs, or meat. Now it gets too little of an essential amino acid to make its proteins, and the victim develops patchy skin, discolored hair, a bloated look, and falls behind developmentally.

Lipids Are Carried in Lipoprotein Complexes Lipids are insoluble in water, and they are carried in blood by proteins in packages called **lipoprotein complexes.** Defects in this system can cause heart disease, so we'll look more closely at these species.

Lipoproteins are classified according to their densities, and each class has its own functions. They range in densities from 0.93 to 1.21 g/cm^3, and the lipoproteins with the lowest density are called **chylomicrons.** They are only 2% protein or less and are put together in the liver. The accompanying figure shows what happens to them. The numbers that follow refer to the numbers in this figure. Some of the carrier units in the figure have letters, like E, C, B-48, and B-100. They represent polypeptides, and they not only participate as lipid carriers but as the species that receptor proteins of other tissues, like the liver, can recognize. For example, B-100 is the polypeptide recognized by one key receptor at the liver.

Chylomicrons Carry Lipids to the Liver Where lipids from the processes of digestion $\boxed{1}$ enter circulation $\boxed{2}$ chylomicrons form around any that appear: triacylglycerols, cholesterol, and free fatty acids. They transport these lipids $\boxed{3}$ and while they are in capillaries of adipose tissue, they unload some of their triacylglycerols $\boxed{4}$. (A lipoprotein lipase catalyzes the hydrolysis of triacylglycerols. The resulting fatty acids and glycerol are absorbed by adipose tissue and reconstituted as triacylglycerols that are stored until needed.)

This process leaves chylomicron remnants $\boxed{5}$, which are now richer in cholesterol, and the liver has special receptor proteins that recognize and help the liver absorb these remnants $\boxed{6}$.

Liver Cholesterol Is Assigned to Various Needs The liver sees cholesterol from two sources, the diet and its own cholesterol-making cells. It can excrete cholesterol by way of the bile, which goes from the gallbladder into the intestinal tract. It can make bile salts from cholesterol. These salts are needed to help digest lipids and absorb fat-soluble vitamins. Finally, the liver can send cholesterol into circulation to be used in tissues for the making of cell membranes and steroid hormones.

The Liver Packages Cholesterol into VLDL Complexes The cholesterol to be sent into circulation, both as free cholesterol and as fatty acid esters of cholesterol, is organized into another lipoprotein complex designated very low density lipoprotein, or VLDL, for short $\boxed{7}$. This has a slightly higher density than the chylomicrons.

The liver, as we said, can both receive and make cholesterol, and it also receives and makes triacylglycerols. Thus the VLDL complexes that are now put together by the liver carry, besides a small amount of protein, cholesterol, cholesterol esters, and triacylglycerols, regardless of their origins, to adipose tissue and muscles $\boxed{8}$.

These tissues remove most of the triacylglycerols from the VLDL units (again via hydrolysis and reconstitution), which leaves a lipoprotein complex $\boxed{9}$ of slightly higher density. It is mostly cholesterol in one form or another, and the complex is now designated as an intermediate-density lipoprotein complex, or IDL for short.

Liver Receptor Proteins Accept Back About Half of the IDL The liver has special receptor proteins that recognize the B-100 protein in IDL, which helps take IDL packages back into the liver $\boxed{10}$. The IDL that is not reabsorbed here experiences further losses of triacylglycerol. This causes a further increase in its density, and now the particles are classified as low density lipoproteins, or LDL $\boxed{11}$.

A good portion of the LDL is captured by the liver by receptor proteins. In fact, transporting cholesterol back to the liver $\boxed{12}$ is one function of LDL. However, about a third of the LDL reaches peripheral tissues $\boxed{13}$ including the adrenal glands. Thus LDL carries cholesterol wherever cholesterol is needed to make cell membranes or to make steroid hormones. Some of these tissues carry special receptor proteins, but others can obtain cholesterol from LDL by other means.

HDL Units Carry Unused Cholesterol Back to the Liver Any leftover cholesterol must now be removed from the extrahepatic tissues (tissues other than the liver), and this job is handled by the high density lipoprotein complexes, or HDL $\boxed{14}$, which the liver makes for this purpose. The chief function of the HDL is to carry cholesterol back to the liver $\boxed{15}$.

The Absence of Liver Receptor Proteins Causes Atherosclerosis The receptor proteins for the IDL and LDL have a crucial function. If they are reduced in number or are absent, there is little ability by the liver to remove excess cholesterol and export it via the bile. The level of cholesterol in the blood, therefore, becomes too high, and this is a cause of atherosclerosis. (This is a disease in which several substances, including collagen, elastic fibers, triacylglycerols, but chiefly cholesterol and its esters, form plaques on the inside of the arterial wall. It is the chief cause of heart attacks.)

Some people have a genetic defect that bears specifically on the liver receptors. Two genes are involved. Those who carry two mutant genes have *familial hypercholesterolemia,* a genetically caused high level of cholesterol in the blood. Even on a zero-cholesterol diet, the victims have very high cholesterol levels. Their cholesterol slowly comes

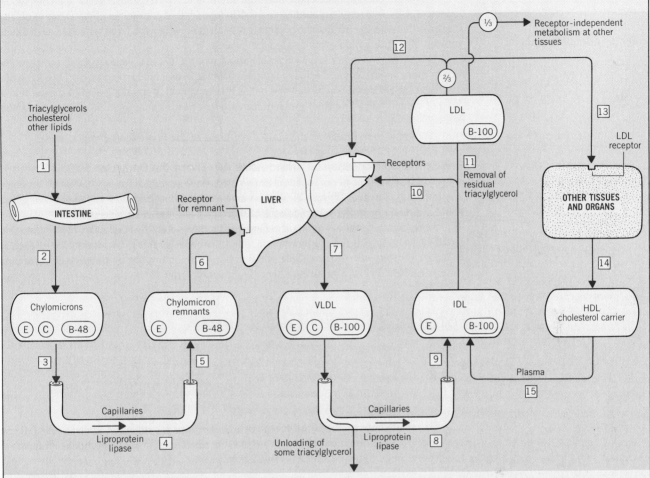

The transport of cholesterol and triacylglycerols by lipoprotein complexes. (Adapted by permission from J. L. Goldstein, T. Kita, and M. S. Brown, *New England Journal of Medicine,* August 4, 1983, page 289.)

out of the blood at valves and other sites, reduces the dimensions of the blood capillaries, and this restricts blood flows. Atherosclerosis has set in.

The heart must now work harder, and eventually arteries and capillaries in the heart itself no longer are able to bring oxygen to heart tissue. The victims generally have their first heart attacks as children and are dead by their early twenties. People with one defective gene and one normal gene for the LDL receptor proteins generally have blood cholesterol levels that are two or three times higher than normal. Although they number only about 0.5% of all adults, they account for 5% of all heart attacks among those younger than 60.

High blood cholesterol levels also occur in many other people, even those with normal genes for the receptor proteins, and the causes have not been fully unraveled. High-cholesterol foods appear to contribute to this problem, and there is some evidence that as the liver receives more and more cholesterol from the diet it loses more and more of the receptor proteins. This forces more and more cholesterol to linger in circulation. Smoking, obesity, and lack of exercise contribute to the cholesterol problem also.

A protein with all essential amino acids in the right proportions is called an **adequate protein.**

To obtain enough of all of the essential amino acids from cassava alone, the child would have to eat 27.5 lb/day, amounting to 16,400 kcal/day of energy, an obvious impossibility. Contrast this with the 35 g of the protein in human milk, which is all that is needed each day to provide all of the amino acids needed for growth.

Vegetarians who are knowledgeable about what to eat and when to eat it can combine two or more vegetables and achieve the right balance in essential amino acids. Thus 43 g of a 1 : 1 mixture of rice (low in lysine) and beans (low in valine) is equal in value to the protein in 35 g of human milk. The chief problem for vegetarians is vitamin B_{12}. No grain or vegetable has it.

Let's now look briefly at the body's synthesis of the nonessential amino acids.

The So-Called Nonessential Amino Acids Are Those the Body Can Make. The pathways for the synthesis of nonessential amino acids have several steps, and we won't examine any in detail. However, Figure 20.6 gives an overview that illustrates in general terms how some are made from the intermediates of glycolysis and the citric acid cycle.

Many of the syntheses outlined in Figure 20.6 depend on the availability of glutamic acid (actually, at body pH, the glutamate ion), and it is made from α-ketoglutarate by a reaction called **reductive amination.** This reaction uses ammonium ion as a source of the amino group and NADPH as a reducing agent. The overall result is

$$\overset{\overset{\textstyle O}{\|}}{^-O_2CCH_2CH_2CCO_2^-} + NH_4^+ + NADPH + H^+ \rightleftharpoons$$
α-Ketoglutarate

$$\overset{\overset{\textstyle NH_3^+}{|}}{^-O_2CCH_2CH_2CHCO_2^-} + NADP^+ + H_2O$$
Glutamate

Glutamate now becomes the source of amino groups for still other amino acids that can be made by a reaction called **transamination,** the transfer of an amino group. We'll illustrate the general case; Figure 20.6 has several examples.

$$\overset{\overset{\textstyle O}{\|}}{G-C-CO_2^-} + \overset{\overset{\textstyle NH_3^+}{|}}{^-O_2CCH_2CH_2CHCO_2^-} \rightleftharpoons \overset{\overset{\textstyle NH_3^+}{|}}{G-CH-CO_2^-} + \overset{\overset{\textstyle O}{\|}}{^-O_2CCH_2CH_2CCO_2^-}$$
A keto acid   Glutamate   An amino acid α-Ketoglutarate

The enzymes for transaminations, the *transaminases,* use a B vitamin, pyridoxal, to make their cofactors.

SUMMARY

Glycogen metabolism The regulation of glycogenesis and glycogenolysis is a part of the machinery for glucose tolerance in the body. Hyperglycemia stimulates the secretion of insulin and somatostatin, and insulin helps cells of adipose tissue to take glucose from the blood. Somatostatin helps to suppress the release of glucagon (which otherwise stimulates glycogenolysis and leads to an increase in the blood sugar level).

When glucose is abundant, the body either replenishes its glycogen reserves, or it makes fat. In danger or other stresses, epinephrine stimulates the release of glucose from glycogen.

When glucose is in short supply, the body makes its own by gluconeogenesis from noncarbohydrate molecules, including several amino acids. In diabetes, some cells that are starved for glucose make their own, also. Such cells are unable to obtain glucose from circula-

tion, so the blood sugar level is hyperglycemic to a glucosuric level. The glucose tolerance test is used to see how well the body handles an overload of glucose. In the management of diabetes, the maintenance of a reasonably steady blood sugar level in the normal range is vital.

Gluconeogenesis Most of the steps in gluconeogenesis are simply the reverse of steps in glycolysis, but there are a few that require rather elaborate bypasses. Special teams of enzymes and supplies of high-energy phosphate energy are used for these. Many amino acids can be used to make glucose by gluconeogenesis.

Lipid absorption and distribution As fatty acids and glycerol migrate out of the digestive tract they become reconstituted as triacylglycerols. The adipose tissue is the principal storage site, and fatty material comes and goes from this tissue according to the energy budget of the body.

Biosynthesis of fatty acids Fatty acids can be made by a repetitive cycle of steps called the lipigenesis cycle. It begins by building one butyryl group from two acetyl groups. The four-carbon butyryl group is attached to an acyl carrier protein that acts as a swinging arm on the enzyme complex. This arm moves the growing fatty acyl unit over first one enzyme and then another as additional two-carbon units are added. The process consumes ATP and NADPH.

Biosynthesis of cholesterol Cholesterol is made from acetyl groups by a long series of reactions. The synthesis of one of the enzymes is inhibited by excess cholesterol, which gives the system a mechanism for keeping its own cholesterol synthesis under control.

Amino acid synthesis Ten of the amino acids are essential amino acids, meaning they must be in the diet. Vegetables and grains are deficient in at least one essential amino acid, but when properly combined, they can furnish a balanced protein diet. Processes of transamination and reductive amination make nonessential amino acids from intermediates of glucose and fatty acid catabolism.

REVIEW EXERCISES

The answers to these review exercises are in the *Study Guide* that accompanies this book.

Glycogen Metabolism and Blood Sugar

20.1 What are the end products of the complete digestion of the carbohydrates in the diet?

20.2 Why can we treat the catabolism of carbohydrates as almost entirely that of glucose?

20.3 What is meant by *blood sugar level?* By *normal fasting level?*

20.4 What is the range of concentrations in mg/dL for the normal fasting level of whole blood?

20.5 What characterizes the following conditions?
(a) Glucosuria   (b) Hypoglycemia
(c) Hyperglycemia   (d) Glycogenolysis
(e) Gluconeogenesis (f) Glycogenesis

20.6 Explain how severe hypoglycemia can lead to disorders of the central nervous system.

Hormones and the Blood Sugar Level

20.7 When epinephrine is secreted, what soon happens to the blood sugar level?

20.8 At which one tissue is epinephrine the most effective?

20.9 What is the end product of glycogenolysis, and what does phosphoglucomutase do to it?

20.10 Why can liver glycogen but not muscle glycogen be used to resupply blood sugar?

20.11 What is glucagon, what does it do? What is its chief target tissue?

20.12 Which is probably better at increasing the blood sugar level, glucagon or epinephrine? Explain.

20.13 How does human growth hormone manage to promote the supply of the energy needed for growth?

20.14 What is insulin? Where is it released? What is its chief target tissue?

20.15 What triggers the release of insulin into circulation?

20.16 If brain cells are not insulin-dependent cells, how can too much insulin cause insulin shock?

20.17 What is somatostatin? Where is it released? What kind of effect does it have on the pancreas?

Glucose Tolerance

20.18 What is meant by *glucose tolerance?*

20.19 For what chief purpose is a glucose tolerance test performed, and how is it carried out?

20.20 Describe what happens when each of the following types has a glucose tolerance test.
(a) A nondiabetic individual.   (b) A diabetic individual.

20.21 Describe a circumstance in which hyperglycemia might arise in a nondiabetic individual.

Gluconeogenesis

20.22 In a period of prolonged fasting or starvation, what does the system do to try to maintain its blood sugar level?

20.23 Amino acids are not excreted, and they are not stored in the same way that glucose residues are stored in a polysaccharide. What probably happens to the excess amino acids in a high-protein diet of an individual who does not exercise much?

20.24 The amino groups of amino acids can be replaced by keto groups. Which amino acids (write their structures) could give the following keto acids that participate in carbohydrate metabolism?
(a) Pyruvic acid   (b) Oxaloacetic acid

Absorption and Distribution of Lipids

20.25 What are the end products of the digestion of triacylglycerols?

20.26 What happens to the products of the digestion of triacylglycerols as they migrate out of the intestinal tract?

Biosynthesis of Fatty Acids and Cholesterol

20.27 Where are the principal sites for each activity in a liver cell?
(a) Fatty acid catabolism   (b) Lipigenesis

20.28 What metabolic pathway in the body is the chief supplier of NADPH for lipigenesis?

20.29 How does cholesterol itself work to inhibit the activity of an enzyme needed for its synthesis?

Biosynthesis of Amino Acids

20.30 What makes some amino acids *essential?*

20.31 Alanine is not on the list of essential amino acids. Why?

20.32 Write the equation for the reductive amination that produces glutamate. Use NADPH as the reducing agent.

20.33 Write the structure of the keto acid that forms when phenylalanine undergoes transamination with α-ketoglutarate.

20.34 When valine and α-ketoglutarate undergo transamination, what new keto acid forms? Write its structure.

Glycogen Storage Diseases (Special Topic 20.1)

20.35 For each of the following diseases, name the defective (or missing) enzyme, and state the biochemical and physiological consequences.
(a) Von Gierke's disease
(b) Cori's disease
(c) McArdle's disease
(d) Andersen's disease

Diabetes Mellitus (Special Topic 20.2)

20.36 What is the biochemical distinction between type I and type II diabetes?

20.37 Juvenile-onset diabetes is usually which type?

20.38 Adult-onset diabetes is usually which type?

20.39 Briefly state the six stages in the onset of type I diabetes.

20.40 Viruses that cause diabetes attack which target cells?

20.41 There is some evidence that diabetes is an autoimmune disease. What does this mean, as applied to diabetes?

20.42 A sustained, elevated blood glucose level causes damage to which specific tissue, damage that might be responsible for other complications?

20.43 How is glucose involved in the formation of a Schiff base? What other kinds of compounds react with glucose in this way?

20.44 When the glucose level in blood drops, what happens to the level of glycosylated hemoglobin? Why?

20.45 What happens to the Schiff bases involving glucose if given enough time? Why is this serious?

20.46 Describe a theory that explains how the hydrogenation of glucose might contribute to blindness.

Lipoprotein Complexes (Special Topic 20.3)

20.47 What are chylomicrons and what is their function?

20.48 What happens to chylomicrons as they move through capillaries of, say, adipose tissue?

20.49 What happens to chylomicrons when they reach the liver?

20.50 What are the two chief sources of cholesterol that the liver exports?

20.51 What do the following symbols stand for?
(a) VLDL   (b) IDL
(c) LDL   (d) HDL

20.52 The loss of what kind of substance from the VLDL converts them into IDL?

20.53 What do IDL release as they change over to LDL?

20.54 What tissue can reabsorb IDL complexes?

20.55 What is the chief constituent of LDL?

20.56 In extrahepatic tissue, what two general uses await delivered cholesterol?

20.57 If the liver lacks the key receptor proteins, which specific lipoprotein complexes can't be reabsorbed?

20.58 Explain the relationship between the liver's receptor proteins for lipoprotein complexes and the control of the cholesterol level of the blood.

20.59 What is the chief job of the HDL?

20.60 Some scientists believe that a relatively high level of HDL provides protection against atherosclerosis, and that one's risk of having heart disease declines when the level of HDL is raised by exercise and losing weight. Why would a low level of HDL tend to promote heart disease?

Chapter 21
Nucleic Acids

The Units of Hereditary
Ribonucleic Acids
mRNA-Directed Polypeptide Synthesis
Viruses

Recombinant DNA Technology and Genetic Engineering
Hereditary Diseases

Dog chow and boy chow, when fully digested, yield essentially the same set of simple chemicals. But after a meal, the different systems use them in ways unique to dogs and boys. The chemistry of heredity tells us how this works.

21.1   THE UNITS OF HEREDITY

Genetic information is carried by the sequence of side-chain bases on deoxyribonucleic acid, DNA.

We have learned that nearly every reaction in an organism requires its special catalyst, and that these are proteins called enzymes. The set of enzymes in one organism is not the same as in another, although many enzymes from different species are similar.

Nucleic Acids Carry Instructions for Making Enzymes.   In reproduction, each organism transmits to its offspring the ability to make a unique set of enzymes. The organism, however, does not duplicate the enzymes themselves and pass them on directly. Rather, it sends on the *instructions* for making the enzymes. It does this by duplicating compounds in a different family, the nucleic acids. These then direct the synthesis of enzymes in the offspring. Our purpose in this chapter is to study how nucleic acids do this.

Genes Are the Units of Heredity.   Each cell nucleus carries intertwined filaments of nucleoprotein called **chromatin.** Chromatin is like a strand of pearls, each pearl made of proteins called *histones,* around which are tightly coiled one of the kinds of nucleic acid, DNA for short. DNA also links the "pearls." Molecules of DNA have sections that constitute individual **genes,** the fundamental units of heredity.

Each "pearl" is called a *nucleosome.*

Each human cell has about 100,000 genes.

Prior to Cell Division, Genes Replicate.   When cell division begins, the chromatin strands thicken and become rodlike bodies called **chromosomes.** The thickening is caused by the synthesis of new nucleic acid (and histone). The new chromatin is an exact copy of the old, if all goes well (as it usually does). During cell division each gene undergoes **replication,** meaning it is reproduced in duplicate. Thus, through the replication of DNA, the genetic message of the first cell is made available to each of the two new cells.

Nucleic Acids Are Polymers Made of Nucleotides.   The **nucleic acids** occur as two types of polymers given the symbols DNA and RNA. **DNA** is **deoxyribonucleic acid,** and **ribonucleic acid** is **RNA.** The monomers for these polymers are called **nucleotides.**

The nucleotides include phosphate ester groups, so they can be hydrolyzed. The hydrolysis of a representative mixture of nucleotides produces three kinds of products: inorganic phosphate, a pentose sugar, and a group of heterocyclic amines called the **bases.** Their structures are given in Figure 21.1, but usually the bases are referred to by their names or their single-letter symbols, as follows:

Bases from DNA	Bases from RNA
Adenine A	Adenine A
Thymine T	Uracil U
Guanine G	Guanine G
Cytosine C	Cytosine C

You can see that three bases are common to both DNA and RNA, and that one base is different.

Another difference between DNA and RNA is the sugar unit. The R in RNA stands for ribose (a pentose) and the D in DNA stands for deoxyribose.

With these hydrolysis products in mind, we can work backwards and see how a nucleic acid is built. First, in Figure 21.2, we look at how a typical nucleotide, adenosine monophosphate (AMP), is made from phosphoric acid, ribose, and its base, adenine. The other nucleotides are put together by using other bases and one or the other of the pentoses.

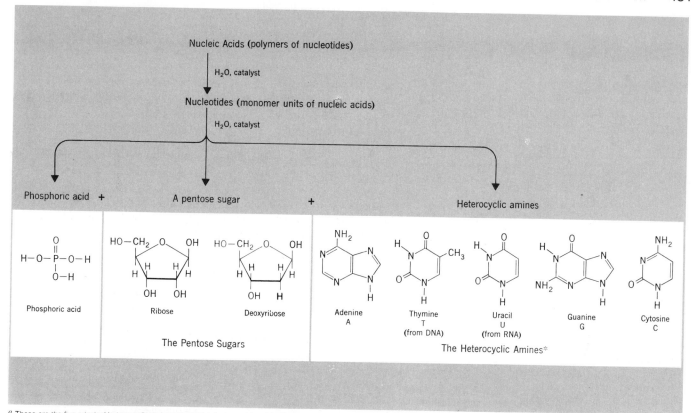

a These are the five principal heterocyclic amines obtainable from nucleic acids. Others, not shown, are known to be present. While they differ slightly in structure they are informationally equivalent to one or another of the five shown here.

FIGURE 21.1
The hydrolysis products of nucleic acids.

The Sequence of Bases Projecting from the Backbones of Nucleic Acids Is How Genetic Information Is Stored. Figure 21.3 shows how a number of nucleotides go together to make a nucleic acid. At least on paper, this is nothing more than the splitting out of water between a phosphate unit of one nucleotide and an alcohol group on the pentose of the next. The result is a phosphodiester system linking sugar unit to sugar unit. Many steps are required, of course, each calling for its own enzyme, but when these are repeated thousands of times, a nucleic acid forms. Thus the pattern for the backbone of a nucleic acid is alternating phosphate and pentose units.

Some 20 enzymes are required, the chief being DNA polymerase, discovered by Arthur Kornberg (Nobel prize, 1959) and his coworkers.

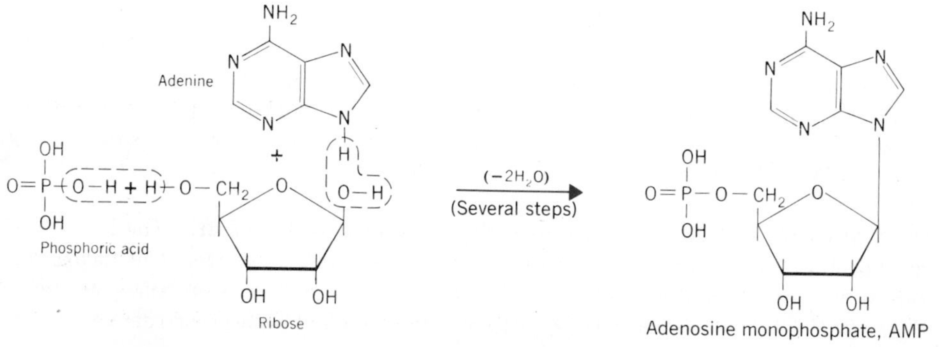

FIGURE 21.2
A typical nucleotide, AMP, and the smaller units from which it is assembled.

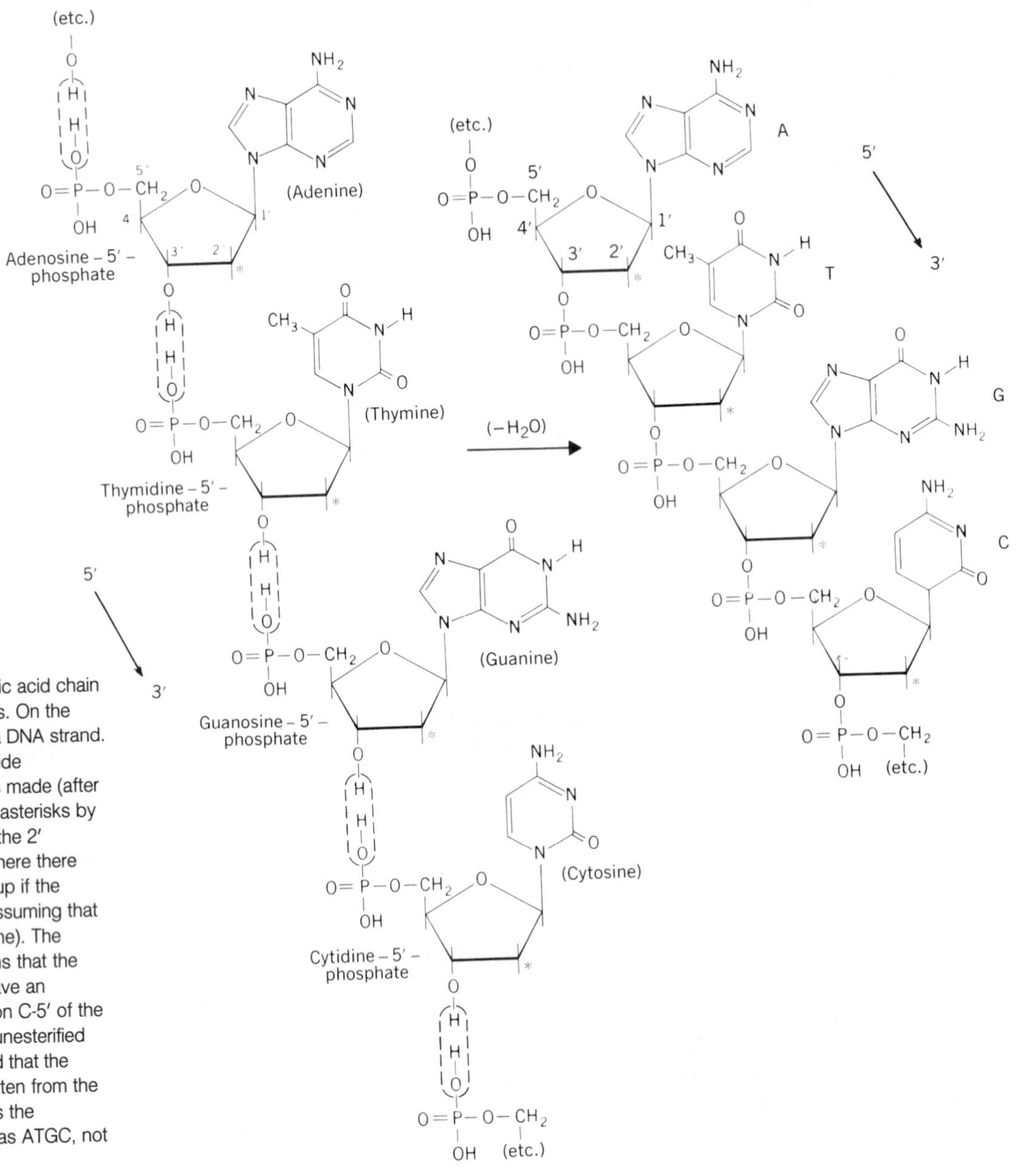

FIGURE 21.3
The relationship of a nucleic acid chain to its nucleotide monomers. On the right is a short section of a DNA strand. On the left are the nucleotide monomers from which it is made (after many steps). The colored asterisks by the pentose units identify the 2' positions of these rings where there would be another OH group if the nucleic acid were RNA (assuming that uracil also replaced thymine). The designation 5' → 3' means that the complete strand would have an unesterified —OH group on C-5' of the first pentose unit and an unesterified C-3' on the other end, and that the sequence of bases is written from the 5' end to the 3' end. Thus the sequence here is written as ATGC, not as CGTA.

Each pentose unit holds one of the four bases, so the bases project at very regular intervals along the backbone. The distinctiveness of a nucleic acid is the *sequence of these bases*. At the molecular level of life, the central genetic fact is that unique genetic information is carried by unique sequences of bases. Knowing this, we can simplify our structures, as shown in Figure 21.4.

Pairs of Bases Are Attracted to Each Other by Hydrogen Bonds. The bases have functional groups so arranged geometrically that **base pairing** occurs. This means that certain pairs of bases are able to fit to each other by means of hydrogen bonds, as seen in Figure 21.5. The geometries and relative locations of their functional groups are such that in

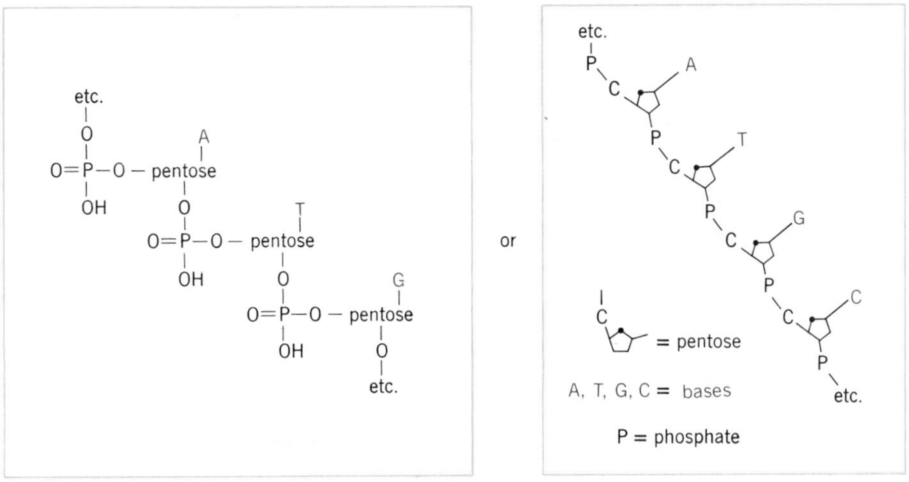

FIGURE 21.4
Condensed structures of nucleic acids. Shown here is a representation of the same short segment of DNA that was given in Figure 21.3. The same representations could be used for RNA if U replaced T, and if the pentose units were understood to be ribose instead of deoxyribose.

DNA G and C form a base pair, and A and T form another. In RNA, G and C always pair, and U and A always pair. The arrangments of hydrogen-bond donor and acceptor sites allow only certain base pairs to exist. Neither G nor C ever pairs with A, T, or U. They can't fit to each other.

DNA Occurs As Two Paired Strands Twisted into a Double Helix. In 1953, Francis Crick of England and James Watson of the United States proposed a **double helix** structure for DNA. Using X-ray data obtained by Rosalind Franklin, they deduced that the long DNA molecules occur in cell nuclei as pairs of complementary strands twisted into right-hand helices.

Being *complementary* means that whenever adenine (A) is on one strand, thymine (T) is opposite it on the other strand; whenever guanine (G) projects from one strand, then cytosine (C) is opposite it on the other. The hydrogen bonds between A and T and between G and C hold the strands together.

A molecular model of the DNA double helix is shown in Figure 21.6. The system resembles a spiral staircase in which the steps, which are perpendicular to the long axis of the spiral, consist of the base pairs. Hydrogen bonds between the pairs are centered around the long axis. Since Crick and Watson's work, a small number of variations in this structure have been found. Figure 21.6 shows what is now called DNA-B, the type Crick and Watson worked with and the most common kind.

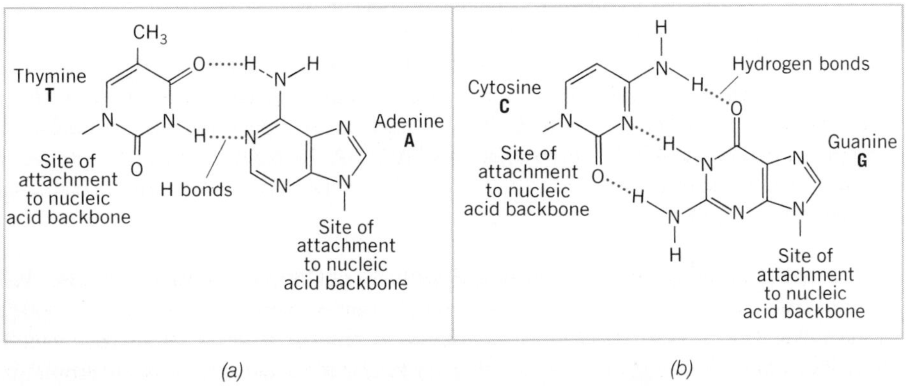

Crick and Watson shared the 1962 Nobel prize in medicine and physiology with Maurice Wilkins.

FIGURE 21.5
Hydrogen bonding between base pairs. (a) Thymine (T) and adenine (A) form one base pair between which are two hydrogen bonds. (b) Cytosine (C) and guanine (G) form another base pair between which are three hydrogen bonds. Adenine can also base pair to uracil (U).

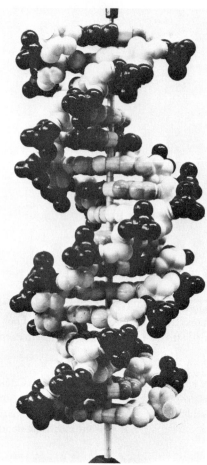

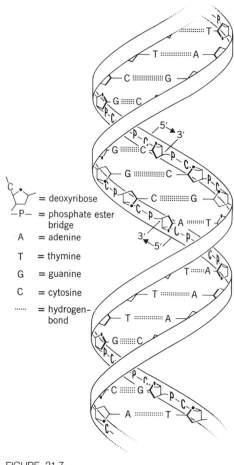

FIGURE 21.6
A molecular model of a short section of a DNA double helix. The darkest parts are phosphate groups. The lightest are deoxyribose units. What appear to be horizontal steps in a spiral staircase are the interchain base pairs.

FIGURE 21.7
A schematic representation of the DNA double helix. The two spiraling strands are held side by side by the hydrogen bonds (rows of colored dots) between base pairs on opposite strands. (See the legend to Figure 21.3 for the meaning of the 5' → 3' designations.)

C = deoxyribose
—P— = phosphate ester bridge
A = adenine
T = thymine
G = guanine
C = cytosine
······ = hydrogen-bond

Figure 21.7 is a schematic representation that points out an additional feature of a double helix: The two DNA strands run in opposite directions. This is indicated in the figure by the 5' → 3' specifications shown by the pentose units.

Figures 21.6 and 21.7 show only short segments of DNA double helices. They do not show that the helices are further twisted and coiled into superhelices. Such coiling and folding is necessary if the cell's DNA is to fit into its nucleus. A typical human cell nucleus, for example, is only about 10^{-7} m across, but if all its DNA double helices were stretched out, they would measure over 1 m, end to end. (The different DNA types concern various ways of twisting and folding.)

There are 3–5 billion base pairs in one nucleus of a human cell.

The nucleotide monomer molecules are present in the nuclear fluid.

In DNA Replication, the Bases Guide the Formation of Complementary Strands. When DNA replicates, the cell makes an exact complementary strand for each of the original strands, and two identical double helices emerge. A number of enzymes are involved. The guarantee that each new strand is a complement to one of the old strands is the requirement

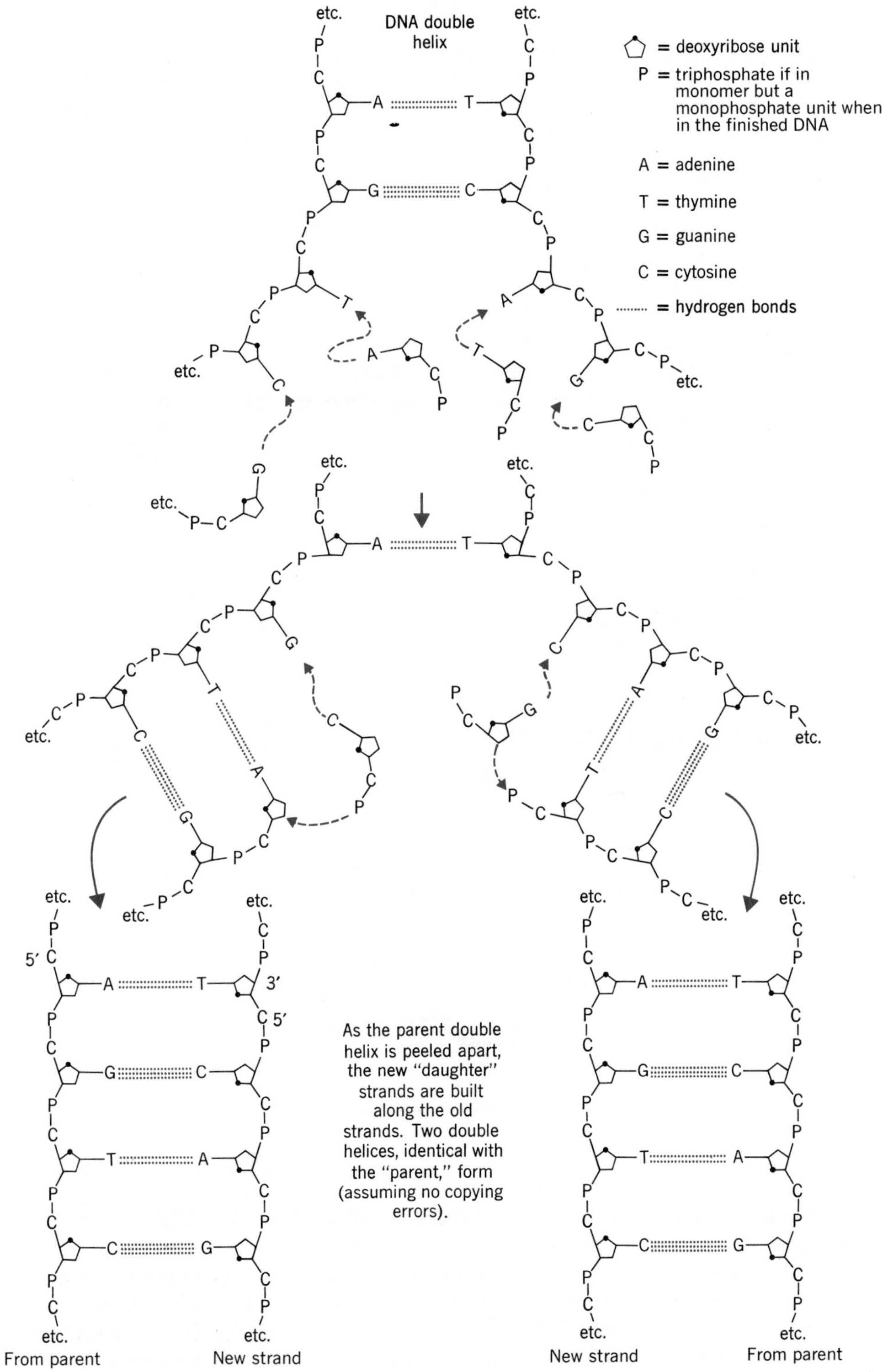

FIGURE 21.8
The accuracy of the replication of DNA is related to the exclusive pairing of A with T and G with C. The two new strands at the bottom are replicas of the original parent strand at the top.

that the only pairings allowed are A to T and C to G. A very general picture of how this works is in Figure 21.8, which explains only one of the many aspects of replication, how base pairing ensures complementary strands.

In the entire set of human genes, the human *genome,* there are many regions consisting of nucleotide sequences repeated in tandem. These regions, called *minisatellites,* all have a common core sequence, but the *number* of repeated sequences in the minisatellites varies from individual to individual. These variations, which can be measured, are so considerable between individuals, that they are the basis of a major new technique in crime prosecution.

Suppose, for example, a sample of semen can be obtained from the sperm left by a rapist. The sample is first hydrolyzed with the use of a specific enzyme, called a *restriction enzyme,* which is able to cut a DNA strand only at specific sites and releases the core segments. The fragments can then be separated and made to bind (by base pairing) to short, radioactively labeled DNA that has been specifically designed to bind to core segments. Now the fragments can be detected by the way in which the atomic radiations affect X-ray film. A series of 30–40 dark bands appears, roughly analogous to the bar codes on groceries that are used for pricing at check out counters.

Each individual has a unique genetic "bar code" or *DNA profile.* Because every cell in the body has the entire genome, a single cut hair of a rape suspect can provide enough material to measure the person's DNA profile and compare it to that obtained from the semen sample. The accompanying figure shows how one suspect was trapped.

When the two DNA profiles match, the jury has evidence as powerful as fingerprints for a conviction. If there is no match, the district attorney looks for another suspect. This kind of evidence is thus as powerful an ally of the innocent as it is an enemy of the guilty. If a crime site specimen is old or has been subjected to harsh environmental conditions and has deteriorated, it will either give a true test or none at all. It does not give a false test that will lead to the conviction of an innocent person.

By the start of the 1990s, several trials involving DNA profiles had taken place in a number of states. Defense attorneys have (properly) issued severe challenges to the validity of DNA profiling. The issue has centered not on the principles involved but on the reliability of the technology used to obtain the profiles. As a result, some improvements have occurred. It can be expected that challenges will continue and that the U. S. Supreme Court will eventually consider this matter.

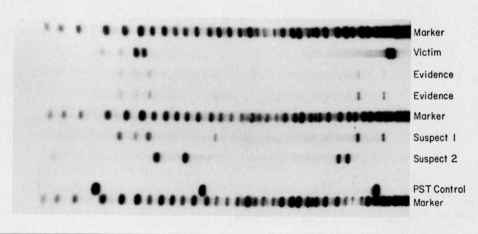

Marker
Victim
Evidence
Evidence
Marker
Suspect 1
Suspect 2
PST Control
Marker

DNA fingerprinting. The banding pattern of the DNA of suspect 1 matches that of the evidence. Neither the rape victim's DNA nor that of suspect 2 does.

In Higher Organisms, Sections of DNA Molecules Called Exons Collectively Carry the Message of One Gene. Human genes were once thought to be *continuous* sequences of nucleotide units, which they are in single-celled species. We now know that genes in higher organisms consist of a series of separated sections of a single strand of DNA. Thus virtually all single genes are divided or split. The sections of a DNA chain that together constitute one gene are called **exons.** Separating the exons are other parts of the DNA chain called **introns,** which have no direct bearing on the genetic message carried by the gene.

We'll see later in this chapter how the exons get their message together and thus give expression to a single gene. For the present, we can consider that a gene is a particular section of a DNA strand minus all of the introns in this section. A gene, in other words, is a specific series of bases strung in a definite sequence along a DNA backbone. In humans, a single gene has between 900 and 3000 bases.

Exon refers to the part that is expressed and *intron* to the segments that *int*errupt the exons.

There are no introns in human genes that code for histones.

Base Sequences in Genes Can Be Determined.   Although the genes of one human being are similar to those of another, there are many small variations. Each person, except for those who are identical twins, has unique genes just as each has unique fingerprints. These differences are behind a powerful new tool for identifying criminals, genetic "fingerprinting," described in Special Topic 21.1.

21.2   RIBONUCLEIC ACIDS

The triplets of bases of the genetic code correlate with individual amino acids.

The general scheme that relates DNA to polypeptides is illustrated in Figure 21.9. Before we can use it we need to learn more about RNA and its various types, particularly the four that participate in expressing genes in higher organisms. Notice, however, how genetic information flows from DNA to a specific polypeptide through four types of RNA.

Ribosomal RNA (rRNA) Is the Most Abundant RNA.   **Ribosomes** are small particles that occur by the thousands on the surfaces of membrane-like tubules that twist and turn throughout a cell outside its nucleus. Each ribosome forms from two subunits, as shown in Figure 21.9, which come together to form a complex with messenger RNA, another type that we'll study soon.

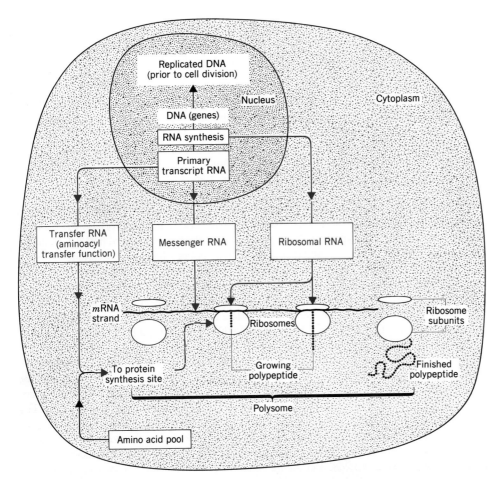

FIGURE 21.9
The relationships of nuclear DNA to the various types of RNA and to the synthesis of polypeptides.

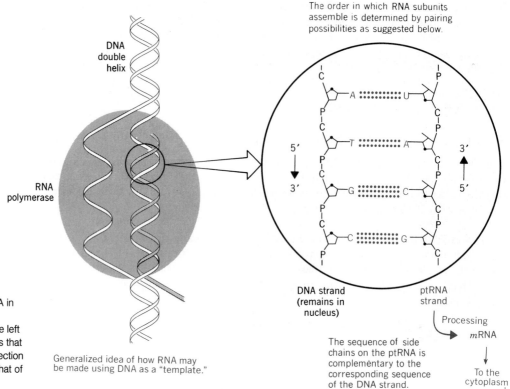

FIGURE 21.10
DNA-directed synthesis of *pt*-RNA in the nucleus of a cell in a higher organism. The shaded oval on the left represents a complex of enzymes that catalyze this step. (Notice the direction of the ptRNA strand is opposite that of the DNA strand.)

Generalized idea of how RNA may be made using DNA as a "template."

The order in which RNA subunits assemble is determined by pairing possibilities as suggested below.

DNA strand (remains in nucleus)

ptRNA strand

The sequence of side chains on the ptRNA is complementary to the corresponding sequence of the DNA strand.

Processing
*m*RNA

To the cytoplasm

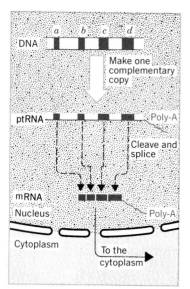

FIGURE 21.11
The RNA made directly at a DNA strand is ptRNA. Only the segments made at sites *a*, *b*, *c*, and *d* — the exons — are needed to carry the genetic message to the cytoplasm. Therefore the ptRNA is processed and its segments that matched the introns of the gene are snipped out. Then the segments that matched the exons are rejoined to make the mRNA strand.

Ribosomes contain proteins and a ribonucleic acid called **ribosomal RNA,** abbreviated **rRNA.** Except in a few viruses, rRNA is single stranded, but its molecules often have hairpin loops in which base pairing occurs.

The precise purpose of rRNA has yet to be worked out, but ribosomes are the sites of polypeptide synthesis. Their rRNA does not itself direct this work. Messenger RNA does this, and it's made from the second kind of RNA of our study.

Primary Transcript RNA (ptRNA) Is Complementary to DNA, Both Exons and Introns.
As indicated in Figure 21.9, when a cell uses a gene to direct the synthesis of a polypeptide, its first step is to use DNA to make **primary transcript RNA,** abbreviated **ptRNA.** A *single* strand of the DNA that bears the polypeptide's gene is used to guide the assemblage of a complementary molecule of ptRNA. Figure 21.10 shows how this takes place. Uracil is now used instead of thymine, so when a DNA strand has an adenine side chain, then uracil not thymine takes the position opposite it on the complementary RNA.

After making ptRNA, the next general step in gene-directed polypeptide synthesis (Figure 21.9) is the processing of ptRNA to form still another kind of RNA, messenger RNA.

Each Messenger RNA (mRNA) Is Complementary to One Gene.   Molecules of ptRNA have large sections complementary to the introns of the DNA, and these sections must be deleted. Special enzymes catalyze reactions that snip these pieces out of ptRNA and splice together just the units corresponding to the exons of the divided gene (see Figure 21.11). The result is a much shorter RNA molecule called **messenger RNA,** or **mRNA.**

In mRNA we have a sequence of bases complementary just to the gene's exons, so this mRNA now carries the unsplit genetic message. We have now moved the genetic message on DNA to a molecule of mRNA, and the name for this overall process is **transcription.**

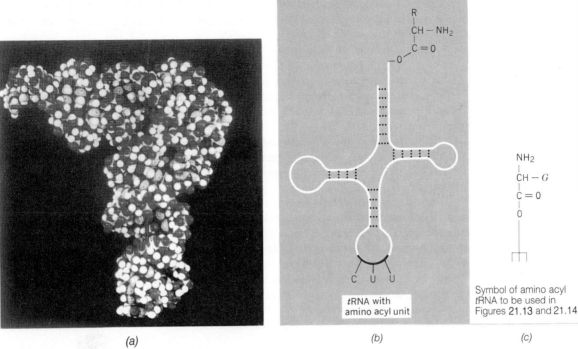

FIGURE 21.12
Transfer RNA (tRNA). *(a)* The tRNA for phenylalanine. Its condon is at the lower tip, and its phenylalanyl residue is attached at the upper left hand point. *(b)* A highly schematic representations of the model to highlight the occurrence of double stranded regions stabilized by the hydrogen bonds. *(c)* The symbol of the aminoacyl—tRNA unit that will be used in succeeding figures.

Triplets of Bases on mRNA Are Genetic Codons. Each group of three adjacent bases on a molecule of mRNA constitutes a unit of genetic information called a **codon** (taken from the word *code*). Thus it is a sequence of *codons* on the mRNA backbone more than a sequence of individual bases that now carries the genetic message. (We'll explain shortly why *three* bases per codon are necessary.)

As indicated in Figure 21.9, once mRNA is made, it moves from the nucleus to the cytoplasm where it attaches ribosomes. Many ribosomes can be strung like beads along one mRNA chain, and such a collection is called a *polysome* (short for *polyribosome*).

Ribosomes are traveling packages of enzymes intimately associated with rRNA. Each ribosome moves along its mRNA chain while the codons on the mRNA guide the synthesis of a polypeptide. To complete this system, we need a way to bring individual amino acids to the polysome's polypeptide assembly sites. For this, the cell uses still another type of RNA.

Transfer RNA (tRNA) Molecules Can Recognize Both Codons and Amino Acids. The substances that carry aminoacyl units to mRNA in the right order for a particular polypeptide are a collection of similar compounds called **transfer RNA** or **tRNA.** Their molecules are small, each typically having only 75 nucleotides. As seen in Figure 21.12, they are single stranded but with hairpin loops stabilized by base pairing.

tRNA Molecules Translate from the Genetic Language to the Polypeptide Language. We're now dealing with the molecular basis of *information,* so we can use language analogies. On a human level, we use language to convey information, and language involves words built from a common alphabet. We are aware that many languages exist among human societies and that the world knows several alphabets. To communicate between languages, we have to translate. The same need for translation occurs at the level of genes and polypeptides. tRNA is the master translator in cells.

tRNA is able to work with two "languages," the genetic and the polypeptide. The genetic language is expressed in an alphabet of 4 letters, the four bases, A, T (or U), G, and C. The polypeptide language has an alphabet of 20 letters, the side chains on the 20 amino acids. To translate from a 4-letter language to a 20-letter language requires that the 4 genetic letters be used in groups of a minimum of 3 letters. Then there will be enough combinations of letters for the larger alphabet of the amino acids. Now there can be at least one genetic "word," built of 3 letters, for each of the 20 amino cids. This is exactly how tRNA is structured for its work with the genetic language. It is able to connect the three-letter codon "words" aligned along an mRNA backbone to a matching alignment of side chains of individual amino acids in a polypeptide being made.

One part of a tRNA molecule can recognize (form base pairs with) a codon because it carries a triplet of bases complementary to the codon. This triplet on tRNA is called an **anticodon.** Each of the 20 unique tRNAs, one for each of the 20 amino acids, carries a particular anticodon. In the tRNA molecule in Figure 21.12, the triplet CUU is its anticodon.

Another part of a tRNA molecule, an OH group at an end ribose unit, can attach a particular aminoacyl unit (by an ester bond). We can use the symbol tRNA-aa for this new compound, where we use "aa" for the aminoacyl group. Twenty such tRNA-aa molecules exist, one for each amino acid, but the anticodon on each is unique. A given tRNA-aa molecule, therefore, can be brought into alignment only with one codon of mRNA at a polysome.

TABLE 21.1
Codon Assignments

First	Second				Third
	U	C	A	G	
U	Phenylalanine	Serine	Tyrosine	Cysteine	U
	Phenylalanine	Serine	Tyrosine	Cysteine	C
	Leucine	Serine	CT[a]	CT	A
	Leucine	Serine	CT	Tryptophan	G
C	Leucine	Proline	Histidine	Arginine	U
	Leucine	Proline	Histidine	Arginine	C
	Leucine	Proline	Glutamine	Arginine	A
	Leucine	Proline	Glutamine	Arginine	G
A	Isoleucine	Threonine	Asparagine	Serine	U
	Isoleucine	Threonine	Asparagine	Serine	C
	Isoleucine	Threonine	Lysine	Arginine	A
	Methionine[b]	Threonine	Lysine	Arginine	G
G	Valine	Alanine	Aspartic acid	Glycine	U
	Valine	Alanine	Aspartic acid	Glycine	C
	Valine	Alanine	Glutamic acid	Glycine	A
	Valine	Alanine	Glutamic acid	Glycine	G

[a] The codon CT is a signal codon for chain termination.

[b] The codon for methionine, AUG, serves also as the codon for N-formylmethionine, the chain-initiating unit in polypeptide synthesis in bacteria and mitochondria.

As each tRNA-aa molecule comes to mRNA, its aminoacyl unit is transferred to a growing polypeptide chain. A unique series of codons *can allow the polypeptide chain to grow only with an equally unique sequence of amino acid residues.* The pairing of the triplets of bases between the codons and anticodons permit only one sequence.

The Genetic Code Is the Correlation between Codons and Amino Acids. Table 21.1 displays the **genetic code,** the known assignments of codons to amino acids. Most amino acids are associated with more than one codon, which apparently minimizes the harmful effects of genetic mutations. (These, in molecular terms, are small changes in the structures of genes.) Phenylalanine, for example, is coded either by UUU or by UUC. (Be sure that you can verify this using Table 21.1.) Alanine is coded by any one of the four triplets: GCU, GCC, GCA, or GCG. Only two amino acids go with single codons, tryptophan (Trp) and methionine (Met).

The Genetic Code Is Almost Universal for All Plants and Animals. A few single-celled species have been found with codon assignments not given in Table 21.1. Moreover, some genes occur in human mitochondria, and they have some unique codons. Apart from these exceptions, the genetic code of Table 21.1 is shared from the lowest to the highest forms of life in both the plant and animal kingdoms. Once again we see a remarkable kinship with nature.

mRNA Codons Also Relate to DNA Triplets. It's important to remember that a codon cannot appear on a strand of mRNA unless a complementary triplet of bases was on an exon unit of the original DNA strand. For example, there could not be the UUC codon on mRNA unless the DNA strand had the triplet AAG, because G pairs with C and A of DNA pairs with U of mRNA.

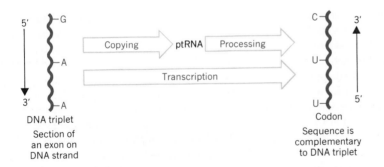

The Direction in Which a Codon Triplet Is Expressed Has Structural Meaning. As shown here, a DNA strand and the RNA strand made directly from it run in opposite directions. To avoid confusion in writing codons on a horizontal line, scientists use the following conventions. The 5′ end of a codon is written on the left end of the three-letter symbol, and the direction, left to right, is 5′ to 3′ (see also Figure 21.3). This is why the codon given above is written as UUC, not as CUU. Opposite this codon is the triplet GAA on the DNA strand, which is also written from the 5′ to 3′ end. To give another example, the complement to the mRNA codon, AAG, is the DNA triplet, CTT.

PRACTICE EXERCISE 1   Using Table 21.1, what amino acids are specified by each of the following codons on an mRNA molecule?

(a) CCU   (b) AGA   (c) GAA   (d) AAG

PRACTICE EXERCISE 2   What amino acids are specified by the following base triads on DNA?

(a) GGA   (b) TCA   (c) TTC   (d) GAT

21.3   mRNA-DIRECTED POLYPEPTIDE SYNTHESIS

tRNA molecules carry aminoacyl groups to places at a mRNA strand where anticodons match codons.

In the previous section we saw how a particular genetic message can be transcribed from the exons of a divided gene to a series of codons on mRNA. We now learn in broad terms how the next general step in Figure 21.9 happens, the mRNA-directed synthesis of a polypeptide.

$$NH_2CHC- \atop G \quad (O=)$$

An aminoacyl unit

$$^+NH_3CHCO_2^- \atop CH_2CH_2SCH_3$$

Methionine (Met)

Genetic *Translation* Follows Transcription.   The steps by which the sequence of codons on mRNA directs the formation of a matching sequence of aminoacyl groups in a polypeptide is called **translation.** As we have already mentioned, there is at least one kind of tRNA molecule for each of the 20 amino acids, and each aminoacyl unit is carried by its own tRNA molecule to the polypeptide assembly site at a ribosome. We will now use the very abbreviated forklike symbol for an aminoacyl-tRNA combination shown in Figure 21.12c. For the rest of our study of translation, we will assume that all needed aminoacyl-tRNA combinations have been made and are waiting like so many spare parts to be used at the assembly line.

The cell begins a polypeptide with an N-terminal methionine residue. After the end of the synthesis, the aminoacyl group of methionine will be left in place only if the polypeptide is supposed to have this as its N-terminal unit. Otherwise, it will be removed, and the second aminoacyl group will be the final N-terminal unit.

The principal steps in making the polypeptide are as follows:

1.   Formation of the elongation complex.

As shown in Figure 21.13, the elongation complex is made of several pieces: two subunits of the ribosome, the first aminoacyl-tRNA unit (which is Met-tRNA$_1$), and the mRNA molecule, beginning at its first codon end. The Met-tRNA$_1$ comes to rest with its anti-codon matched to the first codon and with the bulk of its system in contact with a portion of the ribosome's surface called the P site. This is a site with enzymes that catalyze the transfer of a growing polypeptide chain to a newly arrived amino acyl unit.

The "P" in P site refers to the peptidyl transfer site. The A site is the aminoacyl binding site.

Now the second tRNA unit, tRNA$_2$, which holds the second aminoacyl group, aa$_2$, has to find the mRNA codon that matches its own anticodon, and it does this at another site on the ribosome called the A site. The elongation complex is now complete, and actual chain lengthening can start.

2.   Elongation of the polypeptide chain.

A series of repeating steps now occurs, illustrated in Figure 21.14. The methionine residue transfers from its tRNA to the newly arrived aa$_2$. This makes the first peptide

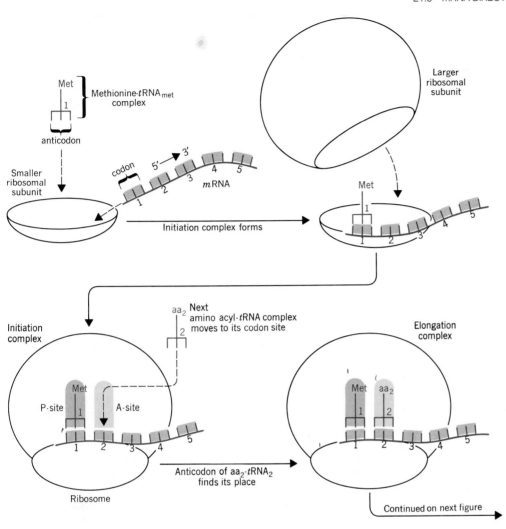

FIGURE 21.13
Formation of the elongation complex at the beginning of the synthesis of a polypeptide.

bond, and it takes place by acyl transfer much as we discussed when we studied the synthesis of amides in Section 13.5.

Methionine unit on its tRNA
$$CH_3SCH_2CH_2CH$$
with NH_2, $O=C$, 1

The next aminoacyl unit on tRNA
$$NH_2$$
$$CH-G$$
$$O=C$$ 2

H^+
1

A dipeptidyl-tRNA unit
$$NH_2$$
$$CH_3SCH_2CH_2CH$$
$$O=C$$
New peptide bond
$$NH$$
$$CH-G$$
$$O=C$$ 2

This translocation of a growing polypeptide to the next aminoacyl unit is blocked by the toxin of the diphtheria bacillus.

The mRNA unit now shifts one codon over, leftward as we have drawn it. (It's actually a relative motion of the mRNA and the ribosome.) This movement positions what is now a *di*peptidyl–tRNA unit over the P site. Now the third aminoacyl–tRNA moves in.

Polypeptide synthesis can occur at several ribosomes moving along the mRNA strand at the same time.

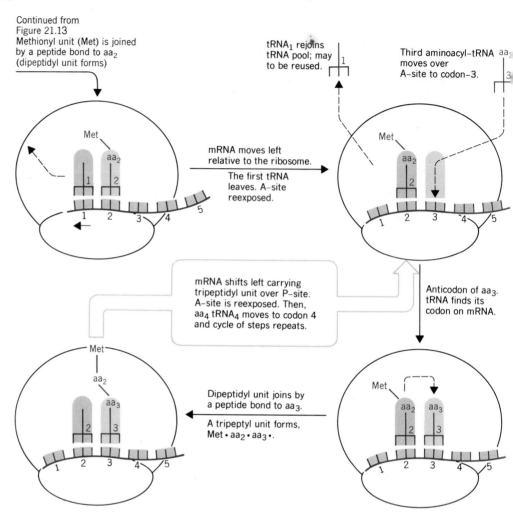

Continued from
Figure 21.13
Methionyl unit (Met) is joined
by a peptide bond to aa_2
(dipeptidyl unit forms)

Met
aa_2

mRNA moves left
relative to the ribosome.

The first tRNA
leaves. A–site
reexposed.

$tRNA_1$ rejoins
tRNA pool; may
to be reused.

Third aminoacyl–tRNA aa_3
moves over
A–site to codon-3.

Met
aa_2

mRNA shifts left carrying
tripeptidyl unit over P–site.
A–site is reexposed. Then,
aa_4 $tRNA_4$ moves to codon 4
and cycle of steps repeats.

Anticodon of aa_3.
tRNA finds its
codon on mRNA.

Met
aa_2
aa_3

Dipeptidyl unit joins by
a peptide bond to aa_3.

A tripeptyl unit forms,
Met·aa_2·aa_3·.

Met
aa_2
aa_3

FIGURE 21.14
The elongation steps in the synthesis
of a polypeptide. The dipeptidyl unit,
Met-aa_2, formed by the process of
Figure 21.13, and it now has a third
amino acid residue, aa_3, added to it.

In mammals, it takes only about
one second to move each amino
acid residue into place in a
growing polypeptide.

Its anticodon finds the matching codon, the third codon of the mRNA strand. Now
tRNA–aa_3 is over the recently vacated A site.

Further elongation occurs; the dipeptide unit transfers to the amino group of aa_3.
Another peptide bond forms, and a tripeptidyl system has been made. The cycle of steps
now takes place again. The mRNA chain first shifts relative to the ribosome so that the
tripeptidyl–tRNA is placed over the P site. The fourth amino acid residue is carried by
tRNA–aa_4 to the mRNA; the tripeptidyl unit transfers to it to make a tetrapeptidyl unit,
and so forth. This cycle of steps continues until a chain-terminating codon is reached.

3.   Termination of polypeptide synthesis.

Once a ribosome has moved to a chain-terminating codon (UAA, UAG, or UGA), the
polypeptide synthesis is complete, and the polypeptide is released. The ribosome can be
reused, and the polypeptide spontaneously acquires its higher levels of structure.

These inhibiting activities render
the *enzymes* for the various steps
inactive.

Inhibition of Polypeptide Synthesis in Bacteria by Antibiotics.   Several antibiotics kill
bacteria by inhibiting their synthesis of polypeptides. Streptomycin, for example, inhibits the
initiation of polypeptide synthesis. Chloramphenicol inhibits the ability to transfer newly
arrived aminoacyl units to the elongating strand. The tetracyclines inhibit the binding of
tRNA–aa units when they arrive at the ribosome. Actinomycin binds tightly to DNA. Erythro-
mycin, puromycin, and cycloheximide interfere with elongation.

21.4   VIRUSES

Viruses take over the genetic machinery of host cells to make more viruses.

They are at the borderline of living systems and are generally regarded only as unique packages of dead chemicals, except when they get inside their host cells. Then they seem to be living things, because they reproduce. They are a family of materials called **viruses.**

Viruses Consist of Nucleic Acids and Proteins.   Viruses are agents of infection made of nucleic acid molecules surrounded by overcoats of protein molecules. Unlike a cell, a virus has either RNA or DNA, but not both. Viruses must use host cells to reproduce because they can neither synthesize polypeptides nor generate their own energy for metabolism. The simplest virus has only four genes and the most complex has about 250.

Viruses are intracellular parasites.

Viruses Have Unique Membrane-Dissolving Enzymes.   The protein overcoat of many viruses includes an enzyme that catalyzes the breakdown of the cell membrane of the host cell. When such a virus particle sticks to the surface of a host cell, its overcoat catalyzes the opening of a hole into the cell. Then the viral nucleic acid squirts into the cell, or the whole virus might move in. Each virus that works this way evidently has its own unique membrane-dissolving enzyme, so any one virus can affect only those cells in the host whose own membranes are susceptible to this enzyme. At least this theory helps us understand how viruses are so unusually selective. A virus that attacks, for example, the nerve cells in the spinal cord has no effect on heart muscle cells. A large number of viruses exist that do not affect any kind of human cell. Many viruses attack only plants.

A complete virus particle outside *its host cell is called a* virion.

Some Viruses Become Silent Genes and Some Replicate.   Once a virus gets inside its host cell one of two possible fates awaits it. It might become turned off and change into a *silent gene;* or it might take over the genetic machinery of the cell and reproduce so much of itself that it bursts out of the host cell. The new virus particles that spill out then infect neighboring host cells, and in this way the infection spreads. A virus that has become a silent gene might later be activated. Some cancer-causing agents, including ultraviolet light, are believed to initiate cancer by this mechanism.

The Retroviruses Can Make DNA Double Helices at the Direction of RNA.   Most viruses are RNA based, not DNA based, and they have special enzymes that enable them to make RNA without their own DNA. The retroviruses are unusual in that they cannot make more RNA without first making double-stranded DNA. To do this, they have an enzyme, *reverse transcriptase,* than can use RNA information to make DNA. This is unusual because it's normally the other way around; DNA information is used to make RNA. But in retroviruses, the flow of information goes in the reverse direction, hence the name, *reverse* transcriptase. (Hence also the *retro,* suggesting reversal or retrograde, in the name retrovirus.)

Cancer-Causing Viruses Transform Normal Genes in Host Cells.   The retroviruses include the only known cancer-causing RNA viruses, technically termed the *oncogenic RNA viruses.* (Several DNA-based viruses also cause cancer.) They transform host cells so that they grow chaotically and continuously. They do this by changing normal genes in the host cell to *oncogenes,* genes that support continued cancerous growth.

Oncogenic means cancer causing.

The Host Cell of the AIDS Virus Is Part of the Human Immune System.   The acquired immunodeficiency syndrome, AIDS, is also caused by a retrovirus, the human immunodeficiency virus or HIV. One reason why this virus is so dangerous is that its host cell, the T4 lymphocyte, is a vital part of the human immune system. By destroying T4 lymphocytes, HIV exposes the body to other infectious diseases, like pneumonia, or to certain rare types of cancer.

Some Viral Infections Produce Interferons, Which Fight Further Infection.   One of the many features of the body's defense against some viral infections is a small family of polypeptides called the interferons. Currently, intensive research is in progress to make individual interferons and to test them in the treatment of viral diseases and cancer. Supplies of interferons are now available by genetic engineering, which we will study next.

21.5   RECOMBINANT DNA TECHNOLOGY AND GENETIC ENGINEERING

Single-celled organisms can be made to manufacture the proteins of higher organisms.

Human insulin, human growth hormone, and human interferons are now being manufactured by a technology that involves the production of *recombinant DNA*. With the aid of Figure 21.15, we'll learn how this technology works. It represents one of the important advances in scientific technology of this century. It has permitted *cloning,* the synthesis of identical copies of a number of genes. The use of recombinant DNA to make genes and the products of such genes is called **genetic engineering.**

Genes Alien to Bacteria Can Be Inserted into Bacterial Plasmids.   Bacteria make polypeptides using the same genetic code as humans. There are some differences in the machinery, however. An *E. coli* bacterium, for example, has DNA not only in its single chromosome but also in **plasmids,** large, circular, supercoiled molecules of DNA. Each plasmid carries just a few genes, but several copies of a plasmid can exist in one bacterial cell. Each plasmid replicates independently of the chromosome.

The plasmids of *E. coli* can be removed and given new DNA material, such as a new gene, with base triplets for directing the synthesis of a particular polypeptide. It can be a gene completely alien to the bacteria, like the subunits of insulin, or growth hormone, or a variety of interferon. The DNA of the plasmids is snipped open by special enzymes called *restriction enzymes* absorbed from the surrounding medium. This medium can also contain naked DNA molecules, such as those of the gene to be cloned. Then, with the aid of a DNA-knitting enzyme called DNA *ligase,* the new DNA combines with the open ends of the plasmid. This recloses the plasmid loops. The DNA of these altered plasmids is called **recombinant DNA.** The altered plasmids are then allowed to be reabsorbed by bacterial cells.

The remarkable feature of bacteria with recombinant DNA is that when they multiply, the plasmids in the offspring also have this new DNA. When these multiply, still more altered plasmids are made.

Between their cell divisions, the bacteria manufacture the proteins for which they are genetically programmed, including the proteins specified by the recombinant DNA. In this way, bacteria can be tricked into making the *human* proteins we have mentioned. The technology isn't limited to bacteria; yeast cells work, too.

Recombinant DNA Can Be Inserted into Cells of Higher Organisms.   Sometimes a desired polypeptide has to be "groomed" by a cell *after* it has been made by genetic translation before it will function. It might have to be attached to a carbohydrate molecule, for example. Bacteria lack the enzymes for such grooming work, so cells of higher organisms are used. When these cells are large enough, the new DNA can be inserted directly into them using glass pipets of extremely small diameters (0.1 μm). Although only a small fraction of such inserted DNA becomes taken up into the cells chromosomes, it can be enough when amplified by successive cell division.

Since viruses are able to get inside cells, some have been used to carry new DNA along. Retroviruses can be customized for this purpose, for example.

Escherichia coli, or *E. coli,* is a one-celled organism found in the human intestinal tract.

The term *cloning* is used for the operation that places new genetic material into a cell where it becomes a part of the cell's gene pool. The new cells that follow this operation are called *clones.*

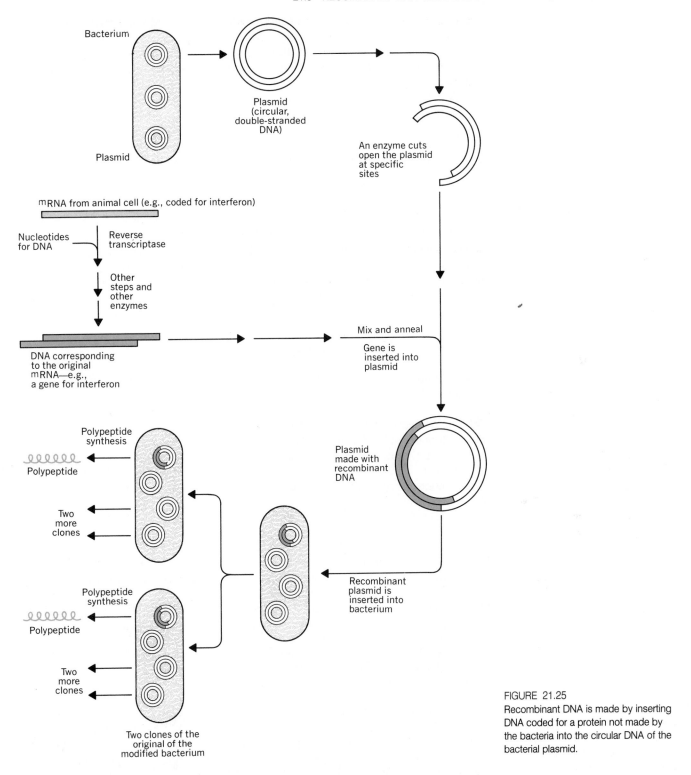

FIGURE 21.25
Recombinant DNA is made by inserting DNA coded for a protein not made by the bacteria into the circular DNA of the bacterial plasmid.

Genetic Engineering Offers Major Advances in Medicine.   One of the hopes of genetic engineering research is to have ways to correct genetic faults. As we will study in the next section, a number of undesirable conditions are caused by flawed or absent genes. Dwarfism, for example, is caused by a lack of growth hormone, a relatively small polypeptide. In experiments with mice, genetic engineering has successfully introduced growth hormone into mice, with dramatic effects on the mouse size.

In another medical application, the smallpox vaccine is being remodeled to provide altered forms that might give immunity to many other diseases, ranging from malaria to influenza.

A polypeptide hormone made by the heart, which reduces blood pressure, can be manufactured by genetic engineering and used by victims of high blood pressure. The clot-dissolving enzyme, tissue plasminogen activator or tPA, has been genetically engineered for use in reducing the damage to heart tissue following a sudden heart attack (Section 17.4).

The alteration of a sticky protein that mussels use to cling to underwater surfaces is being studied to find an adhesive that can be used after surgery.

The list of potential applications of genetic engineering to health problems grows yearly. Hemophiliacs who lack a blood-clotting factor may have it available. Erythropoietin, a polypeptide made in the kidneys that initiates the production of red blood cells, is now made by recombinant DNA technology. Kidney dialysis patients who lack erythropoietin and become anemic need blood transfusions to supply it, but with recombinant erythropoietin they will need fewer transfusions (and in some cases, none at all). The U. S. Food and Drug Administration approved this substance for such use in 1989. People with AIDS, cancer, and those recovering from major surgery will also benefit.

21.6 HEREDITARY DISEASES

About 2500 diseases in humans are caused directly or indirectly by flawed genes.

A Defective Gene Can Make a Defective Enzyme. The victims of cystic fibrosis overproduce a thick mucous in the lungs and the digestive system, which clogs them and which often leads to death in children. About 1 person in 20 carries the defective gene associated with this disease, and it hits about one in every 1000 newborn. The defective gene directs the synthesis of an abnormal enzyme involved in biological oxidations, an enzyme that leads to excessive oxygen consumption and the overproduction of mucous.

Sickle-cell anemia is another disease caused by a defective gene. It was described in Special Topic 16.1.

Albinism, the absence of pigments in the skin and the irises of the eyes, is caused by a defect in a gene that directs the synthesis of an enzyme that is needed to make these pigments. The pigments absorb the ultraviolet rays in sunlight, radiation that can induce cancer, so victims of albinism are more susceptible to skin cancer.

Phenylketonuria, or PKU disease, is a brain-damaging genetic disease in which abnormally high levels of the phenylketoacid called phenylpyruvic acid occur in the blood. This condition causes permanent brain damage in the newborn. Because of a defective gene, an enzyme needed to handle phenylalanine is not made, and this amino acid is increasingly converted to phenylpyruvic acid.

$$C_6H_5CH_2\underset{\underset{NH_3^+}{|}}{C}HCO_2^- \qquad C_6H_5CH_2\overset{\overset{O}{\|}}{C}CO_2H$$

Phenylalanine   Phenylpyruvic acid

PKU can be detected by a simple blood test within 4 or 5 days after birth. If the infant's diet is kept very low in phenylalanine, it can survive the critical danger period and experience no brain damage. The infant's diet should include no aspartame, a low-calorie sweetener (contained in NutraSweet), because it is hydrolyzed by the digestive processes to give phenylalanine.

$$^{+}NH_3CHC-NHCHCOCH_3 + 2H_2O \longrightarrow {}^{+}NH_3CHCO_2{}^{-} + {}^{+}NH_3CHCO_2{}^{-} + CH_3OH$$

CH₂	CH₂	CH₂	CH₂
CO₂⁻	C₆H₅	CO₂⁻	C₆H₅

Aspartame   Aspartic acid   Phenylalanine   Methanol

Maintaining a low phenylalanine diet, of course, is not the easiest and best solution, so genetic engineers are working to correct the fundamental gene defect.

The techniques of genetic engineering have also been used to locate the chromosomes that hold the defective genes for a number of neurologic disorders, like several varieties of dystrophy, familial Alzheimer's disease, manic-depressive illness, and Huntington's chorea. Once the chromosome is known, the gene can be found and its defect analyzed. The hope is that gene therapy will someday be able to correct these disorders.

There is enough phenylalanine in just one slice of bread to be potentially dangerous to a PKU infant.

SUMMARY

Hereditary information The genetic apparatus of a cell is mostly in its nucleus and consists of chromatin, a complex of DNA and proteins. Strands of DNA, a polymer, carry segments that are individual genes. Chromatin replicates prior to cell division, and the duplicates segregate as the cell divides. Each new cell thereby inherits exact copies of the chromatin of the parent cell.

DNA Complete hydrolysis of DNA gives phosphoric acid, deoxyribose, and a set of four heterocyclic amines, the bases adenine (A), thymine (T), guanine (G), and cytosine (C). The molecular backbone of the DNA polymer is a series of deoxyribose units joined by phosphodiester groups. Attached to each deoxyribose is one of the four bases. The order in which triplets of bases occur is the cell's way of storing genetic information.

A gene in higher organisms consists of successive groups of triplets, the exons, separated by introns. Thus the gene is a divided system, not a continuous series of nucleotide units. DNA strands exist in cell nuclei as double helices, and the helices are held near each other by hydrogen bonds that extend from bases on one strand to bases on the other. Base A always pairs with T and C always pairs with G. Using this structure and the faithfulness of base pairing, Crick and Watson explained the accuracy of replication. After replication, each new double helix has one of the parent DNA strands and one new, complementary strand.

RNA RNA is similar to DNA except that in RNA ribose replaces deoxyribose and uracil (U) replaces thymine (T). Four main types of RNA are involved in polypeptide synthesis. One is rRNA, which is in ribosomes. A ribosome contains both rRNA and proteins that have the enzyme activity needed during polypeptide synthesis.

mRNA is the carrier of the genetic message from the nucleus to the site where a polypeptide is assembled. mRNA results from a chemical processing of the longer RNA strand, ptRNA, which is made directly under the supervision of DNA.

tRNA molecules are the smallest RNAs, and their function is to convey aminoacyl units to the polypeptide assembly site. They recognize where they are to go by base pairing between an anticodon on tRNA and its complementary codon on mRNA. Both codon and anticodon consist of a triplet of bases.

Polypeptide synthesis Genetic information is first transcribed when DNA directs the synthesis of mRNA. Each base triplet on the exons of DNA specifies a codon on mRNA. The mRNA moves to the cytoplasm to form an elongation complex with subunits of a ribosome and the first and second tRNA-aa unit to become part of the developing polypeptide.

The ribosome then rolls down the mRNA as tRNA-aa units come to the mRNA codons during the moment when the latter are aligned over the proper enzyme site of a ribosome. Elongation of the polypeptide then proceeds to the end of the mRNA strand or to a chain-terminating codon. After chain termination, the polypeptide strand leaves, and it may be further modified to give it its final N-terminal amino acid residue.

Several antibiotics inhibit bacterial polypeptide synthesis, which causes the bacteria to die.

Viruses Viruses are packages of DNA or RNA that usually are encapsulated by protein. Once they get inside their host cell, virus particles take over the cell's genetic machinery, make enough new virus particles to burst the cell, and then repeat this in neighboring cells. Some viruses become silent genes, which are implicated in human cancer.

Recombinant DNA technology Recombinant DNA is DNA made from bacterial plasmids and DNA, obtained from another source, encoded to direct the synthesis of some desired polypeptide. The altered plasmids are reintroduced into the bacteria, where they become machinery for synthesizing this polypeptide (e.g., insulin, growth hormone, interferon, or erythropoietin). Yeast cells can also be used for this technology.

Hereditary diseases Faulty genes make faulty enzymes, many of which are involved in hereditary diseases. Sometimes the appropriate gene is missing. PKU disease, albinism, and cystic fibrosis are just three of hundreds of hereditary diseases.

REVIEW EXERCISES

The answers to these review exercises are in the *Study Guide* that accompanies this book.

The Hereditary Units

21.1 Chromatin is made up of what kinds of substances?

21.2 What is the name of the chemical that makes up genes?

21.3 What is the relationship between a chromosome and chromatin?

21.4 What is the name of the overall process by which an exact duplicate of a gene is synthesized?

21.5 The duplication of a gene occurs in what part of the cell?

Structural Features of Nucleic Acids

21.6 What is the general name for the chemicals that are most intimately involved in the storage and the transmission of genetic information?

21.7 The monomer units for the nucleic acids have what *general* name?

21.8 What are the names of the two sugars produced by the complete hydrolysis of all of the nucleic acids in a cell?

21.9 What are the names and symbols of the four bases that are liberated by the complete hydrolysis of (a) DNA and (b) RNA?

21.10 How are all DNA molecules structurally alike?

21.11 How do different DNAs differ structurally?

21.12 How are all RNA molecules structurally alike?

21.13 What are the principal structural differences between DNA and RNA?

21.14 When DNA is hydrolyzed, the ratios of A to T and of G to C are each very close to 1 : 1, *regardless of the species*. Explain.

21.15 What is the principal noncovalent force in a DNA double helix?

21.16 What does base pairing mean, in general terms?

21.17 If the AGTCGGA sequence appeared on a DNA strand, what would be the sequence on the DNA strand opposite it in a double helix?

21.18 What does replication of DNA mean, in general terms?

21.19 The accuracy of replication is assured by the operation of what factors?

21.20 What is the relationship between a single molecule of single-stranded DNA and a single gene?

21.21 Suppose that a certain DNA strand has the following groups of nucleotides, where each lowercase letter represents a group several side chains long.

Which sections are likelier to be the introns? Why?

21.22 In general terms only, what particular contribution does a gene make to the structure of a polypeptide?

Ribonucleic Acids

21.23 What is the general composition of a ribosome, and what function does this particle have?

21.24 What is ptRNA, and what role does it have?

21.25 What is a codon, and what kind of nucleic acid is a continuous, uninterrupted series of codons?

21.26 Suppose that sections *x*, *y*, and *z* of the following hypothetical DNA strand are the exons of one gene.

What is the structure of each of the following substances made under its direction?
(a) The ptRNA   (b) The mRNA

21.27 What is an anticodon, and on what kind of RNA is it found?

21.28 Which triplet, ATA or CGC, cannot be a codon? Explain.

21.29 Which amino acids are specified by the following codons?
(a) UUU   (b) UCC   (c) ACA   (d) GAU

21.30 What are the anticodons for the codons of Review Exercise 21.29?

Polypeptide Synthesis

21.31 What is meant by *translation*, as used in this chapter? And what is meant by *transcription*?

21.32 To make the pentapeptide Met · Ala · Try · Ser · Tyr,
(a) What has to be the sequence of bases on the mRNA strand?
(b) What is the anticodon on the first tRNA to move into place?

21.33 How do some of the antibiotics work at the molecular level?

21.34 The genetic code is the key to translating between what two "languages"?

21.35 The genetic code is described as almost universal. What does this mean?

Viruses

21.36 What is a virus made of?

21.37 How do at least some viruses get into their host cells?

21.38 What do viruses do, chemically, in host cells (in general terms)?

21.39 What does the prefix *retro* signify in retrovirus?

21.40 What is the full name of the HIV system?

21.41 What is the host cell of HIV and how does this fact make AIDS so dangerous?

Recombinant DNA

21.42 What is a plasmid, and what is it made of?

21.43 Recombinant DNA is made from the DNA of two different kinds of sources. What are they?

21.44 Recombinant DNA technology is carried out to accomplish the synthesis of what kind of substance (in general terms)?

21.45 What does *genetic engineering* refer to?

Hereditary Diseases

21.46 At the molecular level of life, what kind of defect is the fundamental cause of a hereditary disease?

21.47 What is the molecular defect in PKU, and how does it cause problems for the victims? How is it treated?

DNA Fingerprinting (Special Topic 21.1)

21.48 What fact about cells makes it possible to use them from any part of the body of a suspect for genetic fingerprinting in a rape case for which a semen sample has been obtained?

21.49 A restriction enzyme separates DNA molecules into pieces given what general name? How are they used to give the genetic "bar code?"

Chapter 22
Radioactivity and Health

Atomic Radiations

Ionizing Radiations — Dangers and Precautions

Units to Describe and Measure Radiations

Synthetic Radionuclides

Radiation Technology in Medicine

Because the nuclei of certain elements behave like tiny magnets, magnetic resonance imaging (MRI) is made possible. In this final chapter we study other unusual properties of certain atomic nuclei.

22.1 ATOMIC RADIATIONS

Unstable atomic nuclei eject high-energy radiations as they change to more stable nuclei.

Some atomic nuclei are unstable and the isotopes with such nuclei are **radioactive,** meaning that they emit streams of high-energy radiations. Each radioactive isotope is called a **radionuclide,** and its radiations can cause grave harm to human life. When carefully used, however, their potential benefits outweigh their possible harm. In this chapter we will study what these radiations are, how they can be dangerous, how they can be used wisely, and how they are measured.

Unstable Nuclei Undergo Radioactive Decay, Emit Radiations, and Transmute to Nuclei of Different Elements.   Radioactivity was discovered in 1896 when a French physicist, A. H. Becquerel (1852–1908), happened to store some well-wrapped photographic plates in a drawer that contained samples of uranium ore. The film became fogged, meaning that when developed the picture was like a photograph of fog.

Becquerel might have blamed the accident on faulty film or careless handling, but a mysterious radiation called X rays had recently been discovered by a German scientist, Wilhelm Roentgen (1845–1923). X rays were known to be able to penetrate the packaging of unexposed film and ruin it. What fogged Becquerel's film was a natural radiation that resembled X rays. It was soon found to be emitted by any compound of uranium as well as by uranium metal itself.

Several years later, two British scientists, Ernest Rutherford (1871–1937) and Frederick Soddy (1877–1956), explained radioactivity in terms of events inside unstable atomic nuclei. Such nuclei undergo small disintegrations called **radioactive decay,** and they hurl tiny particles into space or emit a powerful radiation, one which is like X rays but called gamma radiation.

The nuclei that remain after decay almost always are those of an entirely different element, so decay is usually accompanied by the **transmutation** of one element or isotope into another. The natural sources of radiation on our planet emit one or more of three kinds: alpha radiation, beta radiation, and gamma radiation. We receive from the sun and outer space another kind, called *cosmic radiation,* which generates subatomic particles, including neutrons, as it streams into our atmosphere.

Alpha Particles Are the Nuclei of Helium Atoms.   One natural atomic radiation is called **alpha radiation.** It consists of particles called **alpha particles** that move with a velocity almost one-tenth the velocity of light as they leave the atom. Alpha particles are clusters of two protons and two neutrons, so they are actually the nuclei of helium atoms (Figure 22.1). They

Becquerel shared the 1903 Nobel prize in physics with Pierre and Marie Curie.

Roentgen won the 1901 Nobel prize in physics, the first to be awarded.

Nobel prizes in chemistry were awarded to both Rutherford (1908) and Soddy (1921).

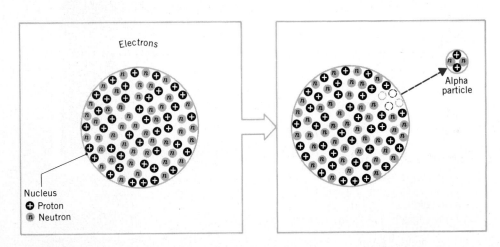

Electrons

Nucleus
⊕ Proton
ⓝ Neutron

Alpha particle

FIGURE 22.1
Emission of an alpha particle.

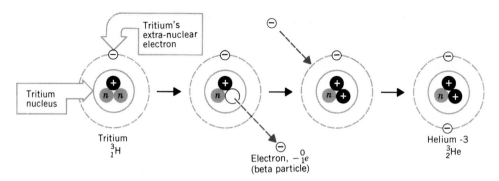

FIGURE 22.2
Emission of a beta particle.

are the largest of the decay particles and have the greatest charge, so when alpha particles travel in air, they soon collide with air molecules and lose their energy (and charge). Alpha particles cannot penetrate even thin cardboard or the outer layer of dead cells on the skin. Exposure to an intense dose of alpha radiation, however, causes a severe burn.

The special symbols for isotopes were introduced in Section 2.3. Thus in $^{238}_{92}U$, 238 is the mass number and 92 is the atomic number.

The most common isotope of uranium, uranium-238 or $^{238}_{92}U$, is an alpha-emitter. When its nucleus ejects an alpha particle, it loses two protons, so the atomic number changes from 92 to 90. It also loses four units of mass number (two protons + two neutrons), so its mass number changes from 238 to 234. The result is that uranium-238 transmutes into an isotope of thorium, $^{234}_{90}Th$.

Beta Radiation Is a Stream of Electrons. Another natural radiation, **beta radiation,** consists of a stream of particles called **beta particles.** These are electrons produced *within* the nucleus (Figure 22.2) and then emitted. With less charge and a much smaller size than alpha particles, beta particles can penetrate matter, including air, more easily than alpha particles. Different sources emit beta particles with different energies, and those of lower energy are unable to penetrate the skin. Those of the highest energy can reach internal organs from outside the body.

Losing one electron from the nucleus is like adding one proton without changing the mass number.

As a nucleus emits a beta particle, a neutron changes into a proton (Figure 22.2). Thus there is no loss in mass number, but the atomic number *increases* by one unit because of the new proton. For example, thorium-234, $^{234}_{90}Th$, is a beta emitter, and when it ejects a beta particle it changes to an isotope of protactinium, $^{234}_{91}Pa$.

Gamma Radiation Often Accompanies Other Radiation. An unstable radionuclide becomes more stable when it emits alpha or beta radiation. But nuclei have different energy states like those of electrons outside the nucleus. Sometimes the emission of an alpha or beta particle leaves the nucleus in one of its higher energy states. When it changes to a lower state, it emits the difference in energy as streams of photons of high-energy electromagnetic radiation. This is **gamma radiation,** and it often accompanies alpha or beta radiation. It is like X rays, only with more energy. Gamma radiation is very penetrating and very dangerous. It easily travels through the entire body.

The composition and symbols of the three radiations studied thus far are summarized in Table 22.1.

TABLE 22.1
Radiations from Naturally Occurring Radionuclides

Radiation	Composition	Mass Number	Electrical Charge	Symbols
Alpha radiation	Helium nuclei	4	2+	^4_2He or α
Beta radiation	Electrons	0	1−	$^0_{-1}e$ or β
Gamma radiation	X-ray type of energy	0	0	$^0_0\gamma$ or γ

Nuclear Equations. In *chemical* reactions no changes in atomic nuclei occur. But nuclear reactions are nearly always accompanied by transmutations. **Nuclear equations,** therefore, differ from chemical equations in important ways. They must, in particular, describe the changes in atomic numbers, mass numbers, and identities of radionuclides.

In nuclear equations the alpha particle is symbolized as ^4_2He, and although it is positively charged, the charge is omitted from the symbol. The particle soon picks up electrons, anyway, from the matter through which it is passing, and it becomes a neutral atom of helium. The beta particle has the symbol $_{-1}^0e$ because its mass number is 0 and its charge is $1-$. Gamma radiation is symbolized simply by γ (or, sometimes it is written as $^0_0\gamma$). Both its mass number and charge are 0.

A nuclear equation is balanced when the sums of the mass numbers on either side of the arrow are equal and when the sums of the atomic numbers are equal. The alpha decay of uranium-238, for example, is represented by the following equation:

$$^{238}_{92}U \longrightarrow ^{234}_{90}Th + ^4_2He$$

Notice that the sums of the atomic numbers agree: $92 = 90 + 2$. And the sums of the mass numbers agree: $238 = 234 + 4$.

The beta decay of thorium-234, which also emits gamma radiation, is represented by the following nuclear equation:

$$^{234}_{90}Th \longrightarrow ^{234}_{91}Pa + _{-1}^0e + \gamma$$

The sums of the atomic numbers agree: $90 = 91 + (-1)$.

It's proper to think of the electron as having an atomic number of -1.

EXAMPLE 22.1   BALANCING NUCLEAR EQUATIONS

Problem: Cesium-137, $^{137}_{55}Cs$, is one of the radioactive wastes that form during fission in a nuclear power plant or an atomic bomb explosion. This radionuclide decays by emitting both beta and gamma radiation. Write the nuclear equation for this decay.

Solution: First we set up as much of the nuclear equation as we can and leave blanks for any information we have to figure out. Thus our partial nuclear equation is

The first thing to do is to figure out the atomic symbol for the isotope that forms. For this we need the atomic number, because then we can look up the symbol in a table. We will find the atomic number by using a step-by-step process, but with just a little experience, you will be able to do this by inspection. We find the atomic number by using the fact that the sum of the atomic numbers on one side of the nuclear equation must equal the sum on the other side. Therefore letting x be the atomic number,

$$55 = -1 + 0 + x$$
$$x = 56$$

Now we can use the periodic table to find that the atomic symbol that goes with element 56 is Ba (barium). Next, to find out which particular isotope of barium forms, we use the fact that the sum of the mass numbers on one side of the nuclear equation must equal the sum on the other side. Letting y equal the mass number of the barium isotope,

$$137 = 0 + 0 + y$$
$$y = 137$$

The balanced nuclear equation therefore is

$$^{137}_{55}\text{Cs} \longrightarrow \, ^{0}_{-1}e + \, ^{0}_{0}\gamma + \, ^{137}_{56}\text{Ba}$$

EXAMPLE 22.2   BALANCING NUCLEAR EQUATIONS

The radium once used in cancer therapy was held in a thin, hollow gold or platinum needle to retain the alpha particles and all the decay products.

Problem: Until the 1950s, radium-226 was widely used as a source of radiation for cancer treatment. It is an alpha emitter and a gamma emitter. Write the equation for its decay.

Solution: We have to look up the atomic number of radium, which turns out to be 88, so the symbol we'll use for this radionuclide is $^{226}_{88}\text{Ra}$. When one of its atoms loses an alpha particle, $^{4}_{2}\text{He}$, it loses 4 units in mass number—from 226 to 222. And it loses 2 units in atomic number—from 88 to 86. Thus the new radionuclide has a mass number of 222 and an atomic number of 86. We have to look up the atomic symbol for element number 86, which turns out to be Rn, for radon. Now we can assemble the nuclear equation.

$$^{226}_{88}\text{Ra} \longrightarrow \, ^{222}_{86}\text{Rn} + \, ^{4}_{2}\text{He} + \gamma$$

PRACTICE EXERCISE 1   Iodine-131 has long been used in treating cancer of the thyroid. This radionuclide emits beta and gamma rays. Write the nuclear equation for this decay.

PRACTICE EXERCISE 2   Plutonium-239 is a by-product of the operation of nuclear power plants. It can be isolated from used uranium fuel and made into fuel itself or into atomic bombs. A powerful alpha and gamma-emitter, it is one of the most dangerous of all known substances. Write the equation for its decay.

A Short Half-Life Means a Rapid Decay and a More Dangerous Radionuclide. Some radionuclides are much more stable than others, and we use their half-lives to describe the differences. The **half-life** or $t_{1/2}$ of a radionuclide is the time it takes for one-half of its atoms

TABLE 22.2
Typical Half-Life Periods

Element	Isotope	Half-Life Period	Radiations
Naturally Occurring Radionuclides			
Potassium	$^{40}_{19}\text{K}$	1.3×10^9 years	Beta, gamma
Neodymium	$^{144}_{60}\text{Nd}$	5×10^{15} years	Alpha
Radon[a]	$^{222}_{86}\text{Rn}$	3.832 days	Alpha
Radium[a]	$^{226}_{88}\text{Ra}$	1590 years	Alpha, gamma
Thorium	$^{230}_{90}\text{Th}$	8×10^4 years	Alpha, gamma
Uranium	$^{238}_{92}\text{U}$	4.51×10^9 years	Alpha
Synthetic Radionuclides			
Hydrogen (tritium)	$^{3}_{1}\text{H}$	12.26 years	Beta
Oxygen	$^{15}_{8}\text{O}$	124 seconds	Positron[b]
Phosphorus	$^{32}_{15}\text{P}$	14.3 days	Beta
Technetium	$^{99m}_{43}\text{Tc}$	6.02 hours	Gamma
Iodine	$^{131}_{53}\text{I}$	8.07 days	Beta
Cesium	$^{137}_{55}\text{Cs}$	30 years	Beta
Strontium	$^{90}_{38}\text{Sr}$	28.1 years	Beta
Americium	$^{243}_{95}\text{Am}$	7.37×10^3 years	Alpha

[a] Although short-lived in relationship to the age of the earth, radon-222 and radium-226 are found in nature because the uranium-238 disintegration series continuously produces them.

[b] The positron has a mass number of 0 and a charge of 1+. (It's sometimes called a positive electron.)

to decay. (Not everything in those that decay vanishes, of course. New isotopes form, and some are also radioactive.) Table 22.2 gives several half-lives. Notice how widely they vary. Some are billions of years, like uranium-238, while others are only days, hours, or even a few seconds long.

Strontium-90, a by-product of nuclear power plants, is a beta emitter with a half-life of 28.1 years. Figure 22.3 shows graphically how a supply of 40 g is reduced successively by units of one-half for each half-life period. At the end of seven half-life periods (196.7 years, from 7×28.1), only 0.3 g of strontium-90 remains in the sample.

The shorter the half-life, the larger the number of decay events per mole per second occurring in the isotope. Mole for mole, it's generally much safer to be near a sample that has a long half-life and thus decays very slowly than to be near one that has a short half-life and decays very rapidly.

A Succession of Decays Occurs in a Radioactive Disintegration Series. As we mentioned, the decay of one radionuclide sometimes produces not a stable isotope but just another radionuclide. This might, in turn, decay to still another radionuclide, with the process repeating until a stable nuclide is finally reached. There are four such series in nature, called **radioactive disintegration series,** and uranium-238 is at the head of one (see Figure 22.4). This series ends in a stable isotope of lead. Notice that the seventh isotope in the series is radon-222 and that its half-life and those following it are short. Their decay produces all three of the radiations we have mentioned. We'll have more to say about radon-222 in Section 22.2.

An extremely long half-life is typical of the radionuclides that head a radioactive disintegration series.

22.2 IONIZING RADIATION—DANGERS AND PRECAUTIONS

Atomic radiations create unstable ions and radicals in tissue, which can lead to cancer, mutations, tumors, or birth defects.

Unstable Ions and Radicals Are Produced in Tissue by Radiations. Alpha and beta particles, as well as X rays and gamma rays, are called **ionizing radiations** because they can

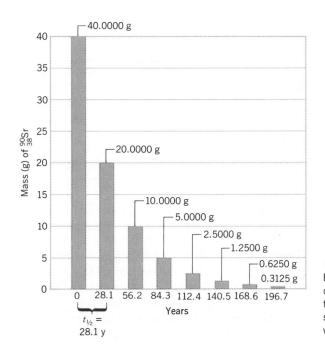

FIGURE 22.3
Each half-life period reduces the quantity of a radionuclide by a factor of two. Shown here is the pattern for strontium-90, a radioactive pollutant with a half-life of 28.1 years.

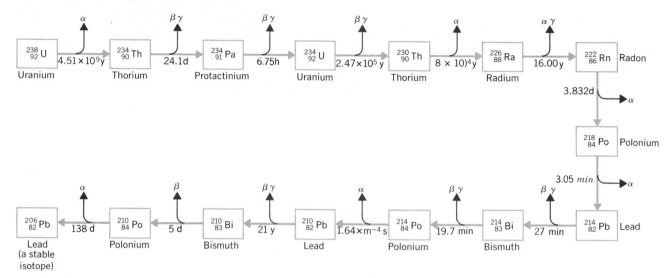

FIGURE 22.4

The uranium-238 radioactive disintegration series. The time given beneath the arrow is the half-life of the preceding isotope (y = year; m = month; h = hour; min = minute; and s = second).

nick electrons from molecules as they strike them and so produce unstable polyatomic ions. Alpha particles, for example, can make ions from water molecules as follows:

$$H—\overset{..}{\underset{..}{O}}—H + \text{high-energy } \alpha \text{ particle} \longrightarrow \left[H—\overset{.}{\underset{..}{O}}—H\right]^+ + e^- + \text{lower-energy } \alpha \text{ particle}$$

The ion, $\left[H—\overset{.}{\underset{..}{O}}—H\right]^+$, is an uncommon, unstable system. Its oxygen atom no longer has an outer octet. Moreover, the collision has left this strange ion with sufficient energy to break up spontaneously as follows:

$$\left[H—\overset{.}{\underset{..}{O}}—H\right]^+ \longrightarrow H^+ + :\overset{.}{\underset{..}{O}}—H$$

Hydroxyl radical

A proton forms plus the hydroxyl radical. This uncommon species is a *neutral* particle with an unpaired electron and without an octet for its oxygen atom. Any particle with an unpaired electron is called a **radical,** and most radicals are very reactive species.

The new ions and radicals produced by ionizing radiations cause unwanted chemical reactions in cells and alter cell compounds in ways foreign to metabolism. If this happens in genes and chromosomes, subsequant reactions could lead to cancer, tumor growth, or a genetic mutation. If they happen in a sperm cell, an ovum, or a fetus, the result might be a birth defect.

Prolonged and repeated exposures to *low* levels of radiation are more likely to induce these problems than bursts of high-level radiation. It depends on whether the injured cell is still able to duplicate itself by cell division. High-energy radiation bursts usually kill a cell outright, or at least render it reproductively dead. For this reason high doses of radiations are used in cancer treatment. But low level radiations that leave a cell reproductively viable can alter the cell contents in ways that are actually passed on to daughter cells.

There Is No Safe Threshold Exposure to Radiations.

All radiations that penetrate the skin or enter the body on food or through the lungs are considered harmful, and *the damage can accumulate* over a working lifetime. Even the ultraviolet radiation in strong sunlight, which barely penetrates the skin, can alter the genetic molecules in skin cells and lead to skin cancer. No "tiny bit of exposure," no **threshold exposure** exists for radiations below which no harm is possible.

Sometimes the term used is *free radical.*

Technical terms are

Carcinogen, a cancer causer

Tumorogen, a tumor causer

Mutagen, a mutation causer

Teratogen, a birth defect causer

In well-run establishments, medical personnel who work with radionuclides, X rays, or emitters of gamma rays are expected to wear devices that automatically record their exposure. The exposure data are periodically logged into a permanent record book, and when the maximum permissible dose has been attained, the worker has to be transferred.

Cells do have a capacity for self-repair. Some exposures carry very low risks, but none is entirely risk-free. The widespread and routine use of X rays for public health screenings has long been curtailed.

No technology in any medical field is entirely risk-free. We take risks when we believe that the benefits outweigh them.

The Collection of Symptoms Caused by Radiations Is Called Radiation Sickness.

Molecules of hereditary materials in the cell's chromosomes are the primary sites of radiation damage. Damage to these leads to all other problems. The first symptoms of exposure to radiation therefore occur in tissues whose cells divide most frequently, for example, the cells in bone marrow. These make white blood cells, so an early sign of radiation damage is a sharp decrease in the blood's white cell count. Cells in the intestinal tract also divide frequently, and even moderate exposures to X rays or gamma rays (as in cobalt-ray therapy for cancer) produces intestinal disorders.

The set of symptoms caused by nonlethal exposures to atomic radiations or X rays is called **radiation sickness.** The symptoms include nausea, vomiting, a drop in the white cell count, diarrhea, dehydration, prostration, hemorrhaging, and the loss of hair. They often appear when sharp bursts of radiations are used to halt the spread of cancer.

Protection from Radiations Is Achieved by Shields, Fast X-Ray Film, and Distance.

Shields have long been used for protection against radiations. (No doubt on a visit to the dentist you have had a lead apron placed over your chest before a dental X ray.) Alpha and beta rays are the easiest to stop as the data in Table 22.3 show. Gamma radiations and X rays

TABLE 22.3
Penetrating Abilities of Some Common Radiations[a]

Type of Radiation	Common Sources	Approximate Depth of Penetration of Radiation into:		
		Dry Air	Tissue	Lead
Alpha rays	Radium-226 Radon-222 Polonium-210	4 cm	0.05 mm[b]	0
Beta rays	Tritium Strontium-90 Iodine-131 Carbon-14	6 to 300 cm[c]	0.06 to 4 mm[b]	0.005 to 0.3 mm
		Thickness to Reduce Initial Intensity by 10%		
Gamma rays	Cobalt-60 Cesium-137 Decay products of radium-226	400 m	50 cm	30 mm
X Rays Diagnostic		120 m	15 cm	0.3 mm
Therapeutic		240 m	30 cm	1.5 mm

[a] Data from J. B. Little. *The New England Journal of Medicine*, Vol. 275, pages 929–938, 1966.

[b] The approximate energy needed to penetrate the protective layer of skin (0.07 mm thick) is 7.5 MeV for alpha particles and 70 keV for beta particles.

[c] The range of β particles in air is about 12 ft/MeV. Thus a 2-MeV β ray has a range of about 24 ft in air.

are stopped effectively only by particularly dense substances. Lead, a very dense metal but still fairly inexpensive, is the most common material used to shield against gamma or X rays. But notice in the data in Table 22.3 that even 30 mm (3.0 cm, a little over an inch) of lead reduces the intensity of gamma radiation by only 10%. A vacuum is least effective, of course, and air isn't much better. Low-density material such as carboard, plastic, and aluminum are poor shielders, but concrete works well if it is thick (and it's much cheaper than lead). Thus by a careful choice of a shielding material, protection can be obtained.

Another protection strategy when taking X rays is to use fast film. With fast film, the *time* of exposure is kept as low as possible.

The simplest self-protection step is to get as far from the source as you can. Radiations, like light from a bulb, move in straight lines spreading out in all the directions open to them from their sources. From any point on the surface of the source, the radiations form a cone of rays, so fewer rays can strike a unit of surface the more distant the surface is from the source. The area of the base of such a cone increases with the square of the distance. Hence, the radiation intensity I on a unit area diminishes with the square of the distance d from the source. This is the **inverse-square law** of radiation intensity.

Inverse-Square Law   The intensity of radiation is inversely proportional to the square of the distance from the source.

$$I \propto \frac{1}{d^2} \tag{22.1}$$

This law holds strictly only in a vacuum, but it holds closely enough when the medium is air to make good estimates. If we move from one location, a, to another location, b, the variation of Equation 22.1 that we can use to compare the intensities at the two different places, I_a and I_b, is given by the following equation:

$$\frac{I_a}{I_b} = \frac{d_b^2}{d_a^2} \tag{22.2}$$

EXAMPLE 22.3   USING THE INVERSE-SQUARE LAW OF RADIATION INTENSITY

Problem: At 1.0 m from a radioactive source the radiation intensity was measured as 30 units. If the operator moves away to a distance of 3.0 m, what will the radiation intensity be?

Solution: As usual, it's a good idea to assemble the data.

$$I_a = 30 \text{ units} \qquad d_a = 1.0 \text{ m}$$
$$I_b = ? \qquad d_b = 3.0 \text{ m}$$

Now we can use Equation 22.2:

$$\frac{30 \text{ units}}{I_b} = \frac{(3.0 \text{ m})^2}{(1.0 \text{ m})^2}$$

Solving for I_b, we get

$$I_b = 30 \text{ units} \times \frac{1.0 \text{ m}^2}{9.0 \text{ m}^2}$$

$$= 3.3 \text{ units}$$

Thus tripling the distance cut the intensity almost by a factor of 10.

Radon-222 is a naturally occurring radionuclide in the family of noble gases produced by the U-238 disintegration series. Chemically, it is as inert as the other noble gases, but radiologically it is a dangerous air pollutant. It is an alpha-emitter and gamma-emitter with a half-life of only four days. Produced in rocks and soil wherever uranium-238 is found, it migrates as a gas into the surrounding air. Basements not fully sealed act like fireplace chimneys to draw radon-222 into homes.

The first indication of how serious radon-222 pollution might be came when an engineer at a nuclear power plant in Pennsylvania set off radiation alarms just by his presence. The problem was traced to his home where the radiation level in the basement was 2700 picocuries per liter of air. In the average home basement, the level is just 1 picocurie/L. (The picocurie is 10^{-12} curie.). The engineer had carried radon-222 and its radioactive decay products on his clothing into his workplace.

As a result of the incident, geologists went looking for unusual concentrations of uranium-238 nearby. They found that the Reading Prong, a formation of bedrock that cuts across Pennsylvania, New Jersey, New York State, and up into the New England states, is relatively rich in U-238.

Radon-222 enters the lungs with breathing, and some decays within the lungs. Several decay products in the series after radon-222 are not gases, like polonium-218 ($t_{1/2}$ 3 min, alpha emitter), lead-214 ($t_{1/2}$ 27 min, beta- and gamma-emitter) and polonium-214 ($t_{1/2}$ 1.6×10^{-4} s, alpha- and gamma-emitter). Left in the lungs, these can cause cancer, and U. S. officials estimate that 7–8% of the country's annual deaths from lung cancer are caused by indoor radiation from radon-222.

The recommended upper limit on radon-222 concentration in home air is 4 picocuries/L. A conservative estimate puts the percentage of homes in the United States with levels above this at 1%.

PRACTICE EXERCISE 3   If the intensity of radiation is 25 units at a distance of 10 m, what does the intensity become if you move to a distance of 0.50 m?

PRACTICE EXERCISE 4   If you are receiving an intensity of 80 units of radiation at a distance of 6.0 m, to what distance would you have to move to reduce this intensity by one-half, to a value of 40 units?

No Escape from the Natural Background Radiation Is Possible. Shielding materials and distance can never completely reduce our exposure. We are constantly exposed to **background radiation,** the radiations given off, for example, by naturally occurring radionuclides in the environment, by radioactive pollutants, cosmic rays, and medical X rays.

About 50 of the roughly 350 isotopes of all elements in nature are radioactive. Natural radiations are in the food we eat, the water we drink, and the air we breathe. Radiations enter our bodies with every X ray taken of us. They come in cosmic-ray showers. On the average, the top 15 cm of soil on our planet has 1 g of radium per square mile. Thus radioactive materials are in the soils and rocks on which we walk and which we use to make building materials.

Radon, a *chemically* inert but radioactive gas and a product of the uranium-238 disintegration series (Figure 22.4), makes the largest contribution to our background radiation (see Special Topic 22.1).

The background radiation varies widely from place to place, and only estimates are possible. Table 22.4 gives the averages of radiations from various sources for the U. S. population. We will discuss the millirem (mrem), the unit of *dose equivalent* used in this table, shortly. For comparison purposes, a dose of 500 rem (500,000 mrem) given to the individuals in a large population would cause the deaths of half of them in 30 days. In relation to this, the intensity of natural background radiation is very small. At higher altitudes, the intensity of background radiation is greater because incoming cosmic rays, which contribute to the background, have had less opportunity to be absorbed and destroyed by the earth's atmosphere.

People became aware of the radon problem only in the early 1980s.

TABLE 22.4

Average Radiation Doses Received Annually by the U. S. Population[a]

Source	Mrem Dose	Percent of Total
Natural radiation—295 mrem, 82%		
Radon[b]	200	55
Cosmic rays	27	8
Rocks and soil	28	8
From inside the body	40	11
Artificial radiation—65 mrem, 18%		
Medical X rays[c]	39	11
Nuclear medicine	14	4
Consumer products	10	3
Others	2	<1
Total[d]	360	

[a] Data from "Ionizing Radiation Exposure of the Population of the United States." Report 93, 1987, National Council on Radiation Protection. These are averages. Individual exposures can vary widely.

[b] See Special Topic 22.1.

[c] A normal chest X ray gives an exposure of 10–20 mrem.

[d] The federal standard for maximum safe occupational exposure in the United States is roughly 5000 mrem/year.

Radioactive Pollutants Are Produced by Atomic Fission. All operating nuclear power plants in the United States today employ atomic fission carried out in huge devices called *reactors*. **Fission** is the disintegration of a large atomic nucleus into small fragments following neutron capture. Fission releases additional neutrons, radioactive isotopes, and enormous yields of heat. In virtually all reactors in use today, the heat generates steam, which drives electric turbines.

The uranium-235 isotope is the only naturally occurring radionuclide that spontaneously undergoes fission when it captures a slow-moving and relatively low-energy neutron. Its nucleus captures the neutron and then splits apart. It can do so in a number of ways to give different products. We'll give the equation for just one mode of splitting. Notice that neutrons are both reactants (initiators) and products.

Neutrons moving too rapidly are poorly captured.

$$^{235}_{92}U + ^{1}_{0}n \xrightarrow[\text{capture}]{\text{neutron}} ^{236}_{92}U \xrightarrow{\text{fission}} ^{139}_{56}Ba + ^{94}_{36}Kr + 3^{1}_{0}n + \gamma + \text{heat}$$

More neutrons are produced than consumed. Hence, one fission event produces enough new initiators (provided they are slowed down) to launch more than one new fission. This enables the continuation of fission among still unused atoms of uranium-235. Because a product is used as a reactant, the reaction is called a **nuclear chain reaction.**

The new isotopes produced by fission are radioactive, and their decay leads to some of the radioactive pollutants of chief concern, principally strontium-90 (a bone-seeking element in the calcium family), iodine-131 (a thyroid-gland-seeker), and cesium-137 (a Group IA radionuclide that goes wherever Na^+ or K^+ can go). The U. S. government has set limits to the release of each radioactive isotope into the air and into the cooling water of nuclear power plants. Those plants that operate in compliance with these standards expose people living near them to an extra dose of no more than 5% of the dose they normally receive from background radiations.

Radioactive Wastes Must Be Stored Away from People. One of the most vexing problems of nuclear energy has been that of permanently storing long-lived radioactive wastes. Most are now in temporary storage at nuclear power plants. They must be sequestered from all human contact for at least 1000 years, and scientists are seeking deep geologic formations as places to store these wastes out of all contact with mining operations, underground water supplies, and future archeologists.

Radioactive Wastes Enter the Food Chain.   During the period of extensive atmospheric tests of nuclear weapons (1954–1962), one of the more vexing wastes released was iodine-131. More recently, some was released from the massive reactor breach at the atomic energy park of Chernobyl, Russia (April 26, 1986). Atmospheric iodine-131 (largely as iodide ion) eventually comes to ground, some of it on pastures grazed by cows. They put it into their milk, which then is sold and so radioactive iodide ion enters humans.

Iodine-131 is a beta and gamma emitter with a short half-life (8 days) and our bodies use iodide ion to make thyroxine, a hormone produced by the thyroid gland. This process, therefore, *concentrates* radioactivity into a small organ. In some people, it will cause cancer.

To put this into perspective, one estimate is that over the next 50 years roughly 1000 extra cancer deaths will occur among the 12 countries affected by the radioactive plume from the Chernobyl accident. That's 20 extra deaths per year, and is statistically negligible. No one desires this, of course, but more deaths will occur during the same time from operations that mine coal and extract oil (to say nothing about the extra deaths associated with the metabolism of ethyl alcohol).

Roughly 8 tons of incandescent, radioactive fuel and wastes were hurled 3200 ft into the night sky at Chernobyl (from steam explosions, not atom bomblike blasts), and the plume drifted over much of northern Europe.

22.3   UNITS TO DESCRIBE AND MEASURE RADIATIONS

Units have been devised to describe the activity of a radioactive sample, the energies of its radiations, and the energies they can deliver to tissue.

Units have been defined for measuring radiations, and each was invented to serve in the answer to a particular question. Note carefully what these questions are and the units will be easier to learn.

The Curie (Ci) Describes How Active a Sample Is.   The curie, the unit of activity most commonly used by U. S. radiologists, was devised to answer the question, "How *active* is a sample of a radionuclide?" It was named after Marie Sklodowska Curie (1867–1934), a Polish scientist, who discovered radium. One **curie, Ci,** is the number of radioactive disintegrations that occur per second in a 1.0-g sample of radium, 3.7×10^{10} disintegrations/second.

Marie Curie is one of two scientists to win two Nobel prizes in a field of science, a share of the physics prize in 1903 and the chemistry prize in 1922.

$$1 \text{ Ci} = 3.7 \times 10^{10} \text{ disintegration/s}$$

This is an intensely active rate, so fractions of the Ci, such as the millicurie (mCi, 10^{-3} Ci), the microcurie (μCi, 10^{-6} Ci), and the picocurie (pCi, 10^{-12} Ci), are often used.

The **becquerel, Bq,** is the SI unit of activity, but it has not yet achieved much popularity among U. S. scientists.

$$1 \text{ Bq} = 1 \text{ disintegration/s}$$

Therefore,

$$1 \text{ Ci} = 3.7 \times 10^{10} \text{ Bq}$$

The Roentgen (R) Describes Exposure to X-Ray or Gamma-Ray Radiation.   The roentgen serves to answer the question, "How *intense* is the exposure to X-ray or gamma-ray radiation?" One **roentgen** of either, when passing through 1 cm^3 of dry air at normal temperature and pressure, generates ions with a total charge of 2.1×10^9 units. If a large population were exposed to 650 roentgens, half of the people would die in 1–4 weeks.

The Rad Describes the Energy Absorbed by Tissue.   The *rad* is commonly used to answer the question, "How much *energy* is *absorbed* by a unit mass of absorbing material?" It is defined in terms of the less used SI unit, the **gray, Gy,** which corresponds to the absorption of 1 joule (J) of energy per kilogram of tissue. (The joule is the SI unit of energy; 1 J = 4.184 cal.)

$$1 \text{ Gy} = 1 \text{ J/kg}$$

Rad comes from *radiation absorbed dose.*

The **rad** is defined as $\frac{1}{100}$th of a gray.

$$1 \text{ rad} = 0.01 \text{ Gy}$$

The joule is an extremely small amount of energy, so both the gray and the rad are also very small quantities. Nevertheless, only 600 rad of gamma radiation would be lethal to most people. We have to remember that it is not the quantity of energy that matters so much as the formation of unstable radicals and ions caused by this energy. A 600-rad dose delivered to water ionizes only one molecule in every 36 million, but the ions or radicals thus produced can begin a cascade of harmful reactions inside a cell.

The roentgen and the rad are close enough in magnitude that they are nearly equivalent from a health standpoint. Thus one roentgen of gamma radiation from a cobalt-60 source, sometimes used in cancer treatment, equals 0.96 rad in muscle tissue and 0.92 rad in compact bone.

The Rem Adjusts Rad Doses for Different Effects in Different Tissues. The rem satisfies the need for a unit of absorbed dose that is additive for different radiations and different target tissues. A dose of one rad of gamma radiation is not *biologically* the same as a dose of 1 rad of beta radiation or of neutrons. Thus the rad is not a good basis for comparison when working with biological effects. The rem fits this need.

Rem comes from *roentgen equivalent for man.*

The **rem** is the unit of *dose equivalent.* To convert rads to rems, we multiply the dose in rads by a factor that takes into account biologically significant properties of the radiation. One rem of any given radiation is the dose that has, in humans, the effect of one roentgen.

The rem, like the rad, is a quantity small in terms of energy but significant in terms of danger. Even millirem quantities of radiation should be avoided, and when this is not possible the workers must wear monitoring devices that allow the day-to-day exposures to be calculated.

The Electron Volt Describes the Energy of X Rays or Gamma Rays. To describe the energies associated with X rays or γ radiation, the **electron volt, eV,** defined as follows, is commonly used.

$$1 \text{ eV} = 1.602 \times 10^{-19} \text{ J}$$

(The electron volt is the energy an electron receives when accelerated by a voltage of 1 V.)

The electron volt is an extremely small amount of energy. Multiples of it are therefore very common, such as the kiloelectron volt (1 keV = 10^3 eV) and the megaelectron volt (MeV = 10^6 eV).

The higher is their energy, the more penetrating are X rays and gamma rays.

Linear accelerators produce radiation for cancer treatment in the range of 6–12 Mev.

X rays used for diagnosis are typically 100 keV or less. The gamma radiations of cobalt-60, used in cancer treatment, have energies of 1.2 and 1.3 MeV. Beta radiation of 70 keV or more penetrates the skin, but alpha particles (which are much larger than beta particles) need energies of more than 7 MeV to do this. Alpha radiation from radium-226 has an energy short of this, 5 MeV. Solar cosmic radiation has energies ranging from 200 MeV to 200 GeV (1 GeV = 1 gigaelectron volt = 10^9 eV).

Film Dosimeters and Geiger Counters Measure Radiations. People who work around radioactive sources cannot avoid some exposure, and they should keep a log of how much they accumulate. A device to measure exposure is called a *dosimeter.* One common type is a film badge that contains photographic film, which becomes fogged by radiations. The degree of fogging, which is related to the exposure, can be measured.

Geiger counters are devices used to measure beta and gamma radiation that has enough energy to penetrate the thin window of a Geiger–Müller tube. Once in the tube, ionizing radiation activates an electrical circuit and a short pulse of electricity is sent to a counting device. The number of counts per second caused by a radioactive source is a measure of the source's activity, at least in beta and gamma radiation.

22.4 SYNTHETIC RADIONUCLIDES

Most radionuclides used in medicine are made by bombarding other atoms with high-energy particles.

Radioactive decay is nature's way of causing transmutations. They can also be caused artificially by bombarding atoms with high-energy particles. Several hundred isotopes that do not occur naturally have been made this way, including several used in medicine.

Various bombarding particles can be used. When alpha particles from a natural radionuclide, for example, are allowed to pass through a tube of nitrogen-14 gas, some high-energy protons and oxygen-17 forms. The alpha particles plow right through the electron clouds of nitrogen-14 atoms and bury themselves in their nuclei. A new nucleus forms, one of fluorine-18, with too much energy to exist for long. Such high-energy, unstable nuclei made by particle capture are called *compound nuclei*. The fluorine-18 ejects a proton and what remains is an atom of oxygen-17. The equation is

> This was the reaction whereby Rutherford discovered artificial transmutation.

$$\underset{\substack{\text{Alpha}\\\text{particle}}}{{}^{4}_{2}\text{He}} + \underset{\substack{\text{Nitrogen}\\\text{nucleus}}}{{}^{14}_{7}\text{N}} \longrightarrow \underset{\substack{\text{Fluorine}\\\text{nucleus}}}{{}^{18}_{9}\text{F}^{*}} \longrightarrow \underset{\substack{\text{Oxygen}\\\text{nucleus}}}{{}^{17}_{8}\text{O}} + \underset{\text{Proton}}{{}^{1}_{1}p}$$

> The protons produced by this nuclear reaction are gaseous protons, true subatomic particles, and not hydronium ions.

(The asterisk by the symbol for fluorine-18 signifies that it contains a compound nucleus.) Oxygen-17 is a rare but nonradioactive isotope of oxygen. Usually, transmutations caused by bombardments produce *radioactive* isotopes of other elements.

Electrically charged bombarding particles, like the alpha particle, are often given greater energy by being accelerated (given higher velocities). By varying the energies and targets, dozens of synthetic radionuclides have been made.

> Remember: The kinetic energy (KE) of a moving object increases with the *square* of its velocity. $KE = \frac{1}{2}mv^2$

Certain isotopes of uranium in atomic reactors eject neutrons, and neutron bombardment is used to make molybdenum-99 from molybdenum-98. The change is nothing more than neutron-capture with γ-ray production.

$$^{98}_{42}\text{Mo} + ^{1}_{0}n \longrightarrow ^{99}_{42}\text{Mo} + \gamma$$

> The symbol for the neutron in a nuclear equation is ${}^{1}_{0}n$.

Molybdenum-99 is radioactive, and its decay leads to one of the radionuclides most commonly used in medicine (Section 22.5).

22.5 RADIATION TECHNOLOGY IN MEDICINE

Both in diagnosis and in cancer treatment, ionizing radiations are used when their benefits are judged to outweigh their harm.

For medical uses, radiations are supplied as X rays, accelerated electron beams, or radionuclides. They are used in diagnosis to locate a cancer or tumor or to assess some organ function, like that of the thyroid gland. Radiations are used in therapy to kill cancer cells. Therapeutic doses are usually of much higher energy than those used for diagnoses. We will look here at the radionuclides used in diagnosis, and discuss beam radiation technologies in Special Topics 22.3 – 22.5.

> *Radiology:* The science of radioactive substances and of X rays.
>
> *Radiologist:* A specialist in radiology who also usually has a medical degree.
>
> *Radiobiology:* The science of the effects of radiations on living things.

To make a dilute solution of technetium-99*m* (in the form of TcO_4^-, a radiologist "milks a molybdenum cow." See the accompanying figure. This device contains molybdenum-99 (in the form of MoO_4^{2-}) mixed with granules of alumina. (Alumina is one of the crystalline forms of Al_2O_3, and it is used here as a support for the molybdenate ion.) The device is charged at a nuclear reactor facility by the neutron bombardment of the molybdenum-98 present (as MoO_4^{2-}) on the alumina granules, as described in Section 22.4. After neutron bombardment, the device is shipped to the hospital, and all the while the molybdenum-99 is decaying to technetium-99*m*.

Each morning, a member of the radiology staff lets a predetermined volume of isotonic salt solution trickle through the bed of granules. Some of the pertechnetate ions dissolve and are leached out. Then the solution so obtained is used for the day's work. After several days, too little molybdenum-99 remains, so the device is shipped back to the reactor for recharging.

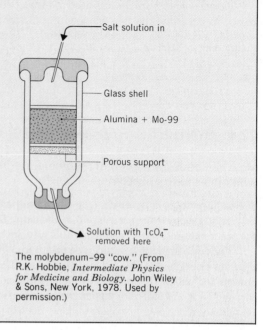

The molybdenum–99 "cow." (From R.K. Hobbie, *Intermediate Physics for Medicine and Biology*. John Wiley & Sons, New York, 1978. Used by permission.)

Both Chemical and Radiological Properties Are Important in the Selection of Radionuclides in Medicine. The *chemical* properties of radionuclides are identical to those of the stable isotopes of the same element. When radionuclides are selected for use in medicine, their chemistry, therefore, has to be considered. They must be *chemically* compatible with the living system even when used just for their radiations. Their *chemistry* is what actually guides them to desired tissues. Any isotope of iodine, for example, is taken up by the thyroid gland to make the hormone thyroxin. This chemical property, therefore, guides any radioactive isotope of iodine, like iodine-131, to the same organ, as we mentioned in Section 22.2.

Minimizing Harm and Maximizing Benefit Guide the Selection of Radionuclides in Medicine. Exposing anyone to any radiation entails some risks. No such exposure is permitted unless the expected benefit from finding and treating a dangerous disease is thought to be greater than the radioactive risk. To minimize risks, the radiologist uses radionuclides with, as much as possible, the following properties.

1. The radionuclide should have a half-life that is short. (Then it will decay *during* the diagnosis when the decay gives some benefit, and as little as possible of the radionuclide will decay later, when the radiations are of no benefit.)

2. The product of the decay of the radionuclide should have little if any radiation of its own and be quickly eliminated. (Either the product should be a stable isotope or have a very long half-life.)

3. The half-life of the radionuclide must be long enough for it to be prepared and administered to the patient.

4. If the radionuclide is to be used for diagnosis, it should decay by penetrating radiation entirely, which means gamma radiation. (Nonpenetrating radiations, such as alpha and beta radiation, add to the risk by causing internal damage without contributing to the detection of the radiation externally. For uses in *therapy*, as in cancer therapy, nonpenetrating radiation is preferred because a radionuclide well placed in cancerous tissue *should* cause damage to such tissue.)

5. The diseased tissue should concentrate the radionuclide, giving a "hot spot" where the diseased area exists, or it should do the opposite and reject the radionuclide, making the diseased area a "cold spot" insofar as external detectors are concerned.

SPECIAL TOPIC 22.3   X RAYS AND CT SCANS

X rays are generated by bombarding a metal surface with high-energy electrons. These can penetrate the metal atom far enough to knock out one of its low-energy-level electrons, like a level 1 electron. This creates a "hole" in the electron configuration, and other electrons at higher levels begin to drop down. In other words, the creation of this "hole" leads to electrons changing their energy levels. The difference between two of the lower levels corresponds to the energy of an X ray which is emitted.

The refinement of X-ray techniques and the development of powerful computers made possible the generation of a diagnostic technology called computerized tomography, or CT for short. A. M. Cormack (United States) and G. N. Hounsfield (England) shared the 1979 Nobel prize in medicine for their work in the development of this technology. The instrument includes a large array of carefully positioned and focused X-ray generators. In the procedure called a CT scan, this array is rotated as a unit around the body or the head of the patient. Extremely brief pulses of X rays are sent in from all angles across one cross section of the patient (see the accompanying photo).

The changes in the X rays that are caused by internal organs or by tumors are sent to a computer, which then processes the data and delivers a picture of the cross section. It's like getting a picture of the inside of a cherry pit without cutting open the cherry. The CT scan is widely used for locating tumors and cancers (see the other accompanying figure).

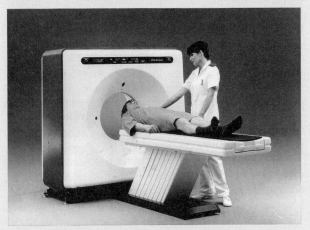

Instrument for the CT scan. Shown here is General Electric's CT/T Total Body Scanner, which can complete an entire scan of the head or body in as little as 4.8 seconds.

A three-dimensional image, based on 63 CT scans, of a section of the vertebrae of a man injured in a motorcycle accident. Both the compression and the twisting are clearly evident.

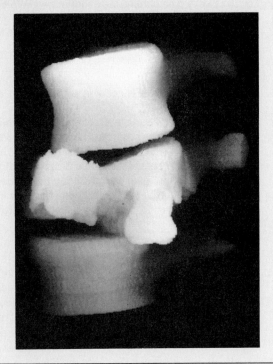

Let us now look at a few of the radionuclides employed in medicine.

Technetium-99*m* **Is the Most Widely Used Radionuclide in Medicine.** Technetium 99*m* is made from the decay of molybdenum-99 (made by neutron bombardment of molybdenum-98, described in the previous section.) See also Special Topic 22.2. As we mentioned earlier, gamma radiation often accompanies other emissions, but with molybdenum-99 the gamma radiation comes after a pause. Molybdenum-99 first emits a beta particle:

$$\mathrm{^{99}_{42}Mo} \longrightarrow \mathrm{^{99m}_{43}Tc} + \mathrm{^{0}_{-1}}e$$

A metastable form of technetium-99 also forms, hence the *m* in 99*m*. It is poised to become more stable, which it does by emitting gamma radiation with an energy of 143 keV:

$$\mathrm{^{99m}_{43}Tc} \longrightarrow \mathrm{^{99}_{43}Tc} + \gamma$$

A number of synthetic radionuclides emit positrons. These are particles that have the same small mass as an electron but carry one unit of *positive* charge. (They're sometimes called a positive electron.) A positron forms by the conversion of a proton into a neutron, as follows:

$$\overset{1}{\underset{1}{p}} \longrightarrow \overset{1}{\underset{0}{n}} + \overset{0}{\underset{1}{e}}$$

Proton   Neutron   Positron
(in atom's (stays in (is emitted)
nucleus)   nucleus)

Positrons, when emitted, last for only a brief interval before they collide with an electron. The two particles annihilate each other, and in so doing their masses convert entirely into energy in the form of two tiny bursts of gamma radiation (511 keV). The gamma radiation formed in this way is called annihilation radiation.

$$\overset{0}{\underset{-1}{e}} + \overset{0}{\underset{1}{e}} \longrightarrow 2\,\overset{0}{\underset{0}{\gamma}}$$

Electron   Positron   Gamma
                         radiation

The two bursts leave the collision site in almost exactly opposite directions.

To make a medically useful technology out of this property of the positron, a positron-emitting nuclide must be part of a molecule with a chemistry that will carry it into the particular tissue to be studied. Once the molecule gets in, the tissue now has a gamma radiator *on the inside.* Thus instead of X rays being sent through the body, as in a CT scan (Special Topic 22.3), the radiation originates right within the site being monitored. The overall procedure is called positron emission tomography, or PET for short.

Three positron emitters are often used: oxygen-15, nitrogen-13, and carbon-11. Glucose, for example, can be made in which one carbon atom is carbon-11 instead of the usual carbon-12. Glucose can cross the blood brain barrier and get inside brain cells. If some part of the brain is experiencing abnormal glucose metabolism, this will be reflected in the way in which positron emitting glucose is handled, and gamma-radiation detectors on the outside can pick up the differences. The use of the PET scan has led to the discovery that glucose metabolism in the brain is altered in schizophrenia and manic depression. PET scanning technology is able to identify extremely small regions in the brain that are in early stages of breakdown and that CT and MRI techniques miss. The PET scan is particularly useful in detecting abnormal brain function in infants, as in early stages of epilepsy.

PET technology is being used to study a number of neuropsychiatric disorders, including Parkinson's disease. When a drug labeled with carbon-11 is used, its molecules go to the parts of the brain with nerve endings that release dopamine. A PET scan then discloses the dopamine-releasing potential of the patient. This potential becomes impaired in Parkinson's disease.

By labeling blood platelets with a positron emitter, scientists can follow the development of atherosclerosis in even the tiniest of human blood vessels. Blood flow in the heart can be monitored without having to insert a catheter.

Technetium-99m almost ideally fits the criteria for a radionuclide intended for diagnostic work. Its half-life is short, 6.02 h. Its decay product, technetium-99, has a very long half-life, 212,000 years, so it has too little activity to be of much concern. (Technetium-99 decays to a stable isotope of ruthenium, $^{99}_{44}\text{Ru}$.) The half-life of technetium-99m, although short, is still long enough to allow time to prepare it and administer it. It decays entirely by gamma radiation, which means that *all* the radiation gets to a detector to signal where the radionuclide is in the body. Finally, a variety of chemically combined forms of technetium-99m have been developed that permit either hot spots or cold spots to form.

One form of technetium-99m is the pertechnetate ion, TcO_4^-. It behaves in the body very much like a halide ion, so it tends to go where chloride ions, for example, go. It is eliminated by the kidneys, so it is used to assess kidney function. Other organs whose functions are also studied by technetium-99m are the liver, spleen, lungs, heart, brain, bones, and the thyroid gland.

Technetium-99m technology has received competition from the CT scan (Special Topic 22.3), the PET scan (Special Topic 22.4), and MRI imaging (Special Topic 22.5). But enough unique applications for technetium-99m remain to ensure that it will continue to be used for some time.

Iodine-131 and Iodine-123. The thyroid gland takes iodide ion and, as we said, makes the hormone thyroxin. The ability of this gland to concentrate iodide ion is so good that if a small dose of iodine-131 is given, nearly 1000 times as much dose concentrates in the thyroid.

When either an underactive or an overactive thyroid is suspected, the patient can be given a drink of flavored water containing radioactive iodine as I⁻. By placing detection

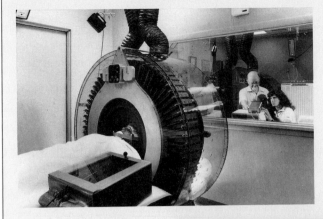

The patient is shown undergoing a brain scan performed by means of PET technology.

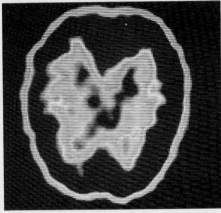
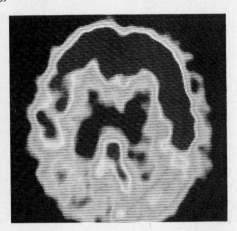

Left: Normal brain scan using PET technology. Right: PET scan of the brain of a patient with Alzheimers disease.

equipment near the thyroid gland, the radiologist can tell how well it takes up circulating iodide ion. For diagnostic purposes, iodine-123 is popular because it has a short half-life (13.3 h) and emits only gamma radiation (159 keV).

Iodine-131 also has a short half-life (8 days), but it emits both beta particles (600 keV) and gamma radiation (mostly 360 keV). The beta particles are of no benefit in diagnosis (they cannot reach the detector), but they are helpful in treating certain types of thyroid cancer.

Technetium-99m as the hydrated pertechnetate ion, TcO_4^-, has about the same radius and charge density as the hydrated iodide ion. Cells of the thyroid gland, therefore, do not distinguish between these two ions at the point where they move inside the cells. When the purpose is to get any kind of detectable radiation inside the thyroid gland in order to detect a tumor or cancer, then Tc-99m, because of its short half-life (6.02 h), is preferred over any radionuclide of iodine.

Cobalt-60 Provides Gamma Radiation above the Megaelectron Volt Level.

Cobalt-60 emits beta radiation (315 keV) and gamma radiation (2.819 MeV); it has a half-life of 5.3 years; and it is used to treat cancer. The sample of cobalt-60 is placed in a lead container many centimeters thick with an opening aimed at the cancerous site. All beta radiations are shielded by a thin piece of aluminum.

Linear Accelerators Make Ultrahigh-Energy X Rays.

Betatrons are linear accelerators that are tending to supplant cobalt-60 therapy for several reasons. These devices generate X rays for therapeutic uses with energies in the range of 6–12 MeV, which are much more powerful and so more penetrating than the gamma rays from cobalt-60. The output of the

Cobalt-60 therapy is being replaced by the radiations from linear accelerators.

The CT scan subjects a patient to large numbers of short bursts of X rays. The PET scan exposes the patient to gamma radiation that is generated on the inside. Thus both technologies carry the usual risks that attend ionizing radiations, and they are used when the potential benefits from correct diagnoses far outweigh such risks. The MRI imaging technology operates without these dangers. At least, none has been discovered thus far. The principal developer of the first hardware for MRI imaging was Raymond Damadian.

MRI stands for magnetic resonance imaging. Atomic nuclei that have odd numbers of protons and neutrons behave as though they were tiny magnets, hence the *magnetic* part of MRI. The nucleus of ordinary hydrogen (which has no neutrons and one proton) constitutes the most abundant nuclear magnet in living systems, because hydrogen atoms are parts of water molecules and all biochemicals. Nuclear magnets spin about an axis much as the electron spins about its axis.

When molecules with spinning nuclear magnets are in a strong magnetic field and are simultaneously bathed with properly tuned radiofrequency radiation (which is of very low energy), the nuclear magnets flip their spins. (This is the *resonance* part of MRI.) As they resonate, they emit electromagnetic energy that is biologically harmless, being of very low energy, and this energy is picked up by detectors. The data are fed into computers, which produce an image much like that of a CT or a PET scan, but without having subjected the patient to ultrahigh energy electromagnetic radiation such as X rays or gamma rays. The MRI images are actually sophisticated plots of the distributions of the spinning nuclear magnets, of hydrogen atoms, for example.

MRI imaging has proven to be especially useful for studying soft tissue, the sort of tissue least well studied by X rays. Different soft tissues have different population densities of water molecules or of fat molecules (which are loaded with hydrogen atoms). And tumors and cancerous tissue have their own water inventories. Calcium ions do not produce any signals to confuse MRI imaging, so bone, which is rich in calcium, is transparent to MRI.

MRI technology has developed during the 1980s as a method superior to the CT scan for diagnosing tumors at the rear and base of the skull and equal to the CT scan for finding other brain tumors. MRI is now the preferred technology for assessing problems in joints (particularly the knees) and in the spinal cord, such as ruptured (herniated, "slipped") disks. Patients with heart pacemakers or embedded shrapnel or surgical clips present problems to the use of MRI because of the powerful magnets used.

CT scans are still better than MRI for the early detection of hemorrhages in the brain, so CT is the method of choice for finding them in potential stroke victims. CT scans are also still preferred for detecting tumors in the kidneys, lungs, pancreas, and the spleen.

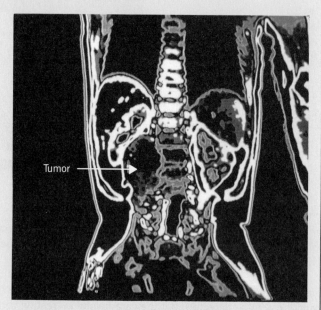

An MRI scan of a seven-month old child revealed a malignant tumor pushing its way into the spinal canal. (It was treated in time.)

machines is stable; the output of cobalt-60 declines over time. The edges of the beams are sharply defined; those from cobalt-60 are less so. This means that X-ray pictures obtained as the therapy is applied are of higher quality. Linear accelerator machines are easily mounted so that they can rotate about a central point, which makes it easier to position the patient and plan the treatment.

Other Medically Useful Radionuclides Are Available for Special Purposes. Indium-111 ($t_{1/2} = 2.8$ days; gamma emitter, 173 and 247 keV) has been found to be a good labeler of blood platelets.

Gallium-67 ($t_{1/2} = 78$ h; gamma emitter; 1.003 MeV) is used in the diagnosis of Hodgkins disease, lymphomas, and bronchogenic carcinoma.

Phosphorus-32 ($t_{1/2} = 14.3$ days; beta emitter; 1.71 MeV) in the form of the phosphate ion has been used to treat a form of leukemia, a cancer of the bone that affects white cells in the blood. Because the phosphate ion is part of the mineral in bone, this ion is a bone seeker.

SUMMARY

Atomic radiations Radionuclides in nature emit alpha radiation (helium nuclei), beta radiation (electrons), and gamma radiation (high-energy X-raylike radiation). This radioactive decay causes transmutation. The penetrating abilities of the radiations are a function of the sizes of the particles, their charges, and the energies with which they are emitted. Gamma radiations, which have no associated mass or charge, are the most penetrating.

Each decay can be described by a nuclear equation in which mass numbers and atomic numbers on either side of the arrow must balance. To describe how stable a radionuclide is we use its half-life, and the shorter this is, the more radioactive is the radionuclide.

The decay of one radionuclide doesn't always produce a stable nuclide. Uranium-238 is at the head of a radioactive disintegration series that involves several intermediate radionuclides until a stable isotope of lead forms.

Ionizing radiations — dangers and precautions When radiations travel in matter, they create unstable ions and radicals that have chemical properties dangerous to health. Intermittent exposure can lead to cancer, tumors, mutations, or birth defects. Intense exposures cause radiation sickness and death. Intense exposures focused on cancer tissue are used in cancer therapy. The use of distance, fast film, and dense shielding material are the best strategies to guard against the hazards of ionizing radiations. According to the inverse-square law, the intensity of radiation falls off with the square of the distance from the source. Complete protection, however, is not possible because of the natural background radiation that now includes traces of radioactive pollutants.

Units of radiation measurement To describe activity we use the curie, Ci, or the becquerel, Bq. To describe the intensity of exposure to X rays or gamma radiation, we use the roentgen. The gray (Gy) or the rad is used to indicate how much energy has been absorbed by a unit mass of tissue (or other matter). Use of the rem puts the damage that different radiations can cause when they have the same values of rads (or grays) on a comparable and additive basis. Finally, to describe the energy possessed by a radiation, we use the electron volt. Diagnostic X rays are on the order of 100 keV. Radiations used in cancer treatment are in the low MeV range. To measure radiations there are devices such as film badges and ionization counters (Geiger–Müller tubes).

Synthetic radionuclides A number of synthetic radionuclides have been made by bombarding various isotopes with alpha radiation, neutrons, or accelerated protons. The target nucleus first accepts the mass, charge, and energy of the bombarding particle, and then it ejects something else to give the new nuclide.

Radionuclides in medicine For diagnostic uses, the radionuclide should have a short half-life (but not so short that it decays before any benefit can be obtained). It should decay by gamma radiation only, and it should be chemically compatible with the organ or tissue so that either a hot spot or a cold spot appears. Its decay products should be as stable as possible, and capable of being eliminated from the body.

Technetium-99m is almost ideal for diagnostic work, particularly for assessing the ability of an organ or tissue to function. Radionuclides of iodine (I-123 and I-131) are used in diagnosing or treating thyroid conditions. Gallium-67, indium-111, and phosphorus-32 are a few of the many other radionuclides used in diagnosis. Cobalt-60 has powerful gamma radiation, and is used in cancer treatment. Linear accelerators also provide high-energy (6–12 MeV) radiation for cancer therapy.

Atomic energy and radioactive pollutants The reactors of most nuclear power plants use U-235 as a fuel. When its atoms capture neutrons, they fission into smaller, usually radioactive atoms as neutrons are released that can cause additional fissions. The heat generated by fission converts water into high-pressure steam that drives electrical turbines.

Radioactive by-products of fission, such as I-131, Sr-90, and Cs-137, cannot be allowed to enter the food supply and must be contained. Some fission products have such long half-lives that radioactive wastes must be kept from human contact for a thousand years.

REVIEW EXERCISES

The answers to the review exercises that require a calculation or the balancing of an equation and that are marked with an asterisk are found in Appendix V. The answers to the other review exercises are found in the *Study Guide* that accompanies this book.

Radioactivity and Radiations

22.1 If a sample is described as *radioactive,* what specifically do we know about it?

22.2 The forerunners of chemists were called alchemists, and in ancient times one of their quests was for a way to change a base metal such as lead into gold. Today what technical word would be used to describe such a change if it were successful?

22.3 The film that Becquerel put in his desk drawer was wrapped in fairly heavy paper. In view of what we know today, his discovery of radioactivity depended on the emission of which radiation from the uranium ore sample?

22.4 When we write nuclear equations, which symbols are used for each of the following?
(a) The alpha particle   (b) The beta particle
(c) A gamma ray

22.5 The energy of an alpha particle is often higher than that of beta or gamma rays. Why is it, then, the least penetrating of the radiations?

22.6 The loss of an alpha particle changes the radionuclide's mass number by how many units? Its atomic number by how many units?

22.7 Why does the loss of a beta particle not change the radionuclide's mass number but *increases* its atomic number?

22.8 What happens to the mass number and to the atomic number of a radionuclide if it is only a gamma emitter?

22.9 Are *all* radioactive decays also transmutations?

22.10 If electrons do not exist in the nucleus, how can one originate in a nucleus in beta decay?

Nuclear Equations

*22.11 Write the symbols of the missing particles in the following nuclear equations.
(a) $^{245}_{96}Cm \rightarrow ^{4}_{2}He +$ _____ (b) $^{22}_{9}F \rightarrow ^{0}_{-1}e +$ _____

22.12 Complete the following nuclear equations by writing the symbols of the missing particles.
(a) $^{220}_{86}Rn \longrightarrow ^{4}_{2}He +$ _____ (b) $^{140}_{56}Ba \rightarrow ^{0}_{-1}e +$ _____

*22.13 Write a balanced nuclear equation for each of the following changes.
(a) Alpha emission from einsteinium-252
(b) Beta emission from magnesium-28
(c) Beta emission from oxygen-20
(d) Alpha and gamma emission from californium-251

22.14 Give the nuclear equation for each of the following radioactive decays.
(a) Beta emission from bismuth-211
(b) Alpha and gamma emission from plutonium-242
(c) Beta emission from aluminum-30
(d) Alpha emission from curium-243

Half-Lives

22.15 Lead-214 is in the uranium-238 disintegration series. Its half-life is 19.7 min. Explain in your own words what being in this series means and what *half-life* means.

22.16 Which would be more dangerous to be near, a radionuclide with a short half-life that decays by alpha emission only or a radionuclide with the same half-life but that decays by beta and gamma emisson? Explain.

*22.17 A 12.00-ng sample of technetium-99*m* will still have how many nanograms of this radionuclide left after 4 half-life periods? (This is about a day.)

22.18 If a patient is given 9.00 ng of iodine-123 (half-life 13.3 h), how many nanograms of this radionuclide remain after 12 half-life periods (about a week)?

Dangers of Ionizing Radiations

22.19 We have ions in every fluid of the body. Why, then, are ionizing radiations dangerous?

22.20 What is a chemical *radical,* and why is it chemically reactive?

22.21 Radiations are teratogenic agents. What does this mean?

22.22 Plutonium-239 is a carcinogen. What does this mean?

22.23 What two properties of ionizing radiations are exploited in strategies for providing radiation protection?

22.24 Atomic radiations are said to have no threshold exposure. What does this mean?

22.25 How is it that the same agent, radiations from a radionuclide, can be used both to cause cancer and to cure it?

22.26 The inverse-square law tells us that if we double the distance from a radioactive source, we will reduce the radiation intensity that we receive by a factor of what number?

22.27 What general property of radiations is behind the inverse-square law?

22.28 List as many factors as you can that contribute to the background radiation.

22.29 Why does a trip in a high-altitude jet plane increase one's exposure to background radiation?

22.30 Name three isotopes made in nuclear power plants that are particularly hazardous to health, and explain in what specific ways they endanger various parts of the body.

22.31 In general terms, how does fission differ from radioactive decay?

22.32 What fundamental aspect of fission makes it possible for it to proceed as a chain reaction?

22.33 The reactors of nuclear power plants are fitted with many movable rods. (Some are made of carbon, for example.) These rods are effective in capturing neutrons. To turn down a reactor or to turn it off, these rods are pushed into the regions where fission occurs. How can their presence control the rate of fissioning?

*22.34 A radiologist discovered that at a distance of 1.80 m from a radioactive source, the intensity of radiation was 140 millirad. How far should the radiologist move away to reduce the exposure to 2 millirad?

22.35 Using a Geiger–Müller counter, a radiologist found that in a 20-min period the dose from a radioactive source would measure 40 millirad at a distance of 10.0 m. How much dose would be received in the same time by moving to a distance of 1.00 m?

Units of Radiation Measurement

22.36 What unit is used to describe the activity of a radioactive sample?

22.37 A hospital purchased a sample of a radionuclide rated at 1.5 mCi. What does this rating mean?

22.38 Give the name of the unit that describes the intensity of an exposure to X rays.

22.39 What is the name and symbol of the unit used in describing how much energy a given mass of tissue receives from exposure to radiation?

22.40 Approximately how many rads would kill half of a large population within 4 weeks, assuming that each individual received this much?

22.41 From a health-protection standpoint, how do the roentgen and the rad compare in their potential danger?

22.42 We cannot add a 1-rad dose to some organ from gamma radiation to a 1-rad dose to the same organ from neutron radiation and say that the total biologically effective dose is 2 rad. Why not?

22.43 How is the problem implied by review exercise 22.42 resolved?

22.44 In units of mrem, what is the average natural background radiation received by the U. S. population, exclusive of artificial radiation?

22.45 What is the name of the energy unit used to describe the energy associated with an X ray or a gamma ray?

22.46 In the unit traditionally used (Review Exercise 22.45), how much energy is associated with diagnostic X rays?

22.47 Why should diagnostic radiations ideally be of much lower energy than radiations used in therapy — in cancer treatment, for example?

22.48 In general terms, how does a film badge dosimeter work?

22.49 Why cannot a Geiger – Müller counter detect alpha radiation?

Synthetic Radionuclides

*22.50 When manganese-55 is bombarded by protons, the neutron is one product. What else is produced? Write a nuclear equation.

22.51 To make indium-111 for diagnostic work, silver-109 is bombarded with alpha particles. What forms if the nucleus of silver-109 captures one alpha particle? Write the nuclear equation for this capture.

*22.52 The compound nucleus that forms when silver-109 captures an alpha particle (Review Exercise 22.51) decays directly to indium-111, plus *two* other identical particles. What are they? Write the nuclear equation for this decay.

22.53 To make gallium-67 for diagnostic work, zinc-66 is bombarded with accelerated protons. When a nucleus of zinc-66 captures a proton, the nucleus of what isotope forms? Write the nuclear equation.

*22.54 The isotope that forms when zinc-66 captures a proton (Review Exercise 22.53) is unstable in a novel way (novel at least to our study). This nucleus is able to capture one of its own electrons. When it does, what new nucleus forms? Write the equation for this kind of nuclear event, called electron capture.

22.55 When fluorine-19 is bombarded by alpha particles, both a neutron and a nucleus of sodium-22 form. Write the nuclear equation, including the compound nucleus that is the intermediate.

*22.56 When boron-10 is bombarded by alpha particles, nitrogen-13 forms and a neutron is released. Write the equation for this reaction.

22.57 When nitrogen-14 is bombarded with deuterons, ^2_1H, oxygen-15 and a neutron form. Write the equation for this reaction.

*22.58 What bombarding particle could change aluminum-27 into phosphorus-32 and a proton? Write the equation.

22.59 What bombarding particle can change sulfur-32 into phosphorus-32 and a proton? Write the equation.

Medical Applications of Radiations

22.60 Why is it desirable to use a radionuclide of short half-life in diagnostic work, when we know that even small samples of such isotopes can be very active?

22.61 The possible dangers related to the high activity of a sample of a radionuclide of short half-life can be minimized by taking advantage of this activity. What can be done with such a radionuclide that could not be done as easily with one of very long half-life?

22.62 We know that gamma radiation is the most penetrating of all natural radiations. Why, then, is a diagnostic radionuclide that emits only gamma radiation preferable to one that gives, say, only alpha radiation?

22.63 Why is iodine-123 better for diagnostic work than iodine-131?

22.64 Why did cobalt-60 replace radium for cancer treatment?

22.65 In what chemical form should phosphorus-32 be used to facilitate its seeking bone tissue? Explain.

Radon in the Environment (Special Topic 22.1)

22.66 How is radon-222 produced in the environment?

22.67 What radiations does it emit?

22.68 Besides its own radiation, what other factors make radon-222 in the lungs particularly hazardous?

22.69 What upper limit on the level of radon-222 in the home is recommended?

Technetium-99m in Medicine (Special Topic 22.2)

22.70 What is done at a nuclear reactor facility to charge a molybdenum "cow"?

22.71 From what radionuclide does Tc-99m form?

22.72 In what chemical form is Tc-99m prepared?

X Rays and CT Scans (Special Topic 22.3)

22.73 In general terms, how are X rays prepared?

22.74 How does the CT scanner differ from an ordinary X-ray machine?

Positron Emission Tomography — The PET Scan
(Special Topic 22.4)

22.75 Compare the positron and electron in terms of mass and charge.

22.76 Describe how a positron forms in a positron-emitting radionuclide.

22.77 What property makes the lifetime of a positron extremely short?

22.78 When a radiologist uses the PET scan, what radiation is converted into an X-raylike picture?

22.79 In general terms, how can a positron-emitting radionuclide be gotten inside a tissue, like the brain?

22.80 Describe an advantage of the PET scan over the CT scan in studying the brain.

Magnetic Resonance Imaging — the MRI Scan
(Special Topic 22.5)

22.81 Why is the MRI less harmful a technique than the CT or PET scan?

22.82 How does MRI complement the use of X rays?

22.83 Why is bone transparent to the MRI?

Appendix I
Mathematical Concepts

I.1 EXPONENTIALS

When numbers are either very large or very small, it's often more convenient to express them in what is called *exponential notation*. Several examples are given in Table I.1, which shows how multiples of 10, such as 10,000 and submultiples of 10, such as 0.0001, can be expressed in exponential notation.

Exponential notation expresses a number as the product of two numbers. The first is a digit between 1 and 10, and this is multiplied by the second, 10 raised to some whole number power or exponent. For example, 55,000,000 is expressed in exponential notation as 5.5×10^7, in which 7 (meaning $+7$) is the exponent.

Exponents can be negative numbers, too. For example, 3.4×10^{-3} is a number with a negative exponent. Now let's learn how to move back and forth between the exponential and the expanded expressions.

Positive Exponents. A positive exponent is a number that tells how many times the number standing before the 10 has to be multiplied by 10 to give the same number in its expanded form. For example:

$$5.5 \times 10^7 = 5.5 \times \underbrace{10 \times 10 \times 10 \times 10 \times 10 \times 10 \times 10}_{10^7}$$

$$6 \times 10^3 = 6 \times 10 \times 10 \times 10 = 6000$$
$$8.576 \times 10^2 = 8.576 \times 10 \times 10 = 857.6$$

TABLE I.1

Number	Exponential Form
1	1×10^0
10	1×10^1
100	1×10^2
1,000	1×10^3
10,000	1×10^4
100,000	1×10^5
1,000,000	1×10^6
0.1	1×10^{-1}
0.01	1×10^{-2}
0.001	1×10^{-3}
0.0001	1×10^{-4}
0.00001	1×10^{-5}
0.000001	1×10^{-6}

The number before the 10 doesn't always have to be a number between 1 and 10. This is just a convention, which we sometimes might find useful to ignore. However, we can't ignore the rules of arithmetic in conversions from one form to another. For example,

$$0.00045 \times 10^5 = 0.00045 \times 10 \times 10 \times 10 \times 10 \times 10$$
$$= 45$$
$$87.5 \times 10^3 = 87.5 \times 10 \times 10 \times 10$$
$$= 87,500$$

In most problem solving situations you find that the problem is given the other way around. You encounter a large number, and (after you've learned the usefulness of exponential notation) you know that the next few minutes of your life could actually be easier if you could quickly restate the number in exponential form. This is very easy to do. Just count the number of places that you have to move the decimal point to the *left* to put it just after the first digit of the given number. For example, you might have to work with a number such as 1500 (as in 1500 mL). You'd have to move the decimal point three places leftward from where it's understood to be (1500 = 1500.) to put it just after the first digit.

$$1\underset{3}{\underbrace{}}5\underset{2}{\underbrace{}}0\underset{1}{\underbrace{}}0$$

Each of these leftward moves counts as one unit for the exponent. Three leftward moves means an exponent of +3. Therefore 1500 can be rewritten as 1.500×10^3. A really huge number and one that you'll certainly meet somewhere during the course is 602,000,000,000,000,000,000,000. It's called Avogadro's number, and you can see that manipulating it would be awkward. (How in the world does one even pronounce it?) In exponential notation, it's written simply as 6.02×10^{23}. Check it out. Do you have to move the decimal point leftward 23 places? (And now you could pronounce it: "six point zero two times ten to the twenty third"—but saying "Avogadro's number" is easier.) Do these exercises for practice.

EXERCISE I.1   Expand each of these exponential numbers.

(a) 5.050×10^6 (b) 0.0000344×10^8 (c) 324.4×10^3

EXERCISE I.2   Write each of these numbers in exponential form.

(a) 422,045   (b) 24,000,000,000,000,000,000   (c) 24.32

Answers to Exercises I.1 and I.2

I.1   (a) 5,050,000   (b) 3440   (c) 324,400
I.2   (a) 4.22045×10^5 (b) 2.4×10^{19} (c) 2.432×10^1

Negative Exponents.   A negative exponent is a number that tells how many times the number standing before the 10 has to be *divided* by 10 to give the number in its expanded form. For example:

$$1 \times 10^{-4} = 1 \div 10 \div 10 \div 10 \div 10$$
$$= \frac{1}{10 \times 10 \times 10 \times 10} = \frac{1}{10,000} = \frac{1}{10^4}$$
$$= 0.0001$$

$$6 \times 10^{-3} = 6 \div 10 \div 10 \div 10$$

$$= \frac{6}{10 \times 10 \times 10} = \frac{6}{1000}$$

$$= 0.006$$

$$8.576 \times 10^{-2} = \frac{8.576}{10 \times 10} = \frac{8.576}{100} = 0.08576$$

You'll see negative exponents often when you study aqueous solutions that have very low concentrations.

Sometimes, you'll want to convert a very small number into its equivalent in exponential notation. This is also easy. This time we count *rightward* the number of times that you have to move the decimal point, one digit at a time, to place the decimal just to the right of the first nonzero digit in the number. For example, if the number is 0.00045, you have to move the decimal four times to the right to place it after the 4.

$$0.\,0\,0\,0\,4\,5$$
$$1\ \ 2\ \ 3\ \ 4$$

Therefore we can write $0.00045 = 4.5 \times 10^{-4}$. Similarly, we can write $0.0012 = 1.2 \times 10^{-3}$. And $0.00000000000000011 = 1.1 \times 10^{-16}$. Now try these exercises.

EXERCISE I.3   Write each number in expanded form.

(a) 4.3×10^{-2} (b) 5.6×10^{-10} (c) 0.00034×10^{-2} (d) 4523.34×10^{-4}

EXERCISE I.4   Write the following numbers in exponential forms.

(a) 0.115 (b) 0.00005000041 (c) 0.000000000000345

Answers to Exercises I.3 and I.4

I.3   (a) 0.043 (b) 0.00000000056 (c) 0.0000034 (d) 0.452334
I.4   (a) 1.15×10^{-1} (b) 5.000041×10^{-5} (c) 3.45×10^{-13}

Now that we can write numbers in exponential notation, let's learn how to manipulate them.

How to Add and Subtract Numbers in Exponential Notation. We'll not spend too much time on this, because it doesn't come up very often. The only rule is that when you add or subtract exponentials, all of the numbers must have the same exponents of 10. If they don't, we have to reexpress them to achieve this condition. Suppose you have to add 4.41×10^3 and 2.20×10^3. The result is simply 6.61×10^3.

$$(4.41 \times 10^3) + (2.20 \times 10^3) = [4.41 + 2.20] \times 10^3$$
$$= 6.61 \times 10^3$$

However, we can't add 4.41×10^3 to 2.20×10^4 without first making the exponents equal. We can do this in either of the following ways. In one, we notice that $2.20 \times 10^4 = 2.20 \times 10 \times 10^3 = 22.0 \times 10^3$, so we have:

$$(4.41 \times 10^3) + (22.0 \times 10^3) = 26.41 \times 10^3 = 2.641 \times 10^4$$

Alternatively, we could notice that $4.41 \times 10^3 = 4.41 \times 10^{-1} \times 10^4 = 0.441 \times 10^4$, so we can do the addition as follows:

$$(0.441 \times 10^4) + (2.20 \times 10^4) = 2.641 \times 10^4$$

The result is the same both ways. The extension of this to subtraction should be obvious.[1]

How to Multiply Numbers Written in Exponential Form. Use the following two steps to multiply numbers that are expressed in exponential forms.

Step 1. Multiply the numbers in front of the 10s.

Step 2. *Add* the exponents of the 10s algebraically.

EXAMPLE I.1

$$(2 \times 10^4) \times (3 \times 10^5) = 2 \times 3 \times 10^{(4+5)}$$
$$= 6 \times 10^9$$

Usually, the problem you want to solve involves very large or very small numbers that aren't yet stated in exponential form. When this happens, convert the given numbers into their exponential forms first, and then carry out the operation. The next example illustrates this and shows how exponentials can make a calculation easier.

EXAMPLE I.2

$$6575 \times 2000 = (6.575 \times 10^3) \times (2.000 \times 10^3)$$
$$= 13.15 \times 10^6$$
$$= 1.315 \times 10^7$$

Work the following exercise to practice.

EXERCISE I.5 Calculate the following products after you have converted large or small numbers to exponential forms.

(a) $6,000,000 \times 0.0000002$ (b) $10^6 \times 10^{-7} \times 10^8 \times 10^{-7}$

(c) $0.003 \times 0.002 \times 0.000001$ (d) $1,500 \times 3,000,000,000,000$

Answers for Exercise I.5

(a) 1.2   (b) 1   (c) 6×10^{-12} (d) 4.5×10^{15}

How to Divide Numbers Written in Exponential Form. To divide numbers expressed in exponential forms, use the following two steps.

Step 1. Divide the numbers that stand in front of the 10s.

Step 2. *Subtract* the exponents of the 10s algebraically.

EXAMPLE I.3

$$(8 \times 10^4) \div (2 \times 10^3) = (8 \div 2) \times 10^{(4-3)}$$
$$= 4 \times 10^1$$

---

[1] Whenever these operations are with pure numbers and not with physical quantities obtained by measurements, we are not concerned about the numbers of significant figures in the answers.

| EXAMPLE I.4 | $(8 \times 10^4) \div (2 \times 10^{-3}) = (8 \div 2) \times 10^{[4-(-3)]}$ |
| | $= 4 \times 10^7$ |

For practice, try the following exercise.

EXERCISE I.6   Do the following calculations using exponential forms of the numbers.

(a) $6{,}000{,}000 \div 1500$

(b) $7460 \div 0.0005$

(c) $\dfrac{3{,}000{,}000 \times 6{,}000{,}000{,}000}{20{,}000}$

(d) $\dfrac{0.016 \times 0.0006}{0.000008}$

(e) $\dfrac{400 \times 500 \times 0.002 \times 500}{2{,}500{,}000}$

Answers for Exercise I.6
(a) 4×10^3 (b) 1.492×10^7 (c) 9×10^{11} (d) 1.2 (e) 8×10^{-2}

The Pocket Calculator and Exponentials. The foregoing was meant to refresh your memory about exponentials, because you almost certainly have studied them in any course in algebra or some earlier course. You probably own a good pocket calculator, at least one that can take numbers in exponential form. Go ahead and use it, but be sure that you understand exponentials well, first. Otherwise, there are a lot of pitfalls.

Most pocket calculators have a key marked *EE* or *EXP*. This is used to enter exponentials. Here is where an ability to *read* exponentials comes in handy. For example, the number 2.1×10^4 reads "two point one times ten to the fourth." The *EE* or *EXP* key on most calculators stands for ". . . times ten to the . . ." Therefore to enter 2.1×10^4, punch the following keys.

$$\boxed{2}\boxed{.}\boxed{1}\boxed{EE}\boxed{4}$$

Try this on your own calculator, and be sure to see that the display is correct. If it isn't, you may have a calculator that works differently than most, so recheck your operations of entering and then check your owner's manual.

To enter an exponential with a negative exponent, you have to use one more key, the $\boxed{+/-}$ key. This switches a positive number to its negative, and you *must* use it rather than the $\boxed{-}$ key in this situation. Thus the number 2.1×10^{-5} enters as follows:

$$\boxed{2}\boxed{.}\boxed{1}\boxed{EE}\boxed{+/-}\boxed{5}$$

Try it and check the display. To see what happens if you use the $\boxed{-}$ key instead of the $\boxed{+/-}$ key, clear the display and enter this number only using the $\boxed{-}$ key instead of the $\boxed{+/-}$ key.

I.2 CROSS-MULTIPLICATION

In this section we will learn how to solve for x in such expressions as

$$\frac{12}{x} = \frac{16}{25} \quad \text{or} \quad \frac{32.0}{11.2} = \frac{6.15x}{13.1}$$

The operation is called *cross-multiplication,* and its object is to get x to stand alone, all by itself, on one side of the $=$ sign, and above any real or understood divisor line.

> To cross-multiply, move a number or a symbol both across the $=$ sign and across a divisor line, and then multiply.

EXAMPLE I.5   ***Problem:*** Solve for x in $\dfrac{25}{x} = 5$

Solution: Notice first the divisor lines; one is understood.

$$\text{divisor line} \qquad \frac{25}{x} = 5 \leftarrow \text{divisor line understood because } 5 = \frac{5}{1}$$

Remember, we want x to stand alone, on top of a divisor line (even if this line is understood). To make this happen, we carry out cross-multiplication as indicated:

$$\frac{25}{x} = \boxed{5}$$

Notice that the arrows show moves that carry the quantities not only across the $=$ sign but also across their respective divisor lines. *It is essential that both crossing-overs be done.* Now we have x standing alone above its (understood) divisor line.

$$\frac{25}{5} = x$$

Now we can do the arithmetic. $x = 5$.

EXAMPLE I.6   ***Problem:*** Solve for x in: $\dfrac{40}{4} = 5x$

Solution: To get x to stand alone, only the 5 has to be moved:

$$\frac{40}{4} = \boxed{5}\, x$$

The result is

$$\frac{40}{4 \times 5} = x \quad \text{or} \quad x = 2$$

EXAMPLE I.7   ***Problem:*** Solve for x in: $\dfrac{25 \times 60}{12} = \dfrac{625}{x}$

Solution: To get x to stand alone, we carry out the following cross-multiplication.

$$\frac{\boxed{25 \times 60}}{\boxed{12}} = \frac{625}{\boxed{x}}$$

The result is

$$x = \frac{625 \times 12}{25 \times 60}$$
$$= 5$$

Now try the following exercise for practice.

EXERCISE I.7   Solve for x in the following:

(a) $\dfrac{12}{x} = \dfrac{16}{25}$ (b) $\dfrac{32.0}{11.2} = \dfrac{6.15x}{13.1}$

Answers for Exercises I.7
(a) $x = 18.75$ (b) $x = 6.085946574$

Nearly always the problem involves physical quantities that have both numbers *and units*. Suppose, for example, you have to solve for vol (volume) in an equation involving a density, 1.357 g/mL. The question might be, for example, what volume is occupied by 46.57 g of a liquid with this density.

$$\frac{1.357 \text{ g}}{\text{mL}} = \frac{46.57 \text{ g}}{\text{vol}}$$

When there are both numbers and units, we have to add one more and very important principle. *We cross-multiply units as well as numbers.*

The result is the following, in which the cancel lines show how the units cancel.

$$\text{vol} = \frac{(46.57 \text{ g}) \times (\text{mL})}{(1.357 \text{ g})} = 34.32 \text{ mL} \quad \text{(correctly rounded)}$$

How to Do Chain Calculations with the Pocket Calculator.   Sometimes the steps in solving a problem lead to something like the following:

$$x = \frac{24.2 \times 30.2 \times 55.6}{2.30 \times 18.2 \times 4.44}$$

Many people will first calculate the value of the numerator and write it down. Then they'll compute the denominator and write it down. Finally, they'll divide the two results to get the final answer. There's no need to do this much work. All you have to do is enter the first number you see in the numerator, 24.2 in our example. Then use the $\boxed{\times}$ key for any number in the numerator and the $\boxed{\div}$ key for any number in the denominator. *Each number in the denominator is entered with the* $\boxed{\div}$ *key.* Any of the following sequences work. Try them.

$$24.2 \times 30.2 \times 55.6 \div 2.30 \div 18.2 \div 4.44 = 218.632 \ldots$$

or

$$24.2 \div 2.30 \times 30.2 \div 18.2 \times 55.6 \div 4.44 = 218.632 \ldots$$

Appendix II
Some Rules for Naming Inorganic Compounds

Only those rules considered sufficient to meet most of the needs of the users of this text are in this appendix. The latest edition of the *Handbook of Chemistry and Physics,* published annually by the CRC Publishing Company, Cleveland, OH, under the general editorship of R. C. Weast, has a section on all of the rules. Virtually all college libraries have this reference.

I. **Binary Compounds** — those made from only two elements
    A. One element is a metal and the other is a nonmetal.
       1. The name of the metal is written first in the name of the compound, and its symbol is placed first in the formula.
       2. The name ending of the nonmetal is changed to *-ide.* Thus the names of the simple ions of Groups VIA and VIIA of the periodic table are

Group VIIA	Group VIA
Fluoride	Oxide
Chloride	Sulfide
Bromide	Selenide
Iodide	Telluride

3. If the metal and the nonmetal each have just one oxidation number, a binary compound of the two is named simply by writing the name of the metal and then that of the nonmetal with its ending modified by *-ide,* as shown above. Greek prefixes such as mono-, di-, tri-, and so on, are not necessary.
   The following are examples:

Some Compounds between Elements of Groups IA and VIIA		Some Compounds between Elements of Groups IA and VIA	
NaF	Sodium fluoride	Na_2O	Sodium oxide (not disodium oxide)
KCl	Potassium chloride	K_2S	Potassium sulfide
LiBr	Lithium bromide	Li_2O	Lithium oxide
RbI	Rubidium iodide	Cs_2S	Cesium sulfide
CsCl	Cesium chloride	Rb_2O	Rubidium oxide

Some Compounds between Elements of Groups IIA and VIIA		Some Compounds between Elements of Groups IIA and VIA	
$BeCl_2$	Beryllium chloride	BeO	Beryllium oxide
$MgBr_2$	Magnesium bromide	MgS	Magnesium sulfide
CaF_2	Calcium fluoride	CaO	Calcium oxide
SrI_2	Strontium iodide	SrS	Strontium sulfide
$BaCl_2$	Barium chloride	BaO	Barium oxide

Some Compounds between Elements of Groups IIA and VIIA		Some Compounds between Elements of Groups IIIA and VIA	
$AlCl_3$	Aluminum chloride	Al_2O_3	Aluminum oxide
AlF_3	Aluminum fluoride	Al_2S_3	Aluminum sulfide

4. If the metal has more than one oxidation number, but the nonmetal has just one, the formal name of the compound includes a roman numeral in parentheses following the name of the metal. This numeral stands for the oxidation number of the metal. Greek prefixes such as mono-, di-, and so on, are not needed.

EXAMPLE 1   Compounds of Iron in Oxidation States of 2+ or 3+

	Formal Name	Common Name
$FeCl_2$	Iron(II) chloride[a]	Ferrous chloride
FeO	Iron(II) oxide	Ferrous oxide
Fe_2O_3	Iron(III) oxide	Ferric oxide
$FeCl_3$	Iron(III) chloride	Ferric chloride

[a] Pronounced "iron two chloride."

EXAMPLE 2   Compounds of Copper in Oxidation States of 1+ and 2+

	Formal Name	Common Name
Cu_2O	Copper(I) oxide	Cuprous oxide
$CuBr$	Copper(I) bromide	Cuprous bromide
$CuCl_2$	Copper(II) chloride	Cupric chloride
CuS	Copper(II) sulfide	Cupric sulfide

5. Molecular compounds of two elements. Greek prefixes such as mono-, di-, and so on, are used, sometimes for *both* elements.
   (a) Oxides of Nonmetals
      (1) Oxides of Carbon
      (2) Oxides of Sulfur

CO	Carbon monoxide	SO_2	Sulfur dioxide
CO_2	Carbon dioxide	SO_3	Sulfur trioxide

   (3) Oxides of Nitrogen (Older Names in Parentheses)

N_2O	Dinitrogen monoxide (nitrous oxide)
NO	Nitrogen oxide (nitric oxide)
N_2O_3	Dinitrogen trioxide
NO_2	Nitrogen dioxide
N_2O_4	Dinitrogen tetroxide
N_2O_5	Dinitrogen pentoxide

   (4) Oxides of Some Halogens

F_2O	Difluorine monoxide
Cl_2O	Dichlorine monoxide
Cl_2O_7	Dichlorine heptoxide

(b) Some Halides of Carbon

CCl_4	Carbon tetrachloride
CBr_4	Carbon tetrabromide

(c) Some Exceptions

H_2O Water NH_3 Ammonia CH_4 Methane

II. Compounds of Three or More Elements

A. A positive and a negative ion are combined.

  1. The name of the positive ion is first followed by the name of the negative ion, just as with binary compounds between metals and nonmetals. Greek prefixes are not needed except where they occur in the name of an ion. (Older names are shown in parentheses.)

Li_2SO_4	Lithium sulfate	$MgSO_4$	Magnesium sulfate
Na_2SO_4	Sodium sulfate	$CaSO_4$	Calcium sulfate
K_2SO_4	Potassium sulfate	$Al_2(SO_4)_3$	Aluminum sulfate
Li_2CO_3	Lithium carbonate	$KMnO_4$	Potassium permanganate
Na_2CO_3	Sodium carbonate	Na_2CrO_4	Sodium chromate
$CaCO_3$	Calcium carbonate	$Mg(NO_3)_2$	Magnesium nitrate
$Al_2(CO_3)_3$	Aluminum carbonate	$NaNO_2$	Sodium nitrite
$LiHCO_3$	Lithium hydrogen carbonate (lithium bicarbonate)[a]		
$NaHCO_3$	Sodium hydrogen carbonate (sodium bicarbonate)		
$NaHSO_4$	Sodium hydrogen sulfate (sodium bisulfate)		
NaH_2PO_4	Sodium dihydrogen phosphate		
K_2HPO_4	Potassium monohydrogen phosphate		
$MgHPO_4$	Magnesium monohydrogen phosphate		
$(NH_4)_2HPO_4$	Ammonium monohydrogen phosphate		
Na_3PO_4	Sodium phosphate		
$Ca_3(PO_4)_2$	Calcium phosphate		

[a] *Bicarbonate* instead of *hydrogen carbonate* is used in this text because it is judged to be the more commonly used name for this ion, particularly among health scientists.

B. Molecular compounds of two or more elements. Most are organic compounds, so their rules of nomenclature are given in the chapters on organic compounds.

III. **Important Inorganic Acids and Their Anions**

Formula	Name	Formula	Name
H_2CO_3	Carbonic acid	HCO_3^-	Bicarbonate ion (hydrogen carbonate ion)
		CO_3^{2-}	Carbonate ion
HNO_3	Nitric acid	NO_3^-	Nitrate ion
HNO_2	Nitrous acid	NO_2^-	Nitrite ion
H_2SO_4	Sulfuric acid	HSO_4^-	Hydrogen sulfate ion (bisulfate ion)
		SO_4^{2-}	Sulfate ion
H_2SO_3	Sulfurous acid	HSO_3^-	Hydrogen sulfite ion (bisulfite ion)
		SO_3^{2-}	Sulfite ion
H_3PO_4	Phosphoric acid (orthophosphoric acid)	$H_2PO_4^-$	Dihydrogen phosphate ion
		HPO_4^{2-}	Monohydrogen phosphate ion
		PO_4^{3-}	Phosphate ion
$HClO_4$	Perchloric acid	ClO_4^-	Perchlorate ion
$HClO_3$	Chloric acid	ClO_3^-	Chlorate ion
$HClO_2$	Chlorous acid	ClO_2^-	Chlorite ion
$HClO$	Hypochlorous acid	ClO^-	Hypochlorite ion
HCl	Hydrochloric acid[a]	Cl^-	Chloride ion

[a] The name of the aqueous solution of gaseous HCl.

Some generalizations about names of acids and their anions.
1. Names of ions from acids whose names end in -ic all end in -ate.
2. When a nonmetal that forms an oxyacid whose name ends in -ic also forms an acid with one fewer oxygen atom, the name of the latter acid ends in -ous (cf. nitric acid, HNO_3, and nitrous acid, HNO_2).
3. When a nonmetal forms an oxyacid with one fewer oxygen atom than are in an -ous acid, then the prefix hypo- is used. (cf. chlorous acid, $HClO_2$, and hypochlorous acid, $HClO$).
4. The binary hydrohalogen acids are called hydrogen halides when they occur as pure gases but are called hydrohalic acids when they occur as aqueous solutions. Thus hydrogen fluoride in water becomes hydrofluoric acid; hydrogen chloride in water becomes hydrochloric acid; and so on.

Appendix III
IUPAC Nomenclature of Common Oxygen Derivatives of Hydrocarbons

ALCOHOLS

Parent Compound. The longest continuous chain of carbons that includes the hydroxyl group. Replace the terminal -e of the name of the corresponding alkane by -ol. For names of diols, triols, and so forth, retain the terminal -e but add -diol, -triol, according to the number of OH groups.

Numbering. Numbering the carbons of the chain to give the carbon with the OH group the lower number, giving its location precedence over all alkyl groups.

CH_3OH	CH_3CH_2OH	$CH_3CH_2CH_2OH$	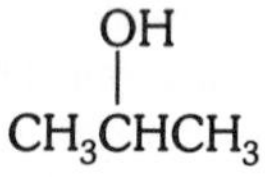 $CH_3\overset{\underset{\mid}{OH}}{C}HCH_3$	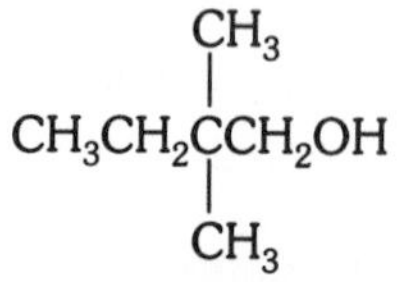 $CH_3CH_2\overset{\underset{\mid}{CH_3}}{\underset{\underset{\mid}{CH_3}}{C}}CH_2OH$
Methanol	Ethanol	1-Propanol	2-Propanol	2,2-Dimethyl-1-butanol

$HOCH_2CH_2OH$	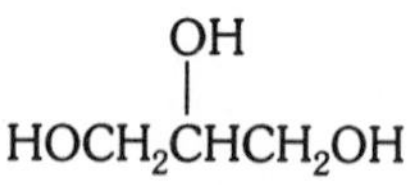 $HOCH_2\overset{\underset{\mid}{OH}}{C}HCH_2OH$	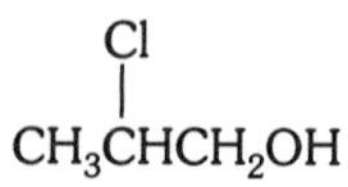 $CH_3\overset{\underset{\mid}{Cl}}{C}HCH_2OH$	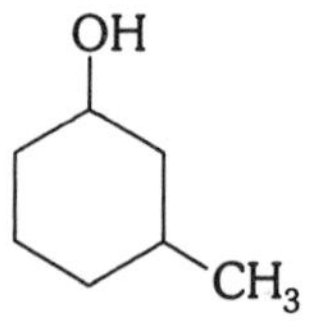 3-Methylcyclohexanol
1,2-Ethanediol	1,2,3-Propanetriol	2-Chloro-1-propanol	

ALDEHYDES

Parent Compound. The longest continuous chain of carbons that includes the carbon of the aldehyde group. Replace the terminal -e of the name of the corresponding alkane by -al.

Numbering. The carbon of the aldehyde group is given number 1. (This gives this carbon precedence over alcohol groups, double bonds, or alkyl groups.)

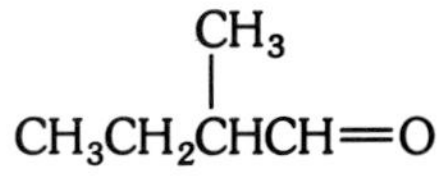

$$CH_3CH_2\overset{\overset{\displaystyle CH_3}{|}}{C}HCH=O \qquad CH_3(CH_2)_4\overset{\overset{\displaystyle CH_3}{|}}{\underset{\underset{\displaystyle CH_3}{|}}{C}}CH_2CH=O$$

2-Methylbutanal 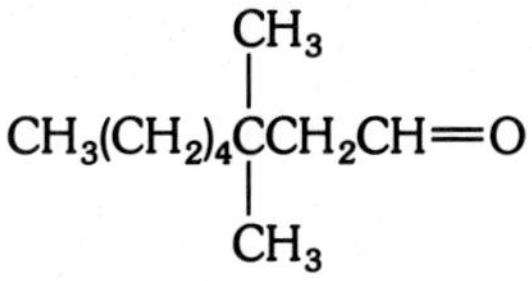   3,3-Dimethyloctanal

KETONES

Parent Compound. The longest continuous chain of carbons that includes the carbon of the keto group. Replace the terminal -e in the name of the corresponding alkane with -one.

Numbering. Number the carbons of the chain to give the carbon of the keto group the lower number. (This gives its location precedence over alkyl and alcohols groups and double bonds.)

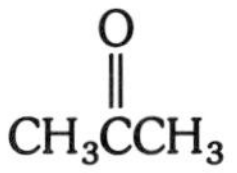

$$CH_3\overset{\overset{\displaystyle O}{||}}{C}CH_3 \qquad CH_3\overset{\overset{\displaystyle O}{||}}{C}CH_2CH_3 \qquad CH_3\overset{\overset{\displaystyle O}{||}}{C}CH_2CH_2CH_3 \qquad CH_3CH_2\overset{\overset{\displaystyle O}{||}}{C}-\overset{\overset{\displaystyle CH_3}{|}}{\underset{\underset{\displaystyle CH_3}{|}}{C}}CH_2CH_2C_6H_5$$

Propanone 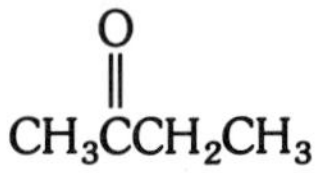   Butanone    2-Pentanone 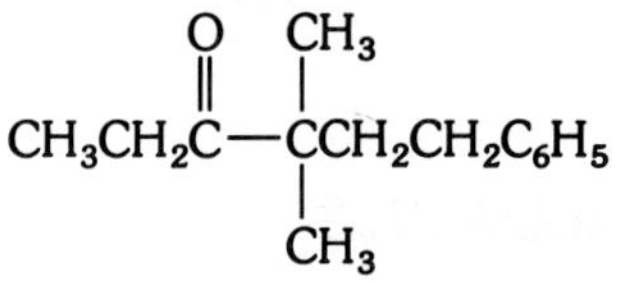   4,4-Dimethyl-6-phenyl-3-hexanone

CARBOXYLIC ACIDS

Parent Compound. The longest continuous chain of carbons that includes the carbon of the carboxyl group. Replace the terminal -e of the corresponding alkane by -oic acid.

Numbering. Number the carbons of the chain to give the carbon of the carboxyl group number 1. (It takes precedence over all other oxygen functions, such as aldehyde and keto groups.)

$$HCO_2H \qquad CH_3CO_2H \qquad CH_3\overset{\overset{\displaystyle CH_3}{|}}{C}HCO_2H$$

Methanoic acid   Ethanoic acid   2-Methylpropanoic acid

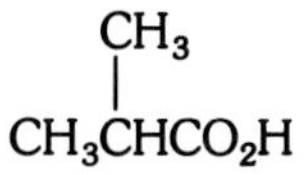

CARBOXYLIC ACID DERIVATIVES

Parent Compound. The name of the parent acid. The change in its name is made according to the type of derivative.

Carboxylic Acid Salts. Change *-ic acid* to *-ate* and precede this word by the name of the other ion (as a separate word).

$$CH_3CO_2^-Na^+$$

$$\underset{\underset{CH_3}{|}}{\overset{\overset{CH_3}{|}}{CH_3CCO_2^-K^+}}$$

Sodium ethanoate   Potassium 2,2-dimethylpropanoate

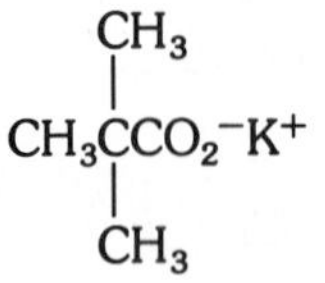

Esters. Change the *-ic acid* of the parent acid to *-ate*, and precede this word by the name of the alkyl group (as a separate word).

$$CH_3CH_2CO_2CH_3$$

$$\overset{\overset{CH_3}{|}}{CH_3CH_2CHCO_2CH_2CH_3}$$

$$\overset{\overset{CH_3}{|}}{C_6H_5CHCO_2C_6H_5}$$

Methyl propanoate   Ethyl 2-methylbutanoate 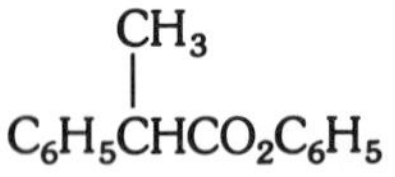   Phenyl 2-phenylpropanoate

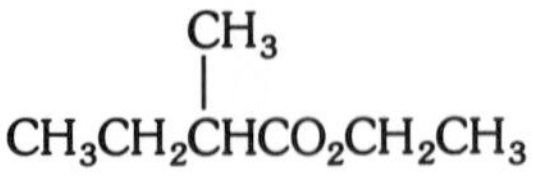

Amides. Change the *-oic acid* to *-amide*.

$$CH_3CONH_2$$

$$\overset{\overset{CH_3}{|}}{CH_3CH_2CH}\overset{\overset{CH_3}{|}}{-CHCONH_2}$$

$$CH_3CH_2CONHCH_3$$

Ethanamide   2,3-Dimethylpentanamide   N-Methylpropanamide

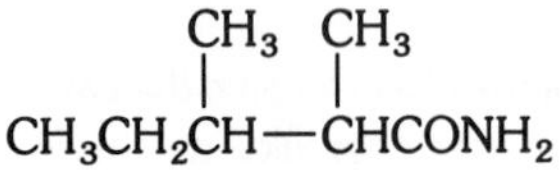

Appendix IV
Some Data Bearing on Human Nutrition

Table IV.1 gives the recommended dietary allowances, the RDAs, of the Food and Nutrition Board, National Academy of Sciences/National Research Council. For a detailed discussion, the reader is referred to *Recommended Dietary Allowances,* 10th ed., 1989, published by the National Academy Press, Washington, D.C.

Table IV.2 gives the chemical structures, good dietary sources, and deficiency diseases of the vitamins.

TABLE IV.1

Recommended Daily Dietary Allowances[a] of the Food and Nutrition Board, National Academy of Sciences National Research Council, Revised 1989

	Age (years)	Weight[b] (kg)	Weight[b] (lb)	Height[b] (cm)	Height[b] (in.)	Protein (g)	Fat-Soluble Vitamins A (μg)[c]	Fat-Soluble Vitamins D (μg)[d]	Fat-Soluble Vitamins E (mg)[e]
Infants	0.0–0.5	6	13	60	24	13	375	75	3
	0.5–1.0	9	20	71	28	14	375	10	4
Children	1–3	13	29	90	34	16	400	10	6
	4–6	20	44	112	44	28	500	10	7
	7–10	28	62	132	54	28	700	10	7
Males	11–14	45	99	157	63	45	1000	10	10
	15–18	66	145	176	69	59	1000	10	10
	19–24	72	160	177	69	58	1000	10	10
	25–50	79	174	176	70	63	1000	5	10
	51+	77	170	173	68	63	1000	5	10
Females	11–14	46	101	157	62	46	800	10	8
	15–18	55	120	163	64	44	800	10	8
	19–24	58	128	164	65	46	800	10	8
	25–50	63	138	163	64	50	800	5	8
	51+	65	143	160	63	50	800	5	8
Pregnant						60	800	10	10
Lactating	First 6 months					65	1300	10	12
	Second 6 months					62	1200	10	11

[a] The allowances are intended to provide for individual variations among most normal persons as they live in the United States under usual environmental stresses. Diets should be based on a variety of common foods in order to provide other nutrients for which human requirements have been less well defined.

[b] These data are the actual medians for the U.S. population for each age range. They are not necessarily ideal.

[c] Retinol equivalents (RE). 1 RE = 1 μg of retinol or 6 μg of β-carotene.

TABLE IV.1   *continued*

Water Soluble Vitamins							Minerals						
Vitamin C (mg)	Folate (μg)	Niacin[f] (mg)	Ribo-flavin (mg)	Thia-mine (mg)	Vitamin B_6 (mg)	Vitamin B_{12} (μg)	Cal-cium (mg)	Phos-phorus (mg)	Iodine (μg)	Iron (mg)	Mag-nesium (mg)	Zinc (mg)	Sele-nium (μg)
30	25	5	0.4	0.3	0.3	0.3	400	300	40	6	40	5	10
35	35	6	0.5	0.4	0.6	0.5	600	500	50	10	60	5	15
40	50	9	0.8	0.7	1.0	0.7	800	800	70	10	80	10	20
45	75	12	1.1	0.9	1.1	1.0	800	800	90	10	120	10	20
45	100	13	1.2	1.0	1.4	1.4	800	800	120	10	170	10	30
50	150	17	1.5	1.3	1.7	2.0	1200	1200	150	12	270	15	40
60	200	20	1.8	1.5	2.0	2.0	1200	1200	150	17	400	15	50
60	200	19	1.7	1.5	2.0	2.0	1200	1200	150	10	350	15	70
60	200	19	1.7	1.5	2.0	2.0	800	800	150	10	350	15	70
60	200	15	1.4	1.2	2.0	2.0	800	800	150	10	350	15	70
50	150	15	1.3	1.1	1.4	2.0	1200	1200	150	15	280	12	45
60	180	15	1.3	1.1	1.5	2.0	1200	1200	150	15	300	12	50
60	180	15	1.3	1.1	1.6	2.0	1200	1200	150	15	280	12	55
60	180	15	1.3	1.1	1.1	2.0	800	800	150	15	280	12	55
60	180	13	1.2	1.0	1.6	2.0	800	1200	150	10	280	15	55
70	400	17	1.6	1.5	2.2	2.2	1200	1200	175	30	320	15	65
95	280	20	1.8	1.6	2.1	2.6	1200	1200	200	15	355	19	75
90	260	20	1.7	1.6	2.1	2.6	1200	1200	200	15	340	16	75

[d] As cholecalciferol. 10 μg of cholecalciferol = 400 IU of vitamin D.

[e] α-Tocopherol equivalents. 1 mg of d-α-tocopherol = 1 α-tocopherol equivalent.

[f] 1 niacin equivalent (NE) equals 1 mg of niacin or 60 mg of dietary tryptophan.

TABLE IV.2

Vitamins Important in Humans[a]

[a] The name beneath a structure is the name recommended by the Joint Commission on Biochemical Nomenclature of the International Union of Pure and Applied Chemistry and the International Union of Biochemistry. More than one compound can sometimes provide the activity of a vitamin, but just one member of any such families is shown here.

Vitamin	Some Good Dietary Sources	Results of Deficiency of Vitamin
Fat-Soluble Vitamins		
A_1 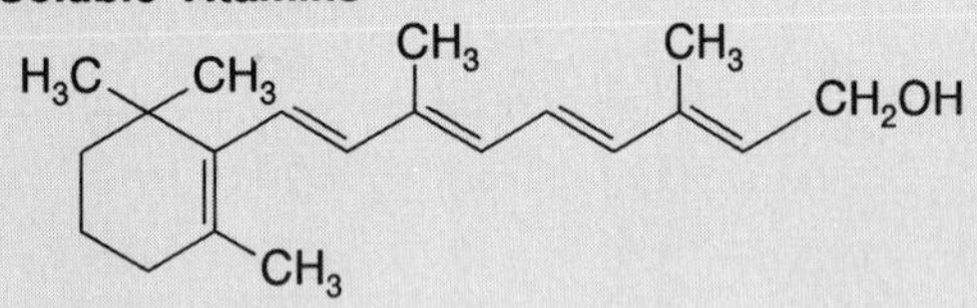 Retinol	Green and yellow vegetables, cod liver oil	Eye disease (xerophthalmia)
D_3 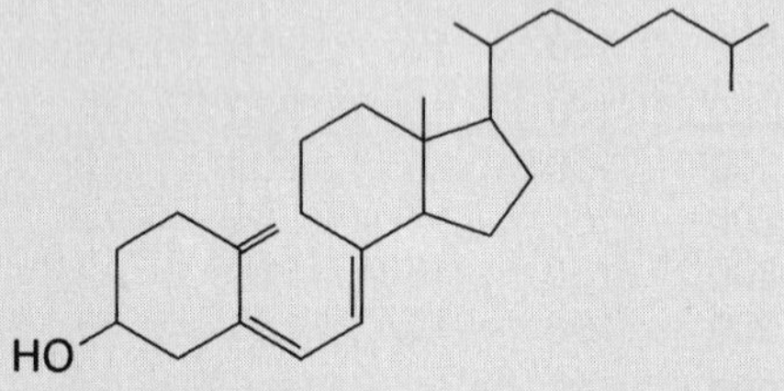 Cholecalciferol	Provitamins in the skin are activated by sunlight. Fish liver oils	Poor use of Ca^{2+}, poor bone formation. Rickets
E 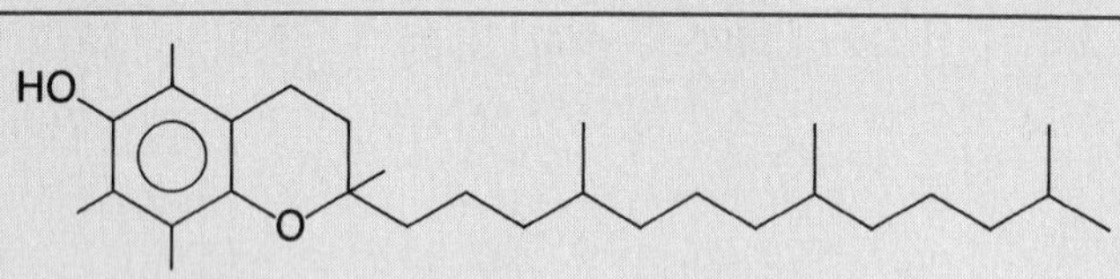 α-Tocopherol	Leafy vegetables. Wheat germ oil. Cottonseed oil	Effects still controversial
K 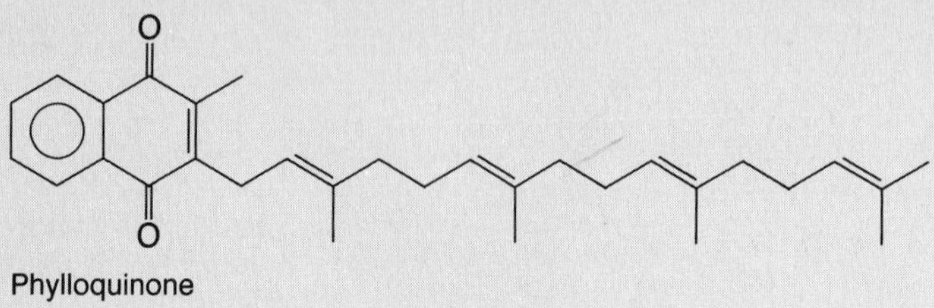 Phylloquinone	Spinach, cabbage. Made in intestinal tract by microorganisms	Poor blood clotting

TABLE IV.2 *continued*

Vitamin	Some Good Dietary Sources	Results of Deficiency of Vitamin
Water-Soluble Vitamins		
C 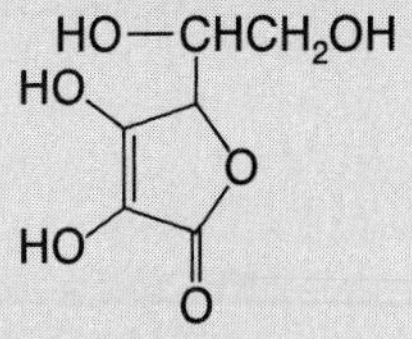 HO—CHCH$_2$OH HO O HO O Ascorbic acid	Citrus fruits, green vegetables, potatoes, berries	Scurvy
Biotin 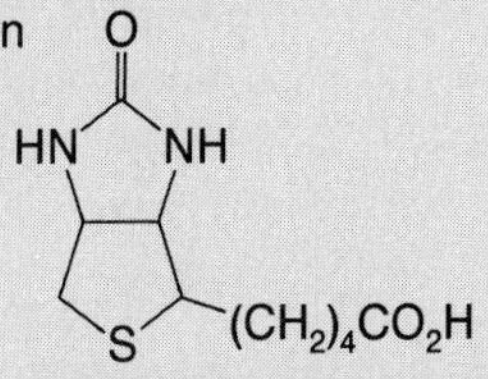 O HN NH S (CH$_2$)$_4$CO$_2$H	Liver, peas, lima beans	Skin disorders
Folic acid 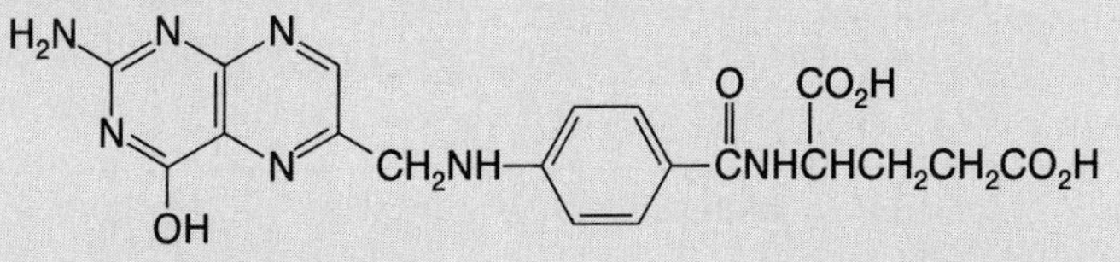 H$_2$N N N N OH N CH$_2$NH O CO$_2$H CNHCHCH$_2$CH$_2$CO$_2$H Pteroylglutamic acid	Amimal organs, yeast wheat germ, chicken, oysters	Gastrointestinal disorders
Pantothenic acid 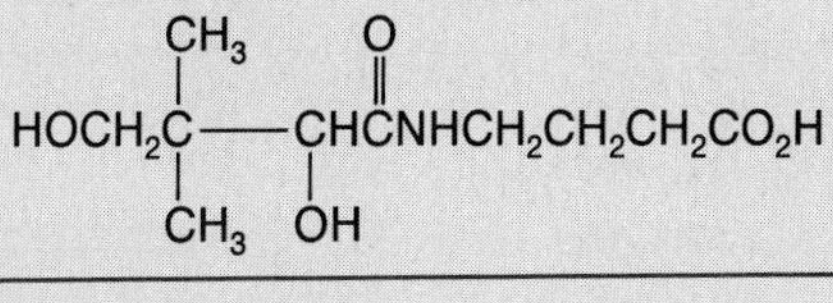 CH$_3$ O HOCH$_2$C——CHCNHCH$_2$CH$_2$CH$_2$CO$_2$H CH$_3$ OH	Egg yolk, kidneys, yeast, liver	Not known for humans
Choline HOCH$_2$CH$_2$N$^+$(CH$_3$)$_3$	Peas, egg yolk, wheat germ, aspargus, spinach	Not known in humans

TABLE IV.2   *continued*

Vitamin	Some Good Dietary Sources	Results of Deficiency of Vitamin
Thiamin 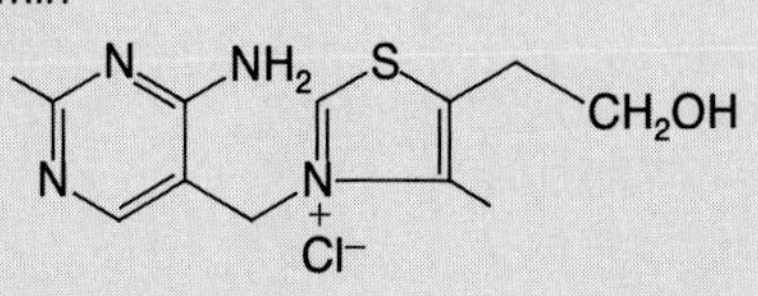	Cereal grains, nuts, liver legumes, heart kidney	Beri beri
Riboflavin 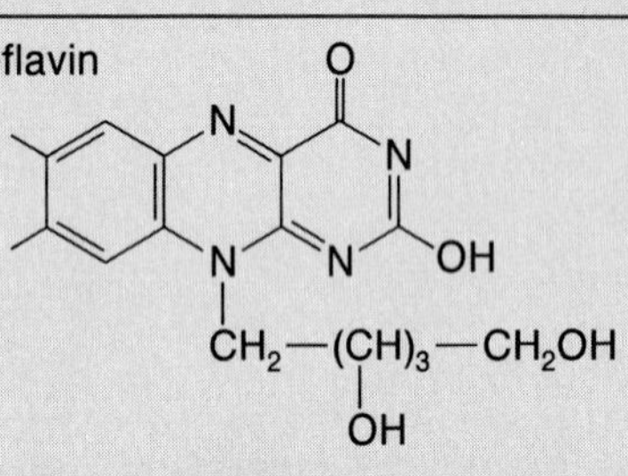	Milk, meat, cheese, eggs, peas, lima beans, wheat germ	Fissures at corners of mouth, reddening of tongue
B_6 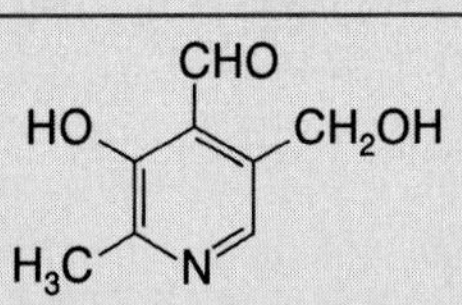	Cereals, liver, legumes, yeast, bananas	Skin disorders, convulsions (in infants)
Nicotinic acid and nicotinamide 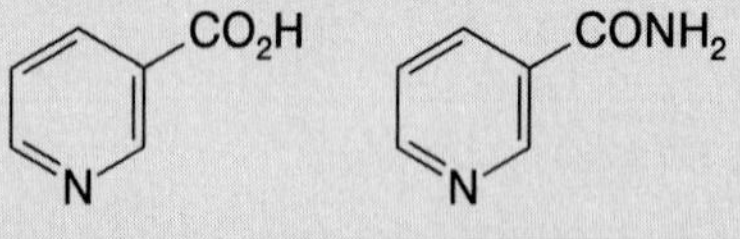	Meats, yeast, wheat germ	Pellagra
B_{12} 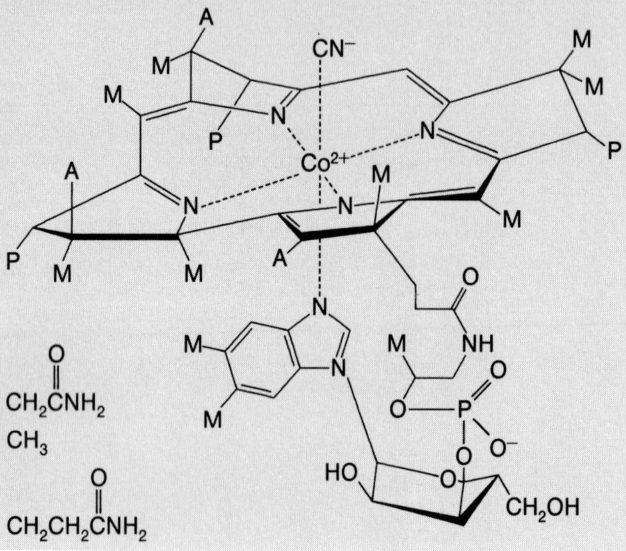	Eggs, meat, liver	Pernicious anemia

Thiamin structure labels: NH_2, S, CH_2OH, N^+, Cl^-

Riboflavin structure labels: O, OH, $CH_2—(CH)_3—CH_2OH$, OH

B_6 structure labels: CHO, HO, CH_2OH, H_3C, N

Nicotinic acid: CO_2H Nicotinamide: $CONH_2$

B_{12} structure labels: A, M, P, CN^-, Co^{2+}, N, NH, O, P, O^-, HO, CH_2OH

$A = CH_2\overset{O}{C}NH_2$

$M = CH_3$

$P = CH_2CH_2\overset{O}{C}NH_2$

Appendix V
Answers to Practice Exercises and Selected Review Exercises

1. 310 K
2. (a) 5.45×10^8 (b) 5.67×10^{12}
   (c) 6.454×10^3 (d) 2.5×10^1
   (e) 3.98×10^{-5} (f) 4.26×10^{-3}
   (g) 1.68×10^{-1} (h) 9.87×10^{-12}
3. (a) 10^{-6} (b) 10^{-9} (c) 10^{-6} (d) 10^3
4. (a) mL   (b) μL   (c) dL   (d) mm
   (e) cm   (f) kg   (g) μg   (h) mg
5. (a) kilogram   (b) centimeter   (c) deciliter
   (d) microgram   (e) milliliter   (f) milligram
   (g) millimeter   (h) microliter
6. (a) 1.5 Mg   (b) 3.45μL   (c) 3.6 mg
   (d) 6.2 mL   (e) 1.68 kg   (f) 5.4 dm
7. (a) 275 kg   (b) 62.5μL   (c) 82 nm or 0.082μm
8. (a) 95   (b) 0.890   (c) 0.0263
   (d) 1.3000   (e) 16.1   (f) 3.8×10^2
   (g) 9.31   (h) 9.1×10^2

9. (a) $\dfrac{1 \text{ g}}{1000 \text{ mg}}$ or $\dfrac{1000 \text{ mg}}{1 \text{ g}}$

   (b) $\dfrac{1 \text{ kg}}{2.205 \text{ lb}}$ or $\dfrac{2.205 \text{ lb}}{1 \text{ kg}}$

10. 0.324 g of aspirin
11. 40 °C
12. 59 °F (quite cool)
13. 20.7 mL
14. 32.1 g

1.27 273 and 373 K
1.29 410 °F
1.31 88 °F
1.33 42.8 °C. No, regardless of the temperature *scale,* it's just as
    hot.

1.51 (a) 192 cm   (b) 111 lb avdp
1.53 16.9 liq oz
1.55 Do not cross; 2.1×10^3 kg $> 1.5 \times 10^3$ kg (the limit)
1.57 3.5 g/pat
1.59 8.8477×10^3 m, 8.8477 km
1.67 23.7 lb avdp of lead
1.69 28.3 mL of acetic acid

1. (a) 15   (b) 24   (c) 24
2. (a) 5   (b) 18   (c) 20
3. (a) 7 *p*, 7 *n*. Electron shells: 2   5
   (b) 13 *p*, 14 *n*. Electron shells: 2   8   3
   (c) 20 *p*, 20 *n*. Electron shells: 2   8   8   2
4. 52.5
5. (a) Sn   (b) Cl   (c) Rb   (d) Mg   (e) Ar
6. (a) 1   (b) 6   (c) 5   (d) 7

2.50 12 times heavier.
2.51 (a) 1.33 times as heavy   (b) 16.0 g of oxygen atoms

1. (a) AgBr   (b) Na_2O (c) Fe_2O_3 (d) $CuCl_2$
2. (a) Copper(II) sulfide; cupric sulfide
   (b) Sodium fluoride
   (c) Iron(II) iodide; ferrous iodide
   (d) Zinc bromide
   (e) Copper(I) oxide; cuprous oxide
3. (a) 3+   (b) 2+   (c) 2+
4. (a) 2   8   8   1.   1+   (b) 2   8   6.   2−
   (c) 2   8   4.   No ion forms
5. (a) 2   8   8   (b) 2   8   8
6. (a) Cs^+ (b) F^- (c) P^{3-} (d) Sr^{2+}

7. (a)
$$Cl-\overset{\displaystyle Cl}{\underset{\displaystyle Cl}{C}}-Cl$$
(b) $H-S-H$

(c) $Cl-\overset{}{\underset{}{N}}-Cl$ (d) $Br-\overset{\displaystyle H}{\underset{\displaystyle Br}{C}}-Br$

8. (a) $NaNO_3$ (b) KOH   (c) $Ca(OH)_2$
   (d) $MgCO_3$ (e) Na_2SO_4 (f) $(NH_4)_3PO_4$
9. (a) Lithium carbonate
   (b) Sodium bicarbonate
   (c) Potassium permanganate
   (d) Sodium dihydrogen phosphate
   (e) Ammonium monohydrogen phoshate

10. (a) $\overset{\delta+}{H}-\overset{\overset{\displaystyle \delta+}{\displaystyle H}}{\underset{\displaystyle \delta-}{N}}-\overset{\delta+}{H}$ (b) $\overset{\delta+}{H}-\overset{\delta-}{F}$

(c) $\overset{\delta+}{H}-\overset{\delta-}{O}-\overset{\delta-}{O}-\overset{\delta+}{H}$

PRACTICE EXERCISES, CHAPTER 4

1. $3O_2 \rightarrow 2O_3$
2. $4Al + 3O_2 \rightarrow 2Al_2O_3$
3. (a) $2Ca + O_2 \rightarrow 2CaO$
   (b) $2KOH + H_2SO_4 \longrightarrow 2H_2O + K_2SO_4$
   (c) $Cu(NO_3)_2 + Na_2S \rightarrow CuS + 2NaNO_3$
   (d) $2AgNO_3 + CaCl_2 \rightarrow 2AgCl + Ca(NO_3)_2$
   (e) $2Al + 3H_2SO_4 \rightarrow Al_2(SO_4)_3 + 3H_2$
   (f) $CH_4 + 2O_2 \rightarrow 2H_2O + CO_2$
4. 8.68×10^{22} atoms of gold per ounce
5. (a) 180   (b) 58.3   (c) 859
6. 0.500 mol of H_2O
7. 4.20 mol of N_2 and 4.20 mol of O_2
8. 450 mol of H_2 and 150 mol of N_2
9. 408 g of NH_3
10. 0.0380 mol of aspirin
11. 11.5 g of O_2
12. 18.4 g of Na and 46.8 g of NaCl
13. (a) 2.45 g of H_2SO_4 (b) 9.00 g of $C_6H_{12}O_6$
14. 156 mL of 0.800 M Na_2CO_3 solution
15. 9.82 mL of 0.112 M H_2SO_4, when calculated step by step with rounding after each step.
    9.84 mL of 0.112 M H_2SO_4, when found by a chain calculation.

REVIEW EXERCISES, CHAPTER 4

4.8 2.01×10^{23} formula units of H_2O
4.10 0.150 g of medication
4.13 (a) 40.0   (b) 100   (c) 98.1
     (d) 106   (e) 158   (f) 128
4.20 (a) 100 g   (b) 250 g   (c) 245 g
     (d) 265 g   (e) 395 g   (f) 320 g
4.22 (a) 1.88 mol   (b) 0.750 mol   (c) 0.765 mol
     (d) 0.708 mol   (e) 0.475 mol   (f) 0.586 mol

4.24 (a) $\dfrac{4 \text{ mol Fe}}{3 \text{ mol } O_2}$ and $\dfrac{3 \text{ mol } O_2}{4 \text{ mol Fe}}$

   (b) $\dfrac{4 \text{ mol Fe}}{2 \text{ mol } Fe_2O_3}$ and $\dfrac{2 \text{ mol } Fe_2O_3}{4 \text{ mol Fe}}$

   (c) $\dfrac{3 \text{ mol } O_2}{2 \text{ mol } Fe_2O_3}$ and $\dfrac{2 \text{ mol } Fe_2O_3}{3 \text{ mol } O_2}$

4.26 (a) 26 mol of O_2 (b) 50 mol of H_2O (c) 26 mol of O_2
4.28 (a) 75.0 mol of H_2 (b) 43 mol of CH_4
4.30 (a) 1.28×10^3 g of NaCl
     (b) 872 g of NaOH (step-wise calculation with rounding after each step)
        873 g of NaOH (when found by a chain calculation)
     (c) 21.8 g of H_2 (using 2.00 g/mol for H_2)
4.32 48.7 g of Na_2CO_3 (step-wise calculation with rounding after each step)
     48.6 g of Na_2CO_3 (when found by a chain calculation)
4.40 (a) 5.85 g of NaCl   (b) 5.63 g of $C_6H_{12}O_6$
     (c) 0.981 g of H_2SO_4 (d) 11.2 g of KOH
4.42 2.5×10^2 mL of 0.10 M HCl solution
4.44 1.0×10^2 mL of 0.010 M $NaHCO_3$ solution
4.46 250 mL of 1.00 M NaOH solution
4.48 167 mL of 0.150 M Na_2SO_4 solution
4.50 540 mL of 0.100 M HCl solution

PRACTICE EXERCISES, CHAPTER 5

1. 1.31×10^3 mL of helium
2. 547 mL of anesthetic
3. 484 atm

REVIEW EXERCISES, CHAPTER 5

5.7   350 mm Hg
5.16 829 mm Hg
5.18 73 mm Hg
5.20 1.84 L of O_2
5.28 (a) 3.55×10^{-2} mol of gas   (b) 32.1   (c) oxygen
5.30 (a) 2.00 mol of H_2 (b) 200 L of H_2
     (c) 1.55×10^5 mL of H_2

PRACTICE EXERCISES, CHAPTER 6

1. 10.2 g of 96% H_2SO_4
2. 1.25 g of glucose and 499 g of water (rounded from 498.75, and assuming that the density of water is 1.00 g/mL)
3. (a) 2.00 g of $KMnO_4$ (b) 0.25 g of NaOH
4. (a) 0.020 Osm   (b) 0.015 Osm
   (c) 0.100 Osm   (d) 0.150 Osm

REVIEW EXERCISES, CHAPTER 6

6.23 $Z \cdot 10H_2O$
6.39 (a) 0.500 g of NaI   (b) 1.25 g of NaBr
     (c) 6.25 g of $C_6H_{12}O_6$ (d) 15.0 g of H_2SO_4
6.41 (a) 12.5 g of $Mg(NO_3)_2$ (b) 5.00 g of NaBr
     (c) 2.50 g of KI   (d) 1.68 g of $Ca(NO_3)_2$
6.43 25.0 mL of ethyl alcohol
6.45 (a) 11.0 mL of KOH solution

(b) 30.0 mL of HCl solution

(c) 650 mL of NaCl solution

(d) 281 mL of KOH solution

6.47 24.7 g of $Na_2SO_4 \cdot 10H_2O$

6.66 10% NaCl, which has 1.7 mmol/100 g solution versus only 0.67 mmol/100 g solution for 10% NaI

6.76 Dissolve 31 mL of 16 M HNO_3 in water and make the final volume equal to 500 mL.

6.79 12.5 g of stock solution, or 11.9 mL of stock solution

6.81 (a) 71.0% (w/w) HNO_3

(b) 24.8 mL of concentrated solution

PRACTICE EXERCISES, CHAPTER 7

1. $HNO_3(aq) + KOH(aq) \rightarrow H_2O + KNO_3(aq)$
$H^+(aq) + NO_3^-(aq) + K^+(aq) + OH^-(aq) \rightarrow$
$$H_2O + K^+(aq) + NO_3^-(aq)$$
$H^+(aq) + OH^-(aq) \rightarrow H_2O$

2. $2NaHCO_3(aq) + H_2SO_4(aq) \rightarrow$
$$2CO_2(g) + 2H_2O + Na_2SO_4(aq)$$
$2Na^+(aq) + 2HCO_3^-(aq) + 2H^+(aq) + SO_4^{2-}(aq) \rightarrow$
$$2CO_2(g) + 2H_2O + 2Na^+(aq) + SO_4^{2-}(aq)$$
$HCO_3^-(aq) + H^+(aq) \rightarrow CO_2(g) + H_2O$

3. $K_2CO_3(aq) + H_2SO_4(aq) \rightarrow CO_2(g) + H_2O + K_2SO_4(aq)$
$2K^+(aq) + CO_3^{2-}(aq) + 2H^+(aq) + SO_4^{2-}(aq) \rightarrow$
$$CO_2(g) + H_2O + 2K^+(aq) + SO_4^{2-}(aq)$$
$CO_3^{2-}(aq) + 2H^+(aq) \rightarrow CO_2(g) + H_2O$

4. $MgCO_3(s) + 2HNO_3(aq) \rightarrow CO_2(g) + H_2O + Mg(NO_3)_2(aq)$
$MgCO_3(s) + 2H^+(aq) + 2NO_3^-(aq) \rightarrow$
$$CO_2(g) + H_2O + Mg^{2+}(aq) + 2NO_3^-(aq)$$
$MgCO_3(s) + 2H^+(aq) \rightarrow CO_2(g) + H_2O + Mg^{2+}(aq)$

5. $Mg(OH)_2(s) + 2HCl(aq) \rightarrow 2H_2O + MgCl_2(aq)$
$Mg(OH)_2(s) + 2H^+(aq) \rightarrow 2H_2O + Mg^{2+}(aq)$

6. (a) $NH_3(aq) + HBr(aq) \rightarrow NH_4Br(aq)$
$NH_3(aq) + H^+(aq) \rightarrow NH_4^+(aq)$
(b) $2NH_3(aq) + H_2SO_4(aq) \rightarrow (NH_4)_2SO_4(aq)$
$NH_3(aq) + H^+(aq) \rightarrow NH_4^+(aq)$

7. $Mg(s) + 2HCl(aq) \rightarrow H_2(g) + MgCl_2(aq)$
$Mg(s) + 2H^+(aq) \rightarrow H_2(g) + Mg^{2+}(aq)$

8. All are weak Brønsted acids.

9. (a) Weak   (b) Weak   (c) Strong   (d) Strong

10. $Cu(NO_3)_2(aq) + Na_2S(aq) \rightarrow CuS(s) + 2NaNO_3(aq)$
$Cu^{2+}(aq) + S^{2-}(aq) \rightarrow CuS(s)$

11. The acetate ion, $C_2H_3O_2^-(aq)$, binds H^+ ions from HCl(aq) because $C_2H_3O_2^-$ is a relatively strong Brønsted base:
$$C_2H_3O_2^-(aq) + H^+(aq) \rightleftharpoons HC_2H_3O_2(aq)$$

12. (a) $Ag^+(aq) + Cl^-(aq) \rightarrow AgCl(s)$
(b) $CaCO_3(s) + 2H^+(aq) \rightarrow Ca^{2+}(aq) + H_2O + CO_2(g)$
(c) No reaction

REVIEW EXERCISES, CHAPTER 7

7.45 0.250 mol of $NaHCO_3$

7.47 6.68 g of Na_2CO_3

7.49 4.91 g of $NaHCO_3$

7.51 29.5 mL of NaOH solution

7.53 $CaCO_3(s) + 2HCl(aq) \rightarrow CO_2(g) + H_2O + CaCl_2(aq)$
$CaCO_3(s) + 2H^+(aq) \rightarrow CO_2(g) + H_2O + Ca^{2+}(aq)$
47.8 g of $CaCO_3$
191 mL of 5.00 M HCl

7.70 3.76 g or 3.76×10^3 mg of Cl^-

7.72 5.01 meq of K^+

7.89 Anion gap $= 10$ meq/L. This is in the normal range of $5 - 14$ meq/L, so no serious disturbance in metabolism is indicated.

PRACTICE EXERCISES, CHAPTER 8

1. (a) 2.5×10^{-6} mol OH^-/L. Basic
(b) 9.1×10^{-8} mol OH^-/L. Acidic
(c) 1.1×10^{-7} mol OH^-/L. Basic

2. (a) 1×10^{-7} to 1×10^{-8} mol H^+/L
(b) Slightly basic
(c) Acidosis

3. $NaHCO_3(aq)$
$$HCO_3^-(aq) + H^+(aq) \rightarrow CO_2(g) + H_2O$$

4. (a) basic   (b) basic   (c) basic
(d) acidic   (e) basic   (f) basic

5. Basic. The acetate ion, a Brønsted base, hydrolyzes.

6. Acidic. The hydrated Cu^{2+} ion hydrolyzes.

7. Yes. Decrease the pH. The NH_4^+ is a weak Brønsted acid.

8. 0.105 M NaOH

9. 0.125 M H_2SO_4

REVIEW EXERCISES, CHAPTER 8

8.4 2.43×10^{-14}

8.37 (a) 0.4000 M HCl   (b) 0.4000 M HBr

8.39 400.0

8.41 (a) 9.115 g of HCl   (b) 5.435 g of HNO_3
(c) 0.3678 g of H_2SO_4

8.43 (a) 0.04457 M Na_2CO_3 (b) 4.723 g of Na_2CO_3/L

PRACTICE EXERCISES, CHAPTER 9

1. (a) $CH_3-CH_2-CH_3$

(b) $CH_3-\underset{\underset{CH_3}{|}}{CH}-CH_3$

(c) $CH_3-\underset{\underset{CH_3}{|}}{\overset{\overset{CH_3}{|}}{C}}-\underset{\overset{CH_3}{|}}{CH}-\underset{\overset{CH_3}{|}}{CH}-CH_3$

2. (a) $CH_3CH_2CH_3$

(b) $CH_3\underset{\overset{CH_3}{|}}{CH}CH_3$

(c) $CH_3\underset{\underset{CH_3}{|}}{\overset{\overset{CH_3}{|}}{C}}-\underset{\overset{CH_3}{|}}{CH}-CHCH_3$

Note that *vertical* bonds are always shown and that sometimes horizontal bonds are left in so that there will be room for other groups.

3. (a) CH_3-CH_3 (structural formula with all H shown)

(b) branched alkane structural formula

(c) branched alkane structural formula

4. Structures (b) and (c) cannot represent real compounds.

5. (a) Identical   (b) Isomers   (c) Identical
(d) Isomers   (e) Different in another way.

PRACTICE EXERCISES, CHAPTER 10

1. The second structure
2. (a) 3-Methylhexane
(b) 4-t-Butyl-2,3-dimethylheptane
(c) 5-sec-Butyl-2,4-dimethylnonane
(d) 2-Bromo-1-chloro-3-iodopropane
3. (a) Ethyl chloride   (b) Butyl bromide
(c) Isobutyl chloride   (d) t-Butyl bromide
4. (a) 2-Methylpropene   (b) 4-Isobutyl-3,6-dimethyl-3-heptene
(c) 1-Chloropropene   (d) 3-Bromopropene
5. (a) cis/trans alkene structures (CH_3CH_2 and CH_3 on $C=C$) and (CH_3CH_2 and H on $C=C$)
(b) Cl $C=C$ Cl structures and H, Cl / Cl, H structures
(c) None   (d) None
6. (a) $CH_3CH_2CH_3$ (b) No reaction
(c) (cyclohexane)$-CH_3$ (d) $CH_3(CH_2)_{16}CO_2H$

7. (a) $CH_3\underset{OH}{CH}CH_2CH_3$ (b) $CH_3\underset{CH_3}{\overset{OH}{C}}CH_3$

(c) $CH_3CH_2\underset{CH_3}{\overset{OH}{C}}$(cyclohexane)

PRACTICE EXERCISES, CHAPTER 11

1. (a) Alcohol   (b) Phenol
(c) Carboxylic acid   (d) Alcohol
(e) Alcohol   (f) Alcohol
2. (a) Monohydric, secondary   (b) Monohydric, secondary
(c) Dihydric, unstable   (d) Dihydric
(e) Monohydric, primary   (f) Monohydric, primary
(g) Monohydric, tertiary   (h) Monohydric, secondary
(i) Trihydric, unstable

3. (a) $CH_3CH=CH_2$ (b) $CH_3CH=CH_2$

(c) $CH_2=\underset{}{\overset{CH_3}{C}}CH_3$ (d) (cyclohexene)

4. (a) $CH_3\overset{CH_3}{CH}CH=O$ and $CH_3\overset{CH_3}{CH}CO_2H$
(b) $C_6H_5-CH=O$ and $C_6H_5-CO_2H$
(c) $CH_2=O$ and HCO_2H

5. (a) $CH_3\overset{O}{\overset{||}{C}}CH_2CH_3$ (b) $C_6H_5-\overset{O}{\overset{||}{C}}CH_3$

(c) (cyclopentanone) $=O$

6. (a) $2CH_3SH$
(b) $(CH_3)_2CH-S-S-CH(CH_3)_2$
(c) $HSCH_2CH_2CH_2CH_2SH$
(d) (cyclopentyl)$-S-S-$(cyclopentyl)

7. (a) CH_3-O-CH_3
(b) $CH_3CH_2CH_2-O-CH_2CH_2CH_3$
(c) (cyclohexyl)$-O-$(cyclohexyl)

8. (a) $C_6H_5-NH_3^+$ (b) $CH_3\overset{+}{N}HCH_3$ with CH_3
(c) $\overset{+}{N}H_3CH_2CH_2\overset{+}{N}H_3$

9. (a) HO, HO (benzene ring) $\underset{}{\overset{OH}{C}}HCH_2NHCH_3$

(b) CH_3O, CH_3O, CH_3O (benzene ring) $-CH_2CH_2NH_2$

PRACTICE EXERCISES, CHAPTER 12

1. (a) $CH_3CH_2\overset{\displaystyle O}{\overset{\|}{C}}CHCH_3$ $\overset{\displaystyle |}{CH_3}$

(b) $CH_3-\overset{\displaystyle O}{\overset{\|}{C}}-C_6H_5$

(c) $CH_3CH_2CH_2\overset{\displaystyle O}{\overset{\|}{C}}CH_2CH_2CH_3$

(d) $CH_3\overset{\displaystyle CH_3}{\underset{\displaystyle CH_3}{C}}-\overset{\displaystyle O}{\overset{\|}{C}}-\overset{\displaystyle CH_3}{\underset{\displaystyle CH_3}{C}}CH_3$

2. (a) $CH_3\overset{\displaystyle OH}{\overset{|}{C}H}CH_2CH_3$

(b) $CH_3\overset{\displaystyle CH_3}{\overset{|}{C}H}CH_2CH_2OH$

(c) cyclohexanol—OH

3. (a) $CH_3\overset{\displaystyle CHOCH_3}{\underset{\displaystyle OH}{}}$

(b) $CH_3CH_2CH_2\overset{\displaystyle CH}{\underset{\displaystyle OH}{}}-O-CH_2CH_3$

(c) $C_6H_5-\overset{\displaystyle OH}{\overset{|}{C}H}-O-CH_2CH_2CH_3$

(d) $HO-CH_2-O-CH_3$

4. (a) $CH_3CH_2CH{=}O + HOCH_3$

(b) $CH_3CH_2OH + O{=}CHCH_2CH_3$

5. (a) $CH_3CH_2\overset{\displaystyle OCH_3}{\underset{\displaystyle OCH_3}{C}H}$

(b) $CH_3\overset{\displaystyle OCH_2CH_2CH_3}{\underset{\displaystyle OCH_2CH_2CH_3}{C}}CH_3$

6. (a) $2CH_3OH + CH_2{=}O$ (b) No reaction

(c) $CH_3\overset{\displaystyle CH_3}{\underset{\displaystyle CH_3}{C}H}C{=}O + 2HOCH_3$

PRACTICE EXERCISES, CHAPTER 13

1. (a) $CH_3CH_2CO_2^-$

(b) CH_3-O-⟨benzene⟩$-CO_2^-$

(c) $CH_3CH{=}CHCO_2^-$

2. (a) CH_3-O-⟨benzene⟩$-CO_2H$

(b) $CH_3CH_2CO_2H$

(c) $CH_3CH{=}CHCO_2H$

3. (a) $CH_3CO_2CH_3$ (b) $CH_3CO_2CH_2CH_2CH_3$

(c) $CH_3CO_2\overset{\displaystyle CHCH_3}{\underset{\displaystyle CH_3}{}}$

4. (a) $HCO_2CH_2CH_3$ (b) $CH_3CH_2CO_2CH_2CH_3$

(c) $C_6H_5CO_2CH_2CH_3$

5. (a) $CH_3CO_2H + CH_3OH$

(b) $CH_3CH_2CO_2H + CH_3\overset{\displaystyle CHCH_3}{\underset{\displaystyle OH}{}}$

(c) $CH_3\overset{\displaystyle CH_3}{\overset{|}{C}H}CO_2H + CH_3CH_2CH_2OH$

6. (a) $C_6H_5-OH + CH_3CO_2^-$

(b) $CH_3OH + CH_3-O-$⟨benzene⟩$-CO_2^-$

7. (a) $CH_3\overset{\displaystyle CH_3}{\overset{|}{C}H}CONHCH_3$ (b) $CH_3CONHC_6H_5$

(c) No amide can form   (d) No amide can form

8. (a) $C_6H_5-CO_2H + NH_2CH_3$

(b) No hydrolysis can occur

(c) $CH_3CO_2H + C_6H_5-NH_2$

(d) $2CH_3CO_2H + NH_2CH_2CH_2NH_2$

PRACTICE EXERCISES, CHAPTER 15

1. $CH_3(CH_2)_{26}\overset{\displaystyle O}{\overset{\|}{C}}O(CH_2)_{25}CH_3$

2. $\underset{H}{CH_3(CH_2)_7}\overset{\displaystyle (CH_2)_7CO_2H}{C{=}C}\underset{H}{}$

3. $\mathbf{1} + 3NaOH \longrightarrow$
   $HOCH_2\overset{\displaystyle CHCH_2OH}{\underset{\displaystyle OH}{}} + NaO_2C(CH_2)_7CH{=}CH(CH_2)_7CH_3$
   $\quad + NaO_2C(CH_2)_{14}CH_3$
   $\quad + NaO_2C(CH_2)_7CH{=}CHCH_2CH{=}CH(CH_2)_4CH_3$

4. $\mathbf{1} + 3H_2 \xrightarrow[\text{heat, pressure}]{\text{catalyst}} CH_2-O-\overset{\displaystyle O}{\overset{\|}{C}}(CH_2)_{16}CH_3$
   $\quad\quad CH-O-\overset{\displaystyle O}{\overset{\|}{C}}(CH_2)_{14}CH_3$
   $\quad\quad CH_2-O-\overset{\displaystyle O}{\overset{\|}{C}}(CH_2)_{16}CH_3$

PRACTICE EXERCISES, CHAPTER 16

1. Glycine: $^+NH_3CH_2CO_2^-$

   Alanine: $^+NH_3CHCO_2^-$
   $|$
   CH_3

   Lysine: $^+NH_3CHCO_2^-$
   $|$
   $(CH_2)_4$
   $|$
   NH_4

   Glutamic acid: $^+NH_3CHCO_2^-$
   $|$
   $(CH_2)_2$
   $|$
   CO_2H

2. (a) $^+NH_3CHCO_2^-$ (b) $^+NH_3CHCO_2^-$
   $|$ $|$
   $CH_2CO_2^-$ CH_2CONH_2

3. $^+NH_3CHCO_2^-$ NH_2^+
   $|$ $\|$
   $CH_2CH_2CH_2NHCNH_2$

4. Hydrophilic; neutral side chain

5. $^+NH_3CHCONHCHCO_2^-$ $^+NH_3CHCONHCHCO_2^-$
   $|$ $|$ $|$ $|$
   CH_3 $CH_2CH_2CO_2H$ CH_2 CH_3
   CH_2CO_2H

PRACTICE EXERCISES, CHAPTER 21

1. (a) Proline   (b) Arginine
   (c) Glutamic acid   (d) Lysine
2. (a) Serine   (b) CT (chain termination)
   (c) Glutamic acid   (d) Isoleucine

PRACTICE EXERCISES, CHAPTER 22

1. $^{131}_{53}I \rightarrow ^{131}_{54}Xe + ^{0}_{-1}e + ^{0}_{0}\gamma$
2. $^{239}_{94}Pu \rightarrow ^{235}_{92}U + ^{4}_{2}He + ^{0}_{0}\gamma$
3. 10,000 units
4. 8.5 m

REVIEW EXERCISES, CHAPTER 22

22.11 (a) $^{241}_{94}Pu$ (b) $^{22}_{10}Ne$
22.13 (a) $^{252}_{99}Es \rightarrow ^{4}_{2}He + ^{248}_{97}Bk$
      (b) $^{28}_{12}Mg \rightarrow ^{0}_{-1}e + ^{28}_{13}Al$
      (c) $^{20}_{8}O \rightarrow ^{0}_{-1}e + ^{20}_{9}F$
      (d) $^{251}_{98}Cf \rightarrow ^{4}_{2}He + ^{247}_{96}Cm + ^{0}_{0}\gamma$
22.17 0.750 ng
22.34 15.1 m
22.50 Iron-55. $^{55}_{22}Mn + ^{1}_{1}H \rightarrow ^{1}_{0}n + ^{55}_{26}Fe$
22.52 $^{113}_{49}In \rightarrow ^{111}_{49}In + 2\,^{1}_{0}n$
22.54 $^{67}_{31}Ga + ^{0}_{-1}e \rightarrow ^{67}_{30}Zn$
22.56 $^{10}_{5}B + ^{4}_{2}He \rightarrow ^{13}_{7}N + ^{1}_{0}n$
22.58 $^{27}_{13}Al + ^{6}_{3}Li \rightarrow ^{32}_{15}P + ^{1}_{1}H$

Glossary

Acetyl Coenzyme A (Acetyl CoA) The molecule from which acetyl groups are transferred into the citric acid cycle or into the lipigenesis cycle.

$$CH_3C(=O)—S—CoA$$

Accuracy In science, the degree of conformity to some accepted standard or reference; freedom from error or mistake; correctness.

Acetal Any organic compound in which two ether linkages extend from one CH unit, as in:

$$R—O—CH—O—R'$$

Achiral Not possessing chirality; that quality of a molecule (or other object) that allows it to be superimposed on its mirror image.

Acid Any substance that can donate a proton, H^+. (Brønsted theory)

Acid–Base Neutralization The reaction of an acid with a base.

Anhydride, Phosphoric

$$—P(=O)(OH)—O—P(=O)(OH)—$$

Phosphoric
anhydride system

Acid Derivative Any organic compound that can be made from an organic acid or that can be changed back to the acid by hydrolysis. (Examples are salts, esters, and amides.)

Acid–Base Indicator (see *Indicator*)

Acidic Solution A solution in which the molar concentration of hydronium ions is greater than that of hydroxide ions.

Acidosis A condition in which the pH of the blood is below normal. *Metabolic acidosis* is brought on by a defect in some metabolic pathway. *Respiratory acidosis* is caused by a defect in the respiratory centers or in the mechanisms of breathing.

Active Site That region of an enzyme molecule most directly responsible for the catalytic effect of the enzyme.

Active Transport The movement of a substance through a biological membrane against a concentration gradient and caused by energy-consuming chemical changes that involve parts of the membrane.

Activity Series A list of elements (or other substances) in the order of the ease with which they release electrons under standard conditions and become oxidized.

Acyl Group

$$R—C(=O)—$$

Acyl Group-Transfer Reaction Any reaction in which an acyl group transfers from a donor to an acceptor.

Addition Reaction Any reaction in which two parts of a reactant molecule add to a double or a triple bond.

Adenosine Diphosphate (ADP) A high-energy diphosphate ester obtained from adenosine triphosphate (ATP) when part of the chemical energy in ATP is tapped for some purpose in a cell.

Adenosine Monophosphate (AMP) A low-energy phosphate ester that can be obtained by the hydrolysis of ATP or ADP; a monomer for the biosynthesis of nuclei acids.

Adenosine Triphosphate (ATP) A high-energy triphosphate ester used in living systems to provide chemical energy for metabolic needs.

Adequate Protein A protein that, when digested, makes available all of the essential amino acids in suitable proportions to satisfy both the amino acid and total nitrogen requirements of good nutrition without providing excessive calories.

ADP (see *Adenosine Diphosphate*)

Aerobic Sequence An oxygen-consuming sequence of catabolism that starts with glucose or with glucose units in glycogen, and proceeds through glycolysis, the citric acid cycle, and the respiratory chain.

Agonist A compound whose molecules can bind to a receptor on a cell membrane and cause a response by the cell.

Albumin One of a family of globular proteins that tend to dissolve in water, and that in blood contribute to the blood's colloidal osmotic pressure and aid in the transport of metal ions, fatty acids, cholesterol, triacylglycerols, and other water-insoluble substances.

Alcohol Any organic compound whose molecules have the OH group attached to a satuated carbon; ROH.

Alcohol Group The OH group when it is joined to a saturated carbon.

Aldehyde An organic compound that has a carbonyl group joined to H on one side and C on the other; $R—CH=O$.

Aldehyde Group $—CH=O$

Aldohexose A monosaccharide whose molecules have six carbon atoms and an aldehyde group.

Aldose A monosaccharide whose molecules have an aldehyde group.

Aldosterone A steroid hormone, made in the adrenal cortex, secreted into the bloodstream when the sodium ion level is low, and that signals the kidneys to leave sodium ions in the bloodstream.

Aliphatic Compound Any organic compound whose molecules lack a benzene ring or a similar structural feature.

Alkali A strongly basic substance such as sodium hydroxide or potassium hydroxide.

Alkali Metals The elements of Group IA of the periodic table — lithium, sodium, potassium, rubidium, cesium, and francium.

Alkaline Earth Metals The elements of Group IIA of the periodic table: beryllium, magnesium, calcium, strontium, barium, and radium.

Alkaloid A physiologically active, heterocyclic amine isolated from plants.

Alkalosis A condition in which the pH of the blood is above normal. *Metabolic alkalosis* is caused by a defect in metabolism. *Respiratory alkalosis* is caused by a defect in the respiratory centers of the brain or in the apparatus of breathing.

Alkane Any saturated hydrocarbon, one that has only single bonds. A *normal alkane* is any whose molecules have straight chains.

Alkene Any hydrocarbon whose molecules have one or more double bonds.

Alkyl Group A substituent group that is an alkane minus one H atom.

Allosteric Activation The activation of an enzyme's catalytic site by the binding of some molecule at a position elsewhere on the enzyme.

Alloy A mixture of two or more metals made by mixing them in their molten states.

Alpha (α) Particle The nucleus of a helium atom; $^4_2$He.

Alpha (α) Radiation A stream of high-energy alpha particles.

Amide Any organic compound whose molecules have a carbonyl-nitrogen unit:

$$-\overset{\overset{\displaystyle O}{\|}}{C}-\overset{\displaystyle |}{N}-$$

Amide Bond The single bond that holds the carbonyl group to the nitrogen atom in an amide.

Amine Any organic compound whose molecules have a trivalent nitrogen atom, as in $R-NH_2$, $R-NH-R$, or R_3N.

Amine Salt Any organic compound whose molecules have a positively charged, tetravalent, protonated nitrogen atom, as in RNH_3^+, $R_2NH_2^+$, or R_3NH^+.

Amino Acid Any organic compound whose molecules have both an amino group and a carboxyl group.

Amino Acid Residue A structural unit in a polypeptide, $-NH-CH-CO-$ furnished by an amino acid, where G is a side-chain group.
$$|\\G$$

AMP (see *Adenosine Monophosphate*)

Amphipathic Compound A substance whose molecules have both hydrophilic and hydrophobic groups.

Amylopectin A polymer found in starch in which linear amylose chains have joined to each other by $\alpha(1\rightarrow6)$ acetal links to give a branched polymer of α-glucose.

Amylose A linear polymer of glucose in which glucose units are joined by $\alpha(1\rightarrow4)$ acetal links.

Anaerobic Sequence The oxygen-independent catabolism of glucose or of glucose units in glycogen to lactate ion.

Anhydrous Without water.

Anion A negatively charged ion.

Anion Gap

$$\text{anion gap} = \frac{\text{meq of Na}^+}{L} - \left(\frac{\text{meq Cl}^-}{L} + \frac{\text{meq HCO}_3^-}{L}\right)$$

Anode The positive electrode to which negatively charged ions (anions) are attracted during electrolysis.

Anoxia A condition of a tissue in which it receives no oxygen.

Antagonist A compound that can bind to a membrane receptor but not cause any response by the cell.

Anticodon A sequence of three adjacent side-chain bases on a molecule of tRNA that is complementary to a codon and that fits to its codon on an mRNA chain during polypeptide synthesis.

Antimetabolite A substance that inhibits the growth of bacteria.

Apoenzyme The wholly polypeptide part of an enzyme.

Aromatic Compound Any organic compound whose molecules have a benzene ring (or a feature very similar to this).

Atmosphere, Standard (see *Standard Atmosphere*)

Atom A small particle with one nucleus and zero charge; the smallest particle of a given element that bears the chemical properties of the element.

Atomic Mass Number (see *Mass Number*)

Atomic Mass Unit (amu) 1.6606×10^{-24} g. A mass very close to that of a proton or a neutron.

Atomic Number The positive charge on an atom's nucleus; the number of protons in an atom's nucleus.

Atomic Weight The average mass, in amu, of the atoms of the isotopes of a given element as they occur naturally.

ATP (see *Adenosine Triphosphate*)

Avogadro's Law Equal volumes of gases contain equal numbers of moles when they are compared at identical temperatures and pressures.

Avogadro's Number 6.023×10^{23}. The number of formula units in one mole of any element or compound.

Background Radiation Cosmic rays plus the natural atomic radiation emitted by the traces of radioactive isotopes in soils and rocks.

Balanced Equation (see *Equation, Balanced*)

Base A proton acceptor; a compound that neutralizes hydrogen ions (Brønsted theory).

Base, Heterocyclic A heterocyclic amine obtained from the hydrolysis of nucleic acids: adenine, thymine, guanine, cytosine, or uracil.

Base Pairing In nucleic acid chemistry, the association by means of hydrogen bonds of two heterocyclic, side-chain bases: adenine with thymine (or uracil) and guanine with cytosine.

Base Quantity A fundamental quantity of physical measurement such as mass, length, and time; a quantity used to define derived quantities such as mass/volume for density.

Base Unit A fundamental unit of measurement for a base quantity such as the kilogram for mass, the meter for length, the second for time, the kelvin for temperature degree, and the mole for quantity of chemical substance; a unit to which derived units of measurement are related.

Basic Solution A solution in which the molar concentration of hydroxide ions is greater than that of hydronium ions.

Benedict's Reagent A solution of copper(II) sulfate, sodium citrate, and sodium carbonate that is used in the Benedict's test.

Benedict's Test The use of Benedict's reagent to detect the presence of any compound whose molecules have easily oxidized functional groups—α-hydroxyaldehydes and α-hydroxyketones—such as those present in monosaccharides. In a positive test the intensely blue color of the reagent disappears and a reddish precipitate of copper(I) oxide separates.

Beta Oxidation The fatty acid cycle of catabolism.

Beta (β) Particle A high energy electron emitted from a nucleus, $_{-1}^{0}e$.

Beta (β) Radiation A stream of high-energy electrons.

Bile A secretion of the gallbladder that empties into the upper intestine and furnishes bile salts; a route of excretion for cholesterol and bile pigments.

Bile Pigment Colored products of the partial catabolism of heme that are transferred from the liver to the gallbladder for secretion via the bile.

Bile Salts Steroid-based detergents in bile that emulsify fats and oils during digestion.

Binding Site That part of an enzyme molecule that holds the substrate molecule and positions it over the active site.

Biochemistry The study of the structures and properties of substances found in living systems.

Blood Sugar The carbohydrates, mostly glucose, present in blood.

Blood Sugar Level The concentration of carbohydrate, mostly glucose, in the blood; usually stated in units of mg/dL.

Binary Compound A compound made from just two elements.

Bohr Model of the Atom The solar system model of the structure of an atom, proposed by Niels Bohr, that pictures the electrons circling the nucleus in discrete energy states called orbits.

Boiling The turbulent behavior in a liquid when its vapor pressure equals the atmospheric pressure and when the liquid absorbs heat while experiencing no rise in temperature.

Boiling Point, Normal The temperature at which a substance boils when the atmospheric pressure is 760 mm Hg (1 atm).

Bond, Chemical A net electrical force of attraction that holds atomic nuclei near each other within compounds.

Boyle's Law (see *Pressure–Volume Law*)

Branched Chain A sequence of atoms to which additional atoms are attached at points other than the ends.

Brønsted Theory An acid is a proton donor and a base is a proton acceptor.

Brownian Movement The random, chaotic movements of particles in a colloidal dispersion that can be seen with a microscope.

Buffer A combination of solutes that holds the pH of a solution relatively constant even if small amounts of acids or bases are added.

Calorie The amount of heat that raises the temperature of 1 g of water by 1 degree Celsius from 14.5 to 15.5 °C.

Carbaminohemoglobin Hemoglobin that carries chemically bound carbon dioxide.

Carbohydrate Any naturally occurring substance whose molecules are polyhydroxyaldehydes or polyhydroxyketones or can be hydrolyzed to such compounds.

Carbon Family The Group IVA elements in the periodic table: carbon, silicon, germanium, tin, and lead.

Carbonate Buffer A mixture or a solution that includes bicarbonate ions and carbonic acid (dissolved carbon dioxide) in which the bicarbonate ion can neutralize added acid and carbonic acid can neutralize added base.

Carboxylic Acid A compound whose molecules have the carboxyl group, CO_2H.

Carcinogen A chemical or physical agent that induces the onset of cancer or the formation of a tumor that might or might not become cancerous.

Catabolism The reactions of metabolism that break molecules down.

Catalysis The phenomenon of an increase in the rate of a chemical reaction brought about by a relatively small amount of a chemical, the catalyst, that is not permanently changed by the reaction.

Catalyst A substance that is able, in relatively low concentrations, to accelerate the rate of a chemical reaction without itself being permanently changed. (In living systems, the catalysts are called enzymes.)

Cathode The negative electrode to which positively charged ions, cations, are attracted during electrolysis.

Cation A positively charged ion.

Centimeter (cm) A length equal to one hundredth of the meter. 1 cm = 0.01 m = 0.394 in.

Charles' Law (see *Temperature–Volume Law*)

Chemical Bond (see *Bond, Chemical*)

Chemical Energy The potential energy that substances have because their arrangements of electrons and atomic nuclei are not as stable as are alternative arrangements that become possible in chemical reactions.

Chemical Equation A shorthand representation of a chemical reaction that uses formulas instead of names for reactants and products; that separates reactant formulas from product formulas by an arrow; that separates formulas on either side of the arrow by plus signs; and that expresses the mole proportions of the chemicals by simple numbers (coefficients) placed before the formulas.

Chemical Property Any chemical reaction that a substance can undergo and the ability to undergo such a reaction.

Chemical Reaction Any event in which substances change into different chemical substances.

Chemiosmotic Theory An explanation of how oxidative phosphorylation is related to a flow of protons in a proton gradient that is established by the respiratory chain, and that extends across the inner membrane of a mitochondrion.

Chiral Molecule A molecule having handedness in its molecular structure (see also *Chirality*)

Chirality The quality of handedness that a molecular structure has that prevents this structure from being superimposable on its mirror image.

Chloride Shift An interchange of chloride ions and bicarbonate ions between a red blood cell and the surrounding blood serum.

Chromatin A nucleoprotein in a cell nucleus made of histones and DNA.

Chromosome Small threadlike bodies in a cell nucleus that carry genes in a linear array and that are microscopically visible during cell division.

Citric Acid Cycle A series of reactions that dismantle acetyl units and send electrons (and protons) into the respiratory chain; a major source of metabolites for the respiratory chain.

Codon A sequence of three adjacent side-chain bases in a molecule of mRNA that codes for a specific amino acid residue when the mRNA participates in polypeptide synthesis.

Coefficients Numbers placed before formulas in chemical equations to indicate the mole proportions of reactants and products.

Coenzyme An organic compound needed to make a complete enzyme from an apoenzyme.

Cofactor A nonprotein compound or ion that is an essential part of an enzyme.

Collagen The fibrous protein of connective tissue that changes to gelatin in boiling water.

Colloidal Dispersion A relatively stable, uniform distribution in some dispersing medium of colloidal particles — those with at least one dimension between 1 and 1000 nm.

Colloidal Osmotic Pressure The contribution made to the osmotic pressure of a solution by substances colloidally dispersed in it.

Combustion A rapid chemical reaction with oxygen that produces heat, light, a flame, and mostly gaseous products.

Compound A substance made from the atoms of two or more elements that are present in a definite proportion by mass and by atoms.

Concentration The quantity of some component of a mixture in a unit of volume or a unit of mass of the mixture.

Condensed Structure (see *Structural Formula*)

Conduction In the science of heat energy, the transfer of heat from a region of higher temperature to a region of lower temperature by means of the transfer of the kinetic energy of atoms, ions, or molecules to their neighbors. In the science of electrical energy, the movement of electrons in a conductor.

Convection In the science of heat energy, the transfer of heat by the circulation of a warmer fluid throughout the remainder of the fluid.

Conversion Factor A fraction that expresses a relationship between quantities that have different units, such as 2.54 cm/in.

Coordinate Covalent Bond A covalent bond in which both of the electrons of the shared pair originated from one of the atoms involved in the bond.

Cosmic Radiation A stream of ionizing radiations, from the sun and outer space, that consists mostly of protons but also includes alpha particles, electrons, and the nuclei of atoms up to atomic number 28.

Covalent Bond The net force of attraction that arises as two atomic nuclei share a pair of electrons. One pair is shared in a single bond, two pairs in a double bond, and three pairs in a triple bond.

Crenation The shrinkage of red blood cells when they are in contact with a hypertonic solution.

Curie (Ci) A unit of activity of a radioactive source.

$$1 \text{ Ci} = 3.70 \times 10^{10} \text{ disintegrations/s}$$

Dalton's Law (see *Law of Partial Pressures*)

Dalton's Theory A theory that accounts for the laws of chemical combination by postulating that matter consists of indestructible atoms; that all atoms of the same element are identical in mass and other properties; that the atoms of different elements are different in mass and other properties; and that in the formation of a compound, atoms join together in definite, whole-number ratios.

Deamination The removal of an amino group from an amino acid.

Decarboxylation The removal of a carboxyl group.

Degree Celsius One-one hundredth (1/100) of the interval on a thermometer between the freezing point and the boiling point of water.

Degree Fahrenheit One-one hundred and eightieth (1/180) of the interval on a thermometer between the freezing point and the boiling point of water.

Denaturation The loss of the natural shape and form of a protein molecule together with its ability to function biologically, but not necessarily accompanied by the rupture of any of its peptide bonds.

Density The ratio of the mass of an object to its volume; the mass per unit volume. Density = mass/volume (usually expressed in g/mL).

Deoxyribonucleic Acid (DNA) The chemical of a gene; one of a large number of polymers of deoxyribonucleotides, whose sequences of side-chain bases constitute the genetic messages of genes.

Derived Quantity A quantity based on a relationship that involves one or more base quantities of measurement such as volume (length³) or density (mass/volume).

Derived Unit A unit of a derived quantity such as g/mL (density).

Desiccant A substance that combines with water vapor to form a hydrate and thereby reduces the concentration of water vapor in the air space around the substance.

***D* Family; *L* Family** The names of the two optically active families to which substances can belong when they are considered solely according to one kind of molecular handedness or the other.

Diabetes Mellitus A disease in which there is an insufficiency of effective insulin and an impairment of glucose tolerance.

Dialysis The passage through a dialyzing membrane of water and particles in solution, but not of particles that have colloidal size.

Diatomic Molecule A molecule made from two atoms.

Diffusion A physical process whereby particles, by random motions, intermingle and spread out so as to erase concentration gradients.

Digestion The hydrolysis of food molecules that occurs in the digestive tract.

Dipeptide A compound whose molecules have two α-amino acid residues joined by a peptide (amide) bond.

Dipolar Ion A molecule that carries one plus charge and one minus charge, such as an α-amino acid.

Dipole, Electrical A pair of equal but opposite (and usually partial) electrical charges separated by a small distance in a molecule.

Diprotic Acid An acid with two protons available per molecule to neutralize a base, for example, H_2SO_4.

Disaccharide A carbohydrate that can be hydrolyzed into two monosaccharides.

Disulfide Link The sulfur–sulfur covalent bond in polypeptides.

Disulfide System —S—S— as in R—S—S—R.

DNA (see *Deoxyribonucleic Acid*)

Double Bond A covalent bond in which two pairs of electrons are shared.

Double Decomposition A reaction in which a compound is made by the exchange of partner ions between two salts.

Double Helix, DNA A spiral arrangement of two intertwining DNA molecules held together by hydrogen bonds between side-chain bases.

Dynamic Equilibrium (see *Equilibrium*)

Dyspnea Air hunger.

Edema The swelling of tissue caused by the retention of water.

Effector A nonsubstrate molecule or ion that activates an enzyme.

Elastase A digestive enzyme that helps to hydrolyze elastin.

Elastin The fibrous protein of tendons and arteries.

Electrical Balance The condition of a net ionic equation wherein the algebraic sum of the positive and negative charges of the reactants equals that of the products.

Electrode A metal object, usually a wire, suspended in an electrically conducting medium through which electricity passes to or from an external circuit.

Electrolysis A procedure in which an electrical current is passed through a solution that contains ions, or through a molten salt, for the purpose of bringing about a chemical change.

Electrolyte Any substance whose solution in water conducts electricity; or the solution itself of such a substance.

Electrolytes, Blood The ionic substances dissolved in the blood.

Electron A subatomic particle that bears one unit of negative charge and has a mass that is 1/1836 the mass of a proton.

Electron Cloud A mental model that views the one or two rapidly moving electrons as creating a cloudlike distribution of negative charge.

Electron Configuration The most stable arrangement (that is, the arrangement of lowest energy) of the electrons of an atom, ion, or molecule.

Electronegativity The ability of an atom joined to another by a covalent bond to attract the electrons of the bond toward itself.

Electron Sharing The joint attraction of two atomic nuclei toward a pair of electrons situated between the nuclei and between which, therefore, a covalent bond exists.

Electron Shell An alternative name for *principal energy level*.

Electron Volt (eV) A very small unit of energy used to describe the energy of a radiation.

$$1\ eV = 1.6 \times 10^{-19}\ \text{joule}$$
$$1\ eV = 3.8 \times 10^{-20}\ \text{calorie}$$
$$1000\ eV = 1\ KeV\ (1\ \text{kiloelectron volt})$$
$$1000\ keV = 1\ MeV\ (1\ \text{megaelectron volt})$$

Element A substance that cannot be broken down into anything that is both stable and more simple; a substance in which all of the atoms have the same atomic number and the same electron configuration; one of the three broad kinds of matter, the others being compounds and mixtures.

Emulsion A colloidal dispersion of tiny microdroplets of one liquid in another liquid.

Enantiomers Isomers whose molecules are related as nonsuperimposable object and mirror image.

End Point The stage in a titration when the operation is stopped.

Endothermic Describing a change that needs a constant supply of heat energy to happen.

Energy A capacity to cause a change that can, in principle, be harnessed for useful work.

Energy Density The energy per gram of stored glycogen or fat.

Energy Level A principal energy state in which electrons of an atom can be.

Energy of Activation The minimum energy that must be provided by the collision between reactant particles to initiate the rearrangement of electrons relative to nuclei that must happen if the reaction is to occur.

Enteropeptidase An intestinal enzyme that catalyzes the conversion of trypsinogen to trypsin.

Enzyme A catalyst in a living system.

Enzyme–Substrate Complex The temporary combination that an enzyme must form with its substrate before catalysis can occur.

Epinephrine A hormone of the adrenal medulla that activates the enzymes needed to release glucose from glycogen.

Equation, Balanced A chemical equation in which all of the atoms represented in the formulas of the reactants are present in identical numbers among the products, and in which any net electrical charge provided by the reactants equals the same charge indicated by the products (see also *Chemical Equation*).

Equilibrium A situation in which two opposing events occur at identical rates so that no net change occurs.

Equivalence Point The stage in a titration when the reactants have been mixed in the exact molar proportions represented by the balanced equation; in an acid–base titration, the stage when the moles of hydrogen ions furnished by the acid matches the moles of hydroxide ions (or other proton acceptor) supplied by the base.

Equivalent The molar mass of an ion divided by its electrical charge.

Equivalent Weight A synonym for *Equivalent*.

Erythrocyte A red blood cell.

Essential Amino Acid An α-amino acid that the body cannot make from other amino acids and that must be supplied by the diet.

Ester A derivative of an acid and an alcohol that can be hydrolyzed to these parent compounds. Esters of carboxylic acids and phosphoric acid occur in living systems.

System in an ester of a carboxylic acid

System in an ester of phosphoric acid

Esterification The formation of an ester.

Ether An organic compound whose molecules have an oxygen attached by single bonds to separate carbon atoms neither of which is a carbonyl carbon atom: R—O—R'.

Evaporation The conversion of a substance from its liquid to its vapor state at a temperature below the liquid's boiling point.

Exon A segment of a DNA strand that eventually becomes expressed as a corresponding sequence of aminoacyl residues in a polypeptide.

Exothermic Describing a change by which heat energy is released from the system.

Extensive Property Any property whose value is directly proportional to the size of the sample, such as volume or mass.

Extracellular Fluids Body fluids that are outside of cells.

Fact In science, something that has physical existence, that can be experienced or observed, and that can be measured by independent observers.

Factor-Label Method A strategy for solving computational problems that uses conversion factors and the cancellation of the units of physical quantities as an aid in working toward the solution.

Fatty Acid Any carboxylic acid that can be obtained by the hydrolysis of animal fats or vegetable oils.

Fatty Acid Cycle The catabolism of a fatty acid by a series of repeating steps that produce acetyl units (in acetyl CoA).

Feedback Inhibition The competitive inhibition of an enzyme by a product of its own action.

Fibrin The fibrous protein of a blood clot that forms from fibrinogen during clotting.

Fibrinogen A protein in blood that is changed to fibrin during clotting.

Fibrous Proteins Water-insoluble proteins found in fibrous tissues.

Fission The splitting of the nucleus of a heavy atom approximately in half, accompanied by the release of one or a few neutrons and energy.

Formula, Empirical A chemical symbol for a compound that gives just the ratios of the atoms and not necessarily the composition of a complete molecule.

Formula, Molecular A chemical symbol for a substance that gives the composition of a complete molecule.

Formula, Structural A chemical symbol for a substance that uses atomic symbols and lines to describe the pattern in which the atoms are joined together in a molecule.

Formula Unit A small particle—an atom, a molecule, or a set of ions—that has the composition given by the chemical formula of the substance.

Formula Weight The sum of the atomic weights of the atoms represented in a chemical formula (and also known as the molecular weight).

Free Rotation The absence of a barrier to the rotation of two groups with respect to each other when they are joined by a single covalent bond.

Fructose A ketohexose present in honey and one product of the hydrolysis of sucrose; levulose.

Functional Group An atom or a group of atoms in a molecule that is responsible for the particular set of reactions that all compounds with this group have.

Galactose An aldohexose that forms, together with glucose, when lactose (milk sugar) is hydrolyzed.

Gamma (γ) One microgram; 1×10^{-6} g.

Gamma Radiation A natural radiation similar to but more powerful than X rays.

Gas Any substance that must be contained in a wholly closed space and whose shape and volume are determined entirely by the shape and volume of its container; a state of matter.

Gas Constant, Universal (R) The ratio of PV to nT for a gas, where P = the gas pressure, V = volume, n = number of moles, and T = the Kelvin temperature. When P is in mm Hg and V is in mL,

$$R = 6.23 \times 10^4 \text{ mm Hg mL/mol K}$$

Gas Tension The partial pressure of a gas over its solution in some liquid when the system is in equilibrium.

Gastric Juice The digestive juice secreted into the stomach and that contains pepsinogen, hydrochloric acid, and gastric lipase.

Gay-Lussac's Law (see *Pressure–Temperature Law*)

Gel A colloidal dispersion of a solid in a liquid that has adopted a semisolid form.

Gene A unit heredity carried on a cell's chromosome and consisting of DNA.

Genetic Code The correlations between codons on mRNA and side chains of aminoacyl residues on polypeptides made under the direction of mRNA.

Geometric Isomerism Isomerism caused by restricted rotation that gives different geometries to the same structural organization; cis–trans isomerism.

Geometric Isomers Isomers whose molecules have identical atomic organizations but different geometries; cis–trans isomers.

Globular Proteins Proteins that are soluble in water or in water that contains certain dissolved salts.

Globulins Globular proteins in the blood that include γ-globulin, an agent in the body's defense against infectious diseases.

Glucagon A hormone, secreted by the α cells of the pancreas in response to a decrease in the blood sugar level, that stimulates the liver to release glucose from its glycogen stores.

Gluconeogenesis The synthesis of glucose from compounds with smaller molecules or ions.

Glucose An aldohexose whose molecules serve as building blocks for glycogen, starch, cellulose, dextrin, maltose, sucrose, and lactose; the chief carbohydrate in blood; *blood sugar; dextrose.*

Glucose Tolerance The ability of the body to manage the intake of dietary glucose while keeping the blood sugar level from fluctuating widely.

Glucose Tolerance Test A series of measurements of the blood sugar level after the ingestion of a considerable amount of glucose; used to obtain information about an individual's glucose tolerance.

Glucosuria The presence of glucose in urine.

Glycogen The starchlike polymer of α-glucose that serves as an animal's means of storing glucose units.

Glycogenesis The synthesis of glycogen.

Glycogenolysis The breakdown of glycogen to glucose.

Glycol A dihydric alcohol.

Glycolipid A lipid whose molecules include a glucose unit, a galactose unit or some other carbohydrate unit.

Glycolysis A series of chemical reactions that break down glucose or glucose units in glycogen until pyruvate remains (when the series is operated aerobically) or lactate forms (when the conditions are anaerobic).

Gradient The presence of a change in value of some physical quantity with distance, as in a *concentration* gradient in which the concentration of a solute is different in different parts of the system.

Gram (g) A mass equal to one thousandth of the kilogram mass, the SI standard mass.

$$1 \text{ g} = 0.001 \text{ kg} = 1000 \text{ mg}; \qquad 1 \text{ lb} = 454 \text{ g}.$$

Group A vertical column in the periodic table; a family of elements.

Half-Life The time needed for one-half of the atoms in a sample of a particular radioactive isotope to undergo radioactive decay.

Halogens The elements of Group VIIA of the periodic table—fluorine, chlorine, bromine, iodine, and astatine.

Hard Water Water that contains one or more of the metallic ions Mg^{2+}, Ca^{2+}, Fe^{2+}, or Fe^{3+}. The negative ions present are usually Cl^- and SO_4^{2-}. If HCO_3^- is the chief negative ion, the water is said to be *temporary hard water;* otherwise it is *permanent hard water.*

Heat The form of energy that transfers between two objects in contact that have initially different temperatures

Heat of Fusion The quantity of heat that one gram of a substance absorbs when it changes from its solid to its liquid state at its melting point.

Heat of Reaction The net energy difference between the reactants and the products of a reaction.

Heat of Vaporization The quantity of heat that one gram of a substance absorbs when it changes from its liquid to its gaseous state.

α-Helix One kind of secondary structure of a polypeptide in which its molecules are coiled.

Heme The deep red, iron-containing prosthetic group in hemoglobin and myoglobin.

Hemiacetal Any compound whose molecules have both an OH group and an ether linkage coming to a —CH— unit:

Hemiketal Any compound whose molecules have both an OH group and an ether linkage coming to a carbon that bears no H atoms:

Hemoglobin (HHb) The oxygen-carrying protein in red blood cells.

Hemolysis The bursting of a red blood cell.

Heterocyclic Compounds Compounds with rings that include atoms other than carbon.

Heterogeneous Nuclear RNA (hnRNA) (see *Primary Transcript RNA*).

High-Energy Phosphate An organophosphate with a phosphoric anhydride system.

Homeostasis The response of an organism to a stimulus such that the organism is restored to its prestimulated state.

Homogeneous Mixture A mixture in which the composition and properties are uniform throughout.

Homolog Any member of a homologous series of organic compounds.

Homologous Series A series of organic compounds in the same family whose successive members differ by individual CH_2 units.

Hormone A primary chemical messenger made by an endocrine gland and carried by the bloodstream to a target organ where a particular chemical response is initiated.

Human Growth Hormone One of the hormones that affects the blood sugar level; a stimulator of the release of the hormone glucagon.

Hydrate A compound in which intact molecules of water are held in a definite molar proportion to the other components.

Hydration The association of water molecules with dissolved ions or polar molecules.

Hydrocarbon Any organic compound that consists entirely of carbon and hydrogen.

Hydrogen Bond The force of attraction between a $\delta+$ on a hydrogen held by a covalent bond to oxygen or nitrogen (or fluorine) and a $\delta-$ charge on a nearby atom of oxygen or nitrogen (or fluorine).

Hydrolysis of Salts Any reaction in which a cation (other than H^+) or an anion (other than OH^-) changes the ratio of the molar concentrations of hydrogen and hydroxide ions in an aqueous solution.

Hydronium Ion H_3O^+

Hydrophilic Group Any part of a molecular structure that attracts water molecules; a polar or ionic group such as OH, CO_2^-, NH_3^+, or NH_2.

Hydrophobic Group Any part of a molecular structure that has no attraction for water molecules; a nonpolar group such as any alkyl group.

Hydroxide Ion OH^-

Hygroscopic Describing a substance that can reduce the concentration of water vapor in the surrounding air by forming a hydrate.

Hyperglycemia An elevated level of sugar in the blood, above 95 mg/dL in whole blood.

Hyperthermia An elevated body temperature.

Hypertonic Having an osmotic pressure greater than some reference; having a total concentration of all solute particles higher than that of some reference.

Hyperventilation Breathing considerably faster and deeper than normal.

Hypoglycemia A low level of glucose in blood—below 65 mg/dL of whole blood.

Hypothermia A low body temperature.

Hypothesis A conjecture, subject to being disproved, that explains a set of facts in terms of a common cause and that serves as the basis for the design of additional tests or experiments.

Hypotonic Having an osmotic pressure less than some reference; having a total concentration of dissolved solute particles less than that of some reference.

Hypoventilation Breathing more slowly and less deeply than normal; shallow breathing.

Hypoxia A condition of a low supply of oxygen.

Ideal Gas A hypothetical gas that obeys the gas laws exactly.

Indicator A dye that, in solution, has one color below a measured pH range and a different color above this range.

Induced Fit Theory Certain enzymes are induced by their substrate molecules to modify their shapes to accommodate the substrate.

Inertia The resistance of an object to a change in its position or its motion.

Inhibitor A substance that interacts with an enzyme to prevent its acting as a catalyst.

Inner Transition Elements The elements of the lanthanide and actinide series of the periodic table.

Inorganic Compound Any compound that is not an organic compound.

Insensible Perspiration The loss of water from the body with no visible sweating; evaporative losses from the skin and the lungs.

Insulin A protein hormone made by the pancreas, released in response to a rise in the blood sugar level, and used by certain tissues to help them take up glucose from circulation.

Insulin Shock Shock brought on by a drastic reduction in the blood sugar level usually after an overdose of insulin.

Intensive Property Any property whose value is independent of the size of the sample, such as temperature and density.

Internal Environment Everything enclosed within an organism.

International System of Units (SI) The successor to the metric system with new reference standards for the base units but with the same names for the units and the same decimal relationships.

International Union of Pure and Applied Chemistry System (IUPAC System) A set of systematic rules for naming compounds designed to give each compound one unique name and for which only one structure can be drawn; the Geneva system of nomenclature.

Interstitial Fluids Fluids in tissues but not inside cells.

Intron A segment of a DNA strand that separates exons and that does not become expressed as a segment of a polypeptide.

Inverse-Square Law The intensity of radiation varies inversely with the square of the distance from its source.

Invert Sugar A 1 : 1 mixture of glucose and fructose.

Iodine Test A test for starch by which a drop of iodine reagent produces an intensely purple color if starch is present.

Ion An electrically charged atomic or molecular-sized particle; a particle that has one or a few atomic nuclei and either one or two (seldom, three) too many or too few electrons to render the particle electrically neutral.

Ionic Bond The force of attraction between oppositely charged ions in an ionic compound.

Ionic Compound A compound that consists of an orderly aggregation of oppositely charged ions that assemble in whatever ratio ensures overall electrical neutrality.

Ionic Equation A chemical equation that explicitly shows all of the particles—ions, atoms, or molecules—that are involved in a reaction even if some are only spectator particles (see also *Net Ionic Equation; Equation, Balanced*).

Ionization A change, usually involving solvent molecules, whereby molecules change into ions.

Ionizing Radiation Any radiation that can create ions from molecules within the medium that it enters, such as alpha, beta, gamma, X, and cosmic radiation.

Ion Product Constant of Water (K_w) The product of the molar concentrations of hydrogen ions and hydroxide ions in water at a given temperature.

$$K_w = [H^+][OH^-]$$
$$= 1.0 \times 10^{-14} \qquad \text{(at 25 °C)}$$

Isoelectric A condition of a molecule in which it has an equal number of positive and negative sites.

Isoelectric Point (pI) The pH of a solution in which a specified amino acid or a protein is in an isoelectric condition; the pH at which there is no net migration of the amino acid or protein in an electric field.

Isoenzymes Enzymes that have identical catalytic functions but which are made of slightly different polypeptides.

Isohydric Shift In actively metabolizing tissue, the use of a hydrogen ion released from newly formed carbonic acid to react with and liberate oxygen from oxyhemoglobin; in the lungs, the use of hydrogen ion released when hemoglobin oxygenates to combine with bicarbonate ion and liberate carbon dioxide for exhaling.

Isomerism The phenomenon of the existence of two or more compounds with identical molecular formulas but different structures.

Isomers Compounds with identical molecular formulas but different structures.

Isotonic Having an osmotic pressure identical to that of a reference; having a total concentration of all solute particles equivalent to that of the reference.

Isotope A substance in which all of the atoms are identical in atomic number, mass number, and electron configuration.

IUPAC System (see *International Union of Pure and Applied Chemistry System*)

K_w (see *Ion Product Constant of Water*)

Kelvin The SI unit of temperature degree and equal to 1/100th of the interval between the freezing point and the boiling point of water when measured under standard conditions.

Keratin The fibrous protein of hair, fur, fingernails, and hooves.

Ketal A substances whose molecules have two ether linkages joined to a carbon that also holds two hydrocarbon groups as in

$$R_2C(OR')_2$$

Ketoacidosis The acidosis caused by untreated ketonemia.

Keto Group The carbonyl group when it is joined on each side to carbon atoms.

Ketohexose A monosaccharide whose molecules contain six carbon atoms and have a keto group.

Ketone Any compound with a carbonyl group attached to two carbon atoms, as in $R_2C{=}O$.

Ketone Bodies Acetoacetate and β-hydroxybutyrate, or their parent acids, and acetone.

Ketonemia An elevated concentration of ketone bodies in the blood.

Ketonuria An elevated concentration of ketone bodies in the urine.

Ketose A monosaccharide whose molecules have a ketone group.

Ketosis The combination of ketonemia, ketonuria, and acetone breath.

Kilocalorie (kcal) The quantity of heat equal to 1000 cal.

Kilogram (kg) The SI base unit of mass; 1000 g; 2.205 lb.

Kilometer (km) A length equal to 1000 m or 0.621 miles.

Kinetic Energy The energy of an object by virtue of its motion.

$$\text{Kinetic energy} = \tfrac{1}{2}(\text{mass})(\text{velocity})^2$$

Kinetics The science of the study of reaction rates.

Kinetic Theory of Gases A set of postulates about the nature of an ideal gas: that it consists of a large number of very small particles in constant, random motion; that in the collisions the particles lose no frictional energy; that between collisions the particles neither attract nor repel each other; and that the motions and collisions of the particles obey all the laws of physics.

Krebs' Cycle The citric acid cycle.

Lambda (λ) One microliter; 1×10^{-6} L.

Law of Conservation of Energy Energy can be neither created nor destroyed but only transformed from one form to another.

Law of Conservation of Mass Matter is neither created nor destroyed in chemical reactions; the masses of all products equals the masses of all reactants.

Law of Definite Proportions The elements in a compound occur in definite proportions by mass.

Law of Partial Pressures (Dalton's Law) The total pressure of a mixture of gases is the sum of their individual partial pressures.

Le Chatelier's Principle If a system is in equilibrium and a change is made in its conditions, the system will change in whichever way most directly restores equilibrium.

Length The base quantity for expressing distances or how long a thing is.

Like-Dissolves-Like Rule Polar solvents dissolve polar or ionic solutes and nonpolar solvents dissolve nonpolar or weakly polar solutes.

Lipid A plant or animal product that tends to dissolve in such nonpolar solvents as ether, carbon tetrachloride, and benzene.

Lipid Bilayer The sheetlike array of two layers of lipid molecules, interspersed with molecules of cholesterol and proteins, that make up the membranes of cells in animals.

Lipigenesis The synthesis of fatty acids from two-carbon acetyl units.

Lipoprotein Complex A combination of a lipid molecule with a protein molecule that serves as the vehicle for carrying the lipid in the blood stream.

Liquid A state of matter in which a substance's volume but not its shape is independent of the shape of its container.

Liter (L) A volume equal to 1000 cm³ or 1000 mL or 1.057 liquid quart.

Lock-and-Key Theory The specificity of an enzyme for its substrate is caused by the need for the substrate molecule to fit to the enzyme's surface much as a key fits to and turns only one tumbler lock.

Macromolecule Any molecule with a very high formula weight—generally several thousand or more.

Maltose A disaccharide that can be hydrolyzed to two glucose molecules; malt sugar.

Manometer A device for measuring gas pressure.

Markovnikov's Rule In the addition of an unsymmetrical reactant to an unsymmetrical double bond of a simple alkene, the positive part of the reactant molecule (usually H^+) goes to the carbon that has the greater number of hydrogen atoms and the negative part goes to the other carbon of the double bond.

Mass A quantitative measure of inertia based on an artifact at Sèvres, France, called the standard kilogram mass; a measure of the quantity of matter in an object relative to this reference standard.

Mass Number The sum of the numbers of protons and neutrons in one atom of an isotope.

Material Balance The condition of a chemical equation in which all of the atoms present among the reactants are also found in the products.

Matter Anything that occupies space and has mass.

Measurement An operation that compares an unknown physical quantity with a known standard.

Melting Point The temperature at which a solid changes into its liquid form; the temperature at which equilibrium exists between the solid and liquid forms of a substance.

Mercaptan A thioalcohol; RSH.

Messenger RNA (mRNA) RNA that carries the genetic code as a specific series of codons for a specific polypeptide from the cell's nucleus to the cytoplasm.

Metabolism The sum total of all of the chemical reactions that occur in an organism.

Metal Any element that is shiny, conducts electricity well, and (if a solid) can be hammered into sheets and drawn into wires.

Metalloids Elements that have some metallic and some nonmetallic properties.

Meter (m) The base unit of length in the International System of Measurements (SI).

$$1 \text{ m} = 100 \text{ cm} = 39.37 \text{ in.} = 3.280 \text{ ft} = 1.093 \text{ yd}$$

Metric System A decimal system of weights and measures in which the conversion of a base unit of measurement into a multiple or a submultiple is done by moving the decimal point; the predecessor to the International System of Measurements (SI).

Microgram (μg) A mass equal to one thousandth of a milligram.

$$1 \ \mu\text{g} = 0.001 \text{ mg} = 1 \times 10^{-6} \text{ g}$$

(Its symbol is sometimes mcg or γ in pharmaceutical work.)

Microliter (μL) A volume equal to one-thousandth of a milliliter.

$$1 \ \mu\text{L} = 0.001 \text{ mL} = 1 \times 10^{-6} \text{ L}$$

(Its symbol is sometimes given as λ in pharmaceutical work.)

Milliequivalent (meq) A quantity of substance equal to one-thousandth of an equivalent.

Milligram (mg) A mass equal to one-thousandth of a gram.

$$1 \text{ mg} = 0.001 \text{ g} \qquad 1000 \text{ mg} = 1 \text{ g} \qquad 1 \text{ grain} = 64.8 \text{ mg}$$

Milliliter (mL) A volume equal to one-thousandth of a liter.

$$1 \text{ mL} = 0.001 \text{ L} = 16.23 \text{ minim} = 1 \text{ cm}^3$$

$$1 \text{ liquid ounce} = 29.57 \text{ mL} \qquad 1 \text{ liquid quart} = 946.4 \text{ mL}$$

Millimeter (mm) A length equal to one-thousandth of a meter.

$$1 \text{ mm} = 0.001 \text{ m} = 0.0394 \text{ in.}$$

Millimeter of Mercury (mm Hg) A unit of pressure equal to $\frac{1}{760}$ atm.

Millimole (mmol) One-thousandth of a mole. 1000 mmol = 1 mol

Mixture One of the three kinds of matter (together with elements and compounds); any substance made up of two or more elements or compounds combined physically and separable into its component parts by physical means.

Model, Scientific A mental construction, often involving pictures or diagrams, that is used to explain a number of facts.

Molar Concentration (*M*) A solution's concentration in units of moles of solute per liter of solution; molarity.

Molar Mass The number of grams per mole of a substance.

Molar Volume The volume occupied by one mole of a gas under standard conditions of temperature and pressure; 22.4 L at 273 K and 1 atm.

Molarity (see *Molar Concentration*)

Mole (mol) A mass of a compound or of an element that equals its formula weight in grams; Avogadro's number of a substance's formula units.

Molecular Compound A compound whose smallest representative particle is a molecule; a covalent compound.

Molecular Equation An equation that shows the complete formulas of all of the substances present in a mixture undergoing a reaction (see also *Net Ionic Equation; Equation, Balanced*).

Molecular Weight The formula weight of a substance.

Molecule An electrically neutral (but often polar) particle made up of the nuclei and electrons of two or more atoms and held together by covalent bonds; the smallest representative sample of a molecular compound.

Monoamine Oxidase An enzyme that catalyzes the inactivation of neurotransmitters or other amino compounds of the nervous system.

Monomer Any compound that can be used to make a polymer.

Monoprotic Acid An acid with one proton per molecule that can neutralize a base.

Monosaccharide A carbohydrate that cannot be hydrolyzed.

Mutagen Any chemical or physical agent that can induce the mutation of a gene without preventing the gene from replicating.

Myosins Proteins in contractile muscle.

Net Ionic Equation A chemical equation in which all spectator particles are omitted so that only the particles that participate directly are represented.

Neurotransmitter A substance released by one nerve cell to carry a signal to the next nerve cell.

Neutral Solution A solution in which the molar concentration of hydronium ions exactly equals the molar concentration of hydroxide ions.

Neutralization, Acid–Base A reaction between an acid and a base.

Neutralizing Capacity The capacity of a solution or a substance to neutralize an acid or a base—expressed as a molar concentration.

Neutron An electrically neutral subatomic particle with a mass of 1 amu.

Nitrogen Balance A condition of the body in which it excretes as much nitrogen as it receives in the diet.

Nitrogen Family The elements of Group VA of the periodic table: nitrogen, phosphorus, arsenic, antimony, and bismuth.

Nitrogen Pool The sum total of all nitrogen compounds in the body.

Noble Gases The elements of Group 0 of the periodic table: helium, neon, argon, krypton, xenon, and radon.

Nomenclature The system of names and the rules for devising such names, given structures, or for writing structures, given names.

Nonelectrolyte Any substance that cannot furnish ions when dissolved in water or when melted.

Nonfunctional Group A section of an organic molecule that remains unchanged during a chemical reaction at a functional group.

Nonmetal Any element that is not a metal (see *Metal*).

Nonsaponifiable Lipid Any lipid, such as the steroids, that cannot be hydrolyzed or similarly broken down by aqueous alkali.

Nonvolatile Liquid Any liquid with a very low vapor pressure at room temperature that does not readily evaporate.

Normal Fasting Level The normal concentration of something in the blood, such as blood sugar, after about 4 hours without food.

Nuclear Chain Reaction The mechanism of nuclear fission by which one fission event makes enough fission initiators (neutrons) to cause more than one additional fission event.

Nuclear Equation A representation of a nuclear transformation in which the chemical symbols of the reactants and products include mass numbers and atomic numbers.

Nucleic Acid A polymer of nucleotides in which the repeating units are pentose phosphate esters, each pentose unit bearing a side-chain base (one of five heterocyclic amines); polymeric compounds that are involved in the storage, transmission, and expression of genetic messages.

Nucleotide A monomer of a nucleic acid that consists of a pentose phosphate ester in which the pentose unit carries one of five heterocyclic amines as a side-chain base.

Nucleus In chemistry and physics, the subatomic particle that serves as the core of an atom and that is made up of protons and neutrons. In biology, the organelle in a cell that houses DNA.

Octet, Outer A condition of an atom or ion in which its highest occupied energy level has eight electrons—a condition of stability.

Octet Rule The atoms of a reactive representative element tend to undergo those chemical reactions that most directly give them the electron configuration of the noble gas that stands nearest the element in the periodic table (all but one of which have outer octets).

Olefin An alkene.

One-Substance–One-Structure Rule If two samples of matter have identical physical and chemical properties, they have identical molecules.

Optical activity The ability of a substance to rotate the plane of polarization of plane-polarized light.

Optical Isomer One of a set of compounds whose molecules differ only in their chiralities.

Optical Isomerism Isomerism that is caused by chirality.

Organic Compounds Compounds of carbon other than those related to carbonic acid and its salts, or to the oxides of carbon, or to the cyanides.

Osmolarity The molar concentration of all osmotically active solute particles in a solution.

Osmosis The passage of water only, without any solute, from a less concentrated solution (or pure water) to a more concentrated solution when the two solutions are separated by a semipermeable membrane.

Osmotic Membrane A semipermeable membrane that permits only osmosis, not dialysis.

Osmotic Pressure The pressure that would have to be applied to a solution to prevent osmosis if the solution were separated from water by an osmotic membrane.

Outer Octet (see *Octet, Outer*)

Outside Level In an atom the highest principal energy level that holds at least one electron.

Oxidation The loss of one or more electrons from an atom, molecule, or ion; in organic chemistry, the loss of hydrogen or the gain of oxygen.

Oxidation Number For simple monoatomic ions, the quantity and sign of the electrical charge on the ion.

Oxidative Deamination The change of an amino group to a keto group with loss of nitrogen.

Oxidative Phosphorylation The synthesis of high-energy phosphates such as ATP from lower energy phosphates and inorganic phosphate by the reactions that involve the respiratory chain.

Oxidizing Agent A substance that can cause an oxidation.

Oxygen Debt The condition in a tissue when anaerobic glycolysis has operated and lactate has been excessively produced.

Oxygen Family The elements in Group VIA of the periodic table: oxygen, sulfur, selenium, tellurium, and polonium.

P_i Inorganic phosphate ion(s) of whatever mix of PO_4^{3-}, HPO_4^{2-}, $H_2PO_4^-$, and possibly even traces of H_3PO_4 that is possible at the particular pH of the system, but almost entirely HPO_4^{2-} + $H_2PO_4^-$.

Partial Pressure The pressure contributed by an individual gas in a mixture of gases.

Pentose Phosphate Pathway The synthesis of NADPH that uses chemical energy in glucose 6-phosphate and that involves pentoses as intermediates.

Peptide Bond The amide linkage in a protein; a carbonyl-to-nitrogen bond.

Percent (%) A measure of concentration.

Volume/volume (v/v) percent: The number of volumes of solute in 100 volumes of solution.

Weight/weight (w/w) percent: The number of grams of solute in 100 g of the solution.

Weight/volume (w/v) percent: The number of grams of solute in 100 mL of the solution.

Milligram percent: The number of milligrams of the solute in 100 mL of the solution.

Period A horizontal row in the periodic table.

Periodic Law Many properties of the elements are periodic functions of their atomic numbers.

Periodic Table A display of the elements that emphasizes the family relationships.

pH The negative power to which the base 10 must be raised to express the molar concentration of hydrogen ions in an aqueous solution.

$$[H^+] = 1 \times 10^{-pH}$$

Phenol Any organic compound whose molecules have an OH group attached to a benzene ring.

Phenyl Group The benzene ring minus one H atom; C_6H_5—.

Phosphate Buffer Usually a mixture or a solution that contains dihydrogen phosphate ions ($H_2PO_4^-$) to neutralize OH^- and monohydrogen phosphate ions (HPO_4^{2-}) to neutralize H^+.

Phosphoglyceride A phospholipid such as a plasmalogen or a lecithin whose molecules include a glycerol unit.

Phospholipid Lipids such as the phosphoglycerides, the plasmalogens, and the sphingomyelins whose molecules include phosphate ester units.

Photon A package of energy released when an electron in an atom moves from a higher to a lower energy state; a unit of light energy.

Photosynthesis The synthesis in plants of complex compounds from carbon dioxide, water, and minerals with the aid of sunlight captured by the plant's green pigment, chlorophyll.

Physical Change Any change that leaves all substances chemically the same.

Physical Property Any observable characteristic of a substance other than a chemical property, such as color, density, melting point, boiling point, temperature, and quantity.

Physical Quantity A property of something to which we assign both a numerical value and a unit, such as mass, volume, or temperature;

$$physical\ quantity = number \times unit$$

Physiological Saline Solution A solution of sodium chloride with an osmotic pressure equal to that of blood.

pI (see *Isoelectric Point*)

Plasmalogens Glycerol-based phospholipids whose molecules also include an unsaturated fatty alcohol unit.

Plasmid A circular molecule of supercoiled DNA in a bacterial cell.

β-Pleated Sheet A secondary structure for a polypeptide in which the molecules are aligned side by side in a sheetlike array with the sheet partially pleated.

Poison A substance that reacts in some way in the body to cause changes in metabolism that threaten health or life.

Polar Bond A bond at which we can write δ+ at one end and δ− at the other end, the end that has the more electronegative atom.

Polar Molecule A molecule that has sites of partial positive and partial negative charge and a permanent electrical dipole.

Polyatomic Ion Any ion made from two or more atoms, such OH^-, SO_4^{2-}, and CO_3^{2-}.

Polymer Any substance with a very high formula weight whose molecules have a repeating structural unit.

Polymerization A chemical reaction that makes a polymer from a monomer.

Polypeptide A polymer with repeating α-aminoacyl units joined by peptide (amide) bonds.

Polysaccharide A carbohydrate whose molecules are polymers of monosaccharides.

Potential Energy Stored or inactive energy.

PP_i Inorganic diphosphate ion(s).

Precipitate A solid that separates from a solution.

Precipitation The formation and separation of a precipitate.

Precision The fineness of a measurement or the degree to which successive measurements agree with each other when several are taken one after the other (see also *Accuracy*).

Pressure Force per unit area.

Pressure–Temperature Law (Gay-Lussac's Law) The pressure of a gas is directly proportional to its Kelvin temperature when the gas volume is constant.

Pressure–Volume Law (Boyle's Law) The volume of a gas is inversely proportional to its pressure when the temperature is constant.

Primary Alcohol An alcohol in whose molecules an OH group is attached to a primary carbon, as in RCH_2OH.

Primary Carbon In a molecule, a carbon atom that is joined directly to just one other carbon, such as the end carbons in $CH_3CH_2CH_3$.

Primary Structure The sequence of aminoacyl residues held together by peptide bonds in a polypeptide.

Primary Transcript RNA (ptRNA) The RNA made at the direction of DNA and complementary to the DNA, both exons and introns.

Principal Energy Level A space near an atomic nucleus where electrons can reside; an electron shell.

Product A substance that forms in a chemical reaction.

Proenzyme An inactive form of an enzyme; a zymogen.

Property A characteristic of something by means of which we can identify it.

Prosthetic Group A nonprotein molecule joined to a polypeptide to make a biologically active protein.

Protein A naturally occurring polymeric substance made up wholly or mostly of polypeptide molecules

Proton A subatomic particle that bears one unit of positive charge and has a mass of 1 amu.

Quantum A quantity of energy possessed by a photon.

Quaternary Structure An aggregation of two or more polypeptide strands each with its own primary, secondary, and tertiary structure.

Rad One rad equals 100 ergs (1×10^{-5} J) of energy absorbed per gram of tissue as a result of ionizing radiations.

Radiation A process whereby light or heat is emitted; also the emitted light or heat. In atomic physics, the emission of some ray such as an alpha, beta, or gamma ray.

Radiation Sickness The set of symptoms that develops following exposure to heavy doses of ionizing radiations.

Radical A particle with one or more unpaired electrons.

Radioactive The property of unstable atomic nuclei whereby they emit alpha, beta, or gamma rays.

Radioactive Decay The change of a radioactive isotope into another isotope by the emission of alpha rays or beta rays.

Radioactive Disintegration Series A series of isotopes selected and arranged such that each isotope except the first is produced by the radioactive decay of the preceding isotope and the last isotope is nonradioactive.

Radioactive Element An element that emits dangerous radiation(s).

Radiomimetic Substance A substance whose chemical effect in a cell mimics the effect of ionizing radiation.

Radionuclide A radioactive isotope.

Rate of Reaction The number of successful (product forming) collisions that occur each second in each unit of volume of the reacting mixture.

Ratio A relationship between two numbers or two physical quantities expressed as a fraction.

Reactant One of the substances that reacts in a chemical reaction.

Reagent Any mixture of chemicals, usually a solution, that is used to carry out a chemical test or other reaction.

Receptor Molecule A molecule of a protein built into a cell membrane that can accept a molecule of a hormone or a neurotransmitter.

Recombinant DNA DNA made by combining the natural DNA of plasmids in bacteria or the natural DNA in yeasts with DNA from external sources, such as the DNA for human insulin, and made as a step in a process that uses altered bacteria or yeasts to make specific proteins (e.g., interferons, human growth hormone, or insulin).

Redox Reaction Abbreviation of *reduction–oxidation;* a reaction in which electrons transfer between reactants.

Reducing Agent A substance that can cause another to be reduced.

Reducing Carbohydrate A carbohydrate that gives a positive Benedict's test.

Reduction The gain of one or more electrons by an atom, ion, or molecule; in organic chemistry, the gain of hydrogen or the loss of oxygen.

Reductive Amination The conversion of a keto group to an amino group by the action of ammonia and a reducing agent.

Rem One rem is the quantity of a radiation that produces the same effect in humans as one roentgen of X rays or gamma rays.

Renal Threshold That concentration of a substance in blood above which it appears in the urine.

Replication The reproductive duplication of a DNA double helix.

Representative Element Any element in any A group of the periodic table; any element in Groups IA–VIIA and those in Group 0.

Respiration The intake and chemical use of oxygen by the body and the release of carbon dioxide.

Respiratory Chain The reactions that transfer electrons from the intermediates made by other pathways to oxygen; the mechanism that creates a proton gradient across the inner membrane of a mitochondrion and that leads to ATP synthesis; the enzymes that handle these reactions.

Respiratory Enzymes The enzymes of the respiratory chain.

Respiratory Gases Oxygen and carbon dioxide.

Ribonucleic Acids (RNA) Polymers of ribonucleotides that participate in the transcription and the translation of the genetic messages into polypeptides (see also *Primary Transcript RNA, Messenger RNA, Ribosomal RNA,* and *Transfer RNA*).

Ribosomal RNA (rRNa) RNA that is incorporated into cytoplasmic bodies called ribosomes.

Ribosome A granular complex of rRNA that becomes attached to a mRNA strand and that supplies some of the enzymes for mRNA-directed polypeptide synthesis.

Ring Compound A compound whose molecules contain three or more atoms joined in a ring.

RNA (see *Ribonucleic Acid*)

Roentgen One roentgen is the quantity of X rays or gamma radiation that generates ions with an aggregate of 2.1×10^9 units of charge in 1 mL of dry air at normal pressure and temperature.

Salt Any crystalline compound that consists of oppositely charged ions (other than H^+ or OH^-).

Salt Bridge A force of attraction between (+) and (−) sites on polypeptide molecules.

Saponifiable Lipid Any lipid with ester groups.

Saponification The reaction of an ester with sodium or potassium hydroxide to give an alcohol and the salt of an acid.

Saturated Compound A compound whose molecules have only single bonds.

Saturated Solution A solution into which no more solute can be dissolved at the given temperature; a solution in which dynamic equilibrium exists between the dissolved and the undissolved solute.

Scientific Method A method of solving a problem that uses facts to devise a hypothesis to explain the facts and to suggest further tests or experiments designed to discover if the hypothesis is true or false.

Scientific Notation The method of writing a number as the product of two numbers, one being 10^x, where x is some positive or negative whole number.

Second (s) The SI unit of time; $\frac{1}{60}$th min.

Secondary Alcohol An alcohol in whose molecules an OH group is attached to a secondary carbon atom; R_2CH—OH.

Secondary Carbon Any carbon atom in an organic molecule that has two and only two bonds to other carbon atoms, such as the middle carbon atom in $CH_3CH_2CH_3$.

Secondary Structure A shape, such as the α helix or a unit in a β-pleated sheet, that all or a large part of a polypeptide molecule adopts under the influence of hydrogen bonds or salt bridges after its peptide bonds have been made.

Semipermeable Descriptive of a membrane that permits only certain kinds of molecules to pass through and not others.

Sensible Perspiration Visible perspiration released by the sweat glands.

Shock, Traumatic A medical emergency in which relatively large volumes of blood fluid leave the vascular compartment and enter the interstitial spaces.

Side Chain An organic group that can be appended to a main chain or to a ring.

Significant Figures The number of digits in a numerical measurement or in the result of a calculation that are known with certainty to be accurate plus one more digit.

Simple Lipid A triacylglycerol; a triglyceride.

Simple Salt A salt that consists of just one kind of cation and one kind of anion.

Simple Sugar Any monosaccharide.

Soap A detergent that consists of the sodium or potassium salts of long-chain fatty acids.

Soft Water Water with little if any of the hardness ions: Mg^{2+}, Ca^{2+}, Fe^{2+}, or Fe^{3+}.

Sol A colloidal dispersion of tiny particles of a solid in a liquid.

Solid A state of matter in which the visible particles of the substance have both definite shapes and definite volumes.

Solubility The extent to which a substance dissolves in a fixed volume or weight of a solvent at a given temperature.

Solute The component of a solution that is understood to be dissolved in or dispersed in a continuous solvent.

Solution A homogeneous mixture of two or more substances that are at the smallest levels of their states of subdivision — at the ion, atom, or molecule level.

Solution, Aqueous A solution in which water is the solvent.

Solution, Concentrated A solution with a high ratio of solute to solvent.

Solution, Dilute A solution with a low ratio of solute to solvent.

Solution, Saturated (see *Saturated Solution*)

Solution, Supersaturated An unstable solution that has a higher concentration of solute than that of the saturated solution.

Solution, Unsaturated A solution into which more solute could be dissolved without changing the temperature.

Solvent That component of a solution into which the solutes are considered to have dissolved; the component that is present as a continuous phase.

Somatostatin A hormone of the hypothalamus that inhibits or slows the release of glucagon and insulin from the pancreas.

Specific Gravity The ratio of the density of an object to the density of water.

Sphingolipid A lipid that, when hydrolyzed, gives sphingosine instead of glycerol, plus fatty acids, phosphoric acid, and a small alcohol or a monosaccharide; sphingomyelins and cerebrosides.

Standard A physical description or embodiment of a base unit of measurement, such as the standard meter or the standard kilogram mass.

Standard Atmosphere (atm) The pressure that supports a column of mercury 760 mm high when the mercury has a temperature of 0 °C.

Standard Conditions of Temperature and Pressure (STP) 0 °C (or 273 K) and 1 atm (or 760 mm Hg).

Standard Solution Any solution for which the concentration is accurately known.

Starch A naturally occurring mixture, obtained from plants, of amylose and amylopectin.

States of Matter The three possible physical conditions of aggregation of matter: solid, liquid, and gas.

Stereoisomer One of a set of isomers whose molecules have the same atom-to-atom sequences but different geometric arrangements; a geometric (cis−trans) or optical isomer.

Steroids Nonsaponifiable lipids such as cholesterol and several sex hormones whose molecules have the four fused rings of the steroid nucleus.

Stoichiometry The branch of chemistry that deals with the mole proportions of chemicals in reactions.

Straight Chain A continuous, open sequence of covalently bound carbon atoms from which no additional carbon atoms are attached at interior locations of the sequence.

Strong Acid An acid with a high percentage ionization. Any species, molecular or ionic, that has a strong tendency to donate a proton to some acceptor.

Strong Base A metal hydroxide with a high percentage ionization in solution. Any species, molecular or ionic, that accepts and strongly binds a proton.

Strong Electrolyte Any substance that has a high percentage ionization in solution.

Structural Formula A formula that uses lines representing covalent bonds to connect the atomic symbols in the pattern that occurs in one molecule of a compound.

Structural Isomer One of a set of isomers whose molecules differ in their atom-to-atom sequence.

Structure Synonym for structural formula (see *Structural Formula*).

Subatomic Particle An electron, a proton, or a neutron; the atomic nucleus as a whole is also a subatomic particle.

Subscripts Numbers placed to the right and a half space below the atomic symbols in a chemical formula.

Substitution Reaction A reaction in which one atom or group replaces another atom or group in a molecule.

Substrate The substance on which an enzyme performs its catalytic work.

Superimposition An operation to see whether one molecular model can be made to blend simultaneously at exactly every point with another model.

Supersaturated Describing an unstable condition of a solution in which more solute is in solution than could be if there were equilibrium between the undissolved and dissolved states of the solute.

Surface-Active Agent (see *Surfactant*)

Surface Tension The quality of a liquid's surface by which it behaves as if it were a thin, invisible, elastic membrane.

Surfactant A substance, such as a detergent, that reduces the surface tension of water.

Suspension A homogeneous mixture in which the particles of at least one component have average diameters greater than 1000 nm.

Synapse The fluid-filled gap between the end of the axon of one nerve cell and the next nerve cell.

Syndet A synthetic detergent that works in hard water.

Target Cell A cell at which a hormone molecule finds a site where it can become attached and then cause some action that is associated with the hormone.

Target Organ The organ whose cells are recognizable by the molecules of a particular hormone.

Temperature The measure of the hotness or coldness of an object. *Degrees* of temperature, such as those of the Celsius, Fahrenheit, or Kelvin scales, are intervals of equal separation on the thermometer.

Temperature–Volume Law (Charles' Law) The volume of a gas is directly proportional to its Kelvin temperature when the pressure is kept constant.

Teratogen Any chemical or physical agent that can cause birth defects in a fetus other than those defects that the fetus has inherited.

Tertiary Alcohol An alcohol in whose molecules an OH group is held by a carbon from which three bonds extend to other carbon atoms; R_3C—OH.

Tertiary Carbon Any carbon in an organic molecule that has three and only three bonds to adjacent *carbon* atoms.

Tertiary Structure The shape of a polypeptide molecule that arises from further folding or coiling of secondary structures.

Tetrahedral Descriptive of the geometry of bonds at a central atom in which the bonds project to the corners of a regular tetrahedron.

Theory An explanation for a large number of facts, observations, and hypotheses in terms of one or a few fundamental assumptions of what the world (or some part of the world) is like.

Thioalcohol A compound whose molecules have the SH group attached to a saturated carbon atom; a mercaptan.

Threshold Exposure The level of exposure to some toxic agent below which no harm is done.

Time A period during which something endures, exists, or continues.

Tollens' Reagent A slightly alkaline solution of the diammine complex of the silver ion, $Ag(NH_3)_2^+$, in water.

Tollens' Test The use of Tollens' reagent to detect an easily oxidized group such as the aldehyde group.

Torr A unit of pressure; 1 torr = 1 mm Hg; 1 atm = 760 torr

Transamination The transfer of an amino group from an amino acid to a receiver with a keto group such that the keto group changes to an amino group.

Transcription The synthesis of messenger RNA under the direction of DNA.

Transfer RNA (tRNA) RNA that serves to carry an aminoacyl group to a specific acceptor site of a mRNA molecule at a ribosome where the aminoacyl group is placed into a growing polypeptide chain.

Transition Elements The elements between those of Groups IIA and IIIA in the long periods of the periodic table; a metallic element other than one in Groups IA or IIA or in the actinide or lanthanide families.

Translation The synthesis of a polypeptide under the direction of messenger RNA.

Transmutation The change of an isotope of one element into an isotope of a different element.

Triacylglycerol A lipid that can be hydrolyzed to glycerol and fatty acids; a triglyceride; sometimes, simply called a glyceride.

Triglyceride (see *Triacylglycerol*)

Triple Helix The quaternary structure of tropocollagen in which three polypeptide chains are twisted together.

Triprotic Acid An acid that can supply three protons per molecule.

Tyndall Effect The scattering of light by colloidal sized particles in a colloidal disperion.

Universal Gas Law $PV = nRT$ (see also *Gas Constant, Universal*).

Unsaturated Compound Any compound whose molecules have a double or a triple bond.

Vacuum An enclosed space in which there is no matter.

Vapor The gaseous form of a liquid

Vapor Pressure The pressure exerted by the vapor that is in equilibrium with its liquid state at a given temperature.

Vaporization The change of a liquid into its vapor.

Vascular Compartment The entire network of blood vessels and their contents.

Vasopressin A hypophysis hormone that acts at the kidneys to help regulate the concentrations of solutes in the blood by instructing the kidneys to retain water (if the blood is too concentrated) or to excrete water (if the blood is too dilute).

Virus One of a large number of substances that consist of nucleic acid (usually RNA) surrounded (usually) by a protein overcoat and that can enter host cells, multiply, and destroy the host.

Vital Force Theory A discarded theory that organic compounds could be made in the laboratory only if the chemicals possessed a vital force contributed by some living thing.

Vitamin An organic substance that must be in the diet; whose absence causes a deficiency disease; which is present in foods in trace concentrations; and that isn't a carbohydrate, lipid, protein, or amino acid.

Volatile Liquid A liquid that has a high vapor pressure and readily evaporates at room temperature.

Volume The capacity of an object to occupy space.

Water of Hydration Water molecules held in a hydrate in some definite mole ratio to the rest of the compound.

Wax A lipid whose molecules are esters of long-chain monohydric alcohols and long-chain fatty acids.

Weak Acid An acid with a low percentage ionization in solution. Any species, molecule or ion, that has a weak tendency to donate a proton and serves poorly as a proton donor.

Weak Base A base with a low percentage ionization in solution. Any species, molecule or ion, that weakly holds an accepted proton and serves poorly as a proton acceptor.

Weak Electrolyte Any electrolyte that has a low percentage ionization in solution.

Weight The gravitational force of attraction on an object as compared to that of some reference.

Zymogen A polypeptide that is changed into an enzyme by the loss of a few amino acid residues or by some other change in its structure; a proenzyme.

Photo Credits

Chapter 13
Opener: Bill Ross/Woodfin Camp & Associates. Page 270: Courtesy of The Department of Surgery, Baylor University, College of Medicine.

Chapter 14
Opener: Bernard Pierre Wolff/Photo Researchers.

Chapter 15
Opener: Ira Kirschenbaum/Stock, Boston.

Chapter 16
Opener: Maureen Fennelli/Comstock. Page 335: (left) Courtesy of François Morel, from *Journal of Cell Biology,* 48, 91–100, 1971; (right) Courtesy Springer-Verlag Publishers, from *Corpuscles,* by Marcel Bessis.

Chapter 17
Opener: Alvin E. Staffan/Photo Researchers.

Chapter 18
Opener: Spencer Grant/Photo Researchers. Page 367: © 1984 United Feature Syndicate.

Chapter 19
Opener: Dilip Mehta/Woodfin Camp & Associates.

Chapter 20
Opener: David S. Strickler/Picture Cube.

Chapter 21
Opener: F. W. Binzen/Photo Researchers. Page 436: Courtesy Life Codes Corporation. Page 439: S. H. Kim, Duke University Medical Center.

Chapter 22
Opener: Spencer Grant/Stock, Boston. Page 467: (left) Courtesy of General Electric, Medical Systems Group; (right) Courtesy of Dimensional Medicine, Inc. Minnetonka, MN. Page 469: (top) Brookhaven National Laboratory; (center, left and right) Dan McCoy/Rainbow. Page 470: Howard Sochurer/Woodfin Camp & Associates.

Index